Modelling of Energy Systems

Tangellapalli Srinivas
Dr B R Ambedkar National Institute of Technology
Jalandhar, India

CRC Press
Taylor & Francis Group
Boca Raton London New York

CRC Press is an imprint of the
Taylor & Francis Group, an **informa** business
A SCIENCE PUBLISHERS BOOK

First edition published 2026
by CRC Press
2385 NW Executive Center Drive, Suite 320, Boca Raton FL 33431

and by CRC Press
4 Park Square, Milton Park, Abingdon, Oxon, OX14 4RN

CRC Press is an imprint of Taylor & Francis Group, LLC

Library of Congress Cataloging-in-Publication Data (applied for)

ISBN: 978-1-032-64248-2 (hbk)
ISBN: 978-1-032-64249-9 (pbk)
ISBN: 978-1-032-64250-5 (ebk)

DOI: 10.1201/9781032642505

Typeset in Times New Roman
by Prime Publishing Services

Foreword 1

The book, *Modeling of Energy Systems* by Dr. Tangellapalli Srinivas, is a useful resource and guide for those dealing with energy technologies and systems. The book describes how to identify, understand, analyze and solve problems related to energy systems, and to develop strategies to improve their efficiency and sustainability. Energy system modeling uses a variety of techniques, often based on engineering and economic approaches, to build mathematical models that represent energy-related systems and processes. Such techniques form an important part of the book.

This book, in particular, provides a road map and support to engineers who want to learn modeling and simulation techniques, including those that form the basis of software tools. For instance, the performance of energy systems modeling and the development of solutions using App Designer allows results to be presented with enhanced precision and in extensive detail.

Modeling of Energy Systems has a broad yet focused content relevant to energy systems and their modeling. The book covers relevant fundamentals in the area of energy systems as well as recent advances and developments aimed at addressing challenges and arriving at solutions.

The fundamentals covered start with the modeling of fluid and thermal properties and processes and technologies, which are relevant to many energy systems. Numerous heat sources are possible, including combustion, waste heat recovery and solar thermal energy, and the book delves into these. A focus is placed on energy system components such as heat exchangers, turbomachinery, and solar energy systems, noting their significant roles.

Building on the growing importance and adoption of solar energy technologies, extensive coverage is provided on solar energy and the design of equipment to support its utilization. The material on solar energy systems includes a wide range of aspects such as mathematical modeling, simulation, analysis, optimization, equipment design and heat exchangers, in addition to focused treatment of relevant heat transfer subjects such as conduction, convection and radiation.

With *Modeling of Energy Systems,* undergraduate students, postgraduate students, practitioners and researchers can obtain pertinent and detailed information on energy systems, and learn new ways to solve problems related them. This can be seen clearly in the broad coverage provided by the book in the many areas of relevance: thermal properties, thermal processes, system modeling and design, power generation, refrigeration and air conditioning, heat exchangers and combustion, as well as solar energy and its collection, storage and utilization.

The book should assist readers in utilizing energy system modeling to address challenges related to energy sources, equipment needs, systems design and various other topics, and thereby to help the world move towards energy systems that are more efficient, clean, economic and, ultimately, sustainable.

Dr. Marc A. Rosen

Professor

Ontario Tech University, Canada

Foreword 2

Energy system modeling and simulation is a key tool for designing, controlling, optimizing, and analyzing energy systems. *Modeling of Energy Systems* authored by Dr. T. Srinivas, covers a variety of topics including system properties, system processes, combustion systems, heat exchangers, turbomachinery, and solar energy systems. This book has significant importance today, due to the ongoing changes in the energy systems and the growing concerns about climate policy, energy security, and economic development. Energy modeling can help businesses and organizations save money on energy bills by identifying areas of inefficiency. It also supports organizations in assessing risks associated with energy use, such as price volatility and supply disruptions. The performance prediction of energy systems provides outputs such as a system's feasibility, greenhouse gas emissions, and energy efficiency. The study of this book can be used in the design of energy systems, technology development, system operations, and energy policy development. The current challenges in operation and maintenance of energy systems can be solved using these computational tools before implementing them. The energy supply sector is the largest contributor to global greenhouse gas emissions. Simulation and modeling can help practitioners optimize energy system operations. Undergraduates, postgraduates, research scholars, practicing engineers, and industrialists can use this book as reference or as a textbook for the design and development of energy systems. I hope the book helps not only understand concepts but also supports the development of solutions and technologies.

Prof. Binod Kumar Kanaujia

Director

Dr B R Ambedkar National Institute of Technology Jalandhar, Punjab, India

Preface

Dr APJ Abdul Kalam encouraged dreams through his statement, "A dream is not that which you see while sleeping it is something that does not let you sleep." Converting dreams, i.e., ideas, into reality in the field of technology needs modeling, simulation, analysis, design, and optimization. Energy is an essential commodity in our day-to-day lives. The conversion of energy from one form to another needs best practices for optimum use and preservation for future generations without harming the environment. The development of energy systems involves many stages, such as discovery (idea generation), definition, development, and delivery. These stages involve technical and non-technical activities. Among technical activities, planning plays an important role in all stages of energy system development in order to avoid failures. Planning involves system modeling, simulation, design, and optimization. System modeling consists of understanding concepts and applying them to achieve the desired outcomes. Modeling and simulation have a great role to play in conducting research and development for product development systems.

This book covers the fundamental requirements of energy system modeling. This modeling can be applied to conventional energy systems and non-conventional (renewable) energy systems. The essential needs of energy system modeling are thermodynamic properties, process solutions, and simulation of components. Thermal models are detailed for the generation of thermodynamic properties, ideal gas properties, real gas properties, compressibility charts for properties, and binary fluid properties.

Process information is equally important for modeling an energy system. The processes related to the power industry, cooling, process industry, etc. are elaborated. The essential energy components, such as boilers, nozzles, turbines, compressors, heat exchangers, etc., are detailed with modeling and simulation. They can be extended to assess the total system, such as a power plant, automobile, aeronautical, air conditioning plant, cement factory, etc. Solar energy systems (renewable energy systems), viz., flat plate collector, air heater, and solar concentrating collector, are detailed with energy modeling and simulation. The basics of solar energy that are required to evaluate solar energy systems are solar geometry and the assessment of solar radiation. This book elaborates on these two topics as well from the basics for easy understanding.

Tangellapalli Srinivas

Acknowledgements

This book is mainly the outcome of the teaching-learning process, through interaction with faculty and students, their expectations and feedback. Towards this, I sincerely acknowledge the role played by my school and college teachers, Ph.D. supervisor, Dr. AVSSKS Gupta, Professor, JNT University, Hyderabad; currently Director, IIIT, Edupulapaya, Andhra Pradesh; postdoctoral supervisor, Dr. B. V. Reddy, Professor, Ontario Tech University, Canada; and research advisor, Dr. PK Nag, Professor, IIT Kharagpur.

One chapter of this book is related to the modelling of solar tracking systems. The Department of Science and Technology (DST), New Delhi, India, supported the idea of the development of a tracking unit without using a torque motor through a financial grant with the title 'Concentrated Solar Air Heater with Passive Tracking Mechanism and Corrugated Receiver' (DST/TMD/CERI/RES/2020/38(G)). My heartfelt thanks to DST and the project co-investigator, Dr. Rajan Kumar, for the successful completion and for supporting every stage of the development of this technology.

My sincere thanks to Director, NIT Jalandhar, Prof. Binod Kumar Kanaujia, for encouraging excellence in teaching, research, and consulting. I thank all my colleagues, both teaching and non-teaching, UG/PG students, research scholars, and all members of Dr. B. R. Ambedkar National Institute of Technology, Jalandhar, Punjab, India. Thanks to my earlier associations, AANM & VVRSR Polytechnic, Gudlavalleru Engineering College, Andhra Pradesh, Vellore Institute of Technology, Vellore, Tamil Nadu, and SK Machine Tools, Hyderabad, Telangana, for supporting my ideas and shaping them into outcomes.

Industry visits, industrial consultancy and associations take the teaching-learning process to another level.. My thanks to Oorja Energy Engineering Pvt. Limited, Hyderabad, Green Builders, Jalandhar, GMR Energy, Vemagiri, Altra Tech Cements, Tadipatri, Sagar Cements, Mattampally, Vijayawada Thermal Power Station, Kindapalli, LANCO, Kondapalli, and many others for providing practical and technical support.

I would like to express my deep sense of gratitude and respect to my parents, who molded my interest in the field of engineering from the beginning. My heart felt gratitude to my wife, Kavitha Devi, elder son Rahul, and younger son Jignesh; without whose constant support and encouragement, I could not have done this work.

Finally, thanks to all who were involved directly or indirectly, in giving shape this book.

Thanking you
Best Regards

Tangellapalli Srinivas

Contents

Chapter 1
Modeling and Simulation

1.1 Introduction

The demand for energy is increasing with the increase in living standards. The development of energy systems needs the essential step of modeling and simulation. Energy systems modeling is a tool used to analyze and optimize energy production, distribution, and consumption. It involves the creation of mathematical models that represent different components of an energy system, from generation to consumption, including conventional and non-conventional energy sources, infrastructure, and markets. The primary processes in the system development are modeling and simulation. The system design, optimization, analysis, and development follow based on the modeling. The modeling helps policymakers, researchers, and industry professionals make informed decisions about energy production, consumption, and sustainability goals. It allows the simulation of different future scenarios to assess the impact of various technologies, policies, and trends on energy systems. Energy system modeling is used to find cost-effective and efficient ways to meet energy demands while minimizing environmental impact and sustainable energy development. The energy supply includes generation sources like fossil fuels (coal, natural gas, oil), nuclear power, and renewable energy (wind, solar, hydro, etc.). Many energy models are increasingly focused on pathways to decarbonize energy systems and reach net-zero emissions goals. As renewable energy sources become more prevalent, modeling efforts are increasingly focused on integrating intermittent power sources (like wind and solar) with reliable grid systems. In essence, energy systems modeling is a crucial tool for navigating the evolving energy landscape and achieving sustainable, resilient energy futures. Polygeneration is one of the most efficient energy systems, which needs system modeling to justify the ideas of integration before the execution of production processes.

1.2 Technologies and methods of energy conversion

Any thermal or energy system is designed for the conversion of one form of energy to another. In such a system, it is possible to see or measure the input and output of the system. For the quantification of these two intermediate processes, information is required. The various technologies and methods of energy conversion are: the kinetic energy of wind into mechanical and electrical form in a wind turbine plant; the chemical energy of fuel into different energies; and electricity in a thermal power plant. Solar radiation is converted into thermal and electricity in a solar thermal power plant; solar to electricity in a photovoltaic cell; chemical to electricity in a fuel cell; pressure energy to electricity in hydropower; nuclear energy to electricity in a nuclear reactor plant, etc. To assess these energy conversions, the first requirement is understanding the thermodynamic properties of the working fluid moving in the system. Only after evaluating thermodynamic properties, it is possible to derive the mass and energy or heat balances to frame the formulae to solve the processes and/or components involved in the total plant. Energy systems modeling and analysis studies provide knowledge in equations framing, thermodynamic properties analysis, simulation techniques, and optimization methods to refine a system.

DOI: 10.1201/9781032642505-1

Thermal power cycles generate power from heat sources, and their operation has a Carnot limitation. It results in energy conversion efficiency, or thermal efficiency, as a function of source and sink temperatures. It works on the heat engine working principle. Examples of power cycles are the Atkine cycle, the Brayton cycle, the combined cycle, the diesel cycle, Ericson cycle, Kalina cycle, Otto cycle, Rankine cycle, Stirling cycle, etc. The stages in power cycle solutions are property solutions, mass and energy balance equation framing, designing, simulation, analysis, and optimization. The cycles can be evaluated in light of the first law of thermodynamics and also in light of the light of the second law of thermodynamics for qualitative analysis. These cycles can be grouped into vapor power cycles and gas power cycles. In gas power, if the mass of fuel is not considered, they can be treated as air standard cycles. Thermodynamic properties and energies at each and every point, i.e., pressure, temperature, volume, entropy, flow rate, exergy, etc., are to be evaluated by applying the energy conversion at each and every component. There are linked equations to be solved in the cycle. By properly substituting these equations, performance formulas can be generated.

For example, the solution needs the fuel, gas, and steam properties to evaluate the complete plant. The mass and energy balance result from the data at the fixed capacity of the plant. The best location of the deaerator can be obtained by analyzing the concerned parameter and changing its pressure between the condenser and boiler. After finalizing the complete set of operational parameters, the sizing of the plant can be completed. The best location of the deaerator can be decided by the optimization process by maximizing the efficiency of the plant. This can be achieved by varying the deaerator pressure between the condenser pressure and boiler pressure.

1.3 Modelling, simulation, analysis, design and optimization

Energy systems modeling plays an important role in the simulation, analysis, optimization, and design of an energy system. Modeling is a powerful tool that can be used to make the energy system effective. This knowledge is widely used in process industries for the design of new plants, debottlenecking, optimizing the performance of units, and even planning for day-to-day production changes.

Modeling is the preparation of a mechanism to predict the performance of a system. The development of derivations to deduce the formula to solve a problem is an example of a mathematical model. Modeling is a fundamental activity for all other activities such as simulation, design, analysis, and optimization. A model is a virtual representation of a problem using a computer application.

Simulation enables the prediction of system performance and its results.. The model is not the real system but it will give an answer that is close to the real one. Simulation involves applying the model and obtaining results (implementation of the model). Therefore, it can be stated that the simulation is similar to an imitation of a real system.

Analysis is concerned with determining the behavior of an existing system or a trial system being designed for a given task. Analysis is finding the behavior of a system with known parameters but unknown performance. The analysis can be carried out in two ways. As mentioned, analysis can be carried out on an existing model with the same parameters, or analysis can also be done without the same parameters. For example, in thermodynamic analysis, analysis can be done to find the specific results, even without knowing the parameters. In this category, the flow may be taken as a unit, i.e., kg/s. The performance results are specific, i.e., specific power, specific cooling, etc.

Design is the process of creating the physical dimensions of a system for a given performance; it helps determine the sizes and shapes of the various parts of the system to meet performance requirements. For example, to manufacture a heat exchanger at a given heat load of 10 kW, the number of tubes, diameter of tubes, length of tube, shell diameter, baffles, material, etc. first need to be designed. It is a planning step as per expectations or requirements.

Optimization tells us how to do things in a better way. An optimum system is much better than a workable system. The modeling, simulation, and analysis can be carried out before designing a product and also after the design process. The complete method of engineering design procedure is outlined in Fig. 1.1.

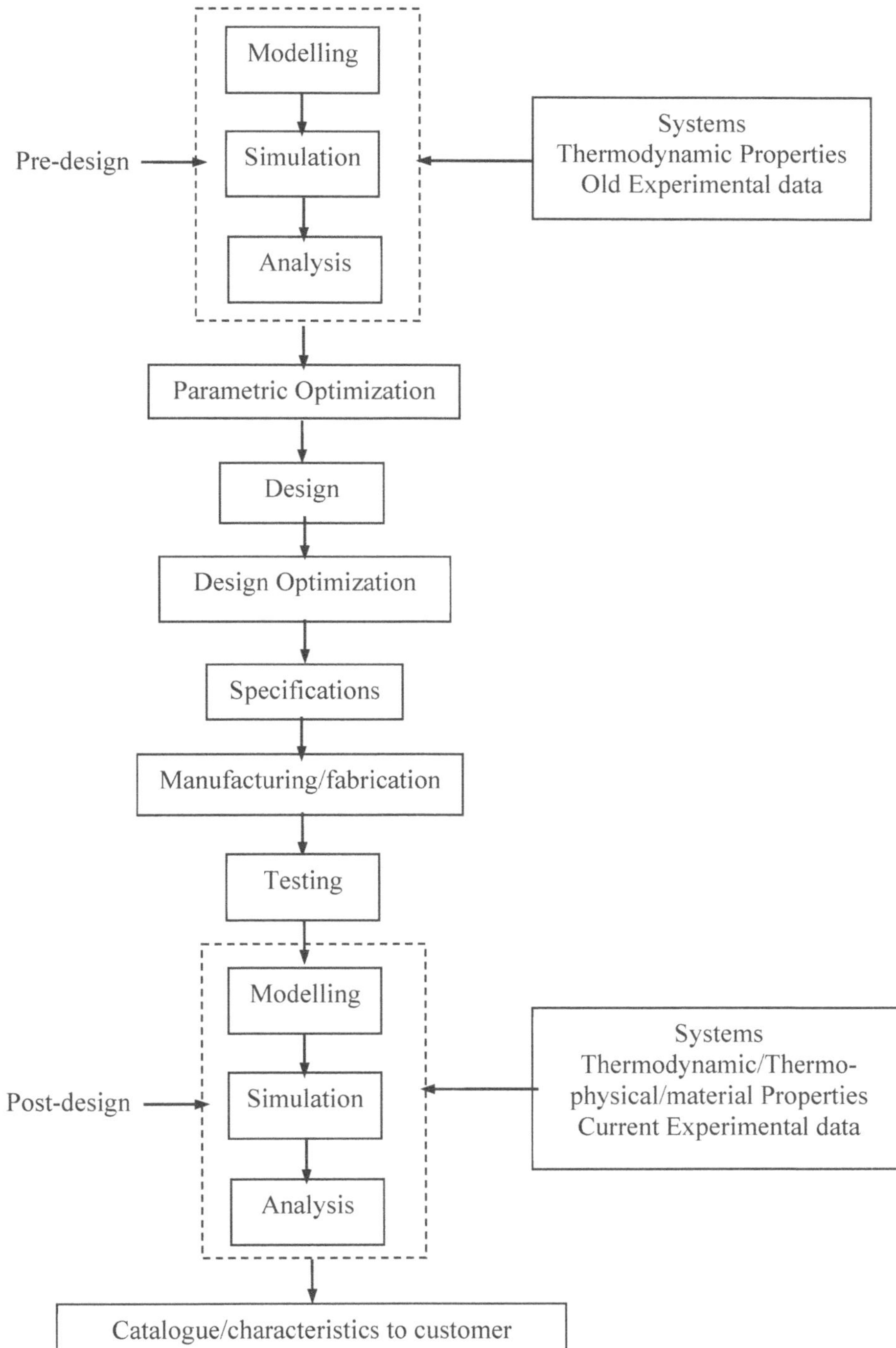

Fig. 1.1. Engineering design of a product/process.

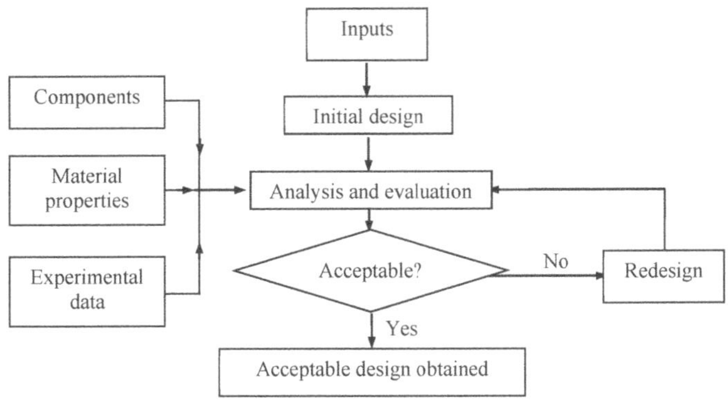

Fig. 1.2. Design of a product with synthesis.

Synthesis is a key element in the design process, as several components and their corresponding analyses are brought together to yield the characteristics of the overall system. Results from various areas have to be linked to synthesize in order to include all the important concerns. Figure 1.2 shows the synthesis of the design.

The development of a product or process needs systematic planning and organization for a successful launch into the market. Decision-making is an important activity that influences the quality of the output. Figure 1.3 arranges the relevant processes in a sequential order that is

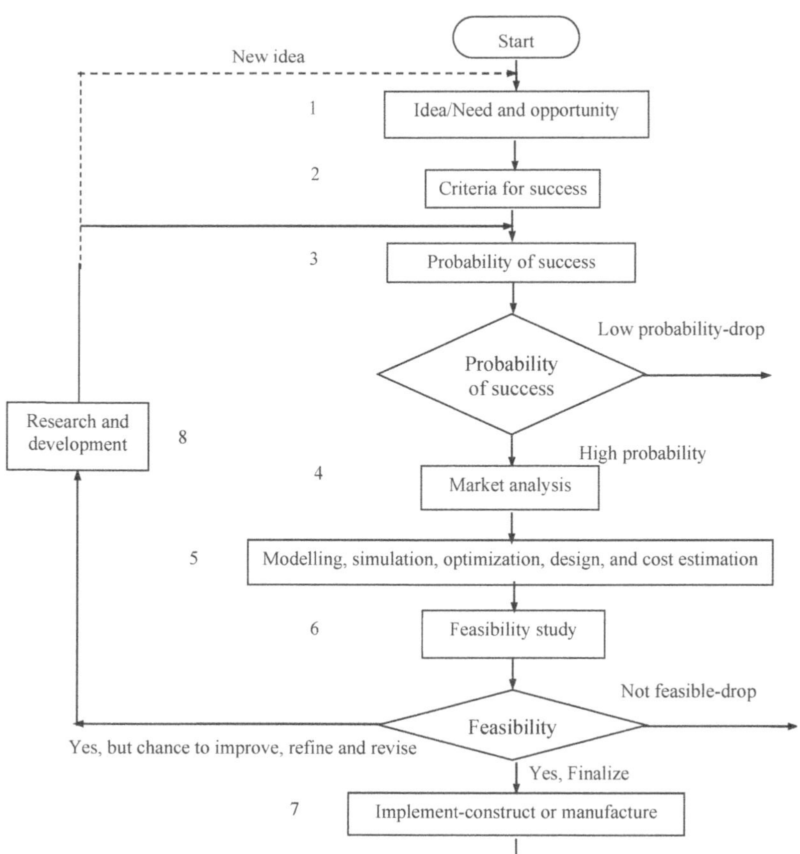

Fig. 1.3. Possible flow diagram in evaluating and planning an engineering undertaking.

universally applicable. First, we need to plan the work and then execute the plan as per the design. The flow chart shown in Fig. 1.3 shows the activities (rectangles) and the decisions (rhombus) at the critical times. The engineering activities are given at 5 and 7, related to design, research, and development. The non-design activities are also equally important for the completeness and commercialization of the product.

At the beginning, we need to specify the need for the work, which gives us the opportunity to provide employment and provides solutions to the identified needs or challenges. The need drives everything in an engineering undertaking.

1.4 Workable system and optimum system

The systems in the technology can be categorized into two: workable and optimum systems. A workable system is designed only to fulfill its function without minding its optimization (Stoecker, 1989). For example, a workable system supplies the required amount of electricity, heating, cooling, fuel, etc. as per demand. It satisfies all the constraints, such as size, weight, operational conditions, properties, etc. Whereas the optimum system involves rigorous research and development process to get the best possible design. But before making an optimization, one should verify various options to fulfill the specified need. Out of the available alternative options, the best one can be selected during the preliminary optimization process or comparative analysis. Later, the optimization process can be extended to the selected technology or product. The optimization also depends on the specified constraints. For example, the constraints for automobiles differ from those of aeronautic technologies.

To differentiate between the workable system and the optimum system, let us consider a pump and piping system to lift water from ground level to the top floor. Assume a water flow rate of 3 kg/s with a horizontal piping length of 250 m and a vertical height of 8 m. Figure 1.4 develops a techno-economic solution to select the best size of pipe for the minimum total cost of the pump and piping system. If we select a pipe size arbitrarily, it is a workable but not an optimum system.

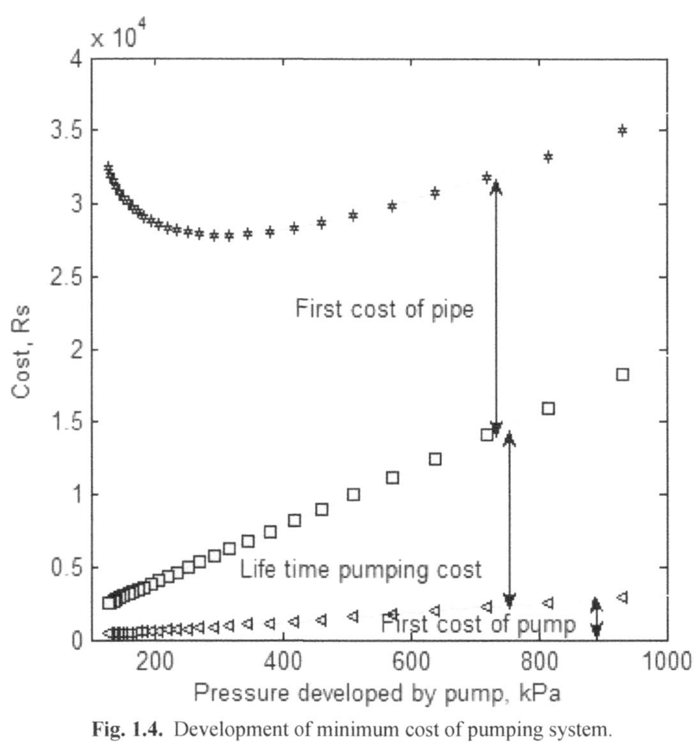

Fig. 1.4. Development of minimum cost of pumping system.

Similarly, the workable system and the optimum system can be differentiated with the example of the use of feedwater heaters in a thermal power plant. The thermal efficiency of a typical thermal power plant increases with an increase in feedwater heater incorporation, both open feedwater heaters and closed feedwater heaters. But the cost also increases with the incorporation of heaters. Studies revealed that with the increase in efficiency, the number of feedwater heaters is diminishes. There is a trade-off between the number of heaters and the total cost of electricity generation. The minimum cost determines the optimum number of heaters. The design of a thermal power plant at a randomly chosen number gives the workable system results. The selection of the right number of heaters for maintaining both thermal efficiency and lower cost gives the optimum system.

A steam-injected gas turbine (STIG) is the latest technology used in a combined cycle power plant, as shown in Fig. 1.5. Some portion of steam from the high-pressure turbine is injected into the combustion chamber. The combustion chamber receives compressed air, steam from the turbine, and fuel. The addition of steam in the combustion chamber increases the mass and energy value of the working fluid that expands in a gas turbine. Therefore, it generates additional electricity. On the other side, the steam injection drops the steam turbine output and increases the latent heat loss in the exhaust gas. Therefore, there is a need to find the optimum amount of steam injection to balance these two (Srinivas et al., 2008).

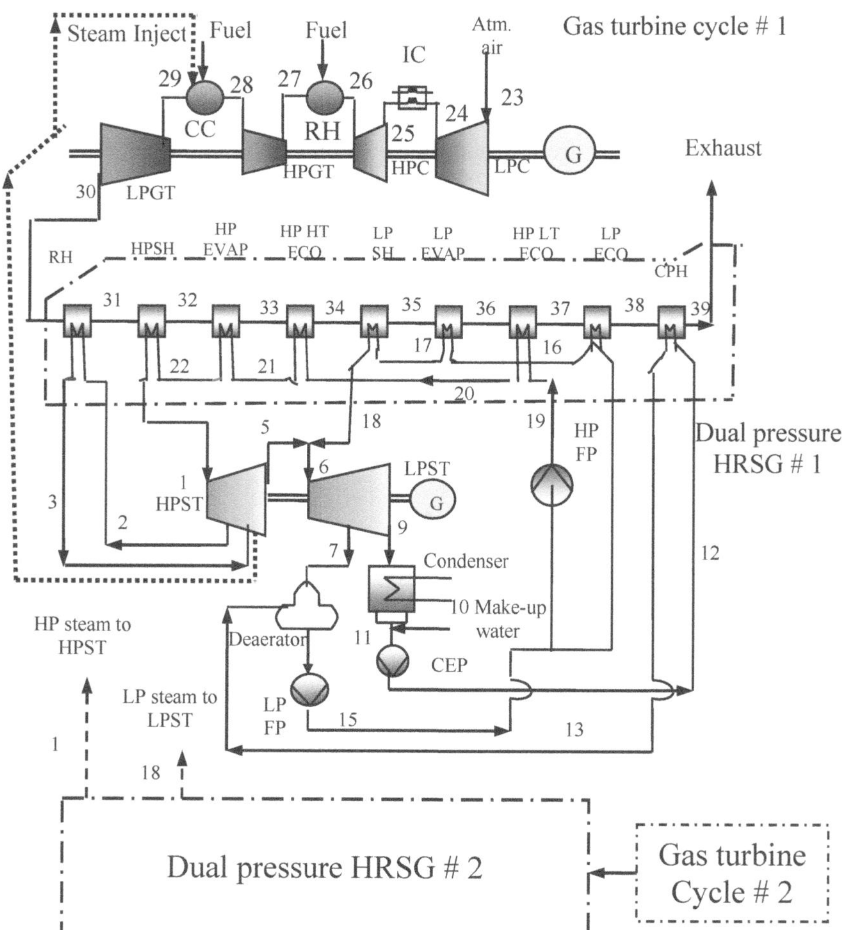

Fig. 1.5. STIG based combined cycle power plant with dual pressure HRSG.

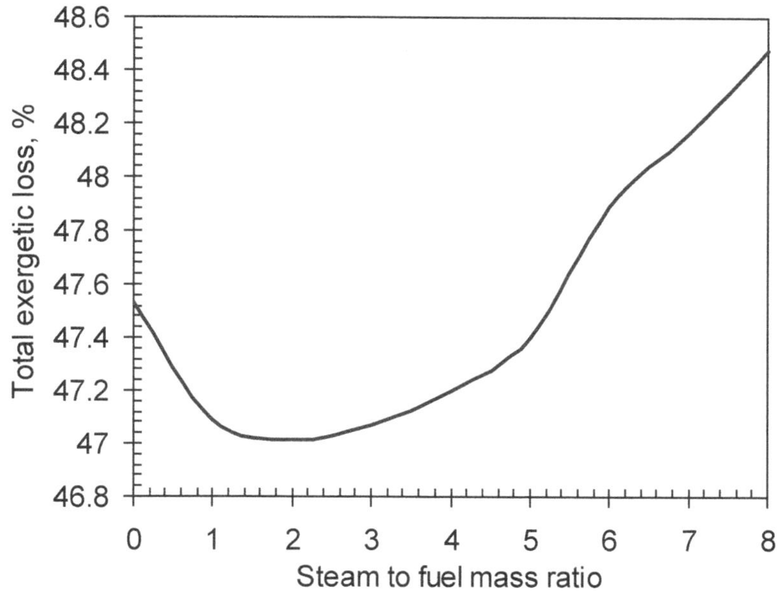

Fig. 1.6. Effect of steam injection mass ratio on total exergetic loss of the combined cycle as a % of exergy input of fuel.

Figure 1.6 shows the total exergetic loss in the combined cycle with the steam injection for different steam injection mass ratios. Steam injection has two effects on the exergetic losses of the cycle components. The exergetic loss of the components for the gas turbine, HRSG, exhaust, and deaerator increases, whereas for the components of the compressor, combustion chamber, steam turbines, and condenser, this loss decreases with an increase in the steam injection. At a low steam injection mass ratio, the decreasing rate of the exergetic loss is greater than the increasing rate. This results in a minimum exergetic loss corresponding to the steam injection mass ratio. It is observed as 2.0 from the plot, which is the optimum steam injection mass ratio.

1.5 Equation fitting

The problem statement or the data can be converted into mathematical equations for further processing. The data may be limited or unlimited. The equation fitting has two categories: exact fit and best fit methods (Fig. 1.7). If a small number of numerical data points are available, an exact fit is suitable, and for a large group of data points, the point with the best fit is suitable. The exact fit, or perfect fit, includes polynomial fit, Lagrange interpolation, etc. The least-squares method is the best-fit solution.

1.5.1 Polynomial representations

Polynomial representation is a simple and easy representation of equation fitting. It can be used for both linear and non-linear equations.

If y is a function of x, then we can write an equation as follows:

$$y = c_1 + c_2x + c_3x^2 + \text{-----} + c_{n+1}x^n \tag{1.1}$$

In this equation, c_1 to c_{n+1} are constants. The value 'n' represents the degree of the equation, which is the highest exponent of x. For example, the mileage of an engine is a function of the speed of the vehicle. A function may also be formed with two variables. An example of a two-variable equation is the real gas equation, in which the pressure is a function of temperature and volume. The

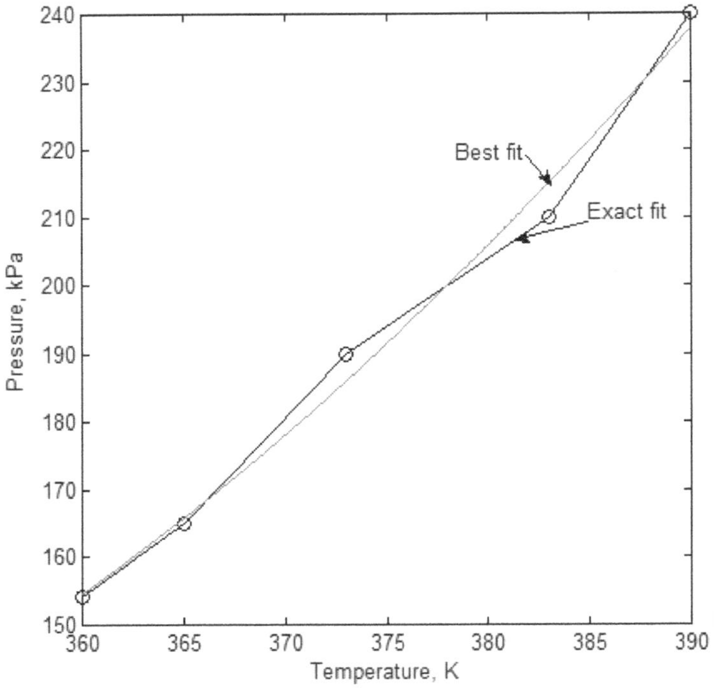

Fig. 1.7. Exact fit method and best fit method for equation development.

number of data points required in the polynomial equation is n + 1. If the data points are more than this, the best fit method is suggested.

For a first-degree equation (linear equation), two data points are sufficient (Fig. 1.8). For example,

$$y = c_1 + c_2x \tag{1.2}$$

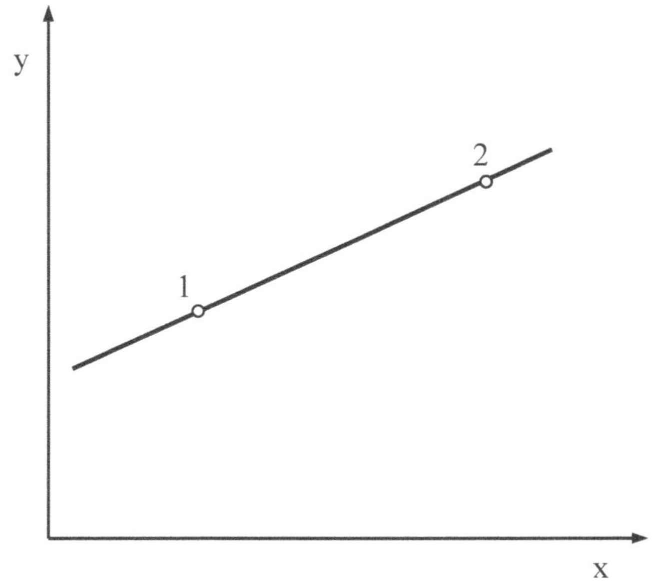

Fig. 1.8. A linear equation with two states (points).

To solve the constants c_1 and c_2, the two known points, (x_1, y_1) and (x_2, y_2) have to be substitutes in the equation.

$$y_1 = c_1 + c_2 x_1 \quad (1.3)$$

$$y_2 = c_1 + c_2 x_2 \quad (1.4)$$

To solve a quadratic equation or second degree equation, we need three points (Fig. 1.9).

$$y = c_1 + c_2 x + c_3 x^2 \quad (1.5)$$

The resultant three equations after substituting these three points are

$$y_1 = c_1 + c_2 x_1 + c_3 x_1^2 \quad (1.6)$$

$$y_2 = c_1 + c_2 x_2 + c_3 x_2^2 \quad (1.7)$$

$$y_3 = c_1 + c_2 x_3 + c_3 x_3^2 \quad (1.8)$$

$$\begin{bmatrix} 1 & x_1 & x_1^2 \\ 1 & x_2 & x_2^2 \\ 1 & x_3 & x_3^2 \end{bmatrix} \begin{bmatrix} a_1 \\ a_2 \\ a_3 \end{bmatrix} = \begin{bmatrix} y_1 \\ y_2 \\ y_3 \end{bmatrix} \quad (1.9)$$

After solving these three simultaneous equations, the constants c_1, c_2 and c_3 will result.

The value of the coefficient decreases from c_1 to c_3 in this quadratic equation (Eq. 1.5). For a large degree equation, the coefficients of higher degree terms are too small. To avoid the difficulty of handling too small values, it is a good practice to write the equation in the following style:

$$y = c_1 + c_2 x + \ldots + c_6 x^5 + c_7 x^6 \quad (1.10)$$

$$y = c_1 + c_2 \frac{x}{100} + \ldots + c_6 \left(\frac{x}{100}\right)^5 + c_{76} \left(\frac{x}{100}\right)^6 \quad (1.11)$$

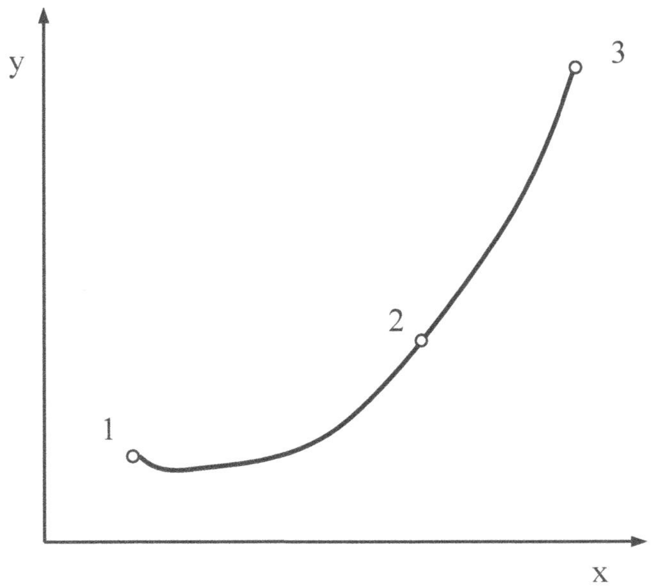

Fig. 1.9. A second degree nonlinear equation with three points.

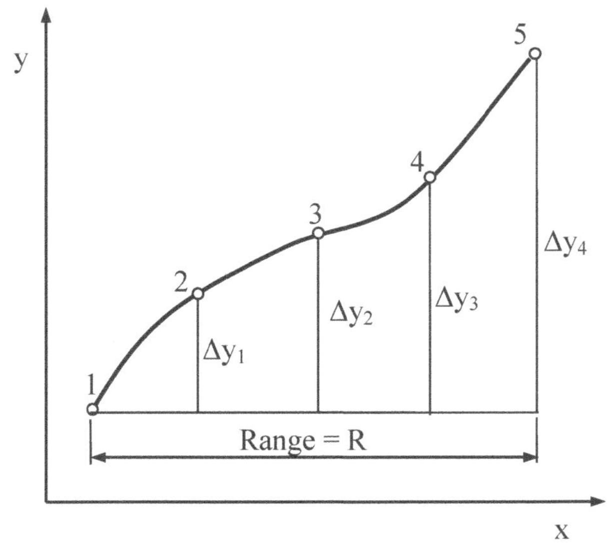

Fig. 1.10. Nonlinear equation with equal spacing in x axis.

If the change in the x axis is equal, the problem may be simplified with further substitution as follows (Fig. 1.10).

For example, if the solar radiation, I (W/m^2) is the function of time, (s), and the equation is expressed at the fourth degree level, and it needs five points. The time period is constant, i.e., $t_2 - t_1 = t_3 - t_2 = t_4 - t_3 = t_5 - t_4$.

The range of t, $t_5 - t_1$, is R, and the solar radiations are $\Delta I_1 = I_2 - I_1$, $\Delta I_2 = I_3 - I_1$, etc.

The nonlinear equation can be expressed with the change in I and t values as follows

$$I - I_1 = c_1\left[\frac{n}{R}(t-t_1)\right] + c_2\left[\frac{n}{R}(t-t_1)\right]^2 + c_3\left[\frac{n}{R}(t-t_1)\right]^3 + c_4\left[\frac{n}{R}(t-t_1)\right]^4 \qquad (1.12)$$

To find c_1 to c_4, first substitute the (t_2, I_2) pair in the above equation.

$$\Delta I_1 = c_1\frac{4(t_2-t_1)}{R} + c_2\left[\frac{4(t_2-t_1)}{R}\right]^2 + c_3\left[\frac{4(t_2-t_1)}{R}\right]^3 + a_4\left[\frac{4(t_2-t_1)}{R}\right]^4 \qquad (1.13)$$

Since the space is uniform in time, n $(t_2 - t_1)/R = 1$, Therefore

$$\Delta I_1 = c_1 + c_2 + c_3 + c_4 \qquad (1.14)$$

Similarly, with (t_3, I_3) point, n $(t_3 - t_1)/R = 2$,

$$\Delta I_2 = 2c_1 + 4c_2 + 8c_3 + 16c_4 \qquad (1.15)$$

and at (t_4, I_4) and (t_5, I_5)

$$\Delta I_3 = 3c_1 + 9c_2 + 27c_3 + 81c_4 \qquad (1.16)$$

$$\Delta I_4 = 4c_1 + 16c_2 + 64c_3 + 256c_4 \qquad (1.17)$$

The coefficients, c_1 to c_4 can be determined from these four equations arranged in the following Table 1.1 with different equations.

Table 1.1. Constants in equation.

Equation	c_4	c_3	c_2	c_1
Fourth degree	$\frac{1}{24}\begin{pmatrix}\Delta I_4 + 4\Delta I_3 \\ +6\Delta I_2 - 4\Delta I_2\end{pmatrix}$	$\frac{\Delta I_4}{6} - \frac{\Delta I_2}{2}$ $+\frac{\Delta I_1}{2} - 6c_4$	$\frac{\Delta I_2}{2} - \Delta I_1$ $-3c_3 - 7c_4$	$\Delta I_1 - c_2 - c_3$ $- c_4$
Cubic		$\frac{1}{6}\begin{pmatrix}3\Delta I_1 + \Delta I_3 \\ -4\Delta I_2\end{pmatrix}$	$\frac{1}{2}(\Delta I_2 - 2\Delta I_1)$ $-3c_2$	$\Delta I_1 - c_2 - c_3$
Quadratic			$\frac{1}{2}(\Delta I_2 - 2\Delta I_1)$	$\Delta I_1 - c_2$
Linear				ΔI_1

In the above equations, the number of variables is one. But most of the engineering solutions consist of multiple variables. If z is a function of x and y, i.e., z = f(x, y), as shown in Fig. 1.11, in polynomial form, the relation between x, y, and z can be expressed as follows:

$$z = a + bx \tag{1.18}$$

where a, b and c are the function of y. Therefore, a and b can be expanded as follows.

$$z = (A_1 + A_2y) + (B_1 + B_2y)x \tag{1.19}$$

at y_1

$$z_{1,1} = a_1 + b_1x_1 \quad \text{at } x_1 \tag{1.20}$$

$$z_{2,1} = a_1 + b_1x_2 \quad \text{at } x_2 \tag{1.21}$$

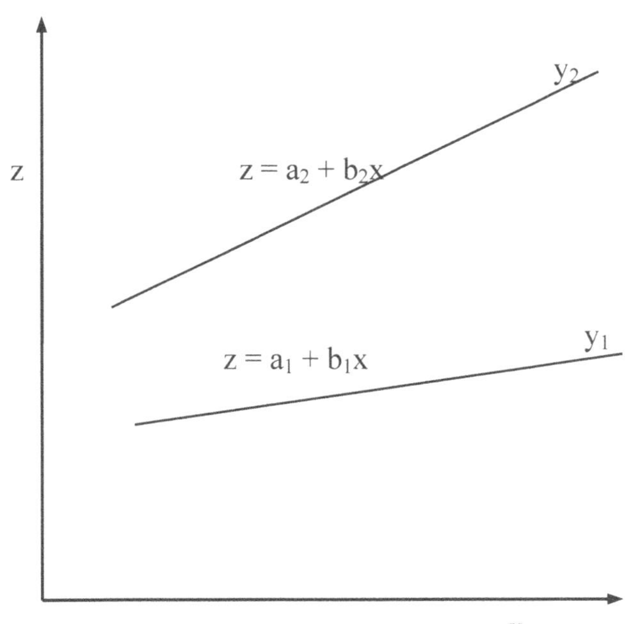

Fig. 1.11. z is the function of x and y.

The solution for a_1 and b_1 is

$$\begin{bmatrix} 1 & x_1 \\ 1 & x_2 \end{bmatrix} \times \begin{bmatrix} a_1 \\ b_1 \end{bmatrix} = \begin{bmatrix} z_{1,\,1} \\ z_{2,\,1} \end{bmatrix} \tag{1.22}$$

Substituting x and y points and solving, a_1, and b_1 can be determined

Therefore,

$$z = a_1 + b_1 x \tag{1.23}$$

Similarly,

For y_2

$$z_{1,2} = a_2 + b_2 x_1 \qquad \text{at } x_1 \tag{1.24}$$

$$z_{2,2} = a_2 + b_2 x_2 \qquad \text{at } x_2 \tag{1.25}$$

The solution for a_2, and b_2 is

$$\begin{bmatrix} 1 & x_1 \\ 1 & x_2 \end{bmatrix} \times \begin{bmatrix} a_2 \\ b_2 \end{bmatrix} = \begin{bmatrix} z_{1,\,2} \\ z_{2,\,2} \end{bmatrix} \tag{1.26}$$

Substituting x and y points and solving, a_2 and b_2 can be determined.

$$z = a_2 + b_2 x \tag{1.27}$$

Next, the constant terms a_1 and a_2 can be expressed by a second-degree equation of y,

at x_1,

$$a = A_1 + A_2 y \tag{1.28}$$

$$a_1 = A_1 + A_2 y_1 \qquad \text{at } y_1 \tag{1.29}$$

$$a_2 = A_1 + A_2\, y_2 \qquad \text{at } y_2 \tag{1.30}$$

the solution for A_0 and A_1 is

$$\begin{bmatrix} 1 & y_1 \\ 1 & y_2 \end{bmatrix} \times \begin{bmatrix} A_1 \\ A_2 \end{bmatrix} = \begin{bmatrix} a_1 \\ a_2 \end{bmatrix} \tag{1.31}$$

Substituting the points of y and a_i and solving, A_0 and A_1 can be determined.

The coefficients of x can also be expressed by equations in terms of the y.

at x_2,

$$b = B_1 + B_2 y \tag{1.32}$$

$$b_1 = B_1 + B_2 y_1 \qquad \text{at } y_1 \tag{1.33}$$

$$b_2 = B_1 + B_2 y_2 \qquad \text{at } y_2 \tag{1.34}$$

The solution for B_1 and B_2 is

$$\begin{bmatrix} 1 & y_1 \\ 1 & y_2 \end{bmatrix} \times \begin{bmatrix} B_1 \\ B_2 \end{bmatrix} = \begin{bmatrix} b_1 \\ b_2 \end{bmatrix} \tag{1.35}$$

Substituting the points of y and b_i and solving, B_1, and B_2 can be determined

Then provide the complete equation

$z = (A_1 + A_2y) + (B_1 + B_2y)x$

Similarly, the following section describes the three-degree-level solution with the polynomial method for two variables.

A performance variable of a component is often a function of two other variables, not just one. If z is a function of x and y, i.e., z = f(x, y), as shown in Fig. 1.12, in polynomial form, the relation between x, y, and z can be expressed as follows:

$$z = a + bx + cx^2 \tag{1.36}$$

where a, b and c are the function of y. Therefore, a, b and c can be expanded as follows.

$$z = (A_1 + A_2y + A_3y^2) + (B_1 + B_2y + B_3y^2)x + (C_1 + C_2y + C_3y^2)x^2 \tag{1.37}$$

at y_1

$$z_{1,1} = a_1 + b_1x_1 + c_1x_1^2 \quad \text{at } x_1 \tag{1.38}$$

$$z_{2,1} = a_1 + b_1x_2 + c_1x_2^2 \quad \text{at } x_2 \tag{1.39}$$

$$z_{3,1} = a_1 + b_1x_3 + c_1x_3^2 \quad \text{at } x_3 \tag{1.40}$$

The solution for a_1, b_1 and c_1 is

$$\begin{bmatrix} 1 & x_1 & x_1^2 \\ 1 & x_2 & x_2^2 \\ 1 & x_3 & x_3^2 \end{bmatrix} \times \begin{bmatrix} a_1 \\ b_1 \\ c_1 \end{bmatrix} = \begin{bmatrix} z_{1,1} \\ z_{2,1} \\ z_{3,1} \end{bmatrix} \tag{1.41}$$

Substituting x and y points and solving, a_1, b_1 and c_1 can be determined.

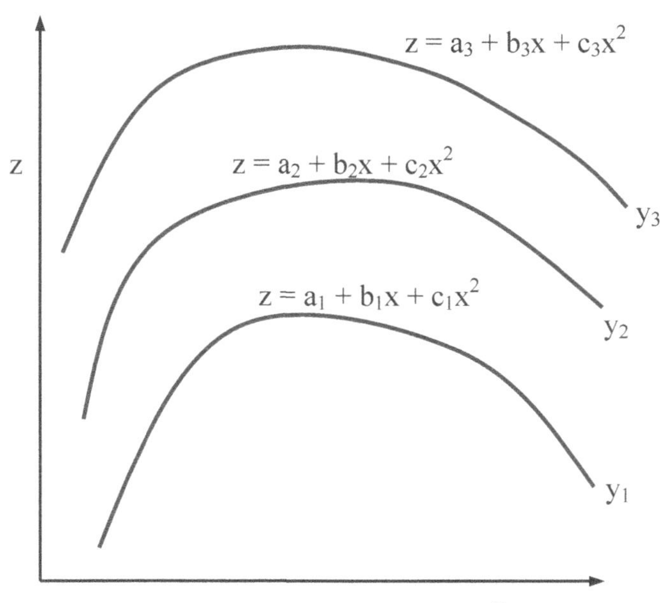

Fig. 1.12. z is the function of x and y.

Therefore, $z = a_1 + b_1x + c_1x^2$ (1.42)

Similarly,

For y_2

$z_{1,2} = a_2 + b_2x_1 + c_2x_1^2$ at x_1 (1.43)

$z_{2,2} = a_2 + b_2x_2 + c_2x_2^2$ at x_2 (1.44)

$z_{3,2} = a_2 + b_2x_3 + c_2x_3^2$ at x_3 (1.45)

The solution for a_2, b_2 and c_2 is

$$\begin{bmatrix} 1 & x_1 & x_1^2 \\ 1 & x_2 & x_2^2 \\ 1 & x_3 & x_3^2 \end{bmatrix} \times \begin{bmatrix} a_2 \\ b_2 \\ c_2 \end{bmatrix} = \begin{bmatrix} z_{1,2} \\ z_{2,2} \\ z_{3,2} \end{bmatrix} \quad (1.46)$$

Substituting x and y points and solving, a_2, b_2 and c_2 can be determined.

$z = a_2 + b_2x + c_2\,x^2$ (1.47)

and for y_3

$z_{1,3} = a_3 + b_3x_1 + c_3x_1^2$ at x_1 (1.48)

$z_{2,3} = a_3 + b_3x_2 + c_3x_2^2$ at x_2 (1.49)

$z_{3,3} = a_3 + b_3x_3 + c_3x_3^2$ at x_3 (1.50)

the solution for a_3, b_3 and c_3 is

$$\begin{bmatrix} 1 & x_1 & x_1^2 \\ 1 & x_2 & x_2^2 \\ 1 & x_3 & x_3^2 \end{bmatrix} \times \begin{bmatrix} a_3 \\ b_3 \\ c_3 \end{bmatrix} = \begin{bmatrix} z_{1,3} \\ z_{2,3} \\ z_{3,3} \end{bmatrix} \quad (1.51)$$

Substituting x and y points and solving, a_3, b_3 and c_3 can be determined.

$z = a_3 + b_3x + c_3\,x^2$ (1.52)

Next, the constant terms a_1, a_2 and a_3 can be expressed by a second-degree equation of y, at x_1,

$a = A_1 + A_2y + A_3y^2$ (1.53)

$a_1 = A_1 + A_2y_1 + A_3y_1^2$ at y_1 (1.54)

$a_2 = A_1 + A_2y_2 + A_3y_2^2$ at y_2 (1.55)

$a_3 = A_1 + A_2\,y_3 + A_3y_3^2$ at y_3 (1.56)

the solution for A_1, A_2 and A_3 is

$$\begin{bmatrix} 1 & y_1 & y_1^2 \\ 1 & y_2 & y_2^2 \\ 1 & y_3 & y_3^2 \end{bmatrix} \times \begin{bmatrix} A_1 \\ A_2 \\ A_3 \end{bmatrix} = \begin{bmatrix} a_1 \\ a_2 \\ a_3 \end{bmatrix} \quad (1.57)$$

Substituting the points of y and a_i and solving, A_1, A_2 and A_3 can be determined.

Therefore, $a = A_1 + A_2y + A_3y^2$

The coefficients of x and x^2 can also be expressed by equations in terms of the y.

at x_2,

$$b = B_1 + B_2y + B_3y^2 \tag{1.58}$$

$$b_1 = B_1 + B_2y_1 + B_3y_1^2 \quad \text{at } y_1 \tag{1.59}$$

$$b_2 = B_1 + B_2y_2 + B_3y_2^2 \quad \text{at } y_2 \tag{1.60}$$

$$b_3 = B_1 + B_2y_3 + B_3y_3^2 \quad \text{at } y_3 \tag{1.61}$$

the solution for B_1, B_2 and B_3 is

$$\begin{bmatrix} 1 & y_1 & y_1^2 \\ 1 & y_2 & y_2^2 \\ 1 & y_3 & y_3^2 \end{bmatrix} \times \begin{bmatrix} B_1 \\ B_2 \\ B_3 \end{bmatrix} = \begin{bmatrix} b_1 \\ b_2 \\ b_3 \end{bmatrix} \tag{1.62}$$

Substituting the points of y and b_i and solving, B_1, B_2 and B_3 can be determined.

Therefore, $b = B_1 + B_2y + B_3y^2$

at x_3,

$$c = C_1 + C_2y + C_3y^2 \tag{1.63}$$

$$c_1 = C_1 + C_2y_1 + C_3y_1^2 \quad \text{at } y_1 \tag{1.64}$$

$$c_2 = C_1 + C_2y_2 + C_3y_2^2 \quad \text{at } y_2 \tag{1.65}$$

$$c_3 = C_1 + C_2y_3 + C_3y_3^2 \quad \text{at } y_3 \tag{1.66}$$

The solution for C_1, C_2 and C_3 is

$$\begin{bmatrix} 1 & y_1 & y_1^2 \\ 1 & y_2 & y_2^2 \\ 1 & y_3 & y_3^2 \end{bmatrix} \times \begin{bmatrix} C_1 \\ C_2 \\ C_3 \end{bmatrix} = \begin{bmatrix} c_1 \\ c_2 \\ c_3 \end{bmatrix} \tag{1.67}$$

Substituting the points of y and c_i and solving, C_1, C_2 and C_3 can be determined.

Therefore, $c = C_1 + C_2y + C_3y^2$

Then the complete equation is

$$z = (A_1 + A_2y + A_3y^2) + (B_1 + B_2y + B_3y^2)x + (C_1 + C_2y + C_3y^2)x^2$$

The A, B and C constants can be computed if nine data points from Fig. 1.12 are available.

1.5.2 Equation fitting with Lagrange interpolation

For arbitrary spacing, alternatively, we can also use the Lagrange interpolation method to find the coefficients without solving the simultaneous equation. The solution is compact compared to the others. But this process is complex compared to the earlier outlined polynomial method. But the complexity is not a big issue due to the development of recent software solutions and programming.

The research or project results can be stored in the form of a correlation equation, which can be used to reproduce the results without repeating the complex simulation or tedious experimental work. The correlation equation can also be used to generate results for any value of inputs that lies in the specified range. Further processes such as simulation, analysis, and optimization, become easier.

Let us consider a quadratic equation,

$$y = c_1 + c_2x + c_3x^2 \tag{1.68}$$

This equation can be rewritten for the Lagrange interpolation as per the figure (Fig. 1.13) thus:

$$y = c_1(x - x_2)(x - x_3) + c_2(x - x_1)(x - x_3) + c_3(x - x_1)(x - x_2) \tag{1.69}$$

The second degree equation needs three points viz. (x_1, y_1), (x_2, y_2) and (x_3, y_3).

To solve the coefficients, substituting (x_1, y_1),

$$y_1 = c_1(x_1 - x_2)(x_1 - x_3)$$

$$c_1 = \frac{y_1}{(x_1 - x_2)(x_1 - x_3)} \tag{1.70}$$

At (x_2, y_2),

$$y_2 = c_2(x_2 - x_1)(x_2 - x_3)$$

$$c_2 = \frac{y_2}{(x_2 - x_1)(x_2 - x_3)} \tag{1.71}$$

Similarly, at (x_3, y_3),

$$y_3 = c_3(x_3 - x_1)(x_3 - x_2)$$

$$c_3 = \frac{y_3}{(x_3 - x_1)(x_3 - x_2)} \tag{1.72}$$

Substituting

$$y = \frac{(x - x_2)(x - x_3)}{(x_1 - x_2)(x_1 - x_3)} y_1 + \frac{(x - x_1)(x - x_3)}{(x_2 - x_1)(x_2 - x_3)} y_2 + \frac{(x - x_1)(x - x_2)}{(x_3 - x_1)(x_3 - x_2)} y_3 \tag{1.73}$$

This equation can be generalized for a n–1 degree as per the following

$$y = \sum_{i=1}^{n} y_i \prod_{j=1}^{n} \frac{(x - x_j) \text{ omitting } (x - x_i)}{(x_i - x_j) \text{ omitting } (x_i - x_i)} \tag{1.74}$$

The symbol pi represents the multiplication.

Lagrange interpolation can be extended to a function of two independent variables. For example, if $z = f(x, y)$, the form can be chosen

$$\begin{aligned} z &= c_{11}(x - x_2)(x - x_3)(y - y_2)(y - y_3) \\ &+ c_{12}(x - x_2)(x - x_3)(y - y_1)(y - y_3) \\ &+ c_{13}(x - x_2)(x - x_3)(y - y_1)(y - y_2) \\ &+ \ldots \\ &+ c_{33}(x - x_1)(x - x_2)(y - y_1)(y - y_2) \end{aligned} \tag{1.75}$$

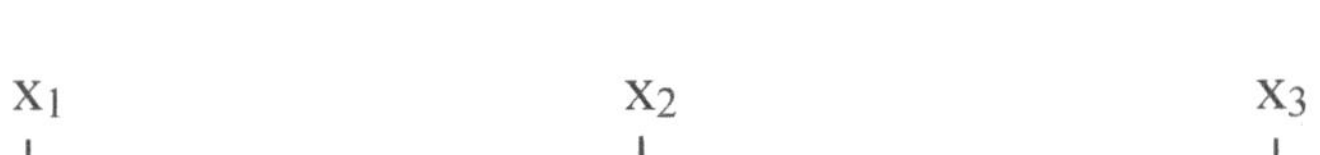

Fig. 1.13. Finding coefficients from the 'x' terms.

1.5.3 Best fit: method of least squares

For a larger number of points, such as an experimental observation, a best-fit method using least squares can be used. In the best fit, the sum of the absolute values of the deviations from the data points is a minimum. In another form of least squares, the sum of the squares of the deviation is a minimum (Fig. 1.14).

For a linear equation,

$$y = c_1 + c_2 x \tag{1.76}$$

If m number of points are known, then the points are (x_1, y_1), (x_1, y_2), -----, (x_m, y_m). The difference between the data point and the equation is $c_1 + c_2 x_i - y_i$.

We can determine the coefficients, c_1 and c_2 to get a minimum deviation

$$\sum_{i=1}^{m}\left(c_1 + c_2 x_i - y_i\right)^2 \rightarrow \min imum \tag{1.77}$$

The partial derivative with respect to c_1 and c_2 and equating to zero result the coefficients.

$$\frac{\partial \sum_{i=1}^{m}\left(c_1 + c_2 x_i - y_i\right)^2}{\partial c_1} = \sum_{i=1}^{m} 2\left(c_1 + c_2 x_i - y_i\right) = 0 \tag{1.78}$$

$$and \quad \frac{\partial \sum_{i=1}^{m}\left(c_1 + c_2 x_i - y_i\right)^2}{\partial c_2} = \sum_{i=1}^{m} 2\left(c_1 + c_2 x_i - y_i\right)x_i = 0 \tag{1.79}$$

Dividing by 2 and separating the above two equations into individual terms gives

$$mc_1 + c_2 \sum x_i = \sum y_i \tag{1.80}$$

$$c_1 \sum x_i + c_2 \sum x_i^2 = \sum x_i y_i \tag{1.81}$$

where m is the number data points.

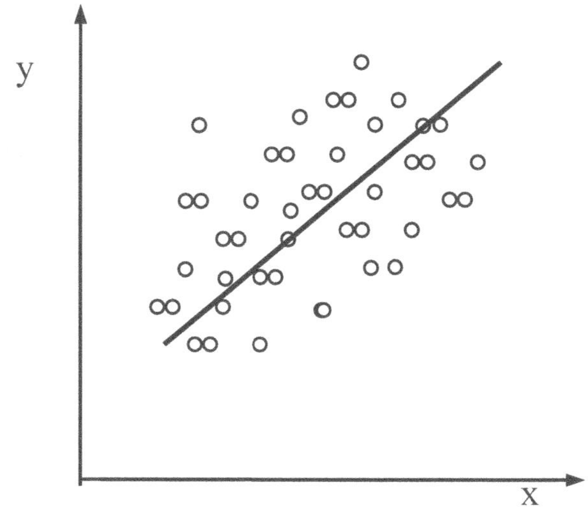

Fig. 1.14. Least square method.

To solve c_1 and c_2, the above two equations can be written in a matrix form

$$\begin{bmatrix} m & \sum x_i \\ \sum x_i & \sum x_i^2 \end{bmatrix} \times \begin{bmatrix} c_1 \\ c_2 \end{bmatrix} = \begin{bmatrix} \sum y_i \\ \sum x_i y_i \end{bmatrix}$$

For non-linear problems, the same procedure will be followed.

$$y = c_1 + c_2\, x + c_3\, x^2 \tag{1.82}$$

For the 'm' number of states, the total can be minimized as

$$\sum_{i=1}^{m} \left(a + bx_i + cx_i^2 - y_i \right)^2 \rightarrow \min imum \tag{1.83}$$

To minimize this function, a partial differentiation with respect to a, b, and c generates three equations, which can be expressed in matrix form.

$$\frac{\partial \sum_{i=1}^{m} \left(c_1 + c_2 x_i + c_3 x_i^2 - y_i \right)^2}{\partial c_1} = \sum 2\left(c_1 + c_2 x_i + c_3 x_i^2 - y_i \right) = 0 \tag{1.84}$$

$$mc_1 + c_2 \sum x_i + c_3 \sum x_i^2 = \sum y_i \tag{1.85}$$

$$\frac{\partial \sum_{i=1}^{m} \left(c_1 + c_2 x_i + c_3 x_i^2 - y_i \right)^2}{\partial c_2} = \sum 2\left(c_1 + c_2 x_i + c_3 x_i^2 - y_i \right) x_i = 0 \tag{1.86}$$

$$c_1 \sum x_i + c_2 \sum x_i^2 + c_3 \sum x_i^3 = \sum x_i y_i$$

$$\frac{\partial \sum_{i=1}^{m} \left(c_1 + c_2 x_i + c_3 x_i^2 - y_i \right)^2}{\partial c_3} = \sum 2\left(c_1 + c_2 x_i + c_3 x_i^2 - y_i \right) x_i^2 = 0 \tag{1.87}$$

$$c_1 \sum x_i^2 + c_2 \sum x_i^3 + c_3 \sum x_i^4 = \sum x_i^2 y_i \tag{1.88}$$

In matrix form,

$$\begin{bmatrix} m & \sum x_i & \sum x_i^2 \\ \sum x_i & \sum x_i^2 & \sum x_i^3 \\ \sum x_i^2 & \sum x_i^3 & \sum x_i^4 \end{bmatrix} \times \begin{bmatrix} c_1 \\ c_2 \\ c_2 \end{bmatrix} = \begin{bmatrix} \sum y_i \\ \sum x_i y_i \\ \sum x_i^2 y_i \end{bmatrix} \tag{1.89}$$

1.6 Simulation

After the completion of the model, it has to be applied to get results. The determination of variables using the model is called simulation. To solve a typical thermal system problem, it needs the thermodynamic and thermophysical properties of the fluids, materials, etc. The performance characteristics of the components are also part of modeling and simulation. In the analysis, the performance or conditions of the system will be analyzed with a change in parameters.

1.6.1 Information flow diagrams

To develop a computer program, an algorithm and flow chart play an important role. In a MATLAB Simulink, the solution consists of the flow diagram. In the simulation, the problem solution can also be expressed in the information flow diagram for easy modeling and simulation. If two components are arranged in series, the results of the first component are supplied as an input to the second component (Fig. 1.15a, b).

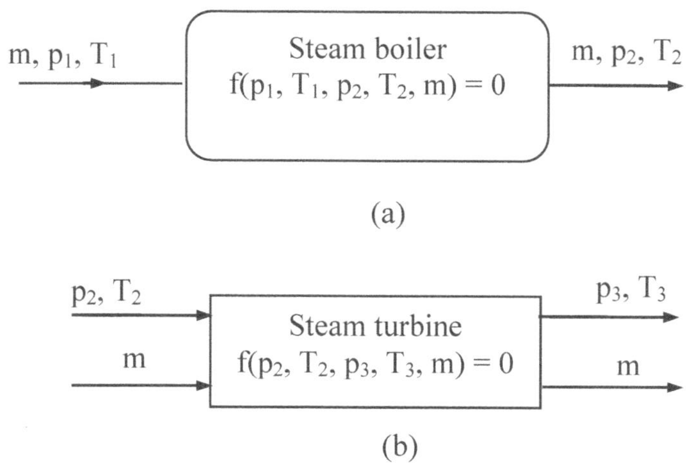

Fig. 1.15. (a) A steam boiler and (b) a steam turbine for information flow diagram.

For example, in a steam power plant, shown in Fig. 1.16, superheated steam has been supplied at a pressure of 100 bar and 500°C to a steam turbine. It is expanded to a condenser pressure of 0.098 bar. The plant has one deaerator located at optimum pressure. To develop an information flow diagram to determine the energy efficiency of the regenerated Rankine cycle by neglecting the pump works and isentropic expansion in the steam turbine, the following formulations can be developed:

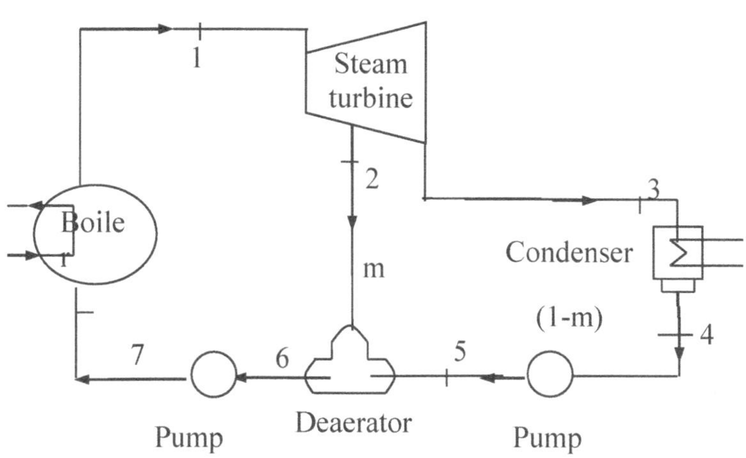

Fig. 1.16. Steam power plant with a deaerator.

$$T_{sat\,2} = T_{sat\,3} + \theta_{deaerator}\,(T_{sat\,1} - T_{sat\,3}) \tag{1.90}$$

Find h_2

Find h_3

From energy balance in deaerator,

$$m = \frac{h_6 - h_5}{h_2 - h_5} \tag{1.91}$$

$$\eta_1 = \frac{(h_1 - h_2) + (1 - m)(h_2 - h_3)}{h_1 - h_6} \tag{1.92}$$

Figure 1.17 shows the information flow diagram for the performance calculation of Rankine cycle.

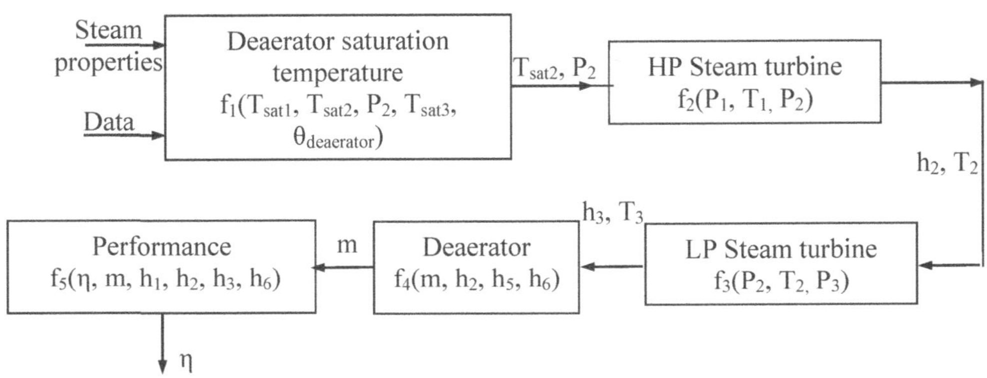

Fig. 1.17. Information flow diagram for the steam power plant.

1.6.2 Successive substitution

In the successive substitution, initially a guess is required to start the iteration. After the completion of the iteration, the final answer has to be the initially guessed variable. The next iteration can be started if it is not this variable. In the next iteration, the initial guess will be substituted, guided by the preceding result. This method is called successive substitution. The iterations will be continued until convergence is reached. The successive substitution method is simple to use and easy to arrive at a solution. But the drawback is that it may not converge for certain cases of initial guess. And it also needs a large number of iterations to reach convergence.

For example, Fig. 1.18 shows a simultaneous heat transfer from the surface of a body through convection and radiation. The temperature difference is shown as ΔT, the difference between the final temperature and the initial temperature. Let us consider Q1, Q2, and Q as convection heat transfer, radiation heat transfer, and total heat transfer, respectively. ΔT is the temperature difference between two bodies where the combined heat transfers from 1 to 2. Following are the formulations required to solve the problem using successive substitution.

$\Delta T = 4\,Q^2 + 200$: Heat transfer and temperature difference
$\Delta T = 400 - 10\,Q_1 - 2\,Q_1^2$: Convection heat transfer
$\Delta T = 450 - 30\,Q_2 - 15\,Q_2^2$: Radiation heat transfer
$Q = Q_1 + Q_2$: Heat balance

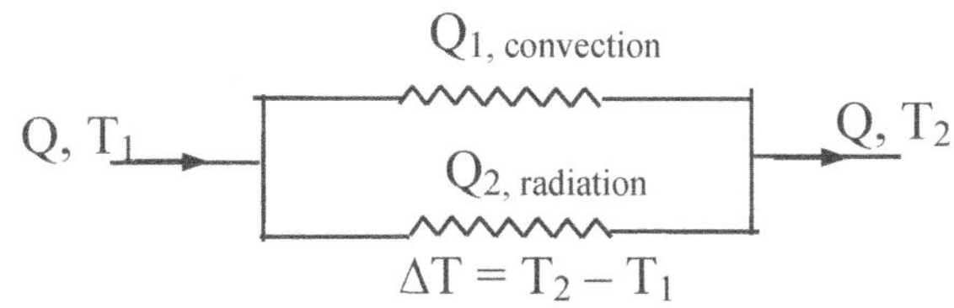

Fig. 1.18. Heat transfer through convection and radiation with a temperature difference between to items.

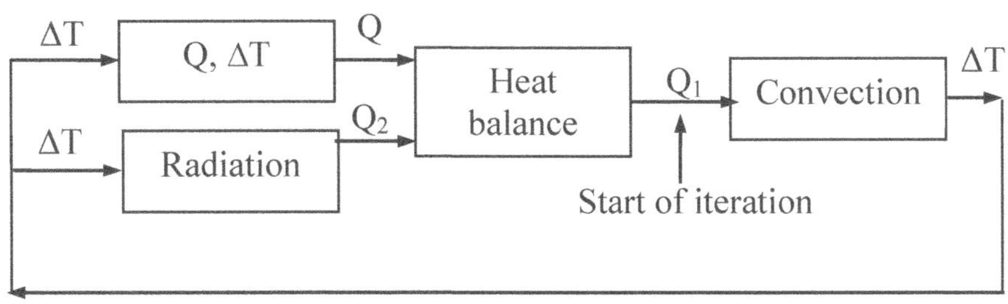

Fig. 1.19. Information flow diagram for option 1 with the initial assumption of Q_1.

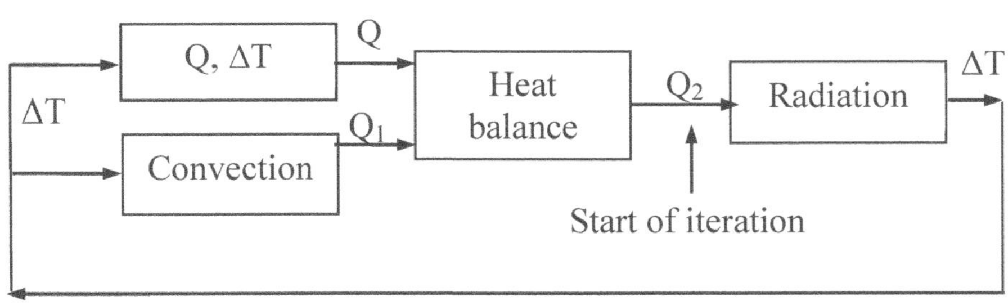

Fig. 1.20. Information flow diagram for option 2.

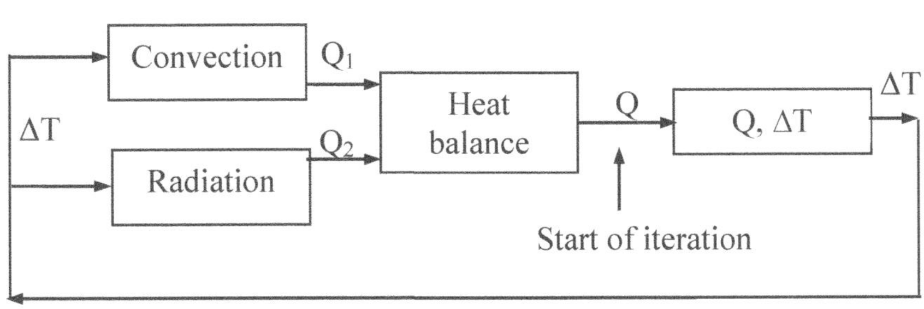

Fig. 1.21. Information flow diagram for option 3.

Figure 1.19 is only one of the possible information flow diagrams that can be generated from the set of equations.

The solution is obtained with option 1 and the results are listed in Table 1.2. In the first option, the problem converges with the given convergence of 0.001. The initial guess for Q_1 is 3. The final result is $\Delta T = 333.0600$, $Q_2 = 1.9659$, $Q = 5.7675$, and $Q_1 = 3.8026$. In addition to this, two more flow diagrams are shown in Figs. 1.20 and 1.21.

Table 1.2. Iterations of successive substitution for option 1 with the initial assumption of Q_1 with the convergence of 0.001.

S. No.	ΔT	Q_2	Q	Q_1
1	352.0000	1.7447	6.1644	**3.0000**
2	316.7300	2.1439	5.4022	4.4197
3	346.1800	1.8144	6.0453	3.2583
4	321.8900	2.0888	5.5202	4.2309
5	342.1400	1.8620	5.9611	3.4314
6	325.4000	2.0506	5.5992	4.0991
7	339.3300	1.8945	5.9019	3.5485
8	327.8100	2.0243	5.6526	4.0074
9	337.3900	1.9168	5.8606	3.6283
10	329.4500	2.0061	5.6889	3.9439
11	336.0500	1.9321	5.8319	3.6828
12	330.5800	1.9935	5.7136	3.8999
13	335.1200	1.9426	5.8121	3.7201
14	331.3600	1.9849	5.7306	3.8695
15	334.4800	1.9498	5.7983	3.7457
16	331.8900	1.9789	5.7422	3.8485
17	334.0400	1.9548	5.7888	3.7633
18	332.2600	1.9748	5.7502	3.8341
19	333.7400	1.9582	5.7823	3.7754
20	332.5100	1.9720	5.7557	3.8241
21	333.5300	1.9605	5.7778	3.7837
22	332.6900	1.9700	5.7595	3.8172
23	333.3900	1.9621	5.7746	3.7894
24	332.8000	1.9687	5.7620	3.8125
25	333.2900	1.9633	5.7725	3.7934
26	332.8900	1.9678	5.7638	3.8092
27	333.2200	1.9640	5.7710	3.7961
28	332.9400	1.9671	5.7651	3.8070
29	333.1700	1.9645	5.7700	3.7979
30	332.9800	1.9667	5.7659	3.8055
31	333.1400	1.9649	5.7693	3.7992
32	333.0100	1.9664	5.7665	3.8044
33	333.1200	1.9652	5.7688	3.8001
34	333.0300	1.9662	5.7669	3.8037
35	333.1000	1.9653	5.7685	3.8007
36	333.0400	1.9660	5.7672	3.8032
37	333.0900	1.9655	5.7683	3.8011
38	333.0500	1.9659	5.7673	3.8028
39	333.0800	1.9655	5.7681	3.8014
40	333.0600	1.9659	5.7675	3.8026

In the second option, a trial value of $Q_2 = 2.0$ has been assumed and the iteration is started as shown in Fig. 1.20. The solution is diverged and the results are tabulated in Table 1.3. In the fifth iteration, an imaginary value resulted, hence there is no convergence.

Finally, in the third option, the successive substitution started with the initial guess of Q as 5, and the order of calculation is shown in Fig. 1.21. But it is also diverged after the completion of the first iteration, as shown in Table 1.4.

Therefore, successive substitutions may not result in a converged value. Therefore, an alternative method, the Newton-Rapson method, is used to get the guaranteed results.

Table 1.3. Iterations of successive substitution for option 2 with the initial assumption of Q_2.

Iteration	ΔT	Q	Q_1	Q_2
1	330.0000	5.7009	3.9226	**2.0000**
2	349.2200	6.1078	3.1250	1.7783
3	227.0600	2.6011	7.1290	2.9828
4	278.3000	4.4245	5.6914	–4.5280
5	463.9300	8.1230	imaginary	–1.2669

Table 1.4. Iterations of successive substitution for option 3 with the initial assumption of Q.

Iteration	ΔT	Q_1	Q_2	Q
1	300.0000	5.0000	2.3166	5.0000
2	414.1300	imaginary		

1.6.3 Taylor-series expansion

This second technique of system simulation is presented by the Newton-Raphson method, which is developed from Taylor-series expansion. In Taylor-series expansion, if z is a function of x and y and it is to be expanded about a point (a, b), i.e., x = a and y = b,

z = const + first degree terms + second degree terms + higher degree terms.

or more specifically,

$$z = c_0 + [c_1(x - a) + c_2(y - b)] + [c_3(x - a)^2 + c_4(x - a)(y - b) + c_5(y - b)^2] + \ldots \tag{1.93}$$

The coefficients in Eq. 1.93 are c_0, c_1, etc. which need solving. If x = a and y = b, all the terms on the right side of the equation becomes to zero except c_0, so that the value of the function at (a, b) is

$$c_0 = z(a, b)$$

To find c_1, partially differentiate Eq. 1.93 with respect to x; then set x = a, and y = b.

$$\frac{\partial z(a,\ b)}{\partial x} = \frac{\partial}{\partial x}\left(c_1 x - c_1 a + c_2(y - b) + c_3(x - a)^2 + c_4(x - a)(y - b) + c_5(y - b)^2 + \ldots\right)$$

The remaining term on right side of above equation is c_1 so

$$c_1 = \frac{\partial z(a,\ b)}{\partial x} \tag{1.94}$$

In a similar manner,

$$c_2 = \frac{\partial z(a,\ b)}{\partial y} \tag{1.95}$$

The constants c_3, c_4 and c_5 are found by partial differentiation twice followed by substitution of x = a, and y = b.

$$\frac{\partial z}{\partial x} = c_1 + 2c_3(x - a) + c_4(y - b)$$

$$\frac{\partial^2 z}{\partial x^2} = 2c_3$$

$$c_3 = \frac{1}{2}\frac{\partial^2 z(a,\ b)}{\partial x^2}$$

$$\frac{\partial\ z}{\partial x \partial y} = c_4$$

$$c_4 = \frac{\partial^2 z(a,\ b)}{\partial x \partial y}$$

$$\frac{\partial z}{\partial y} = c_2 + c_4(x - a) + c_5 2(y - b)$$

$$\frac{\partial^2 z}{\partial y^2} = 2c_5$$

$$c_5 = \frac{1}{2}\frac{\partial^2 z(a,\ b)}{\partial y^2} \tag{1.96}$$

Therefore, the final equation is

$$z(x,\ y) = z(a,\ b) + \frac{\partial z(a,\ b)}{\partial x}(x-a) + \frac{\partial z(a,\ b)}{\partial y}(y-b) + \frac{1}{2}\frac{\partial^2 z(a,\ b)}{\partial x^2}(x-a)^2$$
$$+ \frac{\partial^2 z(a,\ b)}{\partial x \partial y}(x-a)(y-b) + \frac{1}{2}\frac{\partial^2 z(a,\ b)}{\partial y^2}(y-b)^2$$

For the special case where y is a function of one independent variable x, the Taylor-series expansion about the point x = a is

$$y = y(a) + \frac{dy(a)}{dx}(x-a) + \left[\frac{1}{2}\frac{d^2 y(a)}{dx^2}\right](x-a)^2 + \ldots \tag{1.97}$$

The general expression for the Taylor-series expansion if y is a function of n variables x_1, x_2,... x_n around the point ($x_1 = a_1$, $x_2 = a_2$, …, $x_n = a_n$) is

$$y(x_1,\ x_2, \ldots\ldots, x_n) = y(a_1,\ a_2, \ldots\ldots,\ a_n) + \sum_{j=1}^{n} \frac{\partial y(a_1, \ldots., a_n)}{\partial x_i}(x_j - a_j)$$
$$+ \sum_{i=1}^{n}\sum_{j=1}^{n} \frac{\partial^2 y(a_1, \ldots., a_n)}{\partial x_i \partial x_j}(x_i - a_i)(x_j - a_j) + \ldots. \tag{1.98}$$

1.6.3.1 Newton-Raphson with one equation and one unknown

If y = f(x)

At x = a,

$$y = y(a) + \frac{dy(a)}{dx}(x-a) + \left[\frac{1}{2}\frac{d^2y(a)}{dx^2}\right](x-a)^2 + \ldots$$

In the Taylor series expansion of above equation, when x is close to 'a' the higher order terms become negligible.

The equation can be simplified as

$$y = y(a) + \frac{dy(a)}{dx}(x-a)$$

$$y \approx y(a) + [y'(a)]\,(x-a) \tag{1.99}$$

Equation 1.99 is the basis of the Newton-Rahson iterative technique for solving a nonlinear algebraic equation.

For example, the value of x is sought that satisfies the equation

$$x + 2 = e^x \tag{1.100}$$

Define y as

$$y(x) = x + 2 - e^x \tag{1.101}$$

and denote x_c and the correct value of x that solves Eq. 1.99 and makes y = 0

$$y(x_c) = 0 \tag{1.102}$$

The method needs the initial guess of x. If x_t is the temporary value of x, and substituting this in Eq. 1.102, it results in y, but it is not the desired value of y = 0.

Specifically, if $x_t = 2$,

$$y(x_t) = x_t + 2 - e^{x_t}$$
$$= 2 + 2 - 7.39 = -3.39$$

Our trial value of x is incorrect, but now the question is how the value of x should be changed in order to bring y closer to zero.

Returning to the Taylor expansion of Eq. 1.99, express y in terms of x by expanding x_c

$$y(x) \approx y(x_c) + [y'(x_c)](x - x_c) \tag{1.103}$$

For $x = x_t$, Eq. 1.101 becomes

$$y(x_t) \approx y(x_c) + [y'(x_t)](x_t - x_c) \tag{1.104}$$

Equation 1.104 contains the further approximation of evaluating the derivative at x_t rather than at x_c, because the value of x_c is still unknown.

$y(x_c) = 0$, and so it can be solved approximately for the x_c

$$x_c \approx x_t - \frac{y(x_t)}{y'(x_t)}$$

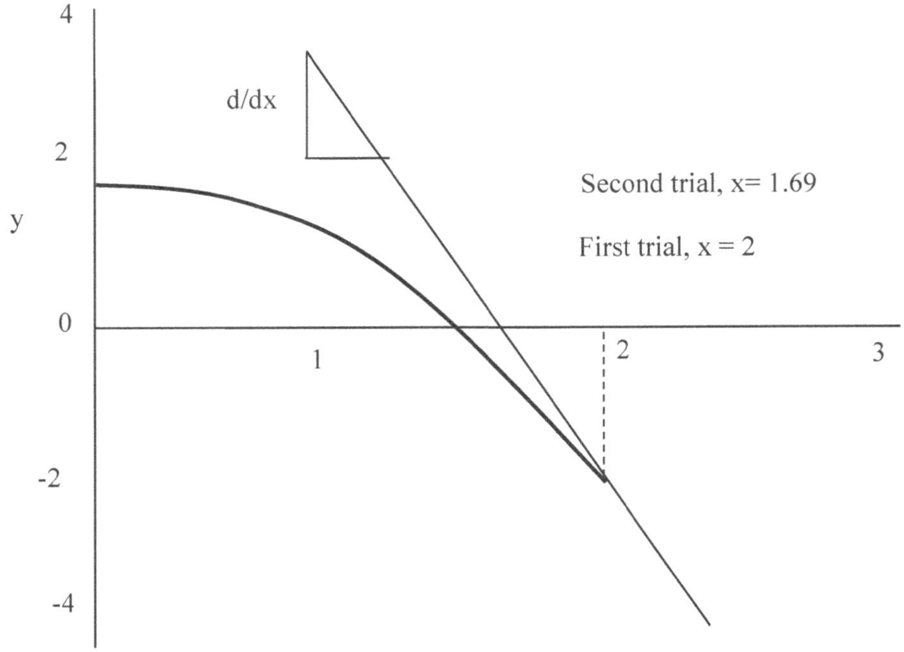

Fig. 1.22. Newton-Raphson iteration.

In the numerical example

$$x_c \approx 2 - \frac{-3.39}{1-e^2} = 1.469$$

The results of x = 1.469 is correct and should be used for the next iteration. The results of the next several iterations are

x	1.469	1.208	1.152
y(x)	–0.876	–0.132	–0.018

From the Fig. 1.22, for the first trial of x = 2, and the deviation of y from zero divided by slope of the curve there suggests a new trial of 1.469.

Alternatively (Balaji, 2021) (Fig. 1.23),

$$f'(x_i) = \frac{f(x_i) - 0}{x_i - x_{i+1}}$$

$$x_i - x_{i+1} = \frac{f(x_i)}{f'(x_i)}$$

$$x_{i+1} = x_i - \frac{f(x_i)}{f'(x_i)}$$

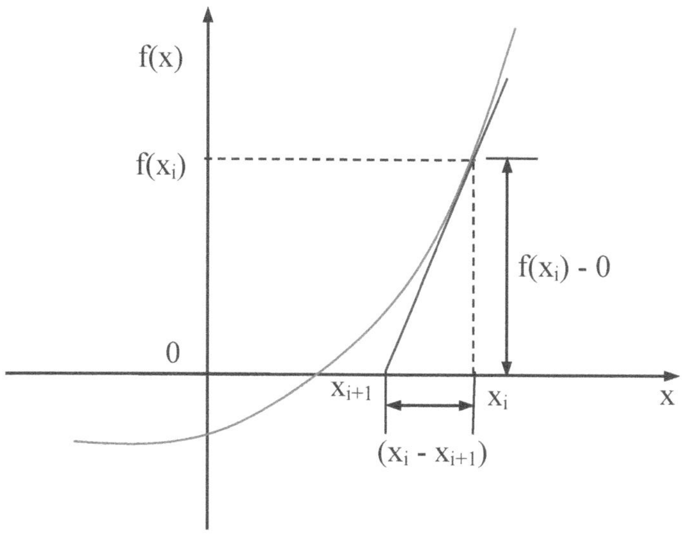

Fig. 1.23. Newton-Raphson method formulation or algorithm.

1.6.3.2 Newton-Raphson method with multiple equations and unknowns

If y is function of x_1 and x_2

$f_1\,(x_1, x_2) = 0$

$f_2\,(x_1, x_2) = 0$

At the beginning, let us assume temporary values of $x_{1,t}$ and $x_{2,t}$. Determine the results of f_1 and f_2 at the temporary values of x_1 and x_2. Calculate the partial derivatives of all functions with respect to all variables. Use the Taylor-series expansion of the form of to establish a set of simultaneous equations.

The Taylor-series expansion equations are

$$f_1(x_{1,t},\ x_{2,t}) \approx f_1(x_{1,c},\ x_{2,c}) + \frac{\partial f_1(x_{1,t},\ x_{2,t})}{\partial x_1}(x_{1,t} - x_{1,c}) + \frac{\partial f_1(x_{1,t},\ x_{2,t})}{\partial x_2}(x_{2,t} - x_{2,c})$$

$$f_2(x_{1,t},\ x_{2,t}) \approx f_2(x_{1,c},\ x_{2,c}) + \frac{\partial f_2(x_{1,t},\ x_{2,t})}{\partial x_1}(x_{1,t} - x_{1,c}) + \frac{\partial f_2(x_{1,t},\ x_{2,t})}{\partial x_2}(x_{2,t} - x_{2,c})$$

The equations in the matrix form is

$$\begin{bmatrix} \dfrac{\partial f_1}{\partial x_1} & \dfrac{\partial f_1}{\partial x_2} \\ \dfrac{\partial f_2}{\partial x_1} & \dfrac{\partial f_{21}}{\partial x_2} \end{bmatrix} \begin{bmatrix} x_{1,t} - x_{1,c} \\ x_{2,t} - x_{2,c} \end{bmatrix} = \begin{bmatrix} f_1 \\ f_2 \end{bmatrix}$$

Solve the set of linear simultaneous equations to determine $x_{i,t} - x_{i,c}$.

Correct the xs

Since $x_{i,\,old} - x_{i,\,new} = x_{i,\,t} - x_{i,\,c}$

$x_{i,\,new} = x_{i,\,old} - (x_{i,\,t} - x_{i,\,c})$

$x_{1,\,new} = x_{1,\,old} - (x_{1,t} - x_{1,c})$

$x_{2,\,new} = x_{2,\,old} - (x_{2,t} - x_{2,c})$

Test for convergence. If the absolute magnitudes of all the fs or all the Δxs are satisfactorily small, terminate; otherwise, repeat the iteration.

Suppose that three nonlinear equations are to be solved for the three unknown variables x_1, x_2 and x_3

$$f_1 (x_1, x_2, x_3) = 0 \tag{1.105}$$

$$f_2 (x_1, x_2, x_3) = 0 \tag{1.106}$$

$$f_3 (x_1, x_2, x_3) = 0 \tag{1.107}$$

Following are the steps for the iteration of the solution.

1. Frame the equations with all the terms on left side of the equality sign, and equate to zero.
2. Take the initial assumption of temporary variables, $x_{1,t}$, $x_{2,t}$, and x_3,t.
3. Determine the values of f_1, f_2 and f_3 at x_1, x_2 and x_3.
4. Calculate the partial derivatives of functions with respect to all variables.
5. Develop the set of simultaneous equation using Taylor-series expansion.

$$f_1(x_{1,t},\ x_{2,t}\ x_{3,t}) \approx f_1(x_{1,c},\ x_{2,c}\ x_{3,c}) + \frac{\partial f_1(x_{1,t},\ x_{2,t}\ x_{3,t})}{\partial x_1}(x_{1,t} - x_{1,c}) + \frac{\partial f_1(x_{1,t},\ x_{2,t}\ x_{3,t})}{\partial x_2}(x_{2,t} - x_{2,c}) + \frac{\partial f_1(x_{1,t},\ x_{2,t}\ x_{3,t})}{\partial x_3}(x_{3,t} - x_{3,c}) \tag{1.108}$$

The set of equations is

$$\begin{bmatrix} \frac{\partial f_1}{\partial x_1} & \frac{\partial f_1}{\partial x_2} & \frac{\partial f_1}{\partial x_3} \\ \frac{\partial f_2}{\partial x_1} & \frac{\partial f_2}{\partial x_2} & \frac{\partial f_2}{\partial x_3} \\ \frac{\partial f_3}{\partial x_1} & \frac{\partial f_3}{\partial x_2} & \frac{\partial f_3}{\partial x_3} \end{bmatrix} \begin{bmatrix} x_{1,t} - x_{1,c} \\ x_{2,t} - x_{2,c} \\ x_{3,t} - x_{3,c} \end{bmatrix} = \begin{bmatrix} f_1 \\ f_2 \\ f_3 \end{bmatrix} \tag{1.109}$$

Solve the set of linear simultaneous equations to determine $x_{i,t} - x_{i,c}$.

Correct the xs

Since $x_{i,\,old} - x_{i,\,new} = x_{i,\,t} - x_{i,\,c}$

$x_{i,\,new} = x_{i,\,old} - (x_{i,\,t} - x_{i,\,c})$

$x_{1,\,new} = x_{1,\,old} - (x_{1,t} - x_{1,c})$

....

$x_{3,\,new} = x_{3,\,old} - (x_{3,t} - x_{3,c})$

Test for convergence. If the absolute magnitudes of all the f's or all the Δxs are satisfactorily small, terminate; otherwise return to step 3.

For example, the biomass is defined by a general formula as $C_{a1}\ H_{a2}\ O_{a3}\ N_{a4}$. For single atom of carbon in fuel, the coefficient a_1 becomes to one. The coefficients of a_2, a_3 and a_4 are the H/C, O/C and N/C mole ratios, respectively. The moisture content in the biomass fuel is neglected. The reactions are solved at thermodynamic equilibrium. The gasification products contain CH_4,

CO, CO_2, H_2, H_2O and N_2. The following is the chemical reaction in biomass gasifier (Srinivas et al., 2009).

$$C_{a_1}H_{a_2}O_{a_3}N_{a_4} + a_5(O_2 + 3.76N_2) + a_6H_2O \Rightarrow b_1CH_4 + b_2CO + b_3CO_2 + b_4H_2 + b_5H_2O + b_6N_2 \quad (1.110)$$

The Eq. (1.110) represents an overall chemical reaction in a gasifier but a number of competing intermediate reactions take place during the process.

Taking atom balances on carbon, hydrogen, oxygen and nitrogen,

C balance: $a_1 = b_1 + b_2 + b_3$ (1.111)

H balance: $a_2 + 2a_6 = 4b_1 + 2b_4 + 2b_5$ (1.112)

O balance: $a_3 + 2a_5 + a_6 = b_2 + 2b_3 + b_5$ (1.113)

N balance: $a_4 + 2\times 3.76a_5 = 2b_6$ (1.114)

There are six unknowns and four equations, (Eq. 1.111–1.114) so there is a need of two equations developed from equilibrium reactions. They are,

Methane reforming $CH_4 + H_2O = CO + 3H_2$ (1.115)

Water shift reaction $CO + H_2O = CO_2 + H_2$ (1.116)

The methane steam reforming reaction (Eq. 1.115) is endothermic because of the stable methane molecule. The C-H bond has a high binding energy (439 kJ/kg mol). Therefore, the dissociation reaction needs a high process temperature to shift the equilibrium composition to the right side. The water gas shift reaction is exothermic and needs a temperature as low as possible in order to shift the equilibrium to the right side. The kinetic rates are in most studies assumed to be very fast compared with the methane steam reforming reaction and therefore to be in equilibrium composition under the conditions of methane steam reforming. All the coefficients are converted in terms of two unknown coefficients, i.e., b_1 and b_2.

$$b_3 = a_1 - b_1 - b_2 \quad (1.117)$$

$$b_4 = 2a_1 + 0.5a_2 - a_3 - 2a_5 - 4b_1 - b_2 \quad (1.118)$$

$$b_5 = a_3 + 2a_5 + a_6 - 2a_1 + 2b_1 + b_2 \quad (1.119)$$

$$b_6 = 0.5a_4 + 3.76\, a_5 \quad (1.120)$$

Total number of species (kg mol) in products,

$$n_t = b_1 + b_2 + b_3 + b_4 + b_5 + b_6 = a_1 + 0.5a_2 + 0.5a_4 + 3.76a_5 + a_6 - 2\,b_1 \quad (1.121)$$

The required two equations results from the equilibrium constants, K_p of reactions. The equilibrium constant $K_{p,1}$ of the reforming reaction can be written as,

$$K_{p1} = \frac{P_{CO}\,P_{H_2}^3}{P_{CH_4}\,P_{H_2O}} = \frac{b_2 b_4^3 P^2}{b_1 b_5 n_t^2}. \quad (1.122)$$

Substituting b_4, b_5 and n_t in Eq. 1.122,

$$K_{p1} = \frac{b_2(2a_1 + 0.5a_2 - a_3 - 2a_5 - 4b_1 - b_2)^3 P^2}{b_1(a_3 + 2a_5 + a_6 - 2\,a_1 + 2b_1 + b_2)(a_1 + 0.5a_2 + 0.5a_4 + 3.76a_5 + a_6 - 2\,b_1)^2} \quad (1.123)$$

where $P = \frac{P_{gasifier}}{P_0}$ (1.124)

The equilibrium constants are given by

$$\ln K_p = \frac{-\Delta G^0}{RT} \quad (1.125)$$

Where, ΔG^0 is the Gibb's free energy (kJ/kg mol), T is the temperature of gasifier in K, assumed initially and R is the universal gas constant in consistent units.

$$\Delta G^0 = \sum h_f^0 + (h_T - h_{298}) - T\,s \quad (1.126)$$

h_f^0 is the enthalpy of formation in kJ/kg mol and s is the entropy measured from 0 K, kJ/kg mol K.

Eq. 1.126 can be written as,

$$b_2 P^2 (2a_1 + a_2 - 2a_3 - 2a_5 - 4b_1 - b_2)^3 - K_{p1} b_1 (2a_3 + 2a_5 + a_6 - 2a_1 + 2b_1 + b_2)$$
$$(a_1 + a_2 + a_4 + 3.76a_5 + a_6 - 2b_1)^2 = 0 = f(b_1, b_2) \quad (1.127)$$

$$K_{p,2} = \frac{P_{CO_2} P_{H_2}}{P_{CO} P_{H_2O}} = \frac{b_3 b_4}{b_2 b_5}. \quad (1.128)$$

Substituting b_3, b_4 and b_5 in Eq. 1.128,

$$K_{p,2} = \frac{(a_1 - b_1 - b_2)(2a_1 + 0.5a_2 - a_3 - 2a_5 - 4b_1 - b_2)}{b_2(a_3 + 2a_5 + a_6 - 2a_1 + 2b_1 + b_2)} \quad (1.129)$$

Eq. 1.129 can be written as,

$$(a_1 - b_1 - b_2 - b_7)(2a_1 + 0.5a_2 - a_3 - 2a_5 - 4b_1 - b_2)$$
$$- K_{p,2} b_2 (a_3 + 2a_5 + a_6 - 2a_1 + 2b_1 + b_2) = 0 = g(b_1, b_2) \quad (1.130)$$

A numerical method is used to solve the b_1 and b_2 from the above two equilibrium constant equations. Therefore, approximate initial values of b_1 and b_2 are taken to start the iteration of the numerical method. By Taylor's series expansion method,

$$f_0 + h\frac{\partial f}{\partial b_1} + k\frac{\partial f}{\partial b_2} = 0 \quad (1.131)$$

$$g_0 + h\frac{\partial g}{\partial b_1} + k\frac{\partial g}{\partial b_2} = 0 \quad (1.132)$$

The values of h and k are the desired degree of accuracy in the coefficients of b_1 and b_2 respectively.

In the above equations of

$f_o = f(b_1, b_2)$
and $g_o = g(b_1, b_2)$ respectively.

The equations are solved to get the values of h and k. If the desired degree of accuracy is not obtained this iteration is repeated with the new values for the coefficients. They are as follows.

$$b_1' = b_1 + h \quad (1.133)$$

$$b_2' = b_2 + k \quad (1.134)$$

This is repeated up to the desired accuracy.

All the six coefficients (b_1 to b_6) in the partial oxidation reaction are calculated by the iteration of the above numerical solution. The outlined procedure is repeated until the temperature value is converged with the energy balance of gasification reaction.

1.7 Optimization

In optimization, the best conditions are determined by the objective function and constraints. In a large system with subcomponents, first the subsystems will be optimized before going to the optimization of the whole system. Optimization can be done in two ways: without using optimization techniques and with the use of optimization techniques. The function can be solved and analyzed by specifying the range for each variable, also known as parametric analysis. After the analysis, the optimum conditions can be selected from the results. In the second category, parametric analysis is not necessary, and optimized results can be generated by solving the optimization technique. The optimization methods are two: the calculus method and the search method. In the calculus method, the constraints must be equality constraints and should be differentiable. The optimization may be with a single variable or with a multivariable function. The problem may be constrained or unconstrained.

In designing a workable system, the system is designed based on arbitrarily assumed parameters. But in the optimization, the parameters are made to float until reaching an optimum solution. The constraints change from case to case. For example, in the design of aircraft engines, the minimum weight is more important than the cost. For a transport engine on the road, the mileage and size of the vehicle have preference compared to the weight. Of course, the minimum cost is a common constraint for all optimization problems. For a thermal system, instead of thermal optimization, thermo-economic optimization results in the best conditions. Figure 1.24 outlines the algorithm to understand the methodology of the optimization process.

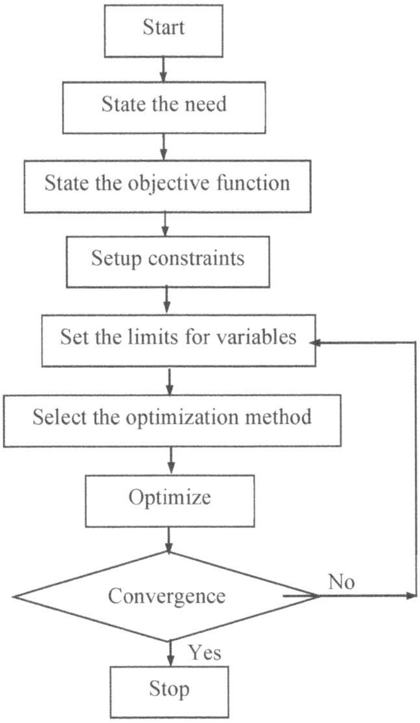

Fig. 1.24. Flow chart showing the optimization procedure.

1.8 Numerical solutions

1. An overhead tank of a big apartment complex has a capacity of 16000 liters as shown in Fig. 1.25. It is desired to select a pump and piping system to transport water from the sump to the tank. The distance between the two is 300 m and the tank is at a level 25 m above the sump. For operational convenience, the time to fill the tank shall be 90 min. Losses in the expansions, contractions, bends and elbows are neglected. Design a workable system for the above and sketch the layout.

Solution

Pressure drop, $\Delta p = \rho g\left(\frac{flv^2}{2gd}\right) = \frac{flv^2\rho}{2d}$

$$\dot{m} = \rho A v = \rho \frac{\pi}{4} d^2 v$$

$$v = \frac{4\dot{m}}{\rho \pi d^2}$$

$$\Delta p = \frac{flv^2\rho}{2d} = \frac{fl\rho}{2d}\left(\frac{4\dot{m}}{\rho\pi d^2}\right)^2 = \frac{8fl\dot{m}^2}{\rho\pi^2 d^5}$$

where f = 0.182 $Re^{-0.2}$

where $Re = \frac{\rho v d}{\mu} = \frac{\rho d}{\mu}\left(\frac{4\dot{m}}{\rho\pi d^2}\right) = \frac{4\dot{m}}{\pi\mu d}$

$$f = 0.182\left(\frac{vd}{\nu}\right)^{-0.2} = 0.182\left(\frac{4\dot{m}}{\rho\pi d^2}\frac{d}{\nu}\right)^{-0.2} = 0.182\left(\frac{4\dot{m}}{\pi d\mu}\right)^{-0.2}$$

$$\dot{m} = \frac{16000}{90\times 60} = 2.963\ kg/s$$

Let d = 1" = 0.0254 m

$\nu = 1.006 \times 10^{-6}\ m^2/s$

$$f = 0.182\left(\frac{4\dot{m}}{\pi d\mu}\right)^{-0.2} = 0.182\left(\frac{4\times 2.963}{\pi 0.0254\times 1000\times 1.006\times 10^{-6}}\right)^{-0.2} = 0.0168$$

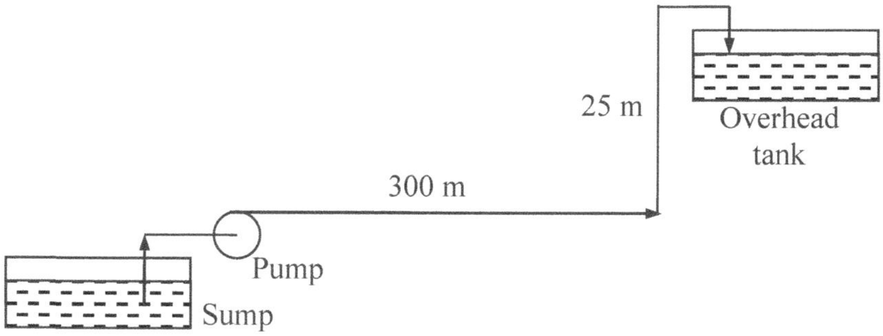

Fig. 1.25. Pumping system.

For check turbulent flow,

$$\text{Re} = \left(\frac{0.182}{f}\right)^5 = \left(\frac{0.182}{0.0168}\right)^5 = 149210 \text{ turbulent}$$

$$\Delta p = \frac{8 f l \dot{m}^2}{\rho \pi^2 d^5} = \frac{8 \times 0.0168 \times 300 \times 2.963^2}{1000 \times \pi^2 \times 0.0254^5} = 3.3925 \times 10^6 \; Pa$$

Total pressure drop = $(3.3925 \times 10^6 + 25 \times 9.81 \times 1000)/1000 = 3637$ kPa

Power = total pressure drop × discharge (m^3/s)
= $3637 \times 2.963 \times 10^{-3} = 10779$ W = 10.779 kW ≈ 15 hp.

It is workable but not optimized.

For example,

Let d = 2" = 0.050 m

Kinetic viscosity, $\nu = 1.006 \times 10^{-6}$ m^2/s

$$f = 0.182\left(\frac{4\dot{m}}{\pi d \mu}\right)^{-0.2} = 0.182\left(\frac{4 \times 2.963}{\pi 0.05 \times 1000 \times 1.006 \times 10^{-6}}\right)^{-0.2} = 0.0193$$

For check turbulent flow,

$$\text{Re} = \left(\frac{0.182}{f}\right)^5 = \left(\frac{0.182}{0.0193}\right)^5 = 74571 \text{ Turbulent}$$

$$\Delta p = \frac{8 f l \dot{m}^2}{\rho \pi^2 d^5} = \frac{8 \times 0.0193 \times 300 \times 2.963^2}{1000 \times \pi^2 \times 0.05^5} = 1.3185 \times 10^5 \; Pa$$

Power = total pressure drop × discharge
= $(1.3185 \times 10^5 + 25 \times 9.81 \times 1000)\, 2.963 \times 10^{-3} = 1117.3$ W = 1.1173 kW ≈ 2 hp.

Still, it is not optimized unless a thermo-economic analysis is done. The problem is analysed in Fig. 1.26.

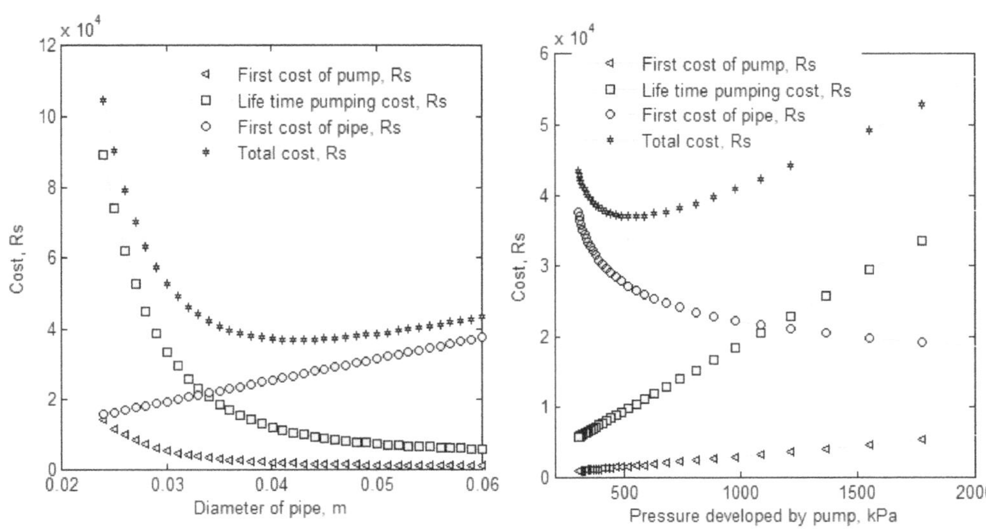

Fig. 1.26. Optimum system.

Table 1.5. Desalinated water with hot water temperature change.

Hot water inlet temperature, °C	Desalination, LPH
52.6	1.66
46.4	0.99
43.6	0.83

2. The desalinated water is increased with increase in hot water temperature in a typical thermal desalination plant as shown in Table 1.5. Compute the constants in the equation $Q = a_0 + a_1t + a_2t^2$ to provide an exact fit using polynomial fit for the following (t, Q) data. Find the desalination at the temperature of 50°C.

Solution

$Q = c_1 + c_2t + c_3t^2$

which gives three equations

$Q_1 = c_1 + c_2t_1 + c_3t_1^2$

$Q_2 = c_1 + c_2t_2 + c_3t_3^2$

$Q_3 = c_1 + c_2t_3 + c_3t_3^2$

$$\begin{bmatrix} 1 & t_1 & t_1^2 \\ 1 & t_2 & t_2^2 \\ 1 & t_3 & t_3^2 \end{bmatrix} \begin{bmatrix} c_1 \\ c_2 \\ c_3 \end{bmatrix} = \begin{bmatrix} Q_1 \\ Q_2 \\ Q_3 \end{bmatrix}$$

$$\begin{bmatrix} 1 & 52.6 & 52.6^2 \\ 1 & 46.4 & 46.4^2 \\ 1 & 43.6 & 43.6^2 \end{bmatrix} \begin{bmatrix} c_1 \\ c_2 \\ c_3 \end{bmatrix} = \begin{bmatrix} 1.66 \\ 0.99 \\ 0.83 \end{bmatrix}$$

After solving the above matrix, $c_1 = 9.7849$, $c_2 = -0.4521$ and $c_3 = 0.0057$

So, the solution is $Q = 9.7849 - 0.4521t + 0.0057t^2$

At t = 50°C,

$Q = 9.7849 - 0.4521 \times 50 + 0.0057 \times 50^2 = 1.326$ LPH

3. Solve the above problem with Lagrange interpolation method.

Solution

With reference to Fig. 1.27, substituting

$$Q = \frac{(t-t_2)(t-t_3)}{(t_1-t_2)(t_1-t_3)}Q_1 + \frac{(t-t_1)(t-t_3)}{(t_2-t_1)(t_2-t_3)}Q_2 + \frac{(t-t_1)(t-t_2)}{(t_3-t_1)(t_3-t_2)}Q_3$$

t	Q
$t_1 = 52.6$	$Q_1 = 1.66$
$t_2 = 46.4$	$Q_2 = 0.99$
$t_3 = 43.6$	$Q_3 = 0.83$

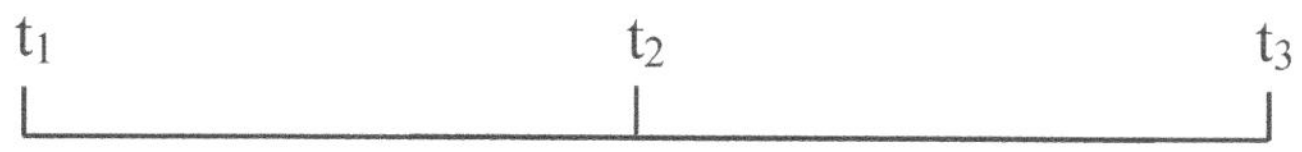

Fig. 1.27. t terms to find the coefficients.

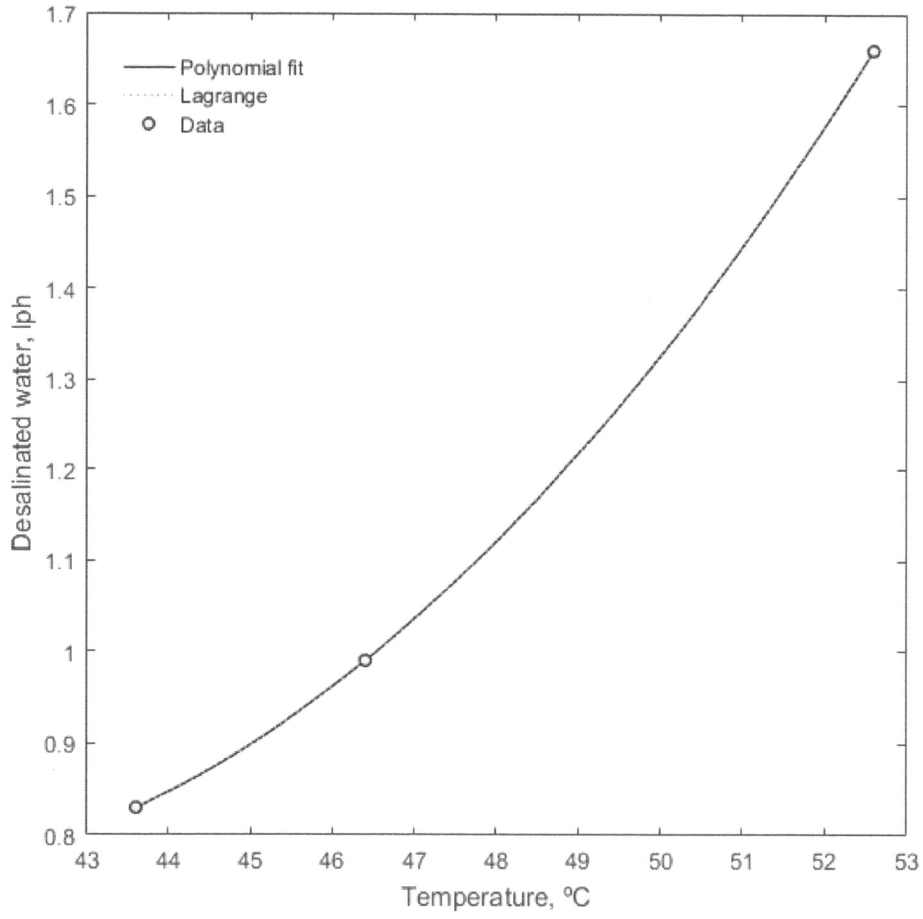

Fig. 1.28. Exact fit solution using polynomial and Lagrange methods.

$$Q = \frac{(t-46.4)(t-43.6)}{(52.6-46.4)(52.6-43.6)} \times 1.66 + \frac{(t-52.6)(t-43.6)}{(46.4-52.6)(46.4-43.6)} \times 0.99$$
$$+ \frac{(t-52.6)(t-46.4)}{(43.6-52.6)(43.6-46.4)} \times 0.83$$

Q = 0.0297(t – 46.4) (t – 43.6) – 0.0589(t – 52.6) (t – 43.6) + 0.0329(t – 52.6) (t – 46.4)

At t = 50,

Q = 0.0297(50 – 46.4) (50 – 43.6) – 0.0589(50 – 52.6) (50 – 43.6) + 0.0329(50 – 52.6) (50 – 46.4)
= 1.326 LPH

Figure 1.28 shows the comparison of results generated by polynomial fit and Lagrange interpolation methods, with the given data.

Table 1.6. Saturation temperature and saturation pressure of a pure substance (steam).

T, K	360	365	373	383	390
P = *f*(T), kPa	154	165	190	210	240

4. Develop an equation fitting to generate the thermodynamic properties using Lagrange interpolation method and MATLAB coding. The following data Table 1.6 gives the data for the steam pressure P vs temperature T: Compute the pressure at T = 375 K by the Lagrange interpolation method.

Solution

Given temperature to find steam pressure, T = 375 K

From Lagrange interpolation

$$y=\sum_{i=1}^{n} y_i \prod_{j=1}^{n} \frac{\left(x-x_j\right)\ omitting\ \left(x-x_i\right)}{\left(x_i-x_j\right)\ omitting\ \left(x_i-x_i\right)}$$

Therefore

$$P=\frac{(T-T_2)(T-T_3)(T-T_4)(T-T_5)}{(T_1-T_2)(T_1-T_3)(T_1-T_4)(T_1-T_5)}P_1+\frac{(T-T_1)(T-T_3)(T-T_4)(T-T_5)}{(T_2-T_1)(T_2-T_3)(T_2-T_4)(T_2-T_5)}P_2$$
$$+\frac{(T-T_1)(T-T_2)(T-T_4)(T-T_5)}{(T_3-T_1)(T_3-T_2)(T_3-T_4)(T_3-T_5)}P_3+\frac{(T-T_1)(T-T_2)(T-T_3)(T-T_5)}{(T_4-T_1)(T_4-T_2)(T_4-T_3)(T_4-T_5)}P_4$$
$$+\frac{(T-T_1)(T-T_2)(T-T_3)(T-T_4)}{(T_5-T_1)(T_5-T_2)(T_5-T_3)(T_5-T_4)}P_5$$

Substituting the values from the Table 1.6, we can get the P at given T

The value of pressure at the temperature, T = 375 K is 194.83 kPa using Lagrange interpolation with the coefficients generated and shown in Fig. 1.29.

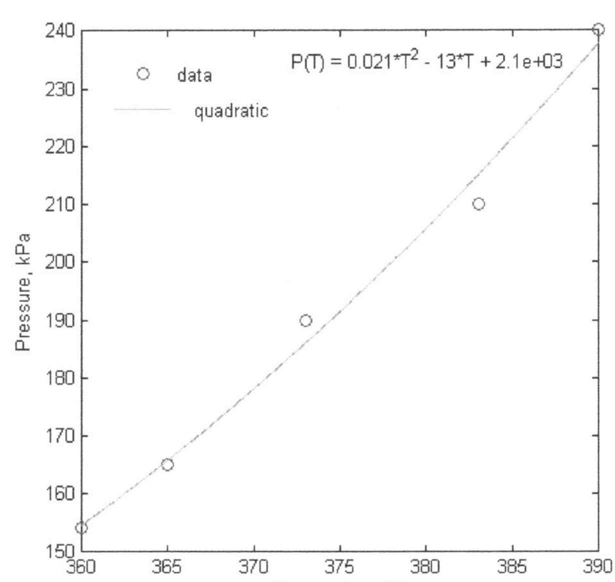

Fig. 1.29. Equation fitting for pressure and temperature relation.

Table 1.7. Exergy efficiency of the combined cycle as a function of steam injection ratio and turbine inlet temperature.

Gas turbine inlet temperature, T, °C (x - axis)	Steam injection ratio, r (y - axis)	
	1	3
1000	42.5	43.3
1200	47.3	48.6

5. In a combined cycle power, plant steam has been taped from the bottoming steam cycle and injected into the gas turbine combustion chamber, the steam injection rate and exergy performance details are given in the Table 1.7 at various gas turbine inlet temperatures. Express the exergy efficiency of combined cycle as a function of steam injection and gas turbine inlet temperature. From the resulting equation fitting find the exergy efficiency of the combined cycle at 1175.4°C and 1.54 steam to fuel mass ratio.

Solution

A first degree polynomial equation in both independent variables will be chosen.

The two pairs of points for r = 1, (1000, 42.5) and (1200, 47.3) can be represented by a straight line as shown in Fig. 1.30.

$\eta = a + bT$

at $r_1 = 1$,

$\eta_{1,1} = a_1 + b_1 T_1$ at $T_1 = 1000°C$

$\eta_{2,1} = a_1 + b_1 T_2$ at $T_2 = 1200°C$

the solution for a_1 and b_1 is

$$\begin{bmatrix} 1 & T_1 \\ 1 & T_2 \end{bmatrix} \times \begin{bmatrix} a_1 \\ b_1 \end{bmatrix} = \begin{bmatrix} \eta_{1,1} \\ \eta_{2,1} \end{bmatrix}$$

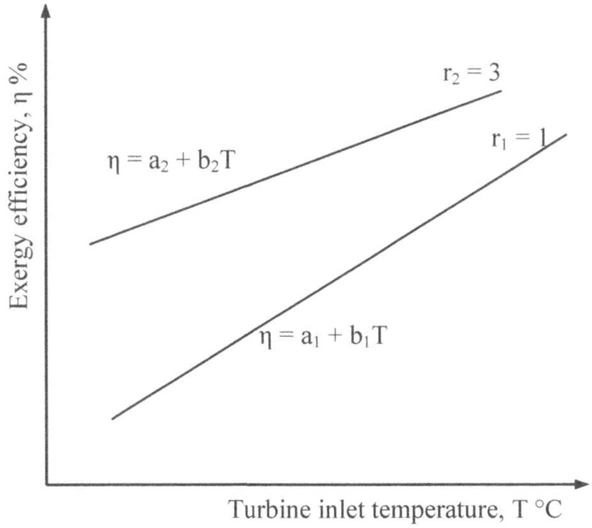

Fig. 1.30. Combined cycle power plant efficiency as a function of turbine inlet temperature and steam injection ratio (steam to fuel mass ratio).

Substituting,

$$\begin{bmatrix} 1 & 1000 \\ 1 & 1200 \end{bmatrix} \times \begin{bmatrix} a_1 \\ b_1 \end{bmatrix} = \begin{bmatrix} 42.5 \\ 47.3 \end{bmatrix}$$

Solving,

$a_1 = 18.5$ and $b_1 = 0.024$

Therefore,

$\eta = 18.5 + 0.024\ T$

Similarly,

For $r = 2$

$\eta_{1,2} = a_2 + b_2 T_1$ at $T_1 = 1000°C$

$\eta_{2,2} = a_2 + b_2 T_2$ at $T_2 = 1200°C$

the solution for a_2 and b_2 is

$$\begin{bmatrix} 1 & T_1 \\ 1 & T_2 \end{bmatrix} \times \begin{bmatrix} a_2 \\ b_2 \end{bmatrix} = \begin{bmatrix} \eta_{1,2} \\ \eta_{2,2} \end{bmatrix}$$

Substituting,

$$\begin{bmatrix} 1 & 1000 \\ 1 & 1200 \end{bmatrix} \times \begin{bmatrix} a_2 \\ b_2 \end{bmatrix} = \begin{bmatrix} 43.3 \\ 48.6 \end{bmatrix}$$

Solving,

$a_2 = 16.8$ and $b_2 = 0.0265$

$\eta = 16.8 + 0.0265\ T$

Next, the constant terms 18.5 and 16.8 can be expressed by a first degree equation of r,

at $T_1 = 1000°C$,

$a = A_1 + A_2 r$

$a_1 = A_1 + A_2 r_1$ at $r_1 = 1$

$a_2 = A_1 + A_2\, r_2$ at $r_2 = 3$

the solution for A_1 and A_2 is

$$\begin{bmatrix} 1 & r_1 \\ 1 & r_2 \end{bmatrix} \times \begin{bmatrix} A_1 \\ A_2 \end{bmatrix} = \begin{bmatrix} a_1 \\ a_2 \end{bmatrix}$$

Substituting,

$$\begin{bmatrix} 1 & 1 \\ 1 & 3 \end{bmatrix} \times \begin{bmatrix} A_1 \\ A_2 \end{bmatrix} = \begin{bmatrix} 18.5 \\ 16.8 \end{bmatrix}$$

Solving,

$A_1 = 19.35$ and $A_2 = -0.85$

Therefore

a = 19.35 – 0.85 r

The second coefficients of T can also be expressed by equations in terms of the m.

at T_2 = 1200°C,

$b = B_1 + B_2 r$

$b_1 = B_1 + B_2 r_1$ at $r_1 = 1$

$b_2 = B_1 + B_2\, r_2$ at $r_2 = 3$

the solution for B_1 and B_2 is

$$\begin{bmatrix} 1 & r_1 \\ 1 & r_2 \end{bmatrix} \times \begin{bmatrix} B_1 \\ B_2 \end{bmatrix} = \begin{bmatrix} b_1 \\ b_2 \end{bmatrix}$$

Substituting,

$$\begin{bmatrix} 1 & 1 \\ 1 & 3 \end{bmatrix} \times \begin{bmatrix} B_1 \\ B_2 \end{bmatrix} = \begin{bmatrix} 0.0240 \\ 0.0265 \end{bmatrix}$$

Solving,

$B_1 = 0.0227$ and $B_2 = 0.0013$

Therefore, $b = 0.0227 + 0.0013\, r$

Then provide the complete equation

$\eta = (19.35 - 0.85\, r) + (0.0227 + 0.0013\, r)\, T$

at T = 1175.4°C and r = 1.54

$\eta = (19.35 - 0.85 \times 1.54) + (0.0227 + 0.0013 \times 1.54) \times 1175.4 = 47.04\ \%$

6. Manufacturers of cooling towers often present catalog data showing the outlet-water temperature as a function of the wet-bulb temperature of the ambient air and the range. The range is the difference between the inlet and outlet temperatures of water. In Table 1.8, for example, when the wet bulb temperature is 20°C and the range is 10°C, the temperature of leaving water is 25.9°C, and so the temperature of the entering water is 25.9 + 10 = 35.9°C. Express the outlet-water temperature, t in Table 1.8 as a function of the wet-bulb temperature (WBT) and range R.

Table 1.8. Outlet-water temperature, °C, of cooling tower.

Range, °C	Wet-bulb temperature, °C		
	20	23	26
10	25.9	27.5	29.4
16	27.0	28.4	30.2
22	28.4	29.6	31.3

Solution

A second degree polynomial equation in both independent variables will be chosen as the form of the equation next, and three different methods for developing the equation will be illustrated here (Fig. 1.31).

Method 1. The three pairs of points for WBT = 20°C, (10, 25.9), (16, 27.0) and (22, 28.4), can be represented by a parabola

$t = a + bR + cR^2$

at $WBT_1 = 20°C$,

$t_{1,1} = a_1 + b_1R_1 + c_1R_1^2$ at $R_1 = 10°C$

$t_{1,2} = a_1 + b_1R_2 + c_1R_2^2$ at $R_2 = 16°C$

$t_{1,3} = a_1 + b_1R_3 + c_1R_3^2$ at $R_3 = 22°C$

the solution for a_1, b_1 and c_1 is

$$\begin{bmatrix} 1 & R_1 & R_1^2 \\ 1 & R_2 & R_2^2 \\ 1 & R_3 & R_3^2 \end{bmatrix} \times \begin{bmatrix} a_1 \\ b_1 \\ c_1 \end{bmatrix} = \begin{bmatrix} t_{1,1} \\ t_{1,2} \\ t_{1,3} \end{bmatrix}$$

Substituting,

$$\begin{bmatrix} 1 & 10 & 10^2 \\ 1 & 16 & 16^2 \\ 1 & 22 & 22^2 \end{bmatrix} \times \begin{bmatrix} a_1 \\ b_1 \\ c_1 \end{bmatrix} = \begin{bmatrix} 25.9 \\ 27.0 \\ 28.4 \end{bmatrix}$$

Solving,

$a_1 = 24.733$, $b_1 = 0.075006$ and $c_1 = 0.004146$

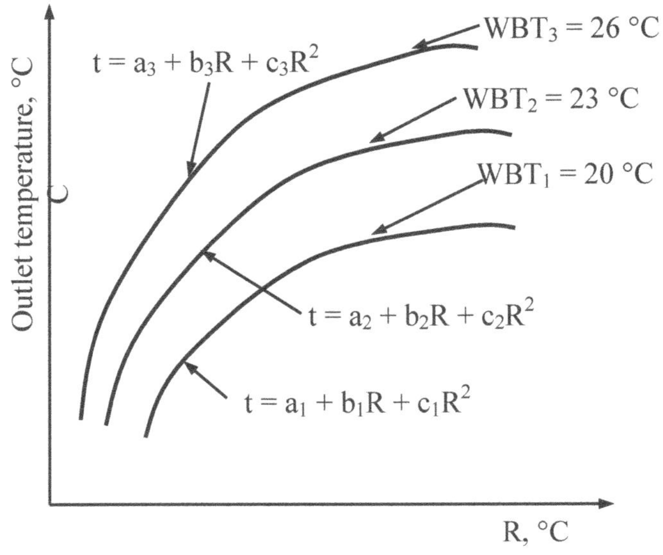

Fig. 1.31. Cooling tower outlet temperature as a function of range and wet bulb temperature of atmospheric air.

Therefore, $t = 24.733 + 0.075006\, R + 0.004146\, R^2$

Similarly,

For WBT = 23°C

$t_{2,1} = a_2 + b_2R_1 + c_2R_1^2$ at $R_1 = 10°C$

$t_{2,2} = a_2 + b_2R_2 + c_2R_2^2$ at $R_2 = 16°C$

$t_{2,3} = a_2 + b_2R_3 + c_2R_3^2$ at $R_3 = 22°C$

the solution for a_2, b_2 and c_2 is

$$\begin{bmatrix} 1 & R_1 & R_1^2 \\ 1 & R_2 & R_2^2 \\ 1 & R_3 & R_3^2 \end{bmatrix} \times \begin{bmatrix} a_2 \\ b_2 \\ c_2 \end{bmatrix} = \begin{bmatrix} t_{2,1} \\ t_{2,2} \\ t_{2,3} \end{bmatrix}$$

Substituting,

$$\begin{bmatrix} 1 & 10 & 10^2 \\ 1 & 16 & 16^2 \\ 1 & 22 & 22^2 \end{bmatrix} \times \begin{bmatrix} a_2 \\ b_2 \\ c_2 \end{bmatrix} = \begin{bmatrix} 27.5 \\ 28.4 \\ 29.6 \end{bmatrix}$$

Solving,

$a_2 = 26.667$, $b_2 = 0.041659$ and $c_2 = 0.0041469$

$t = 26.667 + 0.041659\, R + 0.0041469\, R^2$

and for WBT = 26°C

$t_{3,1} = a_3 + b_3R_1 + c_3R_1^2$ at $R_1 = 10°C$

$t_{3,2} = a_3 + b_3R_2 + c_3R_2^2$ at $R_2 = 16°C$

$t_{3,3} = a_3 + b_3R_3 + c_3R_3^2$ at $R_3 = 22°C$

the solution for a_3, b_3 and c_3 is

$$\begin{bmatrix} 1 & R_1 & R_1^2 \\ 1 & R_2 & R_2^2 \\ 1 & R_3 & R_3^2 \end{bmatrix} \times \begin{bmatrix} a_3 \\ b_3 \\ c_3 \end{bmatrix} = \begin{bmatrix} t_{3,1} \\ t_{3,2} \\ t_{3,3} \end{bmatrix}$$

Substituting,

$$\begin{bmatrix} 1 & 10 & 10^2 \\ 1 & 16 & 16^2 \\ 1 & 22 & 22^2 \end{bmatrix} \times \begin{bmatrix} a_3 \\ b_3 \\ c_3 \end{bmatrix} = \begin{bmatrix} 29.4 \\ 30.2 \\ 31.3 \end{bmatrix}$$

Solving,

$a_3 = 28.733$, $b_3 = 0.024999$ and $c_3 = 0.0041467$

$t = 28.733 + 0.024999\, R + 0.0041467\, R^2$

Next, the constant terms 24.733, 26,667 and 28.733 can be expressed by a second-degree equation of WBT,

at $R_1 = 10°C$,

$a = A_1 + A_2WBT + A_3WBT^2$

$a_1 = A_1 + A_2WBT_1 + A_3 WBT_1^2$ at $WBT_1 = 20°C$

$a_2 = A_1 + A_2 WBT_2 + A_3WBT_2^2$ at $WBT_2 = 23°C$

$a_3 = A_0 + A_1 WBT_3 + A_3WBT_3^2$ at $WBT_3 = 26°C$

the solution for A_1, A_2 and A_3 is

$$\begin{bmatrix} 1 & WBT_1 & WBT_1^2 \\ 1 & WBT_2 & WBT_2^2 \\ 1 & WBT_3 & WBT_3^2 \end{bmatrix} \times \begin{bmatrix} A_1 \\ A_2 \\ A_3 \end{bmatrix} = \begin{bmatrix} a_1 \\ a_2 \\ a_3 \end{bmatrix}$$

Substituting,

$$\begin{bmatrix} 1 & 20 & 20^2 \\ 1 & 23 & 23^2 \\ 1 & 26 & 26^2 \end{bmatrix} \times \begin{bmatrix} A_1 \\ A_2 \\ A_3 \end{bmatrix} = \begin{bmatrix} 24.733 \\ 26.667 \\ 28.733 \end{bmatrix}$$

Solving,

$A_1 = 15.247$, $A_2 = 0.32637$ and $A_3 = 0.007380$

Therefore, $a = 15.247 + 0.32637WBT + 0.007380WBT^2$

The coefficients of R and R^2 can also be expressed by equations in terms of the WBT.

at $R_2 = 16°C$,

$b = B_1 + B_2 WBT + B_3WBT^2$

$b_1 = B_1 + B_2 WBT_1 + B_3 WBT_1^2$ at $WBT_1 = 20°C$

$b_2 = B_1 + B_2 WBT_2 + B_3WBT_2^2$ at $WBT_2 = 23°C$

$b_3 = B_0 + B_2 WBT_3 + B_3WBT_3^2$ at $WBT_3 = 26°C$

the solution for B_1, B_2 and B_3 is

$$\begin{bmatrix} 1 & WBT_1 & WBT_1^2 \\ 1 & WBT_2 & WBT_2^2 \\ 1 & WBT_3 & WBT_3^2 \end{bmatrix} \times \begin{bmatrix} B_1 \\ B_2 \\ B_3 \end{bmatrix} = \begin{bmatrix} b_1 \\ b_2 \\ b_3 \end{bmatrix}$$

Substituting,

$$\begin{bmatrix} 1 & 20 & 20^2 \\ 1 & 23 & 23^2 \\ 1 & 26 & 26^2 \end{bmatrix} \times \begin{bmatrix} B_1 \\ B_2 \\ B_3 \end{bmatrix} = \begin{bmatrix} 0.075006 \\ 0.041659 \\ 0.024999 \end{bmatrix}$$

Solving,

$B_1 = 0.72375$, $B_2 = -0.050978$ and $B_3 = 0.000927$

Therefore, $b = 0.72375 - 0.050978\ WBT + 0.000927\ WBT^2$

at $R_3 = 22°C$,

$c = C_1 + C_2WBT + C_3WBT^2$

$c_1 = C_1 + C_2WBT_1 + C_3\ WBT_1^2$ at $WBT_1 = 20°C$

$c_2 = C_1 + C_2\ WBT_2 + C_3WBT_2^2$ at $WBT_2 = 23°C$

$c_3 = C_0 + C_2\ WBT_3 + C_3WBT_3^2$ at $WBT_3 = 26°C$

the solution for C_1, C_2 and C_3 is

$$\begin{bmatrix} 1 & WBT_1 & WBT_1^2 \\ 1 & WBT_2 & WBT_2^2 \\ 1 & WBT_3 & WBT_3^2 \end{bmatrix} \times \begin{bmatrix} C_1 \\ C_2 \\ C_3 \end{bmatrix} = \begin{bmatrix} c_1 \\ c_2 \\ c_3 \end{bmatrix}$$

Substituting,

$$\begin{bmatrix} 1 & 20 & 20^2 \\ 1 & 23 & 23^2 \\ 1 & 26 & 26^2 \end{bmatrix} \times \begin{bmatrix} C_1 \\ C_2 \\ C_3 \end{bmatrix} = \begin{bmatrix} 0.004146 \\ 0.004169 \\ 0.0041467 \end{bmatrix}$$

Solving,

$C_1 = 0.004147$, $C_2 = 0$ and $C_3 = 0$

Therefore, $c = 0.004147 + 0\ WBT + 0\ WBT^2 = 0.004147$

Then providing the complete equation

$$t = (15.247 + 0.32637WBT + 0.007380WBT^2) + (0.72375 - 0.050978WBT + 0.000927WBT^2)\ R + (0.004147 + 0\ WBT + 0\ WBT^2)\ R^2$$

Method 2. An alternative polynomial form using second-degree expressions for R and WBT is

$$t = c_1 + c_2WBT + c_2WBT^2 + c_4R + c_5R^2 + c_6(R)(WBT) + c_7(WBT^2(R) + c_8(WBT)(R)^2 + c_9(WBT)^2(R)^2$$

The nine sets of t-R-WBT combination expressed in Table 1.8 can be substituted into the above equation to develop nine simultaneous equations, which can be solved for the unknown c_1 to c_9. The c values obtained are

$c_1 = 15.247$	$c_2 = 0.32631$	$c_3 = 0.0073991$
$c_4 = 0.723753$	$c_5 = 0.0041474$	$c_6 = -0.0509782$
$c_7 = 0.00092704$	$c_8 = 0.0$	$c_9 = 0.0$

It is possible to multiply and collect the terms in equation from method 1 to develop the equation of the form of method 2.

Method 3. Lagrange interpolation can be extended to a function of two independent variables. For example, if $z = f(x, y)$, the form can be chosen

$$
\begin{aligned}
z = {} & c_{11}(x - x_2)(x - x_3)(y - y_3)(y - y_3) \\
& + c_{12}(x - x_2)(x - x_3)(y - y_1)(y - y_3) \\
& + c_{13}(x - x_2)(x - x_3)(y - y_1)(y - y_2) \\
& + \dots + c_{33}(x - x_1)(x - x_2)(y - y_1)(y - y_2)
\end{aligned}
$$

To represent the data in Table 1.8, z could refer to the outlet-water temperature, x the WBT, and y the range. In above equation, $x_1 = 20$, $x_2 = 23$, and $x_3 = 26$, while $y_1 = 10$, $y_2 = 16$ and $y_3 = 22$.

To determine the magnitude of c_{12}, for example, values applicable when $x = x_1$ and $y = y_2$ can be substituted in the following equation:

$$c_{12} = \frac{27.5}{(20-23)(20-26)(16-10)(16-22)} = -0.0424$$

7. Determine a_0 and a_1 in the equation $y = a_0 + a_1x$ to provide a best fit in the sense of least-squares deviation to the data points (1, 4.9), (3, 11.2), (4, 13.7) and (6, 20.1).

Solution

The summations to substitute into equations are listed in Table 1.9.

The simultaneous equations to be solved are

$$ma + b\sum x_i = \sum y_i$$

$$a\sum x_i + b\sum x_i^2 = \sum x_i y_i$$

$4a_1 + 14a_2 = 49.9$

$14a_1 + 62a_2 = 213.9$

yielding $a_1 = 1.908$ and $a_2 = 3.019$. Thus

$y = 1.908 + 3.019\,x$

A similar procedure can be followed when fitting a parabola of the form

$y = a + bx + cx^2$

to m data points. The summation to be minimized is

$$\sum_{i=1}^{m}\left(a + bx_i + cx_i^2 - y_i\right)^2 \rightarrow \text{minimum}$$

Table 1.9. Data of x, y and x^2 and xy.

	x_i	y_i	x_i^2	x_iy_i
	1	4.9	1	4.9
	3	11.2	9	33.6
	4	13.7	16	54.8
	6	20.1	36	120.6
Σ	14	49.9	62	213.9

Differentiating partially with respect to a, b and c in turn, results in three linear simultaneous equations expressed in matrix form

$$\begin{bmatrix} m & \sum x_i & \sum x_i^2 \\ \sum x_i & \sum x_i^2 & \sum x_i^3 \\ \sum x_i^2 & \sum x_i^3 & \sum x_i^4 \end{bmatrix} \begin{bmatrix} a \\ b \\ c \end{bmatrix} = \begin{bmatrix} \sum y_i \\ \sum x_i y_i \\ \sum x_i^2 y_i \end{bmatrix}$$

A comparison of the matrix equation with equation (5) and (6) shows a pattern evolving which by analogy permits developing the equations for higher degree polynomials without even differentiating the summation of the squared deviation.

8. Develop the performance curve for a solar flat plate collector from the best fit method of least squares (linear interpolation). The data given in Table 1.10 relates thee data of $\frac{T_{fi} - T_a}{I_T}$ and instantaneous thermal efficiency, η_{th}. Develop the performance equation using the given experimental values. Find the slope and y intercept of the linear equation.

Solution

Consider first the linear equation of the form

y = a + bx

The summations to substitute into the equations are given in Table 1.11.

Table 1.10. The testing data of a solar flat plate collector.

S. No.	$(T_{fi} - T_a)/I_T$	η_{th}, %
1	0.1	10
2	0.095	15
3	0.081	20
4	0.072	25
5	0.043	30
6	0.036	35
7	0.034	40
8	0.032	45
9	0.02	50

Table 1.11. Results to apply in best fit method.

	$[(T_{fi} - T_a)/I_T]_i$	$[\eta_{th}]_i$, %	$[(T_{fi} - T_a)/I_T]_i^2$	$[(T_{fi} - T_a)/I_T]_i\,[\eta_{th}]_i$
	x_i	y_i	x_i^2	$x_i y_i$
	0.1	10	0.0100	1.0000
	0.095	15	0.0090	1.4250
	0.081	20	0.0066	1.6200
	0.072	25	0.0052	1.8000
	0.043	30	0.0018	1.2900
	0.036	35	0.0013	1.2600
	0.034	40	0.0012	1.3600
	0.032	45	0.0010	1.4400
	0.02	50	0.0004	1.0000
Σ	0.5130	270	0.0365	12.195

The simultaneous equations to be solved are

$$ma + b\sum x_i = \sum y_i$$

$$a\sum x_i + b\sum x_i^2 = \sum x_i y_i$$

9a + 0.5130 b = 270

0.5130a + 0.0365b = 12.195

yielding a = 55.1055 and b = –440.4467.

Thus y = 55.1055 – 440.4467 x

The performance equation for a solar flat plate collector has been generated using the best fit method of least squares. The slope of the linear equation is –440.44 and the y axis intercept is 55.10 as shown in Fig. 1.32.

9. Solve the equation using best fit method, $d = cx^n$ as shown in Table 1.12, d = f(x)

Solution

$$\begin{bmatrix} m & \sum \log x_i \\ \sum \log x_i & \sum (\log x_i)^2 \end{bmatrix} \times \begin{bmatrix} \log c \\ n \end{bmatrix} = \begin{bmatrix} \sum (\log d_i) \\ \sum \log x_i \log d_i \end{bmatrix}$$

After substituting the results from Table 1.13,

$$\begin{bmatrix} 5 & -6.72 \\ -6.72 & 10.66 \end{bmatrix} \times \begin{bmatrix} \log c \\ n \end{bmatrix} = \begin{bmatrix} -16.79 \\ 24.69 \end{bmatrix}$$

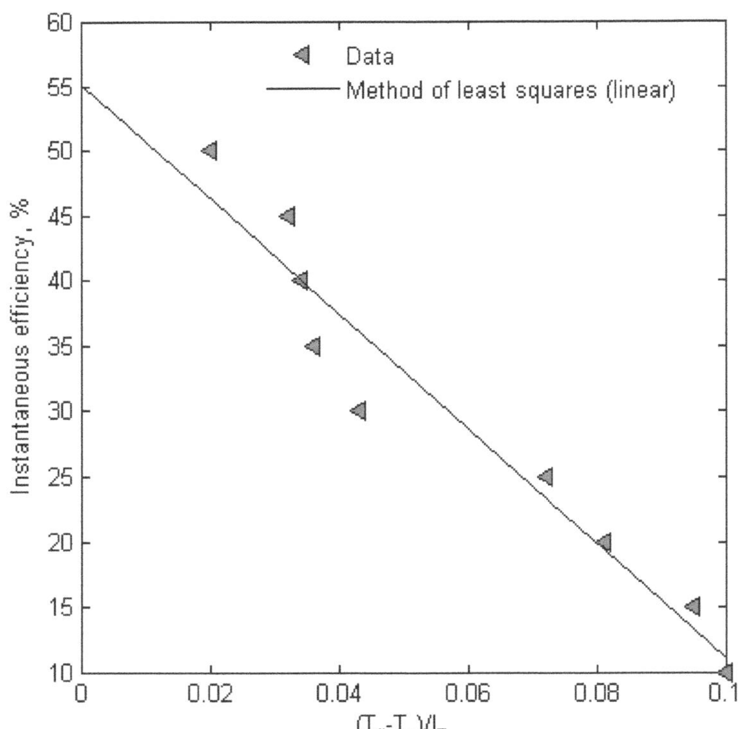

Fig. 1.32. Performance curve for solar flat plate collector using best fit from the method of least squares.

Table 1.12. Values of x and d as per the relation.

S. No.	x	d
1	0.1	0.0100
2	0.2	0.0247
3	0.3	0.0418
4	0.4	0.0608
5	0.5	0.0812

Table 1.13. Generated results for the solution.

S. No.	x	d	log x	log d	(log x)^2	log x log d
1	0.1	0.0100	–2.3026	–4.6052	5.3019	10.6038
2	0.2	0.0247	–1.6094	–3.7010	2.5903	5.9565
3	0.3	0.0418	–1.2040	–3.1749	1.4496	3.8224
4	0.4	0.0608	–0.9163	–2.8002	0.8396	2.5658
5	0.5	0.0812	–0.6931	–2.5108	0.4805	1.7404
		Σ	–6.72	–16.79	10.66	24.69

The above matrix results log c and n as follows

$\log c = -1.6081$

$c = e^{-1.6081} = 0.2$

$n = 1.3$

$d = cx^n$

$d = 0.2\, x^{1.3}$

10. Experiments are carried out in a closed vessel at constant temperature 300 K. Pressure and volume are measured and reported in Table 1.14.
  i) Assume that the data obeys the relations $pv^n = c$. Find the values of n and c using curve fitting method.
  ii) Using the predicted relation, find the volume if the pressure is 2.75 kg/cm^2.

Solution

$pv^n = c$

applying log on both sides,

$\log p + n \log v = \log c$

$\log p = \log c - n \log v$

it is similar to $y = a + bx$

Table 1.14. Data related to pressure and temperature.

p, kg/cm^2	0.5	1	1.5	2	2.5	3
v, liters	1620	1000	750	620	520	460

Table 1.15. Generated results from pressure and volume.

S. No.	p	v	log p	log v	(log v)^2	log v log p
1	0.5	1620	–0.6931	7.3902	54.6148	–5.1225
2	1.0	1000	0.0000	6.9078	47.7171	0.0000
3	1.5	750	0.4055	6.6201	43.8254	2.6842
4	2.0	620	0.6931	6.4297	41.3413	4.4567
5	2.5	520	0.9163	6.2538	39.1104	5.7303
6	3.0	460	1.0986	6.1312	37.5919	6.7358
		Σ	2.4204	39.7328	264.2008	14.4846

so that y = log p, a = logc, b = –n, x = logv

$$\begin{bmatrix} m & \sum x_i \\ \sum x_i & \sum x_i^2 \end{bmatrix} \times \begin{bmatrix} a \\ b \end{bmatrix} = \begin{bmatrix} \sum y_i \\ \sum x_i y_i \end{bmatrix}$$

After converting the terms,

$$\begin{bmatrix} m & \sum \log v_i \\ \sum \log v_i & \sum (\log v_i)^2 \end{bmatrix} \times \begin{bmatrix} \log c \\ -n \end{bmatrix} = \begin{bmatrix} \sum (\log p_i) \\ \sum \log v_i \log p_i \end{bmatrix}$$

Substituting the results from Table 1.15,

$$\begin{bmatrix} 6 & 39.7328 \\ 39.7328 & 264.2008 \end{bmatrix} \times \begin{bmatrix} \log c \\ -n \end{bmatrix} = \begin{bmatrix} 2.4204 \\ 14.4846 \end{bmatrix}$$

The above matrix results log c and n as follows

log c = 9.8218

c = $e^{9.8218}$ = 18432

–n = –1.4223

n = 1.4223

Therefore

$pv^n = c$ becomes to $pv^{1.4223} = 18432$

At p = 2.75

$$v = \left(\frac{c}{p}\right)^{\frac{1}{n}} = \left(\frac{18432}{2.75}\right)^{\frac{1}{1.4223}} = 490.05$$

Figure 1.33 shows the relation between pressure and volume.

11. Prepare the information flow diagram for a desalination plant operating on humidification-dehumidification (HDH) principle as shown in Fig. 1.34. The supplied data is air flow rate (m_a), blower inlet pressure and temperature (p_1, T_1), blower capacity (w_b), surface area of air preheater (A_{aph}), humidifier (A_h) and dehumidifier (A_{dh}), hot water inlet temperatures (T_6, T_8), circulating water inlet temperature ($T_{cw\ in}$). The methodology involved is: finding the blower capacity, air preheater exit temperature, humidifier's air exit temperature, dehumidifier's air exit temperature (Chiranjeevi and Srinivas, 2015).

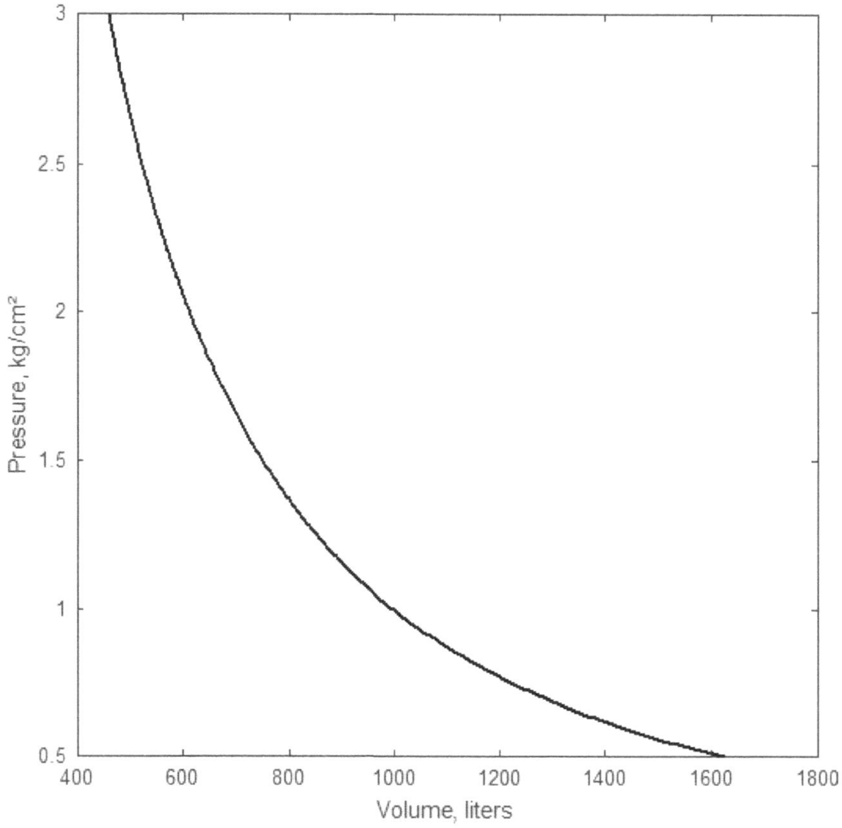

Fig. 1.33. The relation between p and v.

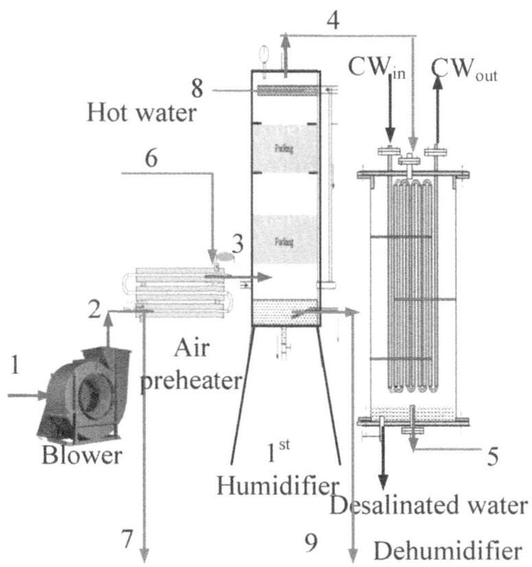

Fig. 1.34. Desalination plant.

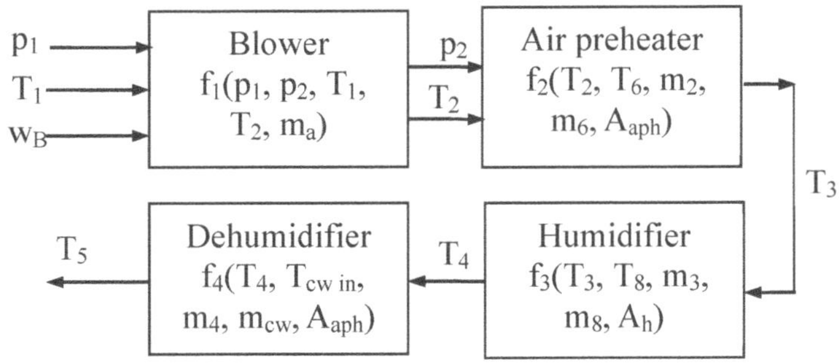

Fig. 1.35. Information flow diagram for desalination plant.

Solution

Figure 1.35 shows the information flow diagram developed for the desalination plant.

12. In a solar tracker, the following mathematical model has been developed (Natarajan and Srinivas, 2017).

Spring load (kg) at known deflection (m), $S = a_1 + a_2\delta + a_3\delta^2 + a_4\delta^3 + a_5\delta^4 + a_6\delta^5$

The constants, $a_1 = 0$, $a_2 = 0.0077E05$, $a_3 = -0.0857E05$, $a_4 = 0.744E05$, $a_5 = -2.6744E05$ and $a_6 = 4.0748E05$.

$\delta = l_{si} - l_s$

Spring angle with collector tilt (ζ), $\xi_i = \tan^{-1}\left(\dfrac{l_s + \Delta l_{noon} + r_t(1-\sin\zeta_i)}{r_t \cos\zeta_i}\right)$

The instantaneous spring length during operation

$$l_{si} = \sqrt{\left(l_s + \Delta l_{noon} + r_t(1-\sin\zeta_i)\right)^2 + \left(r_t \cos\zeta_i\right)^2}$$

The tracking load, $W_{Ei} = 2S_i\left(\dfrac{r_t}{r}\right)\sin(\zeta_i + \xi_i)$

Draw the information flow diagram to find the tracking load. The given data is collector tilt angle, ζ, length of spring, l_s, spring deflection at noon, Δl_{noon}, tracking radius and r_t, sprocket wheel radius, r.

Solution

Figure 1.36 shows the information flow diagram for the solution of the tracking system.

13. Solve the above problem with the following data.

Spring original length = $l_s = 0.5279$ m

Spring deflection at noon = $\Delta l_{noon} = 0.0264$ m

Sprocket wheel radius, r = 0.125 m

Tracking wheel radius, $r_t = 0.07$ m

Instantaneous angle of collector axis to the horizontal, $\zeta_i = 31.82\ °$

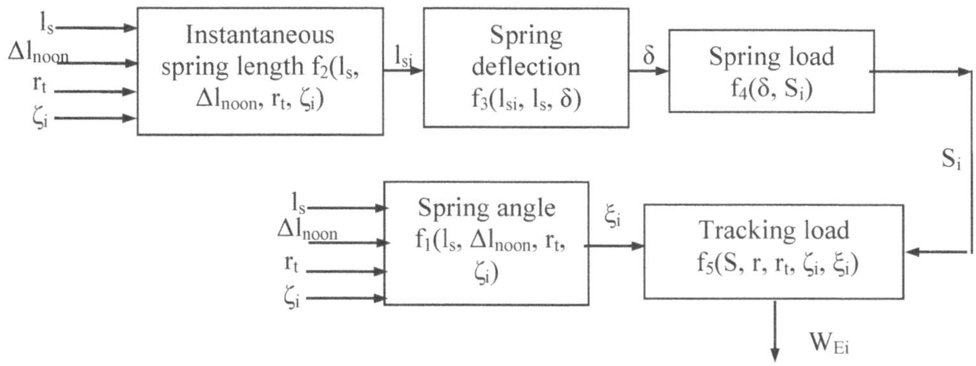

Fig. 1.36. Information flow diagram to solve the tracking mechanism.

Solution

Spring angle with collector tilt (ζ), $\xi_i = \tan^{-1}\left(\dfrac{l_s + \Delta l_{noon} + r_t(1-\sin\zeta_i)}{r_t\cos\zeta_i}\right)$

$$= \tan^{-1}\left(\frac{0.5279 + 0.0264 + 0.007(1-\sin 31.82)}{0.07\cos 31.82}\right)$$

$$= 84.21\,^{\circ}$$

The instantaneous spring length during operation

$$l_{si} = \sqrt{\left(l_s + \Delta l_{noon} + r_t(1-\sin\zeta_i)\right)^2 + \left(r_t\cos\zeta_i\right)^2}$$

$$= \sqrt{\left(0.5279 + 0.0264 + 0.007(1-\sin 31.82)\right)^2 + \left(0.07\cos 31.82\right)^2}$$

= 0.5904 m

$\delta = l_{si} - l_s$

= 0.5904 – 0.5279 = 0.0625 m

Spring load (kg) at known deflection (m), $S = a_1 + a_2\delta + a_3\delta^2 + a_4\delta^3 + a_5\delta^4 + a_6\delta^5$

The constants, $a_1 = 0$, $a_2 = 0.0077E05$, $a_3 = -0.0857E05$, $a_4 = 0.744E05$, $a_5 = -2.6744E05$ and $a_6 = 4.0748E05$.

S = 32.7983 kg

The tracking load, $W_{Ei} = 2S_i\left(\dfrac{r_t}{r}\right)\sin(\zeta_i + \xi_i)$

$$W_{Ei} = 232.7983\left(\frac{0.007}{0.125}\right)\sin(31.82 + 84.21)$$

= 33 kg

14. A fluid is heated in a flat plate collector and parabolic trough collector as shown in Fig. 1.37. Thermal efficiency of the collector is the function of solar radiation (I_T), ambient temperature (T_a) and fluid inlet temperature (T_a) which are grouped into a single operational parameter

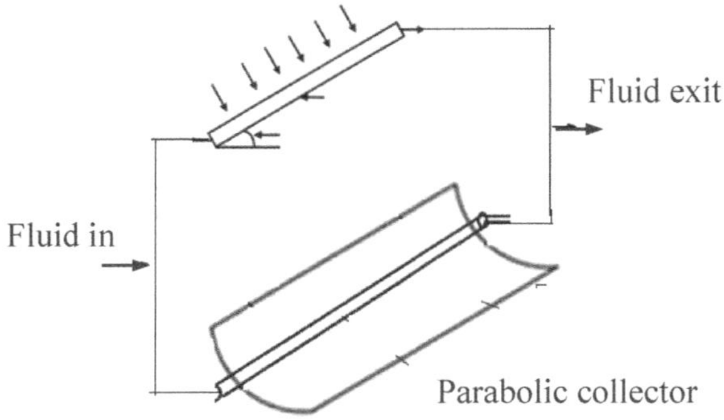

Fig. 1.37. Arrangement of solar thermal collectors for the fluid heating (1-flat plate collector and 2-parabolic collector).

(P). From the testing conditions, the efficiency of the collectors is expressed as a function of these three variables as a group. The relation between these two efficiencies is also modeled. Determine the separate solar collector's efficiencies, total efficiency and operational parameter ($P = (T_{fi}-T_a)/I_T$) using Newton-Raphson method.

Solution

Step 1

$P = 1.03\ \eta^2 + 0.154$

$P = 0.692 - 0.124\ \eta_1 - 0.024\ \eta_1^2$ (parabolic collector)

$P = 0.782 - 0.649\ \eta_2 - 0.0153\ \eta_2^2$ (flat plate collector)

$\eta = (\eta_1 + \eta_2)/2$

$f_1 = P - 1.03\ \eta^2 - 0.154 = 0$

$f_2 = P - 0.692 + 0.124\ \eta_1 + 0.024\ \eta_1^2 = 0$

$f_3 = P - 0.782 + 0.649\ \eta_2 + 0.0153\ \eta_2^2 = 0$

$f_4 = \eta - 0.5\ \eta_1 - 0.5\ \eta_2 = 0$

Step 2. Choose trial values of the variables, which are here selected as $P = 0.3$, $E_1 = 0.7$, $E_2 = 0.4$ and $E = 0.05$.

Step 3. Calculate the magnitudes of the fs at the temporary values of the variables, $f_1 = -0.2248$, $f_2 = -0.2934$, $f_3 = -2210$ and $f_4 = 0.5$.

Step 4. The partial derivatives are shown in Table 1.16.

Table 1.16. Results of partial derivatives.

	$\partial/\partial P$	$\partial/\partial E_1$	$\partial/\partial E_2$	$\partial/\partial E$
$\partial f_1/\partial$	1	0	0	-2.06 E
$\partial f_2/\partial$	1	$0.124 + 0.048\ E_1$	0	0
$\partial f_3/\partial$	1	0	$0649 + 0.0306\ E_2$	0
$\partial f_4/\partial$	0	-0.5	-0.5	1

Step 5. Substituting the temporary values of the variables into the equations for the partial derivatives forms a set of linear simultaneous equations to be solved for the corrections to x:

$$\begin{bmatrix} 1.0 & 0.0 & 0.0 & -1.2360 \\ 1.0 & 0.1576 & 0.0 & 0.0 \\ 1.0 & 0.0 & 0.6612 & 0.0 \\ 0.0 & -0.5 & -0.5 & 1.0 \end{bmatrix} \begin{bmatrix} \Delta x_1 \\ \Delta x_2 \\ \Delta x_3 \\ \Delta x_4 \end{bmatrix} = \begin{bmatrix} -0.22487 \\ -0.2934 \\ 0.2210 \\ 0.050 \end{bmatrix}$$

The solution of the above matrix is

$\Delta x_1 = -0.2596$

$\Delta x_2 = -0.2147$

$\Delta x_3 = 0.0584$

$\Delta x_4 = -0.0282$

where $\Delta x_i = x_{i,t} - x_{i,c}$

Since $x_{i,old} - x_{i,new} = x_{i,t} - x_{i,c}$

$$x_{i,new} = x_{i,old} - (x_{i,t} - x_{i,c})$$

Step 6. Solution of the simultaneous equations

$P = 0.3 - (-0.2596) = 0.5596$

$E_1 = 0.7 - 0.2147 = 0.9147$

$E_2 = 0.4 - 0.0584 = 0.3416$

$E = 0.6 - (-0.0282) = 0.6282$

These values of the variables are returned to step 3 for the next iteration.

The values of the fs and the variables resulting from continued iterations are shown in Table 1.17.

Table 1.17. Results of iterations for solutions of non-linear equations.

After iteration	f_1	f_2	f_3	f_4	P	E_1	E_2	E
1	–0.2248	–0.2934	–0.2210	0.0500	0.5596	0.5596	0.3416	0.6282
2	–8.16e-4	0.001	5.22e-5	0.0000	0.5590	0.9117	0.3424	0.6271
3	–1.23e-6	2.18e-7	1.05e-8	–5.55e-17	0.5590	0.9117	0.3424	0.6271

Chapter 2
Thermal Properties

2.1 Introduction

Modeling and simulation of energy systems demand data on the thermodynamic properties of the working fluids. The processes and the total system simulation are possible with the support of generated thermal properties, which play an important role in the accuracy of the results. Equation fitting is one of the methods to express properties in the form of a mathematical model. A pure substance changes its position from solid to liquid, from liquid to vapor, and vice versa. Information on sensible heat and latent heat, along with their properties, is required to design the energy system with a single fluid system, i.e., a pure substance. A gas with a fixed chemical composition is also a pure substance. The gas cannot be converted into liquid by isothermal compression, but it is possible for the vapor. The properties can be evaluated using the ideal gas model or the real gas model. The ideal gas model is easy to use, but it results in considerable error, which is not suitable for real-time work. Therefore, to develop realistic results with laboratory or experimentation, the real gas model with an acceptable error is more reliable. The analysis of fluid properties is the first and essential step in energy system simulation. The solution is either graphical, such as Mollier charts or compressibility charts, or analytical, through empirical equations or property tables. Thermodynamic properties include specific volume or density, specific internal energy, specific enthalpy, specific entropy, and specific exergy. Thermo-physical properties differ from thermodynamic properties which include thermal conductivity and viscosity.

2.2 Gibbs phase rule

A matter having one fixed chemical composition throughout its mass is called a pure substance. It may exist in one or more phases. In sensible heat transfer, the fluid is in the same state, while in latent heat exchange, the fluid changes state. As per Gibbs phase rule, the properties of a pure substance in a single state are a function of temperature and pressure. During phase change, any one property, either temperature or pressure information (value) is sufficient (Holman and Pilling, 2004).

Gibbs phase rule,

$$F = C + 2 - P \tag{2.1}$$

In the above equation

F = the degree of freedom, i.e., number of variables required to define the state of substance
C = number of components and
P = the number of phases.

DOI: 10.1201/9781032642505-2

For example,

i) Single component without phase change, C = 1, and P = 1

$F = 1 + 2 - 1 = 2$

For sensible heat transfer, two parameters are required to describe the properties of a pure substance. Both temperature and pressure are required.

ii) Single component with phase change: C = 1, and P = 2

Then, $F = 1 + 2 - 2 = 1$.

In solidification, freezing, boiling and condensation, the number of phases is 2. Therefore, any one parameter, i.e., either pressure or temperature, is sufficient to deduce the properties of a pure substance.

iii) Binary fluid system without phase change (aqua ammonia), C = 2, and P = 1

$F = 2 + 2 - 1 = 3$.

In a multicomponent system, three parameters are required to find the other properties. In this case apart from the temperature and pressure, concentration also is required.

iv) Binary fluid system with phase change, C = 2, and P = 2

$F = 2 + 2 - 2 = 2$

Two parameters are sufficient. In this example, if concentration and temperature are known, the pressure can be determined (using saturated temperature). So, any two properties are enough instead of three variables.

v) Single component with three phases, C = 1, P = 3

$F = 1 + 2 - 3 = 0$

In the sublimation process, the triple point is fixed to a fluid. Therefore, the degree of freedom is zero.

The following Table 2.1 summarizes this discussion.

Table 2.1. Gibbs phase rule to find the degree of freedom.

S. No.	Fluid	Phase change	Components, C	Phases, P	Degree of freedom, F = C + 2 - P	Process
1.	Single fluid	Without phase change	1	1	2	Sensible heat transfer
2.	Single fluid	With phase change	1	2	1	Boiling, condensation, freezing and melting
3.	Single fluid	With phase change	1	3	0	Sublimation
4.	Binary fluid	Without phase change	2	1	3	Sensible heat transfer
5.	Binary fluid	With phase change	2	2	2	Boiling, condensation, freezing and melting

2.3 Thermodynamic relations

The need for equations to represent thermodynamic properties arises in the modeling of energy systems. These equations can be used to solve the properties, viz., temperature, pressure, enthalpy, density or specific volume, entropy, and occasionally the internal energy. Generally, thermodynamic properties are a function of temperature and pressure.

The development of thermodynamic properties requires both property relations from classical thermodynamics and experimental data. A frequently used technique to derive the properties is the use of experimental data with the following four equations:

1. Real gas equation of vapour (p-v-T)
2. Specific heat of liquid at reference pressure
3. Saturated pressure and saturated temperature relation during phase change
4. Density of liquid.

These four equations result in non-measurable or derived properties, i.e., heat (internal energy or enthalpy), work, and entropy. These properties can be represented on the property chart shown in Fig. 2.1. Based on the application, a suitable property chart can be used, such as the T-S diagram for power plant cycles, the h-S diagram for the steam Rankine cycle, the p-h chart for the refrigeration cycle, the enthalpy-concentration chart for the binary fluid system, etc.

The primary properties of thermodynamics are pressure, volume, temperature, and entropy. The first three properties are measurable, and the fourth property is derived. The first law of thermodynamics, applied to a closed system consisting of a piston cylinder, such as Fig. 2.2, yields the equation.

$$dq = du + dw \tag{2.2}$$

where q = rate of heat transfers per unit mass in vessel, kJ/kg
u = internal energy, kJ/kg
w = external work per unit mass in vessel, kJ/kg

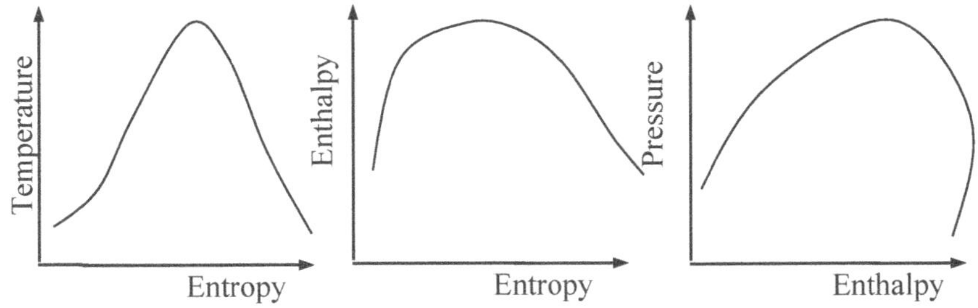

Fig. 2.1. Some widely used property charts.

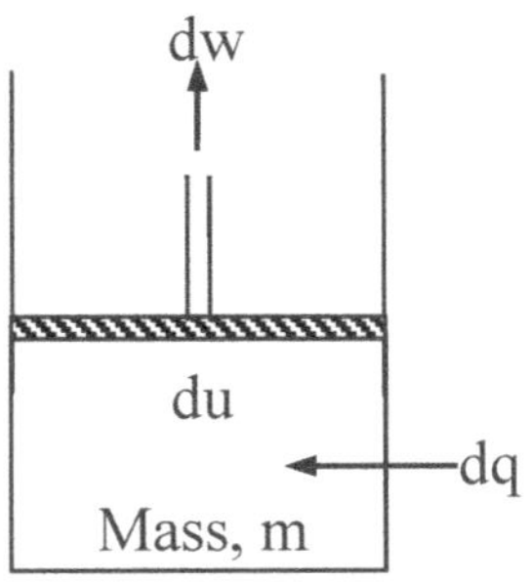

Fig. 2.2. Energy balance in a closed system.

If the heat is transferred to the system in a thermodynamically reversible process,

$$dq_{rev} = Tds \tag{2.3}$$

where T = temperature, K
s = entropy, kJ/kg K

and if the work is performed reversibly

$$dw_{rev} = pdv \tag{2.4}$$

where p = pressure, kPa
v = specific volume, m^3/kg
substitute Eqs. 2.2 and 2.4 into Eq. 2.3 to arrive at

$$Tds = du + pdv \tag{2.5}$$

Even though Eq. 2.5 was arrived at through the use of a non flow process with the assumption of reversibility, thermodynamic logic permits us to state that Eq. 2.5 is perfectly general and is applicable to all processes whether flow or non flow, reversible or irreversible. The reason is that Eq. 2.5 is exclusively a relationship of properties, and the procedure by which we arrived at the equation is immaterial. If v is held constant, dv = 0 and the equation may be written as follows

$$\left.\frac{\partial u}{\partial s}\right|_v = T \tag{2.6}$$

and if s is held constant,

$$\left.\frac{\partial u}{\partial v}\right|_s = -p \tag{2.7}$$

Enthalpy also can be formulated with thermodynamic relations.

$$h = u + pv$$

$$dh = du + pdv + vdp \tag{2.8}$$

In combination with the above equation,

$$dh = Tds + vdp \tag{2.9}$$

From this equation, following two property relations can be formed.

$$\left.\frac{\partial h}{\partial s}\right|_p = T \tag{2.10}$$

$$\left.\frac{\partial h}{\partial p}\right|_s = v \tag{2.11}$$

A set of equations that assists in achieving the goals of developing convenient, yet reasonably accurate, property relations are the Maxwell relations.

The Maxwell relations are

$$\left.\frac{\partial T}{\partial v}\right|_s = -\left.\frac{\partial p}{\partial s}\right|_v \tag{2.12}$$

$$\left.\frac{\partial T}{\partial p}\right|_s = \left.\frac{\partial v}{\partial s}\right|_p \tag{2.13}$$

$$\left.\frac{\partial v}{\partial T}\right|_p = -\left.\frac{\partial s}{\partial p}\right|_T \tag{2.14}$$

$$\left.\frac{\partial p}{\partial T}\right|_v = \left.\frac{\partial s}{\partial v}\right|_T \tag{2.15}$$

The Maxwell relations provide a link between entropy and the properties in the p-v-T equation of state and are derived from four differential equations:

T ds = du + p dv (2.16)

T ds = dh – v dp (2.17)

da = du – T ds – s dT (Helmholtz function) (2.18)

since a = u – Ts

dg = dh – T ds – s dT (Gibbs function) (2.19)

since g = h – Ts

If z if function of x and y, i.e., z = f(x, y)

We can write in the differential form

$$dz = \left.\frac{\partial z}{\partial x}\right| dx + \left.\frac{\partial z}{\partial y}\right| dy$$

and also

$$\frac{\partial^2 z}{\partial x \partial y} = \frac{\partial^2 z}{\partial y \partial x}$$

The same pattern is used in developing each of the Maxwell relations illustrations. The internal energy u can be specified by two independent properties, s and v, so

u = f(s, v) and

$$du = \left.\frac{\partial u}{\partial s}\right|_v ds + \left.\frac{\partial u}{\partial v}\right|_s dv \tag{2.20}$$

Equating the two expressions for du from Eqs. 2.16 and 2.20 gives

$$Tds - pdv = \left.\frac{\partial u}{\partial s}\right|_v ds + \left.\frac{\partial u}{\partial v}\right|_s dv$$

The coefficients of ds and dv must be equal, so

$$T = \left.\frac{\partial u}{\partial s}\right|_v \quad \text{and} \quad -p = \left.\frac{\partial u}{\partial v}\right|_s$$

The second derivatives of u with respect to s and v from each of the two above relations are

$$\frac{\partial^2 u}{\partial s \partial v} = \left.\frac{\partial T}{\partial v}\right|_s = -\left.\frac{\partial p}{\partial s}\right|_v$$

which is the first Maxwell relation.

Similarly, h = f(s, p)

and $dh = \left.\frac{\partial h}{\partial s}\right|_p ds + \left.\frac{\partial h}{\partial p}\right|_s dp$

Equating this expression with dh from Eq. 2.9, T ds = dh – v dp

$$Tds + vdp = \left.\frac{\partial h}{\partial s}\right|_p ds + \left.\frac{\partial h}{\partial p}\right|_s dp$$

The coefficients of ds and dp must be equal, so

$$T = \left.\frac{\partial h}{\partial s}\right|_p \quad \text{and} \quad v = \left.\frac{\partial h}{\partial p}\right|_s$$

The second derivatives of h with respect to s and p from each of the two above relations are

$$\frac{\partial^2 h}{\partial s \partial p} = \left.\frac{\partial T}{\partial p}\right|_s = \left.\frac{\partial v}{\partial s}\right|_p$$

which is the second Maxwell relation.

From Helmholtz function, a = u – Ts

da = du – T ds – s dT

Since T ds = du + p dv

da = T ds – p dv – T ds – s dT

da = –p dv – s dT

i.e., a = f(v, T)

and $da = \left.\frac{\partial a}{\partial v}\right|_T dv + \left.\frac{\partial a}{\partial T}\right|_v dT$

Equating this expression with da, da = – p dv – s dT

$$-pdv - sdT = \left.\frac{\partial a}{\partial v}\right|_T dv + \left.\frac{\partial a}{\partial T}\right|_v dT$$

The coefficients of dv and dT must be equal, so

$$-p = \left.\frac{\partial a}{\partial v}\right|_T \quad \text{and} \quad -s = \left.\frac{\partial a}{\partial T}\right|_v$$

The second derivatives of ***a*** with respect to v and T from each of the two above relations are

$$\frac{\partial^2 a}{\partial v \partial T} = -\left.\frac{\partial p}{\partial T}\right|_v = -\left.\frac{\partial s}{\partial v}\right|_T$$

$$\left.\frac{\partial p}{\partial T}\right|_v = \left.\frac{\partial s}{\partial v}\right|_T$$

which is the third Maxwell relation.

From Gibbs function, g = h – Ts

dg = dh – T ds – s dT

Since T ds = dh – v dp

dg = T ds + v dp – T ds – s dT

dg = v dp – s dT

i.e., g = f(p, T)

and $dg = \left.\frac{\partial g}{\partial p}\right|_T dp + \left.\frac{\partial g}{\partial T}\right|_p dT$

Equating this expression with dg, dg = v dp – s dT

$$vdp - sdT = \left.\frac{\partial g}{\partial p}\right|_T dp + \left.\frac{\partial g}{\partial T}\right|_p dT$$

The coefficients of dp and dT must be equal, so

$$v = \left.\frac{\partial g}{\partial p}\right|_T \quad \text{and} \quad -s = \left.\frac{\partial g}{\partial T}\right|_p$$

The second derivatives of g with respect to p and T from each of the two above relations are

$$\frac{\partial^2 g}{\partial p \partial T} = \left.\frac{\partial v}{\partial T}\right|_p = -\left.\frac{\partial s}{\partial p}\right|_T$$

which is the fourth Maxwell relation.

Summary of the above relations are as follows

$$T = \left.\frac{\partial u}{\partial s}\right|_v = \left.\frac{\partial h}{\partial s}\right|_p$$

$$p = -\left.\frac{\partial u}{\partial v}\right|_s = -\left.\frac{\partial a}{\partial v}\right|_T$$

$$v = \left.\frac{\partial h}{\partial p}\right|_s = \left.\frac{\partial g}{\partial p}\right|_T$$

$$s = -\left.\frac{\partial a}{\partial T}\right|_v = -\left.\frac{\partial g}{\partial T}\right|_p$$

$$\frac{\partial^2 u}{\partial s \partial v} = \left.\frac{\partial T}{\partial v}\right|_s = -\left.\frac{\partial p}{\partial s}\right|_v$$

$$\frac{\partial^2 h}{\partial s \partial p} = \left.\frac{\partial T}{\partial p}\right|_s = \left.\frac{\partial v}{\partial s}\right|_p$$

$$\frac{\partial^2 a}{\partial v \partial T} = -\left.\frac{\partial p}{\partial T}\right|_v = -\left.\frac{\partial s}{\partial v}\right|_T$$

$$\frac{\partial^2 g}{\partial p \partial T} = \left.\frac{\partial v}{\partial T}\right|_p = -\left.\frac{\partial s}{\partial p}\right|_T$$

Diagram for generating the Maxwell relations

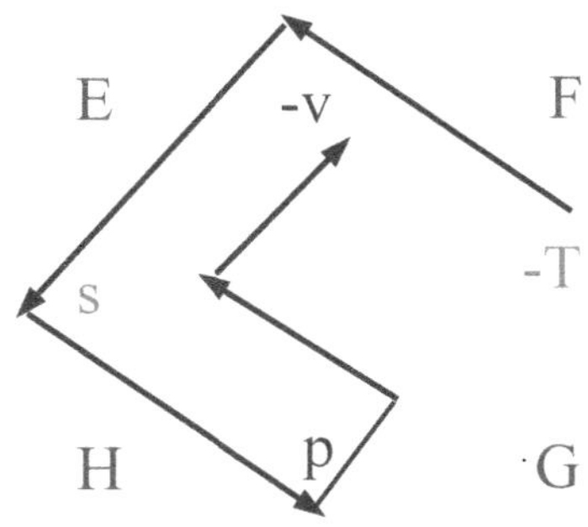

Fig. 2.3. Maxwell relation diagram.

E, F, G and H are in alphabetical order and respectively for internal energy, Helmholtz function, Gibbs function and enthalpy arranged in clockwise direction (Fig. 2.3). The variables s and –T are fixed and –v and p can be changed for flexibility to generate other relations.

From the above diagram, $(-)(-)\left.\frac{\partial T}{\partial v}\right|_s = (-)\left.\frac{\partial p}{\partial s}\right|_v$

2.4 Phase change properties

A useful expression for relating some thermodynamic liquid and vapor properties under saturated conditions is the Clapeyron equation. This equation can be developed from purely thermodynamic principles and is applicable to all pure substances. The development begins with the expression of the Gibbs function:

$g = h - Ts$ (2.21)

In differential form,

$dg = dh - Tds - sdT$ (2.22)

we have $Tds = dh - vdp$,

and so, $dh - Tds = vdp$

$dg = vdp - sdT$ (2.23)

Next, apply Eq. 2.23 to a change between saturated liquid and saturated vapor at the same pressure. In the mixture region, the temperature remains constant in a constant pressure process, so both dT and dp in Eq. 2.23 are zero.

$dg = 0$

So the Gibbs function of a saturated liquid and saturated vapor are equal at the same temperature and pressure, $g_f = g_g$.

The next operation will be to change slightly the pressure and temperature of a sample of liquid/vapor at saturation as shown in Fig. 2.4.

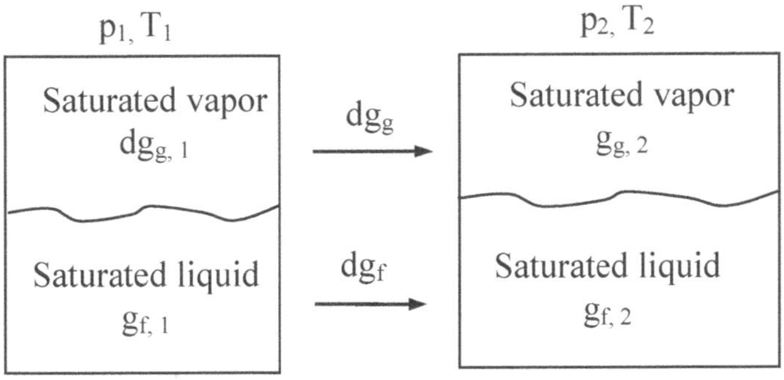

Fig. 2.4. Changing the pressure and temperature of a liquid/vapor mixture.

Since $g_{f,1} = g_{g,1}$, and $g_{f,2} = g_{g,2}$,

$dg_g = dg_f$

$v_g dp - s_g dT = v_f dp - s_f dT$

$dp(v_g - v_f) = dT(s_g - s_f)$

$$\frac{dp}{dT} = \frac{s_g - s_f}{v_g - v_f} = \frac{T(s_g - s_f)}{T(v_g - v_f)} = \frac{h_{fg}}{T(v_g - v_f)} \tag{2.24}$$

Equation 2.24 is the Clapeyron equation and relates the derivative of the saturation pressure with respect to the saturation temperature to certain other properties at saturated conditions.

By making two bold approximations, the Clapeyron equation can be used to suggest the form of an equation to relate the saturation pressure to its corresponding saturation temperature. Assume (1) that $v_f << v_g$, and (2) that the ideal gas equation applies, even for saturated vapor, $pv_g = RT$. Equation 2.24 then becomes

$$\frac{dp}{dT} = \frac{h_{fg}}{T(RT/p)}$$

and $\dfrac{dp}{p} = \dfrac{dT}{T^2}\dfrac{h_{fg}}{R}$

Integrating,

$\ln p = -h_{fg}/(RT) + \text{constant of integration}$

or $\ln p = A + B/T$ (2.25)

where A and B are positive and negative constants, respectively. The form suggested by Eq. 2.25 is widely used and indicates that on a ln p vs 1/T graph a straight line represents the characteristics of a given substance. Antoine equation (1888) introduces one additional parameter,

$\ln p = A + B/(T - C)$ (2.26)

A proposal for the estimate of C is

$C, K = -18 + 0.19\,T$

2.5 Real gas equations

Due to assumptions in an ideal gas model, the following deviations can be observed.

1. There is an attraction between the molecules of the gas
2. Molecules of the gas occupy some space.
3. Perfect gas model is applicable for low pressure and high temperature gases.
4. When a gas is compressed, the molecules are forced into close proximity to one another and ideal gas model is not applicable.

In Van der Waals equation of state, we may consider an ideal gas as one in which:

1. The molecules are very widely spaced and behave as point masses.
2. The molecular forces exert very little influence on collision processes because of the wide molecular spacing.
3. The volume occupied by the molecules is very small compared with the total volume occupied by the gas, again because of the wide molecular spacing.

When a gas is compressed, the molecules are forced into closer proximity to one another and non-ideal-gas behavior will eventually develop. Finally, of course there can be such strong force-field interaction that the gas is converted to a liquid.

Van der Waals proposed an equation of state to account for real gas behavior.

$$\left(P+\frac{a}{v^2}\right)(v-b)=RT$$

a is a constant which takes account of the molecular force of interactions and b is a constant which compensates for the volume occupied by the molecules. The reasoning used to obtain the $\frac{a}{v^2}$ term is as follows, when a group of molecules strikes a surface, the pressure exerted is proportional to the molecular density (molecules per unit volume, $\frac{N}{V}$), because it is this quantity that governs the number of molecular impacts. For molecular impact, the interaction force is reasoned to be proportional to $\left(\frac{N}{V}\right)^2$ since the net effect results from two molecules striking each other. The molecular density is clearly proportional to the macroscopic gas density which in turn is proportional to $\frac{1}{v}$. Thus, the net interaction force is proportional to $\frac{1}{v^2}$ and a is the proportionality constant. The constant b simply takes account of the effective volume occupied by the molecule.

To use the Van der Waals equation, values of the constants a and b must be determined experimentally. Van der Waals equation is not applicable to all ranges of pressures and temperatures.

$$a=\frac{4\bar{R}T_c^2}{P_c e^2}, b=\frac{\bar{R}T_c}{P_c e^2}, \frac{P_c\bar{v}_c}{\bar{R}T_c}=0.271=\frac{2}{e^2}$$

Some other equations of state for real gases are:

Clausius: $P(v-b) = RT$

Berthelot: $\left(P+\frac{a}{Tv^2}\right)(v-b)=RT$

Redlich-Kwong: $\left(P+\frac{a}{\sqrt{T}v(v+b)}\right)(v-b)=RT$

Compared to Van der Waals equation, Redlich-Kwong gives accurate values. A compressibility chart is a simple provision to get the real gas' compressibility factor in an ideal gas equation. To evaluate a real gas, a correction factor in the characteristic gas equation is used, which is called the compressibility factor (Z).

$$Pv = ZRT \rightarrow Z=\frac{Pv}{RT}$$

Different gases at the same reduced pressure and reduced temperature are found to have nearly the same value of the compressibility factor and the same reduced volume.

This is called law of corresponding states.

$$reduced\ pressure,\ \mathrm{Pr}=\frac{pressure,\ P}{critical\ pressure,\ P_c}$$

$$reduced\ temperature,\ Tr=\frac{temperature,\ T}{critical\ temperture,\ T_c}$$

This similarity may be used to determine the volume of a gas at a particular pressure and temperature when its properties in a critical state are known. A generalized chart in which the compressibility factor Z is plotted against the reduced pressure P_r with the reduced temperature T_r as a parameter is called a compressibility chart. From the compressibility chart, $P_r \rightarrow 0$, $Z \rightarrow 1$, and gas behaves like a perfect gas. The value of the compressibility factor Z for any gas may be less than or more than unity. Accuracy within 5% may be obtained by using the chart.

$$\left(p+\frac{a}{v^2}\right)(v-b)=RT \text{ or } p=\frac{RT}{v-b}-\frac{a}{v^2}$$

$$a=\frac{27R^2T_c^2}{64p_c} \text{ and } b=\frac{RT_c}{8p_c}$$

$$\left(p+\frac{27R^2T_c^2}{64p_c}\frac{1}{v^2}\right)\left(v-\frac{RT_c}{8p_c}\right)=RT$$

$$\left(p_rp_c+\frac{27R^2T_c^2}{64p_c}\frac{1}{(v_rv_c)^2}\right)\left(v_rv_c-\frac{v_c}{8}\right)=RT_rT_c$$

$$\left(p_rp_c+\frac{27p_c}{64}\frac{1}{(v_r)^2}\right)\left(v_rv_c-\frac{v_c}{8}\right)=RT_rT_c$$

$$p_cv_c\left(p_r+\frac{27}{64}\frac{1}{(v_r)^2}\right)\left(v_r-\frac{1}{8}\right)=RT_rT_c$$

$$\left(p_r+\frac{27}{64}\frac{1}{(v_r)^2}\right)\left(v_r-\frac{1}{8}\right)=T_r$$

$$\left(p_r + \frac{27}{64}\frac{1}{v_r^{22}}\right)\left(\frac{8v_r - 1}{8}\right) = T_r$$

$$\left(p_r + \frac{27}{64}\frac{1}{v_r^2}\right)(8v_r - 1) = 8T_r$$

Substituting Z,

pv = ZRT

$$p_r p_c v_r v_c = ZRT_r T_c$$

$$p_r = \frac{ZRT_r T_c}{p_c v_r v_c} = \frac{ZT_r}{v_r}$$

Therefore, $p_r\ v_r = ZT_r$

$$\left(p_r + \frac{27}{64}\frac{1}{v_r^2}\right)(8v_r - 1) = 8T_r$$

$$\left(\frac{ZT_r}{v_r} + \frac{27}{64}\frac{p_r^2}{Z^2T_r^2}\right)\left(8\frac{ZT_r}{p_r} - 1\right) = 8T_r$$

$$\left(Z + \frac{27}{64}\frac{v_r}{T_r}\frac{p_r^2}{Z^2T_r^2}\right)\left(\frac{ZT_r}{p_r} - \frac{1}{8}\right) = T_r\frac{v_r}{T_r}$$

$$\left(Z + \frac{27}{64}\frac{p_r}{ZT_r^2}\right)\left(\frac{ZT_r}{p_r} - \frac{1}{8}\right) = v_r$$

$$\left(Z + \frac{27}{64}\frac{p_r}{ZT_r^2}\right)\left(\frac{ZT_r}{p_r} - \frac{1}{8}\right) / \frac{ZT_r}{p_r} = v_r / \frac{ZT_r}{p_r}$$

$$\left(Z + \frac{27}{64}\frac{p_r}{ZT_r^2}\right)\left(1 - \frac{p_r}{8ZT_r}\right) = 1$$

$$\left(\frac{64Z^2T_r^2 + 27p_r}{64ZT_r^2}\right)\left(\frac{8ZT_r - p_r}{8ZT_r}\right) = 1$$

Some algebraic manipulation of the Van der Waals equation along with the relations for the constants (a and b) gives

$$Z^3 - \left(\frac{\text{Pr}}{8Tr} + 1\right)Z^2 + \left(\frac{27\,\text{Pr}}{64Tr^2}\right)Z - \frac{27\,\text{Pr}^2}{512Tr^3} = 0$$

The above equation can be solved for Z explicitly in terms of Pr and Tr and consequently we postulate the existence of relation of the form

$Z = f(P_r, T_r)$ – Law of corresponding states.

The compressibility factor is a complex function of reduced temperature and reduced pressure, and it is not possible to obtain mathematical expressions for Z in terms of T_r and P_r. This is a serious drawback for numerical calculations.

For the generalization, the Redlich-Kwong (RK) equation of state is given by

$$\left(P+\frac{a}{v(v+b)}\right)(v-b)=RT$$

$$P=\frac{RT}{(v-b)}-\frac{a}{v(v+b)}$$

This can be rewritten as

$$\frac{Pv}{RT}=Z=\frac{v}{(v-b)}-\frac{a}{RT(v+b)}=\frac{1}{\left(1-\frac{b}{v}\right)}-\frac{a\left(\frac{b}{v}\right)}{bRT\left(1+\frac{b}{v}\right)}$$

$$Z=\frac{1}{(1-h)}-\frac{ah}{bRT(1+h)}$$

where $h=\frac{b}{v}=\frac{bP}{ZRT}$

Substituting for the parameters a and b in terms of P_c and T_c from these equations, we get

$$a=0.42748R^2\left(\frac{T_c^{2.5}}{p_cT^{1/2}}\right)\ and\ b=0.08664R\left(\frac{T_c}{p_c}\right)$$

$$Z=\frac{1}{1-h}-\left(\frac{h}{1+h}\right)\frac{4.93398}{T_r^{3/2}}$$

$$h=\frac{0.08664P_r}{ZT_r}$$

The above equations can be conveniently used for iterative calculations. The calculations can be initiated with a starting value of Z = 1 in the second equation to obtain the value of h. This value of h can be substituted in the first equation to obtain a revised value of Z. The iterations can be continued until the values of Z do not differ appreciably in successive iterations.

Equations express the compressibility factor Z as a function of T_r and P_r. Any equation of state that expresses Z as a function of T_r and P_r is known as a generalized equation of state. The advantage of a generalized equation of state is that it permits the evaluation of P-v-T data for gases from knowledge of T_c and P_c. Fig. 2.5, Fig. 2.6, and Fig. 2.7 show the values of compressibility factor as a function of reduced temperature and reduced pressure, respectively, for reduced pressure less than 1, up to 10, and 10 to 40.

From PK Nag (Nag, 2008),

$$p=\frac{RT}{v-b}-\frac{a}{\sqrt{T}v(v+b)}$$

$$a=0.42748R^2\left(\frac{T_c^{2.5}}{p_c}\right)\ and\ b=0.08664R\left(\frac{T_c}{p_c}\right)$$

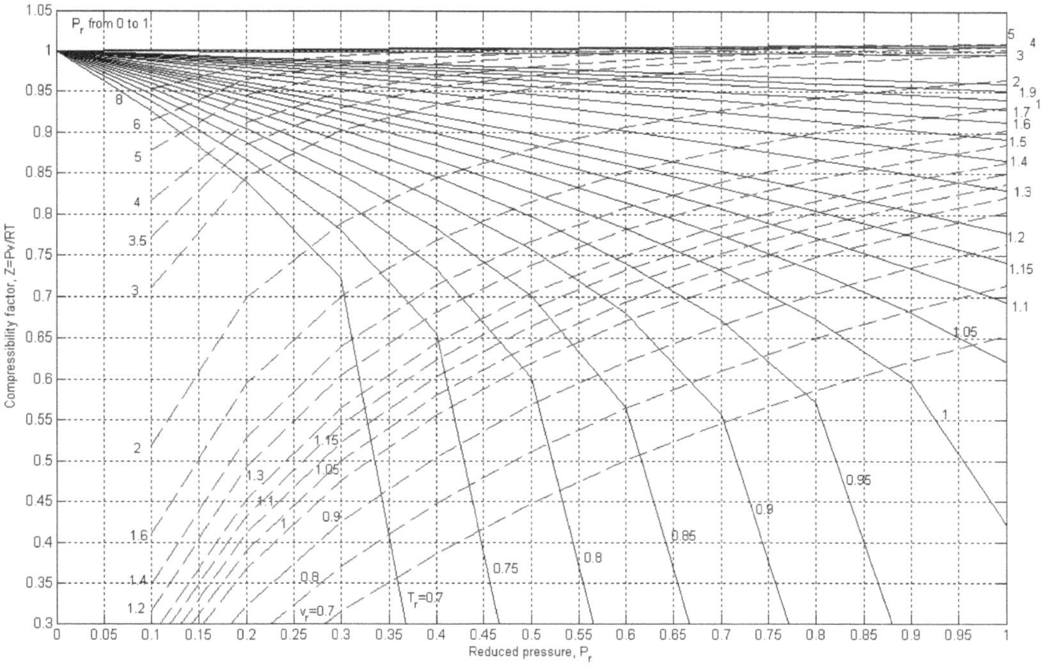

Fig. 2.5. Generalized compressibility chart, $P_r \leq 1$.

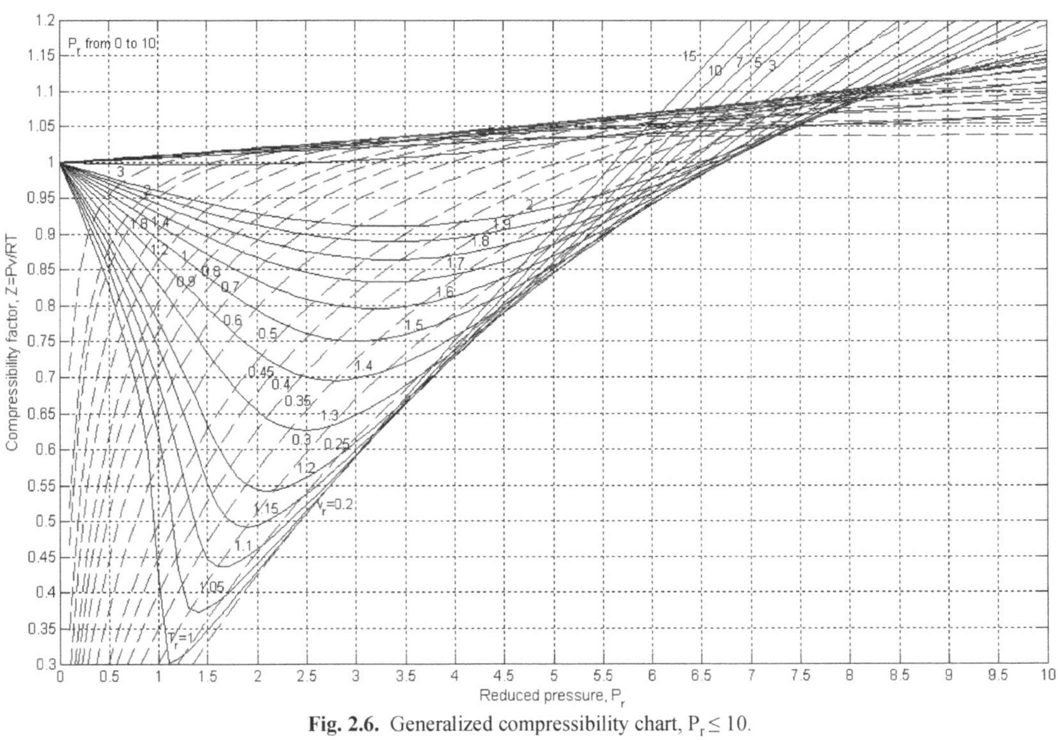

Fig. 2.6. Generalized compressibility chart, $P_r \leq 10$.

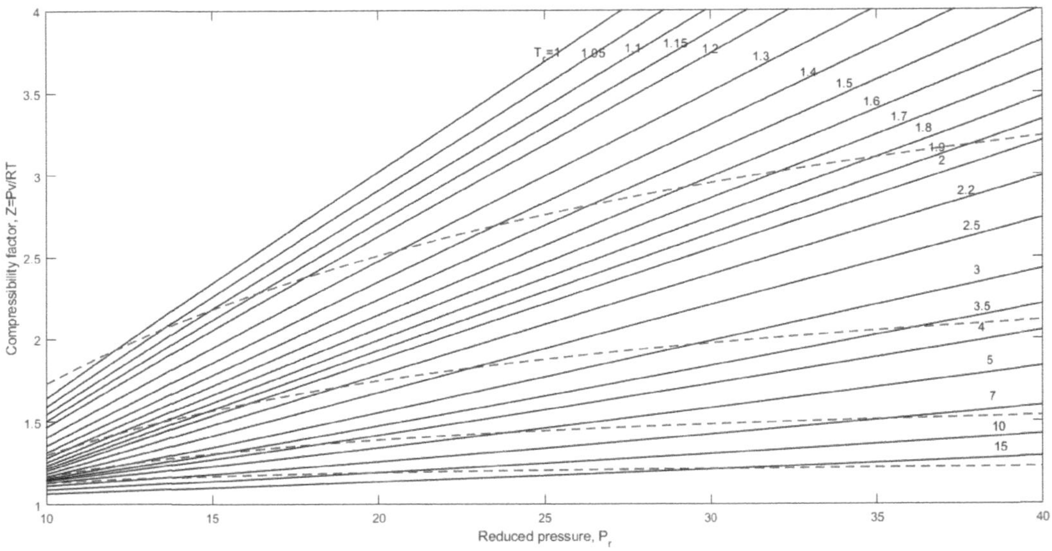

Fig. 2.7. Generalized compressibility chart, $10 \leq P_r \leq 40$.

Compressibility factor, $Z = \dfrac{pv}{RT} = \dfrac{v}{v-b} - \dfrac{a}{RT^{\frac{3}{2}}(v+b)}$

The specific heat of a real gas equation can also be determined as a function of temperature and pressure. One of the tools listed in a section of thermodynamic properties for determining a complete set of thermodynamic properties is specific heat. The two specific heats of interest are those at constant volume and at constant pressure, c_v and c_p, respectively.

Their definitions are

constant volume,

$$c_v = \frac{u}{T}\bigg|_v$$

constant pressure,

$$c_p = \frac{\partial h}{\partial T}\bigg|_p$$

Specific heats are means of computing enthalpies, entropies, and internal energies. Arbitrary base values are always chosen for those properties. In general, specific heats vary both with respect to pressure and temperature, although we shall find that temperature has the dominant influence. The computation of a change in property occurring in a change of pressure and temperature thus entails an integration, for which an equation for the specific heat as a function of p and T would be valuable. The strategy to be followed is to first express the so-called specific heat at zero pressure, c_{p0}, as a function of temperature. Such an expression would come from experimental data, possibly in conjunction with spectrographic measurements. In combination with p-v-T relations, cp0 can be extended into both c_p and c_v at various temperatures and pressures.

An important relation is the partial derivative of cp with respect to pressure at a constant temperature. This relation can be developed by finding the values of *d*s in the equation and showing that for the constant pressure process

$$\left.\frac{1}{T}\frac{\partial h}{\partial T}\right)_p = \left.\frac{\partial s}{\partial T}\right|_p$$

Substituting c_p from equation and differentiating with respect to p at constant T gives

$$\left.\frac{1}{T}\frac{\partial c_p}{\partial p}\right|_T = \frac{\partial^2 s}{\partial T \partial p} = \frac{\partial}{\partial T}\left[\left.\frac{\partial s}{\partial p}\right|_T\right]_p$$

With the substitution of the Maxwell relation, the equation results in

$$\left.\frac{\partial c_p}{\partial p}\right|_T = -T\left.\frac{\partial^2 v}{\partial T^2}\right|_p \tag{2.27}$$

The term $\left.\frac{\partial c_p}{\partial p}\right|_T$ is a function of p-v-T relations in a real gas equation. If we apply it to a perfect gas (pv = RT), the second derivate $\left.\frac{\partial^2 v}{\partial T^2}\right|_p$ becomes zero. Therefore, in a perfect gas, cp is not the function of pressure. In a perfect gas equation, c_p is either constant or a function of temperature. The term cp0 is the specific heat at zero pressure. c_p can be expressed in terms of temperature, and from the real gas equation, c_p can be expressed as a function of temperature and pressure. From the experimental data, the coefficients can be determined to express the function as a polynomial. Table 2.2 shows the coefficients for several common gases. The following are equations to determine the constant pressure of specific heats, specific enthalpy and specific entropy of various ideal gases. In Table 2.2 the constants for the equations are tabulated. The reference point is taken at 25°C and 1.01325 bar (Turns, 1996).

$$\frac{\overline{c}_p}{R} = a_1 + a_2 T + a_3 T^2 + a_4 T^3 + a_5 T^4$$

$$\frac{\overline{h}^0}{RT} = a_1 + \frac{a_2}{2}T + \frac{a_3}{3}T^2 + \frac{a_4}{4}T^3 + \frac{a_5}{5}T^4 + \frac{a_6}{T}$$

$$\frac{\overline{s}^0}{R} = a_1 \ln T + a_2 T + \frac{a_3}{2}T^2 + \frac{a_4}{3}T^3 + \frac{a_5}{4}T^4 + a_7$$

Mean isobaric heat capacity for evaluating enthalpy changes,

$$\overline{c}_p^h = \left[\frac{\overline{h} - \overline{h}^o}{T - T_0}\right]_p = \frac{1}{T - T_0}\int_{T_0}^{T} \overline{c}_p dT$$

Mean molar isobaric exergy capacity for evaluating changes in physical exergy,

$$\overline{c}_p^{\varepsilon} = \left[\frac{\overline{\varepsilon}^{\Delta T}}{T - T_0}\right]_p = \frac{1}{T - T_0}\left[\int_{T_0}^{T} \overline{c}_p dT - T_0 \int_{T_0}^{T} \frac{\overline{c}_p dT}{T}\right]$$

Most useful equations of state implicitly incorporate the specific volume v, so the expression for $\left.\frac{\partial^2 v}{\partial T^2}\right|_p$ is complex. The ultimate use of c_p is likely to be in computing changes in enthalpy or entropy over a change in pressure and temperature which then requires an integration of some

Table 2.2. Coefficients for equations of curve fitting for gas properties (C-H-O-N system).

Species	T(K)	a_1	a_2	a_3	a_4	a_5	a_6	a_7
CO	300–1000	0.03262451E2	0.15119409E-2	−0.03881755E-4	0.05581944E-7	−0.02474951E-10	−0.14310539E5	0.04848897E2
	1000–5000	0.03025078E2	0.14426885E-2	−0.05630827E-5	0.10185813E-9	−0.06910951E-13	−0.14268350E5	0.06108217E2
CO_2	300–1000	0.02275724E2	0.09922072E-1	−0.10409113E-4	0.0686668E-7	−0.0211728E-10	−0.04837314E6	0.10188488E2
	1000–5000	0.04453623E2	0.03140168E-1	−0.12784105E-5	0.0239393996E-8	−0.16690333E-13	−0.0489669E6	−0.09553959E1
H_2	300–1000	0.03298124E2	0.08249441E-2	−0.08143015E-5	0.09475434E-9	0.04134872E-11	−0.10225209E4	−0.03294094E2
	1000–5000	0.02991423E2	0.07000644E-2	−0.05633828E-6	−0.09231578E-10	0.15827519E-14	−0.08350340E4	−0.13551101E1
H_2O	300–1000	0.03386842E2	0.03474982E-1	−0.063546961E-4	0.06968581E-7	−0.02506588E-10	−0.03020811E6	0.02590232E2
	1000–5000	0.02672145E2	0.03056293E-1	−0.0873026E-5	0.12009964E-9	−0.06391618E-13	−0.02989921E6	0.06862817E2
N_2	300–1000	0.03298677E2	0.14082404E-2	−0.03963222E-4	0.05641515E-7	−0.02444854E-10	−0.10208999E4	0.03950372E2
	1000–5000	0.0292664E2	0.14879768E-2	−0.0568476E-5	0.10097038E-9	−0.06753351E-13	−0.09227977E4	0.05980528E2
NO_2	300–1000	0.02670600E2	0.07838500E-1	−0.08063864E-4	0.06161714E-7	−0.02320150E-10	0.02896290E5	0.11612071E2
	1000–5000	0.04682859E2	0.02462429E-1	−0.10422585E-5	0.01976902E-8	−0.13917168E-13	0.02261292E5	0.09885985E1
O_2	300–1000	0.03212936E2	0.11274864E-2	−0.0575615E-5	0.13138773E-8	−0.08768554E-11	−0.10052490E4	0.06034737E2
	1000–5000	0.03697578E2	0.06135197E-2	−0.1258842E-6	0.01775281E-9	−0.11364354E-14	−0.12339301E4	0.03189165E2

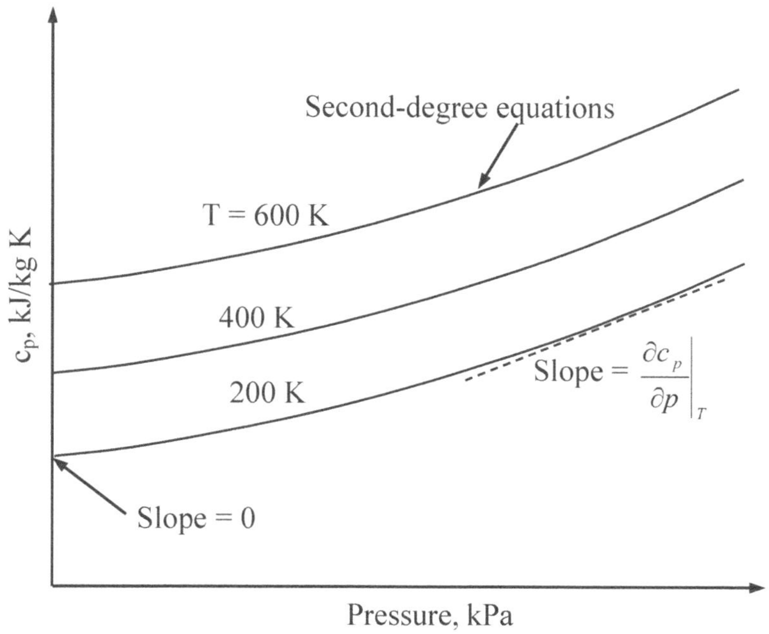

Fig. 2.8. Expression of c_p as a function of temperature and pressure.

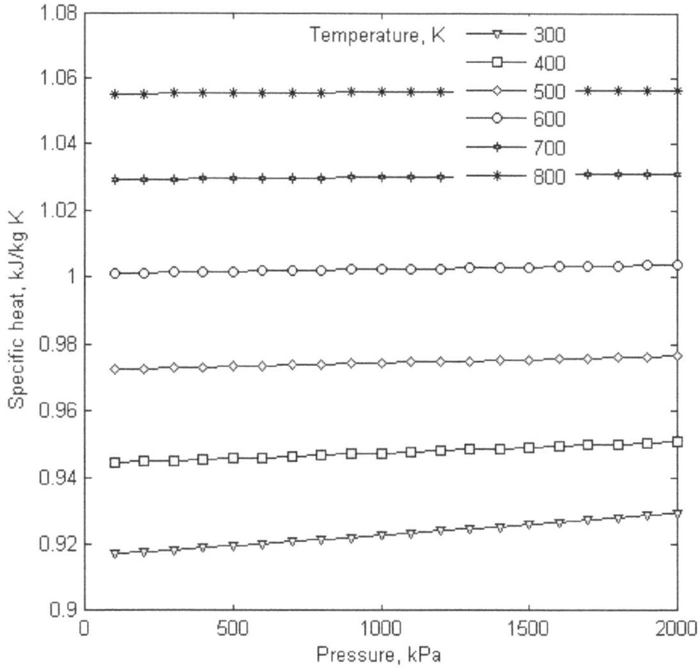

Fig. 2.9. Specific heat is function of pressure and temperature for oxygen.

complex expressions. Figure 2.8 shows that the c_p is the function of temperature and pressure. If the curve for a given temperature is approximated by a second-degree equation, this equation can be developed knowing (1) c_p at zero pressure, (2) that the slope is zero at zero pressure, and (3) the slope at some nonzero pressure. The values of specific heat of oxygen as a function of temperature and pressure are plotted in Fig. 2.9.

For example, for Van der Waals equation of state states,

$$p = \frac{RT}{v-b} - \frac{a}{v^2}$$

The equation is explicit for v

Therefore, $\left.\frac{\partial^2 v}{\partial T^2}\right|_p$ can be solved using numerical derivatives

$$\left.\frac{\partial^2 v}{\partial T^2}\right|_p = \frac{\frac{v_{T+\Delta T} - v_T}{\Delta T} - \frac{v_T - v_{T-\Delta T}}{\Delta T}}{\Delta T}$$

After finding the $\left.\frac{\partial^2 v}{\partial T^2}\right|_p$, c_p can be determined as follows

$c_p = c_{p0} + a_1\, p + a_2\, p^2$

$$\left.\frac{\partial c_p}{\partial p}\right|_T = a_1 + 2a_2 p$$

at p = 0, $\left.\frac{\partial c_p}{\partial p}\right|_T = 0$, therefore, $a_1 = 0$

$$a_2 = \frac{\left.\frac{\partial c_p}{\partial p}\right|_T}{2p}$$

$c_p = c_{p0} + a_2\, p^2$

$$c_p = c_{p0} + \frac{1}{2p}\left.\frac{\partial c_p}{\partial p}\right|_T p^2$$

$$c_p = c_{p0} + \frac{1}{2}\left.\frac{\partial c_p}{\partial p}\right|_T p$$

The simplest and probably most valuable equation of state is the perfect gas equation.

pv = RT

in which the specific gas constant R for the substance can be determined from the universal gas constant.

R = 8.314 kJ/kg K

A refinement of the perfect gas equation incorporates the compressibility factor, Z.

pv = ZRT

If Z is known, the equation gives a precise representation; unfortunately Z is a function of T and p. Graphs are available providing values of Z = f(T, p) for specific substances. For a wide range of temperatures and pressures, the calculation reverts to a cumbersome expression for Z.

An example of a form based heavily on empirical data is the Beattie-Bridgeman equation

$$p = \frac{RT}{v} + \frac{\alpha}{v^2} + \frac{\beta}{v^3} + \frac{\gamma}{v^4}$$

where the values of α, β and γ are determined from best fits of experimental data.

Critical temperatures and pressures of substances are readily available data and the advantage of two equations of state – the Van der Waals and the Redlich-Kwong equations – is that the constants are derivable from the critical properties. At the critical point both the first and second derivatives of p with respect to v are zero.

$$\left.\frac{\partial p}{\partial v}\right|_{T_c} = 0 \text{ and } \left.\frac{\partial^2 p}{\partial v^2}\right|_{T_c} = 0 \tag{2.28}$$

where T_c = critical temperature, K

When Eq. is imposed on the Van der Waals equation

$$p = \frac{RT}{v-b} - \frac{a}{v^2} \tag{2.29}$$

$$a = \frac{27R^2T_c^2}{64p_c} \text{ and } b = \frac{RT_c}{8p_c} \tag{2.30}$$

where p_c = critical pressure, kPa

A still more accurate two-constant equation is the one proposed by Redlich and Kwong.

$$p = \frac{RT}{v-b} - \frac{a}{T^{1/2}v(v+b)} \tag{2.31}$$

Soave made further improvements in accuracy by expressing a constant as a function of temperature. The simplicity of the Redlich-Kwong equation of state is, however, consistent with the approach of this chapter.

$$a = 0.42748\frac{R^2T_c^{5/2}}{p_c} \tag{2.32}$$

$$\text{and } b = 0.08664\frac{RT_c}{p_c} \tag{2.33}$$

The change in enthalpy in the vapor region can be found by integrating an expression for dh which is developed as follows. For vapor, h can be represented as a function of p and T, and then the expression for c_p substituted.

$$dh = \left.\frac{\partial h}{\partial T}\right)_p dT + \left.\frac{\partial h}{\partial p}\right)_T dp = c_p dT + \left.\frac{\partial h}{\partial p}\right)_T dp$$

Also, since T ds = dh – v dp

$$\left.\frac{\partial h}{\partial p}\right)_T = v + T\left.\frac{\partial s}{\partial p}\right)_T$$

Substitute the expression for $(\partial s/\partial p)_T$ from the Maxwell relation, Eq. 25, to obtain

$$dh = c_p dT + \left(v - T \frac{\partial v}{\partial T} \right)_p dp \tag{2.34}$$

For ideal gas, pv = RT,

$$\left(v - T \frac{\partial v}{\partial T} \right)_p dp = \left(v - T \frac{R}{p} \right) dp = \left(v - T \frac{v}{T} \right) dp = 0$$

The expression for a differential change in entropy comes from a parallel development:

$$ds = \left.\frac{\partial s}{\partial T}\right)_p dT + \left.\frac{\partial s}{\partial p}\right)_T dp$$

and from T ds = dh – v dp

$$\left.\frac{\partial s}{\partial T}\right)_p = \frac{1}{T} \left.\frac{\partial h}{\partial T}\right)_p = \frac{c_p}{T}$$

Then $ds = c_p \frac{dT}{T} - \left.\frac{\partial v}{\partial T}\right)_p dp$ (2.35)

For ideal gas, pv = RT

$$ds = c_p \frac{dT}{T} - \frac{R}{p} dp$$

After integration

$$s_2 - s_1 = c_p \ln \frac{T_2}{T_1} - R \ln \frac{p_2}{p_1}$$

Equations 2.34 and 2.35 can be integrated through a change in p and T to find the change in enthalpy and entropy respectively, with c_p treated as a variable with respect to temperature and possibly pressure. In Eq. 2.35 $\partial v/\partial T)_p$ is variable with respect to p. With an ideal gas, for example, $\partial v/\partial T)_p = R/p$, and the integral is $R \ln(p_2/p_1)$. The Redlich-Kwong equation may be used to modify R/p for real gases for the integration.

2.6 Properties of steam

Figure 2.10 shows the p-v diagram for pure water with all its saturated states. If we join all the state points from 'a' to 'f' for all the pressures up to the critical, a vapor dome will be generated as shown. The joining of all the saturation states results in saturation curves. Joining of all the saturated solid states (b), saturated liquid states (c) during melting, saturated liquid states during vaporization (d), and saturated vapor states (e) generates the saturation curves. The region between the saturated liquid curve during vaporization and saturation vapor that appears like a dome is called the vapor dome. The point where the saturated liquid curve and the saturated vapor curve meet in a common state is called the critical point. At the critical point, the latent vaporization is zero. Due to the peculiar property of water, the surface is hidden and shown in dotted lines due to a decrease in specific volume during melting (b-c).

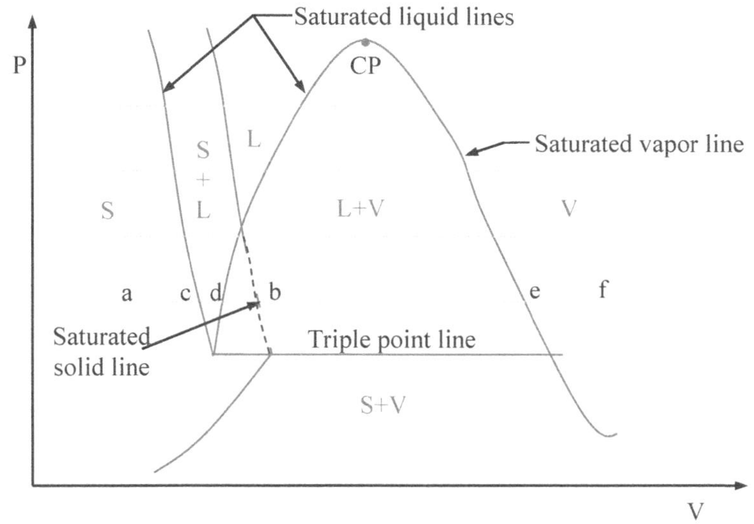

Fig. 2.10. P-v diagram of water which is contracting on melting.

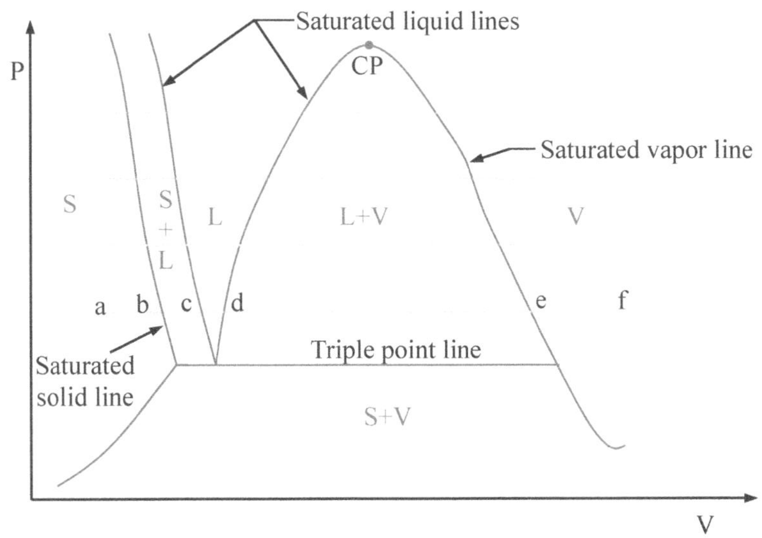

Fig. 2.11. The P-v diagram of a fluid other than water which expands on melting.

Other than water, the specific volume of a fluid always increases during heating and melting, as shown in Fig. 2.11. Compared to water, the process b-c differs, and the rest of the process variations are the same. There is no hidden line on the P-V diagram due to the continuous increment in specific volume with heating. The solid region can be seen on the left side of the saturated liquid line.

Figure 2.12 shows the pressure-volume diagram for liquid and vapor regions. In this diagram, the solid region is omitted as in the thermal power plant evaluation with water as a working fluid; the properties of water and vapor play an important role. The information about solids (ice) is not necessary. The diagram shows the isothermal compression process curves. The curves are spread in subcritical regions and supercritical regions. An isotherm at critical temperature (374.15°C) passes through the critical point. This curve differentiates the vapor from the gas. The vapor can be converted into liquid through isothermal compression, as shown. But the gas cannot be converted into liquid through isothermal compression. Therefore, the vapor and gas can be differentiated with

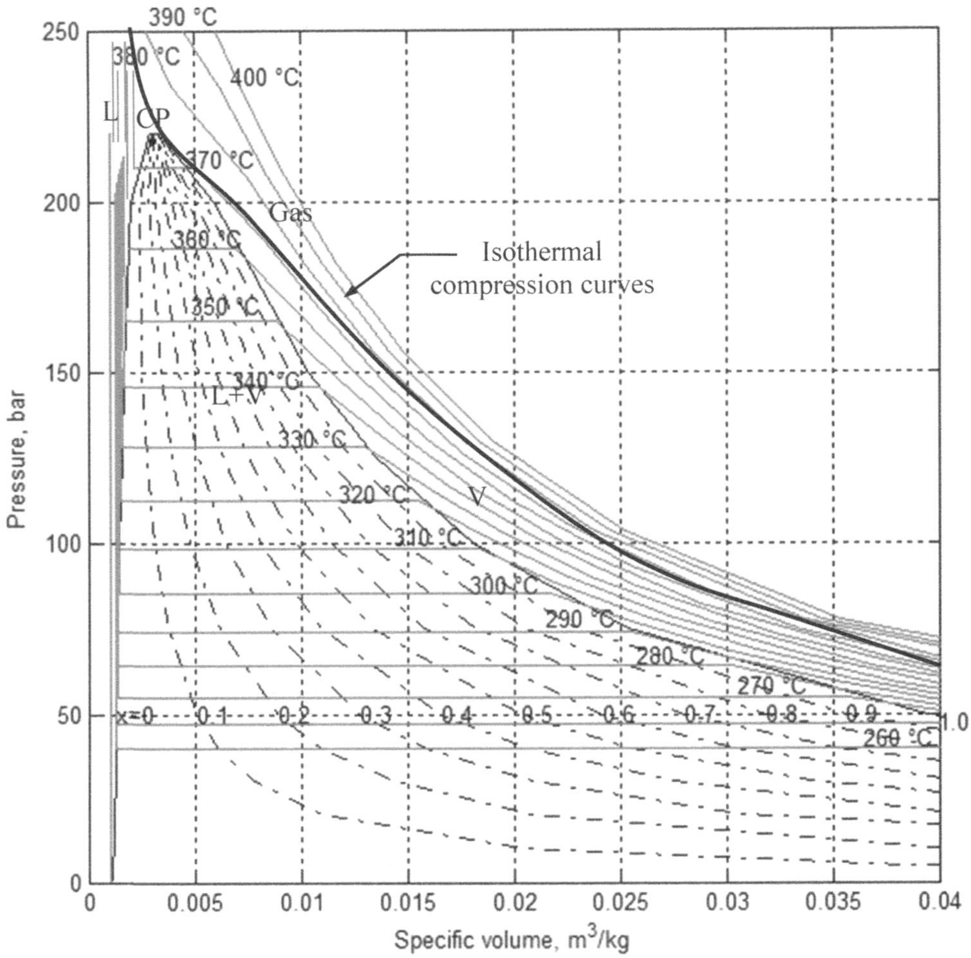

Fig. 2.12. Isotherms (isothermal compression processes) on p-v diagram.

reference to the isothermal compression process passing through the critical point. Since water is an incompressible fluid, only a small amount of volume change can be observed with an increase in pressure. The isotherms are nearly vertical in the liquid region. The latent heat of boiling increases with an increase in pressure. Therefore, cooking in a pressure cooker is easier (with less fuel) compared to cooking in an atmospheric oven. Similarly, the boiling temperature also increases with an increase in pressure. The water in the radiator of the engine is pressurized with a spring-loaded cap at the top. It increases the radiator water pressure. Because of the (small) increase in pressure, the temperature of the boiling point increases. Not only does the boiling point change, but due to the negative slope of the melting curve on the P-T diagram, the freezing temperature decreases. Both of these qualities are desirable for radiator operation. As the water will not boil even when it reaches 100°C, it will not freeze even at 0°C. At the critical point, the latent vaporization is zero. Here suddenly, the liquid flashed completely into vapor without any phase change. For water, the critical pressure (P_c), critical temperature (T_c), and critical volume (v_c) are respectively 221.2 bar, 374.15°C, and 0.00317 m^3/kg.

The vapor pressure differs from the boiling pressure. If a fluid, liquid or solid comes into contact with the vapor, the fluid is in thermodynamic equilibrium with the vapor contact surface temperature (Fig. 2.13). The vapor pressure is directly proportional to the temperature. At atmospheric pressure, the vapor pressure is called the normal boiling point.

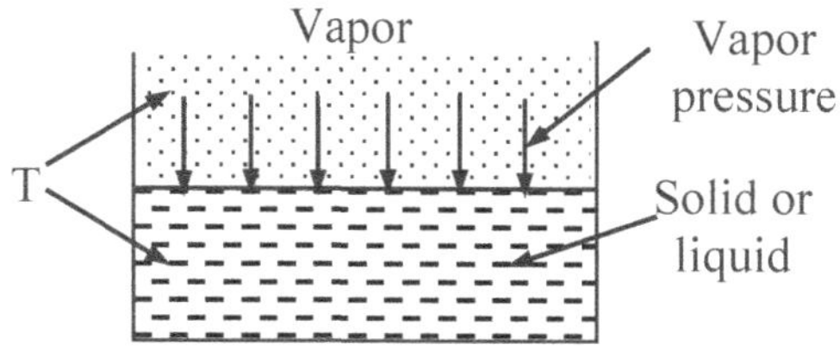

Fig. 2.13. Vapor pressure.

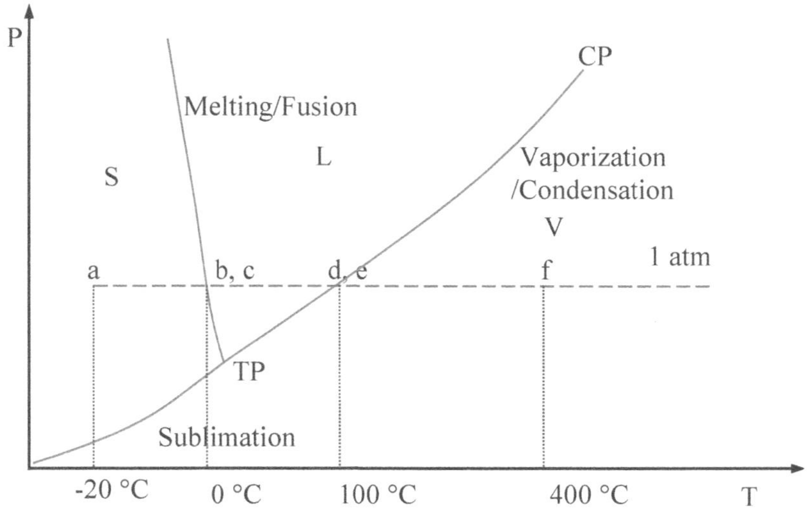

Fig. 2.14. Pressure-temperature diagram of water with three phases.

Figure 2.14 shows the variations in pressure and temperature for water. It shows three curves, and these three meet at a junction point called a triple point. A triple point with three curves divides the region into three regions, viz., solid, liquid, and vapor. As per the earlier discussion developed for the pressure-specific volume diagram, it can be reviewed again to build the pressure-temperature diagram. At a typical atmospheric pressure, the subcooled ice at –20°C is heated to 400°C. It undergoes different processes. They are: a-b: sensible heat transfers of ice; b-c: phase change with melting; c-d: sensible heat transfers of water; d-e: phase change with vaporization; and e-f: sensible heat transfers with superheating. So in these processes, three sensible heat transfers and two phase changes happened. In the reverse direction of heat addition, i.e., cooling, vaporization becomes condensation, and melting becomes freezing. The procedure is repeated at many pressures, and the data is plotted. The slope of the sublimation curve and the vaporization/condensation curve is positive. But the slope of melting or fusion is negative. Due to the peculiar properties of water, the slope of the melting curve is negative.

Similar to Fig. 2.14, Fig. 2.15 also depicts the pressure-temperature diagram, but to study the properties of the fluids applicable to all except water. The slope of all three curves, i.e., sublimation, vaporization, and melting, is positive. The specific volume increases with melting for all fluids except water.

Figure 2.16 shows the P-v-T surfaces of water. The specific volume of water increases during melting or cooling, while the specific volume of water decreases during condensation. It results in a negative slope for the melting curve. Water will contract while cooling, i.e., freezing. Figure 2.17 shows the P-v-T surface for fluids other than water. It results in a positive slope for the melting curve.

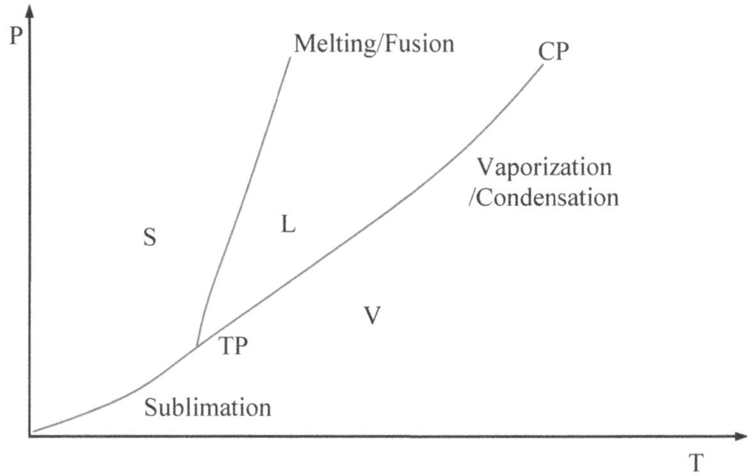

Fig. 2.15. Pressure-temperature diagram of fluids except water with three phases.

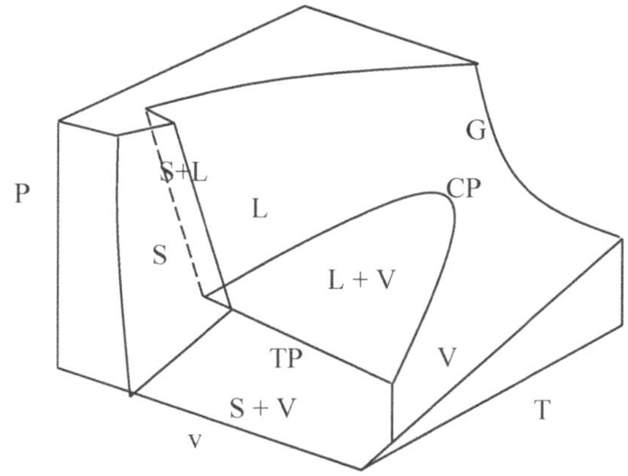

Fig. 2.16. Pressure (P), specific volume (v) and temperature (T) surfaces of water that expands during freezing; CP: critical point, G: gas; L: liquid, S: solid, TP: triple point, V: vapor.

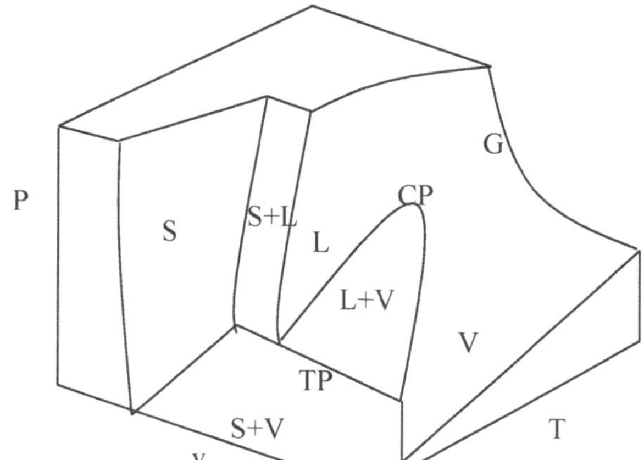

Fig. 2.17. Pressure (P), specific volume (v), temperature (T) surfaces of substance that contracts on freezing (other than water); CP: critical point, G: gas; L: liquid, S: solid, TP: triple point, V: vapor.

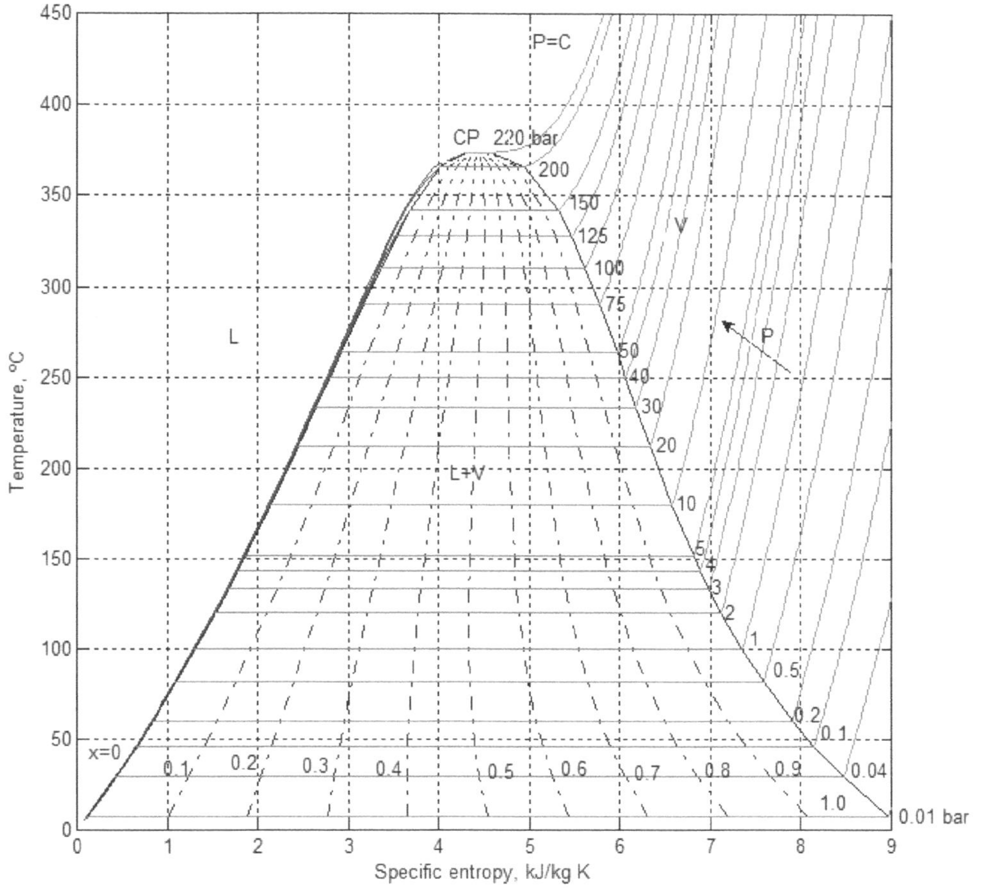

Fig. 2.18. Saturation (or vapor) dome for water.

The pressure-volume-temperature surfaces are depicted for a pure substance other than water. The properties of water have a peculiar characteristic. For other fluids, there is no such issue. The surface is open, and in the front view, all the surfaces are studied. The positive slope of the melting surface has also been depicted.

Figure 2.18 shows the temperature-specific entropy (T-s) diagram for water and steam regions. In this diagram, the properties of ice are skipped. The T-s diagram is more relevant to studying the changes in working fluid states in a thermal power plant.

Figure 2.19 shows the notations used for properties of wet steam. For wet steam, in addition to temperature or pressure, a dryness fraction is also required to describe its properties.

The dryness fraction of steam is defined as the ratio of dry steam mass to the total steam mass. The total mass of wet steam is the sum of water (liquid) and vapor.

$$x = \frac{m_{dry\ steam}}{m_{dry\ steam} + m_{water}} = \frac{m_v}{m_v + m_l} \tag{2.36}$$

The wetness fraction is equal to (1–x).

The 0 dryness indicates a saturated water condition. In this state, the specific volume, specific enthalpy, and specific entropy are v_f, h_f, and s_f. The dryness fraction 1 shows the saturated vapor state. The specific volume, specific enthalpy, and specific entropy are vg, hg, and sg. The difference

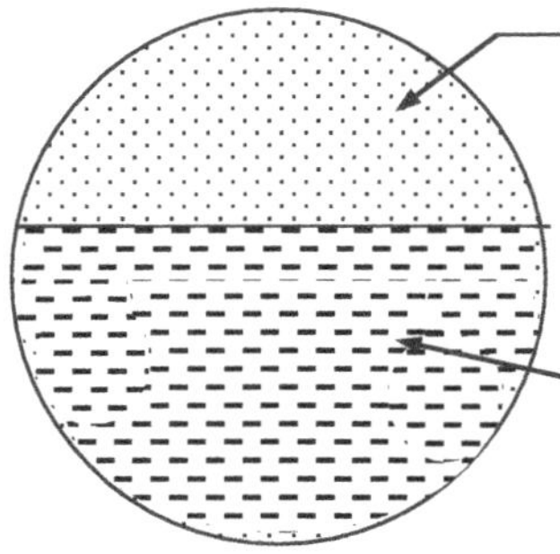

Fig. 2.19. Properties of wet steam.

between these two values is the volume change, latent heat, and entropy change in phase change, as follows:

$$v_{lv} = v_v - v_l \tag{2.37}$$

$$h_{lv} = h_v - h_l \tag{2.38}$$

$$s_{lv} = s_v - s_l \tag{2.39}$$

The total volume, total enthalpy, and total entropy of wet steam are equal to the sum of the properties of water and saturated steam.

Similarly, the total specific properties are also equal to the sum of water and vapor-specific properties.

$$V' = V_l + V_v \tag{2.40}$$

$$mv = m_l\, v_l + m_v\, v_v$$

$$v' = \frac{m_l v_l + m_v v_v}{m}$$

$$= (1-x)\, v_l + x\, v_v$$

$$= v_l - x\, v_l + x\, v_v$$

$$= v_l + x(v_v - v_l)$$

$$v' = v_l + x(v_v - v_l) \tag{2.41}$$

Similarly

$$h' = h_l + x(h_v - h_l) \tag{2.42}$$

$$s' = s_l + x(s_v - s_l) \tag{2.43}$$

The specific volume of water is very low compared to the specific volume of vapor. Therefore, the specific volume of water can be neglected for simple calculations. Then the specific volume of the wet steam is as follows:

$$v' = x v_v \tag{2.44}$$

Figure 2.20 depicts the specific enthalpy-specific entropy (h-s) diagram for pure water. The water, water vapor mixture and superheated vapor regions are shown with isobars, dryness fraction curves and isotherms. The evaluation of a thermal power plant is easy after knowing the enthalpies at each state. Therefore, h-s diagram is much more user friendly compared to the T-s diagram. The

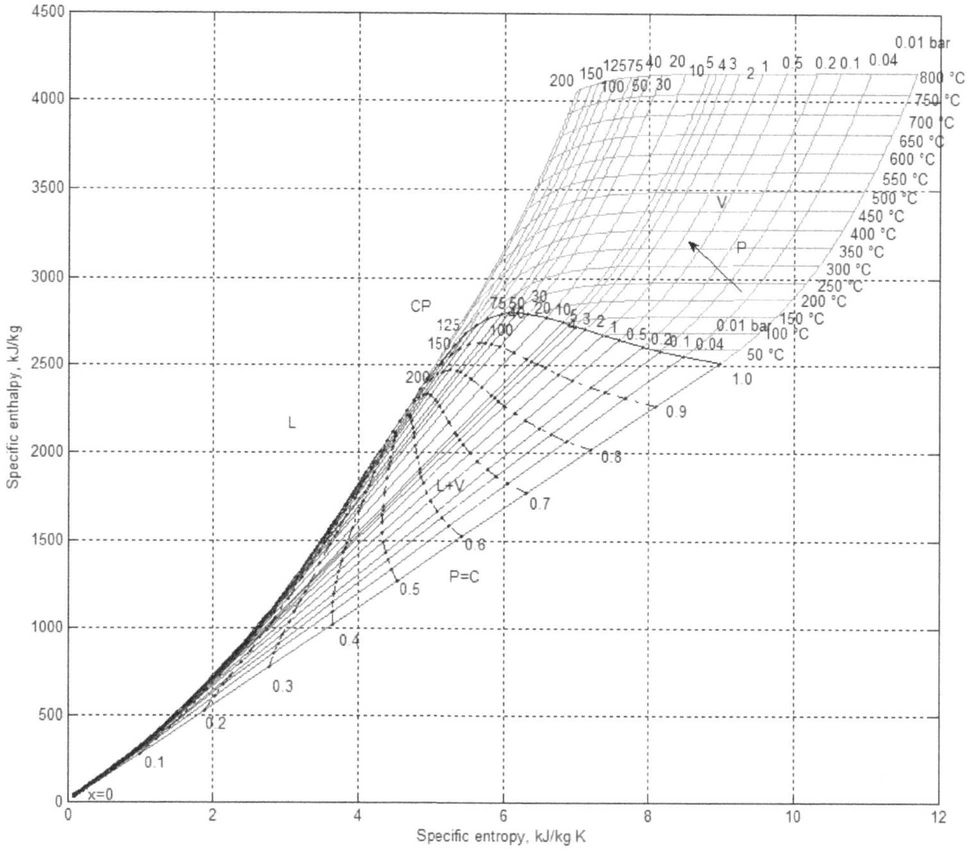

Fig. 2.20. Mollier chart with zero origin.

h-s diagram is also called a Mollier chart. It is more convenient to arrive at the steam related thermal components using this graphical method. A close observation of the curves indicates different slopes of the curves in different regions.

The combined first law of thermodynamics applied to flow process and second law of thermodynamics is as follows

$$Tds = dh - vdp \tag{2.45}$$

Majority of curves in the Mollier chart are at constant pressure. In the above equation, dp = 0 and therefore

$$\left(\frac{\partial h}{\partial s}\right)_P = T \tag{2.46}$$

It indicates that the slope of the isotherm in the Mollier chart is equal to the absolute temperature at the saturation pressure. In the wet steam, the saturation temperature is constant, and therefore, the isobars are straight. In the superheated region, the temperature of the superheated steam continuously increases while it is heating at constant pressure. So, the slope of the isobars in the superheated region increases. With an increase in pressure, the saturation temperature also increases. Therefore, the slope of the curves in the wet mixture and superheated vapor increases from low pressure to high pressure or from low temperature to high temperature. It indicates that the isotherms in the wet region and superheated region are diverging with each other. The isotherms in the superheated region have a near-zero slope or are horizontal.

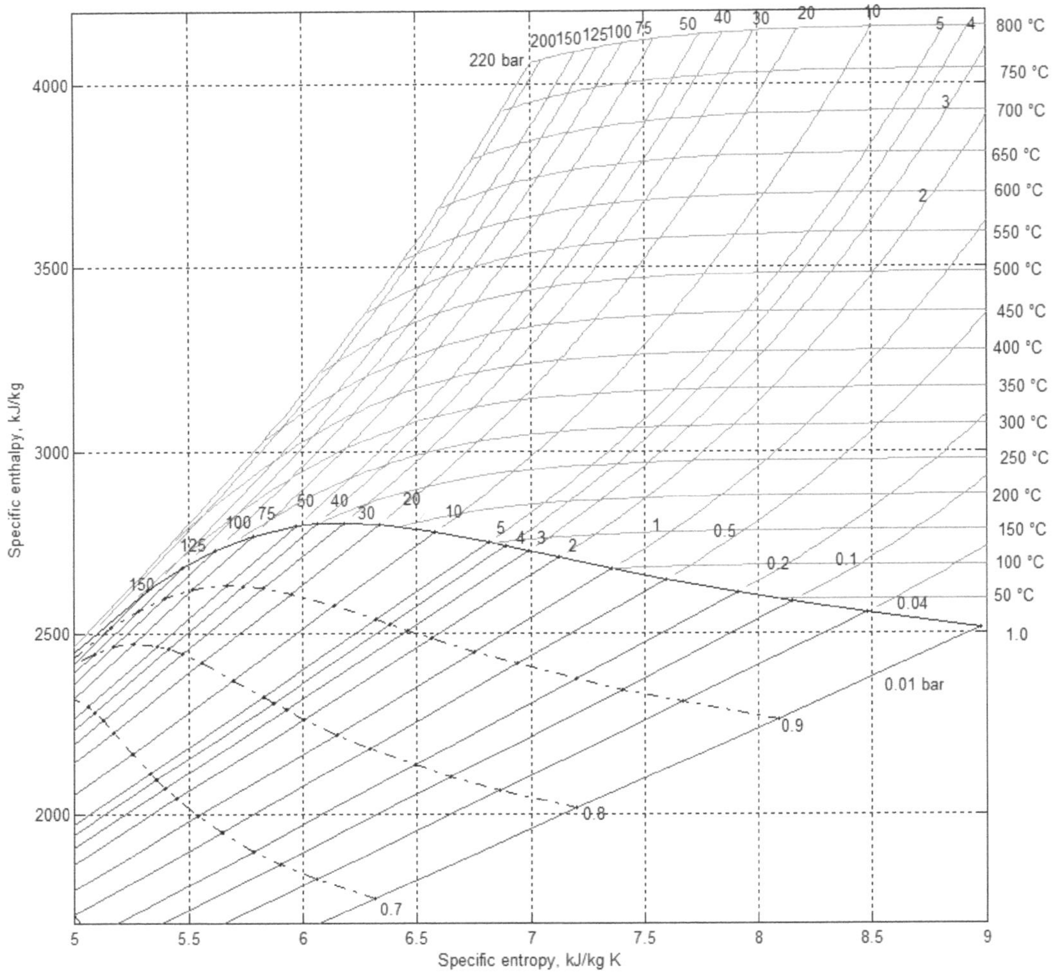

Fig. 2.21. Phase equilibrium diagram on h-s coordinates (Mollier diagram).

Figure 2.21 shows the modified Mollier chart after skipping the solid properties and water properties. The earlier h-s diagram (Fig. 2.20) shows the properties of water. It can be observed that the compressed water properties and too-low dry steam (more wet) curves are concentrated and difficult to read. The saturated steam and superheated steam are reasonably clear. In the modified Mollier chart, the origin is shifted from zero to the right side, and the drier steam, dry saturated steam, and superheated steam properties are focused on. For commercial use of the steam properties, the modified Mollier chart is more reliable and easier to use.

In the analytical method, an exact solution can be obtained. The graphical method gives an approximate solution. Steam tables are used in the analytical solution of the vapor system. Following are the available steam tables based on their region.

1. Saturated water and steam temperature tables
2. Saturated water and steam pressure tables
3. Specific volume of superheated steam
4. Specific enthalpy of superheated steam
5. Specific entropy of superheated steam

Table 2.3 shows the saturation fluid properties for water and steam. The table can be read from temperature or pressure. In cases where the properties are not tabulated at the required pressure or temperature, linear interpolation is recommended. Table 2.4 shows the steam tables developed as a function of pressure.

To read the properties of compressed liquid, it is reasonable to assume water is at saturation. For compressed water, the temperature is below the saturation temperature. The temperature of the compressed water can be referred to as the main variable, and the saturation pressure at this temperature can be identified. Since the influence of pressure is not much on compressed water, the temperature and its saturation temperature are enough to list the properties of water. Compressed water is in a sub-cooled form. The fluid temperature is below its saturation temperature at the same pressure. For example, at atmospheric pressure, if the water temperature is 40°C, the water can be referred to as sub-cooled water or compressed water as its temperature is below the saturation temperature, i.e., 100°C. The fluid properties can be compared at 40°C with its saturation temperature at 40°C. This assumption is reasonable, as the properties of water vary slightly with a change in pressure. To find wet steam properties, the dryness fraction or quality of steam is required. If the property of wet steam is supplied, the dryness fraction can be determined. Three separate superheated steam tables are available to find the specific volume, specific enthalpy, and specific entropy. In addition to the pressure, the superheated steam temperature is also required to determine the state of the superheated steam. Sometimes, instead of supplying the superheated temperature, a degree of superheat may be available. The degree of superheat is defined as the temperature difference between superheated steam and its saturation temperature at the same pressure.

Table 2.3. Properties of saturated steam with temperatures.

t, °C	P, bar	v_l, m³/kg	v_v, m³/kg	h_l. kJ/kg	h_{lv}, kJ/kg	h_v, kJ/kg	s_l, kJ/kg K	s_{lv}, kJ/kg K	s_v, kJ/kg K
0.01	0.006	0.000999	206.160	0.00	2501.57	2501.57	0.00	9.16	9.16
1	0.007	0.000999	192.610	4.17	2499.21	2503.39	0.02	9.12	9.13
2	0.007	0.000999	179.920	8.39	2496.84	2505.22	0.03	9.07	9.10
3	0.008	0.000999	168.170	12.60	2494.46	2507.06	0.05	9.03	9.08
4	0.008	0.000999	157.270	16.80	2492.09	2508.90	0.06	8.99	9.05
5	0.009	0.000999	147.160	21.01	2489.73	2510.73	0.08	8.95	9.03
--	--	--	--	--	--	--	--	--	--

Table 2.4. Saturated steam properties with pressure.

P, bar	t, °C	v_l, m³/kg	v_v, m³/kg	h_l. kJ/kg	h_{lv}, kJ/kg	h_v, kJ/kg	s_l, kJ/kg K	s_{lv}, kJ/kg K	s_v, kJ/kg K
0.0061	0.01	0.000999	206.01	0.03	2501.55	2501.58	0.00	9.16	9.16
0.010	6.98	0.000999	129.21	29.32	2485.04	2514.37	0.11	8.87	8.98
0.015	13.03	0.001000	87.98	54.71	2470.75	2525.46	0.20	8.63	8.83
0.020	17.51	0.001001	67.01	73.45	2460.19	2533.64	0.26	8.46	8.72
0.025	21.10	0.001002	54.26	88.44	2451.74	2540.18	0.31	8.33	8.64
--	--	--	--	--	--	--	--	--	--

2.7 Properties of binary fluid mixtures

The considered binary fluid in this study is an ammonia-water mixture. The mass concentration is defined as the ratio of the mass of ammonia to the total mass of the mixture. The heat addition and rejection occur at a variable temperature with two fluid systems. The addition of ammonia to water permits it to operate at low temperatures (100°C–150°C). The cycle configuration is to be changed to suit the other source temperatures. Water is an ideal work fluid in the power industry. But it is not suitable for low-temperature heat recovery. The ammonia cycle is suitable for low-temperature heat sources. But condensing ammonia in a condenser (absorber) is difficult (except at high pressures) as it needs a very low temperature (below 0°C). Therefore, the addition of ammonia to water (aqua ammonia) requires modifications to the existing Rankine cycle for easy condensation of ammonia in vapor form at the sink temperature. It is diluted with water (a weak mixture) before condensation. This process consists of the absorption of vapor into the liquid. Therefore, the cycle is called the vapor absorption cycle. The changes in layout need additional components such as separators, throttles, and mixtures. The turbine, condenser (absorber), pump, and boiler are the common components in the Rankine cycle and KC. The evaluation of properties for a binary fluid mixture is complex compared to a single fluid system due to the additional property of concentration. The regions of phase with reference to bubble point temperature (BPT) and dew point temperature (DPT) are described to understand the state of the fluid. The Gibbs function is derived to generate properties, and the derived properties are presented with the equation fitting coefficients. A methodology has been developed to identify the region of fluid from the randomly supplied properties, i.e., pressure, temperature, and concentration.

Pressure, temperature, and concentration are required to find the properties of a mixture where the region or phase are not known. The mixture temperature along with the BPT and DPT determines the region of the mixture, that is, the liquid mixture (subcooled or compressed liquid mixture), saturated liquid, liquid-vapor mixture, saturated vapor, and superheated vapor.

Figure 2.22 shows the regions in the binary fluid mixture on the temperature-concentration plot, generated at 40 bar pressure. The BPT curve represents all the starting states of phase change, and the DPT curve represents the completed states of phase change of the mixture at different concentrations. The BPT and DPT curves are also known as saturated liquid curves and saturated vapor curves, respectively. Between these two curves, that is, BPT and DPT, the mixture exists in a

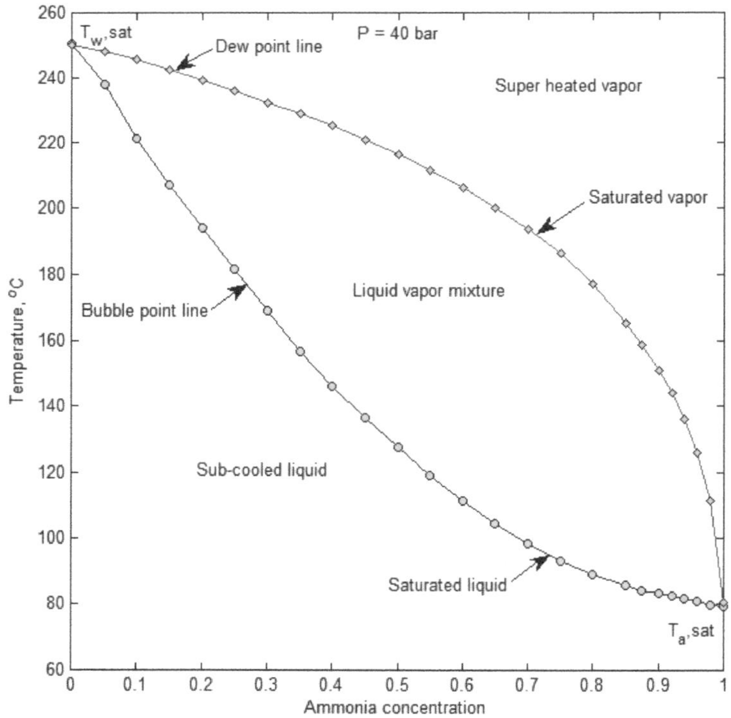

Fig. 2.22. Temperature–concentration diagram with five regions at 40 bar.

liquid-vapor form. Below the BPT, the mixture is completely liquid and is in a subcooled condition. Above the DPT, the mixture is vaporized in a superheated condition. The phase change starts at BPT and ends at DPT at a particular concentration. At zero concentration and 100% concentration, it is a pure substance and shows the respective saturation temperature at the given pressure. The drawing shows five regions, namely, subcooled liquid, saturation liquid, liquid vapor mixture, saturated vapor, and superheated vapor. BPT and DPT are determined from the curve-fit equations.

As per the earlier discussion, the properties are grouped into five regions. The identified regions are subcooled liquid, saturated liquid, liquid-vapor mixture, saturated vapor, and superheated vapor. To find the nature of a mixture, properties such as pressure, temperature, and concentration are required. If the temperature is less than the BPT, it is a subcooled liquid, or if it equals the BPT, then it is a saturated liquid. If the temperature lies between the BPT and DPT, then the state falls between the liquid and vapor conditions, that is, a liquid-vapor mixture. Finding the liquid-vapor mixture properties is complicated compared to either the liquid condition only or the vapor condition only. If the given temperature is greater than the DPT, then it can be concluded that the resulting region is superheated vapor.

Figure 2.23 details the enthalpy-concentration chart to find the properties with the graphical method. It is an enthalpy-concentration plot as a function of ammonia concentration and pressure.

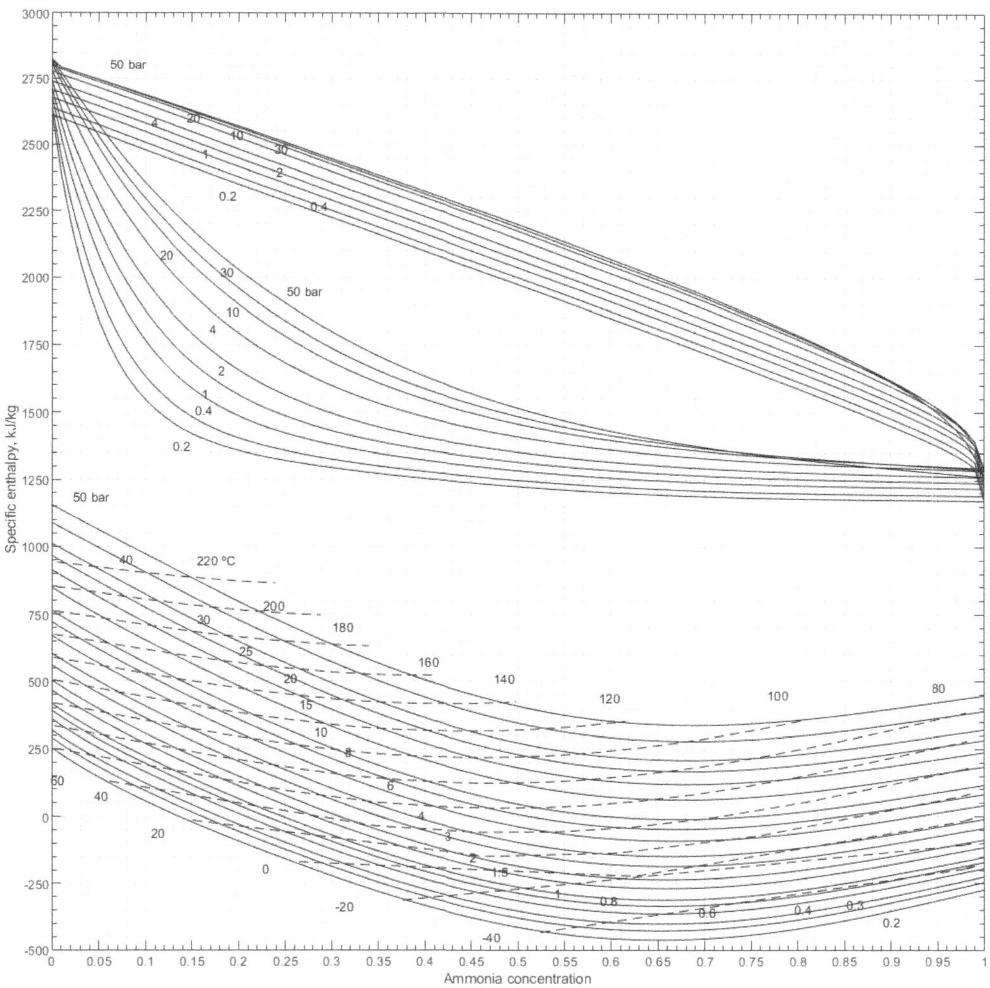

Fig. 2.23. Specific enthalpy-concentration diagram of saturated vapor with concentration and auxiliary curves from 0.2 bar to 50 bar.

The vapor curves consist of saturated vapor and auxiliary curves. The liquid curves are the saturated liquid curves with the BPT information. The properties of a subcooled mixture can also be read from this chart. The enthalpies of liquid and vapor increase with an increase in pressure. The upper curve is the vapor curve, which results from liquid concentration and DPT. The lower curve is the liquid enthalpy plot, which results from the BPT and liquid ammonia concentration. The auxiliary curve results from BPT and vapor ammonia concentration.

2.8 Psychrometric properties

Psychrometric properties are properties of wet air. The atmospheric air is a mixture of dry air and moisture. These properties are required in the modeling and simulation of air-containing units and similar works. Some of the psychrometric properties are dry bulb temperature, wet bulb temperature, relative humidity, specific humidity, specific enthalpy, specific entropy, specific volume, etc. The psychrometric processes can be arrived at using psychrometric properties. The measured temperature by a thermometer with its dry bulb (without wet wick) is called dry bulb temperature (DBT). The measured temperature by a thermometer with its wet wick on the bulb with rotation in air is called the wet bulb temperature (WBT). The rotating wet wick thermometer is called a sling psychrometer. Generally, the air is in an unsaturated condition. It absorbs moisture till the saturation condition. The partial pressure of moisture in the air is determined from Dalton's law of additive pressures.

Figure 2.24 shows the dew point temperature (DPT) and saturation pressure of air at given pressure and temperature. During the cooling process of air at constant pressure, the moisture reaches saturation condition. In this condition, the temperature is called the dew point temperature, as the moisture condenses at this temperature. The saturated air cannot accommodate any further moisture. In this condition, the relative humidity (Ø) is at its maximum, which is 100%. The relative humidity (RH) is defined as the ratio of partial pressure of water vapor to saturated pressure of water at the same temperature. It is also equal to the ratio of the mole fraction of water vapor to the mole fraction of water in a saturated condition at the same temperature and pressure.

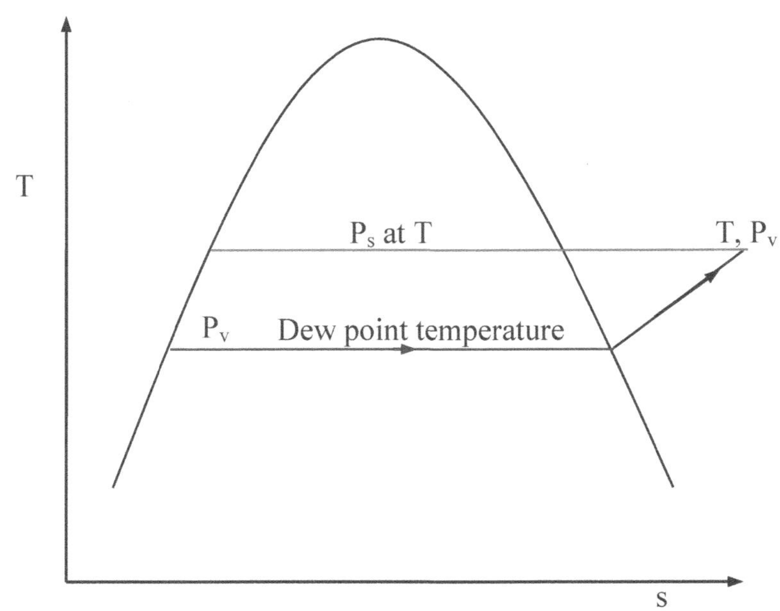

Fig. 2.24. Concept of relative humidity with the function of partial pressure of moisture saturation at the same pressure and temperature.

The following correlation equation results in the saturated pressure of moisture as a function of its saturation temperature.

$$\ln P = \frac{1}{T} + c_2 + c_3 T + c_4 T^2 + c_5 T^3 + c_6 \ln T \tag{2.47}$$

where P_s = saturated vapor pressure of water, kPa
T = temperature, K

The regression coefficients c_1 to c_6 are as follows

$c_1 = -5.80022006 \times 10^3$

$c_2 = -5.516256$

$c_3 = -4.8640239 \times 10^{-2}$

$c_4 = 4.1764768 \times 10^{-5}$

$c_5 = -1.4452093 \times 10^{-8}$

$c_6 = 6.5459673$

Using the perfect gas equation, we can show that

$$RH = \varnothing = \frac{partial\ pressure\ of\ water\ vapour}{saturation\ pressure\ of\ pure\ water\ vapour\ at\ the\ same\ temperature}$$

$$= \frac{P_v}{P_s} = \frac{\left(\frac{n_v \bar{R}}{V}\right)}{\left(\frac{n_s \bar{R}}{V}\right)} = \frac{n_v}{n_s} \tag{2.48}$$

$$= \frac{number\ of\ moles\ vapor}{number\ of\ moles\ vapor\ at\ sat.\ condition}$$

The specific humidity or humidity ratio ω is the mass of water vapor presented in wet air per unit mass of dry air.

$$\omega = \frac{mass\ of\ water\ vapor}{mass\ of\ dry\ air} = \frac{m_v}{m_{da}}$$

$$= \frac{\left(\frac{P_v V}{R_v T}\right)}{\left(\frac{P_a V}{R_a T}\right)} = \frac{P_v R_a}{P_a R_v} = \frac{P_v R_a}{(P_t - P_v) R_v} \tag{2.49}$$

$$= \frac{0.287 \times P_v}{0.4615 \times (P_t - P_v)}$$

Therefore

$$\omega = 0.622 \left(\frac{P_v}{P_t - P_v} \right) \tag{2.50}$$

In the above equation,

P_v = saturation vapor pressure determined at the temperature of the air from the equation or steam tables

P_t = total or atmospheric pressure

Therefore, the specific humidity is the function of partial pressure of water vapor and total atmospheric pressure. As shown in Fig. 2.24, the saturation temperature of water vapor at its partial pressure is the dew point temperature of air.

If the relative humidity and saturation pressure of water vapor is known, the partial pressure of water vapor can be determined from its product as follows

$$P_v = \text{Ø}\, P_s \tag{2.51}$$

Alternatively, the partial pressure of water vapor can also be evaluated from the specific humidity relation given below:

$$\omega = 0.622 \frac{P}{P_t - P_v}$$

$$P_v = \frac{P_t \omega}{0.622 + \omega} \tag{2.52}$$

As shown in Fig. 2.25, the saturation temperature or dew point temperature can be found from the steam tables at the partial pressure of water vapor, P_v.

Alternatively, an approximate equation for dew-point temperature is given by

$$DPT = \frac{4030(DBT + 235)}{4030 - (DBT + 235)\ln\phi} - 235 \tag{2.53}$$

In the above equation, DBT and DPT are in °C.

The degree of saturation is the ratio of the humidity ratio, ω to the humidity ratio of saturated air ω_s at the same temperature and pressure, i.e.

$$\mu = \left.\frac{\omega}{\omega_s}\right|_{T,\,P} \tag{2.54}$$

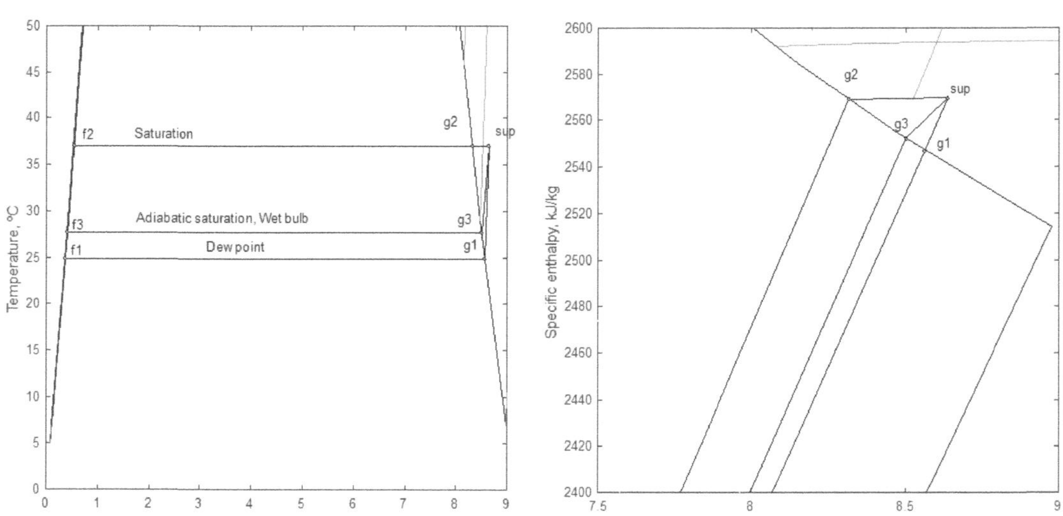

Fig. 2.25. Saturation, wet bulb temperature and dew point temperature on T-s and h-s diagram.

The specific volume is the volume of unit mass of wet air. The specific volume of wet air and dry air are the same.

$$v = \frac{R_a T}{P_a} = \frac{R_a T}{P_t - P_v} \quad m^3/kg\ da \tag{2.55}$$

The specific enthalpy of wet air is the sum of the enthalpy of dry air and the enthalpy of water vapor. Enthalpy values are always based on some reference value. For moist air, the enthalpy of dry air is given a zero value at 0°C, and for water vapor the enthalpy of saturated water is taken as zero at 0°C.

The enthalpy of moist air is given by

$$h = h_a + \omega h_g = c_p t + \omega\,(h_{g\,@\,0°C} + c_{pv} t) \tag{2.56}$$

where c_p = specific heat of dry air at constant pressure, kJ/kg K
t = dry bulb temperature of air-vapor mixture, °C
ω = humidity ratio, kg of water vapor/kg of dry air
h_a = enthalpy of dry air at DBT, kJ/kg
h_g = enthalpy of saturated vapor at 0°C, kJ/kg

The enthalpy of water vapor, h_g at 0°C is 2500 kJ/kg.

$$h = 1.005t + \omega\,(2500 + 1.88t) \tag{2.57}$$

The humid specific heat of moist air can be written as

$$c_{pm} = c_p + \omega c_{pw} \tag{2.58}$$

where c_{pm} = humid specific heat, kJ/kg K
c_p = specific heat of dry air, kJ/kg K
c_{pw} = specific heat of water vapor, kJ/kg
ω = humidity ratio, kg of water vapor/kg of dry air

Since the second term in the above equation ($\omega\ c_{pw}$) is very small compared to the first term, for all practical purposes, the humid specific heat of moist air, c_{pm} can be taken as 1.0216 kJ/kg K dry air.

Alternatively, the enthalpy can also be determined in the following manner.

The enthalpy of air,

$$h = c_p t + \omega h_v$$

$$= c_p t + \omega\,(h_g + c_{pv}\,(t - t_s)) \tag{2.59}$$

Similar to specific enthalpy, the specific entropy of wet air is the sum of the entropy of dry air and the entropy of water vapor. For moist air, the entropy of dry air is given a zero value at 0°C, and for water vapor the entropy of saturated water is taken as zero at 0°C.

The entropy of moist air is given by

$$s = c_p \ln\left(\frac{t + 273.15}{273.15}\right) - R_a \ln\left(\frac{P_a}{P_0}\right) + \omega s_g \tag{2.60}$$

where s_g = specific entropy of steam at partial pressure of water vapor (P_v) and dry bulb temperature, t, °C taken from the steam tables.

The exergy of wet air can be determined from the enthalpy and entropy of the wet air. The properties at the reference state (dead state) are P_0, T_0 and RH_0 respectively 1.01325 bar, 25°C and 100%.

The exergy of moist air,

$$\varepsilon = h - h_0 - T_0 (s - s_0) \tag{2.61}$$

h_0 and s_0 are the specific enthalpy and specific entropy determined at P_0, T_0 and RH_0.

Figure 2.26 shows the psychrometric chart at atmospheric pressure which is used to read the psychrometric properties from the two given properties. The chart consists of DBT lines (vertical lines), specific humidity lines (horizontal towards right), RH curves, WBT lines (inclined and aligned with saturation lines), specific enthalpy lines, specific volume lines, and DPT lines (horizontal towards left).

The straight line law states that "when air is transferring heat and mass (water) to or from a wetted surface, the condition of air shown on a psychrometric chart drives towards the saturation line at the temperature of the wetted surface." For example, as shown in Fig. 2.27, when warm air passes over a wetted surface, its temperature drops from 1 to 2. Also, since the vapor pressure of air at 1 is greater than the saturated vapor pressure at 2, there will be moisture transfer from air to water, i.e., the warm air in contact with a cold wet surface cools and dehumidifies. According to the straight-line law, the final condition of air (i.e., 2) lies on a straight line joining 1 with 2 on the saturation line. This is due to the value of unity of the Lewis number.

Adiabatic saturation temperature is the temperature at which water, by evaporating into air, can bring the air to saturation at the same temperature adiabatically (Fig. 2.28). The thermodynamic wet bulb temperature will be less than the entering air DBT but greater than the dew point temperature.

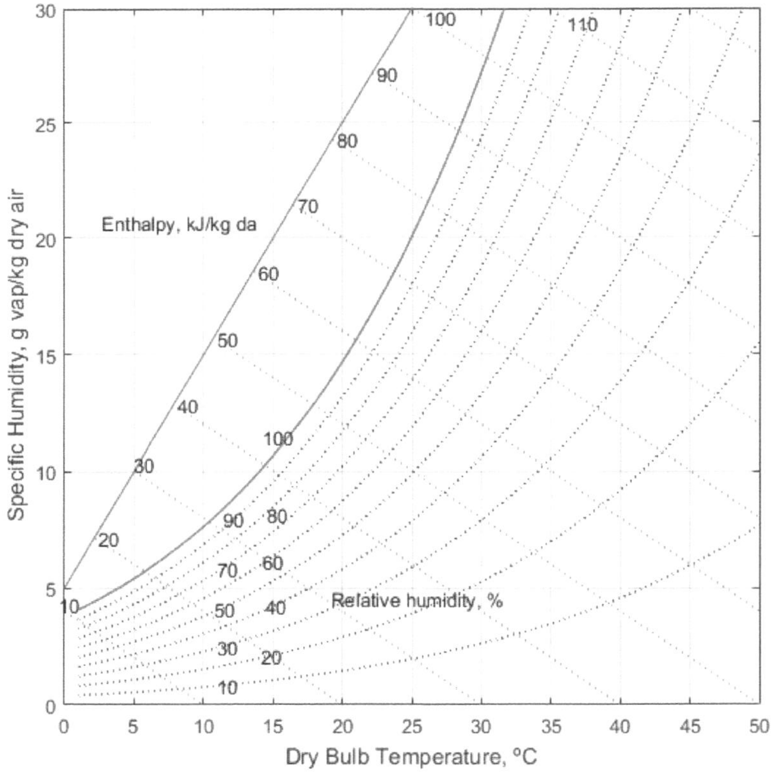

Fig. 2.26. Schematic of a psychrometric chart for a given barometric pressure.

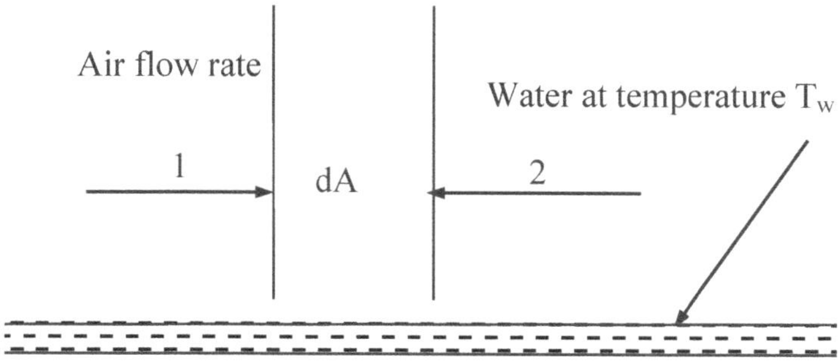

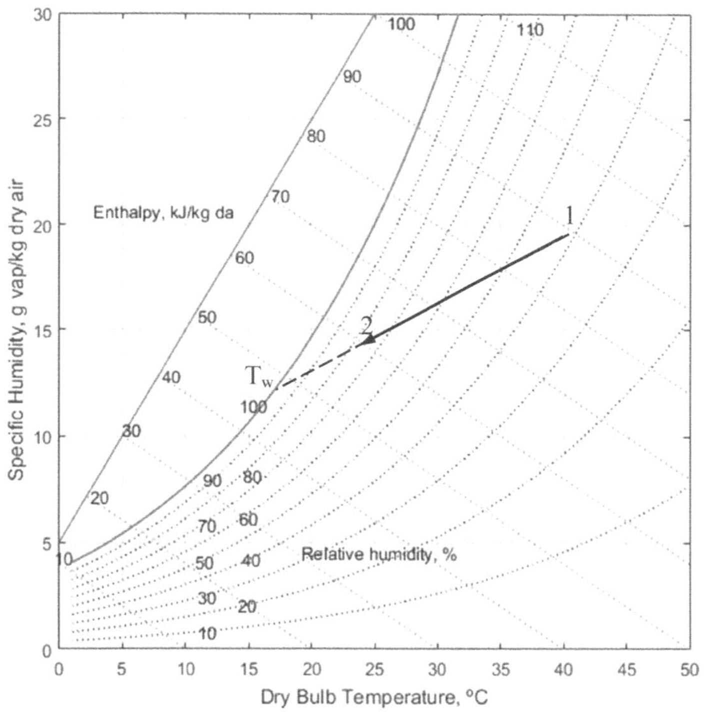

Fig. 2.27. Principle of straight line law for air-water mixtures.

Based on a unit mass flow rate of dry air, this is given by

$$h_1 = h_2 - (\omega_2 - \omega_1)\, h_f \tag{2.62}$$

where h_f is the enthalpy of saturated liquid at the sump or thermodynamic wet bulb temperature, h_1 and h_2 are the enthalpies of air at the inlet and exit of the adiabatic saturator, and ω_1 and ω_2 are the humidity ratio of air at the inlet and exit of the adiabatic saturator, respectively.

Assuming the humid specific heat to be constant, from the enthalpy balance, the thermodynamic wet bulb temperature can be written as

$$t_2 = t_1 - \frac{h_{fg2}}{c_{pm}}(\omega_2 - \omega_1) \tag{2.63}$$

where $h_{fg,2}$ is the latent heat of vaporization at the saturated condition 2. Thus, measuring the dry bulb (t_1) and wet bulb temperature (t_2) one can find the inlet humidity ratio (ω_1) from the above

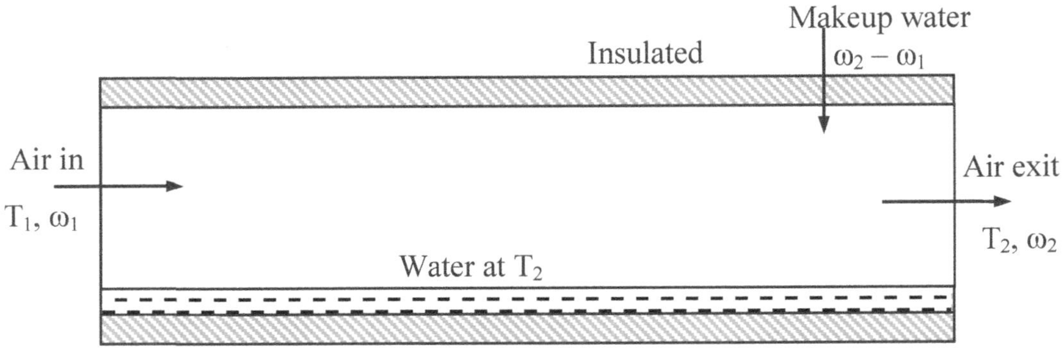

Fig. 2.28. The process of adiabatic saturation of air.

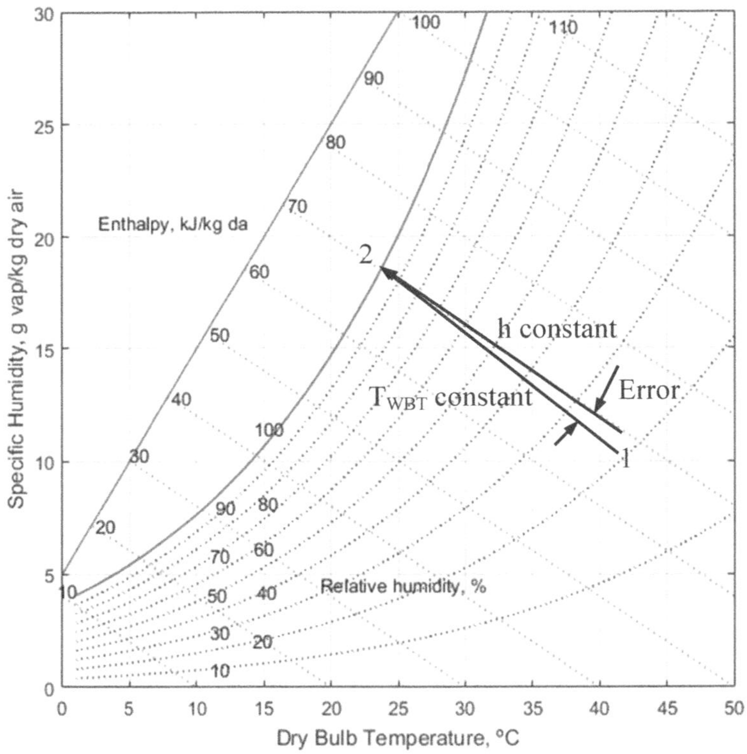

Fig. 2.29. Adiabatic saturation process 1–2 on a psychrometric chart.

expression as the outlet saturated humidity ratio (ω_2) and latent heat of vaporizations are functions of t_2 alone (at fixed barometric pressure).

On the psychrometric chart as shown in Fig. 2.29, point 1 lies below the line of constant enthalpy that passes through the saturation point 2. $t_2 = f(t_1, \omega_1)$ is not a unique function, in the sense that there can be several combinations of t_1 and W_1 which can result in the same sump temperature in the adiabatic saturator. A line passing through all these points is a constant wet bulb temperature line. Thus, all inlet conditions that result in the same sump temperature, for example point 1' have the same wet bulb temperature. The line is a straight line according to the straight-line law. The straight line joining 1 and 2 represents the path of the air as it passes through the adiabatic saturator.

Normally lines of constant wet bulb temperature are shown on the psychrometric chart. The difference between actual enthalpy and the enthalpy obtained by following constant wet bulb temperature is equal to $(\omega_2 - \omega_1)\, h_f$.

In practice, it is not convenient to measure the wet bulb temperature using an adiabatic saturator. Instead, a thermometer with a wetted wick is used to measure the wet bulb. It can be observed that since the area of the wet bulb is finite, the state of air at the exit of the wet bulb will not be saturated; instead, it will be point 2 on the straight line joining 1 and i, provided the temperature of water on the wet bulb is 'i'. It has been shown by Carrier that this is a valid assumption for air-water mixtures. By performing energy balance across the wet bulb, it can be shown that the temperature measured by the wet bulb thermometer is:

$$t_2 = t_1 - \frac{k_w}{h_c} h_{fa}(\omega_l - \omega) \tag{2.64}$$

k_w is the mass transfer coefficient for air-water mixtures, the ratio $(h_c/k_w\, c_{pm})$ = Lewis number is ≈ 1, hence, the wick temperature is approximately equal to the thermodynamic wet bulb temperature. It should be noted that, unlike thermodynamic WBT, the WBT of wet bulb thermometer is not a thermodynamic property as it depends upon the rates of heat and mass transfer between the wick and air.

The following are the empirical relations for the vapor pressure of water in moist air

Modified Apjohn equation,

$$P_v = P_v' - \frac{1.8P(DBT - WBT')}{2700} \tag{2.65}$$

Modified Ferrel's equation,

$$P_v = P_v' - 0.00066(DBT - WBT)\left(1 + \frac{1.8DBT}{1571}\right) \tag{2.66}$$

Carrier equation,

$$P_v = P_v' - \frac{1.8(P - P_v')(DBT - WBT)}{2800 - 1.3(1.8DBT + 32)} \tag{2.67}$$

P_v' is the saturation vapor pressure at wet bulb temperature.

All pressure units in the above equations should be consistent. Once vapor pressure is calculated, all other properties such as relative humidity, humidity ratio, enthalpy, humid volume, etc. can be calculated from the psychrometric equations presented earlier.

The psychrometric properties can be determined from any two given properties at atmospheric pressure. Sometimes, iterations or correlations are used to generate the psychrometric properties from the given data. The following section discusses the various possible supply variables and their solutions. There are many combinations of data supply possibilities. Some of the solutions from the given data are outlined.

The wet air properties can be determined from any of the two following properties:

i. DBT and RH
ii. DBT and WBT
iii. DBT and WBT.
iv. DBT and DPT
v. DBT and DPT
vi. DBT and ω

2.9 Numerical solutions

1. At 50°C and 51°C, respectively, water has the following properties at saturated conditions: v_l = 0.0010121, 0.0010126 m³/kg, v_g = 12.05, 11.50 m³/kg; h_f = 209.26, 213.44 kJ/kg; and h_g = 2592.2, 2593.9 kJ/kg. Use the Clapeyron equation to predict the change in saturation pressure between 50 and 51°C.

Solution:

Substitute the property values at 50.5°C into Eq. 20

$$\frac{dp}{dT} = \frac{s_g - s_f}{v_g - v_f} = \frac{T(s_g - s_f)}{T(v_g - v_f)} = \frac{h_{fg}}{T(v_g - v_f)}$$

$$dp = \frac{2593.1 - 211.4}{(50.5 + 273.15)(11.775 - 0.0010124)}$$

$$= 0.625 \text{ kPa}$$

The steam tables give the following change in the saturation pressure,

12.961 – 12.335 = 0.626 kPa

2. Compute the equation for c_p of oxygen as a function of pressure from 0 to 100 bar at temperature of 500 K if the Van der Waals equation of state applies,

$$p,\ kPa = \frac{RT}{v - b} - \frac{a}{v^2}$$

where R = 0.2598 kJ/kg K
a = 0.1346 kPa(m⁶/kg²)
b = 0.000995 m³/kg

c_{p0} for oxygen = $c_{p0} = \bar{R}\left(a_1 + a_2T + a_3T^2 + a_4T^3 + a_5T^4\right)$, kJ/kg mol K, T in K

a_1	a_2	a_3	a_4	a_5
0.03212936E2	0.11274864E-2	–0.0575615E-5	0.13138773E-8	–0.08768554E-11

Solution:

c_{p0} at 500 K for O_2 = 0.9723 kJ/kg K.

Next compute $\left.\frac{\partial c_p}{\partial p}\right|_T$ at 100000 kPa using Eq. $\left.\frac{\partial c_p}{\partial p}\right|_T = -T\left.\frac{\partial^2 v}{\partial T^2}\right|_p$.

Since the real gas equation is implicit for volume, numerical differentiation can be applied to solve $\left.\frac{\partial^2 v}{\partial T^2}\right|_p$.

$$\left.\frac{\partial^2 v}{\partial T^2}\right|_p = \frac{\frac{v_{T+\Delta T} - v_T}{\Delta T} - \frac{v_T - v_{T-\Delta T}}{\Delta T}}{\Delta T}$$

Extract the second derivative numerically by computing v at 495, 500 and 505 K.

v_{595} = 0.0129 m³/kg

$v_{600} = 0.0130\ m^3/kg$

$v_{605} = 0.0132\ m^3/kg$

The second derivative with respect to temperature at p = 100000 kPa is

$$\frac{\left[\frac{v_{T+5}-v_T}{5}-\frac{v_T-v_{T-5}}{5}\right]}{5} = -7.799\times10^{-9}$$

$$\left.\frac{\partial c_p}{\partial p}\right|_T = 3.899\times10^{-6}\ kJ/kg.K\ \ per\ \ kPa$$

Proposing a second-degree variation with pressure,

At 500 K, $\left.\frac{\partial c_p}{\partial p}\right|_T = 0$ at p = 0

$$\left.\frac{\partial c_p}{\partial p}\right|_T = 3.899\times10^{-6}\ \text{at p = 100000 kPa}$$

Proposing a second-degree variation with pressure,

$c_p = c_{p0} + a_1p + a_2p^2$

$$\left.\frac{\partial c_p}{\partial p}\right|_T = a_1 + 2a_2p$$

where $a_1 = 0$, at p = 0

$$a_2 = \left.\frac{\partial c_p}{\partial p}\right|_T /2p$$

$a_2 = (3.899 \times 10^{-6})/2p$

$= 1.9499 \times 10^{-10}$

Thus, c_p at 500 K and 10000 kPa is

$c_p = 0.9723 + (3.899 \times 10^{-10})\ (10000)^2$

$= 0.9918$ kJ/kg K

3. Steam, initially at 0.5 MPa, 250°C, is cooled at constant volume. At what temperature will the steam become saturated vapor? Find the specific volume using the compressibility chart. Compare the compressibility chart's results (volume and temperature) with the solution obtained with steam tables. What is the % deviation using compressibility chart in finding the volume and temperature? Use steam tables for saturation pressure and temperature in both cases. For water, $P_c = 221.2$ bar and $T_c = 374.15°C$.

Solution:

From the superheated steam tables at P = 5 bar and T = 250°C,

$P_c = 221.2$ bar and $T_c = 374.15°C$.

From steam tables, $v_1 = v_2 = 0.4744$ m³/kg

Using compressibility chart.

$$P_{r1} = \frac{P_1}{P_c} = \frac{5}{221.2} = 0.0226$$

$$T_{r1} = \frac{T_1}{T_c} = \frac{250+273.15}{374.15+273.15} = 0.8082$$

From compressibility chart, $Z_1 = 0.9888$

$$v_{1comp} = \frac{Z_1 R T_1}{P_1} = \frac{0.9888 \times 0.4619 \times (250+273.15)}{5 \times 100} = 0.4799$$

Percentage deviation in volume

$$Deviation = \frac{v_1 - v_{1comp}}{v_1} = \frac{0.4744 - 0.4799}{0.4744} \times 100 = -0.72\%$$

From steam tables, at $v_{1comp} = v_{2comp} = 0.4799$, $T_2 = 142.27$°C

Percentage deviation in temperature

$$Deviation = \frac{T_2 - T_{2comp}}{T_2} = \frac{142.54 - 142.27}{142.54} \times 100 = 0.0657\%$$

4. Compute the specific volume of superheated water vapor at 5000 kPa and 350°C using (a) the perfect gas equation, (b) the Van der Waals equation and (c) the Redlich-Kwong equation.

   Van der Waals constants are a = 1.7048 and b = 0.0016895 for $p = \frac{RT}{v-b} - \frac{a}{v^2}$

   Redlich-Kwong constants are a = 43.951 and b = 0.001171 for $p = \frac{RT}{v-b} - \frac{a}{T^{1/2} v(v+b)}$

Solution:

a) R of water vapor = 8.314/18 = 0.46188 kJ/kg K

$$v = \frac{(0.461889)(350+273.15)}{5000} = 0.057565 \text{ m}^3\text{/kg}$$

b) Using the values of critical pressure and temperature, the Van der Waals constants can be computed

   a = 1.7048 and b = 0.0016895

   The Van der Waals equation

$$p = \frac{RT}{v-b} - \frac{a}{v^2}$$

$$5000 = \frac{0.461889 \times 623.15}{v - 0.0016895} - \frac{1.7048}{v^2}$$

   can be solved for v by an iterative process, such as Newton-Raphson, to find

   v = 0.05303 m³/kg

c) For water, the Redlich-Kwong constants from Eqs. 39 and 40 are

a = 43.951 and b = 0.001171

$$p = \frac{RT}{v-b} - \frac{a}{T^{1/2}v(v+b)}$$

Substituting these constants into Eq.38 and solving for v yields

v = 0.05230 m^3/kg

The values computed in Example 5 may be compared with the specific volume from the steam tables at 5000 kPa and 623.15 K of 0.05194 m^3/kg. The Van der Waals equation showed an appreciable improvement over the perfect gas equation and the Redlich-Kwong equation yielded a result deviating only 0.7% from the tabular value.

5. Starting with a base enthalpy of h = 0 kJ/kg for saturated liquid water at 0°C, compute the enthalpy of superheated vapor at 25 bar and 400°C. Use Redlich-Kwong equation of state, c_{p0} for water vapor (given below) and p-T equation for saturation conditions, $\ln p = 19.335 - \frac{5416}{T}$

The equation for zero-pressure c_{p0}, $\frac{\bar{c}_p}{\bar{R}} = a_1 + a_2T + a_3T^2 + a_4T^3 + a_5T^4$, kJ/kg mol K, T in K

where

a_1	a_2	a_3	a_4	a_5
0.03386842E2	0.03474982E-1	–0.063546961E-4	0.06968581E-7	–0.02506588E-10

Redlich-Kwong constants are a = 43.951 and b = 0.001171 for $p = \frac{RT}{v-b} - \frac{a}{T^{1/2}v(v+b)}$

Solution:

The first stage of the calculation is to move from the base enthalpy to that of saturated vapor at 0°C.

Clapeyron Equation

$$\frac{dp}{dT} = \frac{h_{fg}}{T(v_g - v_f)}$$

With respect to the magnitude of v_g, the specific volume of saturated liquid v_f is negligible.

Therefore, $h_{fg} = \left(\frac{dp}{dT}\right) T v_g$

It is given that, $\ln p = 19.335 - \frac{5416}{T}$

Differentiating both sides, $\frac{1}{p} dp = \frac{5416}{T^2} dT$

$$\frac{dp}{dT} = \frac{5416p}{T^2}$$

At p = 0.6108 kPa, T = 0°C

Therefore, $\frac{dp}{dT} = \frac{5416 \times 0.6108}{(0 + 273.15)^2} = 0.0443\ kPa/K$

It is given that,

The Redlich-Kwong equation is $p = \frac{RT}{v-b} - \frac{a}{T^{1/2}v(v+b)}$

For water, the Redlich-Kwong constants are

a = 43.951 and b = 0.001171

From the Redlich-Kwong equation, when p = 0.6108 and T = 273.15 K the specific volume v_g = 206.36 m³/kg.

Substituting into the Clapeyron equation yields

h_{fg} = (0.0443) (273.15) (206.36) = 2499.2 kJ/kg

h_g at 0°C = 0 + 2499.2 = 2499.2 kJ/kg

The process of moving from saturated vapor at 0°C to superheated vapor at 25 bar and 400 °C comes from an integration of Eq., $dh = c_p dT + \left(v - T\frac{\partial v}{\partial T}\right)_p dp$.

First find the enthalpy at 0.6108 kPa and 500°C by integrating $c_p dT$.

$$\frac{\overline{c_p}}{R} = a_1 + a_2T + a_3T^2 + a_4T^3 + a_5T^4$$

$$h_{0-T} = \frac{R}{M_{H_2O}}\int_0^T \left(a_1 + a_2T + a_3T^2 + a_4T^3 + a_5T^4\right)dT$$

= 779.48 kJ/kg

The integration from 0 to 400°C at 0.6108 kPa of c_{p0} dT is 779.48, so h at 400°C and 0.6108 kPa = 2499.2 + 779.48 = 3278.7 kJ/kg.

The other term to be integrated in the Eq. is

$$\left(v - T\frac{\partial v}{\partial T}\right)_p dp$$

using the Redlich-Kwong equation. Unfortunately, v cannot be solved explicitly in this equation, so a numerical computation of $\partial v/\partial T)_p$ must be made.

For example, at 25 bar the following specific volumes can be computed:

t = 399.9°C	v_a = 0.119872 m³/kg
t = 400.0	v_b = 0.119891 m³/kg
t = 400.1	v_c = 0.119911 m³/kg

Thus

$$\left.v - T\frac{\partial v}{\partial T}\right)_p = v_b - (400 + 273.15)\frac{(v_c - v_a)}{0.2} = -0.0132\ m^3/kg$$

The bracketed term in Eq. $\left(v - T\frac{\partial v}{\partial T}\right)_p dp$ is essentially constant at a value of –0.0132 throughout the pressure range of 0.6108 to 25 bar.

So, the $\int \left(v - T \left(\frac{\partial v}{\partial T} \right)_p \right) dp = -0.0132(p_2 - p_1) = -0.0132\,(2500 - 0.6108) = -32.93$ kJ/kg.

Finally, h at 25 bar and 400°C is 3278.7 – 32.93 = 3245.8 kJ/kg.

From steam tables, at 25 bar and 400°C, the enthalpy of superheated steam = 3240.7 kJ/kg.

The error is $\frac{3240.7 - 3245.8}{3240.7} \times 100 = -0.158\%$

6. Determine the properties of a mixture of ammonia and water which is at 13 bar with 30% of vapor in mixture. The ammonia concentration in the vapor is 0.88.

Solution:

Graphical solution using an enthalpy-concentration chart is shown in Fig. 2.30.
For a graphical solution, locate a state 3 on the vapor curve at 0.88 concentration and 13 bar. Draw a horizontal line till the auxiliary curve and extend to a vertical line from this intersection. The vertical line meets the saturated liquid line at liquid concentration. At state 2 (liquid), both liquid concentration and temperature can be read as 0.28 and 117°C. From the lever rule or locating point with length inversely proportional to the vapor fraction, point 1 can be solved as 0.46. The solution enthalpy from the graph is 770 kJ/kg.

Alternatively, the properties also can be determined as follows:

The dryness fraction (mass of vapor to total mass) is defined by

$$df = \frac{x - x_l}{x_v - x_l} = \frac{x_1 - x_2}{x_3 - x_2}$$

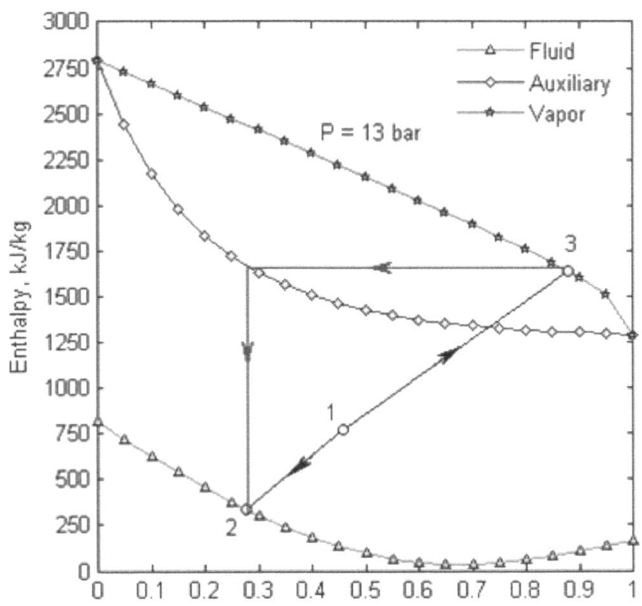

Fig. 2.30. Separation process at known pressure and vapor concentration.

Therefore, solution concentration,

$x = x_1 = x_2 + VF\,(x_3 - x_2)$

$0.28 + 0.3(0.88 - 0.28) = 0.46$

From the chart,

$h_2 = 334$ kJ/kg

$h_3 = 1638$ kJ/kg

$h_1 = (1 - df)\,h_2 + df\,h_3 = (1 - 0.3)\,334 + 0.3 \times 1638 = 725.2$ kJ/kg

7. On a particular day the weather forecast states that the dry bulb temperature is 37°C, while the relative humidity is 50% and the barometric pressure is 101.325 kPa. Find the humidity ratio, dew point temperature, enthalpy, entropy and exergy of moist air on this day.

Solution:

From the steam tables, the saturation pressure at 37°C is

$P_s = 0.06279$ bar

Since the relative humidity, RH = 50%,

The partial pressure of water vapor,

$P_v = RH \times P_s = 0.5 \times 0.06279 = 0.03139$ bar

The specific humidity,

$$\omega = 0.622\frac{P_v}{P_t - P_v} = 0.622\frac{0.03139}{1.01325 - 0.03139} = 0.01989\ kg/kg\ da$$

Dew point temperature is the saturation temperature at P_v

Therefore, from steam tables, the saturation temperature or dew point temperature at 0.0314 bar is 24.85°C.

The specific enthalpy of wet air,

$h = c_p t + \omega h_v$

$= c_p t + \omega\,(h_g + c_{pv}\,(t - t_s))$

$= 1.005 \times 37 + 0.01989 \times (2547.0 + 1.88\,(37 - 24.85))$

$= 88.25$ kJ/kg da

Alternatively,

$h = 1.005t + \omega\,(2500 + 1.88t)$

$= 1.005 \times 37 + 0.01989 \times (2500 + 1.88 \times 37)$

$= 88.25$ kJ/kg da

Specific entropy of wet air,

$$s = c_p \ln\left(\frac{t + 273.15}{273.15}\right) - R_a \ln\left(\frac{P_a}{P_0}\right) + \omega s_g \ \text{(from 0°C)}$$

Therefore, the specific entropy of air,

$$s = 1.005 \times \ln\left(\frac{37 + 273.15}{273.15}\right) - 0.287 \times \ln\left(\frac{1.01325 - 0.03139}{1.01325}\right) + 0.01989 \times 8.63$$

= 0.3083 kJ/kg da K

Exergy of wet air:

The exergy of wet air at the reference state, 25°C DBT, 1.01325 bar and 100% RH,

From the steam tables, the saturation pressure at 25°C is

P_s = 0.0317 bar

Since the relative humidity, RH = 100%,

The partial pressure of water vapor,

$P_v = RH \times P_s = 1 \times 0.0317 = 0.0317$ bar

The specific humidity,

$$\omega = 0.622 \frac{P_v}{P_t - P_v} = 0.622 \frac{0.0317}{1.01325 - 0.0317} = 0.0201 \ kg / kg \ da$$

Dew point temperature is the saturation temperature at P_v

Therefore, from the steam tables, the saturation temperature or dew point temperature at 0.0317 bar is 25°C.

The specific enthalpy of wet air,

$h_0 = c_p t + \omega h_v$

$= c_p t + \omega (h_g + c_{pv} (t - t_s))$ (since air is saturated, there is no degree of superheat)

$= 1.005 \times 25 + 0.0201 \times 2547.3$

= 76.23 kJ/kg da

Specific entropy of wet air,

$$s_0 = c_p \ln\left(\frac{t_0 + 273.15}{273.15}\right) - R_a \ln\left(\frac{P_a}{P_0}\right) + \omega s_g \ \text{(from 0°C)}$$

$$= 1.005 \times \ln\left(\frac{25 + 273.15}{273.15}\right) - 0.287 \times \ln\left(\frac{1.01325 - 0.0317}{1.01325}\right) + 0.0201 \times 8.56$$

= 0.2688 kJ/kg da K

Specific exergy of wet air,

$\varepsilon = h - h_0 - T_0 (s - s_0)$

$= 88.25 - 76.23 - 298.15 \times (0.3083 - 0.2688)$

= 0.24 kJ/kg da

The psychrometric properties are depicted on Fig. 2.31.

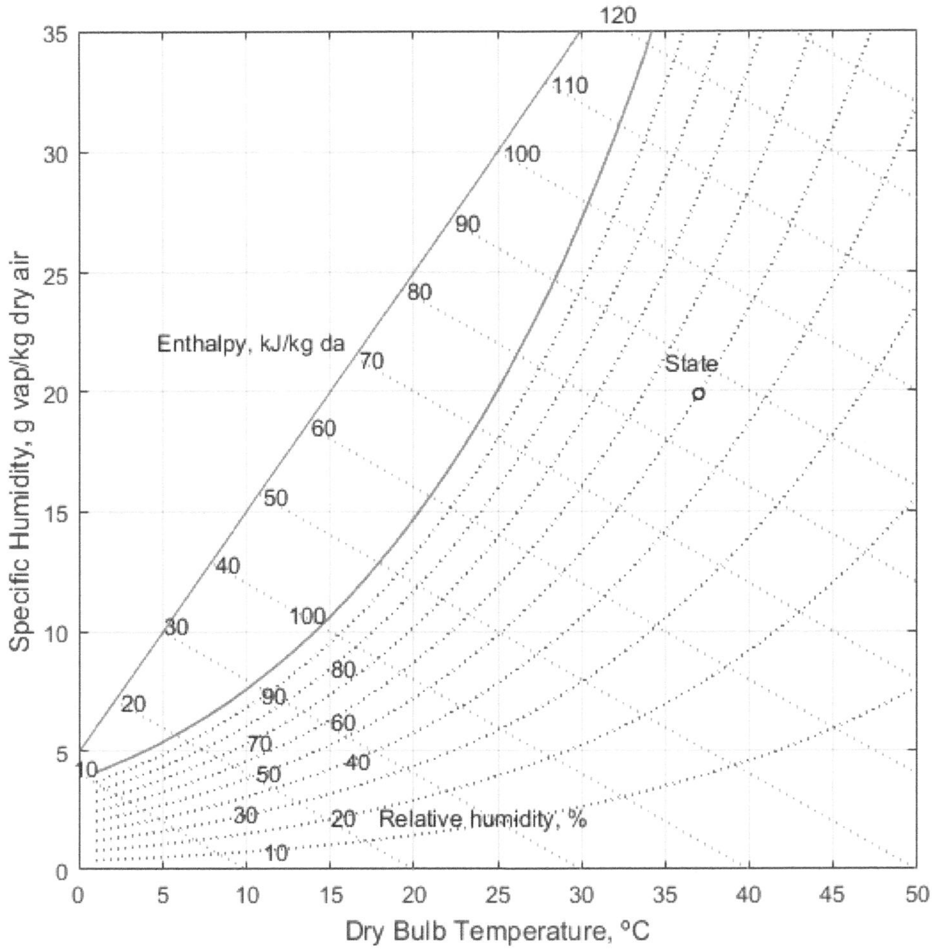

Fig. 2.31. Solution of the problem.

8. Moist air at 1 atmospheric pressure has a dry bulb temperature of 32°C and a wet bulb temperature of 26°C. Calculate (i) the partial pressure of water vapor (ii) humidity ratio (iii) relative humidity (iv) dew point temperature (v) density of air in the mixture (vi) density of water vapor in the mixture (vii) enthalpy of moist air using perfect gas law model and psychrometric equations and (viii) specific entropy.

Solution:

Modified Apjohn equation,

$$P_v = P_v' - \frac{1.8P(DBT - WBT)}{2700}$$

where P_v' is the saturation vapor pressure at wet bulb temperature

From steam tables, the saturation vapor pressure at WBT, $P_v' = 0.0336$ bar

$$P_v = P_v' - \frac{1.8P(DBT - WBT)}{2700}$$

$$= 0.0336 - \frac{1.8 \times 1.01325(32 - 26)}{2700} = 0.02956 \ bar$$

The specific humidity,

$$\omega = 0.622\frac{P_v}{P_t - P_v} = 0.622\frac{0.02956}{1.01325 - 0.02956} = 0.0187 \; kg/kg \; da$$

From steam tables, the saturation pressure at 32°C, P_s = 0.0475 bar

$$RH = \frac{P_v}{P_s} \times 100 = \frac{0.02956}{0.0475} \times 100 = 62.16\%$$

Dew point temperature is the saturation temperature at the P_v

At the saturation state, the dew point temperature is the saturation temperature at P_v = 0.02956 bar

From steam tables,

DPT = 23.85°C

Density of air,

$$\rho_a = \frac{P_a}{R_a T} = \frac{P - P_v}{R_a T} = \frac{(1.01325 - 0.02956) \times 100}{0.287 \times (32 + 273.15)} = 1.1236 \; kg/m^3$$

Similarly, the density of water vapor in air can be approximated from perfect gas equation,

$$\rho_v = \frac{P_v}{R_v T} = \frac{0.02956 \times 100}{0.4615 \times (32 + 273.15)} = 0.021 \; kg/m^3$$

The enthalpy of air,

$h = c_p t + \omega h_v$

$= c_p t + \omega (h_g + c_{pv} (t - t_s))$

$= 1.005 \times 32 + 0.0187 \times (2545.2 + 1.88 (32 - 23.85))$

= 80.05 kJ/kg da

From psychrometric chart it is slightly less than 80 kJ/kg da

Specific entropy of wet air,

$$s = c_p \ln\left(\frac{t + 273.15}{273.15}\right) - R_a \ln\left(\frac{P_a}{P_0}\right) + \omega s_g$$

Specific entropy of steam at 0.02956 bar and 32°C is 8.63 kJ/kg K.

$$= 1.005 \times \ln\left(\frac{32 + 273.15}{273.15}\right) - 0.287 \times \ln\left(\frac{1.01325 - 0.02956}{1.01325}\right) + 0.0187 \times 8.63$$

= 0.2814 kJ/kg da K

The psychrometric properties of the solution are depicted in Fig. 2.32.

9. The atmospheric conditions are 20°C and specific humidity of 0.0095 kg/kg of dry air. Calculate (a) partial pressure of vapor, (b) relative humidity and (c) dew point temperature.

Solution:

From steam tables, the saturation vapor pressure at 20°C, P_s = 0.0234 bar

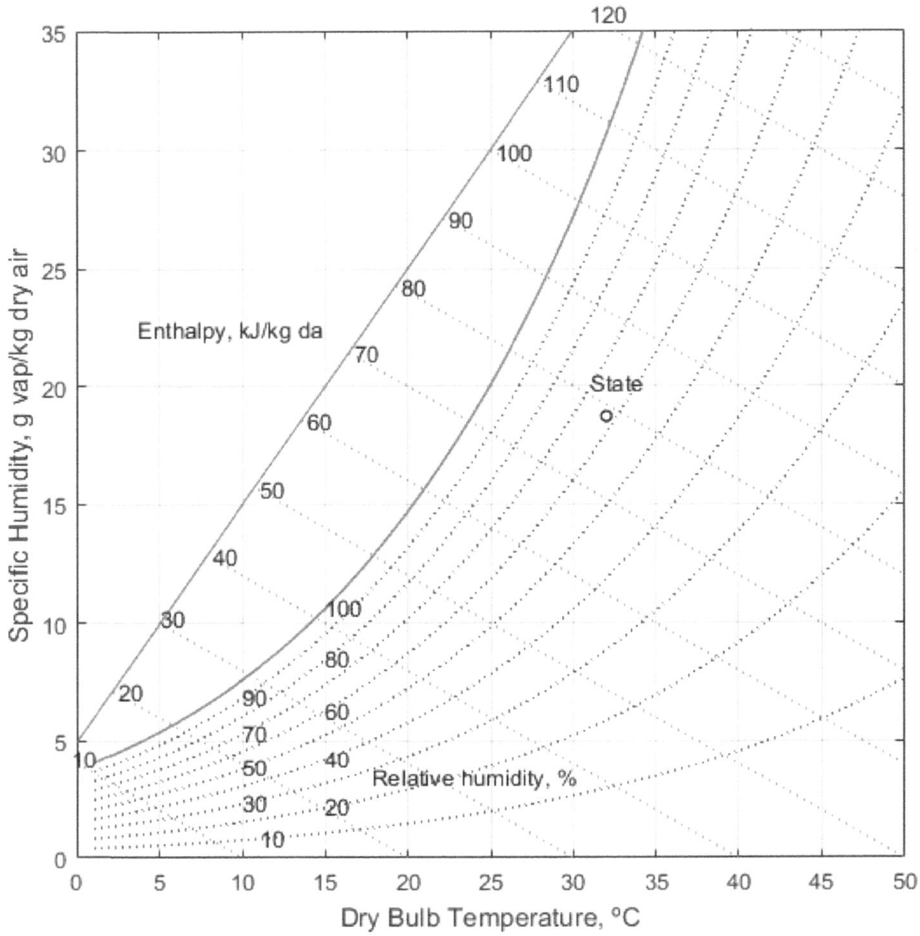

Fig. 2.32. Solution of the psychrometric problem.

The specific humidity,

$$\omega = 0.622\frac{P_v}{P_t - P_v}$$

$$P_t\omega - P_v\omega = 0.622\ P_v$$

$$P_t\omega = (0.622 + \omega)\ P_v$$

$$P_v = \frac{P_t\omega}{0.622 + \omega}$$

$$= \frac{1.01325 \times 0.0095}{0.622 + 0.0095} = 0.01524\ bar$$

$$RH = \frac{P_v}{P_s} \times 100 = \frac{0.01524}{0.0234} \times 100 = 65\%$$

Dew point temperature is the saturation temperature at the P_v

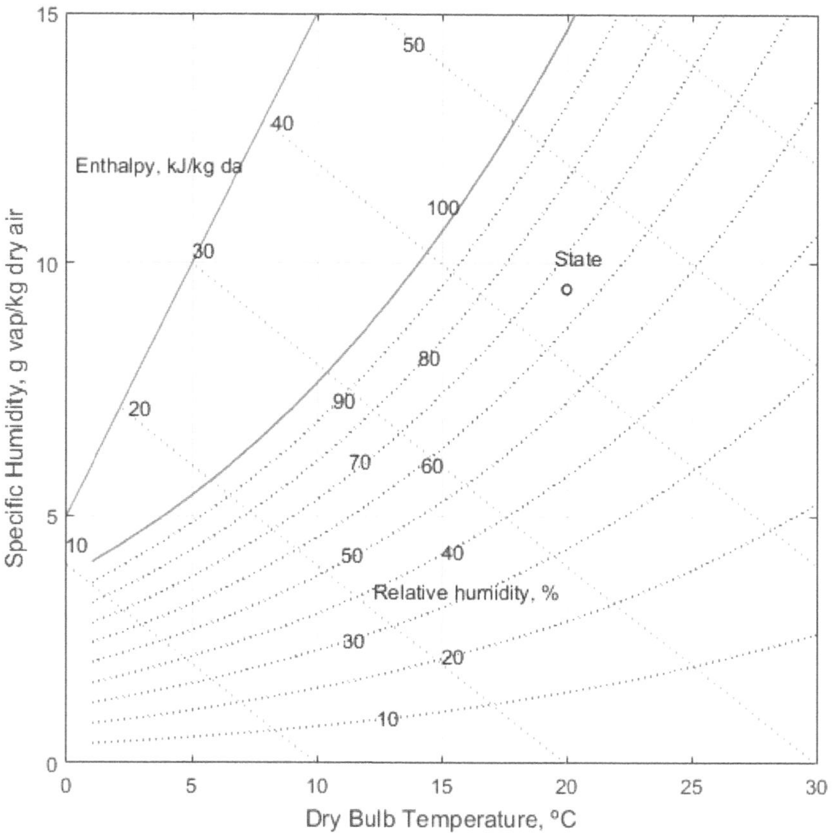

Fig. 2.33. Solution of the psychrometric problem.

At the saturation state, the dew point temperature is the saturation temperature at $P_v = 0.01524$ bar

From steam tables,

DPT = 13.24°C

The psychrometric properties of the solution are depicted on Fig. 2.33.

10. A mixture of dry air and water vapor is at a temperature of 22°C. The dew point temperature is 15°C. Find (a) partial pressure of water vapor, (b) relative humidity (c) specific humidity, (d) enthalpy of air per kg of dry air and (e) specific volume of air per kg of dry air.

Solution:

From steam tables, the saturation vapor pressure at 20°C, $P_s = 0.0264$ bar

P_v is the saturation pressure of water vapor at dew point temperature,

$P_v = 0.0170$ bar

$$RH = \frac{P_v}{P_s} \times 100 = \frac{0.0170}{0.0264} \times 100 = 64.4\%$$

The specific humidity,

$$\omega = 0.622 \frac{P_v}{P_t - P_v}$$

$$= 0.622 \times \frac{0.0170}{1.01325 - 0.0170} = 0.0106 \ \ kg / kg \ \ da$$

The enthalpy of air,

$h = c_p t + \omega h_v$

$= c_p t + \omega (h_g + c_{pv} (t - t_s))$

$= 1.005 \times 22 + 0.0106 \times (2529.1 + 1.88 \times (22 - 15))$

$= 49.05$ kJ/kg da

Specific volume of air

$$v_a = \frac{R_a T}{P_a} = \frac{R_a T}{P - P_v} = \frac{0.287 \times (22 + 273.15)}{(1.01325 - 0.0170) \times 100} = 0.85 \ \ m^3 / kg \ \ da$$

The psychrometric properties of the solution are depicted on Fig. 2.34.

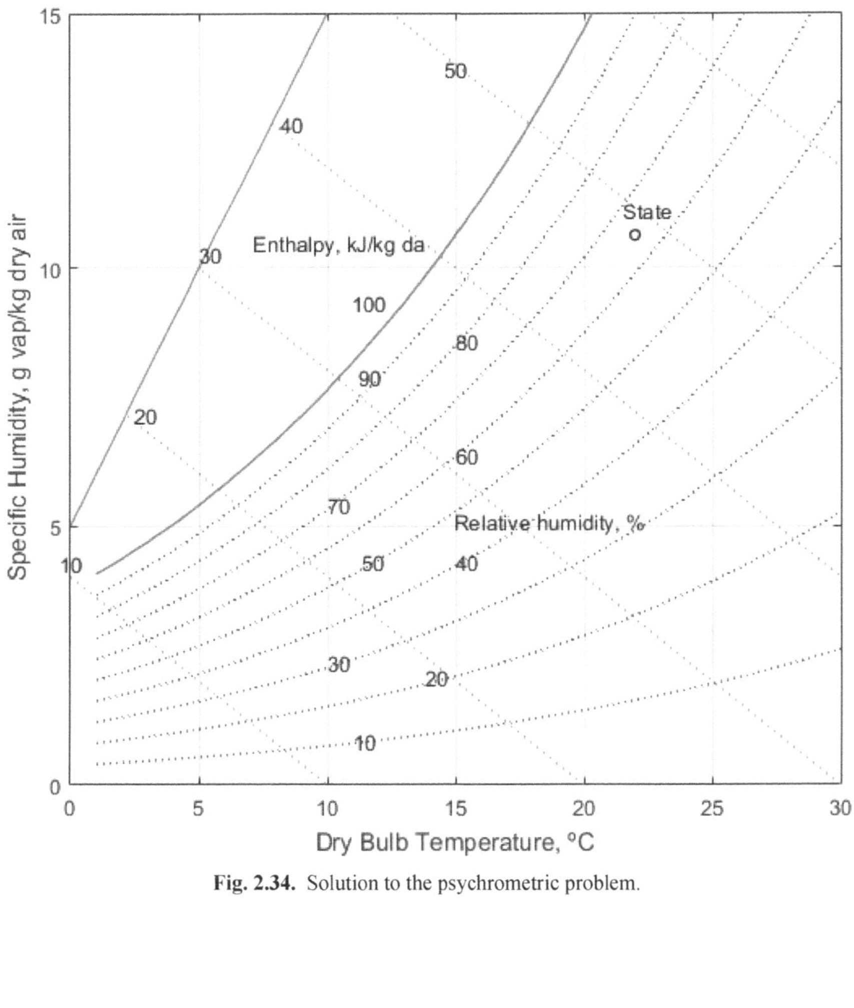

Fig. 2.34. Solution to the psychrometric problem.

11. Moist air has a specific enthalpy of 80 kJ/kg of dry air at 40°C dry-bulb temperature. Find (a) partial pressure of water vapor, (b) relative humidity (c) specific humidity, (d) wet bulb temperature and (e) specific volume of air per kg of dry air.

Solution:

From steam tables, the saturation vapor pressure at 40°C, $P_s = 0.0738$ bar

The enthalpy of air,

$h = c_p t + \omega(h_g + c_{pv} t)$

$h = 1.005t + \omega(2500 + 1.88t)$

h_g is at 0°C is 2500 kJ/kg

$$\omega = \frac{h - c_p t}{h_g + c_{pv} t}$$

$$= \frac{80 - 1.005 \times 40}{2500 + 1.88 \times 40} = 0.0155 \ kg/kg \ da$$

The specific humidity,

$$\omega = 0.622 \frac{P_v}{P_t - P_v}$$

$$P_v = \frac{P_t \omega}{0.622 + \omega}$$

$$= \frac{1.01325 \times 0.0155}{0.622 + 0.0155} = 0.0246 \ bar$$

At 0.0246 bar, the dew point temperature is 20.83°C.

$$RH = \frac{P_v}{P_s} \times 100 = \frac{0.0246}{0.0738} \times 100 = 33.33\%$$

Specific volume of air

$$v_a = \frac{R_a T}{P_a} = \frac{R_a T}{P - P_v} = \frac{0.287 \times (40 + 273.15)}{(1.01325 - 0.0246) \times 100} = 0.91 \ m^3/kg \ da$$

To find the wet bulb temperature, the enthalpy of saturated air at the WBT to be equated to the enthalpy of given air.

$h = c_p t + \omega h_g$ (no superheated vapor)

$80 = 1.005\, t + 0.0106\, h_g$

There is a link between t and h_g can be read from the steam tables

It can be iterated as follows:

For example, if the WBT is selected above the dew point and below the DBT,

WBT = 24°C, $h_g = 2546.06$ kJ/kg

P_v' at WBT = 0.0298 bar

$$\omega' = 0.622\frac{P_v'}{P_t - P_v'} = 0.622 \times \frac{0.0298}{1.01325 - 0.0298} = 0.0188$$

Substituting

$h = c_p t + \omega h_g$

$h = 1.005 \times 24 + 0.0188 \times 2546.06 = 72$ kJ/kg

So, for 80 kJ/kg, the WBT has to be increased to 24°C to refine the solution.

The psychrometric properties of the solution are depicted in Fig. 2.35.

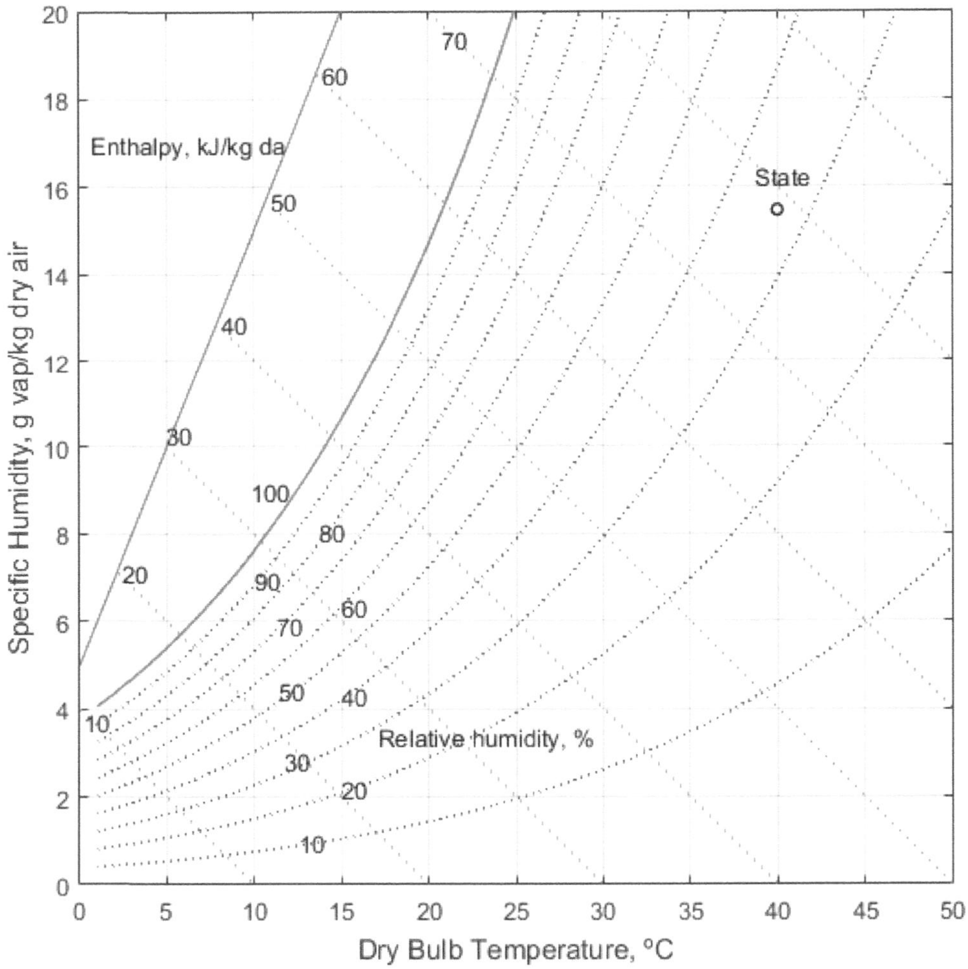

Fig. 2.35. Solution of the psychrometric problem.

Chapter 3
Thermal Processes

3.1 Introduction

Thermodynamic processes are classified as non-flow processes and flow processes based on the type of the system, i.e., closed system and open system respectively. Similarly, with the reference to irreversibilities, the processes are grouped into reversible processes and irreversible processes. The first law of thermodynamics relates heat and work with the system's energy changes in non-flow processes and flow processes. These relations, along with thermodynamic property changes, can be applied to solve the thermodynamic process problems. Entropy plays an important role in distinguishing reversible and irreversible processes. In this chapter, processes involved in different energy systems such as power plants, space technology, aircraft, automobiles, air conditioning, vapor absorption refrigeration, Kalina power cycle, etc. are modelled and presented on the properties charts.

3.2 Ideal gas processes

3.2.1 Isobaric process

In isobaric process of an open system the pressure is maintained constant by changing the volume and temperature.

The following are the energy interactions for work and heat:

$$W = \int -VdP = 0 \tag{3.1}$$

$$Q = H_2 - H_1 = mc_p(T_2 - T_1) \tag{3.2}$$

3.2.2 Isochoric process

In an isochoric process of an open system the volume is maintained constant by changing the pressure and temperature.

The following are the energy interactions:

$$W = \int_1^2 -VdP = -V(P_2 - P_1) = (P_1 - P_2)V \tag{3.3}$$

$$\begin{aligned} Q &= W + (H_2 - H_1) \\ &= (P_1 - P_2)V + H_2 - H_1 \\ &= H_2 - P_2V_2 - (H_1 - P_1V_1) \end{aligned}$$

DOI: 10.1201/9781032642505-3

$= U_2 - U_1$

$$= mc_v (T_2 - T_1) \tag{3.4}$$

3.2.3 *Isothermal process*

In an isothermal process of an open system, the temperature is maintained constant by changing the volume and pressure.

The following are the energy interactions:

$\Delta H = 0$

$Q = W$

3.2.4 *Adiabatic process*

In an adiabatic process of an open system, the boundary is insulated and prevents the heat outflow from the system or inflow to the system from the surroundings.

The following are the energy interactions:

$Q = 0$

$W = H_1 - H_2$

3.2.5 *Expansion in nozzles and diffusers*

With reference to Fig. 3.1, the energy balance equation applied to a steam nozzle,

$$\dot{m}\left(h_1 + \frac{C_1^2}{2} + gZ_1\right) + Q = \dot{m}\left(h_2 + \frac{C_2^2}{2} + gZ_2\right) + W \tag{3.5}$$

$Z_1 = Z_2$ and C_1 can be neglected

Since adiabatic, $Q = 0$, and $W = 0$

$$h_1 = h_2 + \frac{C_2^2}{2}$$

$$C_2 = \sqrt{2(h_1 - h_2)} \tag{3.6}$$

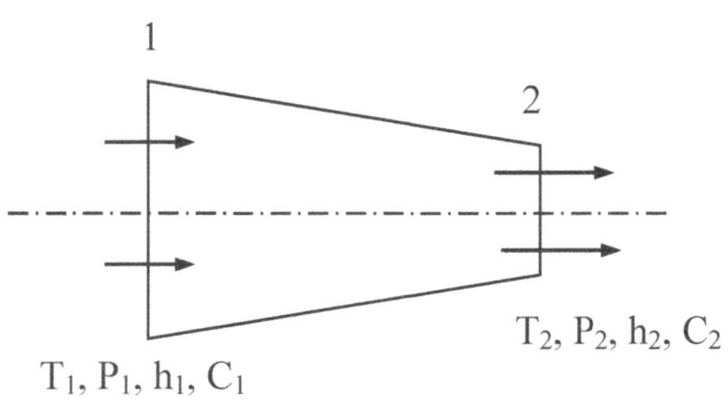

Fig. 3.1. Steam nozzle.

3.2.6 Throttling

The flow through an insulated passage with pressure drop is called throttling (Fig. 3.2). If change in P.E. is neglected, Q = W = 0

$$h_1 + \frac{C_1^2}{2} = h_2 + \frac{C_2^2}{2} \tag{3.7}$$

The pipe velocities in throttling are so slow, therefore K.E. terms can be neglected.

$$h_1 = h_2 \tag{3.8}$$

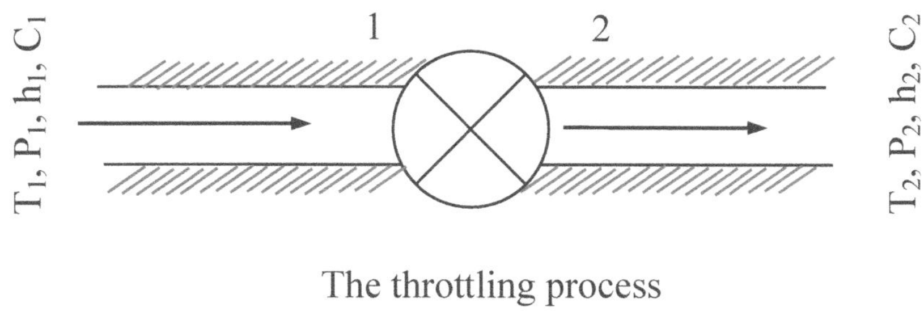

Fig. 3.2. Throttling management.

3.2.7 Expansion in steam turbines, gas turbines and expanders

For the turbine shown in Fig. 3.3, neglecting potential energy (PE) and kinetic energy (KE),

$$\dot{m}\ h_1 = \dot{m}\ h_2 + W$$

$$W = \dot{m}\ (h_1 - h_2) \tag{3.9}$$

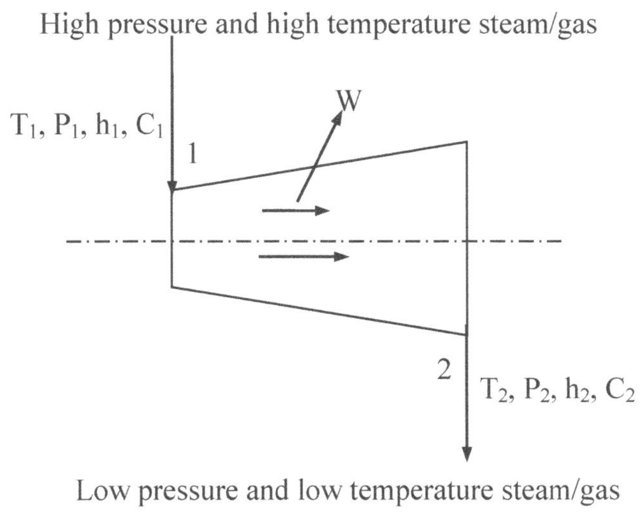

Fig. 3.3. Turbine.

3.2.8 Compression in blowers, pumps and compressors

The energy balance equation applied to a compressor shown in Fig. 3.4,

$$\dot{m}\ h_1 + W = \dot{m}\ h_2$$

$$W = \dot{m}\ (h_2 - h_1) \tag{3.10}$$

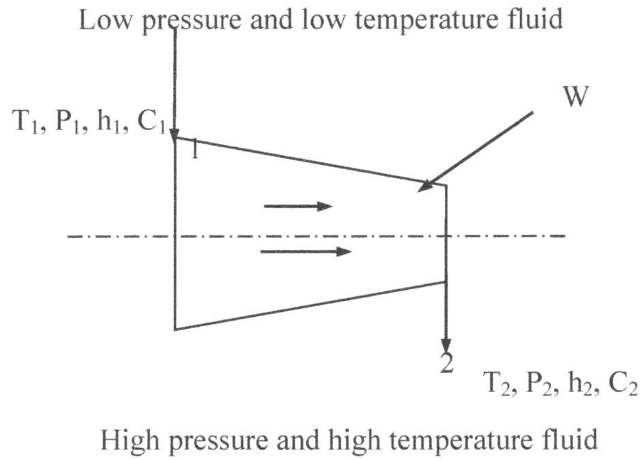

Fig. 3.4. Compressor.

3.2.9 Heat exchanger

Steam condenser is a typical heat exchanger and shown in Fig. 3.5.

Neglecting PE and KE, Q = 0 (to surroundings)

Energy balance

$$m_s h_1 + m_w h_3 = m_s h_2 + m_w h_4$$

$$\text{Heat transfer} = m_s\ (h_1 - h_2) = m_w (h_4 - h_3) \tag{3.11}$$

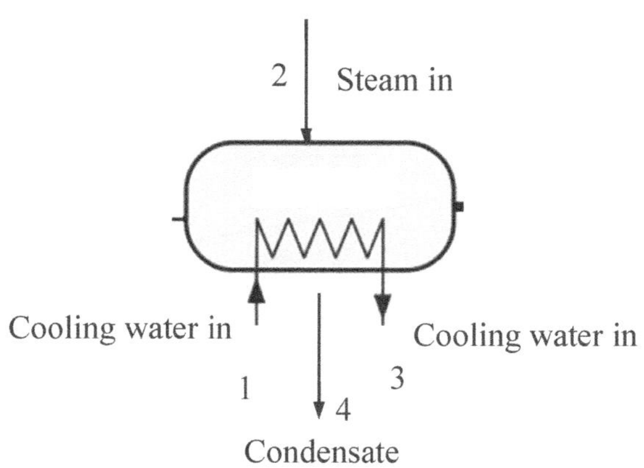

Fig. 3.5. Steam condenser.

3.2.10 Polytropic process in an open system

We know that h = u + Pv (3.12)

dh = du + Pdv + vdP (3.13)

From the first law dq = du + Pdv (3.14)

dh = dq + vdP (3.15)

Since the area under Ts diagram is heat similar to work which is the area under pv diagram,

dh = Tds + vdP (3.16)

dq = Tds = dh – vdP, it is the first law of thermodynamics applied to open system.

In place of W we can found –vdP

Therefore, in open system, $w = \int -vdP$ (3.17)

So the work in polytropic process,

$$w_{1-2} = \int_1^2 -vdp$$

$$= -\int_1^2 \left(\frac{C}{p}\right)^{\frac{1}{n}} dp$$

$$\because pv^n = C$$

$$= -C^{\frac{1}{n}} \left(\frac{p^{-\frac{1}{n}+1}}{-\frac{1}{n}+1} \right)_1^2$$

$$= -C^{\frac{1}{n}} \left(\frac{p_2^{\frac{n-1}{n}} - p_1^{\frac{n-1}{n}}}{\frac{n-1}{n}} \right)$$

$$= \left(\frac{-(p_2 v_2^n)^{\frac{1}{n}} p_2^{\frac{n-1}{n}} + (p_1 v_1^n)^{\frac{1}{n}} p_1^{\frac{n-1}{n}}}{\frac{n-1}{n}} \right)$$

$$= \frac{n(p_1 v_1 - p_2 v_2)}{n-1}$$

$$w = \frac{n(p_1 v_1 - p_2 v_2)}{n-1}$$

$w_{\text{open system}} = n\, w_{\text{closed system}}$

From this equation it is possible to compare the work in a closed system and open system at the same working conditions. Between these two systems is the fluid flow.
i.e.,

$$\frac{W_{open\ system}}{W_{closed\ system}} = n \tag{3.18}$$

If $n < 1$, $W_{open\ system} > W_{closed\ system}$

$n = 1$, $W_{open\ system} = W_{closed\ system}$

$n > 1$, $W_{open\ system} < W_{closed\ system}$

For example, in an isothermal process, the work in an open system and a closed system are the same. In the case of an adiabatic process, the open-system work is less than the closed-system work. The adiabatic expansion from an engine will generate more power than from an adiabatic expansion in a gas turbine at the same capacity. But engines are suitable for small capacity and turbines are good for high capacity output.

Heat transfer, $Q = W + \Delta H = \dfrac{nm(p_1v_1 - p_2v_2)}{n-1} + mc_p(T_2 - T_1)$

$$= \frac{nmR(T_1 - T_2)}{n-1} - mc_p(T_1 - T_2)$$

$$= \frac{mR(T_1 - T_2)}{n-1}\left(n - \frac{(n-1)c_p}{R}\right)$$

$$= W_{closed\ system}\left(\frac{nR - c_p(n-1)}{R}\right)$$

$$= \frac{W_{open\ system}}{n}\left(\frac{nR - c_p(n-1)}{R}\right)$$

$$= W_{open\ system}\left(\frac{nR - c_p(n-1)}{nR}\right)$$

$$Q = W_{open\ system}\left(1 - \frac{c_p(n-1)}{n}\right) \tag{3.19}$$

Change in enthalpy, $\Delta H = Q - W$

$$= W_{closed\ system}\left(\frac{nR - c_p(n-1)}{R}\right) - nW_{closed\ system}$$

$$= W_{closed\ system}\left(\frac{nR - c_p(n-1)}{R} - n\right)$$

$$= W_{closed\ system}\left(\frac{nR - c_p(n-1) - nR}{R}\right)$$

$$= W_{closed\ system}\left(\frac{c_p(1-n)}{R}\right)$$

$$= \frac{W_{open\ system}}{n}\left(\frac{c_p(1-n)}{R}\right)$$

$$\Delta H = W_{open\ system}\left(\frac{c_p}{R}\times\frac{1-n}{n}\right) \quad (3.20)$$

The energy changes in a closed system and an open system are summarised in Table 3.1.

The formulae for various non-flow thermodynamic processes to determine work and heat are listed in Table 3.2.

Table 3.1. Energy transfer in a closed system and an open system.

Energy	**Closed system**	**Open system**
Work, kJ	$\frac{P_1V_1 - P_2V_2}{n-1}$	n W
Heat, kJ	$W\left(\frac{\gamma-n}{\gamma-1}\right)$	$W_{open\ system}\left(1-\frac{c_p(n-1)}{n}\right)$
Change in internal energy, kJ	$\Delta U = W\left(\frac{1-n}{\gamma-1}\right)$	--
Change in enthalpy, kJ	--	$W_{open\ system}\left(\frac{c_p}{R}\times\frac{1-n}{n}\right)$

Table 3.2. Energy transfer in non-flow processes (gas phase).

Process	**Work, kJ**	**Heat, kJ**
Isobaric process	$P(V_2 - V_1)$	$H_2 - H_1$
Isothermal process	$P_1V_1 \ln\left(\frac{V_2}{V_1}\right) = P_1V_1 \ln\left(\frac{P_1}{P_2}\right)$	W
Adiabatic process	$\frac{P_1V_1 - P_2V_2}{\gamma-1}$	0
Isentropic process	$\frac{P_1V_1 - P_2V_2}{\gamma-1}$	0
Polytropic process	$\frac{P_1V_1 - P_2V_2}{n-1}$	$W\left(\frac{\gamma-n}{\gamma-1}\right)$
Isochoric process	0	$U_2 - U_1$

The energy changes in flow processes are summarized in Table 3.3. The expression of energies as a function of polytropic index reduces the complexity in the solution and also avoids the use of many formulae in calculations. The simplification of these formulae reduces the calculation time without affecting the accuracy.

Table 3.3. Energy transfer in flow processes (gas phase).

Process	**Work, kJ**	**Heat, kJ**
Isobaric process	0	$H_2 - H_1$
Isothermal process	$P_1V_1 \ln\left(\frac{V_2}{V_1}\right) = P_1V_1 \ln\left(\frac{P_1}{P_2}\right)$	W
Adiabatic process	$\frac{P_1V_1 - P_2V_2}{\gamma - 1} = H_2 - H_1$	0
Isentropic process	$\frac{P_1V_1 - P_2V_2}{\gamma - 1}$	0
Polytropic process	$n.\frac{P_1V_1 - P_2V_2}{n-1}$	$W_{open\ system}\left(1 - \frac{c_p(n-1)}{n}\right)$
Isochoric process	$(P_1 - P_2)V$	$U_2 - U_1$

3.3 Vapor processes

Similar to the gas processes studied in the earlier section and applied to ideal gas, vapor also undergoes different thermodynamic processes. But the nature of the vapor process is different from the gas process. In the study of pure substances, it has been understood that an isotherm passing through a critical point differentiates the vapor from the gas. Above the isothermal compression process, the fluid is gas, and below this curve, it is vapor before the phase change. The first and second laws of thermodynamics can be applied to vapor processes, but the ideal gas state equation is not applicable to vapor. The real gas equation gives reasonable accuracy to the superheated vapor. The thermodynamic processes applicable to vapor are isobaric processes, isothermal processes, hyperbolic processes, adiabatic processes, isentropic processes, polytropic processes, isochoric processes, and throttling processes. Since the vapor does not follow the characteristic gas equation, the isothermal process differs from the hyperbolic process, whereas in ideal gas both are the same. To solve the vapor process, two conditions are required. The first condition is the type of process, which gives the nature of the process. The second condition is the property at the exit of a process or energy interaction. In the case of property, the pressure, volume, temperature, or entropy at the exit of the process can be stated. Alternatively, heat transfer, work, or change in energy (internal energy or enthalpy) may be supplied. Unlike the gas process, in the vapor process, the exit state may be wet, saturated, or superheated. The exit state will determine the region of the fluid at the exit of the process.

The vapor processes differ from the perfect gas processes as the ideal gas equation is not applicable to the vapor processes. The vapor processes involve the following steps:

1. Finding the properties at the inlet of the process.
2. Identifying the exit region property (wet, dry saturated or superheated vapor) from the reference property.
3. Finding the properties at the exit of the process.
4. Determining the change in property/energy (change in entropy, change in internal energy, change in enthalpy, work done, heat transfer) for the process.
5. Drawing a suitable diagram (P-v, T-s or h-s) to represent the process.

The next section details the following vapor processes.

1. Isobaric process
2. Isothermal process
3. Hyperbolic process
4. Adiabatic/Isentropic process
5. Isochoric process
6. Polytropic process
7. Isenthalpic (throttling) process

3.3.1 Isobaric process

In the isochoric process, the fluid is heated or cooled with constant pressure in a closed system or in a flow process. As per the earlier discussion, in the wet region, the isobars and isotherms are the same due to phase change. In a superheated region, the temperature of steam changes at constant pressure. In a steam boiler, water converts to steam by taking heat from the furnace at constant pressure. The superheated steam is used in the turbine.

The heat transfer in the isobaric process

$$Q = m\ [(h_2 - h_1) \tag{3.21}$$

For a given heating or cooling load in an isobaric process, the exit state of the fluid needs to be simulated. In the isobaric heating process, as shown in Fig. 3.6, the final state may be wet steam, saturated steam or superheated steam depending on the initial condition and heat load. On other hand, on cooling with the isobaric process, the initial state may be wet, saturated or superheated steam.

At the known exit state, the heating or cooling load can be determined directly using the formula after the observations at the initial state and final state. But the supply of heat load or cooling load needs to iterate the exit condition.

The exit enthalpy can be determined as follows:

$$h_2 = h_1 \pm \frac{Q}{m} \tag{3.22}$$

The positive sign is used in heating and the negative in cooling.

The comparison of exit enthalpy with the saturated vapor enthalpy decides the final state. If the exit enthalpy is less than the saturation enthalpy, it is the wet steam. It is equal to saturated enthalpy, then the state is a saturated steam and above this the steam condition is the superheated state. The

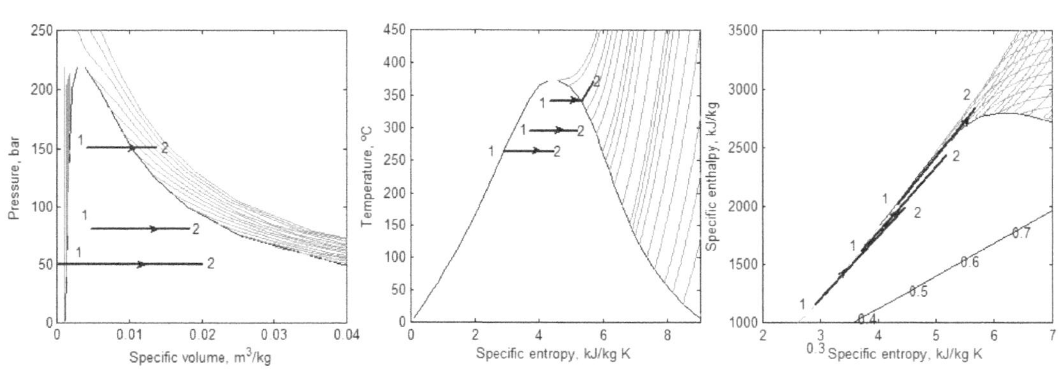

Fig. 3.6. Isobaric heating of wet steam shown on a (a) P-v diagram (b) T-s diagram and (c) h-s diagram.

wet steam condition can be deduced directly through finding the quality of steam without iteration of parameters. The superheated steam needs iteration of the superheated steam temperature till it matches the enthalpy.

For a non-flow process,

$Q = H_2 - H_1$

$W = P (V_2 - V_1)$

$\Delta U = Q - W = H_2 - H_1 - P (V_2 - V_1)$

For a flow process,

$Q = H_2 - H_1$

$W = 0$

$\Delta H = Q - W = H_2 - H_1$

3.3.2 Isothermal process

Figure 3.7 depicts the isothermal process in a wet region and a superheated region. Similar to isobaric, during the phase change process, the temperature and pressure remain constant. After the saturation vapor state, the continuation of the process at a constant temperature changes the pressure and volume. The heating of the vapor in the superheated region at a fixed temperature, expands the fluid and so there is a drop in pressure. This is the isothermal expansion process with heat addition. In the reverse direction, isothermal compression with cooling is possible. Through isothermal compression, it is possible to convert vapor into water below the critical pressure and temperature. But above this state, it is not possible; the vapor remains in the same state and is called gas. At low pressures, the ideal gas equation is applicable. But at high pressures, the ideal gas equation gives a significant error due to the molecule size and its closeness.

In isothermal heating or isothermal cooling, similar to isobaric, the exit condition needs to be iterated. The property at the outlet of the process determines the condition of the steam: either below saturation, at saturation, or above saturation. In the wet region, determination of the quality of steam results in the steam's exit properties. In the superheated region, the exit state requires iteration based on the specified exit condition, whether it is heat transfer or pressure.

For a non-flow process,

$Q = T (S_2 - S_1)$

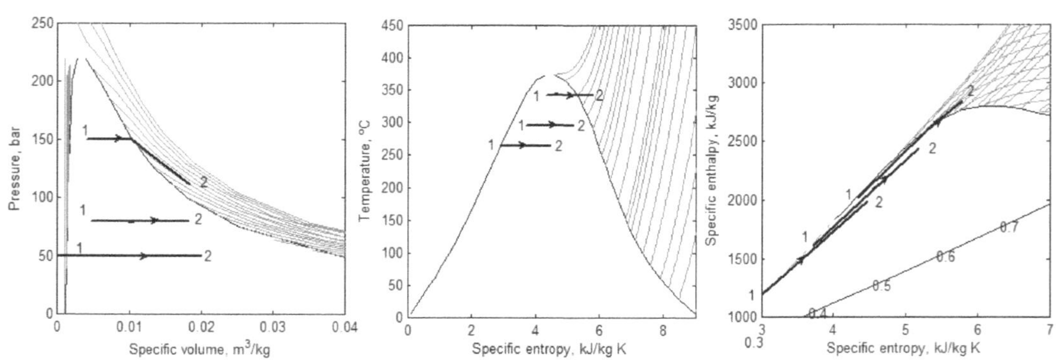

Fig. 3.7. Isothermal heating of wet steam shown on a (a) P-v diagram (b) T-s diagram and (c) h-s diagram.

$W = T(S_2 - S_1) - (H_2 - H_1 - (P_2V_2 - P_1V_1))$

$\Delta U = H_2 - H_1 - (P_2V_2 - P_1V_1)$

For a flow process,

$Q = T(S_2 - S_1)$

$W = T(S_2 - S_1) - (H_2 - H_1)$

$\Delta H = H_2 - H_1$

3.3.3 Hyperbolic process

In the gaseous region with ideal gas approach, the hyperbolic process (Pv = C) and the isothermal process are the same. But in the wet vapor, superheated vapor and real gas models applied to gas these two processes differ as the temperature is not constant even when the product of pressure and volume is constant. Therefore, the hyperbolic process is discussed separately without mixing with isothermal process. The changes in properties in a hyperbolic process are shown in Fig. 3.8.

In a hyperbolic process,

$P_1V_1 = P_2V_2$ or $P_1v_1 = P_2v_2$ in steady state or in a closed system.

For a non-flow process,

$$W = \int_1^2 PdV = mP_1v_1 \ln\left(\frac{v_2}{v_1}\right)$$

$$W = P_1V_1 \ln\left(\frac{V_2}{V_1}\right)$$

$\Delta U = U_2 - U_1 = (H_2 - P_2V_2) - (H_1 - P_1V_1) = H_2 - H_1$

since $P_2V_2 = P_1V_1$

$Q = \Delta U + W$

$$Q = m\left(h_2 - h_1 + P_1v_1 \ln\left(\frac{v_2}{v_1}\right)\right)$$

$$Q = H_2 - H_1 + P_1V_1 \ln\left(\frac{V_2}{V_1}\right) \tag{3.23}$$

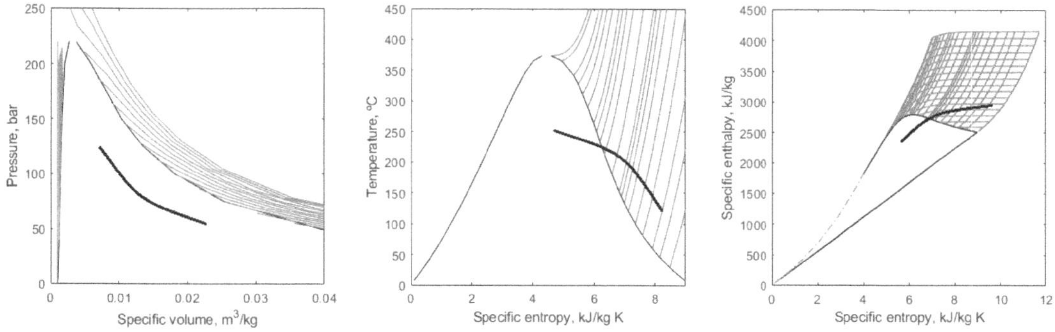

Fig. 3.8. Hyperbolic process on (a) P-v diagram (b) T-s diagram and (c) h-s diagram.

For the flow process,

$W_{\text{open system}} = n\, W_{\text{closed system}}$

Since n = 1 in hyperbolic process, the work in the open system and the closed system are same. The change in energy is also the same as $U_2 - U_1 = H_2 - H_1$ (a special case in hyperbolic process). Therefore, the heat transfer is also the same in the closed system.

In the heating process or cooling process, initially, the saturation condition at the exit pressure can be noted to decide the actual final state. It can be done through the iteration of pressure till matching the product of pressure and volume at the two states, i.e., initial state and saturation state.

3.3.4 Isentropic process/Adiabatic process

In both the isentropic process and the adiabatic process, the heat transfer is zero. In the isentropic process, the entropy is constant as shown in Fig. 3.9. In the adiabatic process, entropy increases on both compression and expansion as shown in Fig. 3.10. That means, the internal dissipative effects such as friction are not considered in the isentropic process, therefore, not entropy generation.

In the isentropic process, the exit condition can be simulated, by equating the inlet and outlet entropy. At the exit of the process, the inlet or outlet entropy may be compared with the saturated liquid entropy and saturated vapor entropy at the exit pressure. Depending on the value of exit entropy, it can be decided whether the exit fluid is wet, dry saturated or superheated heated fluid.

Following are the formulations given for the isentropic process and adiabatic process: Even to use the adiabatic process, at the first level, the isentropic process should be determined; Based on the result of isentropic process, the subsequent adiabatic process can be arrived at. Based on the isentropic efficiency, the adiabatic process can be formulated after finding the solution of the isentropic process. For compression, the isentropic efficiency is defined as the ratio of isentropic enthalpy rise to the actual enthalpy rise.

In compression, the isentropic efficiency of compression,

$$\eta_c = \frac{h_2' - h_1}{h_2 - h_1}$$

The enthalpy of fluid at the exit of the process after the actual compression,

$$h_2 = h_1 - \frac{h_2' - h_1}{\eta_c}$$

From the enthalpy, h2, other properties at state 2 can be determined.

Similarly, in expansion, the isentropic efficiency of expansion,

$$\eta_e = \frac{h_1 - h_2}{h_1 - h_2'}$$

The enthalpy of fluid at the exit of the process after the actual expansion,

$h_2 = \eta_e\,(h_1 - h_2')$

The exit condition, h_2 results in the exit state of adiabatic expansion process.

For a non-flow process,

$Q = 0$

$$W = (H_1 - H_2 - (P_1V_1 - P_2V_2)) \tag{3.24}$$

$$\Delta U = H_2 - H_1 - (P_2V_2 - P_1V_1) \tag{3.25}$$

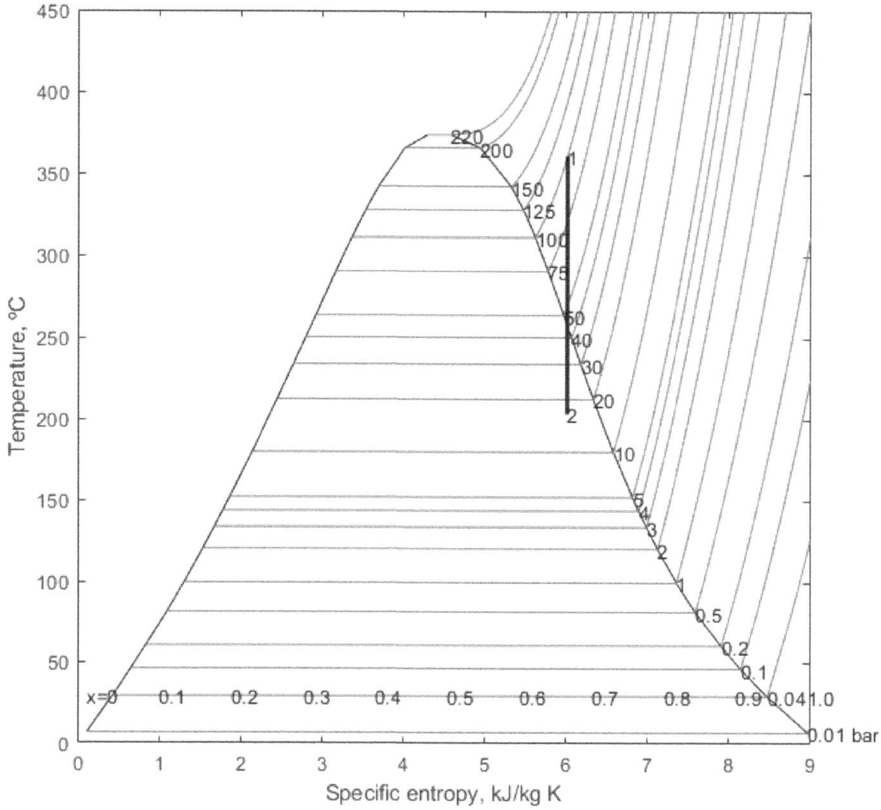

Fig. 3.9. Isentropic process on a T-s diagram.

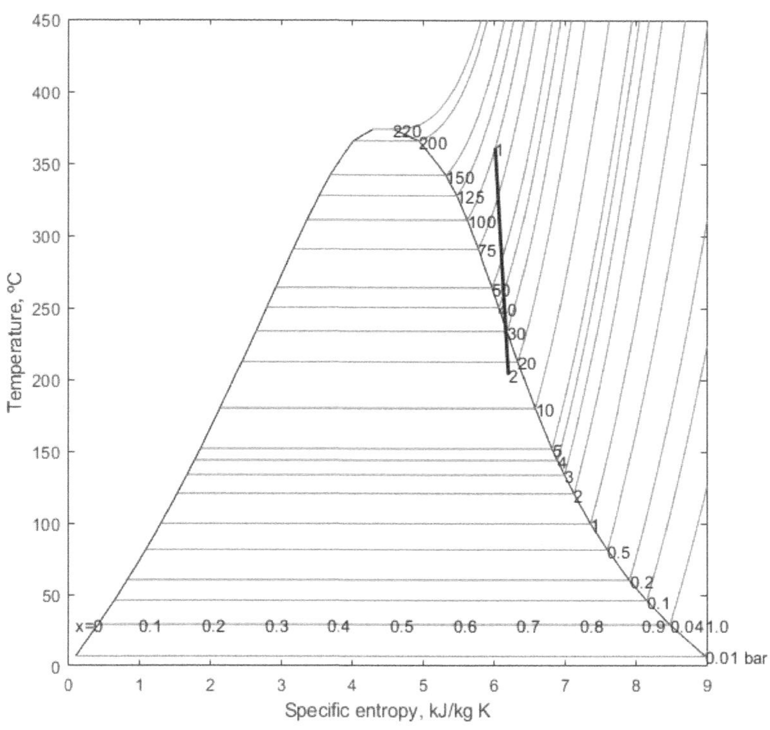

Fig. 3.10. Adiabatic process on a T-s diagram.

For a flow process,

$Q = 0$

$W = H_1 - H_2$

$\Delta H = H_2 - H_1$

3.3.5 Polytropic process

In this process, the vapor follows the relation PV^n = Constant. The vapor does not follow the gas equation, PV = mRT. The polytropic process is a statement of pressure-volume relationship during a reversible process (Fig. 3.11).

In a non-flow process,

$Q = \Delta U + W$

$$W = \frac{P_1V_1 - P_2V_2}{n-1} \tag{3.26}$$

$\Delta U = H_2 - H_1 - (P_2V_2 - P_1V_1)$

In a flow process,

$Q = \Delta H + W$

$$W = n\frac{P_1V_1 - P_2V_2}{n-1} \tag{3.27}$$

$\Delta H = H_2 - H_1$

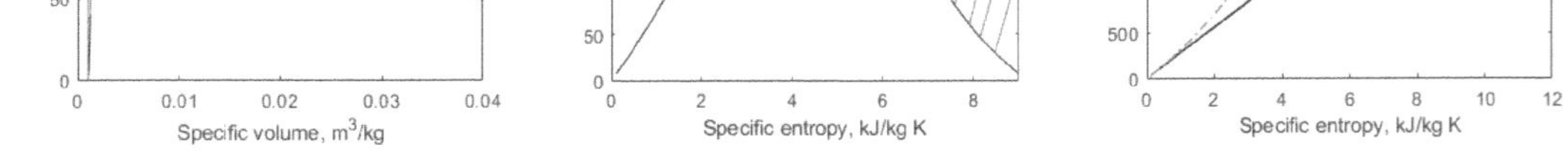

Fig. 3.11. Polytropic process on a T-s diagram.

3.3.6 Isochoric process

In the isochoric process, fluid is heated or cooled with a constant volume or in a fixed container. The constant volume process is either a heating or cooling process. Since the volume is constant in a closed system, the workdone is zero. With reference to the P-v diagram it can be noticed that to reach the critical state through the isochoric process, the initial state should be wet steam and the specific volume should be equal to the critical specific volume. If the initial condition of steam at the beginning of heating is either water or superheated steam, it cannot reach the critical state.

Figure 3.12 shows the cooling of wet steam, saturated steam and superheated steam at constant volume condition. In the P-v diagram it is the vertical; at other T-s and h-s, the process line is inclined. On heating, the entropy increases and on cooling, results drop in entropy as shown.

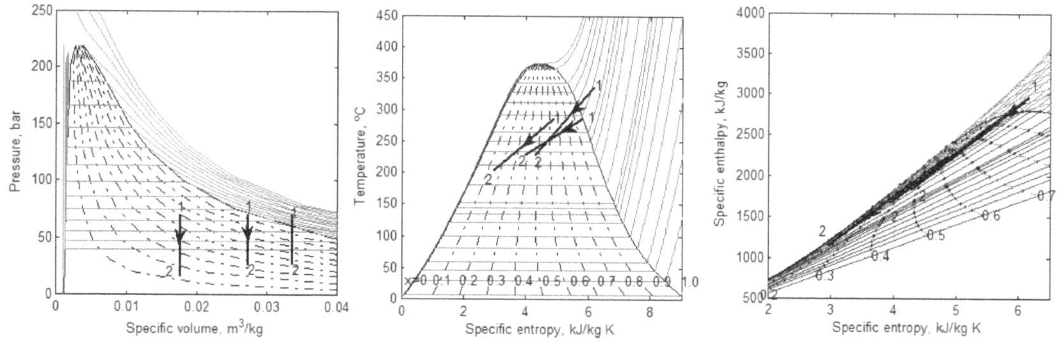

Fig. 3.12. Isochoric cooling of wet steam, saturated steam and superheated steam on a (a) P-v diagram (b) T-s diagram and (c) h-s diagram.

The mass of wet steam

$$v' = xv_v \tag{3.28}$$

The mass, total volume and specific volume during the process are constant.

If the initial and final states of the process are wet steam

$m\ x_1\ v_{v1} = m\ x_2 v_{v2}$

$$x_2 = \frac{x_1 v_{v1}}{v_{v2}} \tag{3.29}$$

For a non-flow process,

Since the work is zero, the heat supplied is equal to the change in internal energy

$$Q = U_2 - U_1 = m\ [(h_2 - p_2 v_2) - (h_1 - p_1 v_1)] \tag{3.30}$$

For a flow process, $W = -V(P_2 - P_1)$

$$\text{Change in energy, } \Delta H = H_2 - H_1 \tag{3.31}$$

$$Q = \Delta H + W = H_2 - H_1 - V\ (P_2 - P_1) \tag{3.32}$$

If the exit condition is given - i.e., pressure after heating or cooling - the final state can be found without any iteration. In the case of supply of heating or cooling capacity, the exit pressure is to be iterated using steam tables or Mollier chart. The use of ideal gas equation and its process relations avoids the use of this iteration but errors are possible.

3.3.7 Throttling process

The throttling process is also called the isenthalpic process due to constant enthalpy during the process (Fig. 3.13). The fluid is passed through a restricted passage for the throttling process. During this expansion, the temperature of the fluid decreases. This is a critical process in the application of refrigeration and air conditioning. Based on the fixed enthalpy before and after the throttling process, the exit condition can be simulated.

$h_1 = h_2$

Based on the known specific enthalpy at the exit pressure, the temperature at the exit of throttling can be iterated.

Tables 3.4 and 3.5 summarizes the above processes for a non-flow process and flow process respectively.

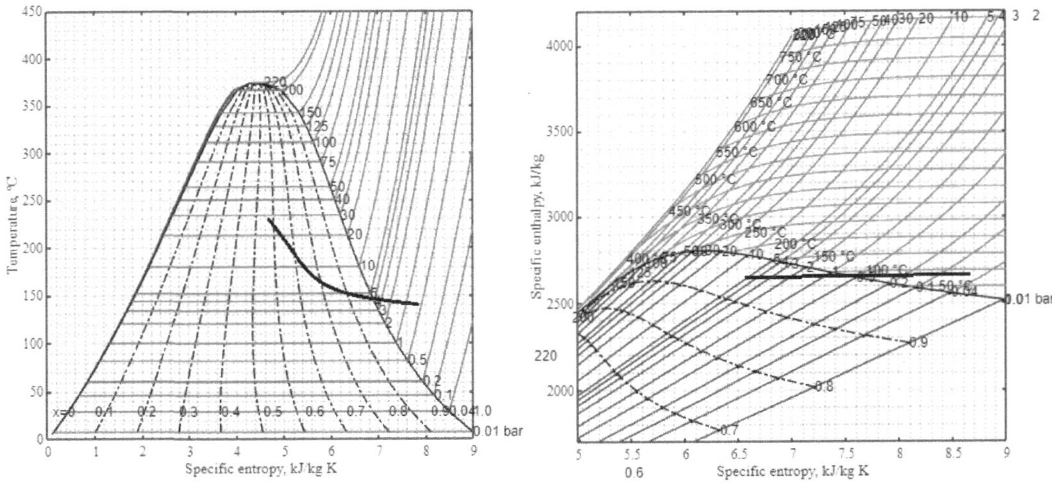

Fig. 3.13. Throttling process on (a)T-s diagram and (b) h-s diagram.

Table 3.4. Energy transfer in non-flow processes.

Process	Change in Energy, kJ	Work, kJ	Heat, kJ
Isobaric process	$H_2 - H_1 - P(V_2 - V_1)$	$P(V_2 - V_1)$	$H_2 - H_1$
Isothermal process	$H_2 - H_1 - (P_2 V_2 - P_1 V_1)$	$T(S_2 - S_1) - (H_2 - H_1 - (P_2V_2 - P_1 V_1))$	$T(S_2 - S_1)$
Hyperbolic process	$H_2 - H_1$	$P_1V_1 \ln\left(\frac{V_2}{V_1}\right)$	$H_2 - H_1 + P_1V_1 \ln\left(\frac{V_2}{V_1}\right)$
Isentropic process	$H_2 - H_1 - (P_2V_2 - P_1V_1)$	$H_1 - H_2 - (P_1V_1 - P_2V_2)$	0
Adiabatic process	$H_2 - H_1 - (P_2V_2 - P_1V_1)$	$H_1 - H_2 - (P_1V_1 - P_2V_2)$	0
Polytropic process	$H_2 - H_1 - (P_2V_2 - P_1V_1)$	$\frac{P_1V_1 - P_2V_2}{n-1}$	$W + \Delta U$
Isochoric process	$H_2 - H_1 - V(P_2 - P_1)$	0	$H_2 - H_1 - V(P_2 - P_1)$

Table 3.5. Energy transfer in a flow process.

Process	Change in Energy, kJ	Work, kJ	Heat, kJ
Isobaric process	$H_2 - H_1$	0	$H_2 - H_1$
Isothermal process	$H_2 - H_1$	$T(S_2 - S_1) - (H_2 - H_1)$	$T(S_2 - S_1)$
Hyperbolic process	$H_2 - H_1$	$nP_1V_1 \ln\left(\frac{V_2}{V_1}\right)$	$H_2 - H_1 + nP_1V_1 \ln\left(\frac{V_2}{V_1}\right)$
Isentropic process	$H_2 - H_1$	$H_1 - H_2$	0
Adiabatic process	$H_2 - H_1$	$H_1 - H_2$	0
Polytropic process	$H_2 - H_1$	$n\frac{P_1V_1 - P_2V_2}{n-1}$	$W + \Delta H$
Isochoric process	$H_2 - H_1$	$V(P_1 - P_2)$	$H_2 - H_1 + V(P_1 - P_2)$

3.4 Binary fluid processes

The processes are formulated using mass balance and energy balance equations. In addition to the properties of pressure, volume and temperature, the additional property involved in the binary fluid processes is concentration. The study of these binary fluid processes equips us to model the power, cooling and its integrations using a binary fluid system (Srinivas, 2021).

The main processes associated with the binary fluid system are as follows:

1. Separation
2. Adiabatic mixing
3. Diabatic mixing
4. Throttling
5. Distillation

3.4.1 Separation

In this separation process, a mixture consists of liquid and vapor. The liquid and vapor are separated from each other and the properties after the separation are found using process equations. After the separation of liquid and vapor, the liquid concentration and vapor concentration differ from the mixture concentration. The liquid concentration is less than the vapor concentration. The mixture concentration lies between the liquid concentration and vapor concentration. As per the assumption, the pressure of the liquid, vapor and mixture are same.

Figure 3.14 shows a separation process (2-1-3) on h-x diagram. The mixture (1) is separated into saturated liquid (2) and saturated vapor (3).

The mass balance in the separation process is

$$m_1 = m_2 + m_3 \tag{3.33}$$

Ammonia mass balance,

$$m_1 x_1 = m_2 x_2 + m_3 x_3 \tag{3.34}$$

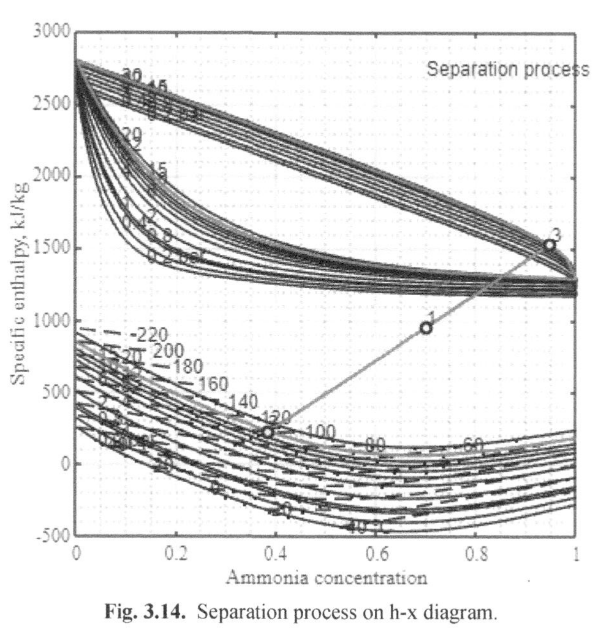

Fig. 3.14. Separation process on h-x diagram.

Either m_2 or m_3 may be eliminated from this equation for the simplification.

Eliminating m_2,

$$m_1x_1 = (m_1 - m_3)\,x_2 + m_3x_3$$

$$m_1x_1 = m_1x_2 - m_3x_2 + m_3x_3$$

$$m_1(x_1 - x_2) = m_3(x_3 - x_2)$$

The dryness fraction,

$$d = \frac{m_3}{m_1} = \frac{x - x_l}{x_v - x_l} = \frac{x_1 - x_2}{x_3 - x_2} \tag{3.35}$$

Similarly, by eliminating m_2 in the ammonia balance, wetness fraction can be obtained as follows:

$$w = \frac{m_2}{m_1} = \frac{x_v - x}{x_v - x_l} = \frac{x_3 - x_1}{x_3 - x_2} \tag{3.36}$$

Equation 3.35 shows that the dryness fraction or vapor fraction is inversely proportional to the distance between liquid and mixture. Similarly, Eq. 3.36 indicates that liquid fraction or wetness fraction is inversely proportional to the distance between the vapor and mixture. It is called the lever rule.

From the energy balance,

$$m_1h_1 = m_2h_2 + m_3h_3 \tag{3.37}$$

Eliminating m_2,

$$m_1h_1 = (m_1 - m_3)h_2 + m_3h_3$$

$$m_1h_1 = m_1h_2 - m_3h_2 + m_3h_3$$

$$m_1(h_1 - h_2) = m_3(h_3 - h_2)$$

Similar to the earlier dryness fraction, which is a function of concentration, the dryness fraction also can be expressed as a function of enthalpies:

$$d = \frac{m_3}{m_1} = \frac{h - h_l}{h_v - h_l} = \frac{h_1 - h_2}{h_3 - h_2} \tag{3.38}$$

Similarly, by eliminating m_2 in the energy balance to obtain the wetness fraction,

$$w = \frac{m_2}{m_1} = \frac{h_v - h}{h_v - h_l} = \frac{h_3 - h_1}{h_3 - h_2} \tag{3.39}$$

Therefore, mixture concentration,

$$x_1 = x_2 + \mathrm{d}\,(x_3 - x_2) \tag{3.40}$$

3.4.2 Adiabatic mixing

Mixing is a process with or without heat transfer. Mixing without heat transfer is called adiabatic mixing and mixing with heat transfer is called diabatic mixing (Refer Fig. 3.15, adiabatic mixing line 1-3-2).

Mass balance,

$$m_1 + m_2 = m_3 \tag{3.41}$$

$$m_1x_1 + m_2x_2 = m_3x_3 \tag{3.42}$$

$$m_1h_1 + m_2h_2 = m_3h_3 \tag{3.43}$$

Eliminating m_1,

$$x_3 = \frac{m_1x_1 + m_2x_2}{m_3} = \frac{(m_3 - m_2)x_1 + m_2x_2}{m_3} = \frac{m_3x_1 - m_2x_1 + m_2x_2}{m_3}$$

$$\therefore x_3 = x_1 + \frac{m_2}{m_3}(x_2 - x_1)$$

$$\therefore \frac{x_3 - x_1}{x_2 - x_1} = \frac{m_2}{m_3} \tag{3.44}$$

Similarly,

$$h_3 = h_1 + \frac{m_2}{m_3}(h_2 - h_1)$$

$$\therefore \frac{h_3 - h_1}{h_2 - h_1} = \frac{m_2}{m_3} \tag{3.45}$$

It shows that as per the lever rule, the length of 3-1 is proportional to m_2 (opposite side) and the total length 1-2 length is proportional to the total mass, m_3.

The outlet concentration and enthalpy can also be calculated using the graphical method. The mixture temperature can be iterated from the enthalpy h_3.

Alternatively, by eliminating, m_2,

$$x_3 = \frac{m_1x_1 + m_2x_2}{m_3} = \frac{m_1x_1 + (m_3 - m_1)x_2}{m_3} = \frac{m_1x_1 + m_3x_2 - m_1x_2}{m_3}$$

$$x_3 = x_2 + \frac{m_1}{m_3}(x_1 - x_2)$$

$$\therefore \frac{x_3 - x_2}{x_1 - x_2} = \frac{m_1}{m_3} \tag{3.46}$$

Similarly,

$$\frac{h_3 - h_2}{h_1 - h_2} = \frac{m_1}{m_3} \tag{3.47}$$

The equation simplified to concentration and enthalpy, represents that the length of 3-2 on concentration–enthalpy chart is proportional to m_1 and the total length 1-2 is proportional to the total mass, m_3 as per the lever rule.

Therefore, the state after mixing can be determined as follows:

Concentration after mixing,

$$x_3 = x_1 + \frac{m_1}{m_3}(x_2 - x_1) = x_2 + \frac{m_2}{m_3}(x_1 - x_2) \tag{3.48}$$

The specific enthalpy after mixing,

$$h_3 = h_1 + \frac{m_1}{m_3}(h_2 - h_1) = h_2 + \frac{m_2}{m_3}(h_1 - h_2) \tag{3.49}$$

3.4.3 Diabatic mixing (Mixing with heat transfer)

Diabatic mixing adds heat transfer from or to the surroundings.

Figure 3.15 depicts diabatic mixing in two stages. In stage 1, the adiabatic mixing is expressed by the process 1-3-2. The second stage involves only heat transfer which is represented by a line 3-4.

Mass balance equations,

$$m_1 + m_2 = m_3 = m_4 \tag{3.50}$$

$$m_1x_1 + m_2x_2 = m_3x_3 = m_4x_4 \tag{3.51}$$

$$m_1h_1 + m_2h_2 = m_3h_3 = m_4h_4 + Q \tag{3.52}$$

Similar to the earlier adiabatic mixing process,

$$\frac{m_1}{m_3} = \frac{x_3 - x_2}{x_1 - x_2} = \frac{h_3 - h_2}{h_1 - h_2} \neq \frac{h_4 - h_2}{h_1 - h_2} \tag{3.53}$$

$$\frac{m_2}{m_3} = \frac{x_3 - x_1}{x_2 - x_1} = \frac{h_3 - h_1}{h_2 - h_1} \neq \frac{h_4 - h_1}{h_2 - h_1} \tag{3.54}$$

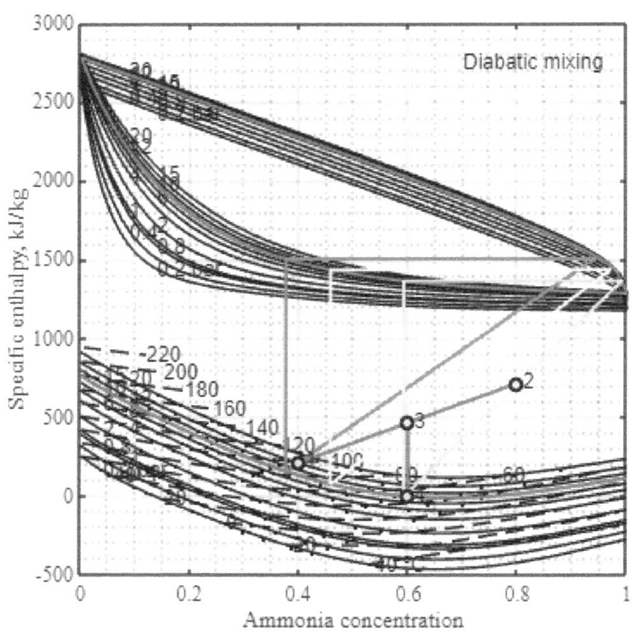

Fig. 3.15. Adiabatic mixing with integrated diabatic mixing in the absorber to give a saturated liquid state.

For example, the process with heat rejection is formulated as follows.

Eliminating m_1,

$(m_4 - m_2)h_1 + m_2h_2 = m_4h_4 + Q$

$m_4h_1 - m_2h_1 + m_2h_2 = m_4h_4 + Q$

$m_4h_1 + m_2(h_2 - h_1) = m_4h_4 + Q$

$$h_4 = \frac{m_4h_1 + m_2(h_2 - h_1) - Q}{m_4}$$

$$= h_1 + \frac{m_2}{m_4}(h_2 - h_1) - \frac{Q}{m_4}$$

Following equation results, the heat-transfer duty in absorber.

$$\frac{Q}{m_4} = h_1 + \frac{m_2}{m_4}(h_2 - h_1) - h_4 \quad (3.55)$$

Eliminating m_2,

$\mathrm{m}_1h_1 + (\mathrm{m}_4 - m_1)h_2 = m_4h_4 + Q$

$m_1h_1 + m_4h_2 - m_1h_2 = m_4h_4 + Q$

$m_1(h_1 - h_2) + m_4h_2 = m_4h_4 + Q$

$$h_4 = \frac{m_1(h_1 - h_2) + m_4h_2 - Q}{m_4}$$

$$= h_2 + \frac{m_1}{m_4}(h_1 - h_2) - \frac{Q}{m_4}$$

The following equation results, the heat-transfer duty in the absorber.

$$\frac{Q}{m_4} = h_2 + \frac{m_1}{m_4}(h_1 - h_2) - h_4 \quad (3.56)$$

Therefore,

$$h_4 = h_1 + \frac{m_2}{m_4}(h_2 - h_1) - \frac{Q}{m_4} = h_2 + \frac{m_1}{m_4}(h_1 - h_2) - \frac{Q}{m_4}$$

or

$$\frac{Q}{m_4} = h_1 + \frac{m_2}{m_4}(h_2 - h_1) - h_4 = h_2 + \frac{m_1}{m_4}(h_1 - h_2) - h_4$$

For convenience, the mixing and heat transfer processes are shown separately. The upward vertical line represents the heat addition to the mixture and the downward vertical line indicates the heat rejection from the mixture. The length of this vertical line is Q/m_4 on a h-x diagram.

To find the enthalpy of the mixture in a diabatic process with heat addition, the equation can be rewritten for h_4.

$$h_4 = h_1 + \frac{m_2}{m_4}(h_2 - h_1) + \frac{Q}{m_4} \quad (3.57)$$

The concentration remains the same and the temperature can be iterated from the saturated liquid condition from a known pressure and concentration.

Alternatively, from the earlier process of adiabatic mixing,

$$h_3 = h_1 + \frac{m_1}{m_3}(h_2 - h_1) = h_2 + \frac{m_2}{m_3}(h_1 - h_2)$$

$$h_4 = h_3 \pm \frac{Q}{m_4}$$

$$h_4 = h_1 + \frac{m_1}{m_3}(h_2 - h_1) \pm \frac{Q}{m_4} = h_2 + \frac{m_2}{m_3}(h_1 - h_2) \pm \frac{Q}{m_4}$$

3.4.4 Throttling

Throttling is an important process in a refrigeration plant to generate cooling. The Kalina cycle allows the absorption of vapor into a weak solution at low pressure. A throttling device changes the weak solution pressure from high pressure to low pressure. Where the fluid temperature decreases after throttling, the process is irreversible. For throttling, expansion valves or capillary tubes are used. The throttling valves separate the high-pressure lines and low-pressure lines. It is assumed that the work and heat transfer in throttling are negligible. The resulting process is an isenthalpic process.

Figure 3.16 shows a typical throttling process in a binary fluid system. The fluid is converted from high pressure to low pressure. The state 1 (before throttling) and state 2 (after throttling) are the same on the enthalpy-concentration chart as there is no change in enthalpy or concentration during the throttling. As per the assumptions, the pressure and temperature of the fluid decrease after throttling. The condition of the fluid before throttling (pressure, temperature, and concentration) is a known state. As per the exit pressure, the exit temperature to be simulated through either analytical method or graphical method. With reference to high pressure, state 1 is a saturated liquid state. The same point when referred to as state 2, it is an unsaturated fluid at low pressure. Since the

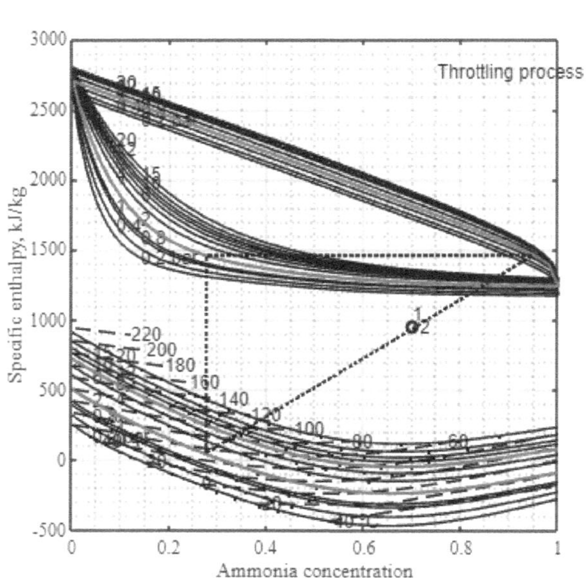

Fig. 3.16. Effect of concentration on throttling exit temperature.

exit temperature is to be simulated at state 2, an iteration method can be applied for the graphical solution.

In the above solution, let us check the exit temperature of throttling if the concentration is shifted from 0.96 to 0.75 without affecting the initial pressure and temperature.

3.4.5 Distillation

The refrigerant concentration is maintained high to result in low temperatures in a refrigeration plant (Fig. 3.17). In a low-concentration mixture, a high content of moisture will exist. We need to condense this moisture to remove it from the mixture and increase concentration. This process increases the concentration of the working fluid. The process of distillation of the working fluid to increase its concentration is called depletion. The equipment used is called a dephlegmator. A dephlegmator is a heat exchanger in which circulating water flows to cool the working fluid. The cooling of the working fluid removes the moisture and so increases concentration. The heat from the working fluid is transferred to the circulating water or some other fluid to improve the performance of the refrigerator. The boiling or condensing temperature of the water is higher than that of ammonia. First, the moisture in the dephlegmator condenses without condensing the ammonia. By gravity, the condensed moisture drips into to the boiler. To condense ammonia, a very low temperature is required, and therefore ammonia remains in vapor form in the dephlegmator. The process in the dephlegmator is a steady-state process, and the process pressure is assumed to be constant. The dephlegmator is compatible the refrigeration plant. But it does not favor the vapor absorption power plant, that is, the Kalina power plant's performance. The use of a dephlegmator in a power plant facilitates operating the boiler at a relative low pressure. With the use of a dephlegmator, a low-capacity or low-cost vapor absorption power plant can be designed. After passing through the dephlegmator, the mixer is sent to a separator. In the separator, the liquid and vapor are separated physically. As per the assumption, distillation and separation occur at a constant pressure. It is assumed that the entire process takes place under constant pressure. In the LiBr vapor absorption system, a dephlegmator is necessary as LiBr is in the powder (solid) form and the water (refrigerant) is in the liquid/vapor form.

The dry fraction (mass of vapor to the total mass) in the separator after the generator,

$$d_1 = \frac{x_4 - x_5}{x_6 - x_5} \tag{3.58}$$

Vapor separated in the generator from m_4,

$$m_6 = \mathrm{d}_1 m_4 \tag{3.59}$$

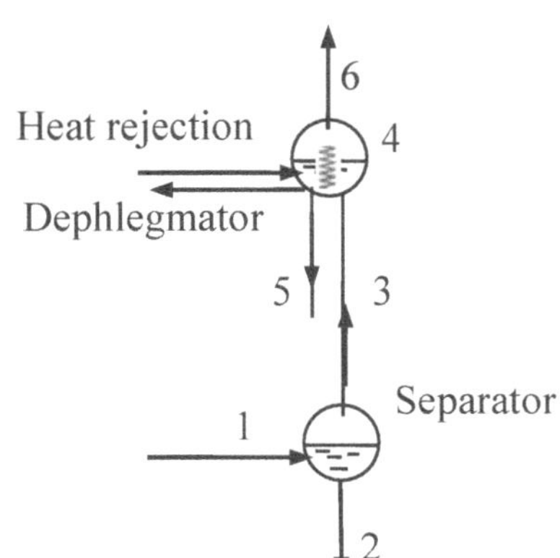

Fig. 3.17. Generator and dephlegmator in a typical vapor absorption system.

Liquid separated in the generator from m_4,

$$m_5 = (1 - d_1)\, m_4 \tag{3.60}$$

Similar to the separator, the vapor–mass ratio in the dephlegmator,

$$d_2 = \frac{x_7 - x_8}{x_9 - x_8} \tag{3.61}$$

Vapor separated in the dephlegmator from m_7,

$$m_9 = d_2 m_7 \tag{3.62}$$

Liquid separated in the dephlegmator from m_7,

$$m_8 = (1 - d_2)\, m_7 \tag{3.63}$$

The removal of heat from the mixture in the dephlegmator results in the condensation of water from the vapor, as the boiling point of water is higher than the boiling point of ammonia (Fig. 3.18). It increases the concentration of vapor, which is the desired effect. The maximum limit of heat removal can also be seen in the concentration-enthalpy chart. The maximum limit for gaining the concertation boost is 100%. The actual resulting concentration can be determined from the effectiveness of distillation. The effectiveness of the dephlegmator is defined as the ratio of the actual concentration increase to the maximum possible concentration increase. It is the function of inlet concentration and outlet concentration (Eq. 3.64).

Effectiveness of dephlegmator,

$$\varepsilon_{dephlegmator} = \frac{x_{exit} - x_{in}}{1 - x_{in}} \tag{3.64}$$

$$h_7 = h_6 - \frac{Q_d}{m_6} \tag{3.65}$$

where Q_d is the heat loss in the dephlegmator. The heat loss can be determined from the required status (concentration) at the exit and the enthalpies.

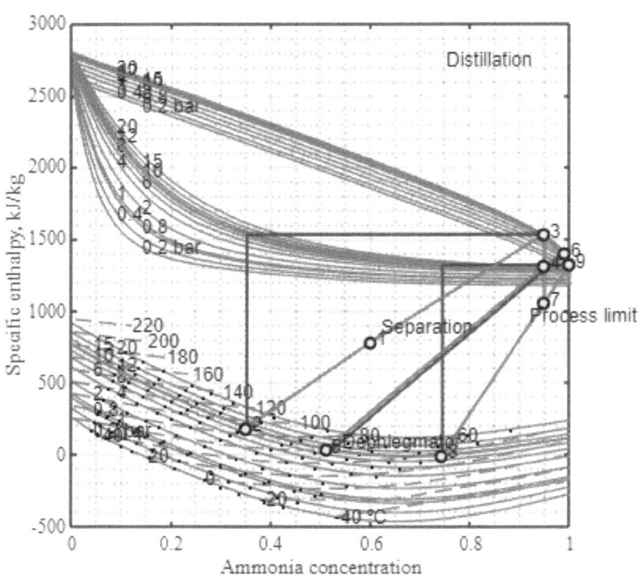

Fig. 3.18. The actual dephlegmator process and the maximum dephlegmator process to define the effectiveness of the process.

3.5 Psychrometric processes

An air washer is a device for conditioning air. In an air washer air comes in direct contact with a spray of water and there will be an exchange of heat and mass (water vapor) between air and water. The outlet condition of air depends upon the temperature of water sprayed in the air washer. Hence, by controlling the water temperature externally, it is possible to control the outlet conditions of air, which then can be used for air conditioning purposes (Stoecker and Jones, 1981).

In the air washer, the mean temperature of water droplets in contact with air decides the direction of heat and mass transfer. As a consequence of the 2nd law, the heat transfer between air and water droplets will be in the direction of the decreasing temperature gradient. Similarly, the mass transfer will be in the direction of the decreasing vapor pressure gradient.

Sensible heating or cooling of wet air takes place without changing the specific humidity or temperature. Figure 3.19 depicts the fluid's sensible heating.

In the air washer, if the water temperature is less than the dew point temperature of supply air, the resultant process is cooling and dehumidification. The temperature and humidity of air decrease by rejecting the heat and condensation as shown in Fig. 3.20. The heat transfer is the sum of the sensible heat transfer and the latent heat transfer.

If the water temperature is equal to the wet-bulb temperature of supply air, the resulted process is adiabatic saturation (Fig. 3.21). Here the sensible heat transfer from air to water is exactly equal to the latent heat transfer from water to air. Hence, no external cooling or heating of water is required. That is, this is a case of pure water recirculation. This is a process that takes place in a perfectly insulated evaporative cooler.

If the water supply temperature is in between the dew point temperature and wet-bulb temperature, it results in a cooling and humidification process (Fig. 3.22). Here the sensible heat transfer is from air to water and latent heat transfer is from water to air, but the total heat transfer is from air to water, hence, the water has to be cooled externally.

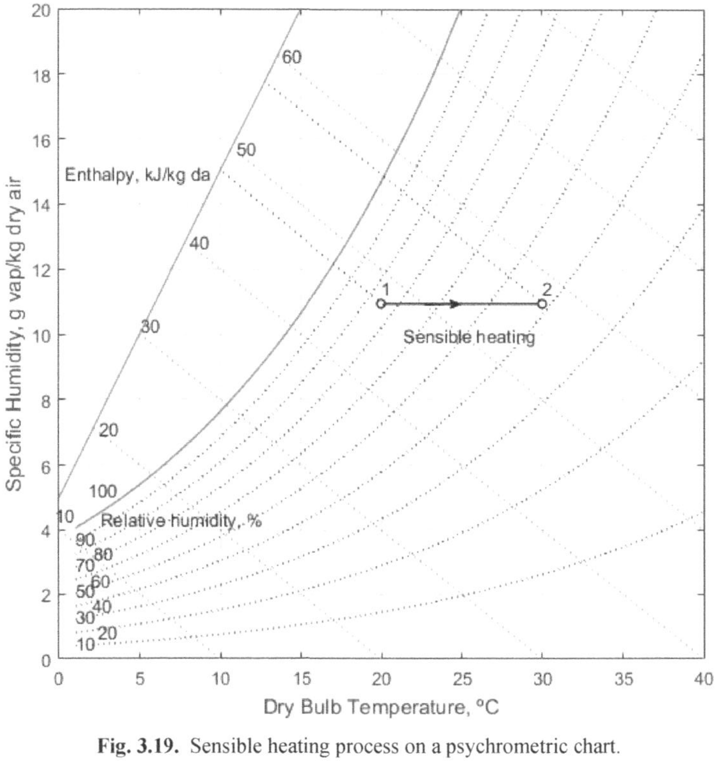

Fig. 3.19. Sensible heating process on a psychrometric chart.

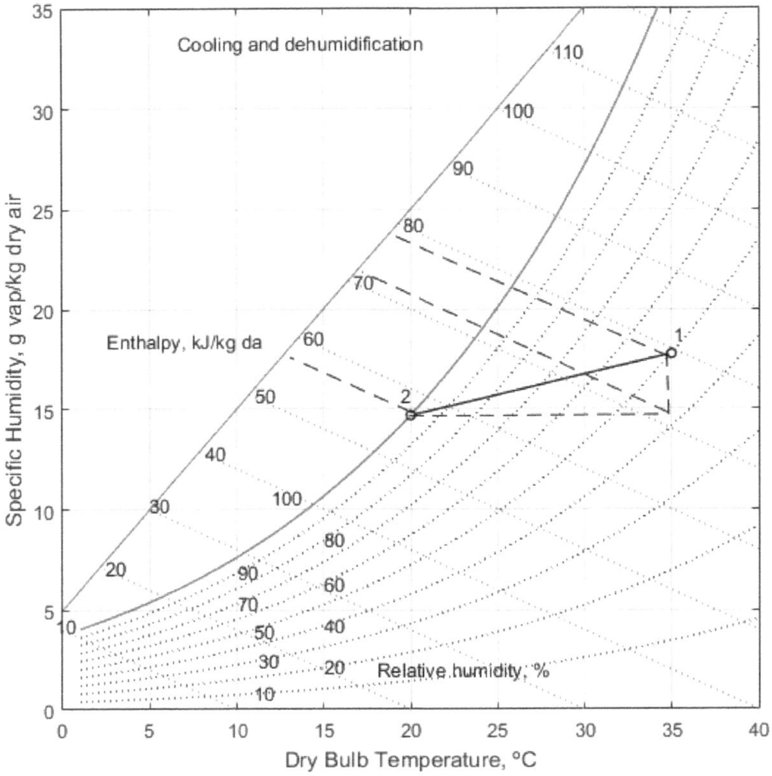

Fig. 3.20. Solution of cooling and dehumidification on psychrometric chart.

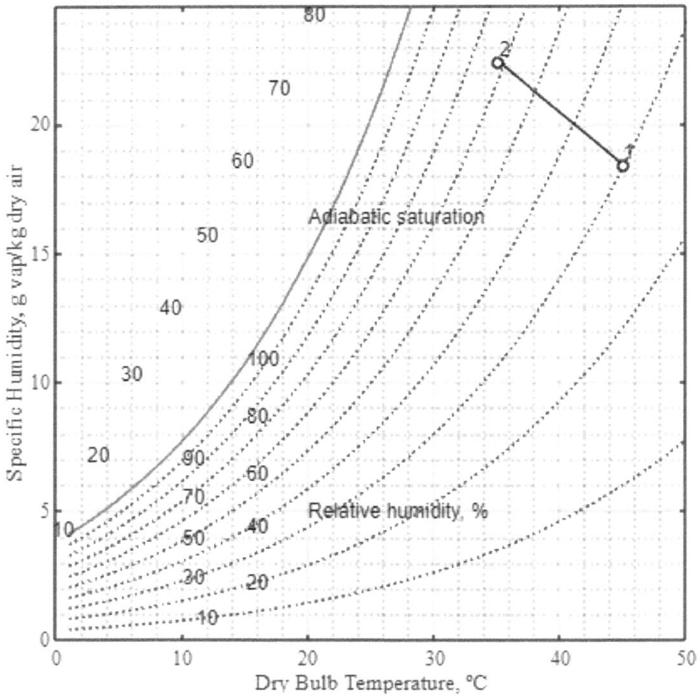

Fig. 3.21. Adiabatic saturation process.

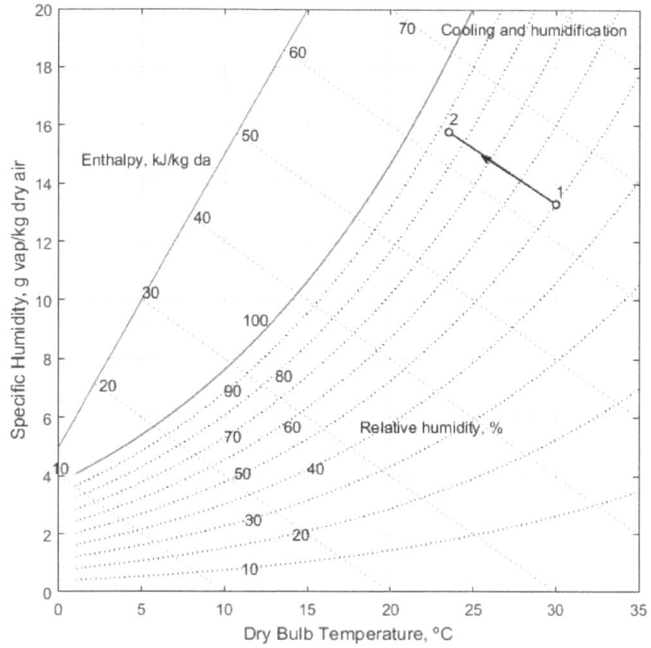

Fig. 3.22. Cooling and humidification on a psychrometric chart.

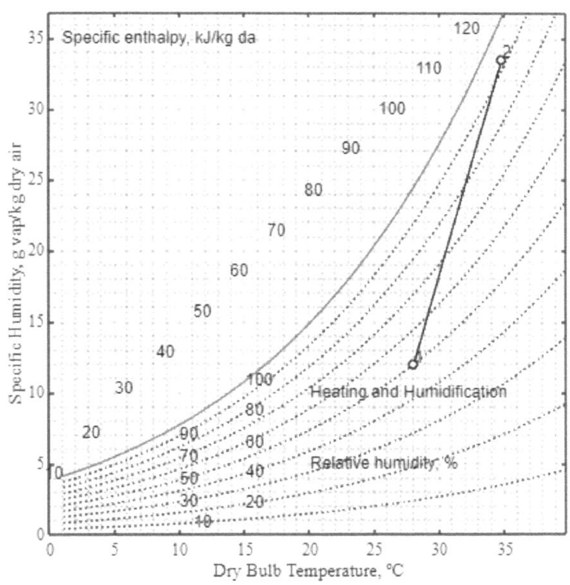

Fig. 3.23. Heating and humidification process.

Similarly, if the water supply temperature is between the wet-bulb temperature and dew point temperature, it also results in a cooling and humidification process (Fig 3.22). Here the sensible heat transfer is from air to water and latent heat transfer is from water to air, but the total heat transfer is from water to air, hence, the water has to be heated externally. This is the process that takes place in a cooling tower. The air stream extracts heat from the hot water coming from the condenser, and the cooled water is sent back to the condenser.

If the water supply temperature is greater than the dry-bulb temperature of supply air, the resultant process is heating and humidification (Fig. 3.23). In this process, both sensible and latent heat transfers are from water to air, hence, water has to be heated externally. Thus, it can be seen that an air washer works as a year-round air conditioning system. Though air washer is an extremely useful simple device, it is not commonly used for comfort air conditioning applications due to concerns about health resulting from bacterial or fungal growth on the wetted surfaces. However, it can be used in industrial applications.

3.6 Numerical examples

1. Find the polytropic non-flow processes for in-air compression and expansion with index values 0, 0.5, 1, 1.4, 2 and ∞. Consider the initial condition at 5 bar and 1 m³. The volume ratio in compression and expansion respectively are 0.5 and 2 (volume ratio = higher volume/lower volume). At constant volume heat addition (compression), consider the final pressure as 25 bar. Similarly, at constant volume heat rejection (expansion), the final pressure is 1 bar. Analyse your result. Find the work in each case.

Solution:

Compression processes:

$V_2 = 0.5V_1 = 0.5 \times 1 = 0.5\ m^3$

i) n = 0, constant pressure process

$$W_{1-2} = \frac{P_1V_1 - P_2V_2}{n-1} = P_2V_2 - P_1V_1 = P(V_2 - V_1)$$

$$= 500\ (0.5 - 1) = -250\ kJ$$

$$Q = \frac{mR(T_1 - T_2)}{(n-1)}\left(\frac{\gamma - n}{\gamma - 1}\right)$$

$$= W\left(\frac{\gamma}{\gamma - 1}\right) = -250 \times \frac{1.4}{1.4 - 1} = -875\ kJ$$

$$\Delta U = U_2 - U_1 = mv_v(T_2 - T_1) = Q - W = -875 - (-250) = -625\ kJ$$

ii) n = 0.5, polytropic process

$$P_2 = P_1\left(\frac{V_1}{V_2}\right)^{0.5} = 500\left(\frac{1}{0.5}\right)^{0.5} = 707.1\ kPa$$

$$W_{1-2} = \frac{P_1V_1 - P_2V_2}{n-1} = \frac{500 \times 1 - 707.1 \times 0.5}{0.5 - 1} = -292.89\ kJ$$

$$Q = W\left(\frac{\gamma - n}{\gamma - 1}\right) = -292.89 \times \frac{1.4 - 0.5}{1.4 - 1} = -659\ kJ$$

$$\Delta U = Q - W = -659 - (-292.89) = -366.1\ kJ$$

iii) n = 1, isothermal process

$$W_{1-2} = P_1V_1 \ln\left(\frac{P_1}{P_2}\right) = P_1V_1 \ln\left(\frac{V_2}{V_1}\right) = 500 \times 1 \times \ln\left(\frac{0.5}{1}\right) = -346.57\ kJ$$

$$Q = W\left(\frac{\gamma - n}{\gamma - 1}\right) = -346.57 \times \frac{1.4 - 1}{1.4 - 1} = -346.57\ kJ$$

$$\Delta U = Q - W = -346.57 - (-346.57) = 0$$

iv) n = γ = 1.4, adiabatic process

$$P_2 = P_1\left(\frac{V_1}{V_2}\right)^{\gamma} = 500\left(\frac{1}{0.5}\right)^{1.4} = 1319.5\ kPa$$

$$W_{1-2} = \frac{P_1V_1 - P_2V_2}{\gamma - 1} = \frac{P_1V_1 - P_2V_2}{\gamma - 1} = \frac{500 \times 1 - 1319.5 \times 0.5}{1.4 - 1} = -302.67\ kJ$$

$$Q = \frac{mR(T_1 - T_2)}{(n-1)}\left(\frac{\gamma - n}{\gamma - 1}\right) = W\left(\frac{\gamma - \gamma}{\gamma - 1}\right) = 0$$

$$\Delta U = Q - W = 0 - (-302.67) = 302.67\ kJ$$

v) n = 2, polytropic process

$$P_2 = P_1\left(\frac{V_1}{V_2}\right)^{2} = 500\left(\frac{1}{0.5}\right)^{2} = 2000\ kPa$$

$$W_{1-2} = \frac{P_1V_1 - P_2V_2}{n - 1} = \frac{500 \times 1 - 2000 \times 0.5}{2 - 1} = -500\ kJ$$

$$Q = W\left(\frac{\gamma - n}{\gamma - 1}\right) = -500 \times \frac{1.4 - 2}{1.4 - 1} = 750\ kJ$$

$$\Delta U = Q - W = 750 - (-500) = 1250\ kJ$$

vi) n = ∞, constant volume process

$$W_{1-2} = 0$$

$$Q = U_2 - U_1 = mc_v(T_2 - T_1) = m\frac{R}{\gamma - 1}(T_2 - T_1) = \frac{(P_2 - P_1)V}{\gamma - 1}$$

$$= \frac{(2500 - 500) \times 1}{1.4 - 1} = 5000\ kJ$$

$$\Delta U = Q - W = 5000 - 0 = 5000\ kJ$$

Expansion processes:

$V_2 = 2V_1 = 2\ m^3$

i) n = 0, Constant pressure process

$$W_{1-2} = \frac{P_1V_1 - P_2V_2}{n - 1} = P_2V_2 - P_1V_1 = P(V_2 - V_1)$$

= 500 (2 – 1) = 500 kJ

$$Q = \frac{mR(T_1 - T_2)}{(n-1)}\left(\frac{\gamma - n}{\gamma - 1}\right)$$

$$= W\left(\frac{\gamma}{\gamma - 1}\right) = 500 \times \frac{1.4}{1.4 - 1} = 1750 \ kJ$$

$$\Delta U = U_2 - U_1 = mv_v\left(T_2 - T_1\right) = Q - W = 1750 - 500 = 1250 \ kJ$$

ii) n = 0.5, polytropic process

$$P_2 = P_1\left(\frac{V_1}{V_2}\right)^{0.5} = 500\left(\frac{1}{2}\right)^{0.5} = 535.55 \ kPa$$

$$W_{1-2} = \frac{P_1V_1 - P_2V_2}{n-1} = \frac{500 \times 1 - 535.55 \times 2}{0.5 - 1} = 414.21 \ kJ$$

$$Q = W\left(\frac{\gamma - n}{\gamma - 1}\right) = 414.21 \times \frac{1.4 - 0.5}{1.4 - 1} = 932 \ kJ$$

$$\Delta U = Q - W = 932 - 414.21 = 517.8 \ kJ$$

iii) n = 1, isothermal process

$$W_{1-2} = P_1V_1 \ln\left(\frac{P_1}{P_2}\right) = P_1V_1 \ln\left(\frac{V_2}{V_1}\right) = 500 \times 1 \times \ln\left(\frac{2}{1}\right) = 346.57 \ kJ$$

$$Q = W\left(\frac{\gamma - n}{\gamma - 1}\right) = 346.57 \times \frac{1.4 - 1}{1.4 - 1} = 346.57 \ kJ$$

$$\Delta U = Q - W = 346.57 - 346.57 = 0$$

iv) n = γ = 1.4, adiabatic process

$$P_2 = P_1\left(\frac{V_1}{V_2}\right)^{\gamma} = 500\left(\frac{1}{2}\right)^{1,4} = 189.46 \ kPa$$

$$W_{1-2} = \frac{P_1V_1 - P_2V_2}{\gamma - 1} = \frac{P_1V_1 - P_2V_2}{\gamma - 1} = \frac{500 \times 1 - 189.46 \times 2}{1.4 - 1} = 302.67 \ kJ$$

$$Q = \frac{mR(T_1 - T_2)}{(n-1)}\left(\frac{\gamma - n}{\gamma - 1}\right) = W\left(\frac{\gamma - \gamma}{\gamma - 1}\right) = 0$$

$$\Delta U = 0 - 302.67 = -302.67 \ kJ$$

v) n = 2, polytropic process

$$P_2 = P_1\left(\frac{V_1}{V_2}\right)^{2} = 500\left(\frac{1}{2}\right)^{2} = 125 \ kPa$$

$$W_{1-2} = \frac{P_1V_1 - P_2V_2}{n-1} = \frac{500 \times 1 - 125 \times 2}{2 - 1} = 250 \ kJ$$

$$Q = W\left(\frac{\gamma - n}{\gamma - 1}\right) = 250 \times \frac{1.4 - 2}{1.4 - 1} = -375 \ kJ$$

$$\Delta U = Q - W = -375 - 250 = -625 \ kJ$$

vi) n = ∞, constant volume process

$$W_{1\text{-}2} = 0$$

$$Q = U_2 - U_1 = mc_v(T_2 - T_1) = m\frac{R}{\gamma - 1}(T_2 - T_1) = \frac{(P_2 - P_1)V}{\gamma - 1}$$

$$= \frac{(100 - 500)1}{1.4 - 1} = -1000 \ kJ$$

$$\Delta U = Q - W = -1000 - 0 = -1000 \ kJ$$

Figure 3.24 is a diagram that plots p-v values to show all the compression and expansion processes. The polytropic index is changed from 0 to ∞ for both compression and expansion. The slope of the curve in compression processes is higher than the slope in expansion processes. That means more pressure rise with a small volume change in compression and a minor pressure change with a large volume change in expansion. It shows that the size of turbine is bigger than the compressor at the same pressure ratio.

Table 3.6 and Table 3.7 show the energy interactions in compressions and expansions respectively. In the compression region, the work is increasing and is at zero with a change in polytropic index from 0 to ∞. Therefore, to minimize the work input to compressor, the index should be low but not equal to zero. Isothermal process results in a minimum work input. In compression, the heat is negative with $n < \gamma$ and positive with $n > \gamma$. Similarly, in compression, the change in internal energy is negative for $n < 1$ and positive for $n > 1$. The work output from expansion is higher with a lower polytropic index. The isothermal process results in low compression work and high

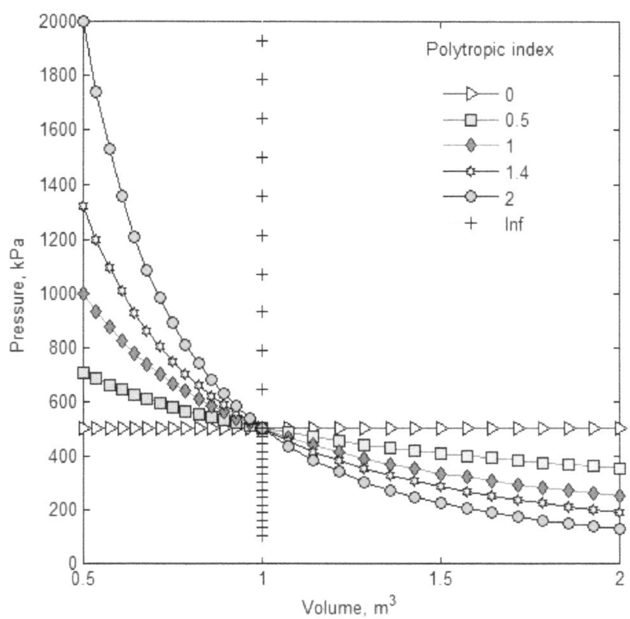

Fig. 3.24. Polytropic processes in compression and expansion at constant volume ratio.

Table 3.6. Energy interactions in compression processes.

Process	Polytropic index	Work, kJ	Heat, kJ	Internal energy, kJ
Constant pressure process	0.00	–250.00	–875.00	–625.00
Polytropic process, n < 1	0.50	–292.89	–659.01	–366.12
Isothermal process	1.00	–346.57	–346.57	0.00
Adiabatic process	1.40	–399.38	0.00	399.38
Polytropic process, n > γ	2.00	–500.00	750.00	1250.00
Constant volume process	∞	0.00	5000.00	5000.00

Table 3.7. Energy interactions in expansion processes.

Process	Polytropic index	Work, kJ	Heat, kJ	Internal energy, kJ
Constant pressure process	0.00	500.00	1750.00	1250.00
Polytropic process, n < 1	0.50	414.21	931.98	517.77
Isothermal process	1.00	346.57	346.57	0.00
Adiabatic process	1.40	302.68	0.00	–302.68
Polytropic process, n > γ	2.00	250.00	–375.00	–625.00
Constant volume process	∞	0.00	–1000.00	–1000.00

expansion work. The heat interaction in expansion is positive with $n < \gamma$ and vice-versa. The internal energy change is also positive with $n < 1$ and vice-versa.

2. Find the polytropic non-flow processes in air compression and expansion with index values 0, 0.5, 1, 1.4, 2 and ∞. Consider the initial condition at 5 bar and 1 m^3. Take the pressure ratio in compression and expansion as 2 (pressure ratio = high pressure/low pressure). Take the volume ratio in constant pressure heat rejection (compression), and constant pressure heat addition (expansion) as 5 (volume ratio = higher volume/lower volume). Find work, heat and change in internal energy at each case. Analyse your results.

Solution:

Compression processes:

$P_2 = r_{pc}\, P_1 = 2 \times 500 = 1000$ kPa

i) n = 0, Constant pressure process

r_{vc} 0.2 and $r_{ve} = 1/0.2 = 5$

Given the final volume, $V_2 = 0.2V_1 = 0.2$ m^3

$$W_{1-2} = \frac{P_1V_1 - P_2V_2}{n-1} = P_2V_2 - P_1V_1 = P\left(V_2 - V_1\right)$$

$= 500\ (0.2 - 1) = -400$ kJ

$$Q = \frac{mR(T_1 - T_2)}{(n-1)}\left(\frac{\gamma - n}{\gamma - 1}\right)$$

$$= W\left(\frac{\gamma}{\gamma - 1}\right) = -400 \times \frac{1.4}{1.4-1} = -1400\ kJ$$

$$\Delta U = U_2 - U_1 = mv_v\left(T_2 - T_1\right) = Q - W = -1400 - (-400) = -1000\ kJ$$

ii) n = 0.5, polytropic process

$P_2 = r_p P_1 = 2 \times 5 = 10$ bar

$$V_2 = V_1\left(\frac{P_1}{P_2}\right)^{1/n} = 1\left(\frac{5}{10}\right)^{1/0.5} = 0.25\ m^3$$

$$W_{1-2} = \frac{P_1V_1 - P_2V_2}{n-1} = \frac{500\times 1 - 1000\times 2.23}{0.5-1} = -500\ kJ$$

$$Q = W\left(\frac{\gamma-n}{\gamma-1}\right) = -500\times\frac{1.4-0.5}{1.4-1} = -1125\ kJ$$

$$\Delta U = Q - W = -1125-(-500) = -625\ kJ$$

iii) n = 1, isothermal process

$$V_2 = V_1\left(\frac{P_1}{P_2}\right) = 1\left(\frac{5}{10}\right) = 0.5\ m^3$$

$$W_{1-2} = P_1V_1 \ln\left(\frac{P_1}{P_2}\right) = P_1V_1 \ln\left(\frac{V_2}{V_1}\right) = 500\times 1\times \ln\left(\frac{0.5}{1}\right) = -346.57\ kJ$$

$$Q = W\left(\frac{\gamma-n}{\gamma-1}\right) = -346.57\times\frac{1.4-1}{1.4-1} = -346.57\ kJ$$

$$\Delta U = Q - W = -346.57-(-346.57) = 0$$

iv) n = γ = 1.4, adiabatic process,

$$V_2 = V_1\left(\frac{P_1}{P_2}\right)^{1/n} = 1\left(\frac{5}{10}\right)^{1/1.4} = 0.61\ m^3$$

$$W_{1-2} = \frac{P_1V_1 - P_2V_2}{\gamma-1} = \frac{P_1V_1 - P_2V_2}{\gamma-1} = \frac{500\times 1 - 1000\times 0.61}{1.4-1} = -273.77\ kJ$$

$$Q = \frac{mR(T_1-T_2)}{(n-1)}\left(\frac{\gamma-n}{\gamma-1}\right) = W\left(\frac{\gamma-\gamma}{\gamma-1}\right) = 0$$

$$\Delta U = Q - W = 0-(-273.77) = 273.77\ kJ$$

v) n = 2, polytropic process,

$$V_2 = V_1\left(\frac{P_1}{P_2}\right)^{1/n} = 1\left(\frac{5}{10}\right)^{1/2} = 0.71\ m^3$$

$$W_{1-2} = \frac{P_1V_1 - P_2V_2}{n-1} = \frac{500\times 14 - 1000\times 0.71}{2-1} = -207.11\ kJ$$

$$Q = W\left(\frac{\gamma-n}{\gamma-1}\right) = -207.11\times\frac{1.4-2}{1.4-1} = 310.66\ kJ$$

$$\Delta U = Q - W = 310.66-(-207.11) = 517.77\ kJ$$

vi) n = ∞, constant volume process,

$$W_{1-2} = 0$$

$$Q = U_2 - U_1 = mc_v(T_2 - T_1) = m\frac{R}{\gamma - 1}(T_2 - T_1) = \frac{(P_2 - P_1)V}{\gamma - 1}$$

$$= \frac{(1000 - 500)1}{1.4 - 1} = 1250\ kJ$$

$$\Delta U = Q - W = 1250 - 0 = 1250\ kJ$$

Expansion process:

$P_2 = P_1/r_e = 500/2 = 250$ kPa

i) n = 0, Constant pressure process

Given the final volume, $V_2 = 5V_1 = 5\ m^3$

$$W_{1-2} = \frac{P_1V_1 - P_2V_2}{n-1} = P_2V_2 - P_1V_1 = P(V_2 - V_1)$$

= 500 (5 – 1) = 2000 kJ

$$Q = \frac{mR(T_1 - T_2)}{(n-1)}\left(\frac{\gamma - n}{\gamma - 1}\right)$$

$$= W\left(\frac{\gamma}{\gamma - 1}\right) = 2000 \times \frac{1.4}{1.4 - 1} = 7000\ kJ$$

$$\Delta U = U_2 - U_1 = mv_v(T_2 - T_1) = Q - W = 7000 - 2000 = 5000\ kJ$$

ii) n = 0.5, polytropic process

$$V_2 = V_1\left(\frac{P_1}{P_2}\right)^{1/n} = 1\left(\frac{500}{250}\right)^{1/0.5} = 4\ m^3$$

$$W_{1-2} = \frac{P_1V_1 - P_2V_2}{n-1} = \frac{500 \times 1 - 250 \times 4}{0.5 - 1} = 1000\ kJ$$

$$Q = W\left(\frac{\gamma - n}{\gamma - 1}\right) = 1000 \times \frac{1.4 - 0.5}{1.4 - 1} = 2250\ kJ$$

$$\Delta U = Q - W = 2250 - 1000 = 1250\ kJ$$

iii) n = 1, isothermal process

$$V_2 = V_1\left(\frac{P_1}{P_2}\right) = 1\left(\frac{500}{250}\right) = 2\ m^3$$

$$W_{1-2} = P_1V_1 \ln\left(\frac{P_1}{P_2}\right) = P_1V_1 \ln\left(\frac{V_2}{V_1}\right) = 500 \times 1 \times \ln\left(\frac{2}{01}\right) = 346.57\ kJ$$

$$Q = W\left(\frac{\gamma - n}{\gamma - 1}\right) = 346.57 \times \frac{1.4 - 1}{1.4 - 1} = 346.57\ kJ$$

$$\Delta U = Q - W = 346.57 - 346.57 = 0$$

iv) n = γ = 1.4, adiabatic process

$$V_2 = V_1\left(\frac{P_1}{P_2}\right)^{1/n} = 1\left(\frac{500}{250}\right)^{1/1.4} = 1.64\ m^3$$

$$W_{1-2} = \frac{P_1V_1 - P_2V_2}{\gamma - 1} = \frac{P_1V_1 - P_2V_2}{\gamma - 1} = \frac{500\times 1 - 250\times 1.64}{1.4 - 1} = 224.58\ kJ$$

$$Q = \frac{mR(T_1 - T_2)}{(n-1)}\left(\frac{\gamma - n}{\gamma - 1}\right) = W\left(\frac{\gamma - \gamma}{\gamma - 1}\right) = 0$$

$$\Delta U = 0 - 224.58 = -224.58\ kJ$$

v) n = 2, polytropic process

$$V_2 = V_1\left(\frac{P_1}{P_2}\right)^{1/n} = 1\left(\frac{500}{250}\right)^{1/2} = 1.41\ m^3$$

$$W_{1-2} = \frac{P_1V_1 - P_2V_2}{n - 1} = \frac{500\times 1 - 250\times 1.41}{2 - 1} = 146.45\ kJ$$

$$Q = W\left(\frac{\gamma - n}{\gamma - 1}\right) = 111.11\times\frac{1.4 - 2}{1.4 - 1} = -166.67\ kJ$$

$$\Delta U = Q - W = -166.67 - 146.45 = -366.12\ kJ$$

vi) n = ∞, constant volume process

$$W_{1-2} = 0$$

$$Q = U_2 - U_1 = mc_v(T_2 - T_1) = m\frac{R}{\gamma - 1}(T_2 - T_1) = \frac{(P_2 - P_1)V}{\gamma - 1}$$

$$= \frac{(250 - 500)1}{1.4 - 1} = -625\ kJ$$

$$\Delta U = Q - W = -625 - 0 = -625\ kJ$$

Figure 3.25 is a diagram plotted for p-v values to show all the compression and expansion processes. The polytropic index is changed from 0 to ∞ for both compression and expansion. The slope of the curve in compression is much more than in the expansion processes. That means there is more pressure rise with a small volume change during compression and a minor pressure change with large volume change during expansion. It shows that the size of turbine is bigger than the compressor at the same pressure ratio. Table 3.8 and Table 3.9 show the energy interactions during compressions and expansions respectively. In compression, the heat is negative with $n < \gamma$ and positive with $n > \gamma$. Similarly, in compression, the change in internal energy is negative for $n < 1$ and positive for $n \geq \gamma$. The work output from expansion is higher with a lower polytropic index. The isothermal process results in low compression work and high expansion work. The heat interactions in expansion are positive with $n < \gamma$ and vice-versa. The internal energy change is also positive with $n < 1$ and vice-versa.

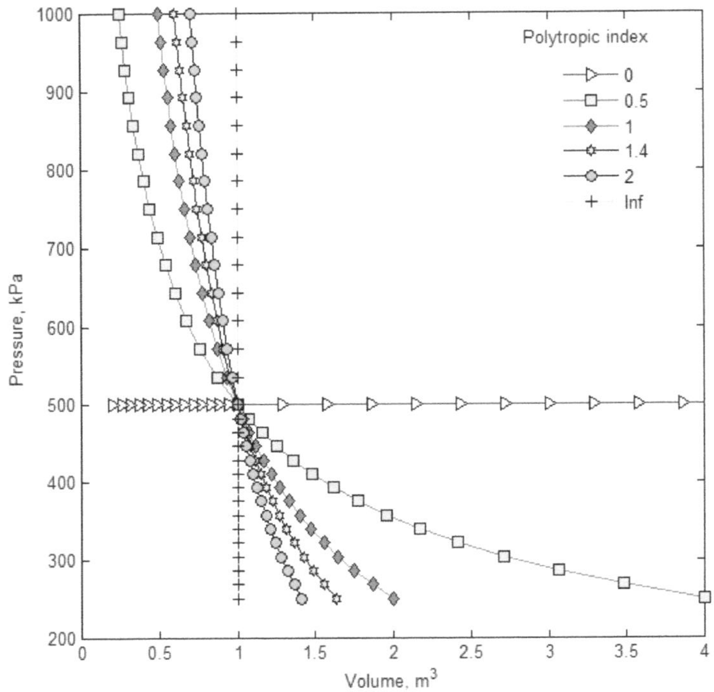

Fig. 3.25. Polytropic processes in compression and expansion under same pressure ratio.

Table 3.8. Energy interactions in compression processes under same pressure ratio (non-flow process).

Process	Polytropic index	Work, kJ	Heat, kJ	Internal energy, kJ
Constant pressure process	0.00	–400.00	–1400.00	–1000.00
Polytropic process, n < 1	0.50	–500.00	–1125.00	–625.00
Isothermal process	1.00	–346.57	–346.57	0.00
Adiabatic process	1.40	–273.77	0.00	273.77
Polytropic process, n > γ	2.00	–207.11	310.66	517.77
Constant volume process	∞	0.00	1250.00	1250.00

Table 3.9. Energy interactions in expansion processes under same pressure ratio (flow process).

Process	Polytropic index	Work, kJ	Heat, kJ	Internal energy, kJ
Constant pressure process	0.00	2000.00	7000.00	5000.00
Polytropic process, n < 1	0.50	1000.00	2250.00	1250.00
Isothermal process	1.00	346.57	346.57	0.00
Adiabatic process	1.40	224.58	0.00	–224.58
Polytropic process, n >1	2.00	146.45	–219.67	–366.12
Constant volume process	∞	0.00	–625.00	–625.00

3. Find the polytropic processes for flow process in air compression and expansion with index values 0, 0.5, 1, 1.4, 2 and ∞. Consider the initial condition at 5 bar and 1 m^3. Take the pressure ratio in compression and expansion as 2 (pressure ratio = high pressure/low pressure). Take the volume ratio in constant pressure heat rejection (compression), and constant pressure heat addition (expansion) as 0.2 and 5 respectively. Find work, heat and change in internal energy in each case. Analyse your results.

Solution:

Compression processes:

$P_2 = r_p P_1 = 2 \times 500 = 1000$ kPa

i) n = 0, Constant pressure process

Given the final volume, $V_2 = 0.2V_1 = 0.2\ m^3$

$$W_{1-2} = \frac{P_1V_1 - P_2V_2}{n-1} = P_2V_2 - P_1V_1 = P(V_2 - V_1)$$

$$= 500\ (0.2 - 1) = -400\ kJ$$

$$Q = \frac{mR(T_1 - T_2)}{(n-1)}\left(\frac{\gamma - n}{\gamma - 1}\right)$$

$$= W\left(\frac{\gamma}{\gamma - 1}\right) = -400 \times \frac{1.4}{1.4 - 1} = -1400\ kJ$$

$$\Delta U = U_2 - U_1 = mv_v(T_2 - T_1) = Q - W = -1400 - (-400) = -1000\ kJ$$

ii) n = 0.5, polytropic process
$P_2 = r_pP_1 = 2 \times 5 = 10$ bar

$$V_2 = V_1\left(\frac{P_1}{P_2}\right)^{1/n} = 1\left(\frac{5}{10}\right)^{1/0.5} = 0.25\ m^3$$

$$W_{1-2} = \frac{P_1V_1 - P_2V_2}{n-1} = \frac{500 \times 1 - 1000 \times 2.23}{0.5 - 1} = -500\ kJ$$

$$Q = W\left(\frac{\gamma - n}{\gamma - 1}\right) = -500 \times \frac{1.4 - 0.5}{1.4 - 1} = -1125\ kJ$$

$$\Delta U = Q - W = -1125 - (-500) = -625\ kJ$$

iii) n = 1, isothermal process,

$$V_2 = V_1\left(\frac{P_1}{P_2}\right) = 1\left(\frac{5}{10}\right) = 0.5\ m^3$$

$$W_{1-2} = P_1V_1 \ln\left(\frac{P_1}{P_2}\right) = P_1V_1 \ln\left(\frac{V_2}{V_1}\right) = 500 \times 1 \times \ln\left(\frac{0.5}{1}\right) = -346.57\ kJ$$

$$Q = W\left(\frac{\gamma - n}{\gamma - 1}\right) = -346.57 \times \frac{1.4 - 1}{1.4 - 1} = -346.57\ kJ$$

$$\Delta U = Q - W = -346.57 - (-346.57) = 0$$

iv) $n = \gamma = 1.4$, adiabatic process

$$V_2 = V_1\left(\frac{P_1}{P_2}\right)^{1/n} = 1\left(\frac{5}{10}\right)^{1/1.4} = 0.61\ m^3$$

$$W_{1-2} = \frac{P_1V_1 - P_2V_2}{\gamma - 1} = \frac{P_1V_1 - P_2V_2}{\gamma - 1} = \frac{500 \times 1 - 1000 \times 0.61}{1.4 - 1} = -273.77 \ kJ$$

$$Q = \frac{mR(T_1 - T_2)}{(n-1)}\left(\frac{\gamma - n}{\gamma - 1}\right) = W\left(\frac{\gamma - \gamma}{\gamma - 1}\right) = 0$$

$$\Delta U = Q - W = 0 - (-273.77) = 273.77 \ kJ$$

v) n = 2, polytropic process

$$V_2 = V_1\left(\frac{P_1}{P_2}\right)^{1/n} = 1\left(\frac{5}{10}\right)^{1/2} = 0.71 \ m^3$$

$$W_{1-2} = \frac{P_1V_1 - P_2V_2}{n - 1} = \frac{500 \times 14 - 1000 \times 0.71}{2 - 1} = -207.11 \ kJ$$

$$Q = W\left(\frac{\gamma - n}{\gamma - 1}\right) = -207.11 \times \frac{1.4 - 2}{1.4 - 1} = 310.66 \ kJ$$

$$\Delta U = Q - W = 310.66 - (-207.11) = 517.77 \ kJ$$

vi) n = ∞, constant volume process

$$W_{1-2} = 0$$

$$Q = U_2 - U_1 = mc_v(T_2 - T_1) = m\frac{R}{\gamma - 1}(T_2 - T_1) = \frac{(P_2 - P_1)V}{\gamma - 1}$$

$$= \frac{(1000 - 500)1}{1.4 - 1} = 1250 \ kJ$$

$$\Delta U = Q - W = 1250 - 0 = 1250 \ kJ$$

Expansion processes:

$P_2 = P_1/r_e = 500/2 = 250$ kPa

i) n = 0, Constant pressure process

Given the final volume, $V_2 = 5V_1 = 5$ m^3

$$W_{1-2} = \frac{P_1V_1 - P_2V_2}{n - 1} = P_2V_2 - P_1V_1 = P(V_2 - V_1)$$

= 500 (5 – 1) = 2000 kJ

$$Q = \frac{mR(T_1 - T_2)}{(n-1)}\left(\frac{\gamma - n}{\gamma - 1}\right)$$

$$= W\left(\frac{\gamma}{\gamma - 1}\right) = 2000 \times \frac{1.4}{1.4 - 1} = 7000 \ kJ$$

$$\Delta U = U_2 - U_1 = mv_v(T_2 - T_1) = Q - W = 7000 - 2000 = 5000 \ kJ$$

ii) n = 0.5, polytropic process

$$V_2 = V_1\left(\frac{P_1}{P_2}\right)^{1/n} = 1\left(\frac{500}{250}\right)^{1/0.5} = 4\ m^3$$

$$W_{1-2} = \frac{P_1V_1 - P_2V_2}{n-1} = \frac{500\times 1 - 250\times 4}{0.5-1} = 1000\ kJ$$

$$Q = W\left(\frac{\gamma - n}{\gamma - 1}\right) = 1000\times\frac{1.4-0.5}{1.4-1} = 2250\ kJ$$

$$\Delta U = Q - W = 2250 - 1000 = 1250\ kJ$$

iii) n = 1, isothermal process

$$V_2 = V_1\left(\frac{P_1}{P_2}\right) = 1\left(\frac{500}{250}\right) = 2\ m^3$$

$$W_{1-2} = P_1V_1\ln\left(\frac{P_1}{P_2}\right) = P_1V_1\ln\left(\frac{V_2}{V_1}\right) = 500\times 1\times\ln\left(\frac{2}{01}\right) = 346.57\ kJ$$

$$Q = W\left(\frac{\gamma - n}{\gamma - 1}\right) = 346.57\times\frac{1.4-1}{1.4-1} = 346.57\ kJ$$

$$\Delta U = Q - W = 346.57 - 346.57 = 0$$

iv) n = γ = 1.4, adiabatic process

$$V_2 = V_1\left(\frac{P_1}{P_2}\right)^{1/n} = 1\left(\frac{500}{250}\right)^{1/1.4} = 1.64\ m^3$$

$$W_{1-2} = \frac{P_1V_1 - P_2V_2}{\gamma-1} = \frac{P_1V_1 - P_2V_2}{\gamma-1} = \frac{500\times 1 - 250\times 1.64}{1.4-1} = 224.58\ kJ$$

$$Q = \frac{mR(T_1 - T_2)}{(n-1)}\left(\frac{\gamma - n}{\gamma - 1}\right) = W\left(\frac{\gamma - \gamma}{\gamma - 1}\right) = 0$$

$$\Delta U = 0 - 224.58 = -224.58\ kJ$$

v) n = 2, polytropic process

$$V_2 = V_1\left(\frac{P_1}{P_2}\right)^{1/n} = 1\left(\frac{500}{250}\right)^{1/2} = 1.41\ m^3$$

$$W_{1-2} = \frac{P_1V_1 - P_2V_2}{n-1} = \frac{500\times 1 - 250\times 1.41}{2-1} = 146.45\ kJ$$

$$Q = W\left(\frac{\gamma - n}{\gamma - 1}\right) = 111.11\times\frac{1.4-2}{1.4-1} = -166.67\ kJ$$

$$\Delta U = Q - W = -166.67 - 146.45 = -366.12\ kJ$$

vi) n = ∞, constant volume process

$$W_{1-2} = 0$$

$$Q = U_2 - U_1 = mc_v(T_2 - T_1) = m\frac{R}{\gamma - 1}(T_2 - T_1) = \frac{(P_2 - P_1)V}{\gamma - 1}a$$

$$= \frac{(250 - 500)1}{1.4 - 1} = -625\ kJ$$

$$\Delta U = Q - W = -625 - 0 = -625\ kJ$$

Figure 3.26 is a diagram plotted for p-v values to show all the compression and expansion processes. The polytropic index is changed from 0 to ∞ for both compression and expansion. The slope of the curve in compression is much more than the slope for the expansion processes. That means there will be more pressure rise with a small volume change in compression and a minor pressure change with a large volume change in expansions. It shows that the size of turbine is bigger than the compressor at the same pressure ratio. Table 3.10 and Table 3.11 show the energy interactions in compressions and expansions respectively. In compression, the heat is negative with $n < \gamma$ and positive with $n > \gamma$. Similarly, in compression, the change in internal energy is negative for $n < 1$ and positive for $n \geq \gamma$. The work output from expansion is higher with a lower polytropic index. The isothermal process results in low compression work and high expansion work. The heat interactions in expansion are positive with $n < \gamma$ and vice-versa. The internal energy change is also positive with $n < 1$ and vice-versa.

4. Steam at pressure of 15 bar and 0.95 dry is generated in a boiler and is made to pass through its superheater where additional quantity of heat is supplied at constant pressure. Consequently, the temperature of steam increases to 350°C. Determine the following (a) change in entropy (b) change in enthalpy and (c) heat supplied

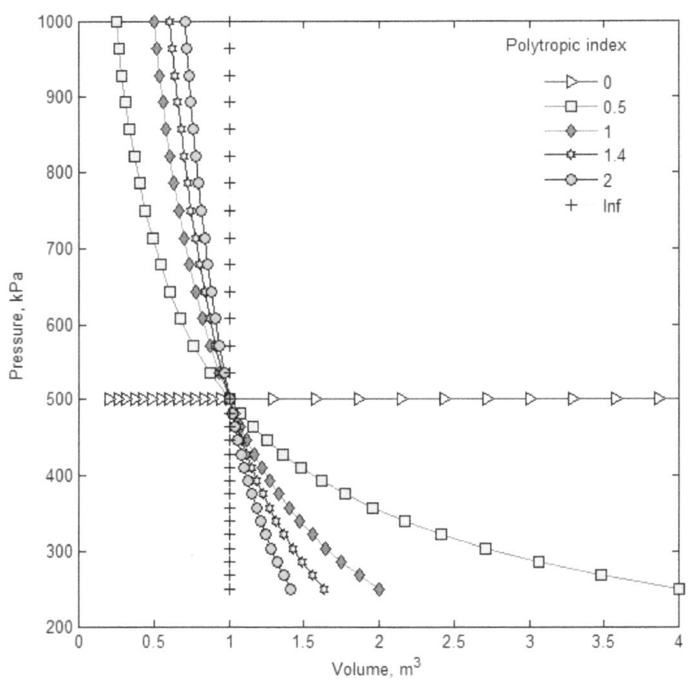

Fig. 3.26. Polytropic processes in compression and expansion under same pressure ratio.

Table 3.10. Energy interactions in compression processes under the same pressure ratio.

Process	Polytropic index	Work, kJ	Heat, kJ	Internal energy, kJ
Constant pressure process	0.00	–400.00	–1400.00	–1000.00
Polytropic process, n < 1	0.50	–500.00	–1125.00	–625.00
Isothermal process	1.00	–346.57	–346.57	0.00
Adiabatic process	1.40	–273.77	0.00	273.77
Polytropic process, n > γ	2.00	–207.11	310.66	517.77
Constant volume process	∞	0.00	1250.00	1250.00

Table 3.11. Energy interactions in expansion processes under the same pressure ratio.

Process	Polytropic index	Work, kJ	Heat, kJ	Internal energy, kJ
Constant pressure process	0.00	2000.00	7000.00	5000.00
Polytropic process, n < 1	0.50	1000.00	2250.00	1250.00
Isothermal process	1.00	346.57	346.57	0.00
Adiabatic process	1.40	224.58	0.00	–224.58
Polytropic process, n >1	2.00	146.45	–219.67	–366.12
Constant volume process	∞	0.00	–625.00	–625.00

Solution:

$P_1 = 15$ bar

$v_1 = 0.125131$ m³/kg

$T_1 = 198.29°C$

$x_1 = 0.95$

$h_1 = 2692.63$ kJ/kg

$s_1 = 6.2343$ kJ/kg K

$P_2 = 15$ bar

$v_2 = 0.186533$ m³/kg

$T_2 = 350°C$

$h_2 = 3148.7225$ kJ/kg

$s_2 = 7.1044$ kJ/kg K

$\Delta S = m\,(s_2 - s_1) = 0.8701$ kJ/K

$\Delta H = m\,(h_2 - h_1) = 456.09$ kJ

$Q = \Delta H = 456.09$ kJ (since $W = 0$)

5. Steam at a pressure of 5.4 bar and dryness fraction 0.8 expands in a cylinder reversibly and isothermally to a pressure of 1 bar. Find (a) the final condition of steam, (b) change in entropy, (c) change in internal energy, (d) work done, and (d) heat transferred per kg of steam.

Solution:

$P_1 = 5.4$ bar

$v_1 = 0.2790$ m³/kg

$T_1 = 154.76°C$

$x_1 = 0.8$

$u_1 = 2180.6132$ kJ/kg

$h_1 = 2331.2650$ kJ/kg

$s_1 = 5.8125$ kJ/kg K

$P_2 = 1$ bar

$v_2 = 1.9589$ m³/kg

$T_2 = 154.76°C$

x_2 = superheated steam

$u_2 = 2589.8863$ kJ/kg

$h_2 = 2785.779498$ kJ/kg

$s_2 =7.6359$ kJ/kg K

$\Delta S = m\,(s_2 - s_1) = 1.8234$ kJ/K

$\Delta U = m\,(u_2 - u_1) = 409.27$ kJ

$Q = m\,T\,(s_2 - s_1) = 780.27$ kJ

$W = Q - \Delta U = 371.00$ kJ

6. Steam at a pressure of 10 bar and 0.9 dry expands to atmospheric pressure hyperbolically. Find (a) the final condition of steam, (b) change in entropy, (c) change in internal energy, (d) work done, and (d) heat transferred per kg of steam.

Solution

P_1=10 bar

$v_1 = 0.1749$ m³/kg

$T_1 = 179.88°C$

$x_1 = 0.9$

$h_1 = h_f + x_1(h_g - h_f) = 2574.8038$ kJ/kg

$u_1 = h_1 - 100\,P_1 v_1 = 2399.8274$ kJ/kg

$s_1 = s_f + x_1(s_g - s_f) = 6.1383$ kJ/kg K

$P_2 = 1.01325$ bar

For hyperbolic process, Pv = Constant, $P_1 v_1 = P_2 v_2$

$$v_2 = \frac{P_1 v_1}{P_2}$$

$$= \frac{10 \times 0.1749}{1.01325}$$

= 1.7268 m³/kg

since $v_2 > v_{sat}$ at 1.01325 bar,

x_2 = superheated

Linear interpolation of temperature from the specific volume of superheated steam tables between 100°C and 110°C,

$T_2 = 110.9937°C$

From the specific enthalpy of superheated steam tables,

$h_2 = 2698.2289$ kJ/kg

$u_2 = h_2 - 100\ P_2v_2 = 2523.25$ kJ/kg

From the specific entropy of superheated steam tables,

$s_2 = 7.4140$ kJ/kg K

Change in entropy, $\Delta S = m\ (s_2 - s_1) = 1.2757$ kJ/K

Change in internal energy, $\Delta U = m\ (u_2 - u_1) = 123.42$ kJ $= \Delta H$

Work done, W = 4.00 kJ

Heat transfer = Q = W + ΔU or ΔH = 127.4310 kJ

7. Two kg of steam at 125 bar is cooled by rejecting 5000 kJ of heat through the constant volume method. Find the final state of steam if the initial state is (a) 75% dry (b) dry saturated steam and (c) at 50°C degree of superheat. Then solve the energy and exergy differences.

Solution:

Q = 30 kJ

a) The initial condition of steam is 1 with 75% quality

$$Q = U_2 - U_1 = m\ [(h_2 - p_2v_2) - (h_1 - p_1v_1)]$$

$$Q = m\ [(h_2 - h_1) - (p_2v_2 - p_1v_1)]$$

8. Binary fluid separation: Separate the liquid and vapor from a mixture at 15 bar, 100°C, 0.70 of ammonia concentration and 1 kg/s of fluid flow.

Solution:

Ref. Fig. 3.14 for separation process

Pressure, P_1 = 15 bar

Temperature, T_1 = 100°C

Concentration, $x_1 = 0.7$

xl = 0.384260

xv = 0.949002

hsep, kJ/kg = 956.707475

hl, kJ/kg = 224.202978

hv, kJ/kg = 1534.381326

From the concentration chart, by referring to the saturated liquid curve, at 15 bar and 100°C,

liquid concentration, $x_2 = 0.38$

With a vertical line from saturated liquid state till the auxiliary curve and then a horizontal line to the vapor curve gives the vapor concentration.

vapor concentration, $x_3 = 0.95$

The dryness fraction can be determined from these three concentrations,

Dryness fraction using the lever rule,

$$d = \frac{m_3}{m_2 + m_3} = \frac{x_1 - x_2}{x_3 - x_2} = \frac{0.7 - 0.38}{0.95 - 0.38} = 0.56$$

Mixture mass, $m_1 = 1$ kg/s $(m_2 + m_3)$

Mass of liquid, $m_2 = (1–d)\ m_1 = 1 – 0.56 = 0.44$ kg/s

Mass of vapor, $m_3 = d\ m_1 = 0.56$ kg/s

From the chart,

$h_1 = 956.70$ kJ/kg

$h_2 = 224.20$ kJ/kg

$h_3 = 1534.38$ kJ/kg

9. Binary fluids mix diabatically as per the following particulars: A fluid at 10 bar, 85°C, 0.4 concentration, 10 kg/s mixes with another fluid at same pressure and 50ºC, 0.8 concentration and 10 kg/s. The mixture is cooled to condense the vapor into the saturated liquid form. Determine the exit properties and heat transfer.

Solution:

Ref. Fig. 3.15 for the solution

Mixing and heat transfer occurs at the constant pressure. Therefore,

Pressure $= P_1 = P_2 = P_3 = P_4 = 10$ bar

$T_1 = 85$°C

$x_1 = 0.4$

$m_1 = 10$ kg/s

$T_2 = 50$°C

$x_2 = 0.8$

$m_2 = 10$ kg/s

State 1 and state 2 can be located on the specific enthalpy-concentration chart using the pressure, temperature and concentration at these two states. First, draw a vertical line from the saturated liquid curve by referring pressure and temperature, draw a vertical line till the auxiliary curve, next draw a horizontal line to saturated vapor curve, and join the saturated vapor state and saturated liquid state which results in an isotherm. The vertical line of concentration cuts this isotherm to locate states 1 and 2.

State 3 can be located from the lever rule, i.e., mass ratio. Since the mass of fluid 1 and fluid 2 are equal, the length of line 1-3 and 3-2 are equal. From state 3, draw a vertical line down till it meets the saturated liquid curve. This is the final state 4. Therefore, line 1-3-2 represents adiabatic mixing and line 3-4 shows the heat transfer process separately.

The temperature, T_3 after the adiabatic mixing can be found from the trial and error method.

Temperature after mixing, $T_3 = 69.77$ºC

Temperature after the condensation, $T_4 = 49.57$ºC

From the graph, the specific enthalpies.

$h_1 = 211.36$ kJ/kg

$h_2 = 717.21$ kJ/kg

$h_3 = 464.29$ kJ/kg

$h_4 = -1.56$ kJ/kg

dryness fraction at state 3, $d_3 = 0.27$

Q can be calculated from the properties collected from the h-x graph.

Q = 9317 kW

10. A binary fluid is expanded through a restricted passage (isenthalpic) to result in a low temperature fluid. Find the exit temperature if the fluid is expanding from 8 bar to 2 bar. The concentration of the fluid is 0.7. The fluid flow is 10 kg/s. The temperature before throttling is 80°C.

Solution:

Refer Fig. 3.16 for the solution.

Using h-x diagram, state 1 can be located using the trial and error method as it is the mixture of liquid and vapor as shown from the pressure, temperature and concentration.

$h_1 = 955.90$ kJ/kg

The concentration and enthalpy remain same after expansion. Therefore, state 1 and state 2 are on the same point on h-x diagram. But the temperature after expansion decreases.

Similar to the inlet state, the exit temperature also can be iterated but at low pressure.

The resulting exit temperature,

$T_2 = 51.51$°C

$h_2 = 955.90$ kJ/kg

The dryness fraction at the exit,

$d_2 = 0.62$

11. A binary fluid is distilled (increased concentration) after the boiling with the following data: The pressure is 10 bar, The temperature and concentration before distillation are 90°C and 0.6 respectively. The fluid supply is 10 kg/s. Consider the effectiveness of distiller or dephlegmator is 0.8. Simulate the exit conditions of the dephlegmator at these operational conditions.

Solution:

Refer Fig. 3.18 for the solution

The given state is a mixture of liquid and vapor, the vapor goes to the dephlegmator after the separation to boost the concentration.

As per the earlier discussion on findings of the separation process, carry forwarded to the dephlegmator for finding the process conditions.

Total mass before separation is 10 kg/s

$x_2 = 0.35$

$x_3 = 0.95$

$x_4 = 0.95$

$x_5 = 0.51$

$x_6 = 0.99$

$x_7 = 0.95$

$x_8 = 0.74$

$x_9 = 1.0$

$h_2 = 185.30$ kJ/kg

$h_3 = 1529.91$ kJ/kg

$h_4 = 1308.06$ kJ/kg

$h_5 = 36.65$ kJ/kg

$h_6 = 1401.67$ kJ/kg

$h_7 = 1057.44$ kJ/kg

$h_8 = -4.77$ kJ/kg

$h_9 = 1317.53$ kJ/kg

Saturated liquid after separation = $m_2 = 5.83$ kg/s

Saturated vapor after separation = $m_3 = 4.17$ kg/s

In dephlegmator, separated liquid = 0.36 kg/s

In dephlegmator, separated vapor = 3.81 kg/s

dryness fraction in separator = 0.41

dryness fraction in dephlegmator = 0.91

Heat rejection in dephlegmator = 535.01 kW

12. Sensible heating: 90 m^3 of air per minute at 20°C and 75% RH is heated until its temperature becomes 30°C, without changing its specific humidity.

a) Calculate RH of heated air and heat addition to air. Hot water has been used for sensible heating of air.

b) The hot water inlet and outlet temperature in the heat exchanger are 60°C and 52°C respectively. Using psychrometric properties, find the exergy loss and exergy efficiency of the sensible heating process.

Solution:

(a) Energy evaluation:

For air at 20°C and 75% RH,

$P_{s1} = 0.0234$ bar (from steam tables at 20°C)

$P_{v1} = \text{Ø}\, P_{s1} = 0.75 \times 0.0234 = 0.01755$ bar

Specific humidity,

$$\omega_1 = 0.622\frac{P_{v1}}{P_t - P_{v1}} = 0.622\frac{0.01755}{1.01325 - 0.01755} = 0.0109\ kg/kg\ da$$

Enthalpy,

$$h_1 = c_p t_1 + \omega_1 h_{v1}$$

$$= c_p t_1 + \omega_1 (h_{g1} + c_{pv} (t_1 - t_{s1}))$$

$$= 1.005 \times 20 + 0.0109 \times (2538.1 + 1.88\,(20\text{–}15.5))$$

$$= 47.85 \text{ kJ/kg da}$$

At 30°C, $P_s = 0.0425$ bar

In sensible heating or cooling P_v is constant,

$$\phi_2 = \frac{P_{v2}}{P_{s2}} = \frac{0.01755}{0.0425} = 41.2\%$$

$\omega_2 = \omega_1 = 0.0109$ kg/kg da

Enthalpy,

$h_2 = c_p t_2 + \omega_2 h_{v2}$

$= c_p t_2 + \omega_2 (h_{g2} + c_{pv} (t_2 - t_{s2}))$

$= 1.005 \times 30 + 0.0109 \times (2538.1 + 1.88 (30 - 15.5))$

$= 58.11$ kJ/kg da

Mass of dry air in 90 m^3 of air supplied,

$$m_a = \frac{PV}{RT} = \frac{(P_t - P_v)V}{RT} = \frac{(1.01325 - 0.01755) \times 100 \times 90}{0.287(20 + 273.15)60} = 1.775 \ kg/s$$

Heat supplied,

$Q = m_{da} (h_2 - h_1)$

$= 1.775 (58.11 - 47.85) = 18.21$ kW

(b) Exergy analysis:

$Q = m_{hw} c_{pw} (T_{hi} - T_{ho}) = m_{da} (h_2 - h_1)$

$$m_{hw} = \frac{m_{da}(h_2 - h_1)}{c_{pw}(T_{hi} - T_{ho})} = \frac{18.21}{4.18 \times (60 - 52)} = 0.54 \ kg/s$$

From the psychrometric properties,

At 20°C DBT and 75% RH

Specific exergy of wet air,

$\varepsilon_1 = 0.42$ kJ/kg da

The specific humidity at this state, $\omega_1 = 10.95$ g/kg da $= \omega_2$

At 30°C DBT and 10.95 g/kg da of specific humidity,

Specific exergy of wet air,

$\varepsilon_2 = 0.4185$ kJ/kg da

Specific exergy of hot water supply,

$\varepsilon_{hi} = h_{wi} - T_0 s_{wi}$

$$\varepsilon_{wi} = c_{pw}\left((T_{wi} - T_0) - T_0 \log\left(\frac{T_{wi}}{T_0}\right)\right)$$

$$= 4.18(60 - 25) - 298.15 \times 4.18 \log\left(\frac{60 + 273.15}{298.15}\right)$$

$= 7.97$ kJ/kg

Specific exergy of hot water at exit,

$\varepsilon_{wo} = h_{wi} - T_0\, s_{wi}$

$$\varepsilon_{wo} = c_{pw}(T_{wo} - T_0) - T_0 c_{pw} \log\frac{T_{wo}}{T_0}$$

$$= 4.18(52-25) - 298.15 \times 4.18 \log\left(\frac{52+273.15}{298.15}\right)$$

$= 4.82$ kJ/kg

Exergy balance of the process

$\varepsilon_{wi} + \varepsilon_1 = \varepsilon_{wo} + \varepsilon_2 + I_{SH}$

$I_{SH} = m_{wi}\,\varepsilon_{wi} + m_a\,\varepsilon_1 - m_{wo}\,\varepsilon_{wo} + m_a\,\varepsilon_2$

$= 0.54 \times 7.97 + 1.775 \times 0.42 - 0.54 \times 4.82 - 1.775 \times 0.4185$

$= 1.7$ kJ/kg

Exergy efficiency of sensible heating process,

$$\eta_{exergy,\ SH} = 1 - \left(\frac{I_{SH}}{m_{wi}\varepsilon_{wi} + m_a\varepsilon_1}\right) \times 100$$

$$= 1 - \left(\frac{1.7}{0.54 \times 7.97 + 1.775 \times 0.42}\right) \times 100$$

$= 66.25\%$

13. Cooling and dehumidification: 120 m^3 of air per minute at 35°C DBT and 50% RH is cooled to 20°C DBT by passing through a cooling coil. Determine (a) RH of out coming air and its WBT, (b) capacity of cooling coil in TR of refrigeration (c) amount of water vapor removed per hour (d) sensible heat factor and slope of line.

If R134a is the refrigerant in the cooling coil with the temperature and pressure of –2°C and 2.7217 bar respectively, find the irreversibility and exergy efficiency of the cooling and dehumidification process. The refrigeration condensate state is 10.166 bar and 40°C.

Solution:

At 35°C DBT and 50% RH,

$P_{s1} = 0.0563$ (from steam tables at 35°C)

$P_{v1} = 0.5 \times 0.0563 = 0.02815$ bar.

Specific humidity,

$$\omega_1 = 0.622\frac{P_{v1}}{P_t - P_{v1}} = 0.622\frac{0.02815}{1.01325 - 0.02815} = 0.0177\ kg/kg\ da$$

Enthalpy,

$h_1 = c_p t_1 + \omega_1 h_{v1}$

$= c_p t_1 + \omega_1\,(h_{g1} + c_{pv}\,(t_1 - t_{s1}))$

$= 1.005 \times 35 + 0.0177 \times (2543.6 + 1.88\,(35\text{–}23))$

(The dew point temperature at 0.02815 bar is 23°C)

= 80.59 kJ/kg da

At 20°C, P_{s2} = 0.0234 bar. It is less than the partial vapor pressure of 0.02815 bar. Therefore, condensation results, and air will saturate at 20°C. RH_2 = 100%

In the saturated condition, the wet-bulb temperature is equal to the dry-bulb temperature.

Specific humidity,

$$\omega_2 = 0.622\frac{P_{v2}}{P_t - P_{v2}} = 0.622\frac{0.0234}{1.01325 - 0.0234} = 0.0147 \ kg/kg \ da$$

Enthalpy,

$h_2 = c_p t_2 + \omega_2 h_{v2}$

$= c_p t_2 + \omega_2 (h_{g2} + c_{pv}(t_2 - t_{s2}))$

$= 1.005 \times 20 + 0.0147 \times (2538.1 + 1.88\,(20–20))$

= 57.41 kJ/kg da

The weight of water vapor removed,

= 0.0177 – 0.0147 = 0.003 kg/kg da

Heat removed per kg of dry air

$= h_1 - h_2$ = 80.59 – 57.41 = 23.57 kJ/kg da

Mass of dry air passing per minute,

$$m_a = \frac{P_a V}{R_a T} = \frac{(P_t - P_v)V}{R_a T} = \frac{(1.01325 - 0.02815)\times 10^2 \times 120}{0.287(35 + 273.15)} = 133.7 \ kg/\min$$

= 2.2278 kg/s

Capacity of cooling coil in TR of refrigeration,

$$Q_{cooling\ coil} = \frac{m_a(h_1 - h_2)}{3.51\times 60} = \frac{133.7\times 23.57}{3.51\times 60} = 13.5 \ TR$$

Amount of water vapor removed per hour

$= m_a(\omega_1 - \omega_2) = 133.7 \times 60\,(0.0177 - 0.00147) = 24.506$ kg/h

We have, $\tan\theta = \frac{\Delta\omega}{\Delta T} = \frac{1}{2451}\left(\frac{1 - SHF}{SHF}\right)$

$$\frac{1}{SHF} - 1 = 2451\times\frac{\Delta\omega}{\Delta T}$$

$$SHF = \frac{1}{1 + 2451\times\left(\frac{\Delta\omega}{\Delta T}\right)}$$

$$SHF = \frac{1}{1 + 2451\times\left(\frac{0.0177 - 0.0147}{35 - 20}\right)} = 0.66$$

Slope of line, $\tan\theta = \frac{\Delta\omega}{\Delta T} = \frac{0.0177 - 0.0147}{35 - 20} = 0.0002\ \text{kg/°C}$

Exergy analysis using psychrometric properties:

35°C DBT and 50% RH

$h_1 = 80.71$ kJ/kg da, $\omega_1 = 17.75$ g/kg da, $s_1 = 0.28$ kJ/kg K and $e_1 = 0.20$ kJ/kg

20°C DBT and 100% RH (from the energy solution)

$h_2 = 57.37$ kJ/kg da, $\omega_2 = 14.68$ g/kg da, $s_2 = 0.20$ kJ/kg K and $e_2 = 0.18$ kJ/kg

From the refrigeration tables of R134a, at 25°C and 1.01325 bar

$h_{r0} = 423.33$ kJ/kg da, $s_{r0} = 1.8964$ kJ/kg K (superheated vapor)

From the refrigeration tables of R134a, at –2°C and 2.7217 bar

$h_{co} = 397.43$ kJ/kg da, $s_{co} = 1.7282$ kJ/kg K (saturated vapor)

$e_{co} = h_{co} - h_{r0} - T_0 (s_{co} - s_{r0})$

$= 397.43 - 423.33 - 298.15\ (1.7282 - 1.8964)$

$= 24.24$ kJ/kg

From the refrigeration tables of R134a, at 40°C and 10.166 bar

$h_{ci} = 256.41$ kJ/kg da, $s_{ci} = 1.1905$ kJ/kg K (saturated liquid)

$e_{ci} = h_{ci} - h_{r0} - T_0 (s_{ci} - s_{r0})$

$= 256.41 - 423.33 - 298.15\ (1.1905 - 1.8964)$

$= 43.54$ kJ/kg

$m_a = 2.2278$ kg/s

Mass of condensate, $m_w = m_a (\omega_1 - \omega_2) = (133.7/60) \times (17.75 - 14.68) = 6.8410$ g/s

Heat balance in the cooling and dehumidifier,

$m_{ref} (h_{co} - h_{ci}) = m_a (h_1 - h_2)$

$$m_{ref} = \frac{m_a (h_1 - h_2)}{(h_{co} - h_{ci})} = \frac{0.2278 \times (80.71 - 57.37)}{(397.43 - 256.41)} = 0.3688\ kg/s$$

Exergy of condensate,

$$e_w = c_{pw}\left(T_w - T_0 - T_0 \times \log\left(\frac{T_w}{T_o}\right)\right)$$

$$= 4.18\left(20 - 25 - 298.15 \times \log\left(\frac{20 + 273.15}{298.15}\right)\right) = 0.1772\ kJ/kg$$

The exergy balance applied to the cooling and dehumidification process,

$\varepsilon_{ci} + \varepsilon_1 = \varepsilon_{co} + \varepsilon_2 + \varepsilon_w + I_{CDH}$

$m_{ref}\, e_{ci} + m_a\, e_1 = m_{ref}\, e_{co} + m_a\, e_2 + m_w\, e_w + I_{CDH}$

$I_{CDH} = m_{ref} (e_{ci} - e_{co}) + m_a (e_1 - e_2) - m_w\, e_w$

= 0.3688 (43.54 – 24.24 + 2.2278 (0.2 – 0.18) – 0.0068410 × 0.1772

= 7.17 kW

Exergy efficiency of cooling and dehumidification process,

$$\eta_{exergy,\ CDH} = 1 - \left(\frac{I_{CDH}}{m_{ref}\varepsilon_{ci} + m_a\varepsilon_1} \right) \times 100$$

$$= 1 - \left(\frac{7.17}{0.3688 \times 43.54 + 2.2278 \times 0.20} \right) \times 100$$

= 56.56%

14. Cooling and humidification: Air at 30°C and 50% RH enters an air washer which has a humidifying efficiency of 80%. Determine the final DBT and RH of air washed with re-circulated spray water.

Solution:

At 30°C DBT and 50% RH,

Humidifier efficiency = 80%

The resulted WBT of inlet air, WBT_1 = 22°C

Humidifier efficiency,

$$\eta_{humidifier} = \frac{DBT_1 - DBT_2}{DBT_1 - WBT}$$

$DBT_2 = DBT_1 - \eta_{humidifier} (DBT_1 - WBT)$

= 30 – 0.8 × (30 – 22) = 23.6°C

The result is RH_2 = 86%

Chapter 4
Combustion

4.1 Introduction

Combustion modeling is critical to the design of combustion chambers in internal combustion engines, thermal power plants, aeronautical gas turbine engines, process furnaces, etc. The combustion process involves the formulation of chemical reactions, mass balance, and energy balance equations. As per the air fuel ratio, combustion may be classified into three categories: complete combustion with excess oxygen, combustion without excess air, and incomplete combustion with an insufficient supply of air. Mass balance and energy balance together result in chemical coefficients and combustion temperature at a given air-fuel ratio. Or the air-fuel ratio can be simulated at the given combustion temperature. Complete combustion solutions involve (with or without excess oxygen) the linear equations, and incomplete combustion solutions involve non-linear equations with a dissociation constant.

4.2 Combustion reactions

A combustion process involves the burning of fuel with oxygen or a substance containing oxygen, such as air. The fuel and oxidation are called reactants, and the constituents resulting from the combustion process are called products. The combustion of carbon, for example, involves a simple reaction.

$$C + O_2 \rightarrow CO_2 \quad (4.1)$$

When the combustion process involves the burning of hydrocarbon fuel, the products will normally contain carbon dioxide and water.

$$CH_4 + 2O_2 \rightarrow CO_2 + 2H_2O \quad (4.2)$$

In this chemical reaction, the coefficients shown for the reactants and products are related to stoichiometric combustion, where a minimum amount of air is supplied for the complete combustion. These coefficients are called stoichiometric coefficients. The oxidizer may be atmospheric air or oxygen. The supply of oxygen in place of atmospheric air results in a higher temperature compared to air supply. In the air supply, the amount of nitrogen, which is not participative in combustion, controls the temperature. By volume, air composition is nitrogen: 78%, 21% nitrogen, and 1% argan. In these calculations, the quantity of argon is neglected. Therefore, air is taken as consisting of 79% nitrogen and 21% oxygen by volume.

Consequently, these are 79/21 = 3.76 mole of nitrogen for each mole of oxygen in the air.

Let x be the kg mole of air supplied for combustion.

$$CH_4 + x\,(O_2 + 3.76\,N_2) \rightarrow b_1\,CO_2 + b_2\,H_2O + b_3\,O_2 + b_4\,N_2 \quad (4.3)$$

DOI: 10.1201/9781032642505-4

C balance, $1 = b_1$

H balance, $4 = 2b_2 \rightarrow b_2 = 2$

O balance, $2x = 2\ b_1 + b_2 + 2\ b_3$

$$2x = 2 + 2 + 2\ b_3$$

$$x = 2 + b_3$$

$$b_3 = x - 2$$

N balance, $2 \times 3.76x = 2\ b_4 \rightarrow b_4 = 3.76x$

Therefore $b_1 = 1$
$b_2 = 2$
$b_3 = x - 2$
$b_4 = 3.76x$

The resultant equation is,

$$CH_4 + x\ (O_2 + 3.76\ N_2) \rightarrow CO_2 + 2\ H_2O + (x - 2)\ O_2 + 3.76x\ N_2 \quad (4.4)$$

In the above equation, $x - 2$ decides the nature of combustion.

If $x - 2 < 0$, indicates incomplete combustion or partial oxidation, deficient in oxygen
$x - 2 = 0$, indicates complete combustion, stoichiometric air-fuel ratio, no excess or deficiency of oxygen
$x - 2 > 0$, indicates combustion with excess oxygen

In the above reaction, in place of 4.76 kg mole of air (1 + 3.76), if 1 kg mole of air (0.21 + 0.79) is considered,

$$CH_4 + x\ (0.21\ O_2 + 0.79\ N_2) \rightarrow b_1\ CO_2 + b_2\ H_2O + b_3\ O_2 + b_4\ N_2 \quad (4.5)$$

C balance, $1 = b_1$

H balance, $4 = 2\ b_2 \rightarrow b_2 = 2$

O balance, $0.21 \times 2x = 2b_1 + b_2 + 2\ b_3$

$0.21 \times 2x = 2 + 2 + 2\ b_3$

$0.21x = 2 + b_3$

$b_3 = 0.21x - 2$

N balance, $2 \times 0.79x = 2\ b_4 \rightarrow b_4 = 0.79x$

Therefore $b_1 = 1$
$b_2 = 2$
$b_3 = 0.21x - 2$
$b_4 = 0.79x$

The resultant equation is

$$CH_4 + x\ (0.21O_2 + 0.79N_2) \rightarrow CO_2 + 2\ H_2O + (0.21x - 2)\ O_2 + 0.79x\ N_2 \quad (4.6)$$

A combustion reaction for a general hydrocarbon C_nH_y burning with the theoretical portion of air may be obtained as follows,

$$C_nH_y + x\ (O_2 + 3.76N_2) \rightarrow b_1\ CO_2 + b_2\ H_2O + b_3O_2 + b_4\ N_2 \quad (4.7)$$

C balance, $n = b_1$

H balance, $y = 2\ b_2 \rightarrow b_2 = y/2$

O balance, $2x = 2b_1 + b_2 + 2\ b_3$

$2x = 2n + y/2 + 2\ b_3$

$x = n + y/4 + b_3$

$b_3 = x - n - y/4$

N balance, $2 \times 3.76x = 2\ b_4 \rightarrow b_4 = 3.76x$

Therefore $b_1 = n$
$b_2 = y/2$
$b_3 = x - n - y/4$
$b_4 = 3.76x$

the final balance is then,

$$C_nH_y + x\ (O_2 + 3.76\ N_2) \rightarrow n\ CO_2 + (y/2)\ H_2O + (x - n - y/4)O_2 + 3.76x\ N_2 \quad (4.8)$$

For the burning of methane with minimum air we would have

$$CH_4 + 2O_2 + 2\ (3.76)\ N_2 \rightarrow CO_2 + 2H_2O + 2 \times 3.76\ N_2 \quad (4.9)$$

The coefficients of air shown in Eq. 4.9 are related to the minimum amount of air, which is called stoichiometric air or theoretical air. At the stoichiometric air supply, the resultant temperature is the maximum, which is called the adiabatic flame temperature at ideal conditions. To control this temperature, an excess amount of air is supplied. For example, if the excess air supply is 10%, the chemical equation is as follows:

$$CH_4 + 1.1 \times 2\ O_2 + 1.1 \times 2 \times 3.76\ N_2 \rightarrow CO_2 + 2H_2O + 0.1 \times 2\ O_2 + 1.1 \times 2 \times 3.76\ N_2 \quad (4.10)$$

In case the air supply is insufficient, the oxidation process is not complete, and it is called partial oxidation. The carbon is partially converted into carbon dioxide and carbon monoxide.

For methane an incomplete reaction might appear as

$$CH_4 + x\ O_2 + 3.76x\ N_2 \rightarrow a\ CO_2 + b\ CO + 2\ H_2O + 3.76x\ N_2 \quad (4.11)$$

In this case we would expect that $x < 2$. The oxygen balance in the equation would yield

$$2\ x = 2a + b + 2 \quad (4.12)$$

For partial oxidation, in addition to mass balance, energy balance and chemical equilibrium should be considered to predict the exact split between CO_2 and CO, i.e., the exact values of the coefficients a and b. If enough O_2 were added to the equation to bring it up to stoichiometric proportions, the CO would then react with the oxygen to form additional CO_2 in accordance with the reaction equation.

$$CO + \tfrac{1}{2}\ O_2 \rightarrow CO_2 \quad (4.13)$$

Even when excess air is supplied, real reactions do not usually go to completion of various factors involved in the combustion process. In a practical sense, therefore, it is usually necessary to make experimental measurements of the composition of the products of combustion to balance the equation.

4.3 Enthalpy of formation

The change in enthalpy for the formation of a compound from its constituent elements at the reference state is the standard enthalpy of formation or enthalpy of formation per unit mole. The following sections give the details of performing energy balances in chemical reactions in order to determine the quantity of heat that is liberated in a particular combustion process, the highest temperature attainable for a given fuel when burning with air, etc. Enthalpy of formation and enthalpy at the given temperature are required to find the energy balance of a chemical reaction. Tables that give the properties as a function of temperature are available for various substances. To formulate such tables, however, it is necessary to establish a reference state for the properties that take proper account of the internal chemical energy of the substance. The base (zero level) energy level is chosen so that the energy of all elements is at zero at 25°C and 1 atm pressure. The energy of a compound is not zero at this reference state but is defined in accordance with the following reasoning as applied to CO_2 at 25°C and 1 atm. To form CO_2 in a steady-flow process, we would have:

$$C + O_2 \rightarrow CO_2 \tag{4.14}$$

and the energy balance:

$$H_R + Q = H_P \tag{4.15}$$

Where H_R is the total enthalpy of the reactants and H_p is the total enthalpy of the products. Q is the heat transfer necessary to make the reaction go. Using the above reference state, $H_R = 0$. Since all the reactants are elements and $H_P = Q$. The value of Q may be determined experimentally and has the following value for CO_2.

$Q = H_P = -393766$ kJ/kg mol.

In other words, the reaction in the equation liberates 393766 kJ of energy for each mole of carbon burned. The enthalpy of CO_2 at 25°C and 1 atm is therefore less than the reference state by this amount. This quantity is called the enthalpy of formation and has the symbol, $\bar{h}_f^0$ where the bar designates the mole of the substance. A reaction in which heat is liberated is called exothermic, with the term endothermic being employed to describe reactions that absorb heat. Most of the formation reactions are exothermic (negative enthalpy of formation), but a few cases, such as NO, N_2O_4 are endothermic reactions that need energy for their formation and these are generally unstable. It is now easy to see how we may calculate the enthalpies of substances with respect to the zero-element state of 298 K and 1 atm. We employ the tables to compute the changes from the reference state and add this to the enthalpy of formation. Thus, the total molar enthalpy at temperature T is

$$\bar{h}_{T,\ total} = \bar{h}_f^0 + \bar{h}_T - \bar{h}_{298} \tag{4.16}$$

For the enthalpy of formation of a solid fuel, the following formulation can be used:

Chemical formula of solid fuel, $C_{a1}\ H_{a2}\ O_{a3}\ N_{a4}$

Mass of dry solid fuel per kg mole of fuel, $m_{df} = 12a_1 + a_2 + 16a_3 + 14a_4$

Higher heating value of dry fuel,

$HHV_{df} = m_{df}\,(0.3491C + 1.1783H + 0.10055S - 0.1034O - 0.151N - 0.0211Ash)1000$ kJ

where C, H, S, O, N and Ash are in % weight basis in the dry solid fuel.

From the definition of HHV, the reactants and products are maintained from and at 25°C with stoichiometric air-fuel ratio.

The energy balance equation to find the enthalpy of formation for dry solid fuel,

$h_{f\ fuel\ dry} = a_1\ hf_{CO2} + 0.5a_2\ hfH_2O_L + HHV_{df}$

As we have seen before, it is possible to choose different zero levels, or reference states, for tabulations of thermodynamic properties. For the steam tables, we choose the saturated liquid state at 0°C, while for the air tables, we choose 0 K and a pressure of 1 atm. For use with the above equations, tables are employed using a reference temperature of 0 K as well as the standard state temperature of 298 K. The 0 K state is perhaps more common for English units, while the 298 K reference is employed for SI tabulations. The modern approach is also to use a reference pressure of 0.1 MPa instead of 1 atm (0.10132 MPa). This small difference has essentially no effect on the enthalpy or internal energies.

As with the enthalpy of formation, it is possible to establish a Gibbs function of formation with a zero level set at 298 K and 1 atm for all elements. The values for a number of compounds are given in the table. The values of absolute entropies are set in accordance with the third law of thermodynamics, which states that the entropy of any pure substance in thermodynamic equilibrium approaches zero as the absolute temperature approaches zero. Using this zero level, one may integrate between zero Kelvin and the standard state of 25°C and 1 atm to obtain the values of absolute entropy given in the table. For those substances that follow ideal gas behavior, the entropy variation when pressure is given through an application of equation $s_2 - s_1 = \varphi_2 - \varphi_1 - R \ln (P_2/P_1)$ written in terms of molar properties:

$$\bar{s}_2 - \bar{s}_1 = \bar{\phi}_2 - \bar{\phi}_1 - \bar{R} \ln\left(\frac{P_2}{P_1}\right) \tag{4.17}$$

For ideal gases the absolute entropy at any pressure and temperature can thus be written as

$$\bar{s}^0(P,T) = \bar{s}^0 + \bar{\phi}_T - \bar{\phi}_{298} - \bar{R} \ln\left(\frac{P}{P_0}\right) \tag{4.18}$$

where $\bar{s}^0$ is taken from table and P_0 is the reference pressure of 1 atm. Absolute values for entropy are required for certain second law analyses of combustion reactions. The important point to remember is that reference states must be carefully specified when dealing with combustion reactions.

4.4 Heat of reaction and heating value

The energy liberated in a chemical reaction is given by the following relations.

For constant volume reactions,

$$Q_V = (\Delta U)_V = (U_P - U_R)_V = \bar{u}_{RP} \tag{4.19}$$

and for constant pressure,

$$Q_P = (\Delta H)_P = (H_P - H_R)_P = \bar{h}_{RP} \tag{4.20}$$

The quantities given in the above two equations are called the heats of reaction or the internal energy and enthalpy of combustion. If the heats are expressed per mole of fuel, then the symbols $\bar{u}_{RP}$ and $\bar{h}_{RP}$ are appropriately used to designate the energies indicated in the equation. The term heating value for a combustion process is in wide use and is a synonym for heat of reaction. When a hydrocarbon fuel is burned in a combustion process, water is produced as one of the products. The maximum energy release will be obtained when all the water in the products due to combustion is in the liquid state; in such cases the heat of the reaction is called 'a higher heating value'. Similarly, when all the water in the products due to combustion is in the vapor state, we say that a lower heating value is obtained. In reality, both liquid and vapor will be present in many combustion processes, so that a heating value between the HHV and LHV can be observed.

4.5 Adiabatic flame temperature

In an adiabatic process, the final temperature that the products will attain is called the adiabatic flame temperature. In this instance, all of the heat of the reaction is converted to internal energy and manifested in the form of a high temperature for the products of combustion. For the temperatures involved in most reactions, the combustion process usually will not go to completion, and an equation of chemical equilibrium must be employed to determine the final temperature. To compute the adiabatic flame temperature, an iterative procedure must be employed because of the complex relationship between the product temperature and its constituent enthalpies.

4.6 The equilibrium constant

The equilibrium constant K_P is defined as,

$$\Delta G^0 = -\bar{R}T \ln\left(\frac{P_3^{\nu_3} P_4^{\nu_4}}{P_1^{\nu_1} P_2^{\nu_2}}\right) \tag{4.21}$$

so that,

$$\Delta G^0 = -\bar{R}T \ln\left(K_P\right) \text{ - The law of mass action.} \tag{4.22}$$

The equation is called the law of mass action. For a constant pressure circumstance, ΔG^o is a function of only the properties of the constituents at temperature T, so it follows that the equilibrium constant is also a function of temperature alone. We re-emphasize that these equations are for perfect gases. In this development, the reader should note that:

1. Pressure is expressed in atmospheres.
2. The values of the vs employed for the calculation of the equilibrium constant are the stoichiometric values for that particular reaction.

At equilibrium, the partial pressures of the constituents may be expressed in terms of the mole fractions of these constituents and the total pressure. Thus

$P_i = x_i P_{total}$

and the relation for the equilibrium constant becomes,

$$K_P = \left(\frac{x_3^{\nu_3} x_4^{\nu_4}}{x_1^{\nu_1} x_2^{\nu_2}}\right) P_{total}^{\nu_3+\nu_4-\nu_1-\nu_2} \tag{4.23}$$

For the simple reactions indicated, the equilibrium constant increases with temperature.

4.7 Equilibrium in multiple reactions

One example of such multiple reactions due to the additional gases in the mixture would be the dissociation of H_2O by the two reactions (Holman and Pilling 2004).

$H_2O \leftrightarrow H_2 + ½\, O_2$

$H_2O \leftrightarrow ½\, H_2 + OH$

Another example might be the combined dissociation of CO_2 and N_2:

$CO_2 \leftrightarrow CO + ½\, O_2$

$½\, O_2 + ½\, N_2 \leftrightarrow NO$

We may write for the two general reactions:

$\nu_1 A_1 + \nu_2 A_2 \leftrightarrow \nu_3 A_3 + \nu_4 A_4$

$v_5 A_5 + v_6 A_6 \leftrightarrow v_7 A_7 + v_8 A_8$

where we are taking one reactant A1 as common to both equations but with a possibly different stoichiometric proportion, i.e., v1 does not necessarily equal v5. We shall designate the above equations' reaction 1 and reaction 2, respectively, with their corresponding degrees of reaction ε1 and ε2. Both degrees of reaction are based on the consumption of reactant A1 in each equation. For reaction 1, we have,

$dn_1 = -v_1 \, d\varepsilon_1$

$dn_2 = -v_2 \, d\varepsilon_2$

$dn_3 = v_3 \, d\varepsilon_3$

$dn_4 = v_4 \, d\varepsilon_4$

while for reaction 2

$dn_1 = -v_5 \, d\varepsilon_5$

$dn_6 = -v_6 \, d\varepsilon_6$

$dn_7 = v_7 \, d\varepsilon_7$

$dn_8 = -v_8 \, d\varepsilon_8$

For the combined reactions, the total consumption of reactant A_1 is:

$dn_1 = -v_1 \, d\varepsilon_1 - v_5 \, d\varepsilon_5$

while the other constituents remain independent. We now apply the condition for equilibrium at constant temperature and pressure and write:

$(\Sigma \bar{g} \, dn_i)_{T,P} = 0$

or

$$\left(-\bar{g}_1 v_1 + \bar{g}_2 v_2 + \bar{g}_3 v_3 + \bar{g}_4 v_4\right) d\varepsilon_1 + \left(-\bar{g}_1 v_5 + \bar{g}_6 v_6 + \bar{g}_7 v_7 + \bar{g}_8 v_8\right) d\varepsilon_2 \tag{4.24}$$

Because we are assuming an ideal gas mixture, we can use the following relation to express the Gibbs function at the reference temperature,

$$dg = v \, dp - s \, dT \tag{4.25}$$

$$\bar{g} - \bar{g}^0 = \int_{P=1}^{P} \bar{v} dP - \int_{T_{ref}}^{T} \bar{s} dT \tag{4.26}$$

The Gibbs function at the reference temperature is:

$$\bar{g} - \bar{g}^0 = \int_{P=1}^{P} \bar{v} dP = \int_{P=1}^{P} \bar{R}T \frac{dP}{P} = \bar{R}T \ln P \tag{4.27}$$

We obtain two relations analogous to that for the single reaction.

$$\Delta G_1^0 = -\bar{R}T \ln\left(\frac{P_3^{v_3} P_4^{v_4}}{P_1^{v_1} P_2^{v_2}}\right) \tag{4.28}$$

$$\Delta G_2^0 = -\bar{R}T \ln\left(\frac{P_7^{v_7} P_8^{v_8}}{P_1^{v_5} P_6^{v_6}}\right) \tag{4.29}$$

where the free energies are,

$$\Delta G_1^0 = v_3\overline{g}_3^0 + v_4\overline{g}_4^0 - v_1\overline{g}_1^0 - v_2\overline{g}_2^0 \tag{4.30}$$

$$\Delta G_2^0 = v_7\overline{g}_7^0 + v_8\overline{g}_8^0 - v_5\overline{g}_1^0 - v_6\overline{g}_6^0 \tag{4.31}$$

Equations may be arranged in the equilibrium constant from as,

$$\ln\left(K_{P1}\right) = \frac{-\Delta G_1^0}{\overline{R}T} \tag{4.32}$$

$$\ln\left(K_{P2}\right) = \frac{-\Delta G_2^0}{\overline{R}T} \tag{4.33}$$

where we recognize that both equilibrium constants may be expressed in terms of mole fractions of the constituents in accordance with the equation for equilibrium constant. To obtain the equilibrium concentrations, the two non-linear equations must be solved simultaneously. For systems involving many simultaneous reactions, this becomes a complicated task and is best accomplished with a computer.

The Van't Hoff isobar equation is quite useful in that it enables us to calculate enthalpy change in a reaction from the equilibrium compositions as determined by KP. In a reciprocal way, energy balances in reactions can yield important information on the variation of K_p with temperature.

Let us rearrange equation as a function of 1/T then,

$$\frac{d\left(\ln K_P\right)}{d\left(\dfrac{1}{T}\right)} = \frac{-\Delta H^0(T, P_0)}{\overline{R}} \tag{4.34}$$

The equation indicates that the slope of the curve in a plot of ln K_P versus 1/T will yield the value of $\frac{\Delta H^0}{\overline{R}}$. For many reactions the value of ΔH^0 is very nearly constant over a rather wide range of temperatures. In such cases, the equation can be integrated between two temperatures T_1 and T_2 to yield:

$$\ln\frac{K_{P2}}{K_{P2}} = \frac{\Delta H^0}{\overline{R}}\left(\frac{1}{T_1} - \frac{1}{T_2}\right) \tag{4.35}$$

and if one knows the value of K_{P1} at T_1 it is an easy task to calculate the value of K_{P2} at temperature T_2.

The van't Hoff equation can also be helpful in evaluating the endothermic or exothermic nature of a reaction. If the equilibrium constant increases with temperature, that is, $K_{P2} > K_{P1}$ for $T_2 > T_1$, the equation indicates that ΔH0 must be positive and the reaction is endothermic. If K_P decreases with an increase in temperature, ΔH° will be negative, and the reaction will be exothermic.

4.8 Exergy

Exergy is the maximum work that can be extracted from a system when its state changes from its initial position to a dead state with reference to the surroundings. Exergy is the function of the system and surroundings. It involves internal and external dissipative effects.

Exergy, E has many components. In the absence of nuclear effects, magnetism, electricity, and surface tension, E is,

$$E = E_k + E_p + E_{ph} + E_0 \tag{4.36}$$

where E_k is kinetic exergy, E_p potential exergy, E_{ph} physical exergy and E_0 chemical exergy. E_k and E_p are associated with high grade energy and E_{ph} and E_0 with low grade energy. The above equation in the specific form is,

$$\varepsilon = \varepsilon_k + \varepsilon_p + \varepsilon_{ph} + \varepsilon_0 \tag{4.37}$$

4.8.1 Physical exergy

The exergy changes during the physical process, which is the maximum work deliverable when the system changes its initial state into dead states P_0 and T_0.

The physical exergy is,

$$\varepsilon_{ph} = (h - T_0 s) - (h_0 - T_0 s_0) \tag{4.38}$$

the change in physical exergy is,

$$\varepsilon_{ph1} - \varepsilon_{ph2} = (h_1 - h_2) - T_0 (s_1 - s_2) \tag{4.39}$$

where,

$$s_2 - s_1 = s_{T2} - s_{T1} - R \ln (P_2/P_1) \tag{4.40}$$

Physical exergy of perfect gas,

$$\varepsilon_{ph} = c_p(T - T_0) - T_0\left(c_p \ln\frac{T}{T_0} - R\ln\frac{P}{P_0}\right) \tag{4.41}$$

Using Janaf tables,

The physical exergy of air,

$$e_{ph\ air} = x\left[\sum_k n_k (h_T - h_{298}) - T_0\left(\sum_k n_k (s_T - s_{298}) - n_{air} R \ln\left(\frac{P}{P_0}\right)\right)\right] \tag{4.42}$$

the physical exergy of products,

$$e_{ph\ products} = \left[\sum_k n_k (h_T - h_{298}) - T_0\left(\sum_k n_k (s_T - s_{298}) - n_p R \ln\left(\frac{P}{P_0}\right)\right)\right] \tag{4.43}$$

In case of combustion, enthalpy of formation should be included to the h_T

4.8.2 Chemical exergy

The work obtained from such a process per mole of substance will equal the molar chemical exergy, $\bar{\varepsilon}_0$. For an ideal gas,

$$\bar{\varepsilon}_0 = \bar{R}T_0 \ln\frac{P_0}{P_{00}} \tag{4.44}$$

For chemical exergy of a mixture, the total work of isothermal compression is

$$\sum_i (\bar{w}_{xi})_{rev} = \bar{R}T_0 \sum_i x_i \ln x_i \tag{4.45}$$

where xi is the mole fraction of the ith component in the mixture. Each separated component emerging from its compressor at P0, T0 has at this point, molar exergy, while its flow rate per mole of mixture is xi. As all the processes taking place in the device are reversible, the exergy of the gas mixture is the sum of the exergy of its components minus the work of compression. Hence,

$$\bar{\varepsilon}_{oM} = \sum_i x_i \bar{\varepsilon}_{0i} + \bar{R}T_0 \sum_i x_i \ln x_i \tag{4.46}$$

Since the second term on the RHS of the above equation is always negative, the exergy of the mixture is always less than the sum of the exergies of its components at the pressure and temperature of the mixture. This equation is also applicable to ideal solutions of liquids, i.e., solutions produced without a change in volume or enthalpy. In the case of real solutions, the exergy of a mixture can be obtained if the activity coefficients, γ_i, for the components are known. The more general version of this equation is,

$$\bar{\varepsilon}_{oM} = \sum_i x_i \bar{\varepsilon}_{0i} + \bar{R}T_0 \sum_i x_i \ln \gamma_i x_i \tag{4.47}$$

Activity coefficients may be greater or smaller than unity, and they are unity for ideal solutions. The chemical exergy of real solutions can also be obtained from exergy concentration diagrams. The work of compression, the second term on the RHS of the equation, can be interpreted as the minimum work necessary for separating the components of the mixture and delivering them at the environmental pressure and temperature. This interpretation will be useful in the analysis of the distillation process.

Using Janaf tables, the chemical exergy of air,

$$e_{ch\ air} = 4.76 \text{ x } \{0.21\ \varepsilon^0_{O_2} + 0.79\ \bar{\varepsilon}^0_{N_2} + RT_0\ [0.21 \times \ln(0.21) + 0.79 \times \ln(0.79)]\} \tag{4.48}$$

Chemical exergy of products,

$$e_{ch,\ products} = n_P \begin{bmatrix} x_{CO_2}\varepsilon^0_{CO_2} + x_{H_2O}\varepsilon^0_{H_2O} + x_{O_2}\varepsilon^0_{O_2} + x_{N_2}\varepsilon^0_{N_2} \\ +RT_0\left(x_{CO_2}\ln\left(x_{CO_2}\right) + x_{H_2O}\ln\left(x_{H_2O}\right) + x_{O_2}\ln\left(x_{O_2}\right) + x_{N_2}\ln\left(x_{N_2}\right)\right) \end{bmatrix} \tag{4.49}$$

4.9 Computation of gas property values

Like the steam tables, the properties of energy and exergy can be obtained from the data listed in the tables for every chemical constituent such as CO, CO2, H_2O, etc. At the reference state, the energy, entropy, and exergy are zero due to the dead state. These properties are generated as a function of temperature at atmospheric pressure, as shown in Fig. 4.1. To present the properties as a function of temperature and pressure, the properties have to be extended with the real gas equation and its solution.

Reference levels: Enthalpy and entropy at 298.15 K, and 1 atmosphere.

For ideal gas the specific entropies at other pressures,

$s = (s_T - s_{298}) - R \ln(p/p_0)$

Where p_0 is the reference pressure of 1 atmosphere and $(s_T - s_{298})$ is the entropy at the given temperature T.

The chemical exergy of a combustible substance,

$$\bar{\varepsilon}^0 = -\Delta\bar{h}^0 + T^0\Delta\bar{s}^0 + \bar{R}T^0\left(x_{O_2}\ln\frac{P^{00}_{O_2}}{P^0} - \sum_k x_k \ln\frac{P^{00}_k}{P^0}\right) \tag{4.50}$$

In the above equation, the subscript k is related to the product of combustion and its components such as CO_2, H_2O, O_2, and N_2, etc. This equation of chemical exergy can be used to determine the (chemical) exergy of gaseous fuel through its chemical composition. However, for the solid, liquid,

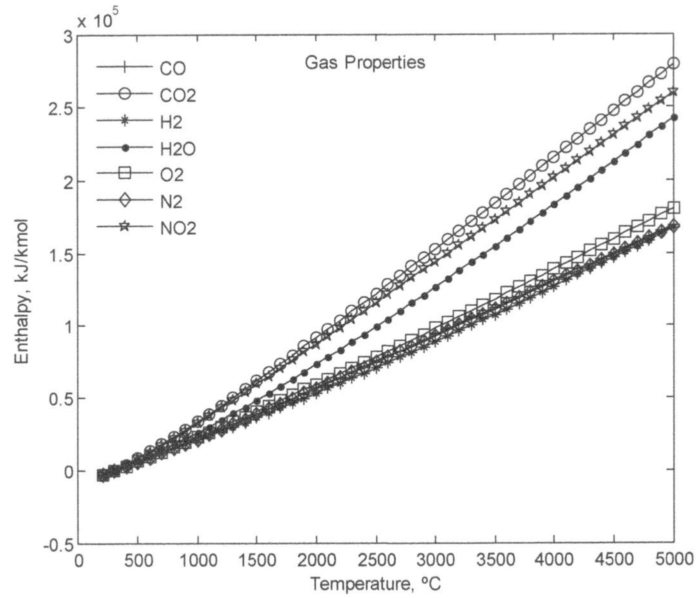

Fig. 4.1. A typical gas enthalpy as a function of temperature.

and industrial fuels, generally there is no readily available chemical formula. From C-H-O-N-S system, the ratio of chemical exergy to the net calorific value for these fuels can be determined as follows.

This ratio of chemical exergy to net calorific value, φ,

$$\phi = \frac{\varepsilon^0}{(NCV)^0} \tag{4.51}$$

4.10 Exergy of fuels

The ratio of chemical exergy to net calorific value for a dry solid fuel can be determined from the following equation with a mass ratio of oxygen to carbon (o/c) less than 0.667 (Kotas, 2012).

$$\phi_{dry} = 1.0437 + 0.1882\frac{h}{c} + 0.0610\frac{o}{c} + 0.0404\frac{n}{c} \tag{4.52}$$

Where c, h, o and n are the mass fractions of C, H, O and N respectively.

For solid fuels with the mass ratio 2.67 > o/c > 0.667,

$$\phi_{dry} = \frac{1.0438 + 0.1882\frac{h}{c} - 0.2509\left(1 + 0.7256\frac{h}{c}\right) + 0.0383\frac{n}{c}}{1 - 0.3035\frac{o}{c}} \tag{4.53}$$

The standard chemical exergy of a solid fuel with moisture,

$$\varepsilon^0\ (\text{kJ/kg}) = [\text{LHV}^0\ (\text{kJ kg}^{-1}) + 2442\ \text{w}]\ \varphi_{dry} + 9417\ \text{S} \tag{4.54}$$

Where w = mass fraction of moisture in the fuel

$h_{fg}^{\,o}$ = latent heat of water at standard temperature, T_0, (25°C) = 2442 kJ/kg steam.

For liquid fuels, the effect of sulfur was included in the correlation giving the expression

$$\phi_{dry} = 1.0401 + 0.1728\frac{h}{c} + 0.0432\frac{o}{c} + 0.2169\frac{s}{c}\left(1 - 2.0628\frac{h}{c}\right) \tag{4.55}$$

The accuracy of this expression is estimated to be ± 0.38%.

For gaseous fuel, the chemical exergy can be determined from the chemical equation. The φ can have values smaller or greater than unity, and for most industrial fuels except peat and wood, this value lies within a small percentage of unity (Table 4.1). The enthalpy of formation for various elements to solve the energy balance equation is listed in Table 4.2.

The exergy efficiency, or second law of efficiency, is defined as a ratio of exergy generation to the exergy value of fuel. The difference between the exergy supply and the total exergy losses (irreversibilities) is equal to the exergy output.

Exergy efficiency,

$$\eta_2 = \left(\frac{\varepsilon^0_{fuel} - i_{total}}{\varepsilon^0_{fuel}}\right) \times 100 \tag{4.56}$$

Table 4.1. Typical values of φ for some industrial fuels and other combustible substances.

S. No.	Fuel	$\varphi = \varepsilon^0/(NCV)^0$
1	Coke	1.05
2	Different type of coal	1.06–1.10
3	Peat	1.16
4	Wood	1.15–1.30
5	Different fuel oils and petrol	1.04–1.08
6	Natural gas	1.04 ± 0.5%
7	Coal gas	1.00 ± 1%
8	Blast furnace gas	0.98 ± 1%
9	Hydrogen	0.985
10	Carbon monoxide	0.973
11	Sulphur (rhombic)	2.017

4.11 Thermo-chemical model of biomass gasifier

Gasifier is a main component in an integrated gasification combined cycle (IGCC). Figure 4.2 outlines the power plant layout with the gasifier. The gasifier may be coal gasifier or biomass gasifier. Due to limitations in coal use, biomass gasifiers are deployed, offering more drive in power plants. To work out the various processes of the power plant, the gas composition of the gasifier is required. Thermochemical model applied to partial oxidation gives the gas composition. The assumptions involved in the gasifier model are as follows.

The heat loss and pressure drop in gasifier	= 5%.
Gasification of solid fuel	= complete
The reference state	= 298.15 K and 1.01325 bar
Polytrophic state efficiency of the air compressor	= 85%
Gasifier pressure	= 12 bar
Steam injection pressure in gasifier	= 12 bar
Pressure of solid fuel supply	= atmospheric pressure
Moisture content in fuel	= neglected

Table 4.2. Enthalpy of formation and chemical exergy of elements at 298.15 K and 1 atmosphere.

Enthalpy of formation, kJ/kg mol		**Standard chemical exergy, kJ/kg mol**	
$\bar{h}^0_{f,\,CO}$	–110541	$\bar{\varepsilon}^0_{CO}$	275430
$\bar{h}^0_{f,\,CO_2}$	–393546	$\bar{\varepsilon}^0_{CO_2}$	20140
$\bar{h}^0_{f,\,H_2O(g)}$	–241845	$\bar{\varepsilon}^0_{H_2O(g)}$	11710
$\bar{h}^0_{f,\,H_2O(l)}$	–285830	$\bar{\varepsilon}^0_{H_2O(l)}$	3120
$\bar{h}^0_{f,\,O_2}$	0	$\bar{\varepsilon}^0_{O_2}$	3970
$\bar{h}^0_{f,\,N_2}$	0	$\bar{\varepsilon}^0_{N_2}$	720
$\bar{h}^0_{f,\,NO_2}$	–33098	$\bar{\varepsilon}^0_{NO_2}$	56220
$\bar{h}^0_{f,\,H_2}$	0	$\bar{\varepsilon}^0_{H_2}$	238490
$\bar{h}^0_{f,\,CH_4}$	–74873	$\bar{\varepsilon}^0_{CH_4}$	836510
$\bar{h}^0_{f,\,C_2H_6}$	–83.8E3	$\bar{\varepsilon}^0_{C_2H_6}$	1504360
$\bar{h}^0_{f,\,C_3H_8}$	–104.7E3	$\bar{\varepsilon}^0_{C_3H_8}$	2163190
$\bar{h}^0_{f,\,C_4H_{10i}}$	–134.2E3	$\bar{\varepsilon}^0_{C_4H_{10i}}$	2818930
$\bar{h}^0_{f,\,C_4H_{10n}}$	–125.6E3	$\bar{\varepsilon}^0_{C_4H_{10n}}$	2818930
$\bar{h}^0_{f,\,C_5H_{12i}}$	–179E3	$\bar{\varepsilon}^0_{C_5H_{12i}}$	3477050
$\bar{h}^0_{f,\,C_5H_{12n}}$	–146.8E3	$\bar{\varepsilon}^0_{5H_{12n}}$	3477050
$\bar{h}^0_{f,\,C_6H_{14}}$	–167.2E3	$\bar{\varepsilon}^0_{C_6H_{14}}$	4134590

The syngas, or producer gas, from the gasifier is used in the gas turbine combustion chamber (GTCC) of the power plant. This gas at high temperature and pressure expands in gas turbines for power generation. The energy in the exhaust gas is recovered in a heat recovery steam generator (HRSG) to generate the steam. The generated steam is expanded in the steam turbine for power generation. A small portion of the steam is supplied to the gasifier to augment the gas yield. It is easy to solve the combustion problem either with complete combustion or partial oxidation by formulating the solid fuel. The proposed formula for a solid fuel is C_{a1}, H_{a2}, O_{a3}, N_{a4}. The coefficient a1 is equal to one for a single carbon atom in fuel. The other coefficients a2, a3, and a4 are determined from H/C, O/C, and N/C mole ratios, respectively. The mass balance, energy balance, and equilibrium correlations lead to the gas composition. CH_4, CO, CO_2, H_2, H_2O, and N_2.

The following is the chemical reaction in a biomass gasifier.

$$C_{a_1}H_{a_2}O_{a_3}N_{a_4} + a_5(O_2 + 3.76N_2) + a_6H_2O$$

$$\Rightarrow b_1CH_4 + b_2CO + b_3CO_2 + b_4H_2 + b_5H_2O + b_6N_2 \quad (4.57)$$

This equation is the total reaction in the gasifier. But it involves many other intermediate reactions, such as methane reforming and shift reactions. The Eq. 4.57 has six unknowns (coefficients related to products) and one variable (air fuel ratio) related to reactants. The balancing of coefficients of C, H, O, and N results in four equations. In addition to these four equations, two more equations of dissociation and air fuel ratio complete the model of partial oxidation. Therefore, at a given air-fuel ratio, the total variables are six; the model needs six equations (four mass balances and two equilibrium equations).

C balance: $a_1 = b_1 + b_2 + b_3$ (4.58)

H balance: $a_2 + 2a_6 = 4b_1 + 2b_4 + 2b_5$ (4.59)

O balance: $a_3 + 2a_5 + a_6 = b_2 + 2b_3 + b_5$ (4.60)

N balance: $a_4 + 2\times3.76a_5 = 2b_6$ (4.61)

The two equilibrium equations are as follows.

Methane reforming $CH_4 + H_2O = CO + 3H_2$ (4.62)

Water shift reaction $CO + H_2O = CO_2 + H_2$ (4.63)

Equation 4.62 is a methane reforming equation, which is an endothermic reaction as methane is a stable molecule. The C-H bond has high bonding energy, i.e., 439 kJ/kg mol. The methane

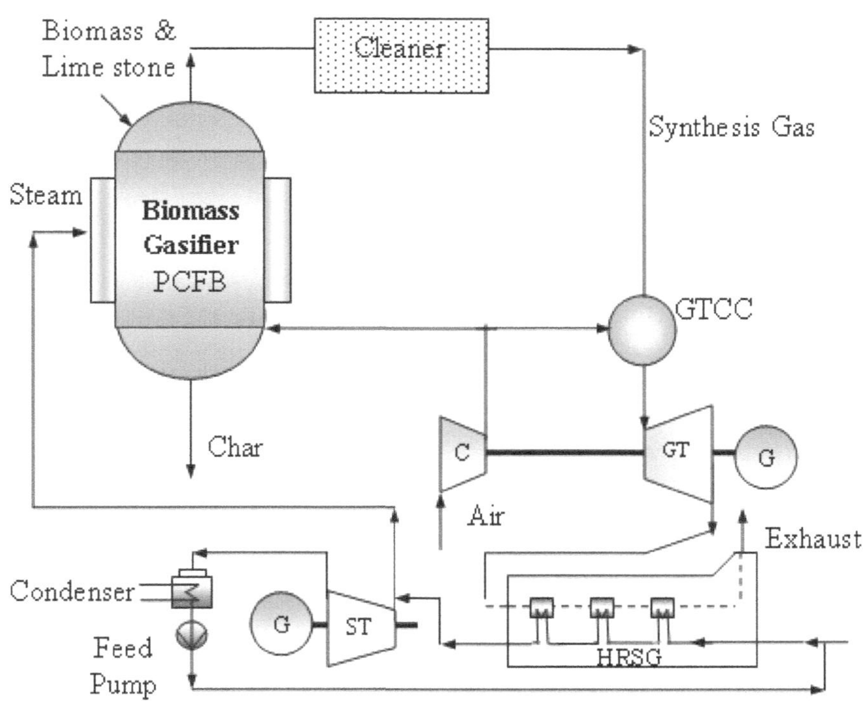

Fig. 4.2. Biomass gasifier in a combined cycle power plant (integrated gasification combined cycle).

reforming chemical process is also used to generate hydrogen from methane. It is a mature process. The dissociation reaction requires a high process temperature to change the equilibrium state to a shift state. On the other side, the second equation, i.e., the water shift reaction is an exothermic reaction. It converts the carbon monoxide and water into carbon dioxide and hydrogen. Its process temperature is low for change of state from equilibrium to shift state. The kinetic rates of the water shift reaction are very fast compared with the methane steam reforming reaction (Srinivas et al., 2012).

The total equation can be condensed with two unknowns (b1 and b2) by the elimination method.

$$b_3 = a_1 - b_1 - b_2 \tag{4.64}$$

$$b_4 = 2a_1 + 0.5a_2 - a_3 - 2a_5 - 4b_1 - b_2 \tag{4.65}$$

$$b_5 = a_3 + 2a_5 + a_6 - 2a_1 + 2b_1 + b_2 \tag{4.66}$$

$$b_6 = 0.5a_4 + 3.76\, a_5 \tag{4.67}$$

Total number of species (kg mol) in products,

$$n_t = b_1 + b_2 + b_3 + b_4 + b_5 + b_6$$

$$= a_1 + 0.5a_2 + 0.5a_4 + 3.76a_5 + a_6 - 2\, b_1 \tag{4.68}$$

The two equations related to the equilibrium are the result equilibrium constants (K_p) of reactions. These are relationships between reactants and products at equilibrium.

The equilibrium constant $K_{p,1}$ of the reforming reaction is as follows:

$$K_{p1} = \frac{P_{CO}\, P_{H_2}^3}{P_{CH_4}\, P_{H_2O}} = \frac{b_2 b_4^3 P^2}{b_1 b_5 n_t^2}. \tag{4.69}$$

Substituting b_4, b_5 and n_t in Eq. 4.69,

$$K_{p1} = \frac{b_2(2a_1 + 0.5a_2 - a_3 - 2a_5 - 4b_1 - b_2)^3 P^2}{b_1(a_3 + 2a_5 + a_6 - 2\, a_1 + 2b_1 + b_2)(a_1 + 0.5a_2 + 0.5a_4 + 3.76a_5 + a_6 - 2\, b_1)^2} \tag{4.70}$$

$$where\ P = \frac{P_{gasifier}}{P_0} \tag{4.71}$$

The equilibrium constants are given by

$$\ln K_p = \frac{-\Delta G^0}{RT} \tag{4.72}$$

In the above equation ΔG^0 is Gibb's free energy, kJ/kg mol, T is the temperature of gasifier K, and R is the universal gas constant, kJ/kg mol K.

$$\Delta G^0 = \sum h_f^0 + (h_T - h_{298}) - T\, s \tag{4.73}$$

h_f^0 is the enthalpy of formation, kJ/kg mol and s is the entropy measured from 0 K, kJ/kg mol K.

Equation 4.70 can be shown as

$$b_2 P^2 (2a_1 + 0.5a_2 - a_3 - 2a_5 - 4b_1 - b_2)^3 - K_{p1} b_1 (a_3 + 2a_5 + a_6 - 2\, a_1 + 2b_1 + b_2)$$

$$(a_1 + 0.5a_2 + 0.5a_4 + 3.76a_5 + a_6 - 2\, b_1)^2 = 0 = f(b_1, b_2) \tag{4.74}$$

$$K_{p,2} = \frac{P_{CO_2} P_{H_2}}{P_{CO} P_{H_2O}} = \frac{b_3 b_4}{b_2 b_5}. \tag{4.75}$$

Substituting the results of b_3, b_4 and b_5 in Eq. 4.75,

$$K_{p,2}=\frac{(a_1-b_1-b_2)(2a_1+0.5a_2-a_3-2a_5-4b_1-b_2)}{b_2(a_3+2a_5+a_6-2a_1+2b_1+b_2)} \tag{4.76}$$

Equation 4.75 can be shown as

$$(a_1-b_1-b_2)(2a_1+0.5a_2-a_3-2a_5-4b_1-b_2)$$

$$-K_{p,2}b_2(a_3+2a_5+a_6-2a_1+2b_1+b_2)=0=g(b_1,b_2) \tag{4.77}$$

In the above two equations (Eqs. 4.73 and 4.76), b_1 and b_2 are the two unknowns. These are the non-linear equations that need iteration with the assumption of b_1 and b_2. A numerical method, Taylor's series expansion method, can be applied to simulate the model of this gasifier.

$$f_0+h\frac{\partial f}{\partial b_1}+k\frac{\partial f}{\partial b_2}=0 \tag{4.78}$$

$$g_0+h\frac{\partial g}{\partial b_1}+k\frac{\partial g}{\partial b_2}=0 \tag{4.79}$$

In Eqs. 4.77 and 4.78, h and k are the desired degree of accuracy for the coefficients b_1 and b2, respectively. These two equations are functions of b_1 and b_2 as follows.

$f_o = f(b_1, b_2)$

and $g_o = g(b_1, b_2)$ respectively.

Now these two equations can be solved for h and k. If the desired degree of accuracy is not satisfied, the new values of b1 and b2 will be considered as follows:

$$b_1' = b_1 + h \tag{4.80}$$

$$b_2' = b_2 + k \tag{4.81}$$

This process is continued until consistency is reached. Finally, all six coefficients (b1 to b6) can be determined after iterations. Chemical equations should be balanced with the energy balance equation, which iterates the gasifier temperature.

The second law evaluation of gasifier involves finding its irreversibility and exergy efficiency.

Chemical exergy of biomass fuel,

$$e_{ch,\ bmf}=a_1\varepsilon_C^0+\frac{1}{2}a_2\varepsilon_{H_2}^0+\frac{1}{2}a_3\varepsilon_{O_2}^0+\frac{1}{2}a_4\varepsilon_{N_2}^0 \tag{4.82}$$

$\varepsilon_{CH_4}^0 = 836510$, $\varepsilon_{CO}^0 = 275430$, $\varepsilon_{CO_2}^0 = 20140$, $\varepsilon_{H_2}^0 = 238490$, $\varepsilon_{H_2O(g)}^0 = 11710$,

$\varepsilon_{O_2}^0 = 3970$, $\bar{\varepsilon}_{N_2}^0 = 690$ and $\varepsilon_C^0 = 410820$

Biomass fuel is supplied at reference temperature; therefore, the physical exergy of biomass has been treated to zero.

Chemical exergy of air supplied to gasifier,

$$e_{ch,\ air}=\sum\nolimits_k n_k\varepsilon_k^0+RT_0\Sigma_k \ln.[P.x_k] \tag{4.83}$$

where x_k is the mole fraction of kth component in the chemical equation.

Similarly, for steam injection the chemical exergy has been determined.

Physical exergy of air supplied to gasifier,

$$e_{ph,\,air} = h_{air} - \Sigma_k T_0 s_k \tag{4.84}$$

Physical exergy of steam injection into gasifier,

$$e_{ph,\,steam} = h_{steam} - \Sigma_k T_0 s_k \tag{4.85}$$

Similarly, the chemical exergy of syn gas ($e_{ch,\,syn\,gas}$) can be determined.

Physical exergy of syn gas,

$$e_{ph,\,syn\,gas} = h_{syn\,gas} - \Sigma_k T_0 s_k \tag{4.86}$$

Total exergy, $e = e_{ch} + e_{ph}$ (4.87)

Irreversibility of gasifier,

$$i_{gasifier} = e_{bmf} + e_{air} + e_{steam} - e_{syn\,gas} \tag{4.88}$$

Exergy efficiency of gasifier,

$$\eta_2 = \frac{\varepsilon_{bmf}^{o} - i_{gasi}}{\varepsilon_{bmf}^{o}} \times 100 \tag{4.89}$$

4.12 Numerical solutions

1. Calculate the amount of air required to burn 1 kg mol of methane at 25°C with the supply of air at 500 K and the products are at 1500 K. Also find the percentage of excess air.

Solution:

$CH_4 + x\,(O_2 + 3.76\,N_2) \rightarrow b_1\,CO_2 + b_2\,H_2O + b_3\,O_2 + b_4\,N_2$

C: balance, $1 = b_1$

H_2: balance, $2 = b_2$

O_2: balance, $x = b_1 + 0.5 \times b_2 + b_3 = 1 + 0.5 \times 2 + b_3$

$b_3 = x - 2$

N_2: balance, $3.76x = b_4$

Therefore,

$CH_4 + x\,(O_2 + 3.76\,N_2) \rightarrow CO_2 + 2\,H_2O + (x-2)\,O_2 + 3.76x\,N_2$

The value of x can be determined from the energy balance of the above reaction equation.

$$\bar{h}_{f,\,CH_4}^{0} + x\{(\bar{h}_{500} - \bar{h}_{298})_{O_2} + 3.76(\bar{h}_{500} - \bar{h}_{298})_{N_2}\}$$
$$= (\bar{h}_f^{0} + \bar{h}_{1500} - \bar{h}_{298})_{CO_2} + 2(\bar{h}_f^{0} + \bar{h}_{1500} - \bar{h}_{298})_{H_2O} + (x-2)(\bar{h}_{1500} - \bar{h}_{298})_{O_2} + 3.76x(\bar{h}_{1500} - \bar{h}_{298})_{N_2}$$

$$\bar{h}_{f,\,CH_4}^{0} - (\bar{h}_f^{0} + \bar{h}_{1500} - \bar{h}_{298})_{CO_2} - 2(\bar{h}_f^{0} + \bar{h}_{1500} - \bar{h}_{298})_{H_2O} + 2(\bar{h}_{1500} - \bar{h}_{298})_{O_2}$$
$$= -x\{(\bar{h}_{500} - \bar{h}_{298})_{O_2} + 3.76(\bar{h}_{500} - \bar{h}_{298})_{N_2}\} + x(\bar{h}_{1500} - \bar{h}_{298})_{O_2} + 3.76x(\bar{h}_{1500} - \bar{h}_{298})_{N_2}$$

$$\bar{h}_{f,\,CH_4}^{0} - (\bar{h}_f^{0} + \bar{h}_{1500} - \bar{h}_{298})_{CO_2} - 2(\bar{h}_f^{0} + \bar{h}_{1500} - \bar{h}_{298})_{H_2O} + 2(\bar{h}_{1500} - \bar{h}_{298})_{O_2}$$
$$= x\left((\bar{h}_{1500} - \bar{h}_{298})_{O_2} - (\bar{h}_{500} - \bar{h}_{298})_{O_2} + 3.76(\bar{h}_{1500} - \bar{h}_{298})_{N_2} - (\bar{h}_{500} - \bar{h}_{298})_{N_2}\right)$$

$$x = \frac{\bar{h}^0_{f,\,CH_4} - (\bar{h}^0_f + \bar{h}_{1500} - \bar{h}_{298})_{CO_2} - 2(\bar{h}^0_f + \bar{h}_{1500} - \bar{h}_{298})_{H_2O} + 2(\bar{h}_{1500} - \bar{h}_{298})_{O_2}}{(\bar{h}_{1500} - \bar{h}_{298})_{O_2} - (\bar{h}_{500} - \bar{h}_{298})_{O_2} + 3.76(\bar{h}_{1500} - \bar{h}_{298})_{N_2} - (\bar{h}_{500} - \bar{h}_{298})_{N_2}}$$

$$= \frac{-74873 - (-393546 + 61681) - 2(-241845 + 48181.15) + 2 \times 40589.83}{40589.83 - 6097.131 + 3.76(38403.9 - 5920.109)}$$

$= 4.631$

air fuel ratio = mass of air/mass of fuel
$= x\,(M_{O2} + 3.76\,M_{N2})/M_{CH4}$
$= 4.631\,(32 + 3.76 \times 28)/16$
$= 39.73$

Stoichiometric air fuel ratio:

$CH_4 + x\,(O_2 + 3.76\,N_2) \rightarrow b_1\,CO_2 + b_2\,H_2O + b_4\,N_2$

After C, H, O and N balance,

$CH_4 + 2\,(O_2 + 3.76\,N_2) \rightarrow CO_2 + 2\,H_2O + 7.52\,N_2$

Now stoichiometric air fuel ratio = mass of air/mass of fuel
$= x\,(M_{O2} + 3.76\,M_{N2})/M_{CH4}$
$= 2\,(32 + 3.76 \times 28)/16$
$= 17.16$

Relative air fuel ratio (Equivalence ratio) = actual air fuel ratio/Stoichiometric air fuel ratio
$= 39.73/17.16 = 2.32$

i.e., the excess air is 132%.

2. Calculate the adiabatic flame temperature for combustion to burn 1 kg mol of methane at 25°C. Consider air temperature at 500 K.

Solution:

Adiabatic flame temperature results at the stoichiometric air fuel ratio, i.e., without deficient or excess amount of air.

$CH_4 + x\,(O_2 + 3.76\,N_2) \rightarrow b_1\,CO_2 + b_2\,H_2O + b_4\,N_2$

C: balance, $1 = b_1$

H_2: balance, $2 = b_2$

O_2: balance, $x = b_1 + 0.5 \times b_2 = 1 + 0.5 \times 2 = 2$

N_2: balance, $3.76x = b_4 = 3.76 \times 2 = 7.52$

Therefore,

$CH_4 + 2\,(O_2 + 3.76\,N_2) \rightarrow CO_2 + 2\,H_2O + 7.52\,N_2$

The adiabatic flame temperature can be iterated with an initial guess and by balancing the energy equation for the above reaction.

Let us assume Adiabatic flame temperature, T = 2500 K

$$H_R = \bar{h}^0_{f,\,CH_4} + 2[(\bar{h}_{500} - \bar{h}_{298})_{O_2} + 3.76(\bar{h}_{500} - \bar{h}_{298})_{N_2}]$$

$= -74873 + 2(6097.1 + 3.76 \times 5920.1) = -18160$ kJ/kg mol

$$H_P = (\bar{h}_f^0 + \bar{h}_{2500} - \bar{h}_{298})_{CO_2} + 2(\bar{h}_f^0 + \bar{h}_{2500} - \bar{h}_{298})_{H_2O} + 7.52(\bar{h}_{2500} - \bar{h}_{298})_{N_2}$$

$$= -393546 + 121940 + 2(241845 + 99112) + 7.52 \times 74305 = 1704.3 \text{ kJ/kg mol}$$

Since $H_P > H_R$, the initial guess of temperature to be decreased.

After the iterations, at T = 2455.5 K,

H_R = –18160 kJ/kg mol (fixed)

H_P = –18159 kJ/kg mol

3. Determine the exergy decay in the combustion process where methane and air are supplied at 25°C and atmospheric pressure. Find the stoichiometric air fuel ratio and adiabatic flame temperature. At this condition, evaluate the irreversibility of the process, i.e., exergy decay in combustion.

Solution:
For stoichiometric combustion,

$$CH_4 + 2\,(O_2 + 3.76\,N_2) \rightarrow CO_2 + 2\,H_2O + 7.52\,N_2$$

Now stoichiometric air fuel ratio = mass of air/mass of fuel
= $2\,(M_{O2} + 3.76\,M_{N2})/M_{CH4}$
= 2 (32 + 3.76 × 28)/16
= 17.16

Since the fuel and air are supplied at 25°C and 1 atmosphere, the enthalpies at this reference state is zero.

Assume adiabatic flame temperature, T = 2500 K,

$$\bar{h}^0_{f,\,CH_4} = (\bar{h}_f^0 + \bar{h}_T - \bar{h}_{298})_{CO_2} + 2(\bar{h}_f^0 + \bar{h}_T - \bar{h}_{298})_{H_2O} + 3.76 \times 2(\bar{h}_T - \bar{h}_{298})_{N_2}$$

LHS = –74873 kJ/kg mole

RHS = (–393546 + 121944.47) + 2(–242845 + 99111.77) + 3.76 × 2 (74304.82)

= 1704.3 kJ/kg mole

The RHS is greater than LHS, therefore adiabatic temperature has to be decreased to match the two values.

After iterations, T = 2327.3 K

The exergy balance of the combustion process is as follows

$$\bar{\varepsilon}^0_{fuel} + \bar{\varepsilon}^0_{air} = \bar{\varepsilon}^0_{gas} + I_{combustion}$$

The exergy of a chemical species is the sum of chemical exergy and physical exergy.

The exergy of fuel,

$$\bar{\varepsilon}^0_{f} = \bar{\varepsilon}^0_{f\,ch} + \bar{\varepsilon}^0_{f\,ph}$$

Since fuel is supplied at dead state (reference state),

$$\bar{\varepsilon}^0_{f\,ph} = 0$$

$$\bar{\varepsilon}^0_{f} = \bar{\varepsilon}^0_{f\,ch} = \bar{\varepsilon}^0_{CH4} = 836510 \ kJ/kgmole$$

Similarly,

The exergy of air,

$$\bar{\varepsilon}_a^0 = \bar{\varepsilon}_{a\,ch}^0 + \bar{\varepsilon}_{a\,ph}^0$$

Air also supplied at the reference state,

$$\bar{\varepsilon}_{a\,ph}^0 = 0$$

$$\bar{\varepsilon}_{a\,ch}^0 = 2\times 4.76(0.21\bar{\varepsilon}_{O2}^0 + 0.79\bar{\varepsilon}_{N2}^0 + RT_0(0.21\ln(0.21) + 0.79\ln(0.79)))$$

$= 2 \times 4.76 (0.21 \times 3970 + 0.79 \times 720 + 8.314 \times 298.15 (0.21 \ln(0.21) + 0.79 \ln(0.79)))$

= 1223.3 kJ/kg mol

$$\bar{\varepsilon}_a^0 = 1223.3 \text{ kJ/kg mole}$$

Total number of moles in products

$n_{total} = n_{CO2} + n_{H20} + n_{N2} = 1 + 2 + 3.76 \times 2 = 10.52$

$$x_{CO2} = \frac{n_{CO2}}{n_{total}} = \frac{1}{10.52} = 0.0951$$

$$x_{H2O} = \frac{n_{H2O}}{n_{total}} = \frac{2}{10.52} = 0.1901$$

$$x_{N2} = \frac{n_{N2}}{n_{total}} = \frac{7.52}{10.52} = 0.7148$$

$$\bar{\varepsilon}_{g\,ch}^0 = n_{total}(x_{CO2}\bar{\varepsilon}_{CO2}^0 + x_{H2O}\bar{\varepsilon}_{H2O}^0 + x_{N2}\bar{\varepsilon}_{N2}^0 + RT_0(x_{CO2}\ln(x_{CO2}) + x_{H2O}\ln(x_{H2O}) + x_{N2}\ln(x_{N2})))$$

$= 10.52 (0.0951 \times 20140 + 0.1901 \times 11710 + 0.7148 \times 720$

$+ 8.314 \times 298.15 (0.0951 \ln(0.0951) + 0.1901 \ln(0.0951) + 0.7148 \ln(0.7148)))$

= 28653 kJ/kg mole

$$h_g = (\bar{h}_T - \bar{h}_{298})_{CO_2} + 2(\bar{h}_T - \bar{h}_{298})_{H_2O} + 7.52(\bar{h}_T - \bar{h}_{298})_{N_2}$$

$= 111340 + 2 \times 89873 + 7.52 \times 67992$

= 802360 kJ/kg mole

$$s_g = (\bar{s}_{ü} - \bar{s}_{298})_{CO_0} + 2(\bar{s} \;\; - \bar{s}_{298})_{H\;O} + 7.52(\bar{s} \;\; - \bar{s}_{298})_N - n_{total}R\ln(P_g / P_o)$$

$= 104.71 + 2 \times 83.8721 + 7.52 \times 65.9643$

= 768.4975 kJ/kg mole K

The exergy of gas, $\bar{\varepsilon}_g^0 = \bar{\varepsilon}_{g\,ch}^0 + \bar{\varepsilon}_{g\,ph}^0$

$$\bar{\varepsilon}_{g\,ph}^0 = h_g - T_0 s_g$$

$= 802360 - 298.15 \times 768.4975$

= 573240 kJ/kg mole

The physical exergy of gas can also be evaluated from the exergy values if available in gas properties without enthalpy and entropy substitutions.

$$\bar{\varepsilon}_g^0 = 28653 + 573240$$

$$= 601890 \text{ kJ/kg mole}$$

Irreversibility of combustion,

$$I_{combustion} = \bar{\varepsilon}_f^0 + \bar{\varepsilon}_a^0 - \bar{\varepsilon}_g^0$$

$$= 836510 + 1223.3 - 601890$$

$$= 235840 \text{ kJ/kg mole}$$

Second law efficiency, or exergy efficiency of the combustion process,

$$\eta_2 = \frac{Exergy\ output}{Exergy\ input} = \frac{\bar{\varepsilon}_f^0 - I_{combustion}}{\bar{\varepsilon}_f^0} \times 100$$

$$= \frac{836510 - 235840}{836510} \times 100$$

$$= 71.8\%$$

4. Following is results of ultimate analysis of coal given on weight basis. C = 82%, H_2 = 4%, S = 1%, O_2 = 5%, N_2 = 1% M = 2% and A = 5%. Find the actual air required for the combustion of coal. The dry flue gas composition is CO_2 = 12%, CO = 1%, O_2 = 7%, and N_2 = 80%. Also determine the stoichiometric and actual air fuel ratio.

Solution:

Theoretical air requirement,

$$a_1C + a_2H_2 + a_3O_2 + a_4N_2 + a_5S + a_6H_2O + a_7(O_2+3.76N_2) = c_1CO_2 + c_2H_2O + c_3N_2 + c_4SO_2$$

From the problem,

$$a_1 = \frac{82}{12} = 6.834$$

$$a_2 = \frac{4}{2} = 2$$

$$a_3 = \frac{5}{32} = 0.15625$$

$$a_4 = \frac{1}{28} = 0.03571$$

$$a_5 = \frac{1}{32} = 0.03125$$

$$a_6 = \frac{2}{18} = 0.1112$$

C balance: $a_1 = c_1$

H_2 balance: $a_2 + a_6 = c_2$

O_2 balance: $a_3 + 0.5a_6 + a_7 = c_1 + ½\, c_2 + c_4$

N_2 balance: $a_4 + 3.76\, a_7 = c_3$

S balance: $a_5 = c_4$

$c_1 = a_1 = 6.834$

$c_2 = a_2 + a_6 = 2 + 0.1112 = 2.1112$

$c_4 = a_5 = 0.03125$

substituting,

$a_3 + 0.5a_6 + a_7 = c_1 + ½\, c_2 + c_4$

$a_7 = c_1 + ½\, c_2 + c_4 - a_3 - 0.5a_6$

$= 6.834 + 0.5 \times 2.1112 + 0.03125 - 0.15625 - 0.5 \times 0.1112$

$= 7.71$ kg mol (stiochiometric)

$c_3 = a_4 + 3.76\, a_7 = 0.03571 + 3.76 \times 7.71 = 29.025$

The combustion reaction with dry flue gas composition can be written as follows:

$$a_1C + a_2H_2 + a_3O_2 + a_4N_2 + a_5S + a_6H_2O + a_7(O_2 + 3.76N_2)$$
$$= p(b_1CO_2 + b_2CO + b_3O_2 + b_4N_2) + b_5H_2O + b_6SO_2$$

From the above equation

C balance: $a_1 = p\,(b_1 + b_2)$

H_2 balance: $a_2 + a_6 = b_5$

O_2 balance: $a_3 + 0.5a_6 + a_7 = p(b_1 + ½\, b_2 + b_3) + ½\, b_5 + b_6$

N_2 balance: $a_4 + 3.76\, a_7 = pb_4$

S balance: $a_5 = b_6$

Solving these 5 equations gives a_7 p, b_5 and b_6

From dry flue gas composition,

$b_1 = 0.12$, $b_2 = 0.01$, $b_3 = 0.07$ and $b_4 = 0.8$

$b_6 = a_5 = 0.03125$

$b_5 = a_2 + a_6 = 2 + 0.1112 = 2.1112$

$$p = \frac{a_1}{b_1 + b_2} = \frac{6.834}{0.12 + 0.01} = 52.56$$

Simplifying and substituting,

$a_7 = p(b_1 + ½\, b_2 + b_3) + ½\, b_5 + b_6 - a_3 - 0.5a_6$

$a_7 = 52.56(0.12 + ½ \times 0.01 + 0.07) + ½ \times 2.1112 + 0.03125 - 0.15625 - 0.5 \times 0.1112$

$= 11.12$

Actual air supply = 11.12 kg mol

$a_4 + 3.76\, a_7 = pb_4$

$$a_7 = \frac{pb_4 - a_4}{3.76} = \frac{52.56 \times 0.8 - 0.03571}{3.76} = 11.17 \text{ (rounding error)}$$

Therefore,

$6.834C + 2H_2 + 0.15625O_2 + 0.03571N_2 + 0.3125S + 0.1112H_2O + 11.12(O_2{+}3.76N_2)$

$= 52.56(0.12CO_2 + 0.01CO + 0.07O_2 + 0.8N_2) + 2.1112H_2O + 0.03125SO_2$

Excess oxygen = (11.12–7.71) × 32 = 109.1200 kg/100 kg coal or 1.09 kg/kg coal

$$AFR_{stoic} = \frac{mass\ of\ air\ (stiochiometric)}{mass\ of\ fuel} = \frac{7.71(32 + 3.76 \times 28)}{100} = 10.58$$

$$AFR_{act} = \frac{mass\ of\ air\ (actual)}{mass\ of\ fuel} = \frac{11.12(32 + 3.76 \times 28)}{100} = 15.27$$

5. Following is the ultimate analysis of a coal. C: 60%, H_2: 4%, S: 3.2%, O_2: 4.8%, N_2: 2%, M: 5% and A: 21%. The exhaust gas has the following volumetric analysis, CO_2: 9%, SO_2: 3%, CO: 2%, O_2: 4% and N_2: 82% Determine the stoichiometric and actual air supply per kg of fuel.

Solution:

Theoretical air requirement,

$a_1C + a_2H_2 + a_3O_2 + a_4N_2 + a_5S + a_6H_2O + a_7(O_2{+}3.76N_2) = c_1CO_2 + c_2H_2O + c_3N_2 + c_4SO_2$

From the problem,

$$a_1 = \frac{82}{12} = 6.834$$

$$a_2 = \frac{4}{2} = 2$$

$$a_3 = \frac{5}{32} = 0.15625$$

$$a_4 = \frac{1}{28} = 0.03571$$

$$a_5 = \frac{1}{32} = 0.03125$$

$$a_6 = \frac{2}{18} = 0.1112$$

C balance: $a_1 = c_1$

H_2 balance: $a_2 + a_6 = c_2$

O_2 balance: $a_3 + 0.5a_6 + a_7 = c_1 + ½\, c_2 + c_4$

N_2 balance: $a_4 + 3.76\, a_7 = c_3$

S balance: $a_5 = c_4$

$c_1 = a_1 = 6.834$

$c_2 = a_2 + a_6 = 2 + 0.1112 = 2.1112$

$c_4 = a_5 = 0.03125$

Substituting,

$a_3 + 0.5a_6 + a_7 = c_1 + ½\, c_2 + c_4$

$a_7 = c_1 + ½\, c_2 + c_4 - a_3 - 0.5a_6$

$= 6.834 + 0.5 \times 2.1112 + 0.03125 - 0.15625 - 0.5 \times 0.1112$

$= 7.71$ kg mol (stiochiometric)

$c_3 = a_4 + 3.76\, a_7 = 0.03571 + 3.76 \times 7.71 = 29.025$

The combustion reaction with dry flue gas composition can be written as follows

$a_1C + a_2H_2 + a_3O_2 + a_4N_2 + a_5S + a_6H_2O + a_7(O_2{+}3.76N_2)$

$= p(b_1CO_2 + b_2CO + b_3O_2 + b_4N_2 + b_5SO_2) + b_6H_2O$

From the above equation:

C balance: $a_1 = p\,(b_1 + b_2)$

H_2 balance: $a_2 + a_6 = b_5$

O_2 balance: $a_3 + 0.5a_6 + a_7 = p(b_1 + ½\, b_2 + b_3 + b_5) + ½\, b_6$

N_2 balance: $a_4 + 3.76\, a_7 = pb_4$

S balance: $a_5 = pb_5$

Solving these 5 equations gives a_7 p, and b_5

From dry flue gas composition,

$b_1 = 0.09$, $b_2 = 0.02$, $b_3 = 0.04$, $b_4 = 0.82$ and $b_5 = 0.03$

$$p = \frac{a_1}{b_1 + b_2} = \frac{6.834}{0.12 + 0.01} = 52.56$$

$b_5 = a_2 + a_6 = 2 + 0.1112 = 2.1112$

Simplifying and substituting,

$a_7 = p(b_1 + ½\, b_2 + b_3 + b_5) + ½\, b_6 - a_3 - 0.5a_6$

$a_7 = 52.56(0.12 + ½ \times 0.01 + 0.07 + 0.03125) + ½ \times 2.1112 - 0.15625 - 0.5 \times 0.1112$

$= 11.12$

Actual air supply = 11.12 kg mol

$a_4 + 3.76\, a_7 = pb_4$

$$a_7 = \frac{pb_4 - a_4}{3.76} = \frac{52.56 \times 0.8 - 0.03571}{3.76} = 11.17 \text{ (rounding error)}$$

Therefore,

$6.834C + 2H_2 + 0.15625O_2 + 0.03571N_2 + 0.3125S + 0.1112H_2O + 11.12(O_2{+}3.76N_2)$

$= 52.56(0.12CO_2 + 0.01CO + 0.07O_2 + 0.8N_2) + 2.1112H_2O + 0.03125SO_2$

Excess oxygen = (11.12–7.71) × 32 = 109.1200 kg/100 kg coal or 1.09 kg/kg coal

$$AFR_{stoic} = \frac{mass\ of\ air\ (stiochiometric)}{mass\ of\ fuel} = \frac{7.71(32+3.76\times 28)}{100} = 10.58$$

$$AFR_{act} = \frac{mass\ of\ air\ (actual)}{mass\ of\ fuel} = \frac{11.12(32+3.76\times 28)}{100} = 15.27$$

6. Following is result of the ultimate analysis of rise husk given on weight basis. C = 36.42%, H = 4.91%, O = 35.88%, N = 0.59% S = 0% and ash = 22.2%. Find the stoichiometric air fuel ratio and adiabatic flame temperature for the rise husk. Solve the problem with a single carbon atom in the fuel.

Solution:

On molar basis,

$$\bar{C} = \frac{C}{12} = \frac{36.42}{12} = 3.035$$

$$\bar{H} = \frac{H}{1} = \frac{4.91}{1} = 4.91$$

$$\bar{O} = \frac{O}{16} = \frac{35.88}{16} = 2.2425$$

$$\bar{N} = \frac{N}{14} = \frac{0.59}{14} = 0.0421$$

$$a_1 = \frac{\bar{C}}{\bar{C}} = 1$$

$$a_2 = \frac{\bar{H}}{\bar{C}} = \frac{4.91}{3.035} = 1.6178$$

$$a_3 = \frac{\bar{O}}{\bar{C}} = \frac{2.2425}{3.035} = 0.7389$$

$$a_4 = \frac{\bar{N}}{\bar{C}} = \frac{0.0421}{3.035} = 0.0139$$

Now the fuel is $C_{a1}\ H_{a2}\ O_{a3}\ N_{a4} = C\ H_{1.6178}\ O_{0.7389}\ N_{0.0139}$

The combustion reaction is

$a_1C + a_2H + a_3O + a_4N + a_5(O_2+3.76N_2) = b_1CO_2 + b_2H_2O + b_3O_2 + b_4N_2$

Higher heating value, HHV

$= m_{dry\ fuel}(0.3491C+1.1783H+0.10055S–0.1034O–0.151N–0.0211Ash)1000$, kJ

where C, H, S, O, N and ash are in % on mass basis in the dry solid fuel.

Mass of dry fuel per kg mole fuel,

$m_{dry\ fuel} = a_1 12 + a_2 1 + a_3 16 + a_4 14$

$= 1 \times 12 + 1.6178 \times 1 + 0.7389 \times 16 + 0.0139 \times 14$
$= 25.6343$ kg

HHV = 25.6343(0.3491× 36.42 + 1.1783 × 4.91 + 0.10055 × 0 – 0.1034 × 35.88

– 0.151 × 0.59 – 0.0211Ash)1000, kJ
= 3,64,830 kJ (fuel is one kg mole of $C_{a1}\ H_{a2}\ O_{a3}\ N_{a4}$)

Enthalpy of formation of fuel,

$h_{f,\ dry\ fuel} = a_1 h_{fCO2} + ½\ a_2 h_{fH2OL} + HHV$

The enthalpy of formation of water condensate is considered fuel for a higher heating value.

$h_{f,dry\ fuel} = 1 \times (-393546) + ½\ 1.6178\ (-285830) + 364830$
$= -259920$ kJ/kg mol fuel

$a_1C + a_2H + a_3O + a_4N + a_5(O_2+3.76N_2) = b_1CO_2 + b_2H_2O + b_3O_2 + b_4N_2$

At stoichiometric air fuel supply, traces of oxygen in products are zero. Therefore, $b_3 = 0$

From the above equation,

C balance: $b_1 = a_1 = 1$

H_2 balance: $b_2 = ½\ a_2 = ½ \times 1.6178 = 0.8089$

O_2 balance: $½\ a_3 + a_5 = b_1 + ½\ b_2$

$a_5 = b_1 + ½\ b_2 - ½\ a_3 = 1 + ½ \times 0.8089 - ½ \times 0.7389 = 1.035$

N_2 balance: $b_4 = ½\ a_4 + 3.76\ a_5$
$= ½ \times 0.0139 + 3.76 \times 1.035 = 3.8985$

$C\ H_{1.6178}\ O_{0.7389}\ N_{0.0139} + 1.035\ (O_2+3.76N_2) = CO_2 + 0.8089\ H_2O + 3.8985\ N_2$

Stoichiometric air fuel ratio = mass of air/mass of fuel
$= a_5\ (M_{O2} + 3.76\ M_{N2})/M_{dry\ fuel}$
$= 1.035\ (32 + 3.76 \times 28)/25.6345$
$= 5.5427$

Adiabatic flame temperature,

$H_R = H_P$

Assume adiabatic flame temperature = 1900 K to start the iteration.

$h_{f,\ fuel} = b_1(h_{f,\ CO2} + h_{CO2,\ T}) + b_2(h_{f,\ H2O} + h_{H2O,\ T}) + b_4\ h_{N2,\ T}$

LHS = –259920

RHS = –393546 + 85392.15 + 0.8089(–241845 + 67725.01) + 3.8985 × 52540.68

R.H.S. = –244170 and it is to be lowered to –259920.

Therefore, the assumed adiabatic temperature should be decreased for the second iteration.
Finally, the adiabatic flame temperature is 1834 K = 1561°C

7. The biomass (solid waste biomass) is defined by a general formula as $C_{a1}\ H_{a2}\ O_{a3}\ N_{a4}$. For single atom of carbon in fuel, the coefficient a_1 becomes one. The coefficients of a_2, a_3 and a_4 are the H/C, O/C and N/C mole ratios, respectively. The reactions are solved at thermodynamic equilibrium. The gasification products contain CH_4, CO, CO_2, H_2, H_2O and N_2. For solid fuel $a_1 = 1$, $a_2 = 1.5933$, $a_3 = 0.5758$, $a_4 = 0.0444$, air supply, $a_5 = 0.5552$ and for air gasification, $a_6 = 0$. Thermodynamic equilibrium constants are $k_{p1} = 0.1506$ and $k_{p2} = 1.9213$. Find the gas composition using Newton-Rapson iteration method or a suitable numerical solution.

Solution:

The following is the chemical reaction in a biomass gasifier.

$$C_{a_1}H_{a_2}O_{a_3}N_{a_4} + a_5(O_2 + 3.76N_2) + a_6H_2O$$
$$\Rightarrow b_1CH_4 + b_2CO + b_3CO_2 + b_4H_2 + b_5H_2O + b_6N_2$$

For typical solid waste biomass fuel, the following equations are observed at the gasifier temperature of 660°C.

C balance: $a_1 = b_1 + b_2 + b_3$

H balance: $a_2 + 2a_6 = 4b_1 + 2b_4 + 2b_5$

O balance: $a_3 + 2a_5 + a_6 = b_2 + 2b_3 + b_5$

N balance: $a_4 + 2\times 3.76a_5 = 2b_6$

Methane reforming $\quad CH_4 + H_2O = CO + 3H_2$

Water shift reaction $\quad CO + H_2O = CO_2 + H_2$

$$k_{p1} = \frac{b_2 b_4^3 P^2}{b_1 b_5 n_t^2}.$$

$$k_{p2} = \frac{b_3 b_4}{b_2 b_5}.$$

where $P = \dfrac{P_{gasifier}}{P_{atm}}$ and $n_t = b_1 + b_2 + b_3 + b_4 + b_5 + b_6$.

The equilibrium constants, k_{p1} and k_{p2} are for methane reforming and a water shift reaction respectively.

$b_3 = a_1 - b_1 - b_2$

$b_5 = a_3 + 2a_5 + a_6 - b_2 - 2(a_1 - b_1 - b_2)$

$\quad a_3 + 2a_5 + a_6 - b_2 - 2a_1 + 2b_1 + 2b_2$

$b_4 = 2a_1 + 0.5a_2 - a_3 - 2a_5 - 4b_1 - b_2$

$b_5 = 0.5a_2 + a_6 - 2b_1 - b_4$

$\quad = 0.5a_2 + a_6 - 2b_1 - (2a_1 + 0.5a_2 - a_3 - 2a_5 - 4b_1 - b_2)$

$\quad = 0.5a_2 + a_6 - 2b_1 - 2a_1 - 0.5a_2 + a_3 + 2a_5 + 4b_1 + b_2$

$b_5 = a_3 + 2a_5 + a_6 - 2a_1 + 2b_1 + b_2$

$b_6 = 0.5a_4 + 3.76\ a_5$

$n_t = b_1 + b_2 + b_3 + b_4 + b_5 + b_6$

$= b_1 + b_2 + a_1 - b_1 - b_2 + 2a_1 + 0.5a_2 - a_3 - 2a_5 - 4b_1 - b_2 + a_3 + 2a_5 + a_6 - 2a_1 + 2b_1 + b_2 + 0.5a_4 + 3.76\ a_5$

$= a_1 + 0.5a_2 + 0.5a_4 + 3.76a_5 + a_6 - 2\ b_1$

$$b_2 P^2 (2a_1 + 0.5a_2 - a_3 - 2a_5 - 4b_1 - b_2)^3 - K_{p1} b_1 (2a_3 + 2a_5 + a_6 - 2\ a_1 + 2b_1 + b_2)$$
$$(a_1 + 0.5a_2 + 0.5a_4 + 3.76a_5 + a_6 - 2\ b_1)^2 = 0$$

$$(a_1 - b_1 - b_2)(2a_1 + 0.5a_2 - a_3 - 2a_5 - 4b_1 - b_2)$$
$$-K_{p,2} b_2 (a_3 + 2a_5 + a_6 - 2a_1 + 2b_1 + b_2) = 0$$

Step 1,

$$f_1 = b_2 P^2 (2a_1 + 0.5a_2 - a_3 - 2a_5 - 4b_1 - b_2)^3 - K_{p1} b_1 (2a_3 + 2a_5 + a_6 - 2\ a_1 + 2b_1 + b_2)$$
$$(a_1 + 0.5a_2 + 0.5a_4 + 3.76a_5 + a_6 - 2\ b_1)^2 = 0$$

$$f_2 = (a_1 - b_1 - b_2)(2a_1 + 0.5a_2 - a_3 - 2a_5 - 4b_1 - b_2)$$
$$-K_{p,2} b_2 (a_3 + 2a_5 + a_6 - 2a_1 + 2b_1 + b_2) = 0$$

Choose trial values of the variables, which are here selected as $b_1 = 0.01$ and $b_2 = 0.2$.

Step 3. Calculate the magnitude of the fs at the temporary values of the variables, $f_1 = 0.2439$ and $f_2 = 0.7237$.

Step 4. The partial derivatives are shown in Table 4.3.

Step 5. Substituting the temporary values of the variables into the equations for the partial derivatives forms a set of linear or nonlinear simultaneous equations to be solved for the corrections to b:

$$\begin{bmatrix} -4.5183 & 0.5518 \\ -4.7990 & -1.8645 \end{bmatrix} \times \begin{bmatrix} \Delta b_1 \\ \Delta b_2 \end{bmatrix} = \begin{bmatrix} 0.2439 \\ 0.7237 \end{bmatrix}$$

where $\Delta b_i = b_i - b_{i+1}$

Step 6. Solution of the simultaneous equations

$b_1 = 0.01 - (-0.0771) = 0.0871$, and $b_2 = 0.2 - (-0.1896) = 0.3896$.

These values of the variables are returned to step 3 for the next iteration.

The values of the fs and the variables resulting from continued iterations are shown in Table 4.4.

$b_3 = a_1 - b_1 - b_2$

$= 1 - 0.514 - 0.4589 = 0.4898$

$b_4 = 2a_1 + 0.5a_2 - a_3 - 2a_5 - 4b_1 - b_2$

$= 2 \times 1 + 0.5 \times 1.5933 - 0.5758 - 2 \times 0.5552 - 4 \times 0.0514 - 0.4589 = 0.4461$

Table 4.3. Partial derivatives of functions.

	$\partial/\partial b_1$	$\partial/\partial b_2$
$\partial f_1/\partial$	–4.5183	0.5518
$\partial f_2/\partial$	–4.7990	–1.8645

Table 4.4. Function values and coefficients.

After iteration	f_1	f_2	b_1	b_2
1	0.2439	0.7237	0.0871	0.3896
2	–0.1439	0.0076	0.0478	0.4665
3	0.0154	–0.0028	0.0513	0.4590
4	1.4113×10^{-4}	-3.4286×10^{-5}	0.0514	0.4589
5	1.2382×10^{-8}	-3.7356×10^{-9}	0.0514	0.4589

$$b_5 = a_3 + 2a_5 + a_6 - 2a_1 + 2b_1 + b_2$$

$$= 0.5758 + 2 \times 0.5552 + 0 - 2 \times 1 + 2 \times 0.0514 + 0.4589 = 0.2478$$

$$b_6 = 0.5a_4 + 3.76\ a_5$$

$$= 0.5 \times 0.0444 + 3.76 \times 0.5552 = 2.1098$$

$$CH_{1.5933}O_{0.5758}N_{0.0444} + 0.5552(O_2 + 3.76N_2)$$

$$\Rightarrow 0.514\ CH_4 + 0.4589\ CO + 0.4898\ CO_2 + 0.4461\ H_2 + 0.2478\ H_2O + 2.1098\ N_2$$

8. One kmol of CO reacts with a theoretical amount of dry air to form an equilibrium mixture of CO_2, CO, O_2, and N_2 at 2500 K, 1 atm. Determine the amount of CO in the equilibrium mixture, in kmol.

Solution

The reaction of CO with the theoretical amount of dry air to form CO_2, CO, O_2, and N_2 is,

$$CO + \frac{1}{2}O_2 + 1.88N_2 \rightarrow n_{CO}CO + n_{O_2}O_2 + n_{CO_2}CO_2 + n_{N_2}N_2$$

Assuming the amount of CO in the equilibrium mixture is z kg mol

C balance: $1 = n_{CO} + n_{CO2} = z + n_{CO2}$, $n_{CO2} = (1–z)$

O balance: $1 + 1 = n_{CO} + 2\ n_{O2} + 2\ n_{CO2} = z + 2\ n_{O2} + 2(1–z) = 2\ n_{O2} + 2 - z$, $n_{O2} = z/2$

N balance: $1.88 = n_{N2}$

Therefore, from C, O and N balanced equations,

$$n_{CO} = z,\ n_{CO_2} = (1-z),\ n_{O_2} = \frac{z}{2}\ and\ n_{N_2} = 1.88$$

$$CO + \frac{1}{2}O_2 + 1.88N_2 \rightarrow zCO + \frac{z}{2}O_2 + (1-z)CO_2 + 1.88N_2$$

where z is the amount of CO, in kmol, present in the equilibrium mixture. The total number of moles n is

$$n = z + \frac{z}{2} + (1-z) + 1.88 = \frac{5.76 + z}{2}$$

At equilibrium $CO_2 \leftrightarrow CO + 1/2\ O_2$; and Equation takes the form

$$K = \left(\frac{P_{CO}P_{O_2}^{1/2}}{P_{CO_2}}\right)$$

From Dalton's law, $\frac{P_{CO}}{P} = \frac{n_{CO}}{n}$

Assuming $P_{ref} = 1$

$$\therefore K = \frac{\frac{n_{CO}}{n} P \left(\frac{n_{O_2}}{n} P \right)^{1/2}}{\frac{n_{CO_2}}{n} P}$$

$$\therefore K = \frac{n_{CO} (n_{O_2})^{1/2}}{n_{CO_2}} \left(\frac{P}{n} \right)^{1/2}$$

$$K = \frac{z(z/2)^{1/2}}{1-z} \left(\frac{P}{(5.76+z)/2} \right)^{1/2}$$

At 2500 K, gives K = 0.0363. Solving iteratively, z = 0.175

9. Biomass is defined by a general formula as $C_{a1} H_{a2} O_{a3} N_{a4}$. For a single atom of carbon in fuel, the coefficient a_1 becomes one. The coefficients of a_2, a_3 and a_4 are the H/C, O/C and N/C mole ratios, respectively. The reactions are solved at thermodynamic equilibrium. The gasification products contain CH_4, CO, CO_2, H_2, H_2O and N_2. The following is the chemical reaction in biomass gasifier.

$$C_{a_1} H_{a_2} O_{a_3} N_{a_4} + a_5(O_2 + 3.76N_2) + a_6H_2O$$
$$\Rightarrow b_1CH_4 + b_2CO + b_3CO_2 + b_4H_2 + b_5H_2O + b_6N_2$$

For a typical solid waste biomass fuel, the following equations result at the gasificier temperature of 660°C.

C balance: $a_1 = b_1 + b_2 + b_3$

H balance: $a_2 + 2a_6 = 4b_1 + 2b_4 + 2b_5$

O balance: $a_3 + 2a_5 + a_6 = b_2 + 2b_3 + b_5$

N balance: $a_4 + 2\times3.76a_5 = 2b_6$

$$k_{p1} = \frac{b_2 b_4^3 P^2}{b_1 b_5 n_t^2}.$$

$$k_{p2} = \frac{b_3 b_4}{b_2 b_5}.$$

where $P = \frac{P_{gasifier}}{P_{atm}}$ and

$n_t = b_1 + b_2 + b_3 + b_4 + b_5 + b_6$.

The equilibrium constants, k_{p1} and k_{p2} are for methane reforming and water shift reaction respectively. For solid fuel $a_1 = 1$, $a_2 = 1.5933$, $a_3 = 0.5758$, $a_4 = 0.0444$, air supply, $a_5 = 0.5552$ and for air gasification, $a_6 = 0$. Thermodynamic equilibrium constants are $k_{p1} = 0.1506$ and $k_{p2} = 1.9213$. Find the gas composition using Newton-Rapson iteration method.

Solution:

There are six unknowns and six equations (five linear and one nonlinear).

Step 1

$f_1 = a_1 - b_1 - b_2 - b_3 = 0$

$f_2 = a_2 + 2a_6 - 4b_1 - 2b_4 - 2b_5 = 0$

$f_3 = a_3 + 2a_5 + a_6 - b_2 - 2b_3 - b_5 = 0$

$f_4 = a_4 + 2 \times 3.76a_5 - 2b_6 = 0$

$$f_5 = k_{p1} - \frac{b_2 b_4^3 P^2}{b_1 b_5 n_t^2}. = 0$$

$$f_6 = k_{p2} - \frac{b_3 b_4}{b_2 b_5}. = 0$$

Step 2. Choose trial values of the variables, which are here selected as b1 = 0.01, $b_2 = 0.2$, $b_3 = 0.2$, $b_4 = 0.5$, $b_5 = 0.5$ and $b_6 = 2$.

Step 3. Calculate the magnitudes of the fs at the temporary values of the variables, $f_1 = 0.59$, $f_2 = -0.4467$, $f_3 = 0.5862$, $f_4 = 3.2195$, $f_5 = -1.22$ and $f_6 = 0.9213$.

Step 4. The partial derivatives are shown in Table 4.5.

Step 5. Substituting the temporary values of the variables into the equations for the partial derivatives forms a set of linear and nonlinear simultaneous equations to be solved for the corrections to b:

$$\begin{bmatrix} -1 & -1 & -1 & 0 & 0 & 0 \\ -4 & 0 & 0 & -2 & -2 & 0 \\ 0 & -1 & -2 & 0 & -1 & 0 \\ 0 & 0 & 0 & 0 & 0 & -2 \\ 137.0576 & -6.8529 & 0 & -8.2235 & 2.7412 & 0 \\ 0 & 5 & -5 & -2 & 2 & 0 \end{bmatrix} \times \begin{bmatrix} \Delta b_1 \\ \Delta b_2 \\ \Delta b_3 \\ \Delta b_4 \\ \Delta b_5 \\ \Delta b_6 \end{bmatrix} = \begin{bmatrix} 0.59 \\ -0.4467 \\ 0.5862 \\ 3.2195 \\ -1.22 \\ 0.9213 \end{bmatrix}$$

where, $\Delta b_i = b_i - b_{i+1}$

Step 6. Solution of the simultaneous equations

$b_1 = 0.01 - (-0.0278) = 0.0378$, $b_2 = 0.4489$, $b_3 = 0.5133$, $b_4 = 0.5105$, $b_5 = 0.2107$ and $b_6 = 2.1098$

Table 4.5. The partial derivatives of functions.

	$\partial/\partial b_1$	$\partial/\partial b_2$	$\partial/\partial b_3$	$\partial/\partial b_4$	$\partial/\partial b_5$	$\partial/\partial b_6$
$\partial f_1/\partial$	−1	−1	−1	0	0	0
$\partial f_2/\partial$	−4	0	0	−2	−2	0
$\partial f_3/\partial$	0	−1	−2	0	−1	0
$\partial f_4/\partial$	0	0	0	0	0	−2
$\partial f_5/\partial$	$(10000b_2b_4^3)/(36481b_1^2b_5)$	$-(10000b_4^3)/(36481b_1b_5)$	0	$-(30000b_2b_4^2)/(36481b_1b_5)$	$(10000b_2b_4^3)/(36481b_1b_5^2)$	0
$\partial f_6/\partial$	0	$(b_3b_4)/(b_2^2b_5)$	$-b_4/(b_2b_5)$	$-b_3/(b_2b_5)$	$(b_3b_4)/(b_2b_5^2)$	0

These values of the variables are returned to step 3 for the next iteration.

The values of the fs and the variables resulting from continued iterations are shown in Table 4.6.

Table 4.6. Iterative solutions.

After iteration	f_1	f_2	f_3	f_4	f_5	f_6	b_1	b_2	b_3	b_4	b_5	b_6
1	-1.11×10^{-16}	0	-2.22×10^{-16}	0	−0.3609	−0.8490	0.0378	0.4489	0.5133	0.5105	0.2107	2.1098
2	-1.11×10^{-16}	-2.22×10^{-16}	-5.55×10^{-17}	0	−0.2791	−0.1327	0.0364	0.4795	0.4841	0.4841	0.4853	2.1098
3	0	-5.55×10^{-16}	1.94×10^{-16}	0	−0.2167	−0.0043	0.0388	0.4814	0.4798	0.4738	0.2452	2.1098

Chapter 5
Heat Exchangers

5.1 Introduction

In the study of energy systems, the design of heat exchangers is significant as these are major components. The simulation of heat exchangers involves both design and off-design conditions. The two methods of modeling heat exchangers are the logarithmic mean temperature difference (LMTD) method and the effectiveness-number of transfer units (ε-NTU) method. The second method is more useful for the simulation of exit conditions, especially for off-design conditions. There are many ways of developing heat exchanger models and their solutions. One approach to engineering system solutions is deriving the variable relations and finally simplifying them into one single equation to find the one variable with the known information. On the other hand, the same problem can also be solved without the derivation or simplification of relations. In this second approach, either successive substitution, the Newton-Raphson method, or a similar method can be used.

5.2 Modeling and system simulation without physical information

System modeling and simulation is taken up in two phases. The first phase, before the erection and implementation of the total system, which has many subsystems, is called pre-design modeling and simulation. The focus here is on operational conditions; the size of the system is not considered. Modeling and simulation in this phase is based on assumptions. The post-design modeling and simulation phase needs information, such as size, of the physical system. The post-design modeling and simulation enables off-design performance predictions. It allows for visualizing problems that can crop up in the future and, thereby plan preventive remedies.

Figure 5.1 shows the modeling details of a heat recovery steam generator (HRSG) used in a typical cogeneration or combined cycle power plant. In the pre-design model, the properties at each defined state can be worked out with a mass and heat balance formulation called modeling. The execution of all these equations to find the result is called the simulation of HRSG. There are three sections in the single-pressure HRSG: the economizer (ECO), the evaporator (EVAP), and the superheater (SH). The assumed equality or inequality constraints can be taken from the design handbook, and they are pinch point (PP), approach point (AP), and terminal temperature difference (TTD). The pinch point is the minimum temperature difference between the hot gas and the steam and water mixture in the evaporator. If the saturated water enters the evaporator, there is a chance of a sudden generation of vapor bubbles at the entry, which causes a pitting action. To avoid this difficulty, the water supply temperature at the inlet of the evaporator section can be fixed below the saturation temperature, which is equal to the approach point. The following are the modeling equations developed to find properties such as temperature T, mass m, and enthalpy h.

In addition to the constraints (PP, AP and TTD), the supplied data is the hot gas inlet temperature (T_1), hot gas supply (m_g), steam pressure (P) and water inlet temperature (T_5).

DOI: 10.1201/9781032642505-5

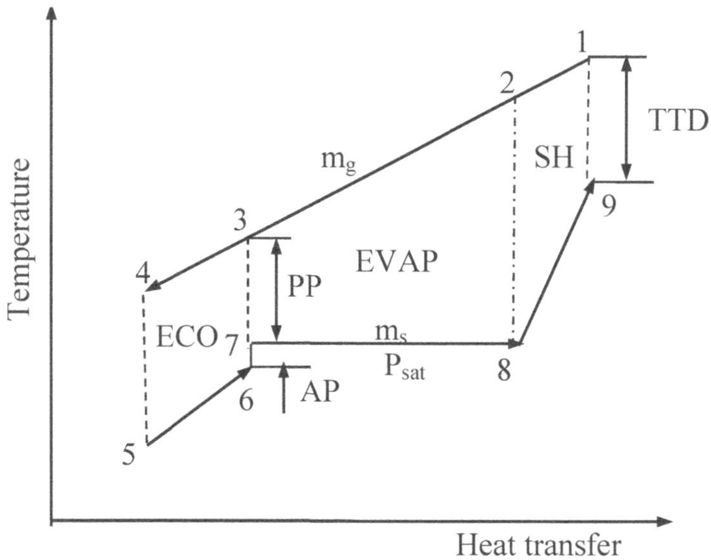

Fig. 5.1. Single pressure heat recovery steam generator (HRSG).

The saturation temperature (T7 and T8) can be taken at the given saturation steam pressure.

$$T_9 = T_1 - TTD \tag{5.1}$$

$$T_3 = T_7 + PP \tag{5.2}$$

$$T_6 = T_7 - AP \tag{5.3}$$

The heat balance in the evaporator and superheater sections (EVAP+SH) results in a steam generation rate as follows.

If the specific heat of gas is assumed constant,

$$m_g c_{pg}(T_1 - T_3) = m_s(h_9 - h_6) \tag{5.4}$$

Therefore,

$$m_s = \frac{m_g c_{pg}(T_1 - T_3)}{(h_9 - h_6)}$$

In the SH section,

$$m_g c_{pg}(T_1 - T_2) = m_s(h_9 - h_8) \tag{5.5}$$

Therefore $T_2 = T_1 - \dfrac{m_s(h_9 - h_8)}{m_g c_{pg}}$

In the ECO section,

$$m_g c_{pg}(T_3 - T_4) = m_s(h_6 - h_5) \tag{5.6}$$

Therefore $T_4 = T_3 - \dfrac{m_s(h_6 - h_5)}{m_g c_{pg}}$

The above formula provides all the gas and steam temperatures in accordance with the heat exchanger constraints. The enthalpy of steam can be obtained from its properties through either a software application or steam tables at a given pressure and temperature. Thus, the complete modeling and simulation of HRSG is done without information (such as size) of the physical system. But it supplies valuable data, for example, the flow rate of hot fluid, the flow rate of cold fluid, and all the inlet and outlet temperatures. Without this preliminary step of pre-design modeling and simulation, it is not possible to proceed further with post-design modeling and simulation.

5.3 Overall heat transfer coefficient

Consider the total thermal resistance to heat flow across a tube between the inside and the outside flow as R (Hewitt et al., 1995).

R = thermal resistance of inside flow + thermal resistance of tube material + thermal resistance outside flow

$$R = \frac{1}{h_i A_i} + \frac{\ln(r_o / r_i)}{2\pi k L} + \frac{1}{h_o A_o} = \frac{1}{h_i 2\pi r_i L} + \frac{\ln(r_o / r_i)}{2\pi k L} + \frac{1}{h_o 2\pi r_o L} \tag{5.7}$$

h_i, h_o = heat transfer coefficient of inside and outside flow
k = thermal conductivity of tube material
R = total thermal resistance

Thermal resistance R can be expressed as an overall heat transfer coefficient based on either the inside or the outside surface of the tube.

$$Q = U_i A_i \Delta T = U_o A_o \Delta T = \Delta T/R \tag{5.8}$$

$$U_o = \frac{1}{A_o R} = \frac{1}{A_0\left(\frac{1}{h_i 2\pi r_i L} + \frac{\ln(r_o / r_i)}{2\pi k L} + \frac{1}{h_o 2\pi r_o L}\right)}$$

$$U_o = \frac{1}{\frac{r_o}{r_i}\frac{1}{h_i} + \frac{r_o}{k}\ln\left(\frac{r_o}{r_i}\right) + \frac{1}{h_o}}$$

$$U_i = \frac{1}{A_i R} = \frac{1}{A_i\left(\frac{1}{h_i 2\pi r_i L} + \frac{\ln(r_o / r_i)}{2\pi k L} + \frac{1}{h_o 2\pi r_o L}\right)}$$

$$U_i = \frac{1}{\frac{1}{h_i} + \frac{r_i}{k}\ln\left(\frac{r_o}{r_i}\right) + \frac{r_i}{r_o}\frac{1}{h_o}} \tag{5.9}$$

If the wall thickness is small $r_i \approx r_o$

$$U_i = U_o = \frac{1}{\frac{1}{h_i} + \frac{1}{h_o}} = \frac{h_i h_o}{h_i + h_o} \tag{5.10}$$

5.4 Fouling factor

The equations developed for the overall heat transfer coefficient are applicable for clean or new heat exchangers without any fouling, mud, or corrosion. With the usage of heat exchangers, the inner surface will corrode and scale with ash, dirt, etc. The formation of rust on heat transfer surfaces results in poor performance of a heat exchanger. It is difficult to predict and evaluate the thickness and thermal conductivity of scale, and hence a fouling factor is introduced to predict the performance of a heat exchanger with scale formation.

Fouling factor (resistance), $F = \frac{1}{U_{dirty}} - \frac{1}{U_{clean}}$ (5.11)

By introducing fouling factor F, m²K/W, to account for fouling on both sides of the inner tube,

$$R = \frac{1}{h_i A_i} + \frac{F_i}{A_i} + \frac{\ln(r_o / r_i)}{2\pi kL} + \frac{F_o}{A_o} + \frac{1}{h_o A_o} \quad (5.12)$$

$$U_o = \frac{1}{A_o R} = \frac{1}{\frac{r_o}{r_i}\frac{1}{h_i} + \frac{r_o}{r_i}F_i + \frac{r_o}{k}\ln\left(\frac{r_o}{r_i}\right) + F_o + \frac{1}{h_o}}$$

$$U_i = \frac{1}{A_i R} = \frac{1}{\frac{1}{h_i} + F_i + \frac{r_i}{k}\ln\left(\frac{r_o}{r_i}\right) + \frac{r_i}{r_o}F_o + \frac{r_i}{r_o}\frac{1}{h_o}} \quad (5.13)$$

The value of the fouling factor can be determined by testing the heat exchangers with and without fouling.

5.5 Logarithmic mean temperature difference (LMTD)

To simplify the heat transfer calculations in a heat exchanger, using the equation Q = UA ΔT, an average value of the temperature difference between the fluids has to be derived. After the derivation, for the mean temperature difference in a heat exchanger, the resultant temperature difference is called the logarithmic mean temperature difference. Therefore, the temperature difference is called the logarithmic mean temperature difference.

The assumptions involved in the performance of a heat exchanger are

i) U, m, c_p are assumed constants

ii) No heat loss to surroundings; perfect insulation.

iii) Steady flow

According to the arrangement of the directions of the two fluids (hot fluid and cold fluid), the heat exchangers can be divided into parallel flow heat exchangers and counter flow heat exchangers. The following sections detail the modeling of these two heat exchangers.

5.5.1 Parallel flow heat exchanger

Figure 5.2 shows the temperature profile along the length of a heat exchanger for a counter flow heat exchanger where the hot fluid and cold fluid flow in the same direction. Let dx be a small element for the development of a mathematical model as shown in Fig. 5.2. The temperature difference in this section between the hot fluid temperature of Th and cold fluid temperature of Tc is ΔT. Assume that the hot fluid temperature and cold fluid temperature in the length of dx remain the same.

Heat flow through an element, $dQ = UdA\,(T_h - T_c)$ (5.14)

Heat balance, $dQ = -m_h\, c_h\, dT_h = m_c\, c_c\, dT_c$

(–ve indicates that the temperature decreases)

$dQ = -C_h\, dT_h = C_c\, dT_c$

Where $C_h = m_h\, c_h$ and $C_c = m_c\, c_c$ → Heat capacity.

$$dT_h = -\frac{dQ}{C_h}, \quad dT_c = \frac{dQ}{C_c}$$

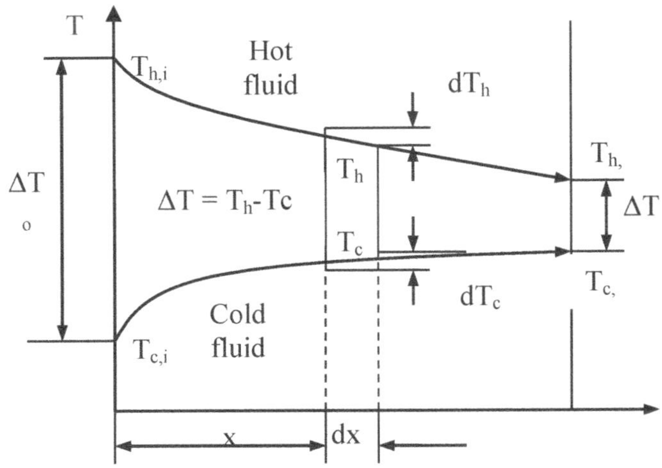

Fig. 5.2. Temperature-length diagram of a parallel flow heat exchanger.

$$dT_h - dT_c = d(T_h - T_c) = -dQ\left(\frac{1}{C_h} + \frac{1}{C_c}\right), \text{ Integrating,} \tag{5.15}$$

$$\int_{\Delta T_o}^{\Delta T_L} d(T_h - T_c) = -\int dQ\left(\frac{1}{C_h} + \frac{1}{C_c}\right)$$

$$\Delta T_L - \Delta T_0 = -Q\left(\frac{1}{C_h} + \frac{1}{C_c}\right)$$

$$\left(\frac{1}{C_h} + \frac{1}{C_c}\right) = -\frac{(\Delta T_L - \Delta T_0)}{Q} \tag{5.16}$$

Combining two equations of heat transfer,

$$dQ = \frac{-d(T_h - T_c)}{\left(\frac{1}{C_h} + \frac{1}{C_c}\right)} = UdA(T_h - T_c) \text{, Separating the variables,}$$

$$\frac{d(T_h - T_c)}{(T_h - T_c)} = -UdA\left(\frac{1}{C_h} + \frac{1}{C_c}\right) \rightarrow \textit{Integrating,}$$

$$\int_{\Delta T_0}^{\Delta T_L} \frac{d(T_h - T_c)}{(T_h - T_c)} = -\int UdA\left(\frac{1}{C_h} + \frac{1}{C_c}\right)$$

$$\ln(T_h - T_c)_{\Delta T_o}^{\Delta T_L} = -UA\left(\frac{1}{C_h} + \frac{1}{C_c}\right)$$

$$\ln\left(\frac{\Delta T_L}{\Delta T_o}\right) = -UA\left(\frac{1}{C_h} + \frac{1}{C_c}\right)$$

$$\because \left(\frac{1}{C_h} + \frac{1}{C_c}\right) = -\left(\frac{\Delta T_L - \Delta T_0}{Q}\right)$$

$$\ln\left(\frac{\Delta T_L}{\Delta T_o}\right) = UA\left(\frac{\Delta T_L - \Delta T_0}{Q}\right)$$

$$Q = UA\left(\frac{\Delta T_L - \Delta T_0}{\ln\left(\frac{\Delta T_L}{\Delta T_0}\right)}\right) = U\ A\ LMTD \tag{5.17}$$

$$LMTD = \frac{\Delta T_L - \Delta T_0}{\ln\left(\frac{\Delta T_L}{\Delta T_0}\right)} \tag{5.18}$$

This can be used for both parallel and counter flow heat exchangers.

5.5.2 Counter flow heat exchanger

For a counter flow heat exchanger, the temperatures of both the fluids decrease in the direction of heat exchanger length (Fig. 5.3). In this case,

Heat flow through an element, $dQ = UdA\,(T_h - T_c)$ (5.19)

$dQ = -C_h\,dT_h = -C_c\,dT_c$

$$dT_h = -\frac{dQ}{C_h},\quad dT_c = -\frac{dQ}{C_c}$$

$$dT_h - dT_c = d(T_h - T_c) = -dQ\left(\frac{1}{C_h} - \frac{1}{C_c}\right) \rightarrow Integrating, \tag{5.20}$$

That means for counter flow the term $\left(\frac{1}{C_h} + \frac{1}{C_c}\right)$ becomes $\left(\frac{1}{C_h} - \frac{1}{C_c}\right)$, the remaining procedure is the same.

$$\Delta T_L - \Delta T_0 = -Q\left(\frac{1}{C_h} - \frac{1}{C_c}\right)$$

$$\left(\frac{1}{C_h} - \frac{1}{C_c}\right) = -\frac{(\Delta T_L - \Delta T_0)}{Q} \tag{5.21}$$

Combining the two equations of heat transfer,

$$dQ = \frac{-d(T_h - T_c)}{\left(\frac{1}{C_h} - \frac{1}{C_c}\right)} = UdA(T_h - T_c)$$, Separating the variables

$$\frac{d(T_h - T_c)}{(T_h - T_c)} = -UdA\left(\frac{1}{C_h} - \frac{1}{C_c}\right) \rightarrow Integrating$$

$$\int_{\Delta T_0}^{\Delta T_L} \frac{d(T_h - T_c)}{(T_h - T_c)} = -\int UdA\left(\frac{1}{C_h} - \frac{1}{C_c}\right)$$

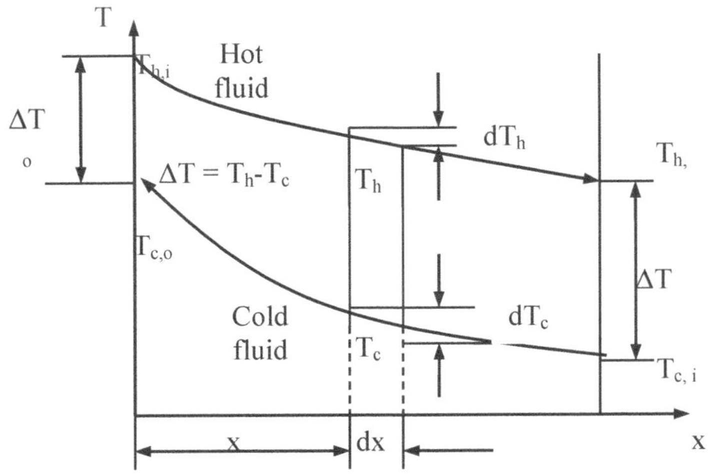

Fig. 5.3. Temperature-length of a heat exchanger for a counter flow heat exchanger.

$$\ln(T_h - T_c)_{\Delta T_o}^{\Delta T_L} = -UA\left(\frac{1}{C_h} - \frac{1}{C_c}\right)$$

$$\ln\left(\frac{\Delta T_L}{\Delta T_o}\right) = -UA\left(\frac{1}{C_h} - \frac{1}{C_c}\right)$$

$$\because \left(\frac{1}{C_h} - \frac{1}{C_c}\right) = -\left(\frac{\Delta T_L - \Delta T_0}{Q}\right)$$

$$\ln\left(\frac{\Delta T_L}{\Delta T_o}\right) = UA\left(\frac{\Delta T_L - \Delta T_0}{Q}\right)$$

$$Q = UA\left(\frac{\Delta T_L - \Delta T_0}{\ln\left(\frac{\Delta T_L}{\Delta T_0}\right)}\right) = U\ A\ LMTD$$

If the capacities of the two fluids are equal in a counter flow arrangement, then it follows from equation that $\Delta T_0 = \Delta T_L$

$$LMTD = \left(\frac{\Delta T_L - \Delta T_0}{\ln\left(\frac{\Delta T_L}{\Delta T_0}\right)}\right) = \frac{0}{\ln(1)} = \frac{0}{0} = indeterminate$$

Applying L'Hospital's Rule:

$$LMTD = \lim_{\Delta T_0 \to \Delta T_L} = \frac{\frac{\partial}{\partial \Delta T_0}(\Delta T_L - \Delta T_0)}{\frac{\partial}{\partial \Delta T_0}\ln\left(\frac{\Delta T_L}{\Delta T_0}\right)} = \frac{1}{(1/\Delta T_0)} = \Delta T_0 = \Delta T_L \tag{5.22}$$

If the temperature difference on both ends of the heat exchanger is the same, the effective temperature difference is equal to ΔT0 and ΔTL. It is observed that the LMTD of a counter flow heat exchanger is greater than the parallel flow heat exchanger under the same conditions. Therefore, the counter flow heat exchanger demands a smaller heat transfer area at the same heat load. Therefore, in the design of a heat exchanger, the usual practice is to select a counter flow heat exchanger for effective heat transfer.

5.6 Correction factor for multi-pass arrangements

The expression of LMTD is used for the evaluation of a single-pass heat exchanger. For a multi-pass heat exchanger, a correction factor is required. The correction factor is multiplied by LMTD of simple pass heat exchanger to get the actual LMTD.

$$Q = F.UA\,(LMTD)_{\text{counter flow}} \tag{5.23}$$

A counter flow heat exchanger is more effective than other heat exchangers. Therefore, the correction factor is less than unity. It also indicates the performance level of a multi-pass heat exchanger at given temperatures.

5.7 Effectiveness – the Number of Transfer Units (NTU) method of heat exchanger analysis

The LMTD method can be applied only with information of all four temperatures. In some cases, where the exit temperatures need to be simulated with the information of inlet temperatures, fluid flow rates, specific heats and the size of the heat exchanger, the LMTD method cannot be used. For the simulation of a heat exchanger with inlet conditions, the use of the effectiveness-number of transfer units (ε-NTU) method is more appropriate. If the inlet and outlet temperatures of the hot and cold fluids, and the overall heat transfer coefficients are specified, the LMTD method, with or without the correction, can be used to solve the rating or sizing problem (Fig. 5.4).

$$Effectiveness\ of\ heat\ exchanger, \varepsilon = \frac{actual\ heat\ transfer}{\max imum\ posible\ heat\ transfer} = \frac{Q_{act}}{Q_{\max}} \tag{5.24}$$

The actual heat transfer is given by the product of the heat capacity rate and temperature difference of hot or cold fluid.

The maximum heat transfer, Q_{max}, is the heat transfer in a counter flow heat exchanger with an infinite area with known inlet temperatures, flow rates, and specific heats. It is also to be noticed that a fluid undergoes a big temperature difference in the heat exchanger.

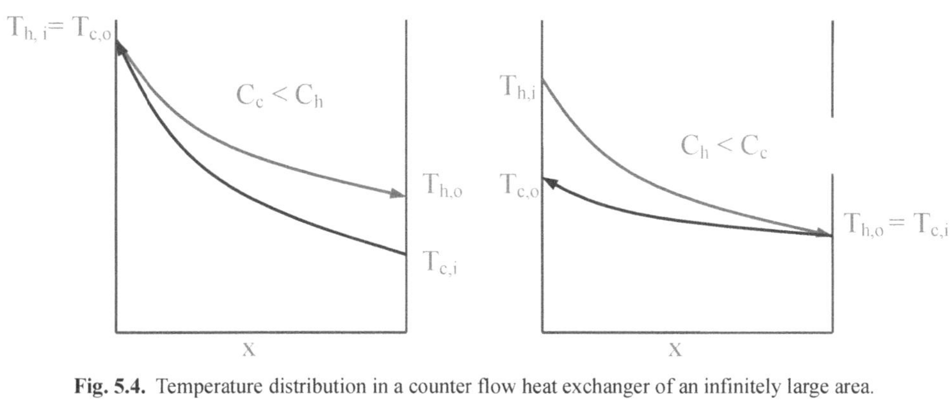

Fig. 5.4. Temperature distribution in a counter flow heat exchanger of an infinitely large area.

i) $C_c < C_h$

For this type of exchanger, with no external heat losses, the outlet temperature of the cold fluid will equal the inlet temperature of the hot fluid (since the area available for heat transfer is infinite). The maximum rate of heat transfer is then given by

$Q_{max} = C_c\,(T_{c,o} - T_{c,i})$

But $T_{c,o} = T_{h,i}$

Therefore, $Q_{max} = C_c\,(T_{h,i} - T_{c,i})$

Also $Q = C_c\,(T_{c,o} - T_{c,i})$

ii) $C_h < C_c$

In this case the outlet temperature of the hot fluid would equal the inlet temperature of the cold fluid. So

$Q_{max} = C_h\,(T_{h,i} - T_{h,o})$

But $T_{h,o} = T_{c,i}$

Therefore, $Q_{max} = C_h\,(T_{h,i} - T_{c,i})$

Also $Q = C_h\,(T_{h,i} - T_{h,o})$

Looking at two equations we may write the general expression

$Q_{max} = C_{min}\,(T_{h,i} - T_{c,i})$

where C_{min} is the smaller of C_c and C_h

$$\varepsilon = \frac{C_h(T_{h,i} - T_{h,o})}{C_{min}(T_{h,i} - T_{c,i})} = \frac{C_c(T_{c,o} - T_{c,i})}{C_{min}(T_{h,i} - T_{c,i})}$$

where C = heat capacity rate = mc

If C_h is minimum, $\varepsilon_h = \dfrac{(T_{h,i} - T_{h,o})}{(T_{h,i} - T_{c,i})}$

If C_c is minimum, $\varepsilon_c = \dfrac{(T_{c,o} - T_{c,i})}{(T_{h,i} - T_{c,i})}$

Number of transfer units, $NTU = \dfrac{UA}{C_{min}}$ (5.25)

5.7.1 Effectiveness for a parallel flow heat exchanger

Heat flow through an element, $dQ = UdA\,(T_h - T_c)$

$dQ = -C_h\,dT_h = C_c\,dT_c$

Where $C_h = m_h\,c_h$ and $C_c = m_c\,c_c$ → Heat capacity.

$$dT_h = -\frac{dQ}{C_h},\quad dT_c = \frac{dQ}{C_c}$$

$$dT_h - dT_c = d(T_h - T_c) = -dQ\left(\frac{1}{C_h} + \frac{1}{C_c}\right)$$

Combining two equations of heat transfer,

$$dQ = \frac{-d\left(T_h - T_c\right)}{\left(\frac{1}{C_h} + \frac{1}{C_c}\right)} = UdA\left(T_h - T_c\right)$$, Separating the variables,

$$or, \frac{d\left(T_h - T_c\right)}{\left(T_h - T_c\right)} = -UdA\left(\frac{1}{C_h} + \frac{1}{C_c}\right) \rightarrow Integrating,$$

$$\int_{\Delta T_o}^{\Delta T_L} \frac{d\left(T_h - T_c\right)}{\left(T_h - T_c\right)} = -\int UdA\left(\frac{1}{C_h} + \frac{1}{C_c}\right)$$

$$\ln\left(T_h - T_c\right)_{\Delta T_o}^{\Delta T_L} = -UA\left(\frac{1}{C_h} + \frac{1}{C_c}\right)$$

$$\ln\left(\frac{\Delta T_L}{\Delta T_o}\right) = -UA\left(\frac{1}{C_h} + \frac{1}{C_c}\right)$$

$$\ln\left(\frac{T_{h,o} - T_{c,o}}{T_{h,i} - T_{c,i}}\right) = -UA\left(\frac{1}{C_h} + \frac{1}{C_c}\right)$$

From the definition of effectiveness,

$$\varepsilon = \frac{C_h\left(T_{h,i} - T_{h,o}\right)}{C_{\min}\left(T_{h,i} - T_{c,i}\right)} = \frac{C_c\left(T_{c,o} - T_{c,i}\right)}{C_{\min}\left(T_{h,i} - T_{c,i}\right)}$$

The values of outlet temperatures $T_{h,o}$ and $T_{c,o}$ are worked out as:

$$T_{h,o} = T_{h,i} - \frac{C_{\min}}{C_h}\left(T_{h,i} - T_{c,i}\right)\varepsilon$$

$$and, T_{c,o} = T_{c,i} + \frac{C_{\min}}{C_c}\left(T_{h,i} - T_{c,i}\right)\varepsilon$$

$$\therefore T_{h,o} - T_{c,o} = \left(T_{h,i} - T_{c,i}\right) - C_{\min}\left(T_{h,i} - T_{c,i}\right)\varepsilon\left(\frac{1}{C_h} + \frac{1}{C_c}\right)$$

$$= \left(T_{h,i} - T_{c,i}\right)\left(1 - \varepsilon C_{\min}\left(\frac{1}{C_h} + \frac{1}{C_c}\right)\right)$$

Substituting this value in logarithmic term,

$$\ln\left(1 - \varepsilon C_{\min}\left(\frac{1}{C_h} + \frac{1}{C_c}\right)\right) = -UA\left(\frac{1}{C_h} + \frac{1}{C_c}\right)$$

$$\left(1 - \varepsilon C_{\min}\left(\frac{1}{C_h} + \frac{1}{C_c}\right)\right) = \exp\left[-UA\left(\frac{1}{C_h} + \frac{1}{C_c}\right)\right]$$

$$\varepsilon = \frac{1 - \exp\left[-UA\left(\frac{1}{C_h} + \frac{1}{C_c}\right)\right]}{C_{\min}\left(\frac{1}{C_h} + \frac{1}{C_c}\right)}$$

$$\varepsilon = \frac{1 - \exp\left[-\frac{UA}{C_h}\left(1 + \frac{C_h}{C_c}\right)\right]}{\frac{C_{\min}}{C_h}\left(1 + \frac{C_h}{C_c}\right)} \tag{5.26}$$

If C_h is assumed to be minimum, then $C_h < C_c$ and therefore $C_h = C_{min}$ and $C_c = C_{max}$. Then

$$\varepsilon = \frac{1 - \exp\left[-\frac{UA}{C_{\min}}\left(1 + \frac{C_{\min}}{C_{\max}}\right)\right]}{\frac{C_{\min}}{C_{\min}}\left(1 + \frac{C_{\min}}{C_{\max}}\right)} \tag{5.27}$$

The dimensionless ratio UA/C_{min} is called the number of transfer units (NTU) and the factor C_{min}/C_{max} represents the capacity ratio (C).

Therefore,

$$\varepsilon = \frac{1 - \exp\left[-NTU(1 + C)\right]}{(1 + C)} \tag{5.28}$$

The same expression can be obtained with the assumption of cold fluid having with a minimum heat capacity rate.

5.7.2 Effectiveness for the counter flow heat exchanger

Heat flow through an element, $dQ = UdA\ d(T_h - T_c)$

$dQ = -C_h\ dT_h = -C_c\ dT_c$

$$dT_h = -\frac{dQ}{C_h},\ dT_c = -\frac{dQ}{C_c}$$

$$dT_h - dT_c = d(T_h - T_c) = -dQ\left(\frac{1}{C_h} - \frac{1}{C_c}\right)$$

Combining the two heat transfer equations,

$$dQ = \frac{-d(T_h - T_c)}{\left(\frac{1}{C_h} - \frac{1}{C_c}\right)} = UdA(T_h - T_c)$$, Separating the variables,

$$or, \frac{d(T_h - T_c)}{(T_h - T_c)} = -UdA\left(\frac{1}{C_h} - \frac{1}{C_c}\right) \rightarrow Integrating,$$

$$\int_{\Delta T_0}^{\Delta T_L} \frac{d(T_h - T_c)}{(T_h - T_c)} = -\int UdA\left(\frac{1}{C_h} - \frac{1}{C_c}\right)$$

$$\ln\left(T_h - T_c\right)_{\Delta T_o}^{\Delta T_L} = -UA\left(\frac{1}{C_h} - \frac{1}{C_c}\right)$$

$$\ln\left(\frac{\Delta T_L}{\Delta T_o}\right) = -UA\left(\frac{1}{C_h} - \frac{1}{C_c}\right)$$

$$\ln\left(\frac{T_{h,o} - T_{c,i}}{T_{h,i} - T_{c,o}}\right) = -UA\left(\frac{1}{C_h} - \frac{1}{C_c}\right)$$

From the definition of effectiveness,

$$\varepsilon = \frac{C_h\left(T_{h,i} - T_{h,o}\right)}{C_{\min}\left(T_{h,i} - T_{c,i}\right)} = \frac{C_c\left(T_{c,o} - T_{c,i}\right)}{C_{\min}\left(T_{h,i} - T_{c,i}\right)}$$

The values of outlet temperatures $T_{h,o}$ and $T_{c,o}$ are worked out as:

$$T_{h,o} = T_{h,i} - \frac{C_{\min}}{C_h}\left(T_{h,i} - T_{c,i}\right)\varepsilon$$

$$and, T_{c,o} = T_{c,i} + \frac{C_{\min}}{C_c}\left(T_{h,i} - T_{c,i}\right)\varepsilon$$

Substituting these values of $T_{h,o}$ and $T_{c,o}$ in logarithmic term,

$$\ln\left(\frac{T_{h,i} - \frac{C_{\min}}{C_h}\left(T_{h,i} - T_{c,i}\right)\varepsilon - T_{c,i}}{T_{h,i} - T_{c,i} - \frac{C_{\min}}{C_c}\left(T_{h,i} - T_{c,i}\right)\varepsilon}\right) = -UA\left(\frac{1}{C_h} - \frac{1}{C_c}\right)$$

$$\left(\frac{\left(T_{h,i} - T_{c,i}\right)\left(1 - \frac{C_{\min}}{C_h}\varepsilon\right)}{\left(T_{h,i} - T_{c,i}\right)\left(1 - \frac{C_{\min}}{C_c}\varepsilon\right)}\right) = \exp\left[-UA\left(\frac{1}{C_h} - \frac{1}{C_c}\right)\right]$$

$$\frac{\left(1 - \frac{C_{\min}}{C_h}\varepsilon\right)}{\left(1 - \frac{C_{\min}}{C_c}\varepsilon\right)} = \exp\left[-UA\left(\frac{1}{C_h} - \frac{1}{C_c}\right)\right]$$

$$1 - \frac{C_{\min}}{C_h}\varepsilon = \exp\left[-UA\left(\frac{1}{C_h} - \frac{1}{C_c}\right)\right] - \frac{C_{\min}}{C_c}\varepsilon \exp\left[-UA\left(\frac{1}{C_h} - \frac{1}{C_c}\right)\right]$$

$$1 - \exp\left[-UA\left(\frac{1}{C_h} - \frac{1}{C_c}\right)\right] = \varepsilon\left\{\frac{C_{\min}}{C_h} - \frac{C_{\min}}{C_c}\exp\left[-UA\left(\frac{1}{C_h} - \frac{1}{C_c}\right)\right]\right\}$$

$$\varepsilon = \frac{1 - \exp\left[-UA\left(\frac{1}{C_h} - \frac{1}{C_c}\right)\right]}{\left\{\frac{C_{\min}}{C_h} - \frac{C_{\min}}{C_c}\exp\left[-UA\left(\frac{1}{C_h} - \frac{1}{C_c}\right)\right]\right\}}$$

$$\varepsilon = \frac{1-\exp\left[-\frac{UA}{C_h}\left(1-\frac{C_h}{C_c}\right)\right]}{\frac{C_{min}}{C_h}\left\{1-\frac{C_h}{C_c}\exp\left[-\frac{UA}{C_h}\left(1-\frac{C_h}{C_c}\right)\right]\right\}} \quad (5.29)$$

If C_h is assumed minimum, then $C_h < C_c$ and therefore $C_h = C_{min}$ and $C_c = C_{max}$. Then

$$\varepsilon = \frac{1-\exp\left[-\frac{UA}{C_{min}}\left(1-\frac{C_{min}}{C_{max}}\right)\right]}{\frac{C_{min}}{C_{min}}\left\{1-\frac{C_{min}}{C_{max}}\exp\left[-\frac{UA}{C_{min}}\left(1-\frac{C_{min}}{C_{max}}\right)\right]\right\}} \quad (5.30)$$

UA/C_{min} is a dimensionless number which is called the number of transfer units (NTU). The ratio C_{min}/C_{max} is the capacity ratio (C).

$$\varepsilon = \frac{1-\exp\left[-NTU(1-C)\right]}{1-C\exp\left[-NTU(1-C)\right]} \quad (5.31)$$

With the cold fluid having the minimum capacity rate, the same equation will result with analysis. The effectiveness is a function of the overall heat transfer coefficient, area, fluid properties, and flow rates.

5.7.3 Limiting values of capacity ratio, C

Two limiting cases of practical interest are:

i) C = 0

In a boiler or condenser, the working fluid temperature is constant. The external fluid temperature changes, i.e., hot fluid in for boiling and cold fluid for condensation. The direction of fluids with each other is not significant due to the constant temperature of the working fluid. By definition, specific heat represents the change of enthalpy with respect to temperature, i.e., $c_p = dh/dT$. With the temperature difference dT being zero, both the effective specific heat and the consequent heat capacity tend to infinity.

In that case $C_{max} = \alpha$ and $C_{min}/C_{max} = 0$.

$$\varepsilon = \frac{1-\exp\left[-NTU(1+C)\right]}{(1+C)} \quad \text{For parallel flow heat exchanger} \quad (5.32)$$

$$\varepsilon = \frac{1-\exp\left[-NTU(1-C)\right]}{1-C\exp\left[-NTU(1-C)\right]} \quad \text{For counter flow heat exchanger} \quad (5.33)$$

Substituting C = 0 in the above two equations of heat exchangers,

The expression for effectiveness then reduces to

$\varepsilon = 1 - \exp(-NTU)$ for both parallel and counter flow configurations (5.34)

ii) C = 1

In a gas turbine power plant, a heat exchanger at the exit of the turbine heats the compressed air to increase the efficiency of the power plant. In this case, both the fluids have approximately equal thermal capacities, and so the capacity ratio, $C = C_{min}/C_{max}$, comes very close to unity.

Then for the parallel flow configuration,

$$\varepsilon = \frac{1-\exp(-2NTU)}{2} \tag{5.35}$$

Obviously, no matter how large the exchanger is or how high the overflow at transfer coefficient, the maximum effectiveness for a parallel flow heat exchanger is only 50%.

For the counter flow arrangement, effectiveness becomes indeterminate and that necessitates fresh analysis. With $C = C_{min}/C_{max} \rightarrow 1$, it is possible to write that

$(1–C) = 0$; $NTU\ (1–C) << 1$

And $\exp[–NTU\ (1–C)] = 1 – NTU\ (1–C)$ - - - - -

Substituting these results in the equation

$$\varepsilon = \frac{1-\left[1-NTU(1-C)\right]}{\left\{1-C\left[1-NTU(1-C)\right]\right\}} = \frac{NTU(1-C)}{\left\{1-\left[C-CNTU(1-C)\right]\right\}}$$

$$= \frac{NTU(1-C)}{\left\{1-C+CNTU(1-C)\right\}} = \frac{NTU(1-C)}{(1-C)(1+CNTU)}$$

$$\varepsilon = \frac{NTU}{(1+NTU)} \tag{5.36}$$

$\because C = 1$

5.8 Numerical solutions

1. Determine the overall coefficient of a heat exchanger with water and oil as fluids. The water flows in a copper tube of 180 mm inside diameter and 15 mm thickness, while the oil flows through the annulus formed by this pipe and an outer concentric pipe. The thermal conductivity of the tube wall is 348 W/m K, and the fouling factors on the oil and waterside are 0.00086 m^2 K/W and 0.000344 m^2 K/W, respectively. The oil and waterside heat transfer coefficients can be taken as 1280 and 4650 W/m^2 K, respectively.

Solution:

Inside diameter of the tube, $D_i = 180$ mm

Thickness, t = 15 mm

Outside diameter of tube, $D_o = 180 + 2\times 15 = 210$ mm

Thermal conductivity of the tube wall, k = 348 W/m K

Fouling factors on outside (oil) $F_o = 0.00086\ m^2k\ /W$

Fouling factors on inside (water) $F_i = 0.000344\ m^2k\ /W$

Outside heat transfer coefficient, $h_o = 1280\ W/m^2K$

Inside heat transfer coefficient, $h_i = 4650\ W/m^2K$

Inner radius, $r_i = 0.09$ m

Outer radius, $r_o = 0.105$ m

Overall heat transfer coefficient,

$$U_o = \frac{1}{A_o R} = \frac{1}{\frac{r_o}{r_i}\frac{1}{h_i} + \frac{r_o}{r_i}F_i + \frac{r_o}{k}\ln\left(\frac{r_o}{r_i}\right) + F_o + \frac{1}{h_o}}$$

$$\text{Or } U_i = \frac{1}{A_i R} = \frac{1}{\frac{1}{h_i} + F_i + \frac{r_i}{k}\ln\left(\frac{r_o}{r_i}\right) + \frac{r_i}{r_o}F_o + \frac{r_i}{r_o}\frac{1}{h_o}}$$

$$\frac{1}{U_o} = \frac{r_o}{r_i}\frac{1}{h_i} + \frac{r_o}{r_i}F_i + \frac{r_o}{k}\ln\left(\frac{r_o}{r_i}\right) + F_o + \frac{1}{h_o}$$

$$= \frac{0.105}{0.09}\frac{1}{4650} + \frac{0.105}{0.09}0.000344 + \frac{0.105}{349}\ln\left(\frac{0.105}{0.09}\right) + 0.00086 + \frac{1}{1280}$$

$= 2.33 \times 10^{-3}$

Uo = 427.37 W/m²K

2. In a chemical plant, 1 kg/s of oil (c_p = 2.8 kJ/kg K) is cooled from 120°C to 50°C in a shell and tube heat exchanger with one shell pass and two tube passes. The heat transfer area of the heat exchanger is 4 m². The overall heat transfer coefficient of the heat exchanger is known to be 800 W/m²K. The cooling water at 20°C enters the heat exchanger at a rate of 2.67 kg/s. Calculate the effectiveness of the heat exchanger and the number of heat transfer units.

Solution:

$m_c c_c = 2.67 \times 4.2 = 11.2$ kJ/K

$m_h c_h = 1 \times 2.8 = 2.8$ kJ/K

NTU = UA/C_{min} = 1.143

$C_{min}/C_{max} = 0.25$

From graph, ε = 60%, at NTU = 1.143 and C = 0.25.

3. Cold water leading to a shower enters a thin-walled double-pipe counter-flow heat exchanger at 15°C at a rate of 0.25 kg/s and is heated to 45°C by hot water that enters at 100°C at a rate of 3 kg/s. If the overall heat transfer coefficient is 950 W/m² °C, determine the rate of heat transfer and the heat transfer surface area of the heat exchanger using the ε–NTU method. The specific heats of cold and hot water are given to be 4.18 and 4.19 kJ/kg.°C, respectively.

Solution:

The heat capacity rates of the hot and cold fluids are,

$$C_h = \dot{m}_h C_{ph} = 0.25 \times 4180 = 1045 \text{ W/°C}$$

$$C_c = \dot{m}_c C_{pc} = 3 \times 4190 = 12{,}570 \text{ W/°C}$$

Therefore, $C_{min} = C_c = 1045$ W/°C

$$\text{and } C = \frac{C_{min}}{C_{max}} = \frac{1045}{12{,}570} = 0.083$$

Then the maximum heat transfer rate becomes,

$$\dot{Q}_{\max} = C_{\min}(T_{h,in} - T_{c,in}) = 1045 \times (100\text{-}15) = 88,825 \text{ W}$$

The actual rate of heat transfer is:

$$\dot{Q} = C_h(T_{h,in} - T_{h,out}) = 1045 \times (45 - 15) = \mathbf{3}1,350 \ W$$

Then the effectiveness of this heat exchanger becomes:

$$\varepsilon = \frac{Q}{Q_{\max}} = \frac{31,350}{88,825} = 0.35$$

The NTU of this heat exchanger is determined using the relation in Table 13-5 to be:

$$NTU = \frac{1}{C-1}\ln\left(\frac{\varepsilon - 1}{\varepsilon C - 1}\right) = \frac{1}{0.083-1}\ln\left(\frac{0.35-1}{0.35\times 0.083-1}\right) = 0.438$$

Then the surface area of the heat exchanger is determined from:

$$NTU = \frac{UA}{C_{\min}} \longrightarrow A = \frac{NTU\ C_{\min}}{U} = \frac{0.438\times 1045}{950} = 0.482\ m^2$$

Chapter 6
Steam Energy Systems

6.1 Introduction

Examples of steam energy systems are steam boiler, steam nozzle, steam turbine, steam condenser, etc. A steam generator or boiler is a closed pressure vessel in which steam is generated from feedwater by the combustion of fuel in a furnace or any equivalent heat course. The generated steam can be used for the production of electricity in steam engines or steam turbines, processing heat in industries such as textile, sugar, chemical, etc., and space heating. The boilers are classified based on orientation, fluid flow, place of furnace, circulation, pressure, size, and number of tubes as horizontal, vertical, or inclined orientation; fire tubes and water tubes; internally fired and externally fired; natural circulation and forced circulation; low pressure and high pressure; portable, stationary, and movable; and single tube and multi-tube boilers.

A steam nozzle is a device where steam expands continuously in a variable cross-sectional area by converting the pressure energy (heat) into kinetic energy. The main objective of a steam nozzle is to generate a high-velocity jet to operate a steam turbine and finally electricity. In the nozzle, the pressure is converted into velocity, and it transforms into mechanical energy, i.e., shaft work in a steam turbine. The steam turbines are either impulse or reaction turbines. In an impulse turbine, all the kinetic energy is converted into mechanical work, and in a reaction turbine, the kinetic energy is converted into fixed blades (nozzles) and moving blades. The cross-sectional area of the nozzle decreases, remains constant, and increases later, respectively, called convergent, throat, and divergent sections. The nozzles may be convergent or convergent-divergent. Convergent-divergent type results in higher velocity compared to convergent nozzles, which are more suitable for thermal power plants.

A steam condenser is a heat exchanger used to condense the steam into liquid by rejecting heat to surroundings through the water-cooling or air-cooling method. The use of a condenser permits the expansion of steam in a turbine below the atmospheric pressure. This additional expansion to vacuum generated more power from the steam turbine. In a non-condensing plant, i.e., a plant without a condenser, the steam expands into atmospheric pressure (back pressure turbine), which results in lower efficiency compared to the condensing power plant. A hot well connected to a condenser collects the saturated liquid, and this feedwater is pumped to the boiler for steam generation. The steam condensers are jet condensers or surface condensers. There are three types of jet condensers, viz., low level, high level, and ejector type jet condensers. The models of surface condensers are down flow, central flow, inverted flow, regenerative flow, and evaporative flow types of condensers.

6.2 Modelling of a steam boiler

The evaporative capacity (kg/unit) of a steam generator indicates the performance of a boiler. Since the working conditions differ from boiler to boiler, a standard feed temperature and working pressure have been adopted to compare the performance of boilers. The feed temperature is 100°C and formed into dry and saturated steam at 100°C at 1.01325 bar (from and at 100°C). It considers

DOI: 10.1201/9781032642505-6

the latent heat of evaporation only at 100°C. The latent heat of evaporation at 100°C is 2257 kJ/kg (Ganapathy, 2002).

Therefore, the equivalent evaporation from and at 100°C,

$$E = \frac{Total\ heat\ required\ to\ evaporate\ feedwater}{2257}, kg/kg\ fuel$$

If the feed water at a temperature of t_1 is heated in a boiler to generate the steam, the total heat supplied in a steam boiler is

$$Q = m_e(h - h_{f1}) \quad (6.1)$$

$m_e = m_s/m_f$
m_s = mass of steam, kg/h
m_f = mass of fuel burnt in furnace, kg/h
m_e = mass of steam per unit mass of fuel, kg/kg fuel
h is the total heat of steam as follows
$h = h_f + x\,h_{fg}$ for wet steam
$h = h_f + h_{fg}$ for dry saturated steam
$h = h_{sup}$ for superheated steam
h_{f1} is the enthalpy or sensible heat of feedwater at T_1.

Therefore, the equivalent evaporation from and at 100°C,

$$E = \frac{Q}{2257} = \frac{m_s(h - h_{f1})}{2257}, kg/kg\ fuel \quad (6.2)$$

The factor, $\frac{h - h_{f1}}{2257}$ is known as factor of evaporation

$$F_e = \frac{h - h_{f1}}{2257} \quad (6.3)$$

Therefore,

$$E = m_s F_e \quad (6.4)$$

The boiler efficiency is the ratio of heat used to generate steam to the heat supplied in the furnace. It is also known as the thermal efficiency of a boiler.

$$\eta_b = \frac{heat\ suppled\ to\ generate\ the\ steam}{Heat\ supplied\ in\ furnace}$$

$$\eta_b = \frac{m_e(h - h_{f1})}{CV} \quad (6.5)$$

m_e = mass of steam generation per unit mass of fuel, kg/kg fuel
CV = calorific value of the fuel, kJ

If a boiler consists of an economizer, evaporator, and superheater and is considered a single unit, its thermal efficiency is the overall thermal efficiency of the boiler.
The performance of a boiler can be evaluated through a boiler trial.

The main objectives of a boiler trial are

1. To find the capacity of the boiler.
2. To calculate the thermal efficiency of the boiler
3. To apply the energy balance and draw the heat balance sheet for the boiler.

The heat losses in the boiler need to be assessed to generate the heat balance sheet. The thermal efficiency of a boiler is the ratio of the output energy of steam to the fuel energy supplied. The difference between the fuel energy supply and generated steam energy is the total heat loss.

Following are the various heat losses related to a steam boiler.

1. Heat loss through the chimney's dry flue gases, kJ/kg fuel
2. Heat loss with moisture in the fuel, kJ/kg fuel
3. Heat loss with steam in products of combustion, kJ/kg fuel
4. Heat loss with unburned carbon in an ash pit, kJ/kg fuel
5. Heat loss with unburned carbon into carbon monoxide (CO), kJ/kg fuel
6. Radiation heat losses (by difference), kJ/kg fuel

1. Heat loss through chimney's dry fluid gases,

   The heat loss with dry flue gases, kJ/kg fuel,

   $Q_{fg} = m_g\, c_{pg}\, (T_g - T_{br})$

   m_g = mass of dry flue gases, kg/kg of fuel
   c_{pg} = specific heat of dry flue gases, kJ/kg K
   T_g = temperature of flue gases leaving chimney, °C and
   T_{br} = temperature of boiler room, °C

   The heat loss with dry flue gas is the maximum loss in a steam boiler.
2. Heat loss with the moisture in fuel, kJ/kg fuel

   The moisture content in the fuel converts into superheated steam in the products of combustion at atmospheric pressure,

   Heat lost in moisture present in fuel,

   $$Q_{m,\,f} = m_m\,(h_{sup} - h_b) = m_m\,(h_g + c_{pv}\,(T_g - T) - h_{br}) \quad (6.6)$$

   $= m_m\,(2676 + c_p\,(t_g - 100) - h_b)$
   From steam tables, corresponding to 1.01325 bar, h_g = 2676 kg/kg and T = 100°C
   m_m = mass of moisture per kg of fuel
   h_{br} = enthalpy or sensible heat of water at boiler room temperature
3. Heat lost to steam formed by combustion of hydrogen, kJ/kg fuel,
   Let H_2 is the mass of hydrogen present, kg/kg fuel.
   Mass of steam formed = 9 H_2, kg/kg fuel
   The heat lost to steam, kJ/kg fuel,

   $$Q_{H2} = 9H_2\,(2676 + c_{pv}(T_g - 100) - h_{br}) \quad (6.7)$$

   Note: Heat lost to steam and moisture, kJ/kg fuel,

   $$Q_{H2} + Q_{moisture,\,f} = (9H_2 + m_m)\,(2676 + c_{pv}(T_g - 100) - h_{br}) \quad (6.8)$$

   where m_m is the mass of moisture, kg/kg fuel.
4. Heat loss with unburned carbon in ash pit, kJ/kg fuel
   The heat lost due to unburned carbon, kJ/kg fuel,

   $$Q_{uc} = m_1\, C_1 \quad (6.9)$$

   where m_1 = mass of carbon in ash pit, kg/kg fuel
   C_1 = calorific value of carbon, kJ/kg carbon

5. Heat loss with unburned carbon into carbon monoxide (CO), kJ/kg fuel
   This loss, generally occurs in a boiler due to insufficient air supply.
   The heat lost due to incomplete combustion,

   $$Q_{CO} = m_2 C_2 \quad (6.10)$$

   where m_2 = mass of carbon monoxide in flue gas per kg of fuel and
   C_2 = calorific value of carbon monoxide
6. Radiation heat losses (by difference), kJ/kg fuel

   The radiation heat loss can be determined by the subtracting the heat utilized for steam generation and heat losses, from the heat supplied with the fuel.

   The heat balance sheet shown in Table 6.1 details the energy supply on the left and the heat consumption on right. The heat supply is the calorific value of the fuel. The right side shows the heat utilization for the steam generation and all other heat losses.
   We know that heat utilized in raising steam per kg of fuel,

   $$Q = m_e (h - h_{fi}) \quad (6.11)$$

The heat balance sheet (Table 6.1) for a boiler trial per kg of fuel is drawn as below:

Table 6.1. Heat balance sheet of a boiler.

Heat supplied	kJ	Heat consumed	kJ	%
Heat supplied by 1 kg of dry fuel	Q	Heat utilized in raising steam	Q_1	-
		Heat loss through chimney's dry flue gases, kJ/kg fuel	Q_2	-
		Heat loss with moisture in the fuel, kJ/kg fuel	Q_3	-
		Heat loss with steam in products of combustion, kJ/kg fuel	Q_4	-
		Heat loss with unburned carbon in the ash pit, kJ/kg fuel	Q_5	-
		Heat loss with unburned carbon turning into carbon monoxide (CO), kJ/kg fuel	Q_6	-
		Radiation heat losses (by difference), kJ/kg fuel	$Q - (Q_1 + Q_2 + -Q_3 + Q_4 + Q_5 + Q_6)$	-
Total	Q		Q	**100**

The density of hot gas in a chimney is less than the density of atmosphere pressure. Therefore, the lighter hot gas expands and floats upwards above the atmospheric air. It creates a draft. The boiler draft is the pressure difference of atmospheric air and hot gases in the furnace (Fig. 6.1). Therefore, the draft works as a fan without using electricity, and it creates the velocity for the fluid.

m_a = mass of air required per kg of fuel burnt
m_g = mass of hot gases = m_a + 1 per kg of fuel
T_a = Atmospheric temperature, K
T_0 = absolute temperature corresponding to 0°C
T_1 = absolute temperature of outside air, K = T_a
T_2 = mean absolute temperature of chimney gases, K = T_g
p_a = atmospheric pressure

Pressure at the grate level of chimney side,

$$p_1 = p_a + \rho_g g H \quad (6.12)$$

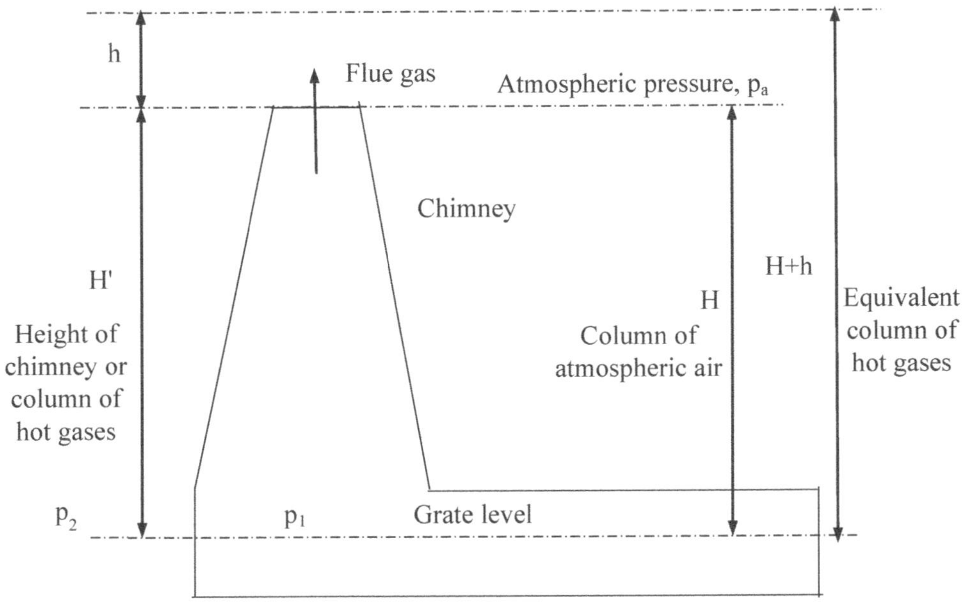

Fig. 6.1. Thermal modelling of a boiler draught.

Pressure at the grate level of outside (open side)

$$p_2 = p_a + \rho_a g H \tag{6.13}$$

Net pressure difference causing the draught

$$\Delta p = p_1 - p_2 = \rho_a g H - \rho_g g H = (\rho_a - \rho_g)\, gH \text{ (static draught)} \tag{6.14}$$

Volume of outside air at NTP (0°C and atmospheric pressure)

$$v_0 = \frac{m_a R T_0}{p_a} = \frac{m_a 0.287 \times 273.15}{101.325} = 0.773 m_a \tag{6.15}$$

Volume of outside air,

$$v_1 = \frac{v_0 T_1}{p_a} = \frac{0.773 m_a T_1}{101.325} = \frac{m_a T_1}{353} \tag{6.16}$$

Density of outside air,

$$\rho_1 = \rho_a = \frac{m_a}{v_1} = \frac{m_a}{\left(\frac{m_a T_1}{353}\right)} = \frac{353}{T_1} \tag{6.17}$$

Similarly, volume of flue gas at T_2,

$v_2 = \dfrac{m_a T_2}{353}$ (According to Avogadro's rule, the flue gas occupies the same volume of air at NTP)

Density of gas at T_2,

$$\rho_2 = \rho_g = \frac{m_a + 1}{v_2} = \frac{m_a + 1}{\left(\frac{m_a T_2}{353}\right)} = \left(\frac{m_a + 1}{m_a}\right)\frac{353}{T_2} = \left(\frac{m_a + 1}{m_a}\right)\frac{353}{T_g} \tag{6.18}$$

$$\Delta p = (\rho_a - \rho_g)\, gH \tag{6.19}$$

$$\Delta p = 353gH\frac{1}{T_a} - 353gH\left(\frac{m_a+1}{m_a}\right)\frac{1}{T_g}$$

$$\Delta p = 353gH\left(\frac{1}{T_a} - \left(\frac{m_a+1}{m_a}\right)\frac{1}{T_g}\right) \tag{6.20}$$

If h_w is the height of water column in mm, which produces the pressure drop, Δp,

$$h_w = 353H\left(\frac{1}{T_a} - \left(\frac{m_a+1}{m_a}\right)\frac{1}{T_g}\right) \quad \text{(1 mm of water = 9.81 Pa)} \tag{6.21}$$

Assuming that the draught pressure, Δp produced is equivalent of H_1 meters of burnt gases,

$$\Delta p = \rho_g g H_1 = 353g\left(\frac{m_a+1}{m_a}\right)\frac{1}{T_g}H_1 \tag{6.22}$$

Equating these two,

$$\Delta p = 353g\left(\frac{m_a+1}{m_a}\right)\frac{1}{T_g}H_1 = 353gH\left(\frac{1}{T_a} - \left(\frac{m_a+1}{m_a}\right)\frac{1}{T_g}\right)$$

$$\left(\frac{m_a+1}{m_a}\right)\frac{1}{T_g}H_1 = H\frac{1}{T_a} - H\left(\frac{m_a+1}{m_a}\right)\frac{1}{T_g}$$

$$H_1 = H\left(\left(\frac{m_a}{m_a+1}\right)\frac{T_g}{T_a} - 1\right) \tag{6.23}$$

Assuming no losses in the chimney, the velocity of gases in the chimney,

$C = \sqrt{2gH_1}$

If the pressure loss in the chimney is h_l meters of hot gas column,

$C = \sqrt{2g(H_1 - h_l)}$

$$C = 4.43\sqrt{H_1\left(1 - \frac{h_l}{H_1}\right)} = 4.43\sqrt{\left(1 - \frac{h_l}{H_1}\right)}\sqrt{H_1} = K\sqrt{H_1} \tag{6.24}$$

where $K = 4.43\sqrt{\left(1 - \frac{h_l}{H_1}\right)}$

The value of K varies according to the construction and material of the chimney.
For steel built chimney, K = 1.1
And for brick built chimney, K = 0.825

The mass of flue gases flowing through any section of the chimney will be given by

$$\dot{m}_g = \rho_g AC$$

$$m_g = \rho_g\left(\frac{\pi}{4}D^2\right)C$$

$$D = \sqrt{\frac{m_g}{\pi\rho_g C}}$$

$$D = 1.128\sqrt{\frac{m_g}{\rho_g C}} \tag{6.25}$$

The chimney draught is a function of hot gas temperature and atmospheric temperature. At certain values of hot gas temperature, atmospheric temperature, and fluid flow rates, the chimney draught may be maximum.

Neglecting the chimney losses,

$$C = \sqrt{2gH_1}$$

Substituting H_1,

$$C = \sqrt{2gH\left(\left(\frac{m_a}{m_a+1}\right)\frac{T_g}{T_a}-1\right)}$$

The mass of hot gas flow rate discharged through the chimney,

$$\dot{m}_g = \rho_g AC$$

The density of hot gas,

$$\rho_g = \frac{p}{RT_g}$$

Inserting the density in the gas flow rate and velocity of gas,

$$\dot{m}_g = \frac{p}{RT_g}A\sqrt{2gH\left(\left(\frac{m_a}{m_a+1}\right)\frac{T_g}{T_a}-1\right)}$$

$$\dot{m}_g = \frac{K}{T_g}\sqrt{\left(\left(\frac{m_a}{m_a+1}\right)\frac{T_g}{T_a}-1\right)} \tag{6.26}$$

where $K = \dfrac{pA\sqrt{2gH}}{R}$

The discharge of gas is maximum at $\dfrac{d\dot{m}_g}{dT_g} = 0$ at a fixed T_a and m_a,

$$\frac{d\dot{m}_g}{dT_g} = \frac{d}{dT_g}\left(\frac{K}{T_g}\sqrt{\left(\left(\frac{m_a}{m_a+1}\right)\frac{T_g}{T_a}-1\right)}\right) = 0$$

$$\frac{d}{dT_g}\left(\frac{1}{T_g}\sqrt{\left(\left(\frac{m_a}{m_a+1}\right)\frac{T_g}{T_a}-1\right)}\right)=0$$

$$\frac{d}{dT_g}\left(\frac{\sqrt{ZT_g-1}}{T_g}\right)=0$$

where $Z=\left(\frac{m_a}{m_a+1}\right)\frac{1}{T_a}$

$$\frac{d}{dT_g}\left(\sqrt{ZT_g-1}\times T_g^{-1}\right)=0$$

$$\left(ZT_g-1\right)^{1/2}(-1)T_g^{-2}+T_g^{-1}\left(ZT_g-1\right)^{-1/2}Z=0$$

$$\left(ZT_g-1\right)^{1/2}(-1)T_g^{-2}+\frac{1}{2}T_g^{-1}\left(ZT_g-1\right)^{-1/2}Z=0$$

$$\frac{-\left(ZT_g-1\right)^{1/2}}{T_g^2}+\frac{Z}{2T_g\left(ZT_g-1\right)^{1/2}}T_g^{-1}=0$$

$$\frac{-2\left(ZT_g-1\right)+ZT_g}{2T_g^2\left(ZT_g-1\right)^{1/2}}=0$$

$$-2(ZT_g-1)+ZT_g=0$$

$$-2ZT_g+2+ZT_g=0$$

$$2-ZT_g=0$$

$$ZT_g=2$$

Substituting Z,

$$\left(\frac{m_a}{m_a+1}\right)\frac{T_g}{T_a}=2$$

$$\frac{T_g}{T_a}=2\left(\frac{m_a+1}{m_a}\right) \tag{6.27}$$

The ratio of absolute temperatures of hot gas temperature and atmospheric temperature has a certain value based on air flow rate into the furnace to influence the draught of the chimney.

Substituting $\frac{T_g}{T_a}$ in discharge, in the draught pressure in a hot gas column, H_1

$$H_1=H\left(\left(\frac{m_a}{m_a+1}\right)\frac{T_g}{T_a}-1\right)$$

$$H_{1,\ \max} = H\left(\left(\frac{m_a}{m_a+1}\right)2\left(\frac{m_a+1}{m_a}\right)-1\right) = H \tag{6.28}$$

At an equal height of the hot gas column and the chimney, maximum discharge, i.e., draught, results.

The draught in mm of water column for maximum discharge can be evaluated by substituting $\frac{T_g}{T_a}$ in $h_{w.}$

$$h_w = 353H\left(\frac{1}{T_a}-\left(\frac{m_a+1}{m_a}\right)\frac{1}{T_g}\right) \qquad \because \frac{T_g}{T_a} = 2\left(\frac{m_a+1}{m_a}\right)$$

$$h_w = 353H\left(\frac{1}{T_a}-\frac{1}{2T_a}\right)$$

$$h_w = \frac{176.5H}{T_a}$$

Finally, the maximum discharge,

$$\dot{m}_g = \frac{p}{RT_g}A\sqrt{2gH\left(\left(\frac{m_a}{m_a+1}\right)\frac{T_g}{T_a}-1\right)} \qquad \because \frac{T_g}{T_a} = 2\left(\frac{m_a+1}{m_a}\right)$$

$$= \frac{pA}{2RT_a}\left(\frac{m_a}{m_a+1}\right)\sqrt{2gH}$$

$$\dot{m}_g = \frac{pA}{2RT_a}\left(\frac{m_a}{m_a+1}\right)\sqrt{2gH} \tag{6.29}$$

The chimney needs a minimum temperature for the creation of the draught. Therefore, the exit temperature of the natural draught is higher than the exit temperature of the artificial draught. It drops the heat utilization for the steam generation and hence a loss in thermal efficiency of the power plant.

Let

$T_{g,\ natural} = T_g$ = absolute temperature of fluid gases with the chimney to create the draught of h_w mm of water

$T_{g,\ artificial}$ = absolute temperature of fluid gases with the artificial draught of h_w mm of water

c_p = mean specific heat of hot gases

The additional heat loss for high temperature gas with natural draught per kg of flue gas

$= c_p(T_{g,natural} - T_{g,\ artificial})$

Draught in hot gas column,

$$H_1 = H\left(\left(\frac{m_a}{m_a+1}\right)\frac{T_g}{T_a}-1\right)$$

The maximum energy of this head per kg of flue gas

$$= H\left(\left(\frac{m_a}{m_a+1}\right)\frac{T_g}{T_a}-1\right)g,\ \ Nm/kg\ \ or\ \ J/kg$$

Efficiency of the chimney,

$$\eta_{chimney} = \frac{H\left(\left(\frac{m_a}{m_a+1}\right)\frac{T_g}{T_a}-1\right)g}{c_p(T_g - T_{artificial})}\times 100 \tag{6.30}$$

The power or capacity of a fan is the work done on the air or gas to deliver the fluid volume.

m_a = mass of air per kg of fuel
m_f = mass of fuel
T_0 = temperature at NTP

$$P_{fan} = \frac{pV}{\eta_{fan}}$$

Forced draught handles the atmospheric air.

Volume of air at T_0,

$$V_0 = \frac{m_a m_f}{\rho_0}$$

$$\because \frac{V_a}{V_0} = \frac{T_a}{T_0}$$

$$V_a = \frac{T_a V_0}{T_0}$$

$$V_a = \frac{T_a m_a m_f}{\rho_0 T_0}$$

Power of the forced draught fan,

$$P_{FD} = \frac{pV_a}{\eta_{fan}} = \frac{pT_a m_a m_f}{\eta_{fan}\rho_0 T_0} \tag{6.31}$$

Induced draught handles the flue gases.

$$\because \frac{V_g}{V_0} = \frac{T_g}{T_0}$$

$$V_g = \frac{T_g V_0}{T_0} \quad \text{(based on Avogadro's rule)}$$

$$V_g = \frac{T_g m_a m_f}{T_0 \rho_0}$$

Power of the forced draught fan,

$$P_{ID} = \frac{pV_g}{\eta_{fan}} = \frac{pT_g m_a m_f}{\eta_{fan}\rho_0 T_0} \tag{6.32}$$

Comparing these two powers,

$$\frac{P_{ID}}{P_{FD}} = \frac{\dfrac{pT_g m_a m_f}{\eta_{fan}\rho_0 T_0}}{\dfrac{pT_a m_a m_f}{\eta_{fan}\rho_0 T_0}} = \frac{T_g}{T_a} > 1$$

Power required for the induced draught fan is greater than the power of the forced draught fan.

6.3 Modelling of a steam nozzle

The steam nozzle converts the pressure energy into kinetic energy by creating high velocity steam to drive the steam turbine. The modelling of a steam nozzle required to find the exit velocity, dimensions of nozzle, and operating conditions is as below:

Let C = Velocity of steam, m/s
h_1 = enthalpy of steam at nozzle inlet, kJ/kg
h_2 = enthalpy of steam at state 2, kJ/kg, and
h_d = the enthalpy drop in the nozzle = $h_1 - h_2$, kJ/kg

Considering a frictionless, adiabatic flow in a steam nozzle with 1 kg/s flow rate,

The increase in kinetic energy = enthalpy drop in a nozzle (without friction)

$$\frac{C^2}{2} = h_d$$

$C = \sqrt{2 \times 1000\ h_d}$, where h_d is in kJ/kg

$$= 44.72\sqrt{h_d}$$

Generally there is a 10 to 15% of heat drop with friction. With friction, the heat drop for kinetic energy will decrease. Let the available heat drop for velocity generation be kh_d.

The velocity, $C = 44.72\sqrt{kh_d}$

The condition for maximum discharge through the nozzle,

Let p_1 = initial pressure of steam,
v_1 = initial volume of 1 kg of steam at pressure p_1, m^3
p_2 = steam pressure at the throat,
v_2 = volume of 1 kg of steam at pressure p_2, m^3
A = cross sectional area of nozzle at throat, m^2 and
C = velocity of steam, m/s

The steam flowing through the nozzle follows approximately the equation given below

Pv^n = constant

where n = 1.135 for saturated steam
n = 1.3 for superheated steam
n = 1.035 + 0.1x for wet steam

With reference to Fig. 6.2,

$$w = \left(\frac{n}{n-1}\right)\left(p_1 v_1 - p_2 v_2\right) \tag{6.33}$$

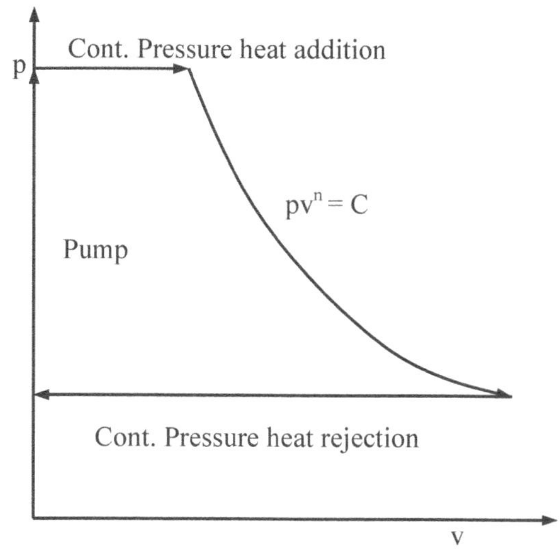

Fig. 6.2. Work done in an open system: area under pv diagram.

and, gain in kinetic energy = adiabatic heat drop = work done during Rankine cycle

$$\frac{C^2}{2} = \left(\frac{n}{n-1}\right) p_1 v_1 \left(1 - \frac{p_2 v_2}{p_1 v_1}\right)$$

$$= \left(\frac{n}{n-1}\right) p_1 v_1 \left(1 - \frac{T_2}{T_1}\right)$$

$$= \left(\frac{n}{n-1}\right) p_1 v_1 \left(1 - \left(\frac{p_2}{p_1}\right)^{\frac{n-1}{n}}\right)$$

$$C^2 = 2\frac{n}{n-1} p_1 v_1 \left(1 - \left(\frac{p_2}{p_1}\right)^{\frac{n-1}{n}}\right)$$

$$C = \sqrt{2\frac{n}{n-1} p_1 v_1 \left(1 - \left(\frac{p_2}{p_1}\right)^{\frac{n-1}{n}}\right)}$$

If m is the mass of steam discharged in kg/s

Then,

$$m = \frac{AC}{v_2}$$

Substituting the value of v2 from the equation

$$m = \frac{AC}{v_1 \left(\frac{p_1}{p_2}\right)^{\frac{1}{n}}}$$

$$m = \frac{A}{v_1\left(\frac{p_1}{p_2}\right)^{\frac{1}{n}}}\sqrt{2\frac{n}{n-1}p_1v_1\left(1-\left(\frac{p_2}{p_1}\right)^{\frac{n-1}{n}}\right)}$$

$$= \frac{A}{v_1}\sqrt{2\frac{n}{n-1}p_1v_1\left(\left(\frac{p_2}{p_1}\right)^{\frac{2}{n}} - \left(\frac{p_2}{p_1}\right)^{\frac{n-1}{n}}\left(\frac{p_2}{p_1}\right)^{\frac{2}{n}}\right)}$$

$$m = \frac{A}{v_1}\sqrt{2\left(\frac{n}{n-1}\right)p_1v_1\left(\left(\frac{p_2}{p_1}\right)^{\frac{2}{n}} - \left(\frac{p_2}{p_1}\right)^{\frac{n+1}{n}}\right)} \tag{6.34}$$

As per Fig. 6.3, at the critical pressure ratio, $\frac{p_2}{p_1}$ the discharge from a nozzle is maximum. It can be determined by differentiating the discharge with respect to the critical pressure ratio and equating to zero. In this differentiation, the critical pressure ratio has only been considered a variable with other conditions as constants.

$$\frac{d}{d\left(\frac{p_2}{p_1}\right)}\left[\left(\frac{p_2}{p_1}\right)^{\frac{2}{n}} - \left(\frac{p_2}{p_1}\right)^{\frac{n+1}{n}}\right] = 0$$

$$\frac{2}{n}\left(\frac{p_2}{p_1}\right)^{\frac{2}{n}-1} - \left(\frac{n+1}{n}\right)\left(\frac{p_2}{p_1}\right)^{\frac{n+1}{n}-1} = 0$$

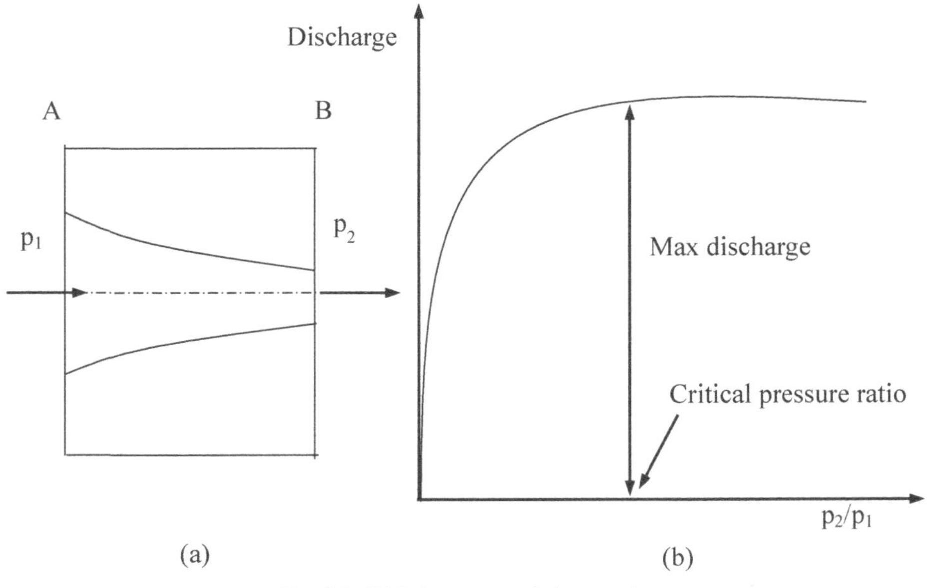

Fig. 6.3. Critical pressure ratio in a nozzle.

$$\left(\frac{p_2}{p_1}\right)^{\frac{2-n}{n}} = \left(\frac{n+1}{2}\right)\left(\frac{p_2}{p_1}\right)^{\frac{1}{n}}$$

$$\left(\frac{p_2}{p_1}\right)^{2-n} = \left(\frac{n+1}{2}\right)^{n}\left(\frac{p_2}{p_1}\right)$$

$$\left(\frac{p_2}{p_1}\right)^{2-n-1} = \left(\frac{n+1}{2}\right)^{n}$$

$$\left(\frac{p_2}{p_1}\right)^{1-n} = \left(\frac{n+1}{2}\right)^{n}$$

$$\left(\frac{p_2}{p_1}\right) = \left(\frac{2}{n+1}\right)^{\frac{n}{n-1}} \tag{6.35}$$

It is the critical pressure ratio to result the maximum discharge from a nozzle.

$$\frac{throat\ \ pressure}{inlet\ \ pressure} = \left(\frac{p_2}{p_1}\right) = \left(\frac{2}{n+1}\right)^{\frac{n}{n-1}}$$

For saturated steam, n = 1.135

$$\left(\frac{p_2}{p_1}\right) = \left(\frac{2}{1.135+1}\right)^{\frac{1.135}{1.135-1}} = 0.58$$

For superheated steam, n = 1.3

$$\left(\frac{p_2}{p_1}\right) = \left(\frac{2}{1.3+1}\right)^{\frac{1.3}{1.3-1}} = 0.546$$

Substituting the value of $\frac{p_2}{p_1}$ from equation, we get the maximum discharge,

$$m_{\max} = \frac{A}{v_1}\sqrt{2\left(\frac{n}{n-1}\right)p_1v_1\left(\left(\left(\frac{2}{n+1}\right)^{\frac{n}{n-1}}\right)^{\frac{2}{n}} - \left(\left(\frac{2}{n+1}\right)^{\frac{n}{n-1}}\right)^{\frac{n+1}{n}}\right)}$$

$$= \frac{A}{v_1}\sqrt{2\left(\frac{n}{n-1}\right)p_1v_1\left(\left(\frac{2}{n+1}\right)^{\frac{2}{n-1}} - \left(\frac{2}{n+1}\right)^{\frac{n+1}{n-1}}\right)}$$

$$= A\sqrt{2\left(\frac{n}{n-1}\right)\frac{p_1}{v_1}\left(\left(\frac{2}{n+1}\right)^{\frac{2}{n-1}} - \left(\frac{2}{n+1}\right)^{\frac{n+1}{n-1}}\right)}$$

$$= A\sqrt{2\left(\frac{n}{n-1}\right)\frac{p_1}{v_1}\left(\frac{2}{n+1}\right)^{\frac{n+1}{n-1}}\left(\left(\frac{2}{n+1}\right)^{\frac{2}{n-1}-\frac{n+1}{n-1}} - 1\right)}$$

$$= A\sqrt{2\left(\frac{n}{n-1}\right)\frac{p_1}{v_1}\left(\frac{2}{n+1}\right)^{\frac{n+1}{n-1}}\left(\left(\frac{2}{n+1}\right)^{\frac{1-n}{n-1}}-1\right)}$$

$$= A\sqrt{2\left(\frac{n}{n-1}\right)\frac{p_1}{v_1}\left(\frac{2}{n+1}\right)^{\frac{n+1}{n-1}}\left(\left(\frac{2}{n+1}\right)^{-1}-1\right)}$$

$$= A\sqrt{n\left(\frac{p_1}{v_1}\right)\left(\frac{2}{n+1}\right)^{\frac{n+1}{n-1}}}$$

$$m_{max} = A\sqrt{n\left(\frac{p_1}{v_1}\right)\left(\frac{2}{n+1}\right)^{\frac{n+1}{n-1}}} \tag{6.36}$$

It shows that the maximum discharge is a function of the throat area, inlet pressure, inlet volume, and polytropic index. The maximum discharge is independent of the exit pressure. The addition of a divergent part to a convergent nozzle does not affect the maximum discharge. The divergent portion increases the exit velocity through the extra expansion of steam. Initially, the discharge increases with a decrease in the exit pressure. But after reaching a maximum value, it remains constant.

The maximum velocity can be determined with this condition of maximum discharge.

$$\frac{p_2}{p_1} = \left(\frac{2}{n+1}\right)^{\frac{n}{n-1}}$$

from the equation into the equation

$$C = \sqrt{2\frac{n}{n-1}p_1v_1\left(1-\left(\frac{p_2}{p_1}\right)^{\frac{n-1}{n}}\right)}$$

$$C_{max} = \sqrt{2\left(\frac{n}{n-1}\right)p_1v_1\left(1-\left(\left(\frac{2}{n+1}\right)^{\frac{n}{n-1}}\right)^{\frac{n-1}{n}}\right)}$$

$$= \sqrt{2\left(\frac{n}{n-1}\right)p_1v_1\left(1-\frac{2}{n+1}\right)}$$

$$= \sqrt{2\left(\frac{n}{n+1}\right)p_1v_1}$$

$$C_{max} = \sqrt{2\left(\frac{n}{n+1}\right)p_1v_1} \tag{6.37}$$

Therefore the maximum velocity is the function of only inlet operational conditions, i.e., p_1 and v_1.

In actual practice, the velocity of steam from the nozzle decreases with the following reasons.

1. Friction between nozzle walls and fluid
2. Fluid internal friction and
3. Losses due to shock waves

Compared to a convergent nozzle, the majority of losses happen in the divergent part of the nozzle. Because of the frictional losses, the following effects result.

1. The process deviates from the isentropic process, i.e., entropy increases.
2. The dryness fraction of steam at the exit of the nozzle increases due to heat generation with friction, with the penalty of kinetic energy.
3. As the dryness fraction increases, the specific volume also increases.

Figure 6.4 represents the influence of friction on nozzle flow with a Mollier diagram.

The process 1 - 2' - 3' shows the isentropic process without friction. The actual process is 1 - 2 - 3. The enthalpy drop in actual expansion is less than the isentropic enthalpic drop. Let it be 1 - 3". Draw a horizontal line from 3" till it touches the actual process, i.e., state 3, which is the final state of expansion. The dryness fraction of steam at 3 is higher than the 3'. For the friction, the coefficient is defined as the ratio of the actual enthalpy drop to the isentropic enthalpy drop.

$$k = \frac{actual\ heat\ drop}{isentropic\ heat\ drop} = \frac{1-3}{1-3'} = \frac{h_1 - h_3}{h_1 - h_3'} \tag{6.38}$$

Sometimes the friction in the convergent portion is neglected as major frictional losses occur in the divergent portion.

The nozzle efficiency is the ratio of the actual enthalpy drop to the isentropic enthalpy drop.

$$Nozzle\ efficiency = \frac{h_1 - h_3}{h_1 - h_3'} \tag{6.39}$$

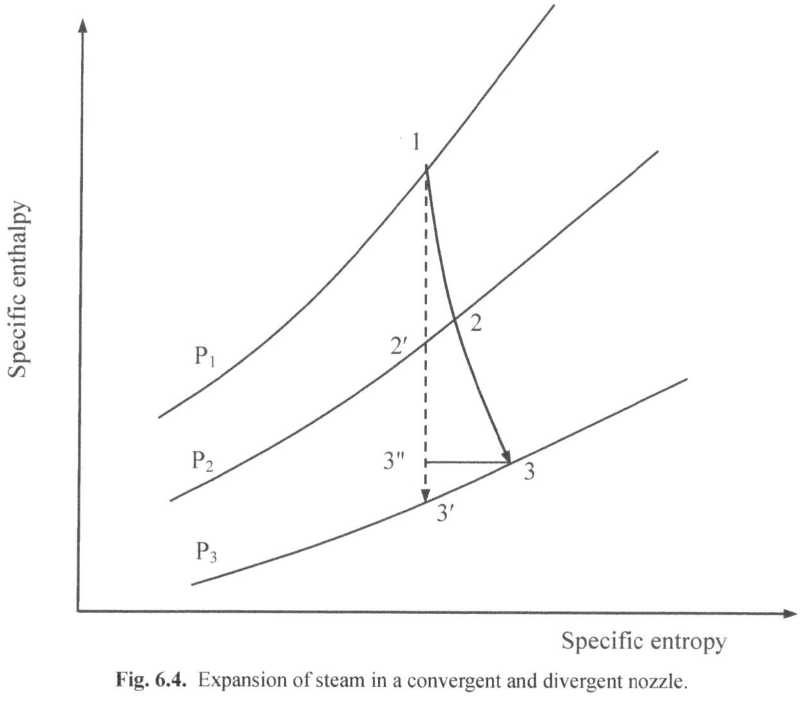

Fig. 6.4. Expansion of steam in a convergent and divergent nozzle.

Let C_3 is the exit velocity in actual expansion and C_3' is the exit velocity after the isentropic expansion.

$$h_1 + \frac{C_1^2}{2} = h_3' + \frac{C_3'^2}{2} \; or \; h_1 - h'_3 = \frac{C_3'^2 - C_1^2}{2} \tag{6.40}$$

and $h_1 + \frac{C_1^2}{2} = h_3 + \frac{C_3^2}{2} \; or \; h_1 - h_3 = \frac{C_3^2 - C_1^2}{2}$

$$\text{Nozzle efficiency} = \frac{C_3^2 - C_1^2}{C_3'^2 - C_1^2} \tag{6.41}$$

If the initial velocity, C_1 is neglected due to a small value,

$$\text{Nozzle efficiency} = \frac{C_3^2}{C_3'^2} \tag{6.42}$$

The velocity coefficient of a nozzle is defined as the ratio of exit velocity of steam in actual expansion and the exit velocity of steam for isentropic expansion.

$$\text{i.e., velocity co-efficient} = \frac{C_3}{C_3'} \tag{6.43}$$

The velocity coefficient is the square root of the nozzle efficiency if the inlet velocity is neglected.

Figure 6.5 shows the discretization of fluid flow to make a general relation between area, velocity and pressure.

Consider an isentropic flow in the nozzle.

Assume

i) Full flow at nozzle.
ii) Uniform velocity over the entire section.

By the continuity equation, $m = \frac{AC}{v} = \frac{(A+\delta A)(C+\delta C)}{(v+\delta v)}$

$v(A + \delta A)(C + \delta C) = AC(v + \delta v)$

$v(AC + A\,\delta C + C\,\delta A + \delta A\,\delta C) = AC\,v + AC\,\delta v$, neglecting $\delta A \delta C$

$A\,\delta C\,v + C\,\delta A\,v = AC\,\delta v$

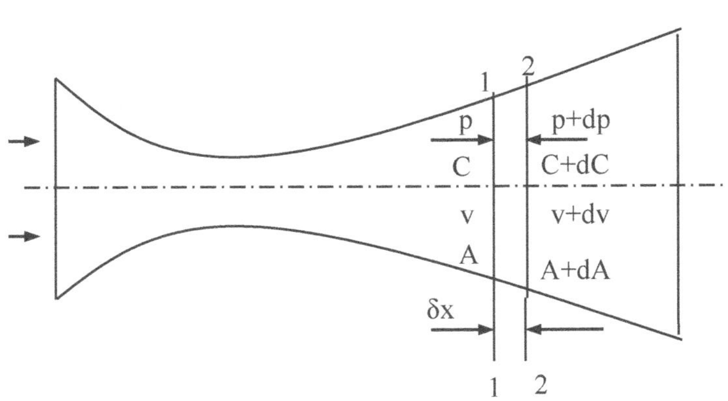

Fig. 6.5. Flow through nozzles.

A δC v + C δA v = AC δv, dividing both sides by ACv

$$\frac{\delta C}{C}+\frac{\delta A}{A}=\frac{\delta v}{v}$$

$$\frac{\delta C}{C}+\frac{\delta A}{A}-\frac{\delta v}{v}=0$$, in the limit δx, δA, δC, δv → 0

$$\frac{dC}{C}+\frac{dA}{A}-\frac{dv}{v}=0 \tag{6.44}$$

Using another method:

$m=\dfrac{AC}{v}=constant$, differentiating

$\dfrac{C}{v}dA+\dfrac{A}{v}dC-\dfrac{AC}{v^2}dv=0$, dividing both sides by $\dfrac{AC}{v}$

$$\frac{dA}{A}+\frac{dC}{C}-\frac{dv}{v}=0$$

We have pv^n = Constant, using the log and differentiating,

log p + n log v = log Const = Const2

$$\frac{dp}{p}+n\frac{dv}{v}=0$$

$$\frac{dv}{v}=-\frac{1}{n}\frac{dp}{p}$$

For the momentum equation,

CdC = –vdp (KE = work), dividing both sides by C^2

$$\frac{dC}{C}=-\frac{v}{C^2}dp \tag{6.45}$$

Substituting these $\dfrac{dv}{v}$ and $\dfrac{dC}{C}$ values in above derived equations in $\dfrac{dA}{A}+\dfrac{dC}{C}-\dfrac{dv}{v}=0$

$$\frac{dA}{A}-\frac{v}{C^2}dp+\frac{1}{n}\frac{dp}{p}=0$$

$$\frac{dA}{A}=\frac{v}{C^2}dp-\frac{1}{n}\frac{dp}{p}=\frac{1}{n}\frac{dp}{p}\left(\frac{vnp}{C^2}-1\right)$$ Since vnp = a^2

$$\frac{dA}{A}=\frac{1}{n}\frac{dp}{p}\left(\frac{a^2}{C^2}-1\right),\qquad Mach\ number=\frac{fluid\ velocity}{sound\ velocity}=\frac{C}{a}$$

$$\frac{dA}{A}=\frac{1}{n}\frac{dp}{p}\left(\frac{1}{M^2}-1\right) \tag{6.46}$$

Accelerated flow (velocity increasing), $\frac{dp}{p}$ is negative → flow through nozzles

a) M < 1, i.e., C < a, $\frac{dA}{A} = (-)(+) = (-)$ → convergent nozzle

b) M = 1, i.e., C = a, $\frac{dA}{A} = 0$ → throat (no change in area)

c) M > 1, i.e., C > a, $\frac{dA}{A} = (-)(-) = (+)$ → divergent nozzle

Retarded flow (velocity decreasing) $\frac{dp}{p}$ is positive → flow through diffusers

a) M < 1, i.e., C < a, $\frac{dA}{A} = (+)(+) = (+)$ → divergent diffuser

b) M = 1, i.e., C = a, $\frac{dA}{A} = 0$ → throat (no change in area)

c) M > 1, i.e., C > a, $\frac{dA}{A} = (+)(-) = (-)$ → convergent diffuser

Generally, in steam turbines, the velocity of steam entering the turbine is very low, but the final velocity required is supersonic because more power needs to be developed, so a convergent-divergent nozzle is used. If the entry velocity of steam at the nozzle is the local sonic velocity, a divergent nozzle will be used. If the exit velocity from the nozzle has to be equal to the sonic velocity, only a convergent nozzle will be used.

The actual discharge from a steam nozzle is higher than the theoretical discharge, which is not expected. It is because of the short length of the convergent nozzle and insufficient time for the expansion. The condensation is not normal; it is metastable, which is a supersaturated state. In this condition, the steam is undercooled to a temperature less than the saturation temperature at the same pressure. Finally, the density of fluid increases, which increases the discharge. Prof. Wilson conducted experiments on the supersaturated flow and showed that dry saturated steam expands suddenly but does not condense until its density is about 8 times that of the saturated vapor of the same pressure.

Figure 6.6 shows the expansion of steam in a nozzle with supersaturated flow. Process 1-2 shows the isentropic expansion without condensation. It also shows a superheat-constant-pressure curve. State 2 is the limit of the supersaturated state. From state 2, it reaches its normal state 3 at constant enthalpy and pressure. Later, the steam expands isentropically to exit pressure to reach state 4. Without supersaturation, the exit state is 4'. Similar to the effect of friction, the supersaturated flow also leads to enthalpy drop and increases the dryness fraction of steam. The Wilson line shows the limit of undercooling at which condensation starts and is assumed to restore the conditions of normal thermal equilibrium. The expansion of supersaturated flow can be shown with the polytropic process, $pv^{1.3} = C$.

Thus, $$\textit{Enthalpy drop} = \frac{n}{n-1} p_1 v_1 \left[1 - \left(\frac{p_2}{p_1}\right)^{\frac{n-1}{n}}\right] \tag{6.47}$$

To find the supercooled temperature the ration between temperature ratio and pressure ratio can be used.

$$\frac{T_2}{T_1} = \left(\frac{p_2}{p_1}\right)^{\frac{n-1}{n}}$$

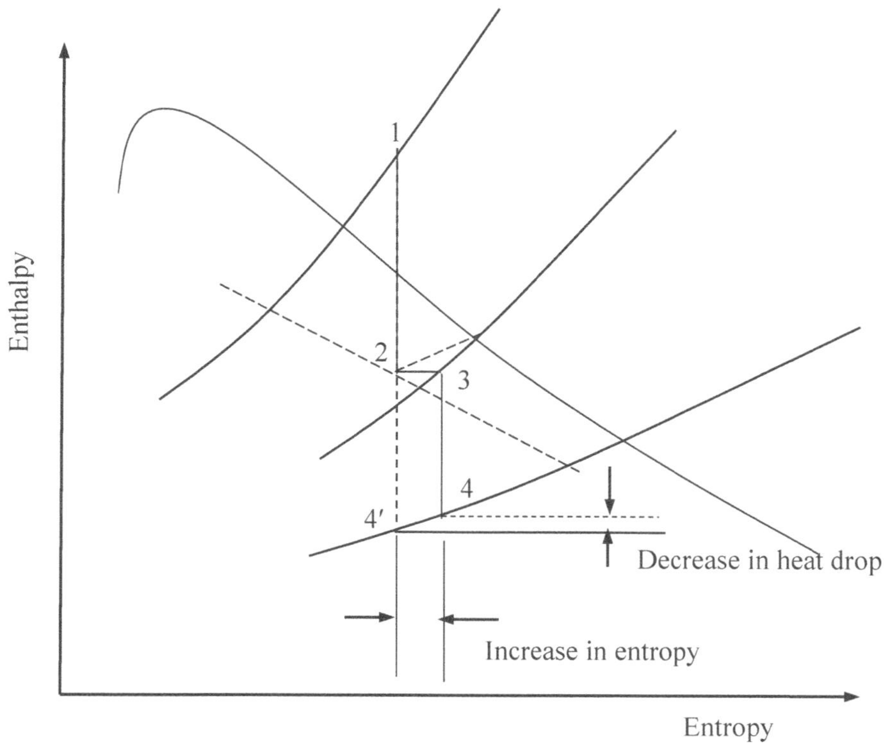

Fig. 6.6. Supersaturated flow of steam.

The degree of undercooling is the temperature difference between the saturation temperature and supercooled temperature during the supersaturation flow.

The effects of supersaturation are as follows:

1. It leads to an increase in entropy and the specific volume of steam.
2. The heat drop decreases compared to the isentropic process; hence, the exit velocity decreases.
3. Due to lack of condensation in supersaturation expansion, the temperature is less than the saturation temperature corresponding to the pressure. Therefore, its density is greater than the density at equilibrium conditions, which increases the discharge.
4. The dryness fraction of steam increases.

The solution of supersaturated flow cannot be arrived at using the Mollier chart without a Wilson line.

The exit velocity of steam,

$$C_2 = \sqrt{2 \times \left(\frac{n}{n-1}\right) p_1 v_1 \left[1 - \left(\frac{p_2}{p_1}\right)^{\frac{n-1}{n}}\right]}$$

Specific volume, $v_2 = v_1 \left(\frac{p_2}{p_1}\right)^{\frac{1}{n}}$

Apparent temperature, $T_2 = T_1\left(\frac{p_2}{p_1}\right)^{\frac{n-1}{n}}$

and $A_2 = \frac{mv_2}{C}$

The temperature difference between the supersaturated temperature and saturation temperature at that pressure is known as the degree of undercooling.

Mathematically, the degree of undercooling = $T_2 - T_2'$

The ratio of pressures corresponding to temperatures T_2 and T_2' is known as the degree of supersaturation. Mathematically, the degree of supersaturation,

$$DoS = \frac{pressure\ corresponding\ to\ T_2}{pressure\ corresponding\ to\ T_2'} = \frac{p_2}{p_2'} \quad (6.48)$$

The following relations may be used to find the supersaturated flow.

$$v = \frac{0.0023(h-1943)}{p} \quad (6.49)$$

$Pv^{1.3}$ = constant and

$$\frac{p}{T^{13/3}} = Contsant \quad (6.50)$$

6.4 Steam condensers

The performance of the condenser is determined by the condenser efficiency and vacuum efficiency. Condenser efficiency is the ratio of the rise in temperature of cooling water to the temperature difference between the saturation temperature of condensed steam and the inlet temperature of cooling water.

Condenser efficiency,

$$\eta_{cond} = \frac{rise\ in\ temperture\ of\ cooling\ water}{saturation\ temperature\ of\ condensed\ steam - inlet\ temperature\ of\ water} \quad (6.51)$$

The pressure below the atmospheric pressure is called a vacuum. It may be expressed in mm of mercury. Vacuum pressure is measured by a vacuum pressure gauge. In calculations, the vacuum gauge reading is corrected to a standard barometric reading of 760 mm of Hg.

$$\eta_{vacuum} = \frac{actual\ vacuum}{barometric\ pressure - absolute\ pressue\ of\ stteam} \quad (6.52)$$

$$= \frac{actual\ vacuum}{\max\ obtainable\ vacuum}$$

Corrected vacuum in mm of Hg = 760 – absolute pressure in mm of Hg
= 760 – actual barometric height – actual vacuum

Also 760 mm of Hg = 1.01325 bar

Therefore, mm of Hg = 1.01325/760 = 0.00133 bar

If the absolute pressure of steam at condensate temperature is equal to the absolute pressure in the condenser, the condenser efficiency is 100%. Due to the presence of some quantity of air in the condenser, which may leak in and be accompanied by the inlet steam, the efficiency is less than

100%. Therefore, the vacuum efficiency depends on the amount of air present in the condenser. The air is removed by a vacuum pump.

The heat balance in a condenser gives the mass of cooling water required to condense the steam.

heat lost by steam = heat gained by cooling water,

$$m_s\,(h_f + xh_{fg} - h_{f1}) = m_w c_{pw}\,(T_{we} - T_{wi}) \tag{6.53}$$

$$m_w = \frac{ms(h_f + xh_{fg} - h_{f1})}{c_{pw}(T_{we} - T_{wi})}$$

m_w = mass of cooling water required, kg/s
m_s = mass of steam, kg/s
T_{we} = cooling water exit temperature, °C
T_{wi} = cooling water inlet temperature, °C
h_{f1} = total heat of condensate leaving the condenser, kJ/kg
x = dryness fraction of steam entering the condenser
h_{fg} = latent of heat of steam entering the condenser, kJ/kg
h_f = total heat of water at the pressure of the steam entering the condenser, kJ/kg

6.5 Numerical solutions

1. As an experiment a steam generator is used for a day with steam generated at 10 bar, dry saturation, from 15 ton of water using 1.5 ton of coal. Coal has 3% moisture and 4% ash. Feedwater temperature is 35°C. Determine, (i) factor of evaporation (ii) the boiler efficiency, and (iii) the equivalent evaporation. Heating value of coal is 30100 kJ/kg.

Solution through MATLAB App Designer:

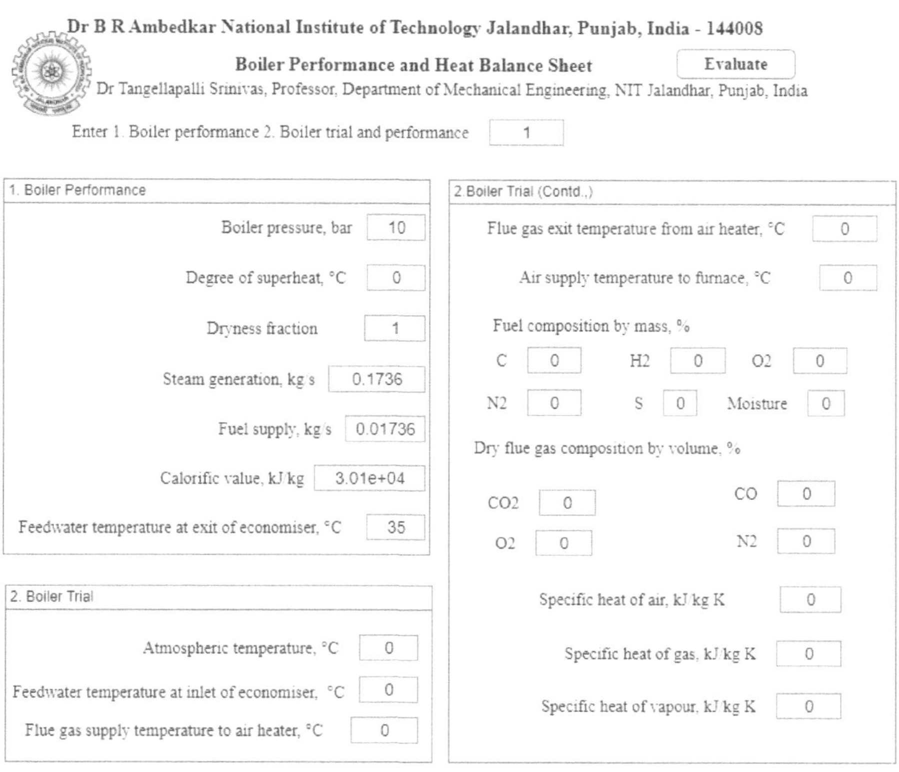

Parameter	Equation	Substitution	Result
Enthalpy of steam at exit of boiler	From steam properties	At 10 bar, dry saturated steam	2776.17 kJ/kg
Enthalpy of feedwater	From steam properties	At 35°C	147.45 kJ/kg
Mass of steam per unit fuel	$m_e = \frac{m_s}{m_{fuel}}$	$\frac{15000}{1500}$	10 kg/kg fuel
Factor of evaporation	$F_e = \frac{h - h_{f1}}{2257}$	$\frac{2776.17 - 147.45}{2257}$	1.16
Equivalent evaporation	$E = \frac{m_e(h - h_{f1})}{2257}$	$\frac{10(2776.17 - 147.45)}{2257}$	11.65 kg/kg fuel
Boiler efficiency	$\eta_b = \frac{m_e(h - h_{f1})}{CV} \times 100$	$\frac{10(2776.17 - 147.45)}{30100} \times 100$	87.33%

2. Draw a boiler performance and heat balance sheet with the following data.

   The boiler has an economiser and air preheater. Atmospheric temperature is 15°C. Steam is generated at 40 bar, 400°C, and 8 kg/kg coal. Feed water temperature inlet and exit temperature at economiser are 27°C and 137°C respectively. Moisture content in fuel is 1.5%. Flue gas temperatures at the inlet and outlet of the air heater are 300°C and 150°C respectively. Temperature of the air entering boiler furnace is 120°C. Dry coal composition by mass is 84% C, 4% H_2, 7% O_2 and the is remainder ash. Dry flue gas composition by volume is 12.5% CO_2, 7.5% O_2, 80% N_2. Datum temperature is 15°C. Calorific value of coal is 32600 kJ/kg. For air and dry flue gas, the specific heat is 1.0032 kJ/kg K. Specific heat vapour is 2.0064 kJ/kg K.

Solution through App Designer:

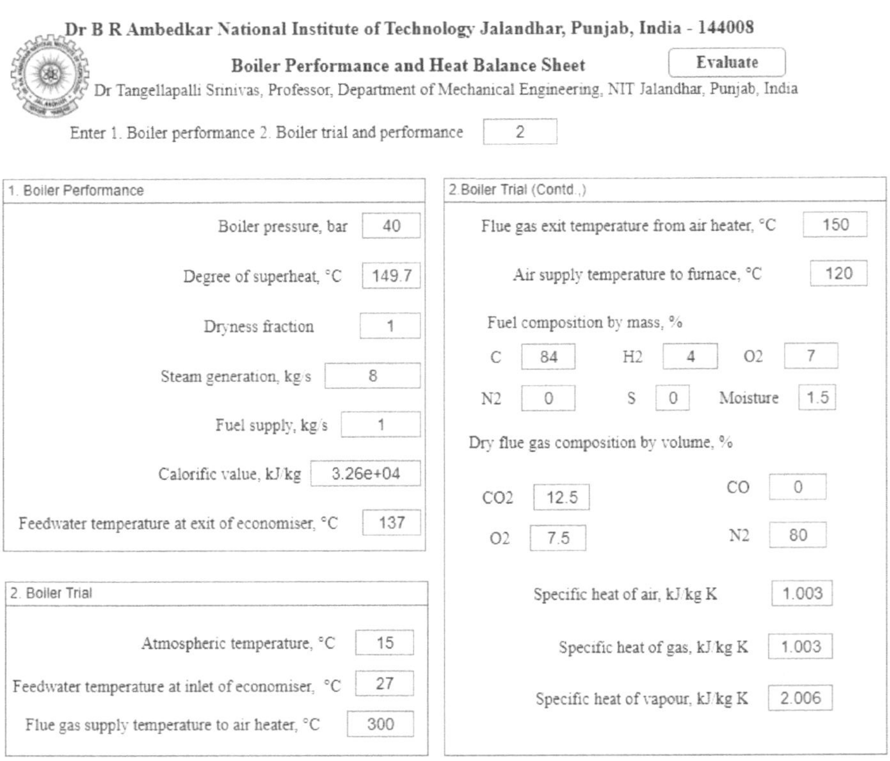

Combustion equation:

$a_1C + a_2H_2 + a_3O_2 + a_4N_2 + a_5S + a_6H2O + a_7(O_2{+}3.76N_2) = p(b_1CO_2 + b_2CO + b_3O_2 + b_4N_2) + b_5H_2O + b_6SO_2$

In the above equation, a_5 and a_6 are zero as per the given data.

Parameter	**Equation**	**Substitution**	**Result**
Enthalpy of steam at exit of boiler	From steam properties	At 40 bar, 400°C	3215.76 kJ/kg
Enthalpy of feedwater	From steam properties	At 137°C	578.63 kJ/kg
Mass of steam per unit fuel	$m_e = \frac{m_s}{m_{fuel}}$	$\frac{8}{1}$	8 kg/kg fuel
Factor of evaporation	$F_e = \frac{h - h_{f1}}{2257}$	$\frac{3215.76 - 578.63}{2257}$	1.17
Equivalent evaporation	$E = \frac{m_e(h - h_{f1})}{2257}$	$\frac{8(3215.76 - 578.63)}{2257}$	9.35 kg/kg fuel
Boiler efficiency	$\eta_b = \frac{m_e(h - h_{f1})}{CV} \times 100$	$\frac{8(3215.76 - 578.63)}{32600} \times 100$	64.71%
Fuel composition: gravimetric to volumetric $a_1C + a_2H_2 + a_3O_2 + a_4N_2 + a_5S + a_6H2O$	$a_1 = \frac{C}{12}, a_2 = \frac{H_2}{2}, a_3 = \frac{O_2}{32}, a_4 = \frac{N_2}{28}, a_6 = \frac{H_2O}{18}$	$a_1 = \frac{0.84}{12}, a_2 = \frac{0.04}{2}, a_3 = \frac{0.07}{32}, a_6 = \frac{0.015}{18}$	$a_1 = 0.07$, $a_2 = 0.02$, $a_3 = 0.0022$, $a_4 = 0$, and $a_6 = 8.3241 \times 10^{-4}$
Water vapor's coefficient in flue gas	$b_5 = a_2 + a_6$	$0.02 + 8.3241 \times 10^{-4}$	0.0208
Fuel coefficient, p	$p = \frac{a_1}{b_1 + b_2}$	$\frac{0.07}{0.0013 + 0}$	56
Coefficient of air supply	$a_7 = p(b_1{+}0.5b_2{+}b_3){+}0.5b_5{+}b_6{-}a_3{-}0.5a_6$	$56(0.0013{+}0{+}7.5{\times}10^4){+}0.5{\times}0.0208{+}0{-}0.0022{-}0.5{\times}8.3241 \times 10^{-4}$	0.1198 kg mol
Coefficient of minimum air supply	$a_{7d} = a_1{+}0.5(a_2{+}a_6){+}a_5{-}a_3{-}0.5a_6$	$7{+}0.5(0.02{+}0.0832){+}0{-}0.2188{-}0.5{\times}0.0832$	7.78 kg mol
Mass of dry flue gas	$m_{dfg} = p(b_1M_{CO2}{+}b_2M_{CO}{+}b_3M_{O2}{+}b_4M_{N2})$	$56(0.013{\times}44{+}0{+}7.5{\times}10^{-4} \times 32{+}0.008 \times 28$	16.97 kg/s
Mass of carried steam	$m_{cs} = 9\ H_2$	9×0.04	0.36 kg/s
Heat from fuel	$Q_f = m_{fuel}\ CV$	1×32600	32600 kW (100%)
Heat used to generate steam	$Q_s = m_s\ (h - h_{f1})$	$8 \times (3215.76 - 578.63)$	2109.7 kW (64.71%)
Heat loss by dry flue gas	$Q_{dfg} = m_{dfg}\ c_{pg}\ (T_{goah} - T_a)$	$6.97 \times 1.0032\ (150{-}15)$	2298.9 kW (7.05%)
Partial pressure of vapour in flue gas	$p_{H2O,g} = \frac{b_5}{p(b_1 + b_2 + b_3 + b_4 + b_6}$	$\frac{0.0208}{56(0.0013 + 0 + 7.5 \times 10^{-4} + 0.008}$	0.0375 bar
Pressure of vapour at T_a	Saturation pressure at T_a	Steam properties at 15°C	0.017 bar
Heat loss with carried steam	$Q_{cs} = m_{cs}\ (h_{H2O,\ g} - h_a)$	$0.36 \times (2783.5 - 62.94)$	979.41 kW (3%)
Heat loss due to radiation (balance)	$Q_r = Q_f - (Q_s + Q_{dfg} + Q_{cs})$	$32600 - (2109.7{+}2298.9{+}979.41)$	8224.7 kW (25.23%)

3. A boiler using 1400 kg of coal per hour needs m kg of air per kg of coal for complete combustion. The temperature of flue gases is 285°C and the temperature of outside air is 25°C. If the draught produced by a 35 m high steel chimney is 18 mm of water column, design and determine the performance of draught.

Solutions through App Designer;

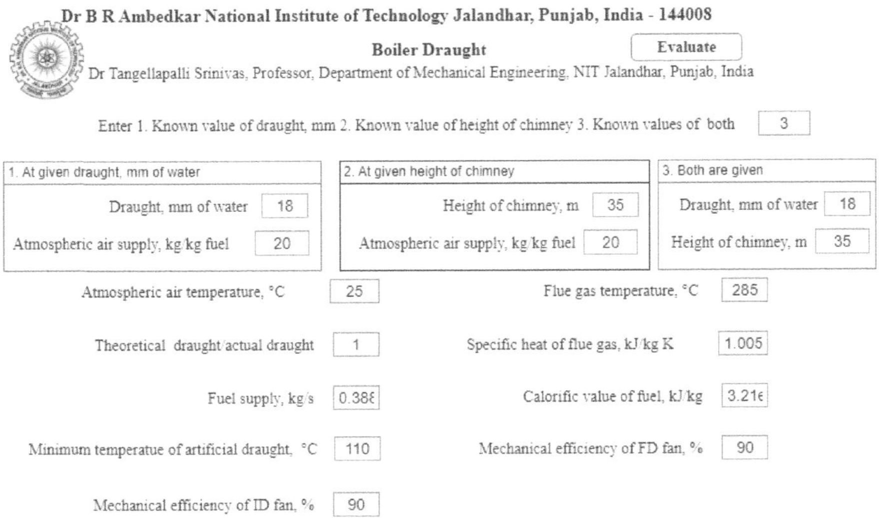

Parameter	**Equation**	**Substitution**	**Result**
Density of air at 0°C	$\rho_a = \frac{p_a}{RT_0}$	$\frac{101.325}{0.287 \times 273.15}$	1.29 kg/m³
Atmospheric air per kg of fuel	$m_e = \frac{1}{\left(\frac{1}{T_a} - \frac{h_w}{353H}\right)T_g - 1}$	$\frac{1}{\left(\frac{1}{298.15} - \frac{18}{353 \times 35}\right)558.15 - 1}$	16.98 kg/kg fuel
Density of gas	$\rho_g = 353\left(\frac{m_a + 1}{m_a}\right)\frac{1}{T_g}$	$353\left(\frac{16.98 + 1}{16.98}\right)\frac{1}{558.15}$	0.67 kg/m³
Draught	$p_{draught} = h_{w\,g}$	18 × 9.81	176.58 N/m²
Height of hot gas column	$H_1 = H\left(\frac{m_a}{m_a + 1}\right)\left(\frac{T_g}{T_a} - 1\right)$	$35\left(\frac{16.98}{16.98 + 1}\right)\left(\frac{558.15}{298.15} - 1\right)$	26.87 m
Velocity of fluid	$C = \sqrt{2gH_1}$	$\sqrt{2 \times 9.81 \times 26.87}$	22.96 m/s
Diameter of chimney	$D = \sqrt{\frac{4m_{t,g}}{\pi C \rho_g}}$	$\sqrt{\frac{4 \times (16.98 + 1)0.3889}{\pi \times 22.96 \times 0.67}}$	0.76 m
Efficiency of chimney	$\eta_{chimney} = \frac{H_1 g}{c_{pg}(T_g - T_{artificial\,d})} \times 100$	$\frac{26.87 \times 9.81}{1.005 \times 1000(285 - 110)} \times 100$	0.15%
Power of FD fan	$P_{FD} = \frac{p_{draught} T_a m_a m_f}{\eta_{fan} \rho_0 T_0}$	$\frac{176.58 \times 298.15 \times 16.98 \times 0.3889}{0.9 \times 1.2925 \times 273.15}$	1.09 kW
Power of ID fan	$P_{ID} = \frac{p_{draught} T_g m_a m_f}{\eta_{fan} \rho_0 T_0}$	$\frac{176.58 \times 558.15 \times 16.98 \times 0.3889}{0.9 \times 1.2925 \times 273.15}$	2.05 kW

4. The convergent-divergent nozzles of a steam turbine are supplied with dry saturated steam at a pressure of 9 bar. The exit pressure is 1 bar. Assuming nozzle efficiency as 90%, design a steam nozzle at 1 kg/s of fluid flow.

Solution through App Designer:

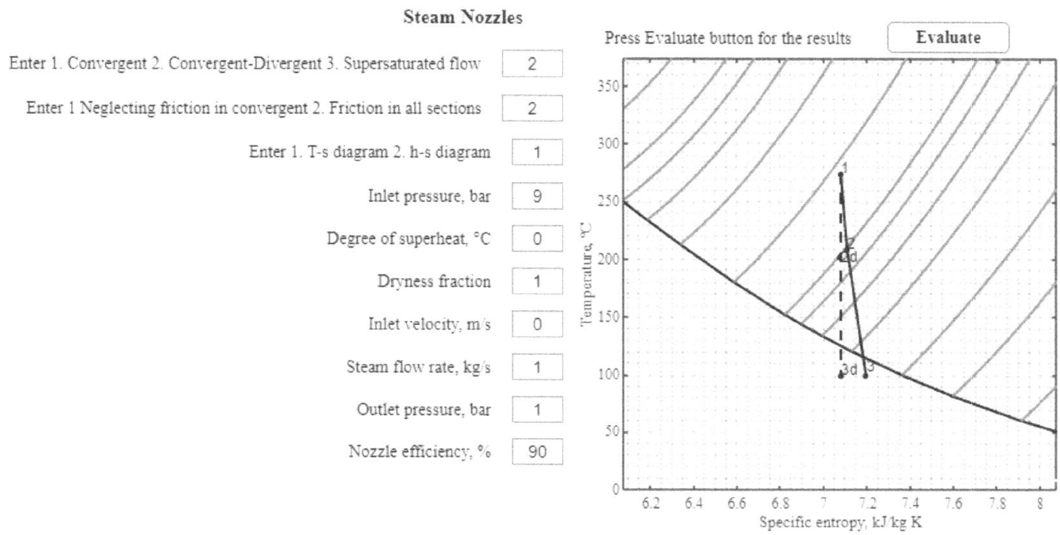

Parameter	Equation	Substitution	Result
Enthalpy and entropy of steam at entry of the nozzle	From steam properties	At 9 bar, dry saturated	2772.1 kJ/kg and 6.62 kJ/kg K
Throat pressure at the maximum discharge condition	$p_2 = p_1\left(\frac{2}{n+1}\right)^{n/(n-1)}$	$9\left(\frac{2}{1.135+1}\right)^{1.135/(1.135-1)}$	5.20 bar
Enthalpy of steam at the throat pressure after isentropic expansion	From steam properties	At 5.2 bar, wet steam (0.96)	2669.2 kJ/kg
Enthalpy of steam at the exit pressure after isentropic expansion	From steam properties	At 1 bar, wet steam (0.88)	2399.3 kJ/kg
Enthalpy of steam at the throat pressure after actual expansion	$h_2 = h_1 + \eta_n(h_1 - h_2')$	2772.1 – 0.9(2772.1–2669.2)	2679.8 kJ/kg
Enthalpy of steam at the exit pressure after actual expansion	$h_3 = h_1 + \eta_n(h_1 - h_3')$	2772.1 – 0.9(2772.1–2399.3)	2436.6 kJ/kg
Velocity of steam at the throat	$C_2 = \sqrt{2000(h_1 - h_2)}$	$\sqrt{2000(2772.1 - 2679.8)}$	429.83 m/s
Velocity of steam at the throat	$C_3 = \sqrt{2000(h_1 - h_3)}$	$\sqrt{2000(2772.1 - 2436.6)}$	819.16 m/s
Throat area	$A_2 = \frac{m_s v_2}{C_2}$	$\frac{1 \times 0.3494}{429.83}$	0.0008 m^2
Exit area	$A_3 = \frac{m_s v_3}{C_3}$	$\frac{1 \times 1.5147}{819.16}$	0.0018 m^2

5. The following observations are results of testing of a steam condenser: Barometer reading is 765 mm of Hg. The condenser vacuum is 685 mm of Hg. The mean condenser temperature is 34°C. The condensate temperature is = 28°C. The condensate is 1750 kg/h. The cooling water used is 78000 kg/h. The cooling water inlet and outlet temperatures are 18°C and 30°C respectively. Model and simulate (i) vacuum corrected to the standard barometer reading, (ii) vacuum efficiency (iii) undercooling of the condensate (iv) condenser efficiency (v) quality of the steam entering the condenser and (vii) mass of air per kg of uncondensed steam

Solution through App Designer:

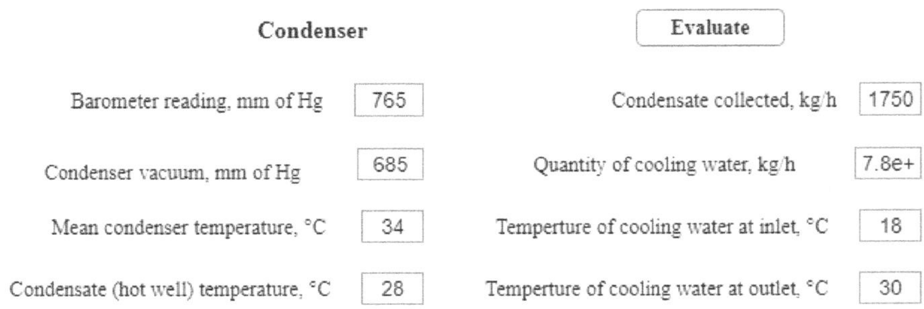

Parameter	Equation	Substitution	Result
Vacuum corrected to the standard barometric reading	$P_{cmm} = 760 - (p_{BRmm} - p_{CVmm})$	760 – (765–685)	680 mm
Absolute saturation pressure at condenser mean temperature (ideal pressure)	From steam properties	At 34°C	0.0532 bar = 39.98 mm
Gauge saturation pressure at condenser mean temperature (ideal pressure)	$P_{abs,\ mm} = p_{BRmm} - p_{satmm}$	765 – 39.98	725.01 mm
Vacuum efficiency	$\eta_{vacuum} = \dfrac{Condenser\ vacuum\,(corrected)}{Ideal\ vacuum}$	$\dfrac{685}{725.01}$	94.48%
Undercooling	$UC = T_{mc} - T_c$	34 – 28	6
Absolute pressure in condenser	$p_{BRmm} - p_{CVmm}$	765 – 685	80 mm = 0.1064 bar
Vacuum temperature at saturation pressure	From steam properties	At 0.1064 bar	47.05 °C
Condenser efficiency	$\eta_{condenser} = \dfrac{T_{wo} - T_{wi}}{T_v - T_{wi}}$	$\dfrac{30-18}{47.05-18}$	41.30%
Specific enthalpies at Vacuum temperature and hot well temperature	From steam properties	At 47.05°C and 28°C	196.92 kJ/kg, 2586.9 kJ/kg, 117.3 kJ/kg
Dryness fraction of steam	$x = \left(\dfrac{\dfrac{m_w c_{pw}(T_{wo} - T_{wi})}{m_s} + h_{f1} - h_f}{h_g - h_f} \right)$	$\left(\dfrac{\dfrac{21.66 \times 4.18(30-18)}{0.4861} + 117.3 - 196.92}{2586.9 - 196.92} \right)$	0.90
Partial pressure of air	$p_a = p_{abs} - p_{sat,\ abs}$	0.1064 – 0.0532	0.0532
Mass of air present per kg of uncondensed vapour	$m_a = \dfrac{p_a v_g}{RT_{mc}}$	$\dfrac{0.0532 \times 100 \times 26.6}{0.287 \times (34 + 273.15)}$	1.61 kg

Chapter 7
Turbomachinery

7.1 Introduction

A turbomachine is a dynamic rotary machine that transfers energy between fluid, and a rotor that has blades. Turbomachines are classified into hydraulic turbomachines and thermal turbomachines. Examples of turbomachinery are pumps, compressors, blowers, fans, and turbines. Compressors, fans, blowers, and pumps need energy (electricity) to drive the machine. This machine increases the fluid power by increasing the pressure and velocity as per requirement. Turbines generate electrical energy from fluid energy.

Similar to the heat exchanger, turbomachinery can also be developed with pre-design and post-design processes. The exit temperature of a turbine (either gas, steam, or any other vapor) can be predicted with modeling and simulation without having information of the physical system but with a set of assumptions, and the thermodynamic (isentropic) process and its relations.

7.2 Basics of turbomachinery

In a turbine, a stage refers to a set of blades that work together to extract energy from a fluid flow. Turbines can be designed with multiple stages to efficiently utilize the fluid's energy across different pressure and temperature ranges (Logan, 1993).

A suitable thermodynamic process for assessing turbomachinery is the adiabatic process. Compared to changes in enthalpy and kinetic energy, the change in potential energy can be neglected (Yahya, 2011).

The steady flow energy equation (SFEE) applied to a compressor,

$$h_1 + \frac{C_1^2}{2} + w_s = h_2 + \frac{C_2^2}{2} \tag{7.1}$$

The shaft work,

$$w_s = \left(h_2 + \frac{C_2^2}{2}\right) - \left(h_1 + \frac{C_1^2}{2}\right) \tag{7.2}$$

$$w_s = h_{02} - h_{01} = c_p\,(T_{02} - T_{01}) \tag{7.3}$$

If the inlet and outlet velocities are neglected,

$$w_s = h_2 - h_1$$

$$h_1 + \frac{C_1^2}{2} = h_{2s} + \frac{C_{2s}^2}{2} = h_2 + \frac{C_2^2}{2}$$

DOI: 10.1201/9781032642505-7

$h_{01} = h_{02s} = h_{02}$

For isentropic expansion, the ideal gas velocity is

$$C_{2s}^2 = 2(h_{01} - h_{2s})$$

For a perfect gas,

$$C_{2s}^2 = 2c_p(T_{01} - T_{2s})$$

For irreversible adiabatic expansion, the actual outlet velocity of the compressor,

$$C_2^2 = 2(h_{02} - h_2) \quad h_{01} = h_{02s} = h_{02}$$

For a perfect gas,

$$C_2^2 = 2c_p(T_{02} - T_2)$$

For the isentropic process,

$P_{01} = P_{02s}$

The stagnation enthalpy is constant.

$h_{01} = h_{02s}$

$$h_1 + \frac{1}{2}C_1^2 = h_{2s} + \frac{1}{2}C_2^2$$

C_{2s} is the outlet velocity at the exit of the diffuser.

For the actual process,

$$\frac{1}{2}(C_1^2 - C_{2s}^2) = \frac{1}{2}(C_1^2 - C_2^2)$$

$C_2 = C_{2s}$

$h_{01} = h_{02} = h_{02s}$

$h_2 = h_{2s}$

$\Delta P_0 = P_{01} - P_{02}$

A small-stage efficiency refers to an infinitesimally small pressure rise (or drop) in a turbine and a true measure of the aerodynamic performance of the compressor. It is an imaginary stage and independent of reheat effect. Let us take a small state between P and P + dP.

$(\Delta h)_s = c_p (\Delta T)_s$

$(\Delta h) = c_p (\Delta T)$

Small or infinitesimal stage efficiency of a compressor,

$$\eta_P = \frac{(dh)_s}{dh} = \frac{dT_s}{dT} \tag{7.4}$$

For an infinitesimal isentropic process,

$$\frac{dT_s}{T} = \frac{dh_s}{c_pT} = \frac{\left(\frac{dP}{\rho}\right)}{c_pT} = \frac{dPRTRT(\gamma-1)}{PR\gamma T} = \frac{dP}{P}\left(\frac{\gamma-1}{\gamma}\right)$$

Therefore, from η_p

$$\frac{dT}{T} = \frac{1}{\eta_p}\frac{dP}{P}\left(\frac{\gamma-1}{\gamma}\right)$$

This equation is for the actual compression process (1-2).

Integrating,

$$\ln p^{\frac{1}{\eta_p}\frac{\gamma-1}{\gamma}} = \ln T + \ln const$$

$$\frac{p^{\frac{1}{\eta_p}\frac{\gamma-1}{\gamma}}}{T} = const$$

It is applicable for the actual compression of a finite-stage or multi-stage machine. The small stage efficiency η_p is constant and can be determined from boundary conditions.

Applying the equation for states 1 and 2,

$$\frac{p_1^{\frac{1}{\eta_p}\frac{\gamma-1}{\gamma}}}{T_1} = \frac{p_2^{\frac{1}{\eta_p}\frac{\gamma-1}{\gamma}}}{T_2}$$

$$\frac{T_2}{T_1} = \left(\frac{p_2}{p_1}\right)^{\frac{1}{\eta_p}\frac{\gamma-1}{\gamma}}$$

Taking logs on both sides and rearranging.

$$\ln\left(\frac{T_2}{T_1}\right) = \frac{1}{\eta_p}\left(\frac{\gamma-1}{\gamma}\right)\ln\left(\frac{P_2}{P_1}\right)$$

$$\eta_p = \frac{\frac{\gamma-1}{\gamma}\ln\left(\frac{P_2}{P_1}\right)}{\ln\left(\frac{T_2}{T_1}\right)} = \frac{\ln\left(\frac{T_{2s}}{T_1}\right)}{\ln\left(\frac{T_2}{T_1}\right)} \tag{7.5}$$

Assuming the irreversible adiabatic compression (1-2) is equal to a polytropic process with index n,

$$\left(\frac{P_2}{P_1}\right)^{\frac{1}{\eta_p}\frac{\gamma-1}{\gamma}} = \left(\frac{P_2}{P_1}\right)^{\frac{n-1}{n}}$$

Equating the indices

$$\eta_p = \frac{\gamma-1}{\gamma}\frac{n}{n-1}$$

$$n = \frac{\gamma\eta_p}{1-\gamma(1-\eta_p)}$$

The efficiency of a finite stage can be related to the small stage efficiency.

The actual temperature rise in the stage,

$$T_2 - T_1 = T_1\left(\frac{T_2}{T_1} - 1\right) = T_1\left(p_r^{\frac{1}{\eta_p}\frac{\gamma-1}{\gamma}} - 1\right)$$

$$\eta_c = \frac{h_{2s} - h_1}{h_2 - h_1} = \frac{T_{2s} - T_1}{T_2 - T_1}$$

$$\eta_c = \frac{T_1\left(\frac{T_{2s}}{T_1} - 1\right)}{T_1\left(\frac{T_2}{T_1} - 1\right)}$$

$$= \frac{\left(\frac{T_{2s}}{T_1} - 1\right)}{\left(\frac{T_2}{T_1} - 1\right)}$$

The total or overall stage efficiency of a compressor,

$$\eta_c = \frac{p_r^{\frac{\gamma-1}{\gamma}} - 1}{p_r^{\frac{1}{\eta_p}\frac{\gamma-1}{\gamma}} - 1} \tag{7.6}$$

For a multi-stage compressor, η_c is replaced by the overall efficiency η_{cm} of the compressor and p_r by the overall pressure ratio p_{rm}.

$$\eta_c = \frac{p_{rm}^{\frac{\gamma-1}{\gamma}} - 1}{p_{rm}^{\frac{1}{\eta_p}\frac{\gamma-1}{\gamma}} - 1}$$

Figure 7.1 shows generalized inlet and outlet velocity triangles to formulate the energy transformation in a turbomachine.

The angular velocity of the rotor,

$$\omega = \frac{2\pi N}{60}$$

The inlet velocity of blade at diameter d_1,

$$C_{b1} = \frac{\pi d_1 N}{60}$$

The outlet velocity of blade,

$$C_{b2} = \frac{\pi d_2 N}{60}$$

The absolute velocity is observed by the observer at rest and the relative velocity is observed by the observer with the rotor.

These three velocities—blade velocity, absolute velocity, and relative velocity—are related as follows:

$$C = C_b + C_r$$

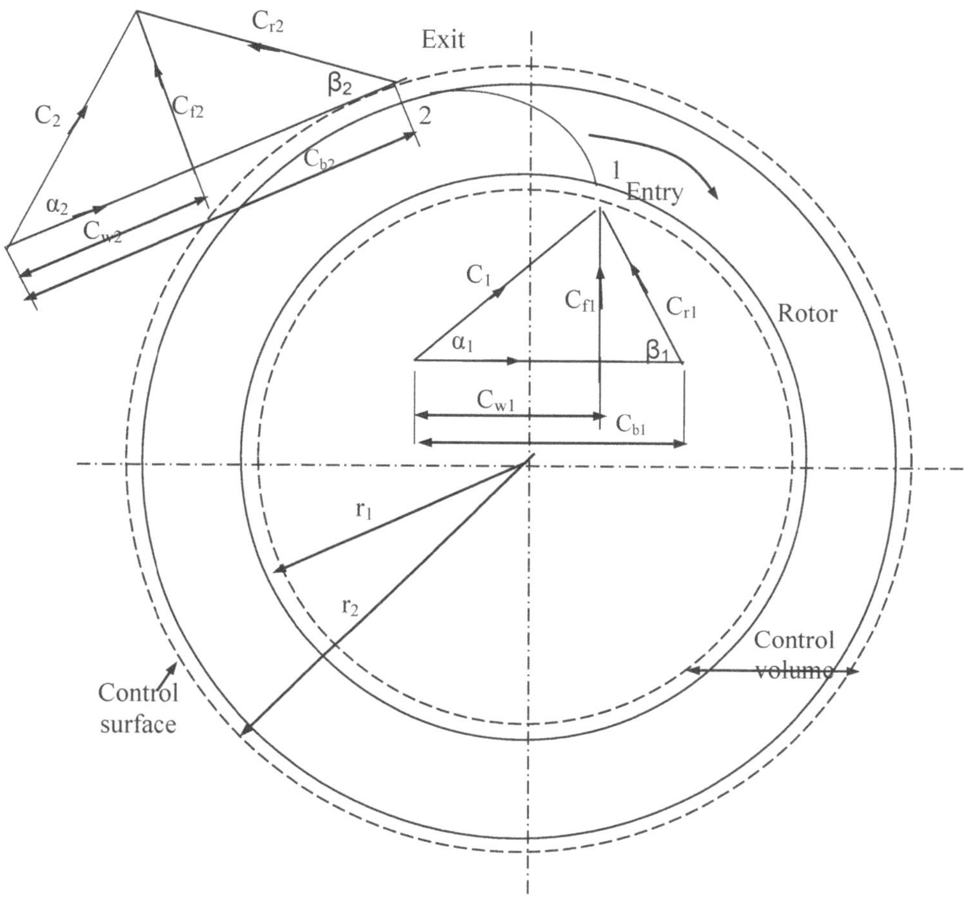

Fig. 7.1. Inlet and outlet velocity triangles of a general turbo machine.

The absolute velocity (C) has two components, one tangential (C_w) and the other, a radial component (C_f). The torques related to the rotor can be determined from Newton's second law of motion.

$$\sum F \times r = \frac{\partial}{\partial t}\int_{CV} \rho C_r dV + \int_{CS} \rho\left(CdA\right)C \times r$$

For steady flow,

$$\frac{\partial}{\partial t}\int_{CV} \rho\left(Cr\right)dV = 0$$

Therefore,

$$\sum F \times r = \int_{CS} \rho\left(CdA\right)C \times r$$

$$\tau = \sum F_\theta r = \int_{exit} \rho C_n dArC_w - \int_{in} \rho C_n dArC_w$$

The infinitesimal flow rate through the control surface,

$$d\dot{m} = \rho C_n dA$$

Therefore,

$$\tau = \sum F_{\theta} r = \int_{exit} rC_w d\dot{m} - \int_{in} rC_w d\dot{m}$$

Keeping 1 and 2 respectively for inlet and outlet,

$$\tau = r_2 C_{w2} \dot{m}_2 - r_1 C_{w1} \dot{m}_1$$

$m_1 = m_2 = m =$ constant for steady flow.

Therefore,

$$\tau_c = \dot{m}\left(r_2 C_{w2} - r_1 C_{w1}\right)$$

The torques exerted by the blades causes change in moment of momentum (angular momentum) at the inlet and exit of the rotor.

The energy from the fluid to the rotor decreases from inlet to outlet for the prime movers.

$r_1 C_{w1} > r_2 C_{w2}$

According to Newton's third law, there is an equal and opposite torque exerted by the fluid on the rotor blades.

$$\tau_r = \dot{m}\left(r_1 C_{w1} - r_2 C_{w2}\right)$$

Work = torque × angular velocity of the rotor

For compressors,

$$W_c = \tau_c \omega = \dot{m}\left(\omega r_2 C_{w2} - \omega r_1 C_{w1}\right)$$

$$W_c = \dot{m}\left(C_{b2} C_{w2} - C_{b1} C_{w1}\right)$$

The specific work is

$$w_c = C_{b2} C_{w2} - C_{b1} C_{w1}$$

Similarly, for turbines,

$$W_T = \tau_T \omega = \dot{m}\left(C_{b1} C_{w1} - C_{b2} C_{w2}\right)$$

$$w_T = C_{b1} C_{w1} - C_{b2} C_{w2}$$

These equations are Euler's equations.

The differential forms of equations,

$$dw_c = d(C_b\, C_w)$$

$$dw_T = -\, d(C_b\, C_w)$$

The resultant force on a turbomachine's blade has three components: tangential, axial, and radial. The tangential force provides a change in the momentum of the fluid in the peripheral direction, which is the driving force for the rotor. The axial thrust causes the static pressure and momentum change in the axial direction. It acts on the thrust bearings and does not cause the motion of the rotor. Similar to the axial force, the radial force results in static pressure and a change in momentum in the axial direction. It also does not participate in the rotation and adds to the thrust bearing. The design engineer has to maximize the tangential force and minimize the axial force and radial force.

The energy transfer equations from inlet velocity triangles,

$$C_{f1}^2 = C_1^2 - C_{w1}^2 = C_{r1}^2 - \left(C_{b1} - C_{w1}\right)^2 \tag{7.7}$$

$$C_1^2 - C_{w1}^2 = C_{r1}^2 - \left(C_{b1}^2 + C_{w1}^2 - 2C_{b1}C_{w1}\right)$$

$$C_1^2 = C_{r1}^2 - C_{b1}^2 + 2C_{b1}C_{w1}$$

$$C_{b1}C_{w1} = \frac{1}{2}\left(C_1^2 + C_{b1}^2 - C_{r1}^2\right) \tag{7.8}$$

From outlet velocity triangles,

$$C_{b2}C_{w2} = \frac{1}{2}\left(C_2^2 + C_{b2}^2 - C_{r2}^2\right) \tag{7.9}$$

Substituting and rewriting the equations with three parts of velocities,

$$w_T = \frac{1}{2}\left(C_1^2 - C_2^2\right) + \frac{1}{2}\left(C_{b1}^2 - C_{b2}^2\right) + \frac{1}{2}\left(C_{r2}^2 - C_{r1}^2\right) \tag{7.10}$$

$$w_c = \frac{1}{2}\left(C_2^2 - C_1^2\right) + \frac{1}{2}\left(C_{b2}^2 - C_{b1}^2\right) + \frac{1}{2}\left(C_{r1}^2 - C_{r2}^2\right) \tag{7.11}$$

In the differential form,

$$dw_T = -d\left(C_b C_w\right) = -d\left(\frac{1}{2}C^2\right) - d\left(\frac{1}{2}C_b^2\right) + d\left(\frac{1}{2}C_r^2\right) \tag{7.12}$$

The energy transfer in a turbine has three components. The value of C_b C_w decreases with a decrease in absolute velocity C_2, blade velocity C_{b2}, and an increase in relative velocity C_r. The differential form of absolute velocity is the change in the kinetic energy of the fluid through the machine. It is the change in the dynamic head of the turbomachine. Similarly, the differential form of blade velocity indicates the change in the centrifugal energy of the fluid, which is due to the rotation of the fluid and causes a change in the radius of rotation of the fluid and a change in the static head of the fluid. The differential form of relative velocity is the change in kinetic energy of the fluid in relative coordinates. It is also due to a change in the static head of fluid in a rotor. The energy transfer in the compressor also occurs in a similar manner.

A turbine's isentropic work represents the machine's maximum output. At these conditions, the tangential whirl velocity components, C_{wl} and C_{w2} are ideal.

Euler's enthalpy drop, $\Delta h_E = w_{ET} = C_{b1}\, C_{w1} - C_{b2}\, C_{w2}$

Euler's pressure drop, $(\Delta p)_{ET} = p_1 - p_2$

The irreversible or actual work generated by the turbine is less than Euler's work.

The isentropic pressure drop,

$(\Delta p)_{sT} = p_1 - p_{2s} = p_1 - p_{2a}$

The actual work can be determined from the actual values of the tangential velocity components.

$w_{aT} = C_{b1}\, C'_{w1} - C_{b2}\, C'_{w2}$

where C'_{w1} and C'_{w2} represent the actual values of the tangential velocity components.

In a compressor, the stage develops a lesser pressure rise than the Euler's pressure rise during an isentropic process. Therefore, in a compressor, the actual work is higher than the isentropic work.

7.3 Compressors

Figure 7.2 shows the classification of compressors. Basically, these are classified into two categories, viz., positive displacement compressors and non-positive compressors. In positive compressors, the pressure rise is due to the positive admission and delivery preventing undesired reversal of flow. The work is transferred by the hydrostatic force. In non-flow compressors, there is no provision to prevent the reversal flow. The fluid undergoes a change in momentum with the centrifugal force. The process in positive and non-positive compressors is considered an open system due to the continuous flow of fluid. The subdivisions can be seen in Fig. 7.2.

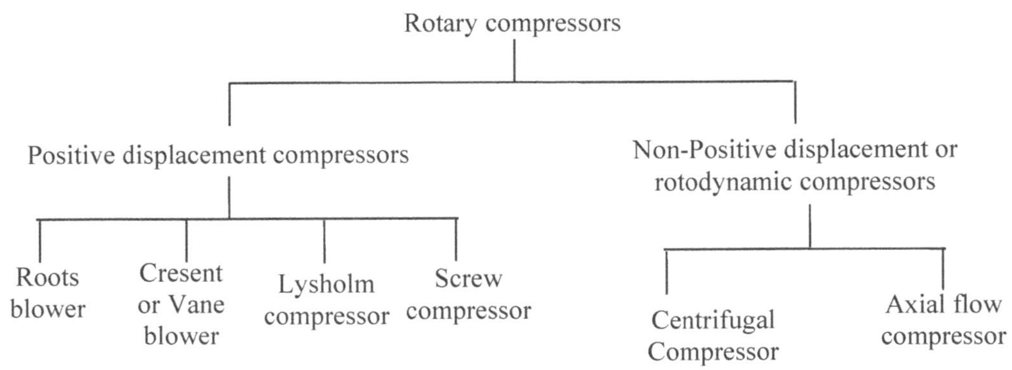

Fig. 7.2. Classification of rotary compressors.

7.3.1 Reciprocating compressors

Figure 7.3 shows the p-v diagram of a reciprocating compressor with the inner dead centre (IDC), clearance volume, and outer dead centre (ODC).

The swept volume or piston displacement,

$$V_p = (V_1 - V_3) = \frac{\pi D^2}{4} L \tag{7.13}$$

where D is the inner diameter and L is the length of stroke (IDC to ODC).

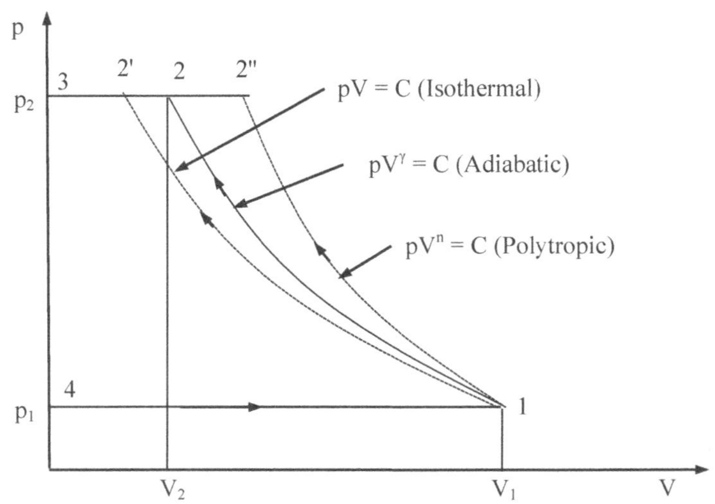

Fig. 7.3. P-v diagram of a reciprocating compressor.

In Fig. 7.3 of p-V diagram, 1 and 2 present the suction and delivery states of the fluid. At the end of the compression, the discharge valve opens. Figure 7.4 shows the p-V diagram of a compressor with the clearance volume.

$$w = p_2 v_2 + \frac{p_2 v_2 - p_1 v_1}{n-1} - p_1 v_1 \tag{7.14}$$

$$= \left(1 + \frac{1}{n-1}\right)(p_2 v_2 - p_1 v_1) = \frac{n}{n-1}(p_2 v_2 - p_1 v_1)$$

$$= \frac{n}{n-1} p_1 v_1 \left(\frac{p_2 v_2}{p_1 v_1} - 1\right)$$

$$= \frac{n}{n-1} p_1 v_1 \left(\frac{T_2}{T_1} - 1\right)$$

$$= \frac{n}{n-1} p_1 v_1 \left(\left(\frac{p_2}{p_1}\right)^{\frac{n-1}{n}} - 1\right)$$

The work with clearance volume is as follows:

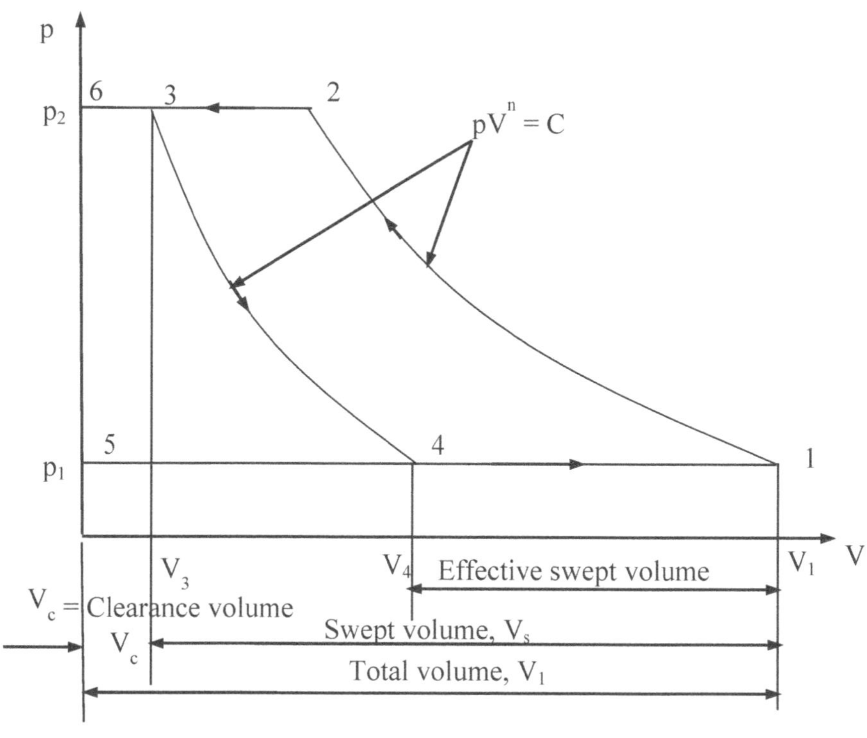

Fig. 7.4. P-V diagram with clearance volume.

The work without considering the clearance volume is the area as shown in Fig. 7.3. With clearance volume, a similar area for the expansion process 3-4 to be deducted (area on y-axis).

Therefore, the work with clearance volume,

$$w = \frac{n}{n-1} p_1 v_1 \left(\left(\frac{p_2}{p_1} \right)^{\frac{n-1}{n}} - 1 \right) - \frac{n}{n-1} p_1 v_4 \left(\left(\frac{p_2}{p_1} \right)^{\frac{n-1}{n}} - 1 \right) \quad \text{(since } p_4 = p_1\text{)}$$

$$w = \frac{n}{n-1} \left(p_1 v_1 - p_1 v_4 \right) \left(\left(\frac{p_2}{p_1} \right)^{\frac{n-1}{n}} - 1 \right) \tag{7.15}$$

7.3.2 Roots blower

Roots blower has two rotors which have separate parallel axes and rotate in opposite directions. The mechanical energy for the blower is taken from an external source and is supplied to one of the rotors while the other one is a gear driven from the first. The air is trapped and is positively contained in the specified boundaries. Roots blowers don't compress the air within the lobes themselves. Instead, they move the air from a low-pressure area to a high-pressure area. When the outlet is opened, there is a rise in pressure due to the resistance from the air in the receiver; therefore, the pressure in blower increases to the level of the pressure in the receiver (Fig. 7.5).

Let the intake pressure be P_1 and receiver pressure be P_2 then the pressure rise will be $(P_2 - P_1)$ at constant volume. If V is the volume between a lobe face and the casing, i.e., in space A, free air delivered will be 4 V and 6 V for a two lobe and 3 lobe rotor respectively per revolution of the compressor.

For a two lobe rotor, work done per revolution of rotor:

$$W = 4\,(P_2 - P_1)\,V \tag{7.16}$$

If Vs (= 4V) is the volume handled per minute at P_1 and T_1, the actual work done per minute will be

$$W_{roots} = (P_2 - P_1)\,V_s \tag{7.17}$$

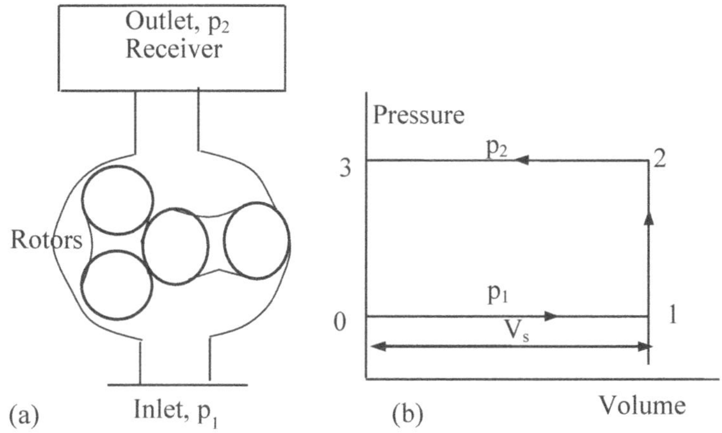

Fig. 7.5. Roots blower and its p-v diagram.

For an isentropic compression process from p_1 to p_2 the work input will be

$$W_{ideal} = \frac{\gamma}{\gamma - 1} P_1 V_s \left(\left(\frac{P_2}{P_1} \right)^{\frac{\gamma-1}{\gamma}} - 1 \right) \tag{7.18}$$

The performance of a roots blower is adjusted on the basis of roots efficiency.

Roots efficiency is the ratio of adiabatic (isentropic) work done to the actual work done on a compressor.

$$\eta_{roots} = \frac{isentropic\ work}{actual\ work} = \frac{W_{ideal}}{W_{roots}} \tag{7.19}$$

$$\eta_{roots} = \frac{\frac{\gamma}{\gamma - 1} P_1 V_s \left(\left(\frac{P_2}{P_1} \right)^{\frac{\gamma-1}{\gamma}} - 1 \right)}{(P_2 - P_1) V_s} \tag{7.20}$$

If r is the pressure ratio P_2/P_1 and substituting,

$$\frac{\gamma}{\gamma - 1} = \frac{c_p}{R}$$

$$\eta_{roots} = \frac{c_p}{R} \frac{\left((r)^{\frac{\gamma-1}{\gamma}} - 1 \right)}{r - 1} \tag{7.21}$$

Observe the roots efficiency is plotted in Fig. 7.6. The equation shows that η_{roots} will decrease when the pressure ratio increases.

Higher pressure ratio may be obtained by placing two or more blowers in series and having an intercooler between each stage. Roots blowers have been designed for delivery capacities ranging from 0.14 m³/min to 1400 m³/min and the pressure ratios of the order of 1 to 2 for a single machine and 1 to 3 for a two stage machine.

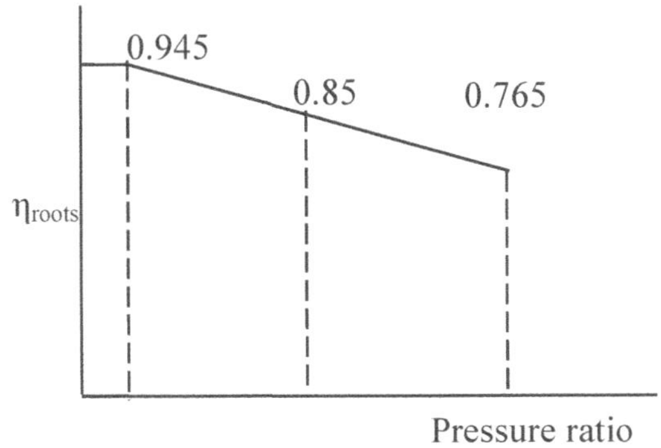

Fig. 7.6. Efficiency of a roots blower with pressure ratio.

7.3.3 Cresent or vane blower

The machine shown in Fig. 7.7, represents outline of a vane blower. It comprises a truly cylindrical outer casing into which a rotor is fitted eccentrically. It has a set of vanes of non-metallic material, usually of fiber or carbon type which can slide radially in the accurately fitted slots of the rotor.

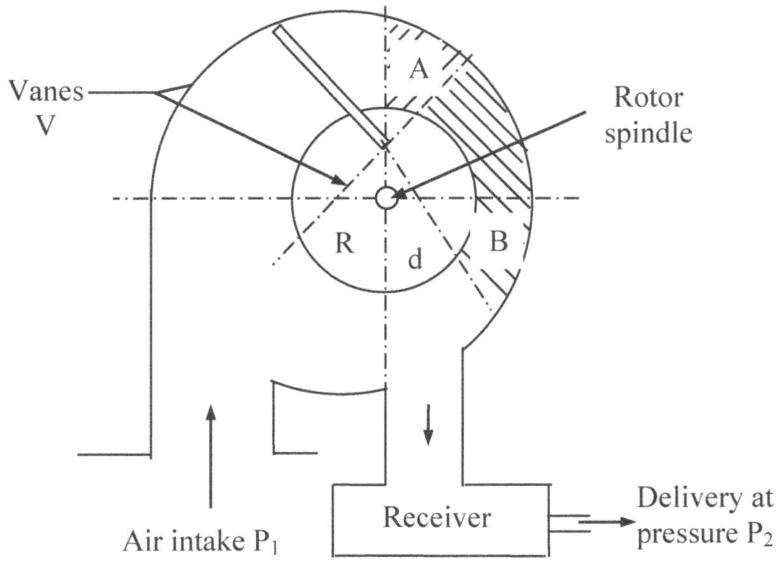

Fig. 7.7. Vane blower.

When the rotor rotates, centrifugal force acts on the vanes which throws the vanes out to make contact with the casing for all angular positions of the rotor. In order to remove the pressure of the blades from the casing, restraining rings having a bore slightly less than that of the outer casing are provided.

The compression of air trapped depends upon the eccentricity of the rotor to the casing and on the intake and delivery ports. The intake side opens to a large area of vanes and delivery opens to a much smaller area as shown in Fig. 7.3.

Assuming a rotor with only a single vane, work input can be calculated as below:

For position 'a' of the vane, the air is trapped at pressure p_1 and vane V continues collecting air till it reaches the point 'b', i.e., the position where there is little gap between the casing and vanes. Due to the eccentricity of the rotor and hence the decreasing space volume between the rotor and the casing, the compression of air starts at b. When the rotor reaches point c and further travels to d the air occupying volume A is compressed into a smaller volume B. Work input can be considered equal to area A but the machine is so designed that half of the total pressure rise takes place due to the momentary back flow action of the high pressure air. The p-v diagram for the compression process of a vane type blower is shown in Fig. 7.8.

The vane blower is used often for water vapour refrigerators and superchargers. It may be mentioned here that for a given air flow and for a given pressure ratio, the vane blower needs less work input than a roots blower.

Total work input = Area A + Area B

$$W = \frac{\gamma}{\gamma - 1} P_1 V_1 \left(\left(\frac{P_c}{P_1} \right)^{\frac{\gamma-1}{\gamma}} - 1 \right) + V_c \left(P_2 - P_1 \right) \tag{7.22}$$

$$p_1 V_1^{\gamma} = p_c V_c^{\gamma}$$

$$V_c = \left(\frac{p_1}{p_c} \right)^{\frac{1}{\gamma}} V_1 \tag{7.23}$$

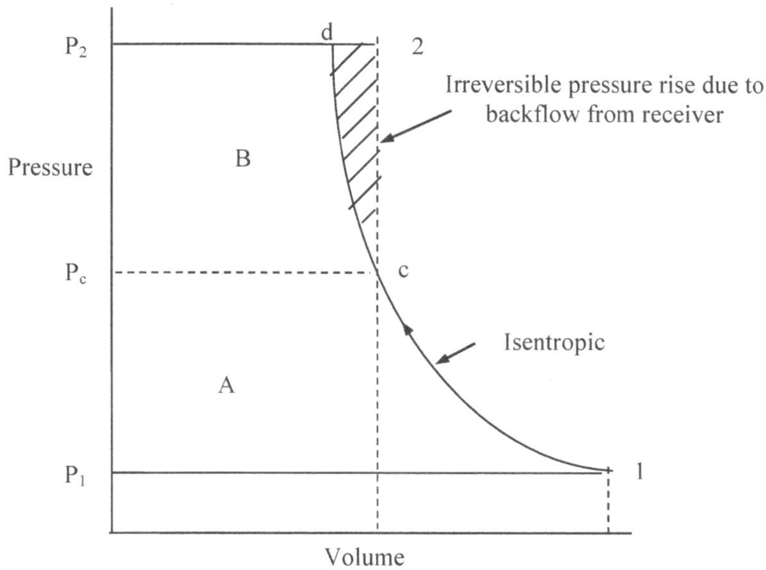

Fig. 7.8. p-v diagram of vane blower.

$$W = \frac{\gamma}{\gamma-1} P_1 V_1 \left(\left(\frac{P_c}{P_1} \right)^{\frac{\gamma-1}{\gamma}} - 1 \right) + \left(\frac{P_1}{P_c} \right)^{\frac{1}{\gamma}} V_1 (P_2 - P_1) \tag{7.24}$$

$$W_{ideal} = \frac{\gamma}{\gamma-1} P_1 V_1 \left(\left(\frac{P_2}{P_1} \right)^{\frac{\gamma-1}{\gamma}} - 1 \right) \tag{7.25}$$

$$\eta_{vane} = \frac{isentropic\ work}{actual\ work} = \frac{W_{1-d}}{W_{1-c-2}} \tag{7.26}$$

$$\eta_{vane} = \frac{\frac{\gamma}{\gamma-1} P_1 V_1 \left(\left(\frac{P_2}{P_1} \right)^{\frac{\gamma-1}{\gamma}} - 1 \right)}{\frac{\gamma}{\gamma-1} P_1 V_1 \left(\left(\frac{P_c}{P_1} \right)^{\frac{\gamma-1}{\gamma}} - 1 \right) + \left(\frac{P_1}{P_c} \right)^{\frac{1}{\gamma}} V_1 (P_2 - P_1)} \tag{7.27}$$

With the information of kinetic energies of fluid at the inlet and outlet, the performance of the rotary compressor can be evaluated with stagnation properties. Due to high speed and if friction is not present, ideal compression of air is represented by frictionless adiabatic pressure (1-2') $PV^{\gamma} = C$ as shown in Fig. 7.9 and Fig. 7.10.

But if friction is present, the actual process is (1-2) $PV^n = C$

If compression takes place isothermally, (1-3) $PV = C$, work input will be minimum.

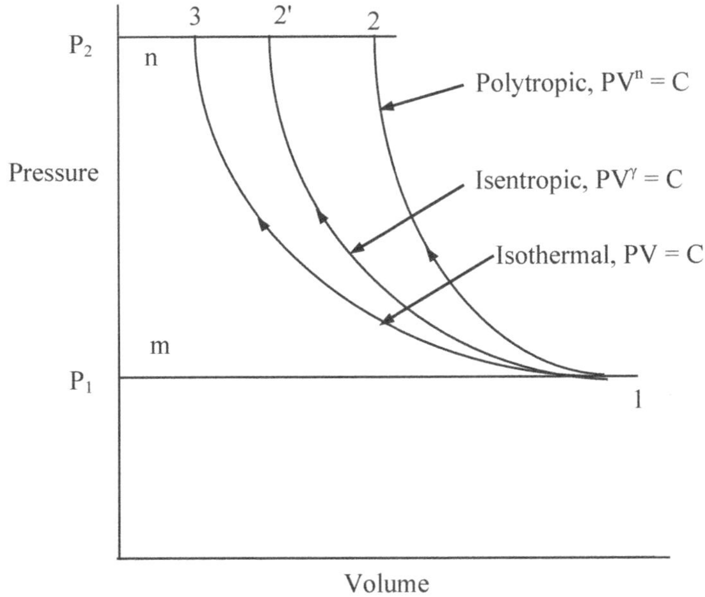

Fig. 7.9. Representation of isothermal, isentropic and polytropic compression on a P-V diagram.

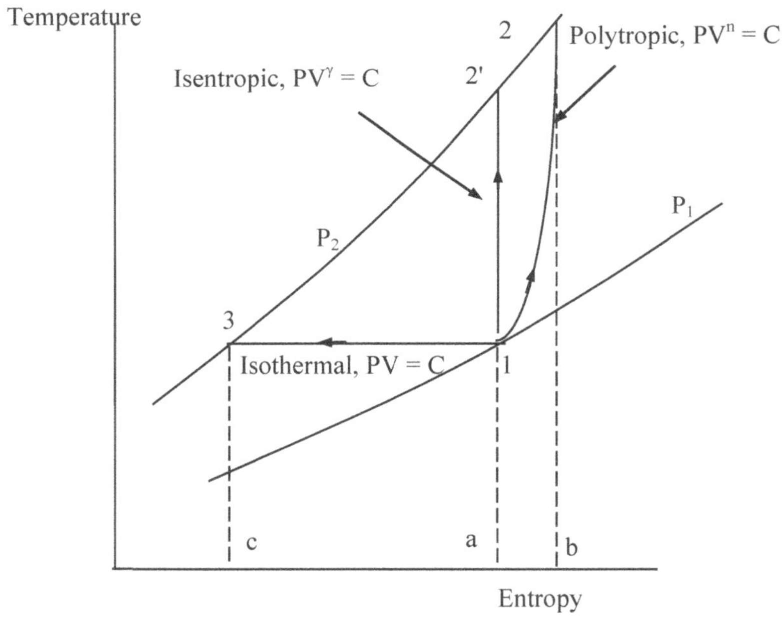

Fig. 7.10. Representation of isothermal, isentropic and polytropic compression on a T-s diagram.

Work of compression of PV diagram by (1-2'),

$$W = -\int VdP$$

$$PV^{\gamma} = C$$

$$V^{\gamma} = \frac{C}{P_0}$$

$$\text{or } V = \left(\frac{C}{P_0}\right)^{\frac{1}{\gamma}}$$

$$W = -\int_1^2 V dP$$

$$= -\int_1^2 \left(\frac{C}{P_0}\right)^{\frac{1}{\gamma}} dP$$

$$= -C^{\frac{1}{\gamma}} \int_1^2 (P_0)^{\frac{-1}{\gamma}} dP$$

$$= -C^{\frac{1}{\gamma}} \left(\frac{P_0^{1-\frac{1}{\gamma}}}{1-\frac{1}{\gamma}}\right)_1^2$$

$$= -\frac{C^{\frac{1}{\gamma}}}{\frac{\gamma-1}{\gamma}} \left(P_{0,2}^{1-\frac{1}{\gamma}} - P_{0,1}^{1-\frac{1}{\gamma}}\right)$$

$$= \frac{-\gamma}{\gamma-1}\left(C^{\frac{1}{\gamma}} P_{0,2}^{1-\frac{1}{\gamma}} - C^{\frac{1}{\gamma}} P_{0,1}^{1-\frac{1}{\gamma}}\right)$$

$$= \frac{-\gamma}{\gamma-1}\left(\left(P_{0,2}V_2'^{\gamma}\right)^{\frac{1}{\gamma}} P_{0,2}^{1-\frac{1}{\gamma}} - \left(P_{0,1}V_1^{\gamma}\right)^{\frac{1}{\gamma}} P_{0,1}^{1-\frac{1}{\gamma}}\right)$$

$$= \frac{-\gamma}{\gamma-1}\left(P_{0,2}V_2' - P_{0,1}V_1\right)$$

$$= \frac{-\gamma}{\gamma-1} P_{0,1}V_1 \left(\frac{P_{0,2}V_2'}{P_{0,1}V_1} - 1\right)$$

We have, P01V = mRT01

and

$$P_{0,1}V_1^{\gamma} = P_{0,2}V_2^{\gamma}$$

$$\left(\frac{V_2'}{V_1}\right)^{\gamma} = \frac{P_{0,1}}{P_{0,2}} = \left(\frac{P_{0,2}}{P_{0,1}}\right)^{-1}$$

$$\left(\frac{V_2'}{V_1}\right) = \left(\frac{P_{0,2}}{P_{0,1}}\right)^{\frac{-1}{\gamma}}$$

$$W = \frac{-\gamma}{\gamma-1} P_{0,1}V_1 \left(\frac{P_{0,2}V_2'}{P_{0,1}V_1} - 1\right)$$

$$= \frac{-\gamma}{\gamma-1} P_{0,1} V_1 \left(\frac{P_{0,2}}{P_{0,1}} \left(\frac{P_{0,2}}{P_{0,1}} \right)^{\frac{-1}{\gamma}} - 1 \right)$$

$$= \frac{-\gamma}{\gamma-1} P_{0,1} V_1 \left(\left(\frac{P_{0,2}}{P_{0,1}} \right)^{\frac{\gamma-1}{\gamma}} - 1 \right)$$

$$W = \frac{-\gamma}{\gamma-1} mRT_{0,1} \left(\left(\frac{P_{0,2}}{P_{0,1}} \right)^{\frac{\gamma-1}{\gamma}} - 1 \right) \tag{7.28}$$

Work done can also be determined by applying the steady flow energy equation.

$$m\left(h_1 + \frac{C_1^2}{2} + gZ_1 \right) + Q = m\left(h_2' + \frac{C_2'^2}{2} + gZ_2 \right) + W$$

With Q = 0 and Z1 = Z2

$$m\left(h_1 + \frac{C_1^2}{2} \right) = m\left(h_2'' + \frac{C_2'^2}{2} \right) + W$$

Considering the stagnation enthalpy

$$mh_{01,} = mh'_{0,2,} + (-W)$$

$$W = m(h'_{0,2} - h_{0,1})$$

$$W = mc_p(T'_{0,2} - T_{0,1})$$

$$W = m\frac{\gamma}{\gamma-1} R(T'_{0,2} - T_{0,1}) \tag{7.29}$$

Work done for a polytropic process (actual work) can be calculated from the steady flow energy equation

$$m\left(h_1 + \frac{C_1^2}{2} \right) + W_a = m\left(h_2 + \frac{C_2^2}{2} \right)$$

$$W_a = m(h_{0,2} - h_{0,1})$$

$$W_a = mc_p(T_{0,2} - T_{0,1})$$

Friction work = Actual work – isentropic work

$$W_f = c_p(T_{0,2} - T_{0,1}) - c_p(T'_{0,2} - T_{0,1})$$

= area m-1-2-n) – area (m-1-2'-n)
= area (1-2-2'-1)

Friction work consists of two parts

i) Heat generated by internal friction (1-2-b-1) and
ii) Increase in $-\int VdP$ work, i.e., (1-2-2') due to increase in volume

Polytropic work,

$$W_p = \frac{n}{n-1}\left(P_{0,2}V_2 - P_{0,1}V_1\right)$$

$$W_p = \frac{n}{n-1} mR\left(T_{0,2} - T_{0,1}\right)$$

$$W_p = \left(\frac{n}{n-1}\right)\left(\frac{\gamma-1}{\gamma}\right)c_p\left(T_{0,2} - T_{0,1}\right) \tag{7.30}$$

7.3.4 Centrifugal compressors

Centrifugal compressors are non-positive displacement compressors widely used for medium discharge and pressure applications. The coordinates in the centrifugal compressors are radial, tangential, and axial. The reference direction for the blade angle measurement is the tangential direction. The absolute and relative blade angles are represented with the notations α_1, α_2, and β_1, β_2, respectively. In axial compressors, the blade velocity is constant. But in the centrifugal compressors, due to changes in radius, the blade velocities are not equal.

$$C_{b1} = \frac{\pi d_1 N}{60} \tag{7.31}$$

$$C_{b2} = \frac{\pi d_2 N}{60} \tag{7.32}$$

Figure 7.11 shows the fluid flow in the inducer section of the centrifugal compressor's impeller without use of inlet guide valves (IGVs). The absolute inlet velocity C_1 is in the axial direction, and therefore, $\alpha_1 = 90°$. The inlet relative velocity C_{r1} changes as per the inlet guide vane's angle β_1 measured from the tangential direction. Therefore, the inlet whirl velocity is zero, $C_{w1} = 0$.

$$\tan\beta_1 = \frac{C_1}{C_{b1}} = \frac{C_{f1}}{C_{b1}} \tag{7.33}$$

Figure 7.12 shows the flow in the inducer section of a centrifugal compressor with inlet guide vanes. In the earlier case, the inducer section of the impeller is curved due to the absence of inlet

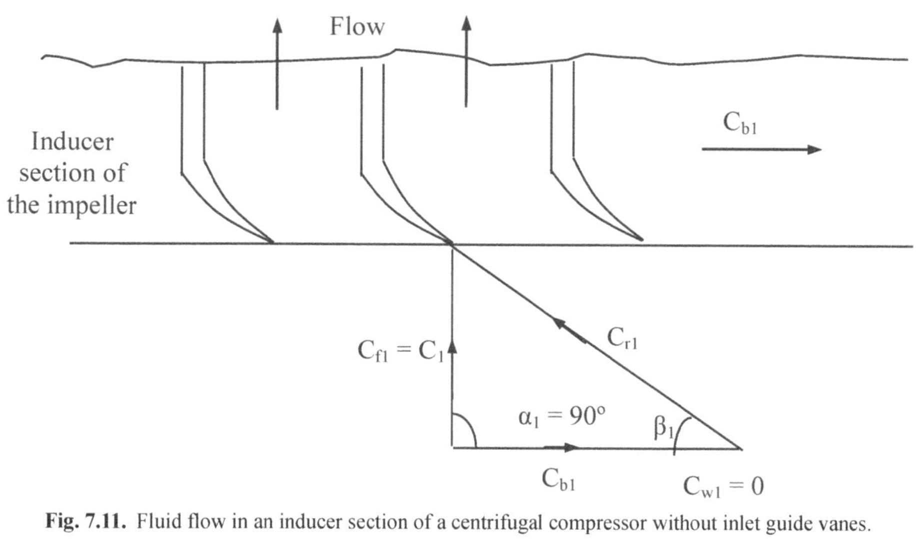

Fig. 7.11. Fluid flow in an inducer section of a centrifugal compressor without inlet guide vanes.

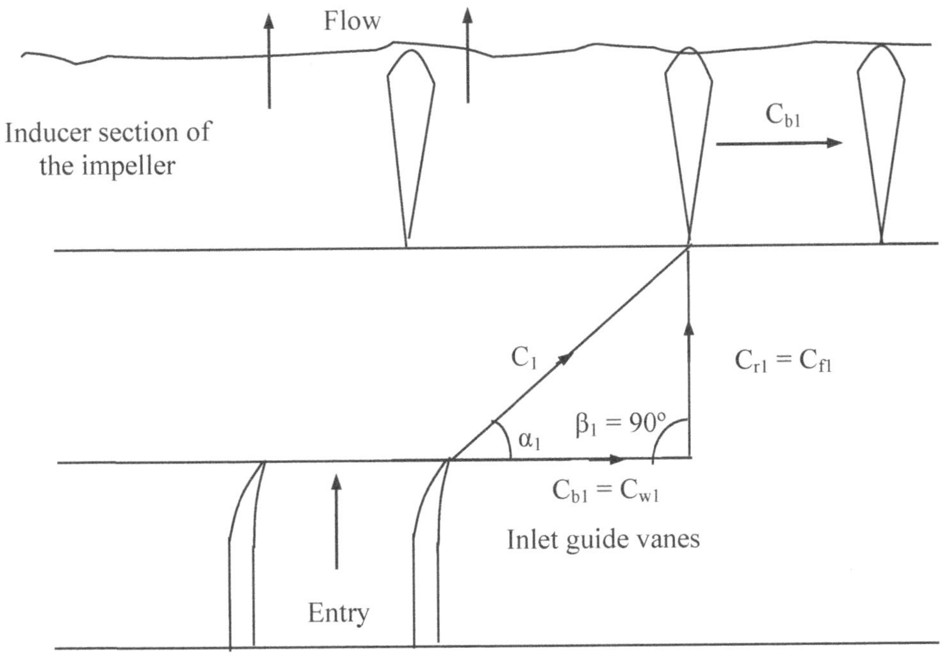

Fig. 7.12. Fluid flow in the inducer section of a centrifugal compressor with inlet guide vanes.

guide vanes. But with the inlet guide valves, the inducer section of a centrifugal compressor's impeller is straight, which is easy to manufacture. The absolute fluid angle α1 at the exit of inlet guide vanes provides an axial entry of the fluid to meet the straight vanes (C_{r1} is axial with $\beta_1 = 90°$).

From the inlet velocity triangles,

$$C_{w1} = C_{b1} \tag{7.34}$$

In this case, the absolute inlet velocity C_1 has a swirl component C_{w1}. Without guide vanes, there is a swirl component, which is called a zero whirl or swirl at the inlet. Unless mentioned, we are assuming zero whirl in all the cases.

Figure 7.13 shows the inlet and outlet velocity triangles with backward swept angles ($\beta_2 < 90°$). The fluid outlet relative velocity is Cr_2 with an angle of β_2. The outlet absolute velocity is C_2 with an angle of α_2. The tangential velocity and radial velocity (flow) are C_{w2} and C_{f2} respectively.

$$C_{f1} = C_1 \sin \alpha_1 = C_{r1} \sin \beta_1 \tag{7.35}$$

$$C_{w1} = C_1 \cos \alpha_1 = C_{f1} \cot \alpha_1 = C_{b1} - C_{f1} \cot \beta_1 \tag{7.36}$$

$$C_{f2} = C_2 \sin \alpha_2 = C_{r2} \sin \beta_2 \tag{7.37}$$

$$C_{w2} = C_2 \cos \alpha_2 = C_{f2} \cot \alpha_2 = C_{b2} - C_{f2} \cot \beta_2 \tag{7.38}$$

Figure 7.14 shows the inlet and outlet velocity triangles of a radial tipped impeller with inducer blades.

The outlet velocity triangles are special blades with radial tipped blades having $\beta_2 = 90°$.

$$C_{f2} = C_{r2} = C_2 \sin \alpha_2 \tag{7.39}$$

$$C_{w2} = C_2 \cos \alpha_2 = C_{f2} \cot \alpha_2 = C_{b2} \tag{7.40}$$

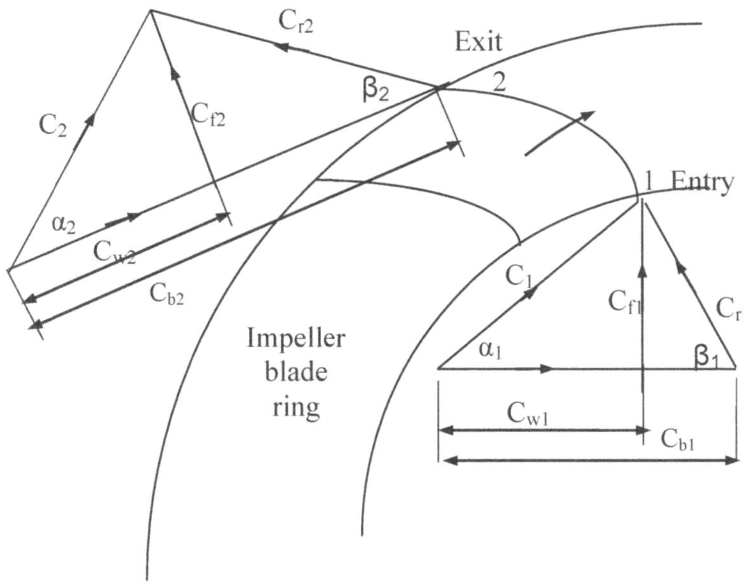

Fig. 7.13. Inlet and outlet blade velocity triangles with backward swept blades at the exit ($\beta_2 < 90°$).

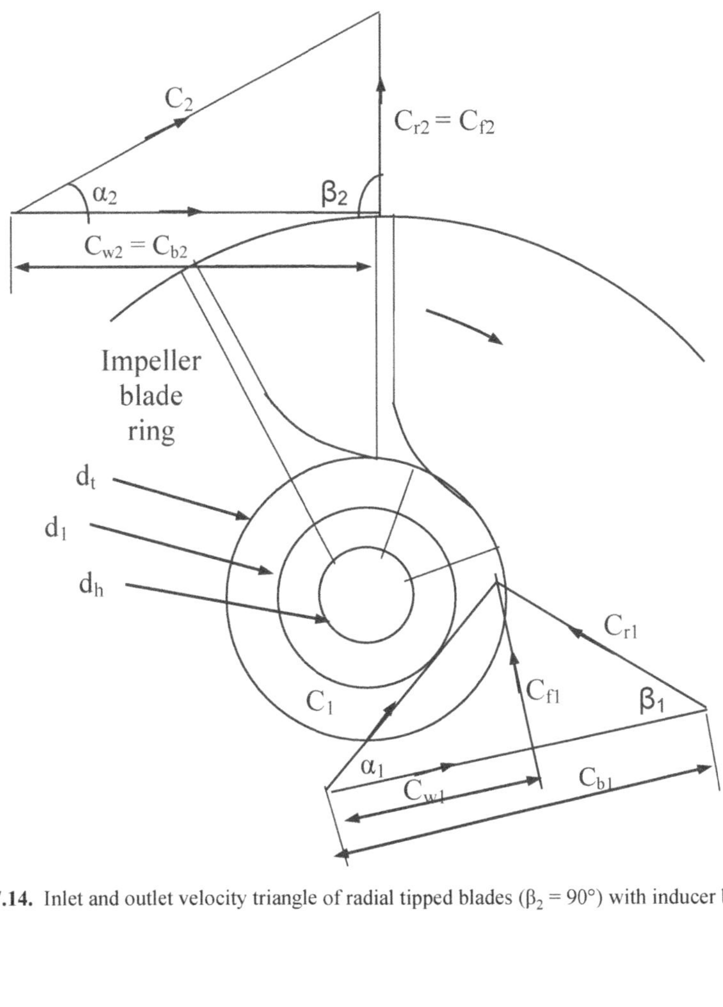

Fig. 7.14. Inlet and outlet velocity triangle of radial tipped blades ($\beta_2 = 90°$) with inducer blades,

The continuity equation,

$$\dot{m} = \rho_1 C_{f1} \pi d_1 b_1 = \rho_2 C_{f2} \pi d_2 b_2 \tag{7.41}$$

$$\dot{m} = \rho_1 C_{f1} \frac{\pi}{4}\left(d_1^2 - d_h^2\right) = \rho_2 C_{f2} \pi d_2 b_2 \tag{7.42}$$

Figure 7.15 shows the inlet and outlet velocity triangles for forward-swept blades ($\beta_2 > 90°$) with zero swirl at the entry. The swirl velocity at the exit of the blade is higher than the blade velocity ($Cw_2 > C_{b2}$) with a large angle of deflection of the fluid. The work and pressure rise increase with this design. But this is not suitable for the high speeds and causes losses. It is used in fans and centrifugal blowers. The backward and radial tipped blades in a centrifugal compressor have stable operating conditions.

With zero swirl at entry,

$C_{w1} = 0$

$C_{f2} = C_2 \sin \alpha_2 = C_{r2} \sin (180 - \beta_2) = C_{r2} \sin \beta_2$

$C_{w2} = C_2 \cos \alpha_2 = C_{f2} \cot \alpha_2 = C_{b2} + C_{f2} \cot (180 - \beta_2) = C_{b2} - C_{f2} \cot \beta_2$

The blade velocities at the inlet and outlet are C_{b1} and C_{b2} respectively.

The work,

$$w = C_{b2}\, C_{w2} - C_{b1}\, C_{w1} \tag{7.43}$$

For zero swirl at inlet, C_{w1} is zero.

$$w = C_{b2}\, C_{w2} \tag{7.44}$$

Substituting C_{w2},

$w = C_{b2} (C_{b2} - C_{f2} \cot \beta_2)$

The flow coefficient,

$$\phi_2 = \frac{C_{f2}}{C_{b2}} \tag{7.45}$$

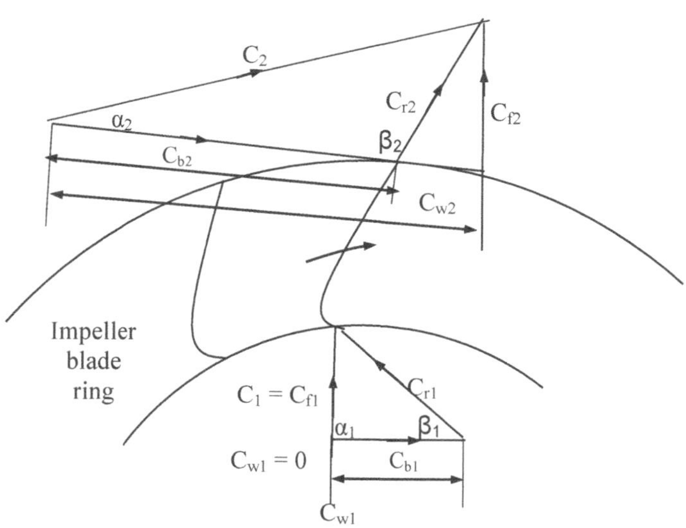

Fig. 7.15. Inlet and outlet velocity triangles for forward swept blades ($\beta_2 > 90°$) with zero swirl at entry.

Therefore,

$$w = C_{b2}^2 (1 - \varphi_2 \cot \beta_2) \tag{7.46}$$

The work can also be expressed in the form of Euler's equation.

$$w = \frac{1}{2}\left(C_2^2 - C_1^2\right) + \frac{1}{2}\left(C_{r1}^2 - C_{r2}^2\right) + \frac{1}{2}\left(C_{b2}^2 - C_{b1}^2\right) \tag{7.47}$$

For a radial-tipped impeller ($\beta_2 = 90°$) with zero swirl at entry ($\alpha_1 = 90°$),

$$w = C_{b2}^2 \tag{7.48}$$

Pressure Coefficient,

$$\psi = \frac{w}{C_{b2}^2} \tag{7.49}$$

$$\psi = \frac{C_{b2}C_{w2}}{C_{b2}^2}$$

$$\psi = \frac{C_{w2}}{C_{b2}} = \frac{C_{b2}(1 - \phi_2 \cot \beta_2)}{C_{b2}} \tag{7.50a}$$

$$\psi = 1 - \varphi_2 \cot \beta_2 \tag{7.50b}$$

Figure 7.16 plots the pressure coefficient - flow coefficient ($\varphi - \psi$) for a centrifugal compressor with forward-swept, radial and backward-swept impeller blades and zero swirl at inlet.

for radial-tipped blades,

$\psi = 1$

The actual work.

$$w_a = h_{04} - h_{01} = C_{b2}^2 \left(1 - \phi_2 \cot \beta_2\right) \tag{7.51a}$$

For a perfect gas,

$$w_a = c_p \left(T_{04} - T_{01}\right) = C_{b2}^2 \left(1 - \phi_2 \cot \beta_2\right) \tag{7.51b}$$

The ideal work between the same static pressures p_1 and p_4

$$w_s = h_{04ss} - h_{01} = c_p \left(T_{04ss} - T_{01}\right) \tag{7.52a}$$

$$w_s = c_p T_{01} \left(\frac{T_{04ss}}{T_{01}} - 1 \right)$$

$$w_s = c_p T_{01} \left(p_{r0}^{\frac{\gamma-1}{\gamma}} - 1 \right) \tag{7.52b}$$

The stagnation pressure ratio,

$$p_{r0} = \frac{p_{04ss}}{p_{01}} \approx \frac{p_{04}}{p_{01}} \tag{7.53}$$

This is valid for incompressible fluid flow.

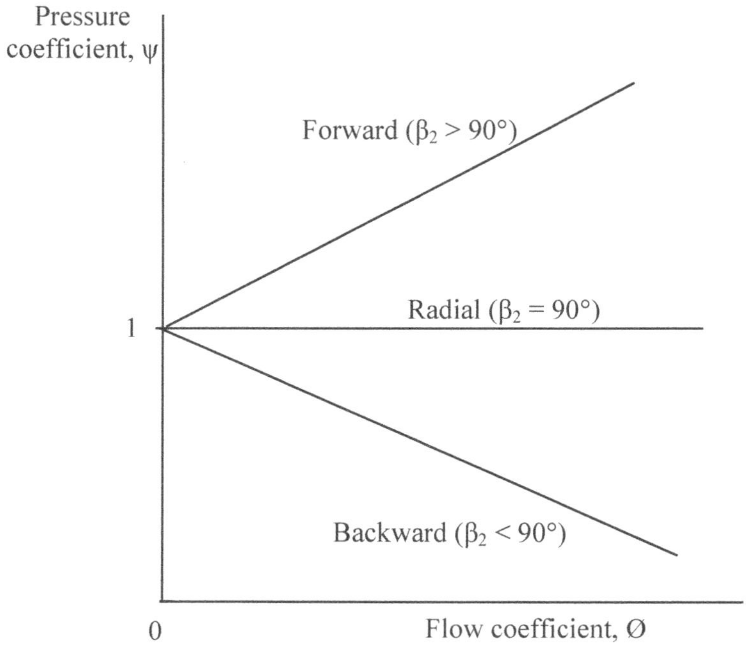

Fig. 7.16. Pressure coefficient and flow coefficient of a centrifugal compressor with zero swirl at inlet ($C_{w1} = 0$) and different blades.

Assume $c_4 \approx c_{4ss}$

The total-to-total efficiency of the stage,

$$\eta_{st} = \frac{w_s}{w_a} = \frac{h_{04ss} - h_{01}}{h_{04} - h_{01}} \tag{7.54a}$$

$$\eta_{st} = \frac{c_p\left(T_{04ss} - T_{01}\right)}{c_{b2}^2\left(1 - \phi_2 \cot \beta_2\right)} \tag{7.54b}$$

$$\eta_{st} = \frac{c_p T_{01}\left(p_{r0}^{\frac{\gamma-1}{\gamma}} - 1\right)}{C_{b2}^2\left(1 - \phi_2 \cot \beta_2\right)} \tag{7.54c}$$

$$p_{r0} = \left(1 + \eta_{st}\left(1 - \phi_2 \cot \beta_2\right)\frac{C_{b2}^2}{c_p T_{01}}\right)^{\frac{\gamma-1}{\gamma}} \tag{7.55}$$

Therefore the pressure ratio is a function of C_{b2}, φ_2 and β_2.
For the reversible stage, $\eta_{st} = 1$.

At the exit of the impeller, the energy is in the form of kinetic energy. It will be converted into pressure energy by the diffuser and volute casing at the exit of the compressor. The sharing of pressure rise in the impeller and the diffuser can be determined from the degree of reaction. The degree of reaction is the ratio of pressure rise in the impeller to pressure rise in stage (rotor (impeller) and its associated diffuser).

Degree of reaction,

$$R = \frac{\text{change in static enthalpy in the impeller}}{\text{change in static enthalpy in the stage}}$$

$$R = \frac{h_2 - h_1}{h_{02} - h_{01}} \tag{7.56}$$

$$h_{02} - h_{01} = \frac{1}{2}\left(C_2^2 - C_1^2\right) + \frac{1}{2}\left(C_{r1}^2 - C_{r2}^2\right) + \frac{1}{2}\left(C_{b2}^2 - C_{b1}^2\right)$$

Therefore,

$$h_2 - h_1 = \frac{1}{2}\left(C_{b2}^2 - C_{b1}^2\right) + \frac{1}{2}\left(C_{r1}^2 - C_{r2}^2\right) \tag{7.57}$$

At zero swirl at the entry, $C_{w1} = 0$

$$h_{02} - h_{01} = C_{b2}\, C_{w2} \tag{7.58}$$

$$R = \frac{\left(C_{b2}^2 - C_{b1}^2\right) + \left(C_{r1}^2 - C_{r2}^2\right)}{2C_{b2}C_{w2}} \tag{7.59}$$

At constant flow velocity (radial)

$$C_1 = C_{f1} = C_{f2}$$

$$C_{r1}^2 - C_{b1}^2 = C_{f1}^2 = C_{f2}^2 \tag{7.60}$$

$$C_{r2}^2 = C_{f2}^2 + \left(C_{b2} - C_{w2}\right)^2 = C_{f2}^2 + C_{b2}^2 + C_{w2}^2 - 2C_{b2}C_{w2}$$

$$C_{b2}^2 - C_{r2}^2 = 2C_{b2}C_{w2} - C_{w2}^2 - C_{f2}^2 \tag{7.61}$$

$$R = \frac{\frac{1}{2}\left(C_{r1}^2 - C_{r2}^2\right) + \frac{1}{2}\left(C_{b2}^2 - C_{b1}^2\right)}{C_{b2}C_{w2}}$$

$$= \frac{C_{r1}^2 - C_{b1}^2 + C_{b2}^2 - C_{r2}^2}{2C_{b2}C_{w2}}$$

$$= \frac{C_{f2}^2 + 2C_{b2}C_{w2} - C_{w2}^2 - C_{f2}^2}{2C_{b2}C_{w2}}$$

$$= \frac{2C_{b2}C_{w2} - C_{w2}^2}{2C_{b2}C_{w2}}$$

$$R = 1 - \frac{1}{2}\left(\frac{C_{w2}}{C_{b2}}\right) \tag{7.62a}$$

Simplifying,

$$R = 1 - \frac{1}{2}\left(1 - \phi_2 \cot \beta_2\right)$$

$$= 1 - \frac{1}{2} + \frac{1}{2}\phi_2 \cot\beta_2$$

$$R = \frac{1}{2} + \frac{1}{2}\phi_2 \cot\beta_2 \tag{7.62b}$$

Figure 7.17 analyses the effect of the flow coefficient and impeller exit angle on the degree of reaction. The degree of reaction (50%) is constant for a radial-tipped impeller ($\beta_2 = 90°$). The degree of reaction increases with an increase in flow coefficient with a backward swept impeller ($\beta_2 < 90°$) and decreases with an increase in flow coefficient with a forward swept impeller ($\beta_2 > 90°$).

$$R = 1 - \frac{1}{2}\psi \tag{7.63a}$$

$$\psi = 2(1 - R) \tag{7.63b}$$

As per this equation and Fig. 7.18, the degree of reaction and pressure coefficient are inversely proportional. The backward swept angle results in a high degree of reaction and a low pressure coefficient. The forward swept angle gives the lower degree of reaction and the high pressure coefficient.

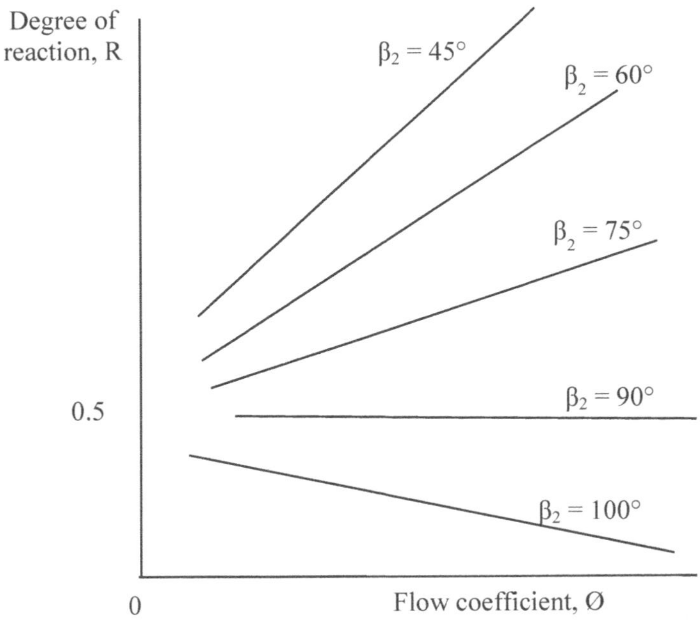

Fig. 7.17. Changes in the degree of reaction with flow coefficient and impeller outlet angle.

7.3.5 Axial flow compressors

Figure 7.19 shows the inlet and outlet velocity triangles for an axial flow compressor. The entry of blade, outlet of blade and the diffuser are numbered as 1, 2, and 3 respectively. The absolute and relative fluid angles are represented by α_1, α_2, α_3 and β_1, β_2 respectively.

As the flow repeats in the next stage,

$C_1 = C_3$ and $\alpha_1 = \alpha_3$

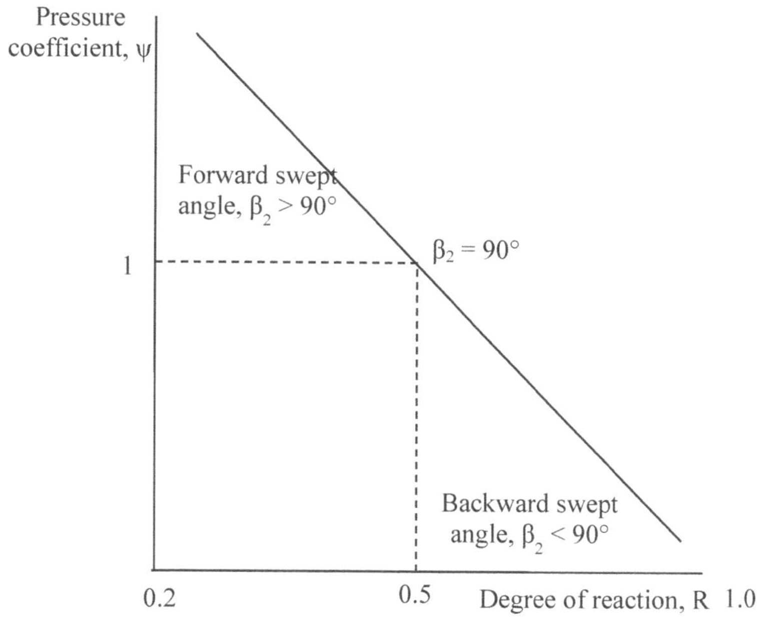

Fig. 7.18. Variation of pressure coefficient with degree of reaction.

The absolute velocities are C_1 and C_2, swirl (tangential) velocities are C_{w1} and C_{w2} and relative velocities are Cr_1 and Cr_2, respectively at the inlet and outlet.

The blade velocities are constant at the inlet and outlet.

$C_b \approx C_{b1} \approx C_{b2}$

From the inlet velocity triangles,

$$C_{f1} = C_1 \cos \alpha_1 = Cr_1 \cos \beta_1 \quad (7.64)$$

$$C_{w1} = C_1 \sin \alpha_1 = C_{f1} \tan \alpha_1 \quad (7.65)$$

$$C_{rw1} = C_{r1} \sin \beta_1 = C_{f1} \tan \beta_1 \quad (7.66)$$

$$C_b = C_{w1} + C_{rw1} \quad (7.67a)$$

$$C_b = C_1 \sin \alpha_1 + C_{r1} \sin \beta_1 \quad (7.67b)$$

$$C_b = C_{f1} (\tan \alpha_1 + \tan \beta_1) \quad (7.67c)$$

From the outlet velocity triangles,

$$C_{f2} = C_2 \cos \alpha_2 = C_{r2} \cos \beta_2 \quad (7.68)$$

$$C_{w2} = C_2 \sin \alpha_2 = C_{f2} \tan \alpha_2 \quad (7.69)$$

$$C_{rw2} = C_{r2} \sin \beta_2 = C_{f2} \tan \beta_2 \quad (7.70)$$

$$C_b = C_{w2} + C_{rw2} \quad (7.71a)$$

$$C_b = C_2 \sin \alpha_2 + C_{r2} \sin \beta_2 \quad (7.71b)$$

$$C_b = C_{f2} (\tan \alpha_2 + \tan \beta_2) \quad (7.71c)$$

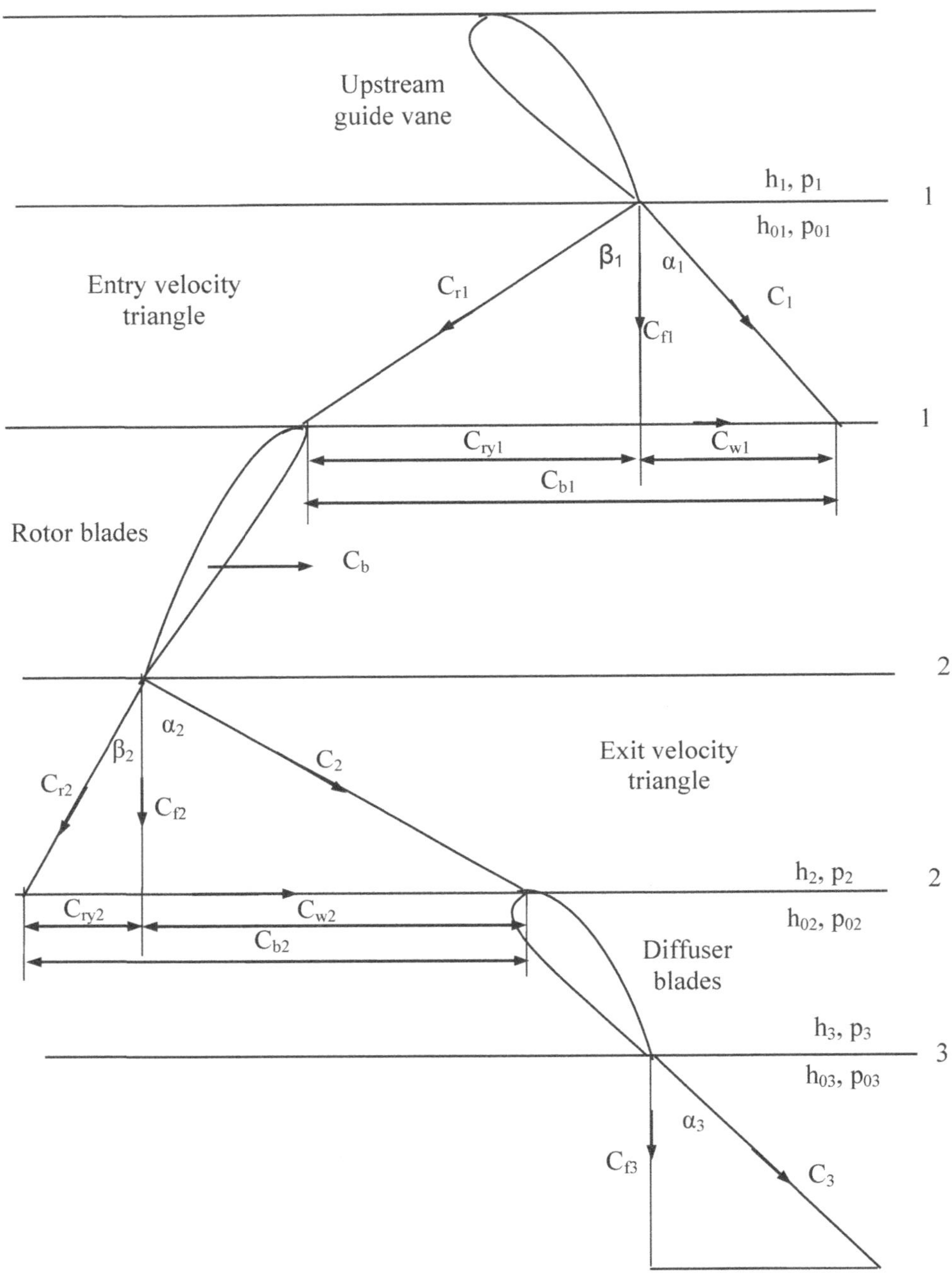

Fig. 7.19. Velocity triangles for an axial compressor stage.

The axial velocity is constant.

$$C_{f1} = C_{f2} = C_{f3} = C_f \tag{7.72}$$

$$C_f = C_1 \cos \alpha_1 = C_{r1} \cos \beta_1 = C_2 \cos \alpha_2 = C_{r2} \cos \beta_2 \tag{7.73}$$

$$C_{w1} + C_{rw1} = C_{w2} + C_{rw2}$$

$$C_{w2} - C_{w1} = C_{rw1} - C_{rw2} \tag{7.74}$$

$$C_f (\tan \alpha_2 - \tan \alpha_1) = C_f (\tan \beta_1 - \tan \beta_2) \tag{7.75}$$

Alternatively,

$$\frac{C_b}{C_f} = \tan\alpha_1 + \tan\beta_1$$

$$\frac{C_b}{C_f} = \tan\alpha_2 + \tan\beta_2$$

$$\tan\alpha_1 + \tan\beta_1 = \tan\alpha_2 + \tan\beta_2$$

The specific work,

$$w = C_b (C_{w2} - C_{w1}) \tag{7.76a}$$

$$= c_p (T_{02} - T_{01})$$

$$C_{w1} = C_{f1} \tan\alpha_1$$

$$C_{w2} = C_{f2} \tan\alpha_2$$

$$w = C_b C_f (\tan\alpha_2 - \tan\alpha_1) = C_b C_f (\tan\beta_1 - \tan\beta_2) \tag{7.76b}$$

Using velocity triangles and cosine rule,

$$C_{r1}^2 = C_{b1}^2 + C_1^2 - 2C_{b1}C_1 \cos(90 - \alpha_1)$$

$$C_{r1}^2 = C_{b1}^2 + C_1^2 - C_{b1}C_1 \sin\alpha_1$$

C1 sinα1 = Cw1

$$C_{r1}^2 = C_{b1}^2 + C_1^2 - 2C_{b1}C_{w1}$$

$$C_{b1}C_{w1} = \frac{C_{b1}^2 + C_1^2 - C_{r1}^2}{2}$$

Similarly, from the exit velocity triangles,

$$C_{b2}C_{w2} = \frac{C_{b2}^2 + C_2^2 - C_{r2}^2}{2}$$

$$w = C_{b2}C_{w2} - C_{b1}C_{w1}$$

$$= \frac{C_{b2}^2 + C_2^2 - C_{r2}^2}{2} - \frac{C_{b1}^2 + C_1^2 - C_{r1}^2}{2}$$

For an axial-flow compressor ($C_b = C_{b1} = C_{b2}$), the specific work equation is

$$w = \frac{C_2^2 - C_1^2}{2} + \frac{C_{r1}^2 + -C_{r2}^2}{2} \tag{7.76c}$$

The absolute velocity term related to the dynamic pressure rise and the relative velocity indicates the static pressure rise.

The blade loading coefficient or pressure coefficient,

$$\psi = \frac{w}{C_b^2} \tag{7.77}$$

It is a number used to compare the stages of differing sizes and speeds.

The flow coefficient,

$$\phi = \frac{C_f}{C_b} \tag{7.78}$$

$$\psi = \frac{C_b\left(C_{w2} - C_{w1}\right)}{C_b^2}$$

$$\psi = \frac{C_{w2}}{C_b} - \frac{C_{w1}}{C_b}$$

$$\psi = \frac{C_f \tan\alpha_2}{C_b} - \frac{C_f \tan\alpha_1}{C_b}$$

$$\psi = \varphi\,(\tan\alpha_2 - \tan\alpha_1) = \varphi\,(\tan\beta_1 - \tan\beta_2) \tag{7.79}$$

Figure 7.20 shows the φ-ψ plot for an axial compressor stage. The main objective of a compressor is to increase the static pressure of fluid, which depends on the geometry and rotor speed. The total static pressure rise is the sum of static pressure rises in the rotor and diffuser. In the derivation, a reversible adiabatic flow with constant axial velocity has been assumed. The flow is assumed to be incompressible due to the small pressure rise.

The Bernoulli equation across the rotor blade row,

$$p_1 + \frac{1}{2}\rho C_{r1}^2 = p_2 + \frac{1}{2}\rho C_{r2}^2$$

$$(\Delta p)_R = p_2 - p_1 = \frac{1}{2}\rho\left(C_{r1}^2 - C_{r2}^2\right) \tag{7.80}$$

Using velocity triangles of Fig. 1b,

$$C_{r1}^2 - C_{r2}^2 = C_{f1}^2 + C_{rw1}^2 - C_{f2}^2 - C_{rw2}^2 = C_{rw1}^2 - C_{rw2}^2$$

$$C_{r1}^2 - C_{r2}^2 = C_f^2\left(\tan^2\beta_1 - \tan^2\beta_2\right) \tag{7.81}$$

Bernoulli's equation across the diffuser blade row,

$$(\Delta p)_D = p_3 - p_2 = \frac{1}{2}\rho\left(C_2^2 - C_3^2\right) \tag{7.82}$$

$$= \frac{1}{2}\rho\left(C_{f2}^2 + C_{w2}^2 - C_{f3}^2 - C_{w3}^2\right)$$

$$= \frac{1}{2}\rho\left(C_{w2}^2 - C_{w3}^2\right)$$

$$= \frac{1}{2}\rho\left(\left(C_{f2}\tan\alpha_2\right)^2 - \left(C_{f3}\tan\alpha_3\right)^2\right)$$

$$= \frac{1}{2}\rho C_{f2}^2\left(\tan^2\alpha_2 - \tan^2\alpha_3\right)$$

Assuming $\alpha_3 = \alpha_1$

$$(\Delta p)_D = \frac{1}{2}\rho C_{f2}^2\left(\tan^2\alpha_2 - \tan^2\alpha_1\right) \tag{7.83}$$

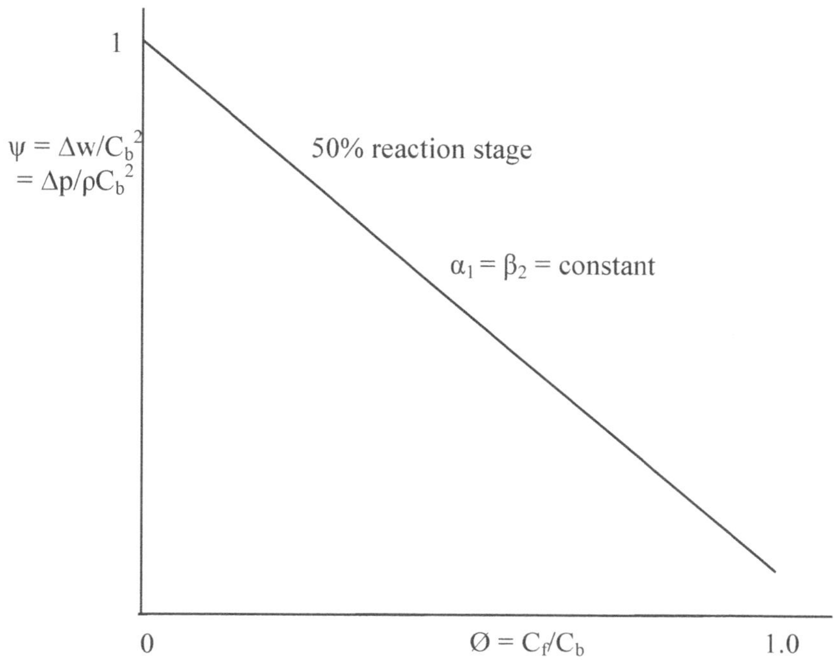

Fig. 7.20. Relation between pressure coefficient and flow coefficient of an axial flow compressor.

The stage pressure rise,

$p_3 - p_1 = \Delta p_R + \Delta p_D$

Substituting,

$$(\Delta p)_{st} = \frac{1}{2}\rho C_f^2\left(\tan^2\beta_1 - \tan^2\beta_2 + \tan^2\alpha_2 - \tan^2\alpha_1\right)$$

$$= \frac{1}{2}\rho C_{f2}^2\left((\tan\beta_1 + \tan\beta_1)(\tan\beta_1 - \tan\beta_1) + (\tan\alpha_2 + \tan\alpha_1)(\tan\alpha_2 - \tan\alpha_1)\right)$$

since,

$$\tan\alpha_1 + \tan\beta_1 = \tan\alpha_2 + \tan\beta_2$$

or

$$\tan\beta_1 - \tan\beta_2 = \tan\alpha_2 - \tan\alpha_1$$

$$(\Delta p)_{st} = \frac{1}{2}\rho C_{f2}^2(\tan\beta_1 - \tan\beta_1)\left((\tan\beta_1 + \tan\beta_1) + (\tan\alpha_2 + \tan\alpha_1)\right)$$

$$= \frac{1}{2}\rho C_{f2}(\tan\beta_1 - \tan\beta_1)\left(C_{f2}(\tan\beta_1 + \tan\beta_1) + C_{f2}(\tan\alpha_2 + \tan\alpha_1)\right)$$

$$= \frac{1}{2}\rho C_{f2}(\tan\beta_1 - \tan\beta_1)\left(C_{b1} + C_{b2}\right)$$

$$= \rho C_f C_2(\tan\beta_1 - \tan\beta_1) \quad (7.84)$$

$$(\Delta p)_{st} = \rho C_f C_b(\tan\alpha_2 - \tan\alpha_1) \quad (7.85)$$

For isentropic flow,

$$\frac{(\Delta p)_{st}}{\rho} = \Delta h_{st} = w \tag{7.86}$$

The ideal work,

$$w_s = h_{03ss} - h_{01} = c_p\,(T_{o3ss} - T_{01}) \tag{7.87}$$

Actual work,

$$w_a = h_{02} - h_{01} = h_{03} - h_{01} = c_p\,(T_{03} - T_{01}) \tag{7.88}$$

Total-to-total efficiency

$$\eta_{tt} = \frac{\textit{ideal stage work between total conditions at entry and exit}}{\textit{actual stage work}}$$

$$\eta_{tt} = \frac{w_s}{w_a} = \frac{h_{03ss} - h_{01}}{h_{03} - h_{01}} = \frac{T_{03ss} - T_{01}}{T_{03} - T_{01}} \tag{7.89}$$

The stage work determined from the actual velocity triangles is equal to the actual work.

$$h_{03} - h_{01} = C_b\,(C_{w2} - C_{wl}) \tag{7.90a}$$

$$h_{03} - h_{01} = C_b\,C_f(\tan\alpha_2 - \tan\alpha_1) = C_b\,C_f\,(\tan\beta_1 - \tan\beta_2) \tag{7.90b}$$

$$h_{03} - h_{01} = \frac{1}{2}\left(C_2^2 - C_1^2\right) + \frac{1}{2}\left(C_{r1}^2 - C_{r2}^2\right) \tag{7.90c}$$

Actual value of the loading coefficient,

$$\psi = \frac{h_{03} - h_{01}}{C_b^2} = \frac{c_p\left(T_{03} - T_{01}\right)}{C_b^2} \tag{7.91}$$

Substituting,

$$\psi = \frac{h_{03ss} - h_{01}}{\eta_{tt} C_b^2} = \frac{c_p\left(T_{03ss} - T_{01}\right)}{\eta_{tt} C_b^2} \tag{7.92}$$

It is not valid for an actual flow as

$$w > \frac{(\Delta p)_{st}}{\rho}$$

For isentropic flow,

$$\frac{(\Delta p)_{st}}{\rho} = w_s = h_{03ss} - h_{01} = c_p(T_{03ss} - T_{01}) \tag{7.93}$$

Stage efficiency,

$$\eta_{tt} = \frac{(\Delta p)_{st}}{\rho(h_{03} - h_{01})} \tag{7.94a}$$

$$\eta_{tt} = \frac{(\Delta p)_{st}}{\rho u(C_{w2} - C_{w1})} \tag{7.94b}$$

$$\eta_{tt} = \frac{(\Delta p)_{st}}{\rho c_p(T_{03} - T_{01})} \tag{7.94c}$$

The stage pressure rise $(\Delta p)_{st}$ can be measured using manometers and the actual work can be measured with torque measuring instrument.

The axial velocity is constant, then for $\alpha_1 = \alpha_3$, $C_1 = C_3$

$$h_{03} - h_{01} = h_3 - h_1 \tag{7.95}$$

$$T_{03} - T_{01} = T_3 - T_1$$

These relations are also valid approximately for small values of C_1 and C_3.

For isentropic flow,

$$h_{03ss} - h_{01} = h_{3ss} - h_1 \tag{7.96}$$

$$T_{03ss} - T_{01} = T_{3ss} - T_1$$

Static-to-static efficiency,

$$\eta_{ss} = \frac{h_{3ss} - h_1}{h_3 - h_1} = \frac{T_{3ss} - T_1}{T_3 - T_1} \tag{7.97}$$

Degree of reaction,

$$R = \frac{\textit{isentropic change of enthalpy in the rotor}}{\textit{isentropic change of enthalpy in the } \text{stage}}$$

This gives

$$R = \frac{\int_1^{2s} dh}{\int_1^{3ss} dh} = \frac{h_{2s} - h_1}{h_{3ss} - h_1} \tag{7.98}$$

For isentropic flow,

$dh = \frac{dp}{\rho}$; therefore,

$$R = \frac{\int_1^{2s} \frac{dp}{\rho}}{\int_1^{3ss} \frac{dp}{\rho}}$$

For incompressible flow assumption due to small pressure rise,

$$R = \frac{\int_1^{2s} dp}{\int_1^{3ss} dp}$$

$$R = \frac{p_2 - p_1}{p_3 - p_1} \tag{7.99}$$

For an actual compressor stage, the degree of reaction,

$$R = \frac{\textit{actual change of enthalpy in the rotor}}{\textit{actual change of enthalpy in the stage}}$$

$$R = \frac{h_2 - h_1}{h_3 - h_1} = \frac{T_2 - T_1}{T_3 - T_1} \tag{7.100}$$

For $C_1 = C_3$, $h_3 = h_1 = h_{03} - h_{01} = C_b (C_{w2} - C_{w1})$

$$h_2 - h_1 = \frac{1}{2}\left(C_{r1}^2 - C_{r2}^2\right)$$

$$R = \frac{h_2 - h_1}{h_{03} - h_{01}} \tag{7.101a}$$

$$R = \frac{C_{r1}^2 - C_{r2}^2}{2C_b\left(C_{w2} - C_{w1}\right)} \tag{7.101b}$$

Using velocity triangles,

$$C_{w1} = C_b - C_f \tan\beta_1$$

$$C_{w2} = C_b - C_f \tan\beta_2$$

$$C_{w2} - C_{w1} = C_f\left(\tan\beta_1 - \tan\beta_2\right)$$

$$C_{r1}^2 = C_f^2 + \left(C_f \tan\beta_1\right)^2$$

$$C_{r2}^2 = C_f^2 + \left(C_f \tan\beta_2\right)^2$$

$$C_{r1}^2 - C_{r2}^2 = C_f^2\left(\tan^2\beta_1 - \tan^2\beta_2\right)$$

Therefore

$$R = \frac{C_f^2\left(\tan^2\beta_1 - \tan^2\beta_2\right)}{2C_bC_f\left(\tan\beta_1 - \tan\beta_2\right)}$$

$$R = \frac{1}{2}\left(\frac{C_f}{C_b}\right)\left(\tan\beta_1 + \tan\beta_2\right) \tag{7.101c}$$

But $\dfrac{C_f}{C_b} = \phi$ *and* $\dfrac{1}{2}\left(\tan\beta_1 + \tan\beta_2\right) = \tan\beta_m$

Therefore

$$R = \phi\tan\beta_m \tag{7.101d}$$

Rearranging,

$$R = \frac{1}{2}\left(\frac{C_f}{C_b}\right)\left(\left(\tan\beta_1 + \tan\alpha_1\right) - \left(\tan\alpha_1 - \tan\beta_2\right)\right)$$

$$\tan\beta_1 + \tan\alpha_1 = \frac{C_b}{C_f}$$

Therefore

$$R = \frac{1}{2} - \frac{1}{2}\left(\frac{C_f}{C_b}\right)(\tan\alpha_1 - \tan\beta_2) \tag{7.102}$$

$C_f \tan\alpha_1 = C_{w1} = C_b - C_{rw1}$

$C_f \tan\beta_2 = C_{rw2} = C_b - C_{w2}$

Therefore, after substituting

$$R = \frac{1}{2} - \frac{1}{2}\left(\frac{C_{w1}}{C_b} - \frac{C_{rw2}}{C_b}\right) \tag{7.103}$$

$$R = \frac{1}{2} - \frac{1}{2C_b}(C_{w2} - C_{rw1}) \tag{7.104}$$

For a low degree of reaction,

$C_{rw1} < C_{w2}$

A low-degree of reaction is not desirable for obtaining higher efficiencies. To decrease the burden on blade rows, the static pressure rise can be divided equally between the rotor and diffuser.

Therefore,

$$h_2 - h_1 = h_3 - h_2 = \frac{1}{2}(h_3 - h_1) \tag{7.105}$$

$$R = \frac{1}{2}(\textit{fifty per cent reaction})$$

Equation for R = ½ gives

$$C_{r1}^2 - C_{r2}^2 = C_2^2 - C_1^2 \tag{7.106}$$

For R = ½,

$$\frac{1}{2} = \frac{C_f(\tan\beta_1 + \tan\beta_2)}{2C_b}$$

or

$$\frac{C_b}{C_f} = \tan\beta_1 + \tan\beta_2$$

from velocity triangles,

$$\frac{C_b}{C_f} = \tan\alpha_1 + \tan\beta_1 = \tan\alpha_2 + \tan\beta_2$$

$$\alpha_1 = \beta_2 \tag{7.107}$$

and

$\alpha_2 = \beta_1$

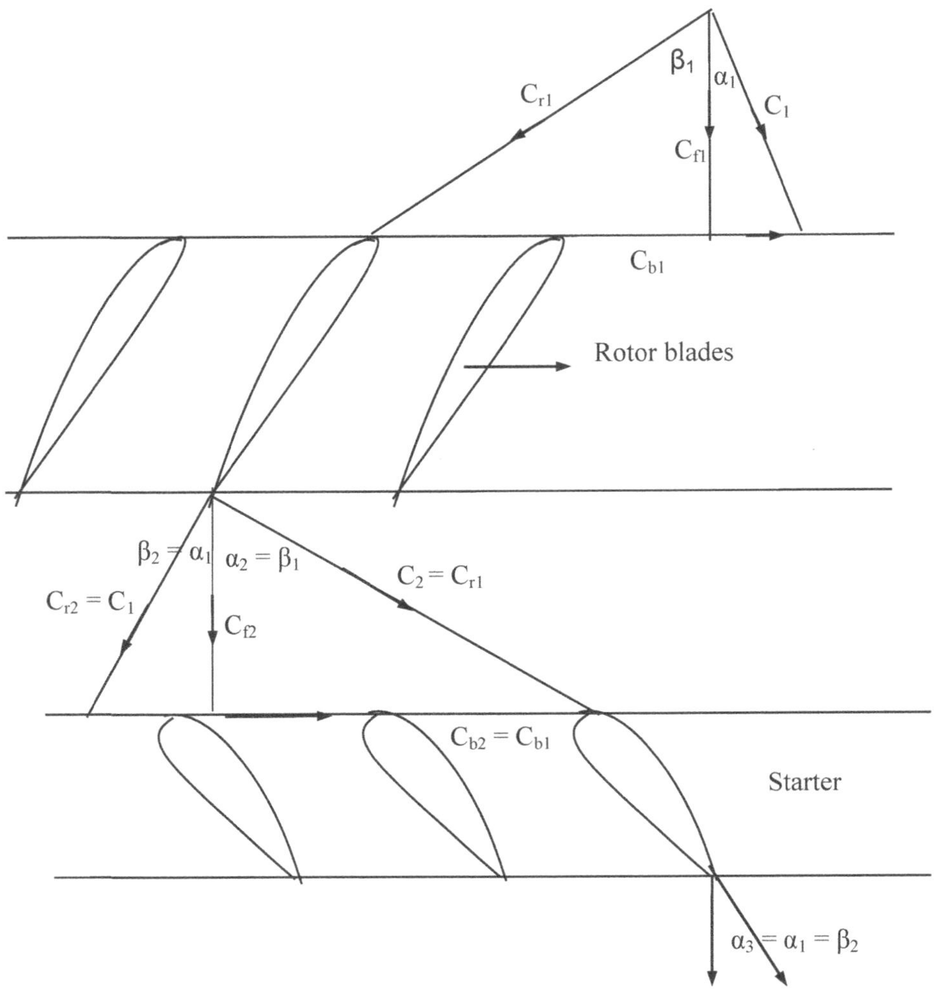

Fig. 7.21. Fifty percent degree of reaction stage compressor.

$$C_{r1} = C_2$$

$$C_{r2} = C_1 \tag{7.108}$$

Figure 7.21 shows the inlet and outlet velocity triangles for a fifty percent degree reaction-staged compressor.

As per the relations, the inlet and outlet velocity triangles are symmetrical as shown in Fig. 7.22. The whirl components at the inlet of the rotor and inlet of diffuser are equal.

$$C_{w1} = C_{rw2}$$

$$C_{rw1} = C_{w2} \tag{7.109}$$

The static pressure rise in the rotor of a high reaction stage is higher compared to that in the diffuser, i.e., $(\Delta p)_R > (\Delta p)_D$. For such a stage the quantity $(\tan \alpha_1 - \tan \beta_2)$ in the equation is negative, giving $R > ½$.

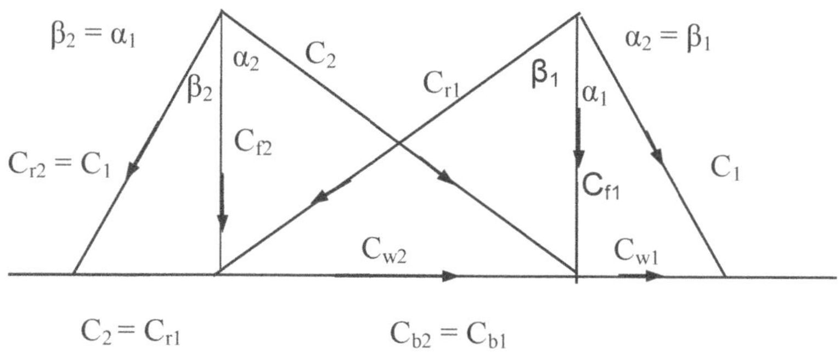

Fig. 7.22. Velocity triangles.

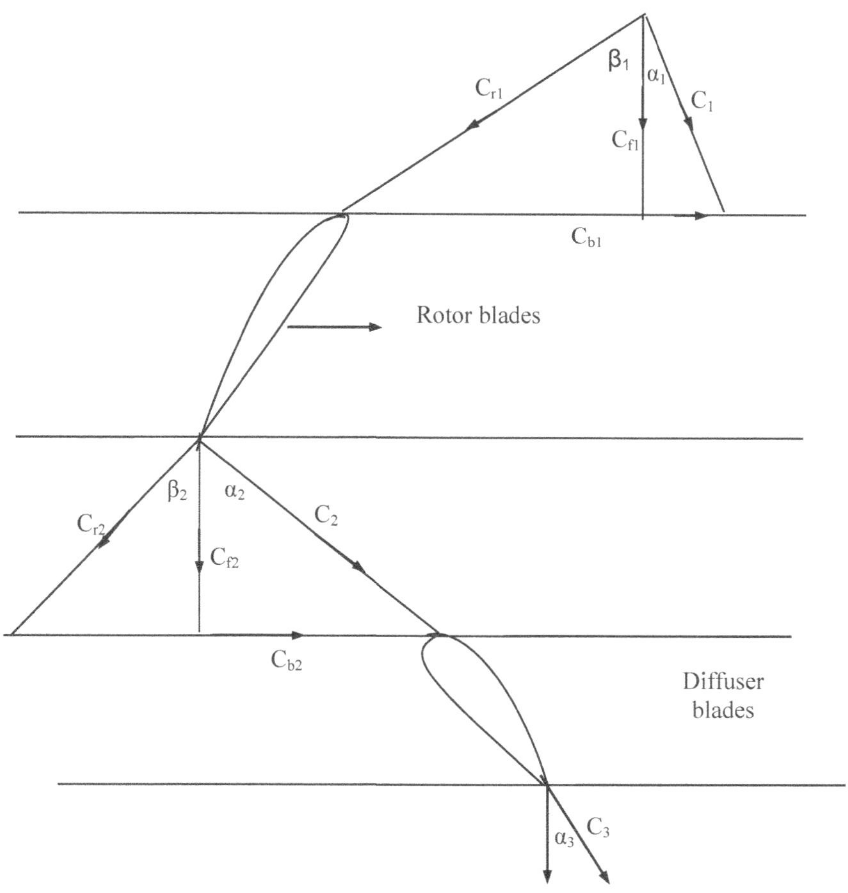

Fig. 7.23. High reaction stage ($R > ½$, $\alpha_1 < \beta_2$).

Therefore, for such a stage

$\beta_2 > \alpha_1$

$C_{r1} > C_2$

Figure 7.23 shows the inlet and outlet velocity triangles for a high-degree reaction-staged compressor. The rotor static pressure rise is high due to the large swirl component at the inlet. The swirl component at the exit is small. Therefore, these stages result in higher efficiency due to the large pressure rise in the rotor.

7.4 Steam turbines

The jet of steam with absolute velocity, C_1 impinges on the moving blade at an angle of α, to the tangent of blade (Fig. 7.24). The tangential component of this jet does work on the blade, because it is in the same direction as the motion of the blade, C_b; this tangential component is known as the velocity of whirl, C_{w1}. The axial component of C_{f1} of velocity C_1 does no work on the blade because it is perpendicular to the direction of the motion of the blade. This component C_{f1} is known as the velocity of flow because it is responsible for the flow through the turbine. The velocity of flow causes an axial thrust on the rotor.

Due to motion of blade u, in the peripheral direction, the entering steam jet with velocity C_1 will have a relative velocity C_{r1} to the blade. This relative velocity may be found by subtracting the vectors of the steam velocity C_1 and the blade velocity, C_b.

As the steam is to glide over the blade without shock, it follows that the surface of the blade at inlet must parallel the relative velocity, C_{r1}. That is, the moving blade at the inlet must be inclined to the tangent of the blade at an angle β_1. This is the principle on which the blade is designed.

A similar vector diagram is shown at the outlet tip of the moving blade. The steam glides off the blade and comes out with a relative velocity C_{r2} inclined at angle β_2 to the tangent. Addition of vector of blade velocity, C_b and relative velocity C_{r2} gives the absolute velocity of the leaving steam C_2. Inclined at angle α_2 to the tangent. After obtaining the vector C_2, its tangential component C_{w2} and axial component C_{f2} may be drawn.

where,

C_b = Blade velocity

C_1 = Inlet velocity of steam

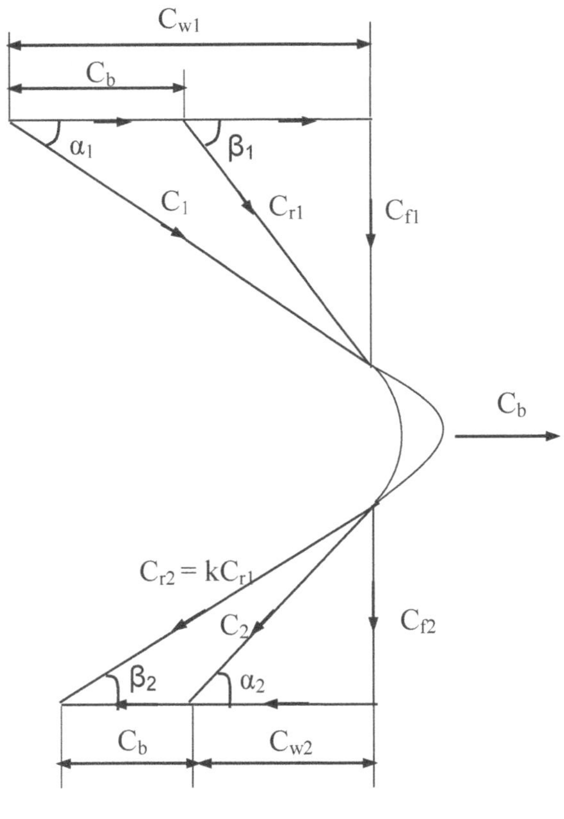

Fig. 7.24. Velocity triangles.

C_2 = Exit velocity of steam
C_{w1} = Whirl velocity at inlet
C_{w2} = Whirl velocity at the exit
C_{f1} = Flow velocity at inlet
C_{f2} = Flow velocity at exit
C_{r1} = Relative velocity at inlet
C_{r2} = Relative velocity at exit
α_1 = Angle of steam inlet. °
α_2 = Angle of steam exit, °
β_1 = Moving blade inlet angle, °
β_2 = Moving blade exit angle, °

In a combined velocity vector diagram, the blade is omitted and with the common blade velocity, the inlet and outlet velocity diagrams are developed as shown in Fig. 7.25.

The force applied by steam on the blade in the direction of blade is given by,

F = Change in momentum in the direction of the blade.

F = $\dot{m}$ (Change in whirl velocity)

The resultant force parallel to BF is inclined at an angle ψ to the plane in which the wheel rotates. The resultant force may be resolved into two components.

1. A useful thrust or tangential force acting in the direction of blade motion
2. An idle thrust or axial force perpendicular to the wheel plane, i.e., parallel to the turbine axis.

Forces on the blade and work done by blades,

From Newton's second law of motion, tangential force on the blade,

F_t = mass × acceleration in the tangential direction
= mass per second × change in velocity in tangential direction

$$F_t = m(C_{w1} - C_{w2}) \tag{7.110}$$

m = mass rate of steam, kg/s

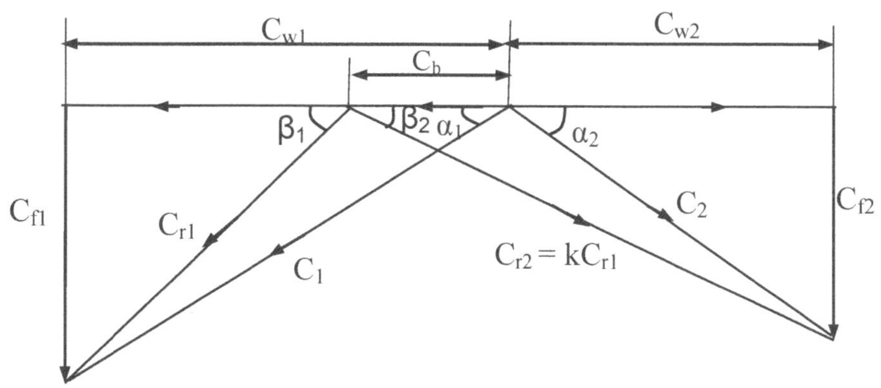

Fig. 7.25. Combined velocity triangles.

It is obvious from the velocity diagram that C_{w2} is opposite in direction to the blade motion. By giving the same direction of C_{w2}, the force becomes,

$$F_t = m(C_{w1} - (-C_{w2})) = F_t = m(C_{w1} + C_{w2}) \tag{7.111}$$

Work done on the blade per second:

W = Tangential force × distance moved in unit time in the direction of force
= tangential force × blade velocity in the direction of force
= $F_t C_b$
= $m(C_{w1} + C_{w2}) C_b$ Nm/s or J/s

$$\text{Power, } P = m\,(C_{w1} + C_{w2})\,C_b \tag{7.112}$$

If there are n number of identical stages,

$$P = n\,m\,(C_{w1} + C_{w2})\,C_b$$

It is the ratio of work done on the blade to the energy entering the blade.

$$\eta_{blade} = \frac{work\ done\ on\ blade/s}{energy\ entering\ the\ blade/s} = \frac{m(C_{w1} + C_{w2})C_b}{\frac{1}{2}mC_1^2}$$

$$\eta_{blade} = \frac{2(C_{w1} + C_{w2})C_b}{C_1^2} \tag{7.113}$$

Axial force is the ratio of mass and the acceleration in the axial direction.

Axis force = mass × acceleration in axial direction

$$F_a = m(C_{f1} - C_{f2})\ \text{N}$$

Where there is no friction in the blade $C_{r1} = C_{r2}$, $\beta_1 = \beta_2$ then $C_{f1} = C_{f2}$

$$F_a = 0$$

State efficiency is the ratio of work done in the stage to the isentropic enthalpy drop in the stage,

$$\eta_{stage} = \frac{work\ done\ in\ stage}{isentropic\ enthalpy\ drop\ in\ stage} = \frac{m(C_{w1} + C_{w2})C_b}{m(h_1 - h_2)} \tag{7.114}$$

$$= \frac{(C_{w1} + C_{w2})C_b}{\frac{1}{2}C_1^2} = \frac{\frac{1}{2}C_1}{\frac{1}{2}m(h_1 - h_2)} = \eta_{blade}\eta_{nozzle} \tag{7.115}$$

η_{stage} or $\eta_o = \eta_{blade}\eta_{nozzle}$

Energy converted to heat by blade friction,

$$Q_{friction} = m\left(\frac{C_{r1}^2 - C_{r2}^2}{2}\right) \tag{7.116}$$

The influence of ratio of blade speed to steam speed on blade efficiency in a single-state impulse turbine:

In an impulse turbine, blades are symmetrical, i.e., $\beta_1 = \beta_2$

We have

$$\eta_{blade} = \frac{2(C_{w1} + C_{w2})C_b}{C_1^2}$$

$$C_{w1} + C_{w2} = C_{r1}\cos\beta_1 + C_{r2}\cos\beta_2 = C_{r1}\cos\beta_1\left(1 + \frac{C_{r2}\cos\beta_2}{C_{r1}\cos\beta_1}\right) \quad (7.117)$$

$$= C_{r1}\cos\beta_1\left(1 + kB\right) where\ B = \frac{\cos\beta_2}{\cos\beta_1}$$

β_1 and β_2 are usually almost equal, for this analysis, β is constant

We have $C_{r1}\cos\beta_1 = C_1\cos\alpha_1 - C_b$

$C_{w1} + C_{w2} = (C_1\cos\alpha_1 - C_b)(1 + aB)$

Blade speed ratio $\rho = \frac{u}{C_1}$, $C_b = \rho C_1$

$$\eta_{blade} = \frac{2(C_1\cos\alpha_1 - C_b)(1 + aB)C_b}{C_1^2}$$

$$= \frac{2(C_1 C_b\cos\alpha_1 - C_b{}^2)(1 + aB)}{C_1^2}$$

$$= \frac{2C_1\left(\frac{C_b}{C_1}\cos\alpha_1 - \frac{C_b{}^2}{C_1^2}\right)(1 + aB)}{C_1^2}$$

$$\eta_{blade} = \left(\frac{C_b}{C_1}\cos\alpha_1 - \frac{C_b{}^2}{C_1^2}\right)(1 + aB)$$

$$\eta_{blade} = \left(\rho\cos\alpha_1 - \rho^2\right)(1 + aB) \text{ parabola equation,} \quad (7.118)$$

$$if\ \rho = 0, \frac{C_b}{C_1} = 0, \eta_{blade} = 0$$

$$if\ \rho = 1, \frac{C_b}{C_1} = 1, Torque = 0, \eta_{blade} = 0$$

ρ	Displacement	Torque	η_{blade}
0 ($C_b = 0$)	0	Maximum	0
1 ($C_b = C_1$)	Maximum	0	0

For maximum blade efficiency,

$$\frac{d\eta_{blade}}{d\rho} = 0$$

$$\frac{d\left(\rho\cos\alpha_1 - \rho^2\right)(1 + aB)}{d\rho} = 0$$

$$\frac{d\left(\rho \cos\alpha_1 - \rho^2\right)}{d\rho} = 0$$

$$\cos\alpha_1 - 2\rho = 0$$

$$\cos\alpha_1 = 2\rho$$

$$\rho_{opt} = \frac{\cos\alpha_1}{2} \tag{7.119}$$

Substituting this condition in blade efficiency,

$$\eta_{blade,\max} = \left(\frac{\cos\alpha_1}{2}\cos\alpha_1 - \frac{\cos^2\alpha_1}{4}\right)(1+aB)$$

$$= \left(\cos^2\alpha_1 - \frac{\cos^2\alpha_1}{2}\right)(1+aB)$$

$$\eta_{blade,\max} = \left(\frac{\cos^2\alpha_1}{2}\right)(1+aB) \tag{7.120}$$

If a = 1 and B = 1

$\eta_{\text{blade, max}} = \cos^2\alpha_1$

Figure 7.26 shows the maximum blade efficiency condition.

Figure 7.27 shows the velocity triangles for a compounded impulse turbine stage. High pressure steam enters the nozzle and the steam leaving nozzle enters the moving blades. Steam from moving blades is supplied to fixed blades which acts as guide vanes fixed to casing. The nozzle, moving blades, and fixed blades together are termed as a stage of velocity compounded impulse turbine.

$C_{r1} = C_{r2}$ and $C_{r3} = C_{r4}$

Across the fixed blade also there shall be no friction loss resulting in,

$C_2 = C_3$

Due to symmetry of moving blades,

$\beta_1 = \beta_2$ and $\beta_3 = \beta_4$

Due to symmetry of fixed blades,

$\alpha_2 = \alpha_3$

Steam should leave the stage axially ($\alpha_4 = 90°$). The whirl at the outlet is zero. These turbines are called as axial discharge turbines.

Total work from this stage = work from first row of moving blades + work from second row of moving blades.

Work from the first row of moving blades can be written as already described for simple impulse stage.

$W_I = m\,(C_{w1} + C_{w2})\,C_b = (C_1\cos\alpha_1 - C_b)\,(1 + aB)$ at a = 1 and B = 1

$W_I = 2\,m\,C_b\,(C_1\cos\alpha_1 - C_b)$

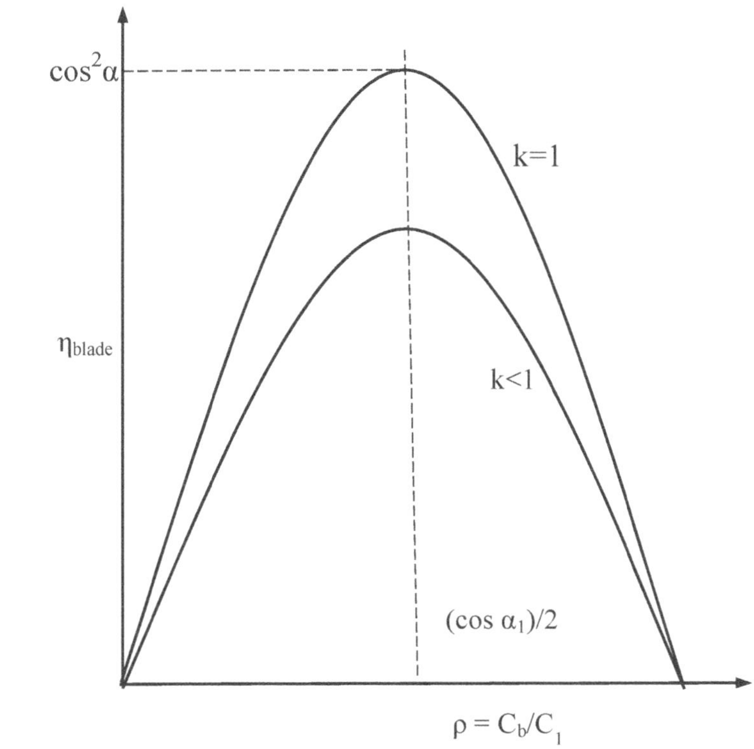

Fig. 7.26. Maximum blade efficiency condition.

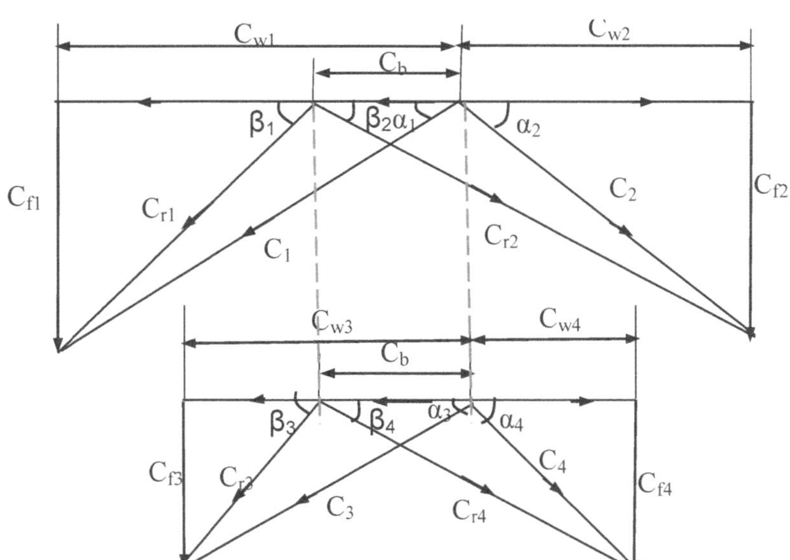

Fig. 7.27. Stage of velocity compounded impulse turbine and velocity diagram for the stage.

Similarly, work from the second row of moving blades,

$W_{II} = 2\ m\ C_b\ (C_3 \cos \alpha_3 - C_b)$

Or $W_{II} = 2\ m\ C_b\ (C_2 \cos\alpha_2 - C_b)$

From the velocity diagram,

$C_2 \cos\alpha_2 = C_{r2} \cos\beta_2 - C_b$

Substituting,

$W_{II} = 2\,m\,C_b\,(C_{r2} \cos\beta_2 - 2\,C_b) = 2\,m\,C_b\,(C_{r1} \cos\beta_1 - 2\,C_b) = 2\,m\,C_b\,(C_1 \cos\alpha_1 - 3\,C_b)$

Therefore,

$W_{II} = 2\,m\,C_b\,(C_1 \cos\alpha_1 - 3\,C_b)$

Total work from stage,

$W_T = W_I + W_{II}$

$W_T = 2m\,C_b\,(C_1 \cos\alpha_1 - C_b) + 2m\,C_b\,(C_1 \cos\alpha_1 - 3\,C_b)$

$W_T = 2m\,C_b\,(2C_1 \cos\alpha_1 - 4\,C_b)$

$$\textit{Diagram efficiency} = \frac{\textit{total work}}{\textit{Energy supplied}}$$

$$\eta_d = \frac{2mC_b(2C_1\cos\alpha_1 - 4C_b)}{\dfrac{mC_1^2}{2}}$$

$$= \frac{2mC_b(2C_1\cos\alpha_1 - 4C_b)}{\dfrac{mC_1^2}{2}}$$

$$\eta_d = \frac{4C_b(2C_1\cos\alpha_1 - 4C_b)}{C_1^2} \tag{7.121}$$

Non dimensional form of velocity $u/C_1 = \rho$, i.e., blade-speed-to-steam-velocity ratio,

Or,

$$\eta_d = 4\rho(2\cos\alpha_1 - 4\rho) \tag{7.122}$$

Differentiating diagram efficiency with respect to ρ and applying conditions to get the optimum value of ρ for maximum diagram efficiency,

$$\frac{d\eta_d}{d\rho} = \frac{d}{d\rho}(8\cos\alpha_1 - 16\rho^2)$$

Substituting,

$$\frac{d\eta_d}{d\rho} = 0$$

$$\rho_{opt} = \frac{\cos\alpha_1}{4} \tag{7.123}$$

Or $4u = C_1 \cos\alpha_1$

Maximum efficiency,

$$\eta_{d,\,max} = \cos^2\alpha_1 \tag{7.124}$$

for the optimum value of ρ_{opt} the work may be estimated as,

Work from I row, $W_I = 6m\ C_b^2$

Work from II row, $W_{II} = 2m\ C_b^2$

Total work from velocity compounded turbine state,

$W_T = W_I + W_{II}$

$W_T = 8m\ C_b^{22}$

Here it is obvious that from the optimum value of ρ, i.e., ρ_{opt} the work available from the second row is 1/4th of the total work from the state. It can be extrapolated from here that if there are more than two rows of moving blades in a velocity compounded turbine state, i.e., there, four or more number of rows, then the optimum blades speed to steam velocity ratio ρ can be generalized for n number of moving blades rows.

$$\rho_{opt} = \frac{\cos\alpha_1}{2n} \tag{7.125}$$

i.e., for three rows,

$$\rho_{opt} = \frac{\cos\alpha_1}{6} \tag{7.126}$$

Similarly, work from the last row of moving blades shall be,

$$W_{last\ row} = \frac{1}{2^n} \text{ of total work.} \tag{7.127}$$

Thus, from a three row stage, work from the last row shall be 1/8th of the total work and from a four row stage, work from the last row shall be 1/16th of the total work. Here it is obvious that as the number of rows increases, the work from last row diminishes. So the number of rows is generally limited to two. If we look at diagram efficiency, then it shows that the efficiency also diminishes with increases in the number of stages. Graphical pattern of the relation between Stage efficiency versus blade-speed-to-steam-velocity ratio ρ is shown in Fig. 7.28.

Steam entering moving blades passes through a converging passage area along with a change in direction (Fig. 7.29). Thus, there is an increase in velocity ($C_{r2} > C_{r1}$) from inlet to exit in moving blades which results in a reaction force. It may be noted that due to shaft rotation being caused by a combination of impulse and reaction forces, these reaction turbines are also termed as impulse-reaction turbine.

$$Degree\ of\ reaction = \frac{Enthalpy\ drop\ in\ moving\ blades}{Total\ enthalpy\ drop\ \text{in the}\ stage} \tag{7.128}$$

Degree of reaction,

$$\varepsilon = \frac{h_1 - h_2}{h_0 - h_2}$$

In case of a 50% degree of reaction,

$\varepsilon = 0.5$

$$h_1 - h_2 = \frac{1}{2}(h_0 - h_2)$$

$$h_0 - h_1 = h_1 - h_2 = \frac{1}{2}(h_0 - h_2)$$

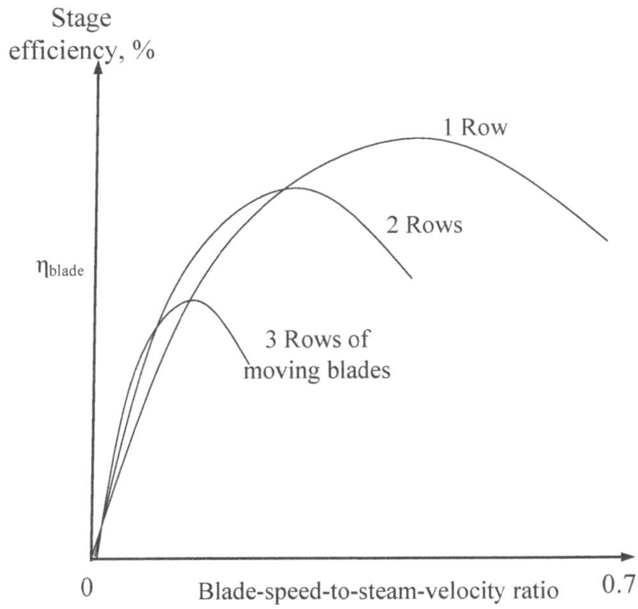

Fig. 7.28. Stage efficiency versus blade-speed-to steam-velocity ratio for velocity compounded impulse turbine.

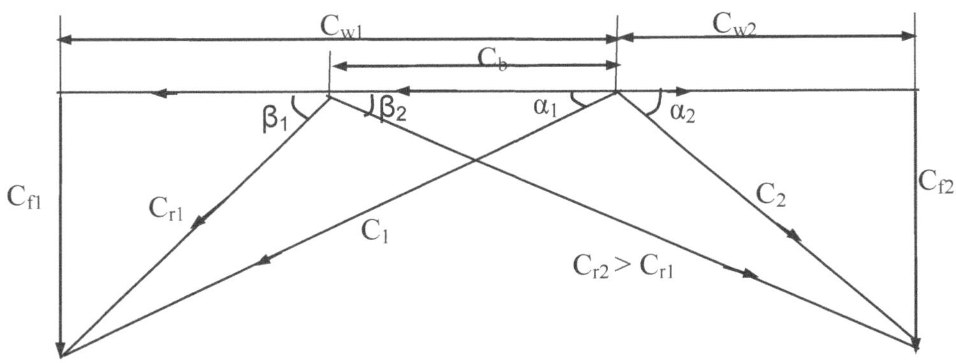

Fig. 7.29. Velocity and force diagram of reaction blading.

Turbines having a 50% degree of reaction are called Parson's turbines. A Parson's turbine has symmetrical blades for moving and stationary blades, i.e., inlet angles of stationary and moving blades are equal and also the exit angles of stationary and moving blades are equal. The term symmetrical blading in a reaction turbine refers to the 50% reaction stage.

A typical degree reaction value of zero refers to the condition when there is no enthalpy drop in moving blades. This refers to the impulse turbine stage where no heat drop occurs in moving blades. In both reaction turbines and impulse turbines, the compounding process is similar for reducing the speed of rotation. Compounding in reaction turbine is where pressure drop gradually occurs in the stationary and moving blade rows. In reaction turbine blading there is a possibility of steam leakage across the tip of the blade and casing as there is large clearance here. This leakage of pressure from the tip clearance is more vulnerable in high pressure regions as there occurs a large pressure drop in high pressure regions for a given enthalpy drop. Due to this fact, the initial stages in a high pressure region are taken to be of impulse type and subsequent stages are reaction stages. The above fact is generally interpreted as inefficient reaction blading in high pressure stages. Generally, modern turbines' first stage is velocity compounded (Curtis stage) followed by pressure compounded stages (Rateau stages) or reaction stages.

For stationary blade row the SFEE gives,

$$h_0 - h_1 = \frac{C_2^2 - C_1^2}{2}$$

For moving blade row the SFEE gives,

$$h_1 - h_2 = \frac{C_{r2}^2 - C_{r1}^2}{2}$$

Total enthalpy drop in this stage, $h_0 - h_2 = (h_0 - h_1) + (h_1 - h_2)$

and for 50% degree of reaction,

$C_1 = C_{r2}$ and $C_{r2} = C_2$

so $h_0 - h_1 = h_1 - h_2$

$h_0 = 2h_1 - h_2$

i.e., $\frac{h_1 - h_2}{h_0 - h_2} = \frac{1}{2}$

Hence it proves that symmetrical blading results in a 50% degree reaction turbine. Rate of work done from reaction stage can be estimated similar to that of the impulse stage.

Work done $= m\ u\ \Delta C_w$

where $\Delta C_w = C_{w2} + C_{w1} = C_1 \cos\alpha_1 + C_2 \cos\alpha_2$

for symmetrical blading,

$C_2 = V_{r1}, \alpha_2 = \beta_1,$

so $\Delta C_w = C_1 \cos \alpha_1 + C_{r1} \cos\beta_1$
$\quad = C_1 \cos \alpha_1 + C_1 \cos\alpha_1 - C_b$

$\Delta C_w = 2C_1 \cos \alpha_1 - C_b$

Therefore,

Work done, $W = m\ C_b\ (2C_1 \cos \alpha_1 - C_b)$

Axial thrust in a turbine can be quantified by the sum of the thrust due to change in axial velocity and due to pressure difference across the moving blade row. Mathematically the axial thrust can be given as:

$F_a = m\ (C_{f1} - C_{f2}) + (p_1A_1 - p_2A_2)$

where A_1 and A_2 are cross-sectional areas available as flow passage at the inlet and exit of moving blade row and p_1 and p_2 are pressures of steam at the inlet and exit sections of moving blades. In case of a 50% reaction turbine $C_{f1} = C_{f2}$, therefore axial thrust is only due to pressure difference across the moving blade row.

Diagram efficiency of a reaction stage can be estimated by knowing the energy input to the moving blades and taking ratio of work done to energy input.

Energy input to moving blades,

$$E_{in} = m\frac{C_1^2}{2} + m\left(\frac{C_{r2}^2 - C_{r1}^2}{2}\right)$$

$$= \frac{m}{2}\left(C_1^2 + C_{r2}^2 - C_{r1}^2\right)$$

$$= \frac{m}{2}\left(C_1^2 + C_1^2 - C_{r1}^2\right) \text{ as } C_1 = C_{r2}$$

$$= m\left(C_1^2 - \frac{C_{r1}^2}{2}\right)$$

From the velocity diagram,

$$C_{r1}^{\ 2} = C_1^{\ 2} + C_b^{\ 2} - 2C_1C_b\cos\alpha_1$$

Substituting value of C_{r1} in the energy equation.

$$E_{in} = m\left(C_1^{\ 2} - \frac{C_1^{\ 2} + C_b^{\ 2} - 2C_1C_b\cos\alpha_1}{2}\right)$$

$$E_{in} = m\left(\frac{C_1^{\ 2} + 2C_1C_b\cos\alpha_1 - C_b^{\ 2}}{2}\right)$$

$$\eta_d = \frac{W}{E_{in}}$$

$$= \frac{mC_b(2C_b\cos\alpha_1 - C_b)}{m\left(\frac{C_1^{\ 2} + 2C_1C_b - C_b^{\ 2}}{2}\right)}$$

$$\eta_d = \frac{2C_b(2C_1\cos\alpha_1 - C_b)}{C_1^{\ 2} + 2C_1C_b\cos\alpha_1 - C_b^{\ 2}}$$

Substituting $\rho = \frac{u}{C_1}$

$$\rho = \frac{C_b}{C_1}$$

$$\eta_d = \frac{2\rho\left(2\cos\alpha_1 - \rho\right)}{1 + 2\rho\cos\alpha_1 - \rho^2} \tag{7.129}$$

The maximum value of diagram efficiency and the optimum value of blade-speed-to-steam-velocity ratio, ρ can be estimated by differentiating with respect to ρ and equating to zero.

$$\frac{d\eta_d}{d\rho} = 0$$

$$2\cos\alpha_1 - 2\rho = 0$$

$$\rho_{opt} = \cos\alpha_1$$

substituting ρ for getting maximum value of diagram efficiency, $\eta_{d,\,max}$

$$\eta_{d,\max} = \frac{2\cos\alpha_1\left(2\cos\alpha_1 - \cos\alpha_1\right)}{1 + 2\cos^2\alpha_1 - \cos^2\alpha_1}$$

$$\eta_{d,\max} = \frac{2\cos^2 \alpha_1}{1+\cos^2 \alpha_1} \tag{7.130}$$

Work output for the condition of maximum diagram efficiency can be obtained by substituting

$\rho = \cos \alpha_1$

or $u = C_1 \cos \alpha_1$

for $\eta_{d,\max}$, $W = m\, C_b\, (2C_1 \cos \alpha_1 - C_b)$

$W = m\, C_1 \cos \alpha_1\, (2C_1 \cos \alpha_1 - C_1 \cos \alpha_1)$

$W = m\, C^2_1 \cos^2 \alpha_1$

$W = mu^2$ for maximum diagram efficiency

Optimum blade-speed-to-steam-velocity ratio yields the condition $C_b = C_1\cos\alpha_1$ which shall modify the velocity diagram for an optimum reaction state as shown in Fig. 7.30.

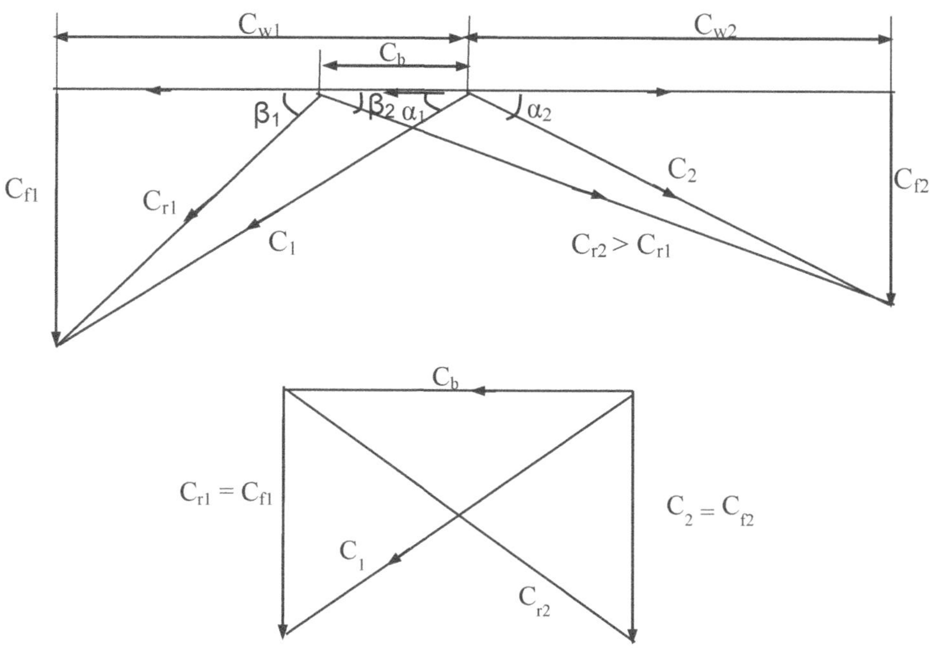

Fig. 7.30. Velocity diagram for an optimum reaction turbine stage having $\rho = \cos \alpha_1$.

7.5 Numerical solutions

1. 20 m³/s of air at 1 bar and 15°C is to be compressed in a centrifugal compressor through a pressure ratio 1.5:1. The compression follows the law pV1.5 = constant. The velocity of flow at the inlet and outlet remains constant and equal to 60 m/s. If the inlet and outlet impeller diameters are, respectively, 0.6 m and 1.2 m, and speed of rotation is 5000 rpm, find (a) the blade angles at the inlet and outlet of the impeller and the angle at which the air enters the impeller casing, and (b) breadth of impeller blade at the inlet and outlet. It may be assumed that no diffuser is fitted and the whole pressure increase occurs in the impeller and that the blades have a negligible thickness. Assume pre-whirl at the inlet to be zero.

Solution through App Designer

$C_{f2} = C_{f1} = C_1 = 60$ m/s

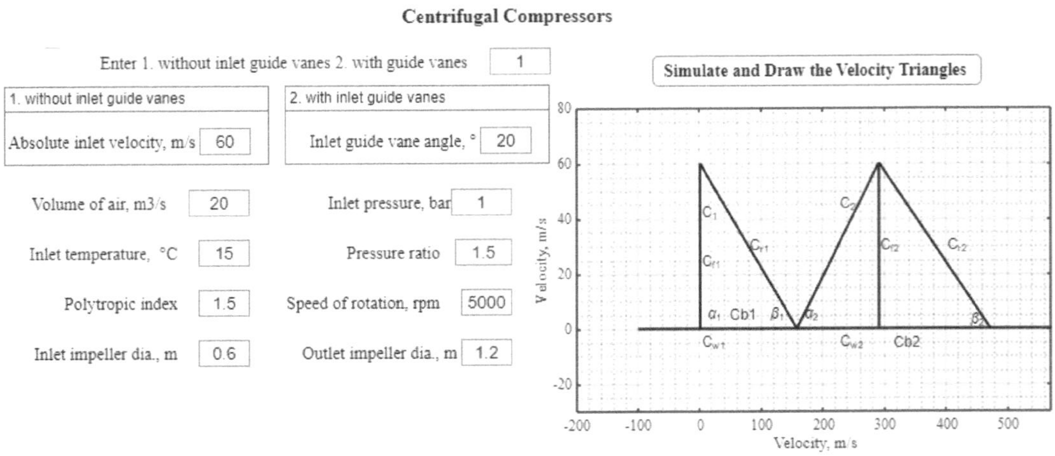

Parameter	Equation	Substitution	Result
Outlet pressure	$p_2 = PR \times p_1$	1.5×1	1.5 bar
Blade inlet velocity	$C_{b1} = \frac{\pi D_1 N}{60}$	$\frac{\pi \times 0.6 \times 5000}{60}$	157.08 m/s
Blade outlet velocity	$C_{b2} = \frac{\pi D_2 N}{60}$	$\frac{\pi \times 1.2 \times 5000}{60}$	314.16 m/s
Relative velocity at inlet	$C_{r1} = \sqrt{C_{b1}^2 + C_{f1}^2}$	$\sqrt{157.08^2 + 60^2}$	168.15 m/s
Blade inlet angle	$\beta_1 = \sin^{-1}\left(\frac{C_{f1}}{C_{r1}}\right)$	$\sin^{-1}\left(\frac{60}{168.15}\right)$	20.95 °
Blade outlet angle	$\beta_2 = \beta_1$		27.25 °
Inlet relative velocity	$C_{r1} = \left(\frac{C_{w1} - u}{\cos(\beta_1)}\right)$	$\left(\frac{927.18 - 200}{\cos(27.25)}\right)$	818 m/s
Outlet temperature from polytropic process	$T_2 = T_1\left(\frac{p_1}{p_1}\right)^{\frac{n-1}{n}}$	$(15 + 273.15)\left(\frac{1.5}{1}\right)^{\frac{1.5-1}{1.5}} - 273.15$	56.70°C
Outlet whirl velocity	$C_{w2} = \frac{c_{pa}(T_2 - T_1)}{C_{b2}}$	$\frac{1005(56.7 - 15)}{314.16}$	133.40 m/s
Blade outlet angle	$\beta_2 = \tan^{-1}\left(\frac{C_{f2}}{C_{b2} - C_{w2}}\right)$	$\tan^{-1}\left(\frac{60}{314.16 - 133.40}\right)$	18.36 °
Exit angle of fluid (diffuser inlet angle)	$\alpha_2 = \tan^{-1}\left(\frac{C_{f2}}{C_{w2}}\right)$	$\tan^{-1}\left(\frac{60}{133.40}\right)$	24.22 °

Table contd. ...

...Table contd.

Parameter	Equation	Substitution	Result
Absolute velocity of fluid at rotor exit	$C_2 = \dfrac{C_{f2}}{\sin^{-1}(\alpha_2)}$	$\dfrac{60}{\sin^{-1}(24.22)}$	146.27 m/s
Relative velocity at exit	$C_{r2} = \dfrac{C_{f2}}{\sin^{-1}(\beta_2)}$	$\dfrac{60}{\sin^{-1}(18.36)}$	190.46 m/s
Blade breadth at inlet	$B_1 = \dfrac{Q_1}{\pi D_1 C_{f1}}$	$\dfrac{20}{\pi 0.6 \times 60}$	0.1768 m
Volume of flow at exit	$Q_2 = \dfrac{p_1 Q_1 T_2}{p_2 T_1}$	$\dfrac{1 \times 20 \times (56.70 + 273.15)}{1.5 \times (15 + 273.15)}$	15.26 m^3
Blade breadth at exit	$B_2 = \dfrac{Q_2}{\pi D_2 C_{f2}}$	$\dfrac{15.26}{\pi 1.2 \times 60}$	0.0675 m

2. A steam jet enters a row of blades with a velocity of 1000 m/s at an angle of 22° with the direction of motion of the moving blades. If the blade speed is 200 m/s, determine the inlet and exit blade angles. Velocity of steam while passing over the blades is reduced by 10 percent. Also determine the power developed by turbine when the rate of flow of steam is 1000 kg per minute.

Solution through App Designer.

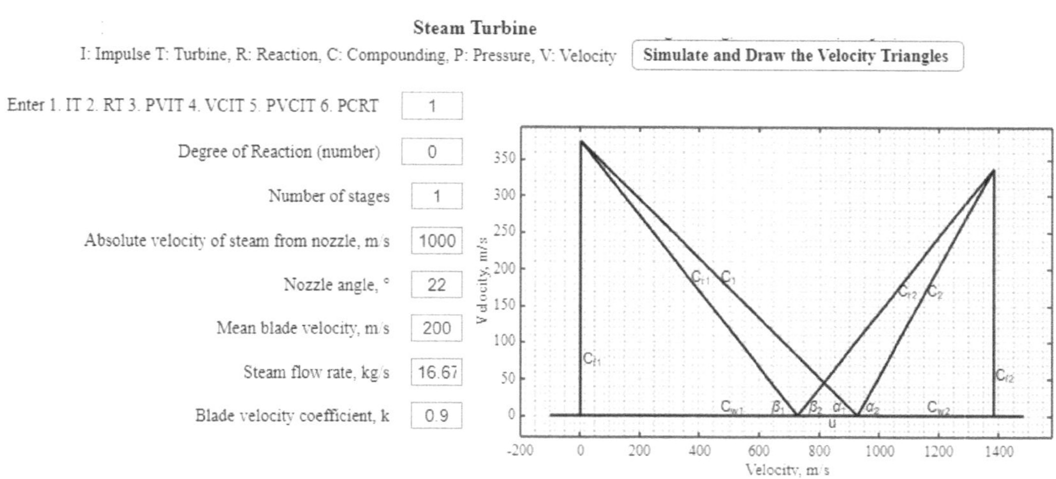

Parameter	Equation	Substitution	Result
Inlet whirl velocity	$C_{w1} = C_1 \cos(\alpha_1)$	1000 cos(22°)	927.18 m/s
Axial flow velocity at inlet	$C_{f1} = C_1 \sin(\alpha_1)$	1000 sin(22°)	374.61 m/s
Blade inlet angle	$\beta_1 = \tan^{-1}\left(\dfrac{C_{f1}}{C_{w1} - C_b}\right)$	$\tan^{-1}\left(\dfrac{374.61}{927.18 - 200}\right)$	27.25 °
Blade outlet angle	$\beta_2 = \beta_1$		27.25 °

Table contd. ...

...Table contd.

Parameter	**Equation**	**Substitution**	**Result**
Inlet relative velocity	$C_r = \left(\frac{C_w - C_b}{\cos(\beta)}\right)$	$\left(\frac{927.18 - 200}{\cos(27.25)}\right)$	818 m/s
Outlet relative velocity	$C_{r2} = 0.9\ C_{r1}$	0.9×818	736.2 m/s
Outlet whirl velocity	$C_{w2} = C_{r2}\cos(\beta_2) - C_b$	$736.2 \times \cos(27.25°) - 200$	454.46 m/s
Axial flow velocity at outlet	$C_{f2} = C_{r2}\sin(\beta_2)$	$736.2\ \sin(27.25°)$	337.15 m/s
Absolute velocity at the exit	$C_2 = \sqrt{C_{f2}^2 - C_{w2}^2}$	$\sqrt{736.2^2 - 454.46^2}$	565.87 m/s
Steam angle at the exit	$\alpha_2 = \tan^{-1}\left(\frac{C_{f2}}{C_{w2}}\right)$	$\tan^{-1}\left(\frac{337.15}{454.46}\right)$	36.57 °
Tangential force	$F_t = m_s(C_{w1} + C_{w2})$	$16.67\ (927.18 + 454.46)$	23032 N
Power developed	$P = Ft\ C_b$	23032×200	4604.2 kW
Axial thrust	$F_a = m_s(C_{f1} - C_{f2})$	$16.67\ (374.61 + 337.15)$	624.47 N
Heat loss to friction	$Q_{friction} = m_s\left(\frac{C_{r2}^2 - C_{r1}^2}{2}\right)$	$16.67\left(\frac{736.2^2 - 818^2}{2}\right)$	1059 kN
Efficiency of blade	$\eta_{blade} = \left(\frac{P}{\frac{1}{2}m_s C_1^2}\right)\times 100$	$16.67\left(\frac{4604.2 \times 1000}{\frac{1}{2}\times 16.67 \times 1000^2}\right)\times 100$	55.26%

Chapter 8
Power Generation Systems

8.1 Introduction

A combined cycle power plant integrates the gas power plant and steam power plant. The gas power plant operates on a thermodynamic cycle known as the Brayton cycle. Steam power plants operate on the Rankine cycle. The basic Brayton cycle for a thermal power plant consists of an air compressor, gas turbine combustion chamber (GTCC) and gas turbine (Fig. 8.1). Many modified Brayton cycles are developed to improve the power and thermal efficiency of the simple cycle. They are the regenerator, gas reheater, supplementary firing, etc. In a gas turbine-steam turbine (GT-ST) a combined cycle power plant, the gas cycle plays an important role on the performance of the combined cycle.

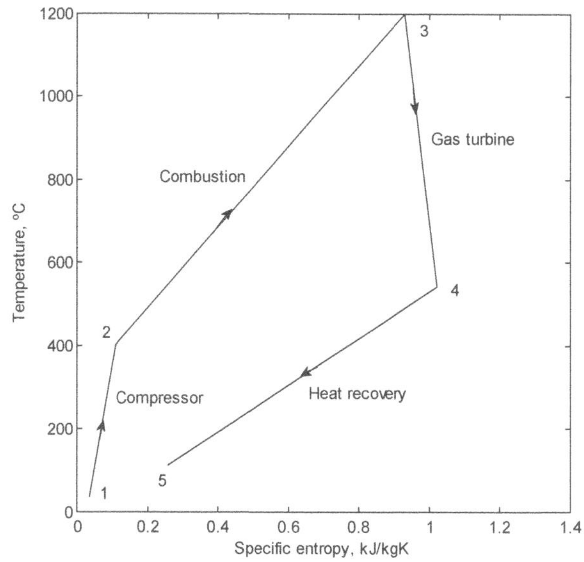

Fig. 8.1. Representation of Brayton cycle on T-s diagram.

8.2 Gas power generation system

Based on the gas properties, the modeling of a gas power plant results in a performance that is per the assumptions. In the energy modeling, the specific heat may be assumed constant in the air standard cycles, specific heat may be determined as a function of temperature, and specific heat may be calculated as a function of temperature and pressure.

DOI: 10.1201/9781032642505-8

For constant c_p of gas,

For a typical gas turbine combustion chamber, the energy balance equation is

$$m_f CV = (m_f + m_a)\, c_{pg}\, (T_{comb} - T_{air}) \tag{8.1}$$

where m_f and m_a are mass flow rates of fuel and air respectively, in kg/s. CV is the calorific value of fuel, kJ/kg. c_{pg} is the specific heat of gas, kJ/kg K. T_{comb} and T_{air} are the temperatures of combustion chamber (products of gas) and compressed air respectively.

The above equation can be simplified as

$$CV = \frac{(m_f + m_a)c_{pg}(T_{comb} - T_{air})}{m_f} = (1 + AFR)c_{pg}(T_{comb} - T_{air}) \tag{8.2}$$

where AFR is the air fuel ratio

Therefore

$$AFR = \frac{CV}{c_{pg}(T_{comb} - T_{air})} - 1 \tag{8.3}$$

finding air fuel ratio at a known combustion temperature,

or

$$T_{comb} = T_{air} + \frac{CV}{(1 + AFR)c_{pg}} \tag{8.4}$$

8.3 Evaluation of gas power plant using JANAF tables

Analyzing the gas power cycle using JANAF tables results in a realistic evaluation compared to basic thermodynamic results. The level of error decreases with this approach but the method of this evaluation involves many iterations. Due to the development of computational tools and methods, it may not be a big issue for solution development. This chapter outlines the procedure to solve problems related to the Brayton cycle for energy performance and exergy performance using gas properties from the JANAF tables.

8.3.1 Air compressor

For an isentropic compression process in the compressor, the relation is given as,

$$0.21\left(s_{O_2,\,2'} - s_{O_2,\,1}\right) + 0.79\left(s_{N_2,\,2'} - s_{N_2,\,1}\right) - R\,\ln\left(\frac{P_2}{P_1}\right) = 0 \tag{8.5}$$

The temperature of the compressed air at state 2' is determined by iteration of the above equation. The actual temperature of the compressed air (T_2) is calculated from the isentropic efficiency of the compressor.

For an isentropic compression process, the change in the entropy for air is zero, i.e., $\Delta s = 0$.

The isentropic efficiency for a compressor and gas turbine is determined from the pressure ratio (r_c), specific heat ratio, γ and polytropic state efficiency, η_∞).

Isentropic efficiency of the compressor,

$$\eta_c = \frac{r_c^{\left(\frac{\gamma-1}{\gamma}\right)} - 1}{r_c^{\left(\frac{\gamma-1}{\gamma\,\eta_\infty}\right)} - 1} \tag{8.6}$$

Initial guess of T'_2

$$T_2' = T_1\left(\frac{P_2}{P_1}\right)^{\frac{\gamma-1}{\gamma}} \tag{8.7}$$

Examples of an iteration for isentropic expansion using JANAF

T_2'	$s_{2'} - s_1$
575.63 (initial guess)	0.3426
575.29	0.3245
574.97	0.3074
- - -	- - -
569.19	0.001

$h_1 = 0.21\ h_{O2,\,T1} + 0.79\ h_{N2,\,T1} = 0$ (for atmospheric condition)

$h_2' = 0.21\ h_{O2,\,T2'} + 0.79\ h_{N2,\,T2'}$

$$\eta_c = \frac{h_2' - h_1}{h_2 - h_1} \tag{8.8}$$

$$h_2 = h_1 + \frac{h_2' - h_1}{\eta_c} \tag{8.9}$$

Initial guess of T_2 based on isentropic efficiency,

$$T_2 = T_1 + \frac{T_2' - T_1}{\eta_c} \tag{8.10}$$

Example of iteration of exit temperature using JANAF tables

T_2	h_2 (= 9163.2 kJ/kg mol)
607.00 (initial guess)	9191.7
606.70	9183.0
606.51	9176.9
- - -	- - -
606.06	9163.8 (consistency 0.01)

$h_1 = 0$ (in case of atmospheric state)

$h_2 = x\ (0.21\ h_{O2,\,T2} + 0.79\ h_{N2,\,T2})$

Work supplied to compressor,

$w_c = h_2 - h_1$

8.3.2 Gas turbine combustion chamber

The energy balance in the combustion chamber at the given temperature results in the air fuel ratio, i.e., the value of 'x' in the chemical reaction. In a reaction equation '*x*' denotes the amount of air in kg mol to be supplied per kg mol of fuel into the GTCC. The amount of air to be compressed to get the required combustion temperature (T_3) is determined from the energy balance of reactants and products.

Let us take the example of natural gas as combustion fuel for heat supply to the gas power plant.

Let us assume natural gas' (NG) composition as follows,

Methane (CH_4), 90.81%, ethane (C_2H_6) 2.96%, propane (C_3H_8) 2.7%, i-butane (C_4H_{10}) 0.54%, n-butane (C_4H_{10}) 0.76%, i-pentane (C_5H_{12}) 0.2%, n-pentane (C_5H_{12}) 0.16%, n-hexane (C_6H_{14}) 0.27%, carbon dioxide (CO_2) 1.49% and nitrogen (N_2) 0.09% (by volume).

The full combustion of NG with compressed air in GTCC is

$$[a_1\, CH_4 + a_2\, C_2H_6 + a_3\, C_3H_8 + a_4\, (C_4H_{10})_i + a_5\, (C_4H_{10})_n + a_6\, (C_5H_{12})_i + a_7\, (C_5H_{12})_n + a_8\, (C_6H_{14})_n$$
$$+ a_9\, CO_2 + a_{10}\, N_2]_{NG} + [x\, (0.21\, O_2 + 0.79N_2)]_{air} \rightarrow [b_1\, CO_2 + b_2\, H_2O + b_3\, O_2 + b_4\, N_2]_{exhaust} \quad (8.11)$$

$a_1 = 90.81\%\ (CH_4) = 0.9081$
$a_2 = 2.96\%\ (C_2H_6) = 0.0296$
$a_3 = 2.7\%\ (C_3H_8) = 0.027$
$a_4 = 0.54\%\ (C_4H_{10}) = 0.0054$
$a_5 = 0.76\%\ (C_4H_{10}) = 0.0076$
$a_6 = 0.2\%\ (C_5H_{12}) = 0.002$
$a_7 = 0.16\%\ (C_5H_{12}) = 0.0016$
$a_8 = 0.26\%\ (C_6H_{14}) = 0.0026$
$a_9 = 1.49\%\ (CO_2) = 0.0149$
$a_{10} = 0.09\%\ (N_2) = 0.0009$

Mass of the fuel,

$$m_{fuel} = a_1\, M_{CH4} + a_2\, M_{C2H6} + a_3\, M_{C3H8} + a_4\, M_{C4H10i} + a_5\, M_{C4H10n} + a_6\, M_{C5H12i} + a_7\, M_{C5H12n} + a_8\, M_{C6H14} + a_9\, M_{CO2} + a_{10}\, M_{N2} \quad (8.12)$$

Fuel energy,

$$HHV = a_1\, h_{fCH4} + a_2\, h_{fC2H6} + a_3\, h_{fC3H8} + a_4\, h_{fC4H10i} + a_5\, h_{fC4H10n} + a_6\, h_{fC5H12i} + a_7\, h_{fC5H12}n + a_8\, h_{fC6H14} - (a_1 + 2a_2 + 3a_3 + 4a_4 + 5a_6 + 5a_7 + 6a_8 + a_9)\, h_{fCO2} - (2a_1 + 3a_2 + 4a_3 + 5a_4 + 5a_5 + 6a_6 + 6a_7 + 7a_8)\, h_{fH2OL} \quad (8.13)$$

Exergy of the fuel,

$$\varepsilon = a_1\, \varepsilon_{CH4} + a_2\, \varepsilon_{C2H6} + a_3\, \varepsilon_{C3H8} + a_4\, \varepsilon_{C4H10} + a_5\, \varepsilon_{C4H10} + a_6\, \varepsilon_{C5H12} + a_7\, \varepsilon_{C5H12} + a_8\, \varepsilon_{C6H14} + a_9\, \varepsilon_{CO2} + a_{10}\, \varepsilon_{N2} + RT_0\, (a_1\, \ln(xa_1) + xa_2\, \ln(xa_2) + xa_3\, \ln(xa_3) + xa_4\, \log(xa_4) + xa_5\, \ln(xa_5) + xa_6\, \ln(xa_6) + xa_7\, \ln(xa_7) + xa_8\, \ln(xa_8) + xa_9\, \ln(xa_9) + xa_{10}\, \ln(xa_{10}))) \quad (8.14)$$

$$[a_1\, CH_4 + a_2\, C_2H_6 + a_3\, C_3H_8 + a_4\, (C_4H_{10})_i + a_5\, (C_4H_{10})_n + a_6\, (C_5H_{12})_i + a_7\, (C_5H_{12})_n + a_8\, (C_6H_{14})_n$$
$$+ a_9\, CO_2 + a_{10}\, N_2]_{NG} + [x\, (0.21\, O_2 + 0.79N_2)]_{air} \rightarrow [b_1\, CO_2 + b_2\, H_2O + b_3\, O_2 + b_4\, N_2]_{exhaust} \quad (8.15)$$

The mass balance of chemical reaction,

C balance:

$$b_1 = a_1 + 2a_2 + 3a_3 + 4a_4 + 4a_5 + 5a_6 + 5a_7 + 6a_8 + a_9 \quad (8.16)$$

H balance:

$$b_2 = 2a_1 + 3a_2 + 4a_3 + 5a_4 + 5a_5 + 6a_6 + 6a_7 + 7a_8 \quad (8.17)$$

O balance:

$$b_3 = a_9 + 0.21x - b_1 - 0.5b_2 \quad (8.18)$$

N balance:

$$b_4 = a_{10} + 0.79x \quad (8.19)$$

$$h_{air} = 0.21\ h_{T,\ O2} + 0.79\ h_{T,\ N2} \quad (8.20)$$

Enthalpy of reactants,

$$HR = h_{fuel} + x\ h_{air} \quad (8.21)$$

Enthalpy of products

$$HP = b_1(h_{fCO2} + h_{T,\ CO2}) + b_2\,(h_{fH2O} + h_{T,\ H2O}) + b_3\ h_{T,\ O2} + b_4\ h_{T,\ N2} \quad (8.22)$$

Example of an iteration of the equation to give the mole value of 'x' using JANAF tables

x	HR - HP
30 (initial guess)	21578
30.21	15925
30.37	11752
30.49	8673.3 - -
- -	- -
30.82	0.0001

8.3.3 Gas turbine

Isentropic efficiency of gas turbine,

$$\eta_{gt} = \frac{1-\left(\dfrac{P_4}{P_3}\right)^{\frac{\eta_{\infty gt}(\gamma-1)}{\gamma}}}{1-\left(\dfrac{P_4}{P_3}\right)^{\frac{(\gamma-1)}{\gamma}}} \quad (8.23)$$

Initial guess of $T_{4'}$

$$T_4' = T_3\left(\frac{P_4}{P_3}\right)^{\frac{\gamma-1}{\gamma}} \quad (8.24)$$

For an isentropic expansion process in the gas turbine, the equation is given as,

$$\Delta s = 0 = s_3 - s_{4'}$$

$$s_3 = [b_1 s_{CO_2} + b_2 s_{H_2O} + b_3 s_{O_2} + b_4 s_{N_2}]_3 - (b_1 + b_2 + b_3 + b_4)\ R\ \ln\left(\frac{P_3}{P_0}\right) \quad (8.25)$$

$$s_{4'} = [b_1 s_{CO_2} + b_2 s_{H_2O} + d_3 s_{O_2} + b_4 s_{N_2}]_{34'} - (b_1 + b_2 + b_3 + b_4)\ R\ \ln\left(\frac{P_4}{P_0}\right) \quad (8.26)$$

The temperature of the gas after expansion in gas turbine (T_4') is estimated from the iteration of Eq. 8.26. The actual temperature of the gas at gas turbine exit (T_4) is determined by the gas turbine isentropic efficiency.

Example of an iteration of the gas turbine's exit temperature in the isentropic process using JANAF tables

T_4'	$s_{4'} - s_3$
742.35 (initial guess)	108.31
753.17	93.30
762.51	80.51
- - -	- - -
823.38	0.01

$$h_3 = b_1\, h_{CO2,\,T3} + b_2\, h_{H2O,\,T3} + b_3\, h_{O2,\,T3} + b_4\, h_{N2,\,T3} \tag{8.27}$$

$$h_{4'} = b_1\, h_{CO2,\,T4'} + b_2\, h_{H2O,\,T4'} + b_3\, h_{O2,\,T4'} + b_4\, h_{N2,\,T4'} \tag{8.28}$$

$$\eta_T = \frac{h_3 - h_4}{h_3 - h_4'} \tag{8.29}$$

$$h_4 = h_3 - \eta_T\,(h_3 - h_4') \tag{8.30}$$

Initial guess of T_4 based on isentropic efficiency,

$$T_4 = T_3 - \eta_T\,(T_3 - T_4') \tag{8.31}$$

Example of an iteration for exit temperature of a gas turbine using JANAF tables with enthalpy reference:

T_4	h_4 (=582980 kJ/kg mol)
877.1 (initial guess)	580250
879.8	583170
879.6	582970
- - -	- - -
879.65	582980

$$h_3 = b_1\, h_{CO2,\,T3} + b_2\, h_{H2O,\,T3} + b_3\, h_{O2,\,T3} + b_4\, h_{N2,\,T3} \tag{8.32}$$

$$h_4 = b_1\, h_{CO2,\,T4} + b_2\, h_{H2O,\,T4} + b_3\, h_{O2,\,T4} + b_4\, h_{N2,\,T4} \tag{8.33}$$

$$w_T = (h_3 - h_4) \tag{8.34}$$

$$w_{net} = w_t - w_c \tag{8.35}$$

But the mass of air is m_a kg/s, therefore the multiplication factor for the chemical reaction is found as follows,

$$m_{a,factor} = \frac{m_a}{x(0.21M_{O_2} + 0.79M_{N_2})} \qquad \because m_a = m_{a,factor}\, x(0.21M_{O_2} + 0.79M_{N_2}) \tag{8.36}$$

Power generation,

$$W_c = \frac{m_{a,factor}\, w_c}{\eta_{m,c}} \tag{8.37}$$

$$W_T = \eta_m\, \eta_G\, m_{a,\,factor}\, w_T \tag{8.38}$$

Net power generation, $W_{net} = W_T - W_C$ (8.39)

Thermal efficiency,

$$\eta_1 = \frac{W_{net}}{m_{a,factor} HHV} \times 100 \tag{8.40}$$

Exergy evaluation

$$s_1 = x\left(0.21 s_{O_2,1} + 0.79 s_{N_2,1} - R \ \ln\left(\frac{P_1}{P_0}\right)\right) \tag{8.41}$$

= 0 (in case of atmospheric condition)

$$s_2 = x\left(0.21 s_{O_2,2} + 0.79 s_{N_2,2} - R \ \ln\left(\frac{P_2}{P_0}\right)\right) \tag{8.42}$$

$$s_3 = [b_1 s_{CO_2} + b_2 s_{H_2O} + b_3 s_{O_2} + b_4 s_{N_2}]_3 - (b_1 + b_2 + b_3 + b_4) \ R \ \ln\left(\frac{P_3}{P_0}\right) \tag{8.43}$$

$$s_4 = [b_1 s_{CO_2} + b_2 s_{H_2O} + b_3 s_{O_2} + b_4 s_{N_2}]_4 - (b_1 + b_2 + b_3 + b_4) \ R \ \ln\left(\frac{P_4}{P_0}\right) \tag{8.44}$$

Mole fractions of combustion gases

$$nt_{total} = b_1 + b_2 + b_3 + b_4 \tag{8.45}$$

$$xb_1 = \frac{b_1}{n_{total}} \tag{8.46}$$

$$xb_2 = \frac{b_2}{n_{total}} \tag{8.47}$$

$$xb_3 = \frac{b_3}{n_{total}} \tag{8.48}$$

$$xb_4 = \frac{b_4}{n_{total}} \tag{8.49}$$

Chemical exergy of air,

$$\text{Chemical exergy, } e_{ch} = \sum_k n_k \overline{\varepsilon^0{}_k} + RT_0 \Sigma_k n_k \ln.(.x_k) \tag{8.50}$$

$$\varepsilon_{ch,\,air} = x\,(0.21\ \varepsilon_{0,\,O2} + 0.79\ \varepsilon_{0,\,N2} + R\ T_0\ (0.21\ \ln(0.21) + 0.79\ \ln(0.79))) \tag{8.51}$$

Physical exergy of air at the inlet of the compressor

$\varepsilon_{ph,\,air\,1} = 0$ (in case of atmospheric condition)

Physical exergy of air at the exit of the compressor

$$\varepsilon_{ph,\,air\,2} = (h - T_0 s)_2 \tag{8.52}$$

Physical exergy of gas at the exit of the combustion chamber/inlet of the gas turbine

$$\varepsilon_{ph,\,gas\,3} = (h - T_0 s)_3 \tag{8.53}$$

Physical exergy of gas at the exit of the gas turbine

$$\varepsilon_{ph,\,gas\,4} = (h - T_0 s)_4 \tag{8.54}$$

Chemical exergy of gas,

$$\varepsilon_{ch,\,gas} = n_{total}\,(xb_1\,\varepsilon_{0,\,CO2} + xb_2\,\varepsilon_{0,\,H2OG} + xb_3\,\varepsilon_{0,\,O2} + xb_4\,\varepsilon_{0N2} + xb_5\,\varepsilon_{0NO2} + R\,T_0\,(xb_1\,\ln(xb_1) + xb_2\,\ln(xb_2) + xb_3\,\ln(xb_3) + xb_4\,\ln(xb_4) + xb_5\,\ln(xb_5))) \quad (8.55)$$

Total exergy of fluid = chemical exergy + physical exergy

Exergy of air at the inlet of the compressor,

$$\varepsilon_1 = \varepsilon_{ch,\,air} + \varepsilon_{ph,\,air\,1} \quad (8.56)$$

Exergy of air at the exit of the compressor/inlet of combustion,

$$\varepsilon_2 = \varepsilon_{ch,\,air} + \varepsilon_{ph,\,air\,2} \quad (8.57)$$

Exergy of gas at the exit of combustion/inlet of the gas turbine,

$$\varepsilon_3 = \varepsilon_{ch,\,gas} + \varepsilon_{ph,\,gas\,3} \quad (8.58)$$

Exergy of gas at the exit of the gas turbine,

$$\varepsilon_4 = \varepsilon_{ch,\,gas} + \varepsilon_{ph,\,gas\,4} \quad (8.59)$$

Irreversibility of the compressor,

$$i_{compressor} = \varepsilon_1 + w_c - \varepsilon_2 \quad (8.60)$$

Irreversibility of the combustion,

$$i_{combustion} = \varepsilon_{fuel} + \varepsilon_2 - \varepsilon_3 \quad (8.61)$$

Irreversibility of gas turbine,

$$i_T = \varepsilon_3 - w_T - \varepsilon_4 \quad (8.62)$$

Exergy efficiency of the gas power plant,

$$\eta_2 = \frac{\varepsilon_{fuel} - \sum i}{\varepsilon_{fuel}} \times 100 \quad (8.63)$$

8.4 Steam Rankine Cycle

Figure 8.2 shows the basic layout of a steam Rankine cycle. The processes are depicted on a T-s diagram (Fig. 8.3). The essential components are a boiler, steam turbine, steam condenser and feed pump. Some of the performance indices in thermal a power plant are steam rate, heat rate, power generation and thermal efficiency.

The components in thermal power plants belong to an open system. The steady flow energy equation (SFEE) can be applied to arrive at the components in the plant.

After applying SFEE to the steam turbine, the gross work from the turbine,

$$W_t = m_s\,(h_1 - h_2) \quad (8.64)$$

The steam flow rate m_s may be supplied or can be determined from the given source after analyzing the boiler and its furnace. The enthalpy h_1 can be read either from the steam tables or Mollier chart. It is the enthalpy at the inlet of turbine and can be read at the boiler pressure and superheated steam temperature. Similarly, h_2 can be found from steam tables or Mollier chart at the condenser pressure and temperature. The analytical solution needs iteration for isentropic expansion. In the graphical solution using Mollier chart, directly by drawing a vertical line (constant entropy line on h-s diagram) the state 2 can be located easily.

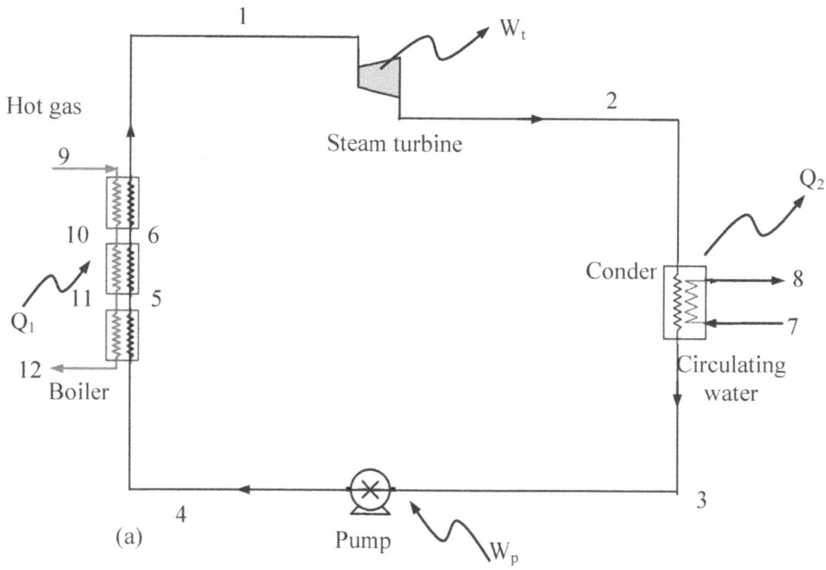

Fig. 8.2. Steam power plant schematic.

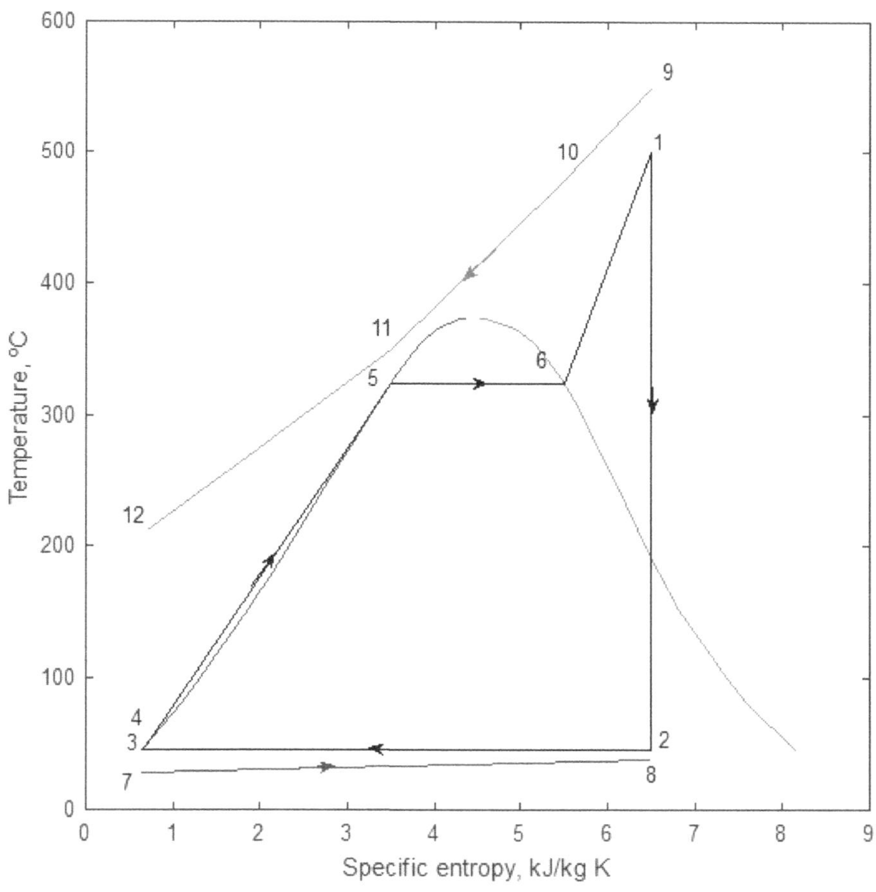

Fig. 8.3. Representation of a thermal power plant on a temperature-entropy plot.

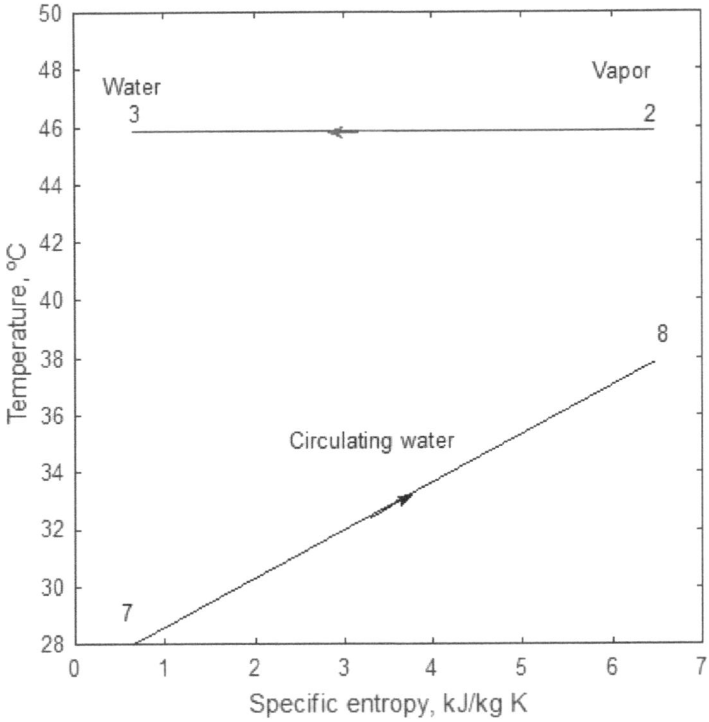

Fig. 8.4. Temperature profile in a steam condenser.

Heat rejection through the condenser to condense the steam into saturated water,

$$Q_2 = m_s (h_2 - h_3) = m_w c_{pw} (T_8 - T_7) \tag{8.65}$$

The enthalpy h_3 is the enthalpy of saturated water at the condenser pressure. It can be obtained from the steam tables but not from the Mollier chart, as the water region is omitted in the modified chart. The circulating water temperature, T_7 and T_8 are below the condensed steam temperature (Fig. 8.4). The above equation also gives the circulating water m_w requirement as per the designed water inlet temperature, T_7 and exit temperature, T_8.

After the condenser, the water flow to pump, the work input to the boiler feed pump,

$$W_p = m_s (h_4 - h_3) \tag{8.66}$$

The enthalpy h_3 can be read from the steam tables at condenser pressure and saturated water condition, h_w. h_4 is to be calculated from the work in a flow process as follows

$$W_p = m\int_3^4 -vdP \tag{8.67}$$

Water is an incompressible fluid and so the density and specific volume are approximately constant.

$$W_p = -mv\int_3^4 dP = -mv\left(P_4 - P_3\right)$$

The boiler pressure, P_4 is greater than the condenser pressure, P_3. The negative sign indicates that the work done on the system to operate the pump.

Therefore, the positive quantity of pump work is:

$$W_p = mv(P_4 - P_3) \tag{8.68}$$

As per the steady flow energy equation, $W_p = m(h_4 - h_3)$.

Therefore,

$$W_p = m(h_4 - h_3) = mv(P_4 - P_3) \tag{8.69}$$

$$h_4 = h_3 + v(P_4 - P_3) \tag{8.70}$$

Finally, in the boiler, SFEE can be used to determine the unknown data.

Heat supplied in the boiler or steam generator,

$$Q_1 = m_s(h_1 - h_4) = m_g c_{pg}(T_9 - T_{12}) \tag{8.71}$$

The above equation gives the gas required for steam generation of m_s or vice-versa. The gas inlet temperature, T_9 and exit temperature, T_{12} are fixed as per the source conditions. Figure 8.5 shows the temperature profile in the boiler for hot fluid (gas) and cold fluid) water/steam. In all the boiler parts or section, the designer has to ensure that the gas temperature is always more than the water or steam temperature. Actually, in a Rankine cycle, a boiler consists of three sections viz. the economizer, evaporator and superheater. The economiser has water tubes where sensible heat transfers when the temperature of water rises to evaporation temperature. In the evaporator, phase change happens and water converts completely into saturated vapor. In the superheater coil, the temperature of the vapor increases till it reaches the turbine inlet temperature.

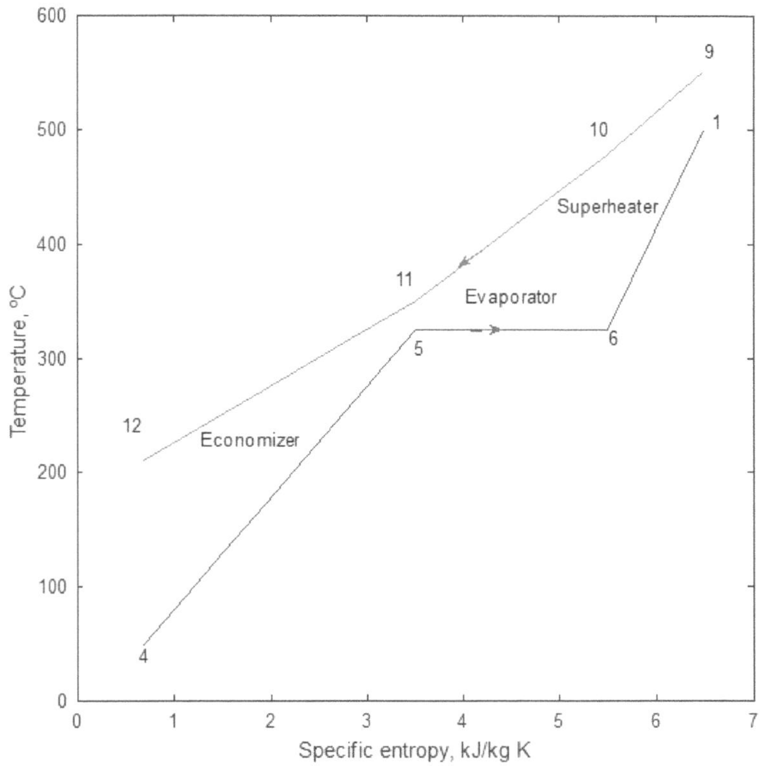

Fig. 8.5. Temperature profile in a steam boiler.

In thermal power plant, steam rate is defined as the amount of steam required to generate unit power production.

$$steam\ rate = \frac{mass\ of\ steam\ (kg/h)}{W_{net}\ (kW)}\ kg/kWh \tag{8.72}$$

Similarly, the heat rate can be defined as the heat supply to generate unit amount of power.

$$heat\ rate = \frac{Q_{\sup ply}\ (kW)}{W_{net}\ (kW)} \tag{8.73}$$

It is the inverse of thermal efficiency and the heat rate number has no units.

Steam rate is useful to express the capacity of the plant, and heat rate is used alternately to represent the thermal efficiency by the heat required to generate unit production.

Thermal efficiency of Rankine cycle,

$$\eta_{thermal} = \frac{W_{net}}{Q_1} \times 100 \tag{8.74}$$

$$= \frac{W_{turbinet} - W_{pump}}{Q_1} \times 100$$

From first law of thermodynamics,

$$\sum Q = \sum W$$

$$Q_1 - Q_2 = W_t - W_p$$

$$\eta_{thermal} = \frac{Q_1 - Q_2}{Q_1} \times 100 = \left(1 - \frac{Q_2}{Q_1}\right) \times 100$$

In general, the performance of the steam Rankine cycle can be assessed from the input and net power generation.

In the reversible Rankine cycle, the steam expansion and pumping are not isentropic processes. They are the adiabatic process. In the adiabatic process, heat transfer is zero but entropy increases due to internal dissipative effects.

Figure 8.6 shows the isentropic expansion (1-2′) and adiabatic expansion (1-2) in a steam turbine. The isentropic process is a hypothetical process used as a reference to the actual process, i.e., adiabatic process. To define the deviation of the actual from the theoretical, isentropic efficiency for a turbine is introduced as follows.

Isentropic efficiency of a steam turbine,

$$\eta_{i\ turbine} = \frac{Actual\ enthalpy\ drop}{Isentropic\ enthalpy\ drop} \times 100$$

In a turbine, the actual enthalpy drop is less than the isentropic enthalpy drops as shown in Fig. 8.6. Therefore,

$$\eta_{i\ turbine} = \frac{h_1 - h_2}{h_1 - h_{2'}} \times 100 \tag{8.75}$$

It is easy to find the $h_{2'}$ by equating the entropy before and after the expansion or alternatively from a Mollier chart. Now, the Eq. 8.75 can be used to find the actual enthalpy after the expansion, h_2, from the known isentropic efficiency of the turbine. The isentropic efficiency can be collected from the manufacturer's manual or from the plant operating data.

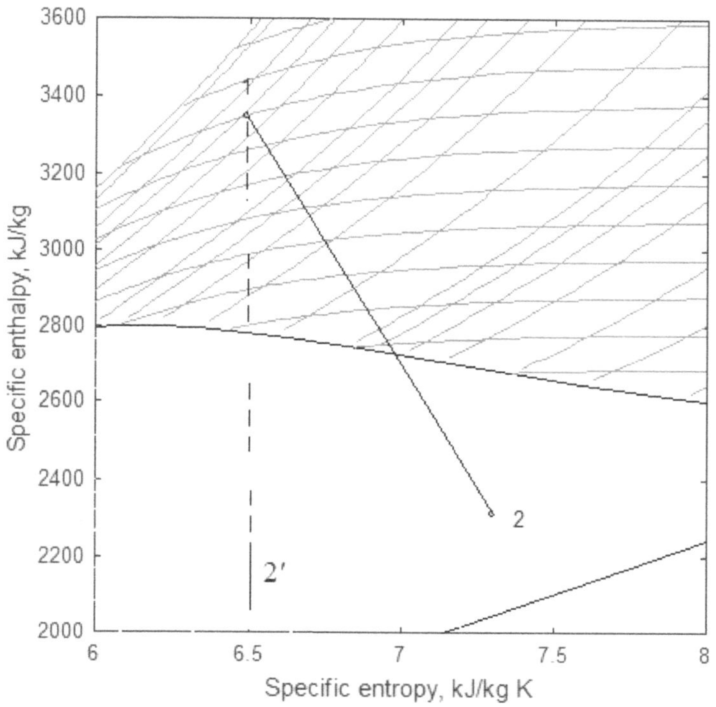

Fig. 8.6. Irreversibility in a steam turbine.

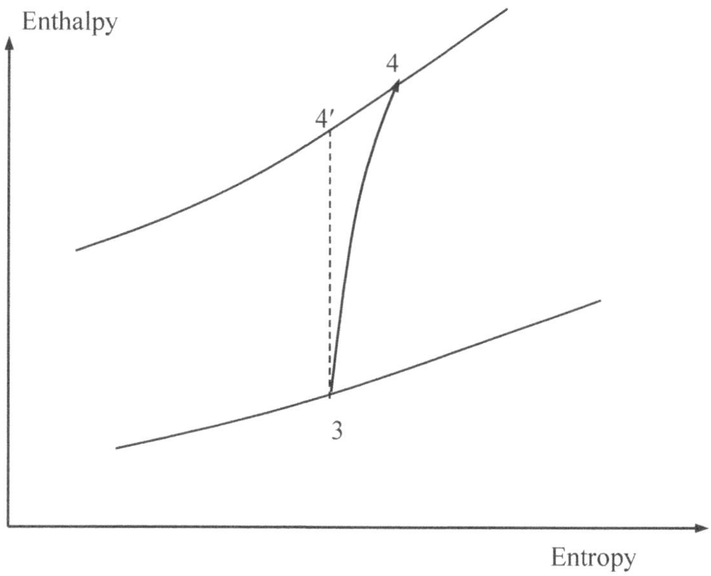

Fig. 8.7. Irreversibility in a boiler feed pump.

Similar to a turbine, a feed pump also has isentropic efficiency which is used to find the actual exit state of the pump. Steam turbines and feed pumps are turbomachines. The efficiency of a turbomachine can be expressed with thermodynamic properties as an isentropic efficiency. Figure 8.7 shows such pumping of water from low pressure (condenser pressure) to high pressure (boiler pressure). In such pumping, the isentropic enthalpy rise is less than the actual enthalpy rise as shown.

$$\eta_{i\ pump} = \frac{Isentropic\ enthalpy\ rise}{Actual\ enthalpy\ rise} \times 100$$

Therefore,

$$\eta_{i\ pump} = \frac{h_{4'} - h_3}{h_4 - h_3} \times 100 \tag{8.76}$$

Enthalpy after pumping can be determined from the SFEE applied to the pump. The actual enthalpy after pumping, h_4 can be determined from the value of isentropic efficiency of the pump.

The Rankine cycle is a base cycle and it offers lower efficiency as no attempts are made to reach the Carnot cycle. Now to improve the Rankine cycle, the available methods are superheating (increasing the source temperature), decreasing the sink temperature, reheating and regeneration through the feedwater heating system. The following sections focus on performance augmentation and analysis of steam power cycles.

T_m = mean temperature of heat addition

$$T_m = \frac{h_1 - h_4}{s_1 - s_4}$$

8.4.1 Reheat Rankine cycle

The reheater in a steam power plant increases the dryness fraction in the turbine and saves its life from corrosion and erosion. The location and its process have been shown in Figs. 8.8 and 8.9, respectively. The location of the reheater depends on its optimum working conditions, as shown in Fig. 8.10.

Thermal efficiency of a reheat Rankine cycle

$$\eta_{thermal} = \frac{(h_1 - h_2) + (h_3 - h_4) - (h_6 - h_5)}{(h_1 - h_6) + (h_3 - h_2)} \times 100 \tag{8.77}$$

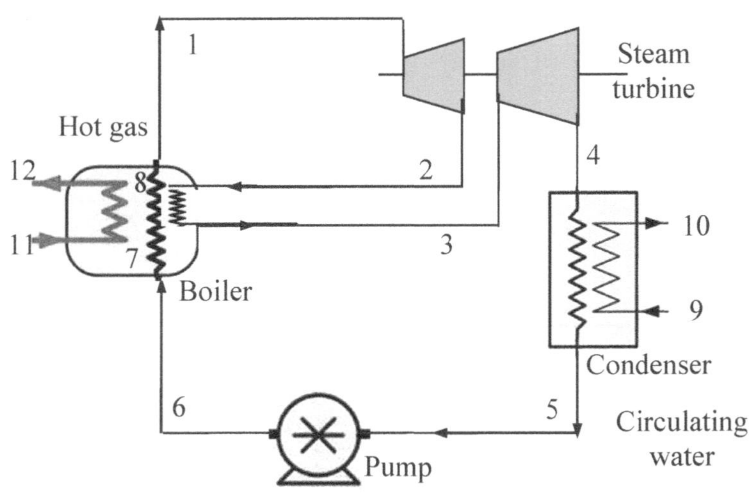

Fig. 8.8. Steam power plant with reheat arrangement.

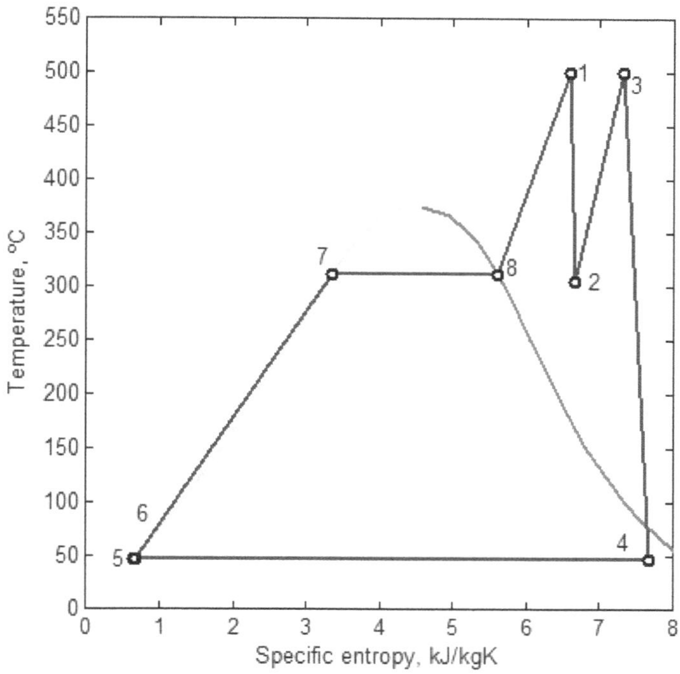

Fig. 8.9. Representation of reheat cycle on a T-s diagram.

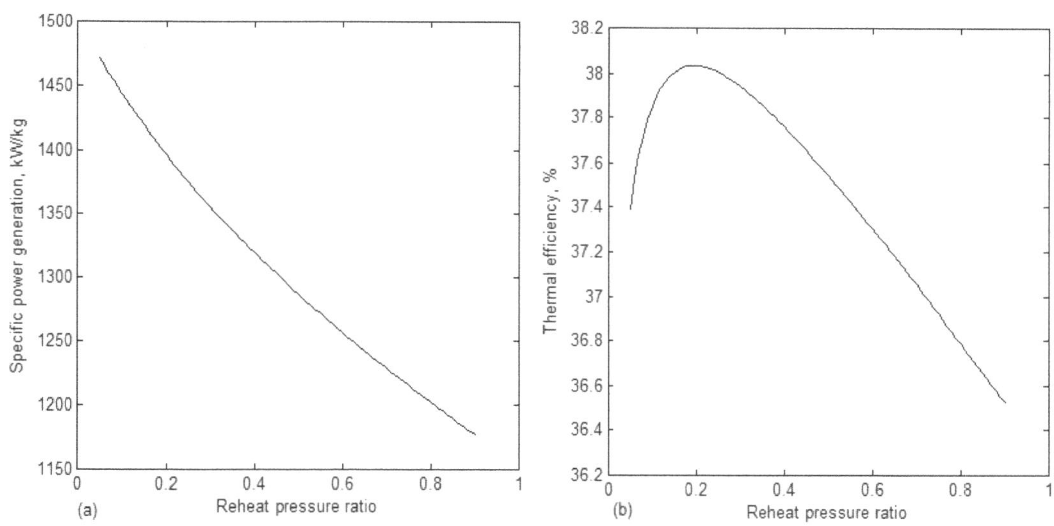

Fig. 8.10. Influence of reheat pressure ratio on reheat Rankine cycle performance.

8.4.2 Regenerative Rankine cycle

8.4.2.1 Open feedwater heater

The open feedwater heater has two functions: (i) removes the dissolved gases from steam and increases the thermal efficiency of the plant by increasing the feedwater temperature at the inlet of the boiler (Fig. 8.11). The bled steam and feedwater directly mix with each other. Figure 8.12 shows the thermodynamic process of feedwater heating on the T-s diagram.

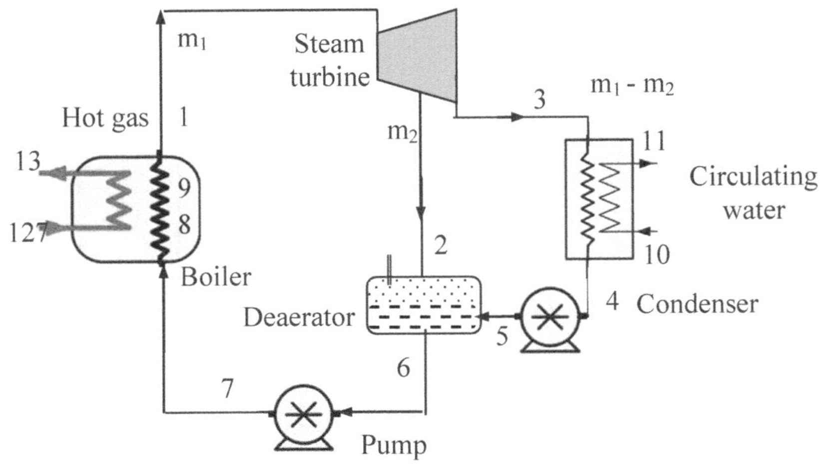

Fig. 8.11. Steam power plant schematic arrangement with a deaerator.

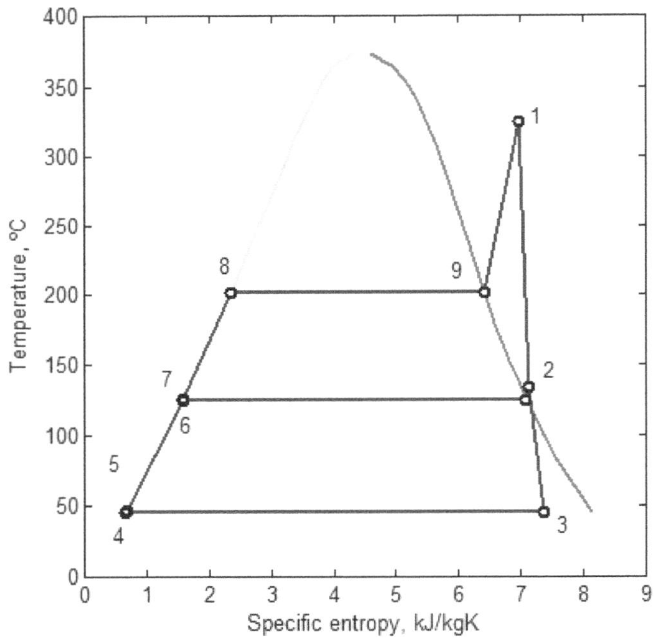

Fig. 8.12. Representation of Rankine cycle with deaerator on a T-s diagram.

The steam generation from the known hot fluid flow rate,

$$m_1 = \frac{m_g c_{pg} \left(T_{13} - T_{12}\right)}{h_1 - h_7} \tag{8.78}$$

Let m_2 be the mass to be removed from the turbine at the steam flow supply of an m_1 in boiler.

The mass balance in an open feedwater heater is,

$$m_3 = m_1 - m_2 \tag{8.79}$$

The heat balance is,

$$m_2 h_2 + (m_1 - m_2) h_5 = m_6 h_6 \tag{8.80}$$

The simplification of this heat balance is as follows,

$$m_2 h_2 + m_1 h_5 - m_2 h_5 = m_1 h_6 \tag{8.81}$$

$$m_2 (h_2 - h_5) = m_1 (h_6 - h_5)$$

$$m_2 = \frac{m_1 (h_6 - h_5)}{h_2 - h_5} \tag{8.82}$$

The gross work from the turbine,

$$W_{ST} = m_1 (h_1 - h_2) + (m_1 - m_2)(h_2 - h_3) \tag{8.83}$$

After the condenser, the water flows to the first feed pump; the work input to the feed pump,

$$W_{p1} = (m_1 - m_2)(h_5 - h_4) \tag{8.84}$$

where,

$$h_5 = h_4 + v(P_5 - P_4) \tag{8.85}$$

Similarly, work supply to the second pump, i.e., boiler feed pump is,

$$W_{p2} = m_1 (h_7 - h_6) \tag{8.86}$$

where,

$$h_7 = h_6 + v(P_7 - P_6) \tag{8.87}$$

$$W_{net} = W_{ST} - W_{p1} - W_{p2} \tag{8.88}$$

Thermal efficiency of the regenerative Rankine cycle,

$$\eta_{thermal} = \frac{m_1 (h_1 - h_2) + (m_1 - m_2)(h_2 - h_3) - (m_1 - m_2)(h_5 - h_4) - m_1 (h_7 - h_6)}{m_1 (h_1 - h_7)} \times 100 \tag{8.89}$$

As per the analysis of deaerator location, at the mid point of boiler saturation temperature and condenser saturation temperature, it results in maximum thermal efficiency. For convenience, the deaerator location is expressed as a dimensionless number, as follows:

$$\theta = \frac{T_{bled\ steam} - T_{co}}{T_{bo,\ sat} - T_{co}} \tag{8.90}$$

8.4.2.2 Single closed feedwater heater

In a closed feedwater heater, there is no direct contact between steam and feedwater. The power plant layout with a single closed feedwater heater and its processes are shown in Fig. 8.13 and Fig. 8.14 respectively.

Steam generation in the boiler from the hot fluid,

$$m_1 = \frac{m_g c_{pg} (T_{15} - T_{14})}{h_1 - h_9} \tag{8.91}$$

In a closed feedwater heater, the heat loss by steam is equal to the heat gain by the feedwater. The exit temperature of the feedwater is below the saturation temperature of bled steam. This temperature difference is called terminal temperature difference (TTD) in the feedwater heater.

Therefore,

$$T_6 = T_7 - TTD \tag{8.92}$$

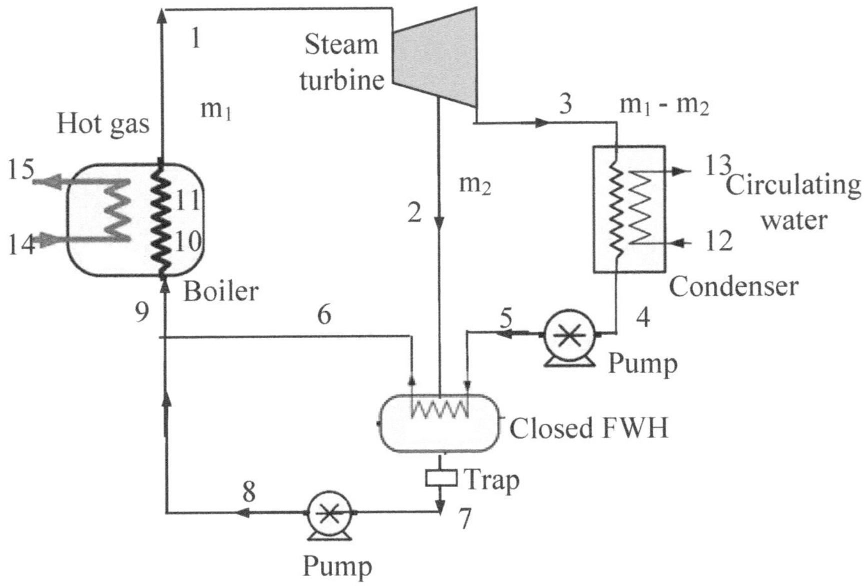

Fig. 8.13. Schematic of flow of regenerative Rankine cycle with closed feedwater heater.

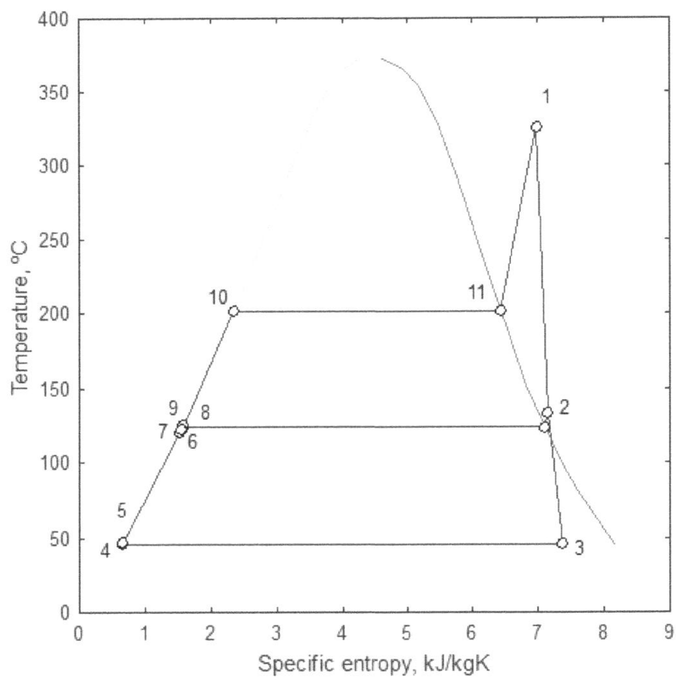

Fig. 8.14. Regenerative Rankine cycle with a single closed feedwater heater on a T-s diagram.

If m_1 is the steam generation in the boiler, m_2 is the bled steam used for the closed feedwater heater, The mass balance in closed feedwater heater is,

$$m_3 = m_1 - m_2 \tag{8.93}$$

The heat balance in feedwater heater (FWH),

$$m_2 h_2 + (m_1 - m_2) h_5 = (m_1 - m_2) h_6 + m_2 h_7 \tag{8.94}$$

Simplification,

$$m_2 h_2 + m_1 h_5 - m_2 h_5 = m_1 h_6 - m_2 h_6 + m_2 h_7$$

$$m_2 (h_2 - h_5 + h_6 - h_7) = m_1 (h_6 - h_5)$$

$$m_2 = \frac{m_1 (h_6 - h_5)}{h_2 + h_6 - h_5 - h_7} \tag{8.95}$$

The gross work from the turbine,

$$W_{ST} = m_1 (h_1 - h_2) + (m_1 - m_2)(h_2 - h_3) \tag{8.96}$$

After the condenser, the water flow to first feed pump, the work input to the feed pump,

$$W_{p1} = (m_1 - m_2)(h_5 - h_4) \tag{8.97}$$

where

$$h_5 = h_4 + v(P_5 - P_4) \tag{8.98}$$

Similarly work supply to second pump, i.e., boiler feed pump is,

$$W_{p2} = m_2 (h_8 - h_7) \tag{8.99}$$

where,

$$h_8 = h_7 + v(P_8 - P_7) \tag{8.100}$$

At the mixing before boiler,

$$h_9 = \frac{(m_1 - m_2) h_6 + m_2 h_8}{m_1} \tag{8.101}$$

Thermal efficiency of a regenerative Rankine cycle,

$$\eta_{thermal} = \frac{m_1 (h_1 - h_2) + (m_1 - m_2)(h_2 - h_3) - (m_1 - m_2)(h_5 - h_4) - m_2 (h_8 - h_7)}{m_1 (h_1 - h_9)} \times 100$$

8.4.2.3 Two closed feedwater heaters

Figure 8.15 shows the arrangement of two closed feedwater heaters in a thermal power plant and their processes are depicted on a T-s diagram (Fig. 8.16).

Steam generation in a boiler,

$$m_1 = \frac{m_g c_{pg} (T_{18} - T_{17})}{h_1 - h_{12}} \tag{8.102}$$

The temperature difference between boiler saturation and condenser saturation,

$$\Delta T_{bc} = T_{boiler} - T_{condenser} \tag{8.103}$$

Saturation temperature at P_2,

$$T_{sat\ 2} = T_{sat\ 4} + \frac{2}{3} \Delta T_{bc} \tag{8.104}$$

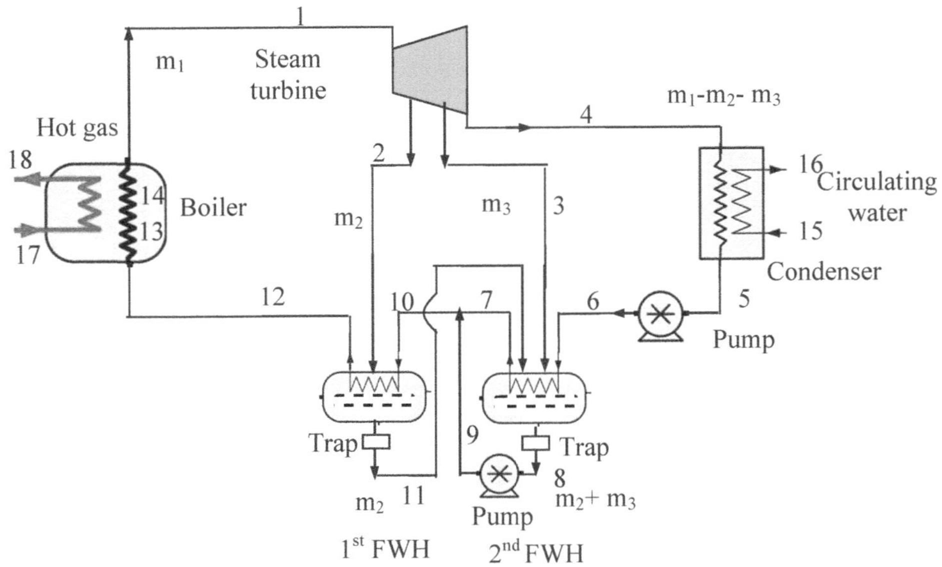

Fig. 8.15. The arrangement of Rankine cycle with two feedwater heaters.

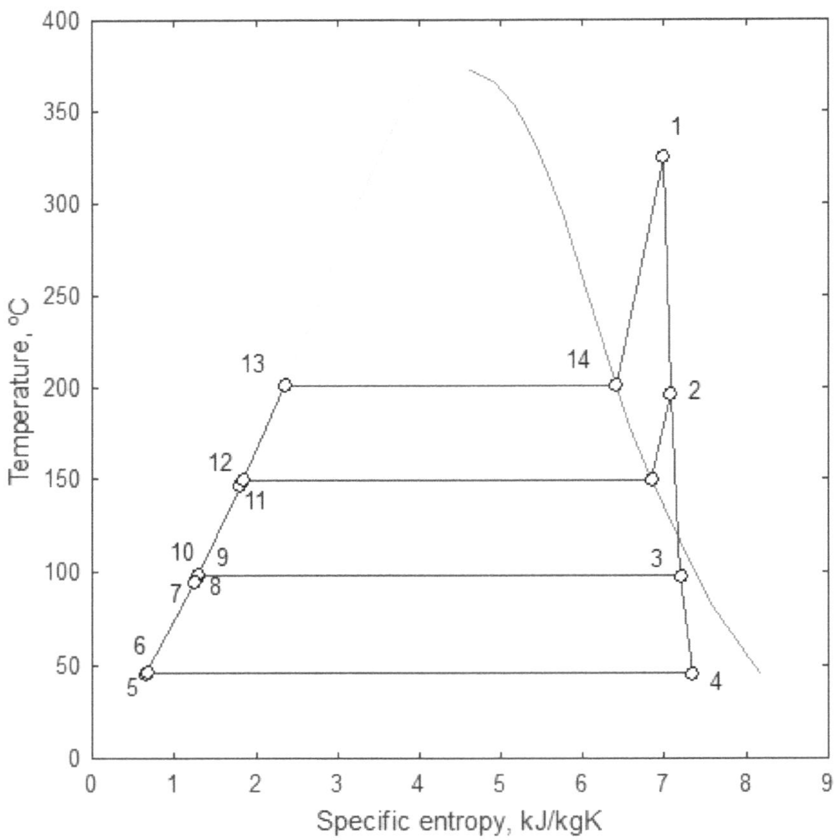

Fig. 8.16. T-s diagram of Rankine cycle with dual feedwater heaters.

Saturation temperature at P_3,

$$T_{sat\ 3} = T_{sat\ 4} + \frac{1}{3}\Delta T_{bc} \tag{8.105}$$

The enthalpy at the exit of the first pump,

$$h_6 = h_5 + v_5 (P_1 - P_4)\ 100 \tag{8.106}$$

The enthalpy at the exit of the second pump,

$$h_9 = h_8 + v_8 (P_1 - P_3)\ 100 \tag{8.107}$$

At the exit of the heat exchanger, there is a temperature difference between the hot fluid and cold fluid. In the deaerator it is zero due to direct mixing. This temperature difference at the terminal of the heat exchanger is called terminal temperature difference (TTD).

The temperature of feedwater at the exit of 2nd FWH,

$$T_7 = T_{sat\ 3} - TTD \tag{8.108}$$

Similarly, the temperature of feedwater at the exit of 1st FWH,

$$T_{12} = T_{sat\ 2} - TTD \tag{8.109}$$

At the exit of 2nd FWH, two jets of steam mix. The enthalpy after mixing,

$$h_{10} = \frac{(m_2 + m_3) h_9 + (m_1 - m_2 - m_3) h_7}{m_1} \tag{8.110}$$

Heat balance in 1st FWH,

$$m_2 (h_2 - h_{11}) = m_1 (h_{12} - h_{10}) \tag{8.111}$$

After simplification,

$$m_2 = \frac{m_1 (h_{12} - h_{10})}{h_2 - h_{11}} \tag{8.112}$$

Similarly at 2nd FWH,

$$m_3 h_3 + m_2 h_{11} = (m_2 + m_3) h_8 + (m_1 - m_2 - m_3)(h_7 - h_6) \tag{8.113}$$

After simplification,

$$m_3 = \frac{(m_1 - m_2)(h_7 - h_6) + m_2 (h_8 - h_{11})}{h_3 - h_8 + h_7 - h_6} \tag{8.114}$$

From the above the equations m_2, m_3 and h_{10} can be determined through iteration. i, If it is assumed $h_{10} = h_7$, in heat balance of a mixing equation, the iteration is not necessary to find the mass flow rates through simultaneous equations.

The net output from the cycle,

$$W_{net} = m_1 (h_1 - h_2) + (m_1 - m_2)(h_2 - h_3) + (m_1 - m_2 - m_3)(h_3 - h_4) - (m_1 - m_2 - m_3)(h_6 - h_5) - (m_2 + m_3)(h_9 - h_8) \tag{8.115}$$

8.4.3 Reheat – Regenerative Rankine Cycle

In actual practice, the reheating and regeneration are used together in a thermal power plant, as shown in Figs. 8.17 and 8.18.

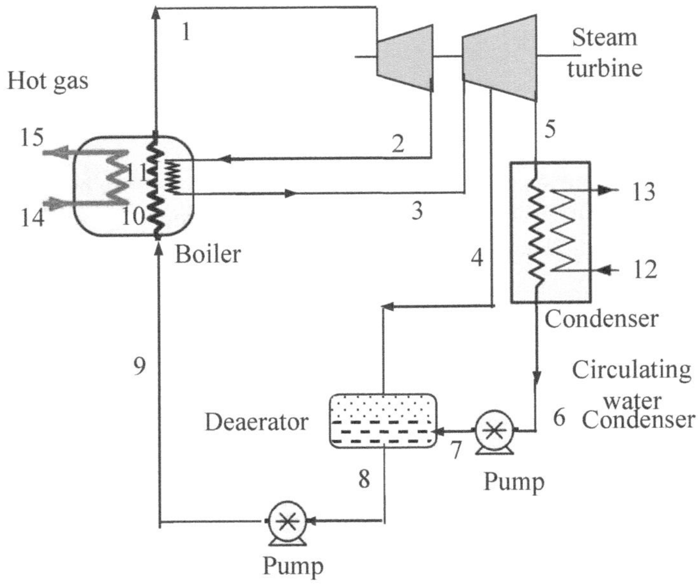

Fig. 8.17. Reheat-regenerative Rankine cycle with deaerator as a feedwater heater.

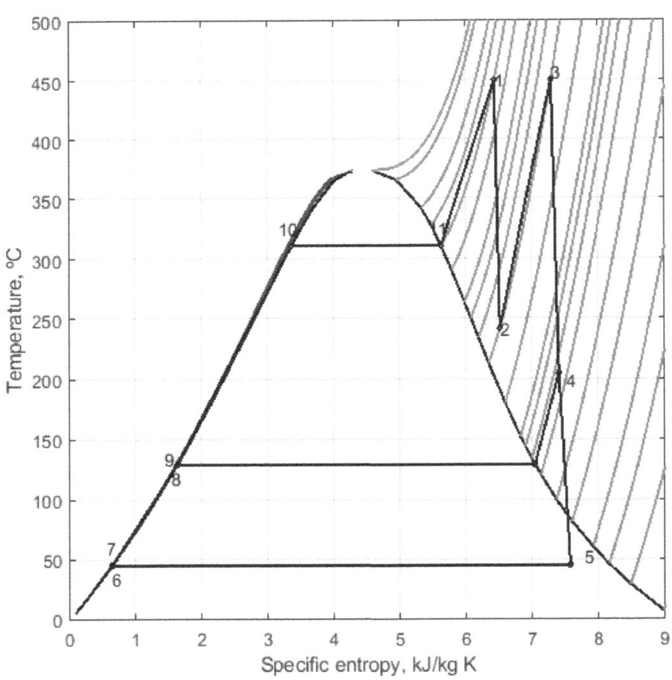

Fig. 8.18. Temperature-specific entropy diagram of Rankine cycle with reheater and regenerator.

8.4.4 Steam Rankine cycle with Waste Heat Recovery

A waste heat recovery process takes the place of the boiler and plays a significant role in controlling carbon dioxide emissions. Figure 8.19 shows the process layout of a thermal power plant from an industrial (cement factory) water heating system. The cycle is shown in Fig. 8.20.

Figure 8.21 shows heat recovery from the hot gas to the plant's working fluid to generate superheated steam from water. The hot gas temperature in the evaporator can be determined from the pinch point (PP) of the heat recovery steam generator (HRSG). Approach point (AP) also maintained in the HRVG to control the sharp transition of water into steam.

The turbine inlet temperature is determined from the hot gas supply temperature and TTD,

$$T_1 = T_{13} - TTD \tag{8.116}$$

where TTD is the temperature difference between hot gas supply temperature and turbine inlet temperature.

The water temperature at the exit of the evaporator,

$$T_8 = T_9 - AP \tag{8.117}$$

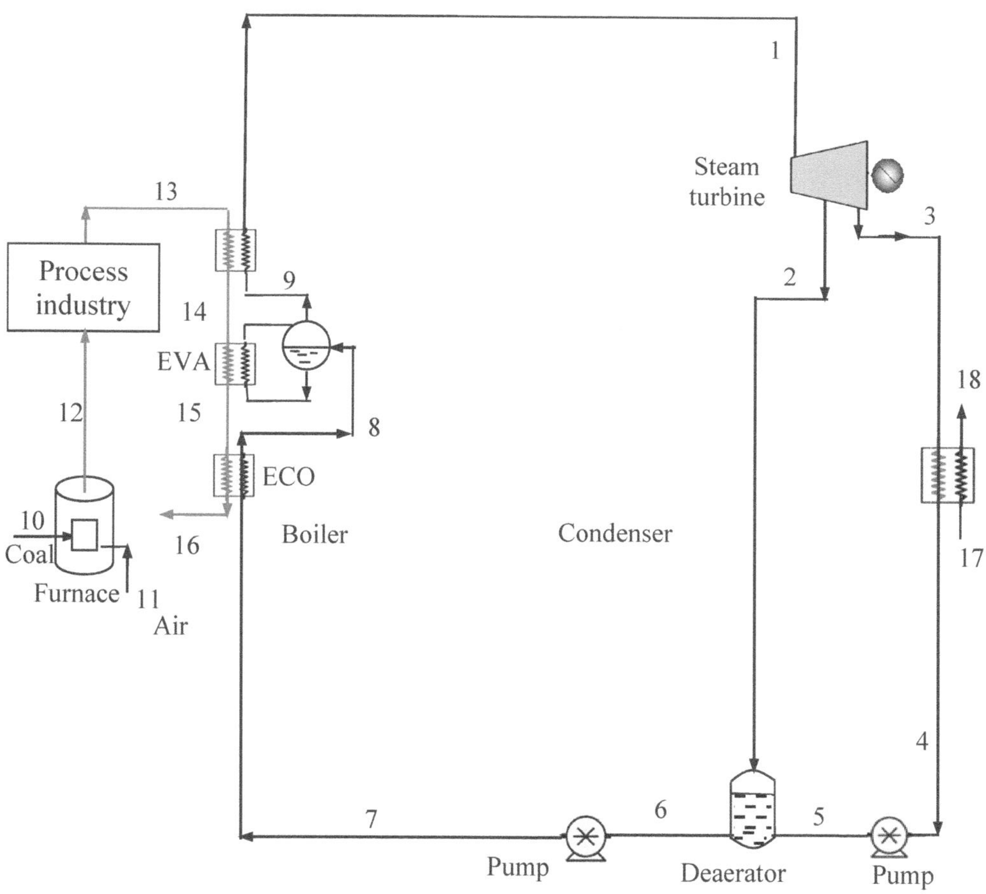

Fig. 8.19. Schematic of a plant layout of SRC with a deaerator.

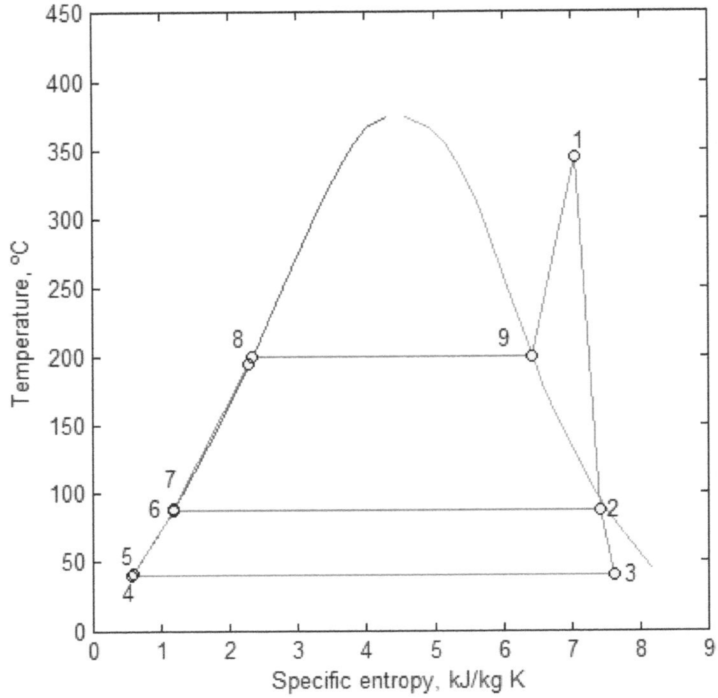

Fig. 8.20. Temperature-entropy diagram for SRC with a deaerator.

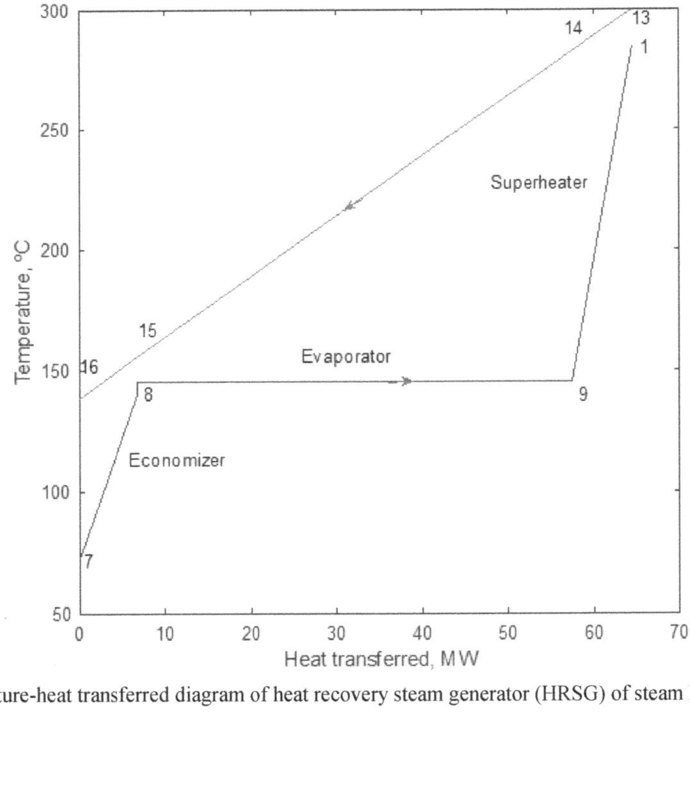

Fig. 8.21. Temperature-heat transferred diagram of heat recovery steam generator (HRSG) of steam Rankine cycle (SRC).

The temperature of hot gas in the evaporator,

$$T_{15} = T_8 + AP + PP \tag{8.118}$$

The local gas temperatures are determined from the heat balance applied to the corresponding heat exchanger.

The deaerator location is selected from its temperature ratio. It is defined as the ratio of the temperature difference between the deaerator and the condenser and the total temperature difference between the boiler saturation and the condenser.

The deaerator temperature ratio,

$$\theta_{deaerator} = \frac{T_{bled\ steam} - T_{condenser}}{T_{HRSG,\ sat} - T_{condenser}} \tag{8.119}$$

The steam capacity from the gas flow and its temperature,

$$m_1 = \frac{m_g c_{pg} \left(T_{13} - T_{15}\right)}{h_1 - h_8} \tag{8.120}$$

The mass of gas is determined from its flow rate (V_g) and molecular weight (M_g).

$$m_g = \frac{V_g M_g}{22.4 \times 3600} \tag{8.121}$$

Let m_2 be the volume of steam to be removed from the turbine at the steam flow supply of m_1 for feedwater heating.

The mass balance in an open feedwater heater (deaerator),

$$m_3 = m_1 - m_2 \tag{8.122}$$

Similarly, the heat balance in a deaerator,

$$m_2\, h_2 + (m_1 - m_2)\, h_5 = m_6\, h_6 \tag{8.123}$$

From the simplification of these two equations,

$$m_2 = \frac{m_1 \left(h_6 - h_5\right)}{h_2 - h_5} \tag{8.124}$$

The gross work from the turbine,

$$W_{ST} = m_1\, (h_1 - h_2) + (m_1 - m_2)\, (h_2 - h_3) \tag{8.125}$$

The work supply to the first pump,

$$W_{p1} = (m_1 - m_2)\, (h_5 - h_4) \tag{8.126}$$

where

$$h_5 = h_4 + v(P_5 - P_4) \tag{8.127}$$

Similarly work supply to the second pump, i.e., the boiler feed pump,

$$W_{p2} = m_1\, (h_7 - h_6) \tag{8.128}$$

where

$$h_7 = h_6 + v(P_7 - P_6) \tag{8.129}$$

The net output from the thermal power plant,

$$W_{net} = W_{ST} - W_{p1} - W_{p2} \tag{8.130}$$

The heat supply to the power plant,

$$Q_{supply} = m_1 (h_1 - h_7) \tag{8.131}$$

Thermal efficiency of steam Rankine cycle (SRC),

$$\eta_{thermal} = \frac{W_{net}}{Q_{sup\,ply}} \times 100 \tag{8.132}$$

For exergy analysis, the irreversibilities are evaluated from the exergy balance of a plant's components.

Irreversibility of HRSG,

$$I_{HRSG} = m_g(\varepsilon_{13} - \varepsilon_{16}) + m_1(\varepsilon_1 - \varepsilon_7) \tag{8.133}$$

where ε (= h – T_0s) is the specific exergy, kJ/kg.

Irreversibility in a steam turbine,

$$I_{ST} = m_1(\varepsilon_1 - \varepsilon_2) + m_3(\varepsilon_3 - \varepsilon_4) - W_{ST} \tag{8.134}$$

Irreversibility in a deaerator,

$$I_{deaerator} = m_2\varepsilon_2 + m_5\varepsilon_5 - m_6\varepsilon_6 \tag{8.135}$$

From the energy balance in steam condenser, the circulating cooling water,

$$m_w = \frac{m_3(h_3 - h_4)}{c_{pw}(T_{18} - T_{17})} \tag{8.136}$$

The irreversibility of a water-cooled condenser

$$I_{condenser} = m_3\left(\varepsilon_3 - \varepsilon_4\right) + m_{17}\left(\varepsilon_{17} - \varepsilon_{18}\right) \tag{8.137}$$

Irreversibility of two pumps,

$$I_{pump} = m_4\left(\varepsilon_4 - \varepsilon_5\right) + m_6\left(\varepsilon_6 - \varepsilon_7\right) + W_p \tag{8.138}$$

The irreversibility of exhaust gas is equal to the exergy loss by exhaust gas.

$$I_{exhaust} = m_g(\varepsilon_{12} - \varepsilon_0) \tag{8.139}$$

where ε_0 is the exergy at the reference state.

Similarly, irreversibility of hot water from the condenser is equal to its exergy loss.

$$I_{hot\ water} = m_{18}(\varepsilon_{18} - \varepsilon_0) \tag{8.140}$$

Total irreversibility losses in SRC,

$$I_{total} = I_{HRSG} + I_{ST} + I_{deaerator} + I_{condenser} + I_{pump} + I_{exhaust} + I_{hot\ water} \tag{8.141}$$

Exergy of hot gas source,

$$E_g = m_{38}(\varepsilon_{13} - \varepsilon_0) \tag{8.142}$$

Exergy efficiency according to the second law of thermodynamics,

$$\eta_2 = \frac{E_g - I_{total}}{E_g} \times 100 \tag{8.143}$$

8.5 Combined cycles

Combined cycle power plants are gaining wider acceptance due to increasing availability of natural gas, higher overall thermal efficiencies, peaking and two-shifting capabilities, fast start-up capabilities, and lower cooling water requirements. Optimization of heat recovery steam generators (HRSG) is particularly interesting for combined plant design in order to maximize work obtained in the vapor cycle (Horlock, 2013). Figure 8.22 shows the combined cycle layout with steam Rankine cycle with the support of T-s diagram (Fig. 8.23) and heat recovery diagram (Fig. 8.24). Similarly, Figs. 8.25, 8.26, and 8.27 detail the combined cycle power plant with a flasher. Another set of Figs. 8.28, 8.29, and 8.30 shows the combined cycle power plant with two flashers. The two flashers generate more steam compared to a single flasher.

A case study has been conducted in a combined cycle power plant with the layout depicted in Fig. 8.31. Natural gas (NG) with the following composition was collected from the GMR energy power plant, India: methane (CH_4), 90.81%, ethane (C_2H_6) 2.96%, propane (C_3H_8) 2.7%, i-butane (C_4H_{10}) 0.54%, n-butane (C_4H_{10}) 0.76%, i-pentane (C_5H_{12}) 0.2%, n-pentane (C_5H_{12}) 0.16%, n-hexane (C_6H_{14}) 0.26%, carbon dioxide (CO_2) 1.49% and nitrogen (N_2) 0.09%.

The full combustion of NG with compressed air in GTCC was

$$[a_1CH_4 + a_2C_2H_6 + a_3C_3H_8 + a_4(C_4H_{10})_i + a_5(C_4H_{10})_n + a_6(C_5H_{12})_i + a_7(C_5H_{12})_n + a_8(C_6H_{14})_n + a_9CO_2 + a_{10}N_2]_{NG} + [x\,(O_2+3.76N_2)]_{air} \rightarrow [d_1CO_2 + d_2H_2O + d_3O_2 + d_4N_2]_{exhaust} \quad (8.144)$$

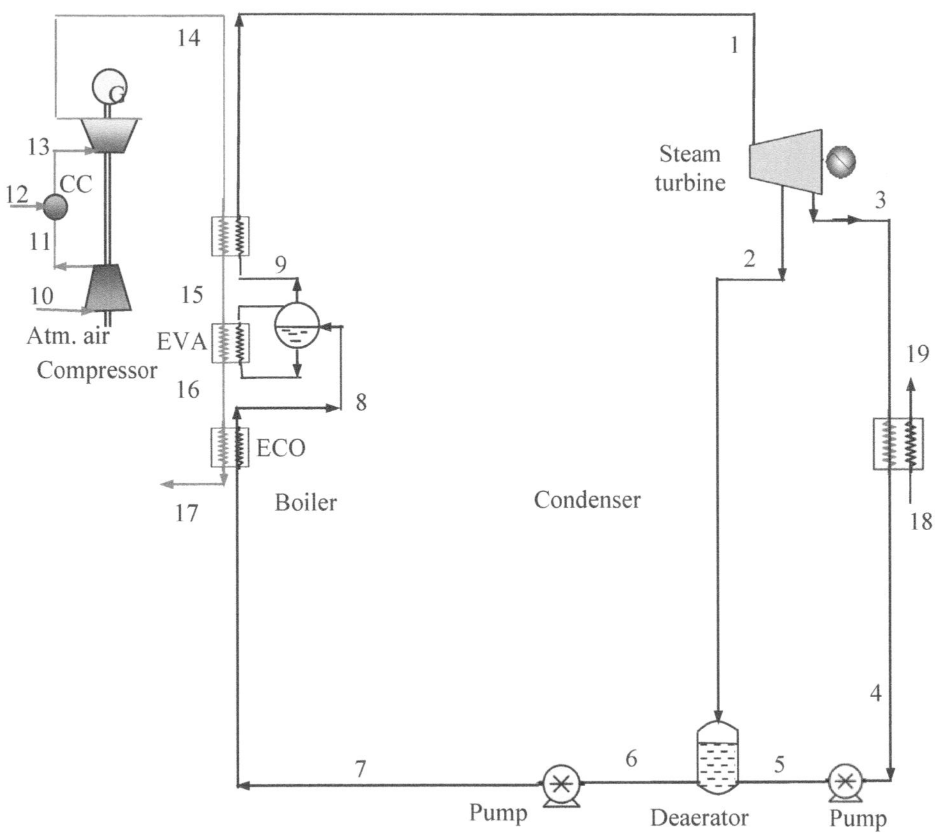

Fig. 8.22. Combined cycle power plant with steam Rankine cycle as a bottoming cycle.

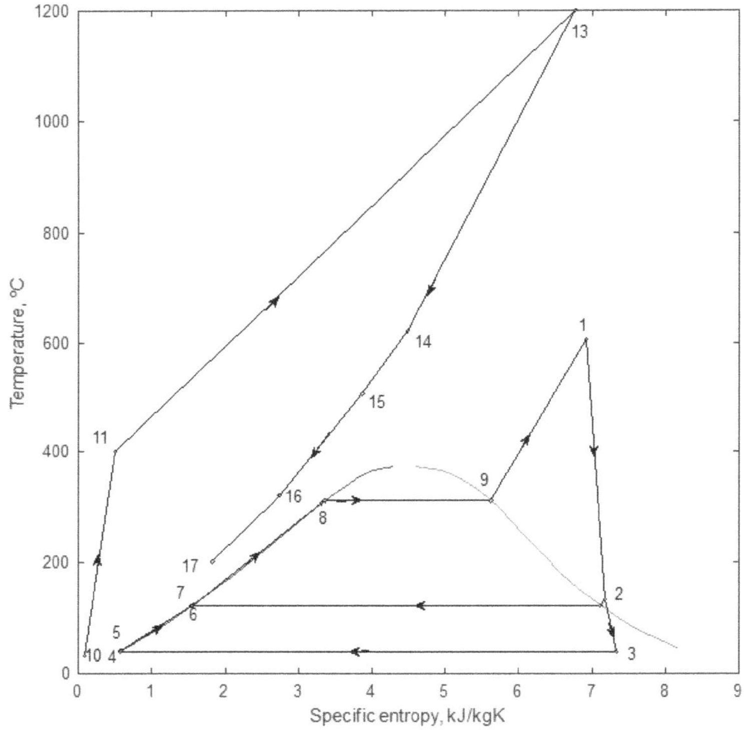

Fig. 8.23. Diagram of a Combined cycle power plant on temperature-specific entropy with steam Rankine cycle as a bottoming cycle.

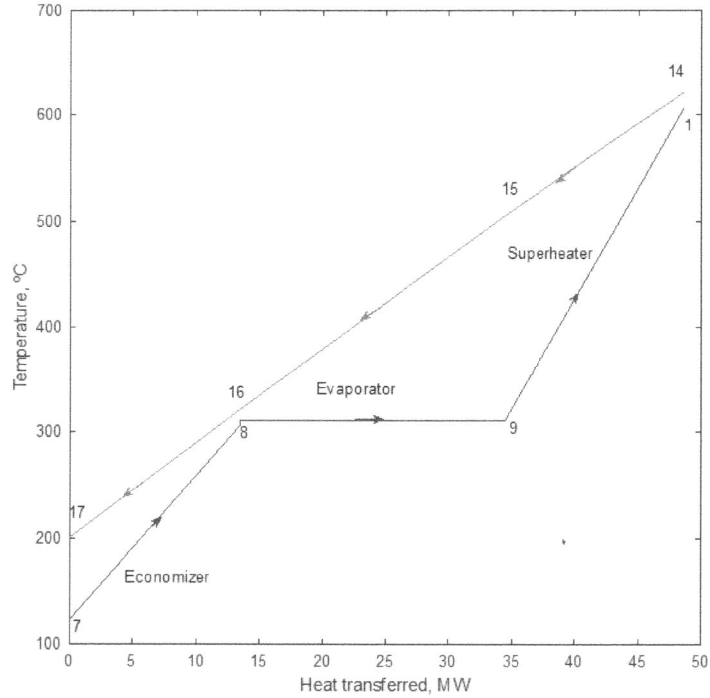

Fig. 8.24. Diagram of a temperature-heat transferred HRSG with steam Rankine cycle as a bottoming cycle.

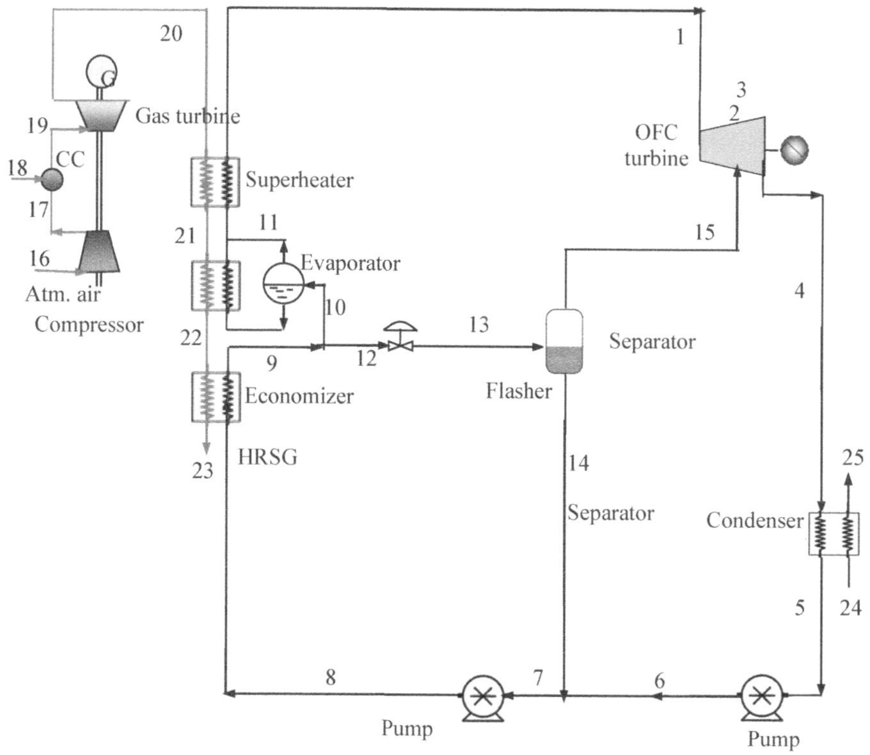

Fig. 8.25. Combined cycle power plant with a single steam flash cycle as a bottoming cycle.

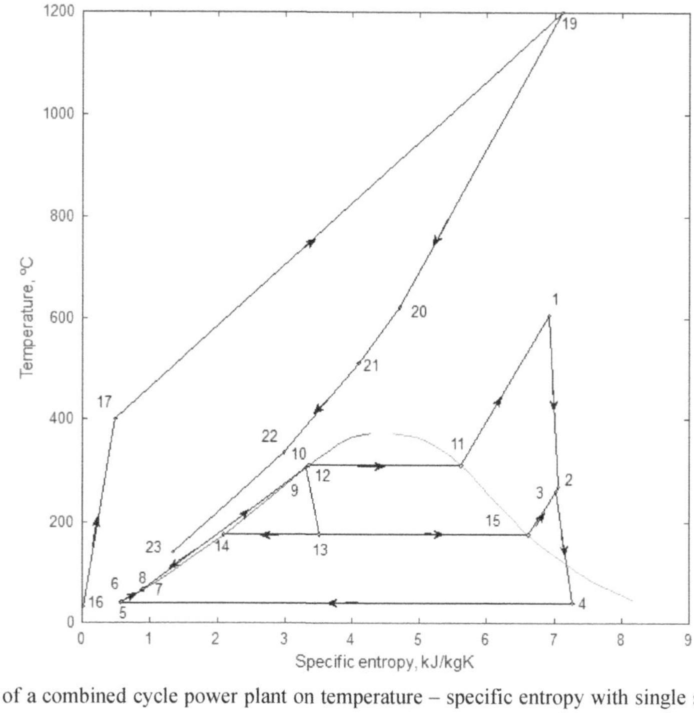

Fig. 8.26. Diagram of a combined cycle power plant on temperature – specific entropy with single steam flash cycle as a bottoming cycle.

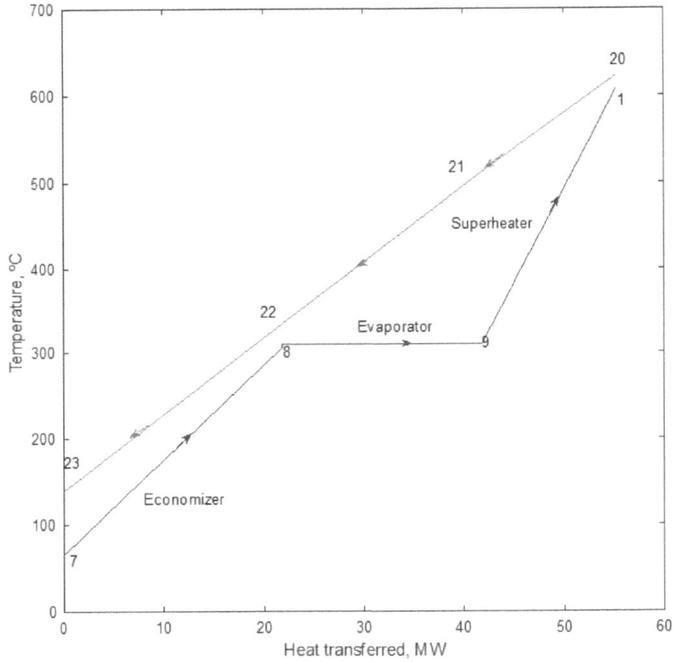

Fig. 8.27. Diagram of a temperature–heat transferred HRSG with single steam flash cycle in a combined cycle power plant.

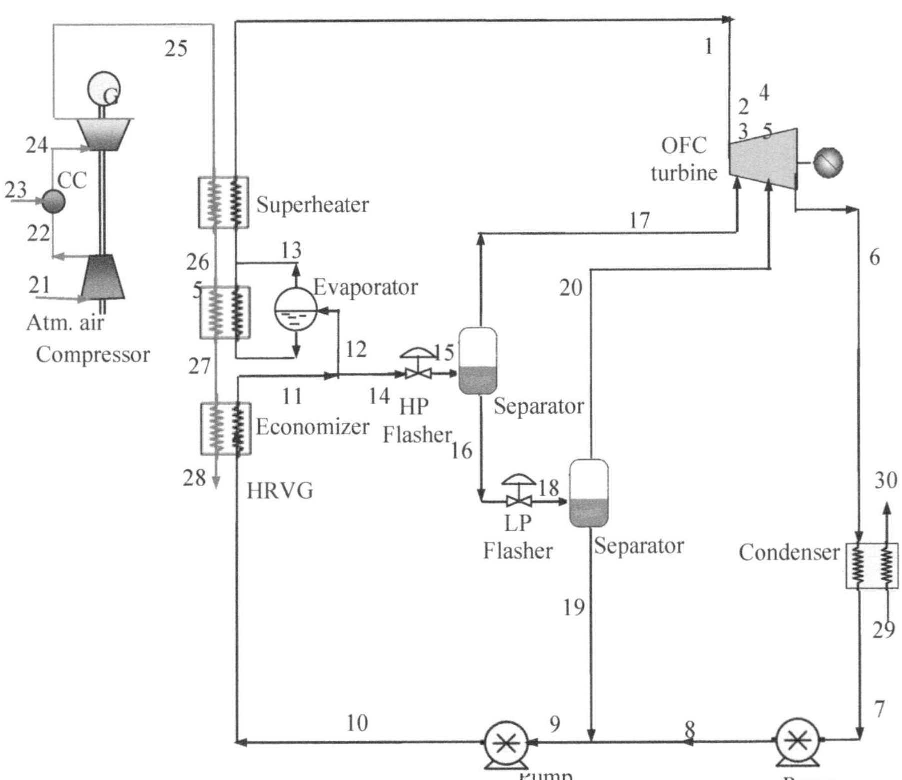

Fig. 8.28. Combined cycle power plant with double steam flash cycle as a bottoming cycle.

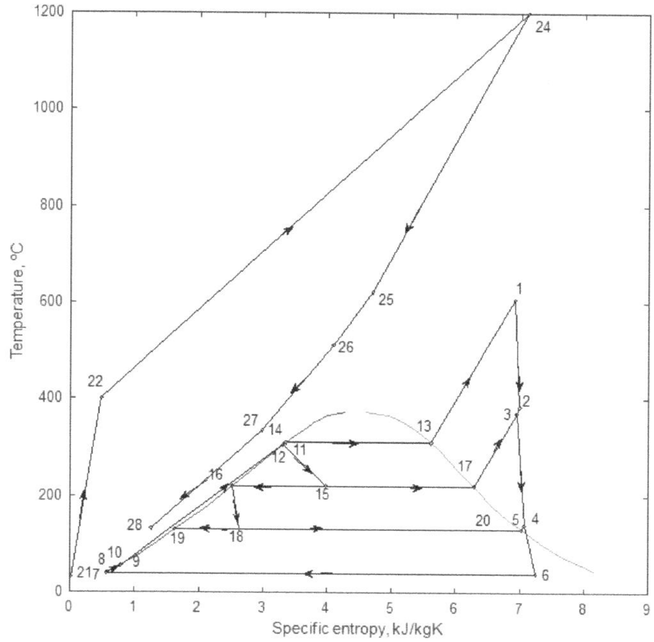

Fig. 8.29. Diagram of a combined cycle on temperature–specific entropy with double steam flash cycle as a bottoming cycle.

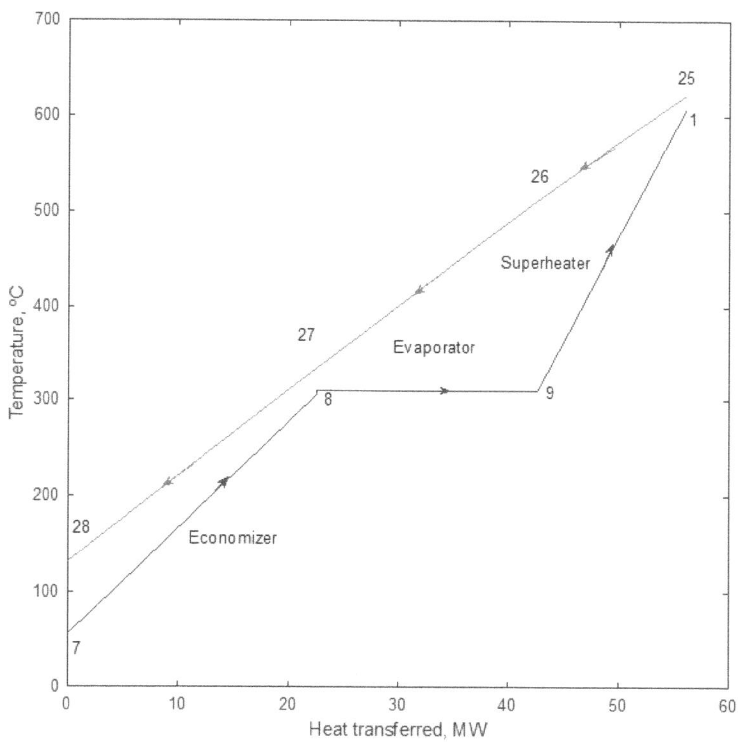

Fig. 8.30. Temperature-heat transferred diagram of HRSG with double steam flash cycle in a combined cycle power plant.

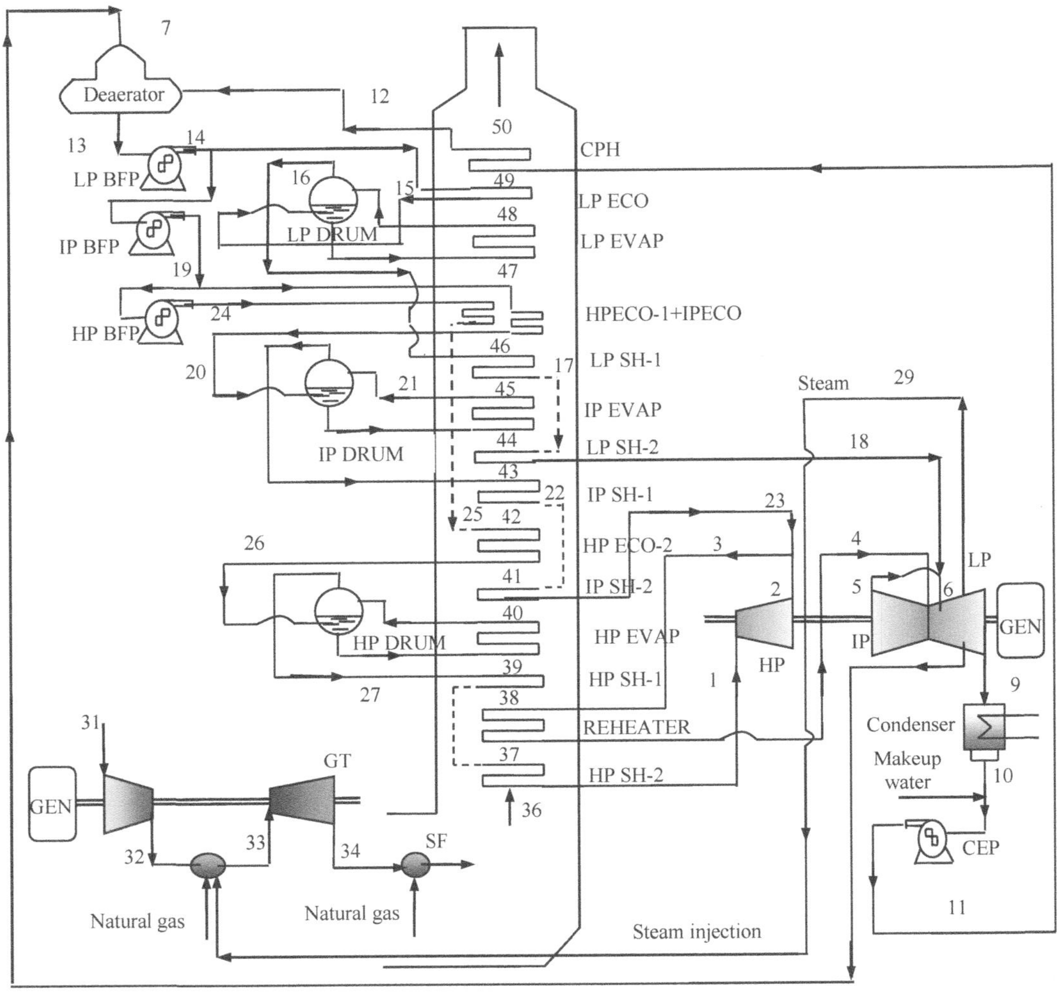

Fig. 8.31. Flow arrangements for fuel, air, gas and water/steam in a steam injected CC plant, BFP: boiler feed pump; CEP: condensate extraction pump; CPH: condensate preheater; ECO: economizer; EVAP: evaporator; GT: gas turbine; GTCC: gas turbine combustion chamber; HP: high pressure; HRSG: heat recovery steam generator; IP: intermediate pressure; LP: low pressure; RH: reheater; SF: supplementary firing; SH: superheater; ST: steam turbine.

In Eq. (8.144) 'x' denotes the amount of air in kg mol supplied per kg mol of NG ($a_1 + a_2 + \text{---} + a_{10}$) into the GTCC. The amount of air to be compressed to get the required combustion temperature (T_{33}) is determined from the energy balance of reactants and products.

The combustion equation at pressure ratio of 10 and 1200°C temperature is:

$$[0.908CH_4 + 0.0297\ C_2H_6 + 0.027\ C_3H_8 + 0.0054\ (C_4H_{10})_i + 0.0076\ (C_4H_{10})_n + 0.002\ (C_5H_{12})_i + 0.0016\ (C_5H_{12})_n + 0.0027\ (C_6H_{14})_n + 0.0149\ CO_2 + 0.0009\ N_2]_{NG} + [5.92\ (O_2{+}3.76N_2)]_{air} \rightarrow [1.15\ CO_2 + 2.12\ H_2O + 3.65\ O_2 + 22.22\ N_2]_{exhaust} \quad (8.145)$$

It results in flue gas with the composition of 3.9% CO_2, 7.2% H_2O, 12.5% O_2 and 76.06% N_2. The air fuel ratio at this condition is 38.94.

For an isentropic compression process in the compressor, the relation is given as:

$$s_{O_2,\ 32'} - s_{O_2,\ 31} + 3.76(s_{N_2,\ 32'} - s_{N_2,\ 31}) - 4.76\ R\ \ln\left(\frac{P_{32}}{P_{31}}\right) = 0 \quad (8.146)$$

The temperature of the compressed air at state 32' is determined by the iteration of the above equation. The actual temperature of the compressed air (T_{32}) is calculated from the isentropic efficiency of the compressor.

For an isentropic expansion process in the gas turbine, the equation is given as

$$\Delta s = 0 = s_{33} - s_{34'} \tag{8.147}$$

$$s_{33} = [d_1 s_{CO_2} + d_2 s_{H_2O} + d_3 s_{O_2} + d_4 s_{N_2}]_{33} - (d_1 + d_2 + d_3 + d_4)\ R\ \ln\left(\frac{P_{33}}{P_0}\right) \tag{8.148}$$

$$s_{34'} = [d_1 s_{CO_2} + d_2 s_{H_2O} + d_3 s_{O_2} + d_4 s_{N_2}]_{34'} - (d_1 + d_2 + d_3 + d_4)\ R\ \ln\left(\frac{P_{34}}{P_0}\right) \tag{8.149}$$

The temperature of the gas after expansion in gas turbine (T_{34}') is estimated from the iteration of Eq. (8.149). The actual temperature of the gas at the gas turbine exit (T_{34}) is determined by the gas turbine's isentropic efficiency. The steam generation rates can be obtained from the mass and energy balance equations in HRSG.

8.6 Numerical solutions

1. Determine the energy and exergy performance of a Brayton cycle with the following particulars: Efficiency of electrical generator = 98%. Mechanical efficiency of gas turbine = 95%. Mechanical efficiency of air compressor = 98%. Atmospheric condition is taken as 25°C and 1.01325 bar. Air flow rate in a compressor is 1 kg/s. The plant pressure ratio and turbine inlet temperature are 10 and 1400 K respectively. The products of combustion expand through the turbine to a pressure of 1.1 bar. Polytrophic efficiency of compressor and turbine is taken as 91% and 87.5% respectively. Pressure drop in the combustion chamber is neglected. Heat losses in the combustion chamber and turbines are not accounted for. The natural gas (NG) composition collected from the GMR energy power plant, India has the following volume fraction: methane (CH_4), 90.81%, ethane (C_2H_6) 2.96%, propane (C_3H_8) 2.7%, i-butane (C_4H_{10}) 0.54%, n-butane (C_4H_{10}) 0.76%, i-pentane (C_5H_{12}) 0.2%, n-pentane (C_5H_{12}) 0.16%, n-hexane (C_6H_{14}) 0.27%, carbon dioxide (CO_2) 1.49% and nitrogen (N_2) 0.09% (by volume).

Solution:

Refer Fig. 8.32 for the notations of a gas power plant.

Consider the combustion as follows

$$[a_1CH_4 + a_2C_2H_6 + a_3C_3H_8 + a_4(C_4H_{10})_i + a_5(C_4H_{10})_n + a_6(C_5H_{12})_i + a_7(C_5H_{12})_n + a_8(C_6H_{14})_n$$
$$+ a_9CO_2 + a_{10}N_2]_{NG} + [x\ (0.21\ O_2 + 0.79\ N_2)]_{air} \rightarrow [b_1CO_2 + b_2H_2O + b_3O_2 + b_4N_2]_{exhaust}$$

Air compressor:

For an isentropic compression process in the compressor, the relation is given as,

$$0.21\left(s_{O_2,\,2'} - s_{O_2,\,1}\right) + 0.79\left(s_{N_2,\,2'} - s_{N_2,\,1}\right)\ R\ \ln\left(\frac{P_2}{P_1}\right) = 0$$

The temperature of the compressed air at state 2' is determined by the iteration of the above equation. The actual temperature of the compressed air (T_2) is calculated from the isentropic efficiency of the compressor.

For an isentropic compression process, the change in the entropy for air is zero, i.e., $\Delta s = 0$. The isentropic efficiency for a compressor and gas turbine is determined from the pressure ratio (r_c), specific heat ratio (γ) and polytropic state efficiency (η_∞).

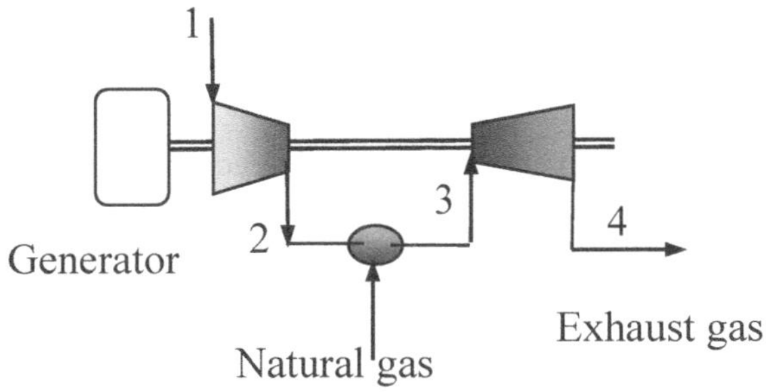

Fig. 8.32. Gas power plant.

Isentropic efficiency of a compressor, $\eta_c = \dfrac{r_c^{\left(\frac{\gamma-1}{\gamma}\right)} - 1}{r_c^{\left(\frac{\gamma-1}{\gamma\,\eta_{\infty c}}\right)} - 1}$

$$= \frac{10^{\left(\frac{1.4-1}{1.4}\right)} - 1}{r_c^{\left(\frac{1.4-1}{1.4\,\times 0.91}\right)} - 1} = 0.88$$

Initial guess of $T'_{2'}$,

$$T_2' = T_1\left(\frac{P_2}{P_1}\right)^{\frac{\gamma-1}{\gamma}}$$

$$= 298.15(10)^{\frac{1.4-1}{1.4}}$$

$$= 575.63\ K$$

Iteration for the exit temperature of compressed air after isentropic compression,

T_2'	$s_{2'} - s_1$
575.63 (initial guess)	0.3426
575.29	0.3245
574.97	0.3074
- - -	- - -
569.19	0.001

$T_2' = 569.19$ K

$h_1 = 0.21\ h_{O2,\,T1} + 0.79\ h_{N2,\,T1} = 0$ kJ/kg mol

$h_2' = 0.21\ h_{O2,\,T2} + 0.79\ h_{N2,\,T2} = 8041.6$ kJ/kg mol

$$\eta_c = \frac{h_2' - h_1}{h_2 - h_1}$$

$$h_2 = h_1 + \frac{h_2' - h_1}{\eta_c}$$

$$= 0 + \frac{8041.6 - 0}{0.88} = 9163.2 \ kJ/kgmole$$

Initial guess of T_2 based on isentropic efficiency,

$$T_2 = T_1 + \frac{T_2' - T_1}{\eta_c}$$

$$= 298.15 + \frac{569.19 - 298.15}{0.88} = 607.00 \ K$$

Iteration to arrive at the actual exit temperature of compressed air after adiabatic compression:

T_2	h_2 (= 9163.2 kJ/kg mol)
607.00 (initial guess)	9191.7
606.70	9183.0
606.51	9176.9
- - -	- - -
606.06	9163.8 (consistency 0.01)

Therefore, the exit temperature of compressed air is 606.06 K

$h_1 = 0$

$h_2 = (0.21\ h_{O2,\ T2} + 0.79\ h_{N2,\ T2})$

$= 9163.8$ kJ/kg mol

Work supplied to the compressor,

$w_c = h_2 - h_1 = 9163.8$ kJ/kg mol

Combustion:

The energy balance in the combustion chamber at the given temperature gives the air fuel ratio, i.e., the value of 'x' in the chemical reaction. In reaction equation 'x' denotes the amount of air in kg mol to be supplied per kg mol of NG ($a_1 + a_2 +$ --- $+ a_{10}$) into the GTCC. The amount of air to be compressed to get the required combustion temperature (T_3) is determined from the energy balance of reactants and products.

The full combustion of NG with compressed air in GTCC is

$$[a_1\, CH_4 + a_2\, C_2H_6 + a_3\, C_3H_8 + a_4\, (C_4H_{10})_i + a_5\, (C_4H_{10})_n + a_6\, (C_5H_{12})_i + a_7\, (C_5H_{12})_n + a_8\, (C_6H_{14})_n$$
$$+ a_9\, CO_2 + a_{10}\, N_2]_{NG} + [x\, (0.21\, O_2 + 0.79\, N_2)]_{air} \rightarrow [b_1\, CO_2 + b_2\, H_2O + b_3\, O_2 + b_4\, N_2]_{exhaust}$$

$a_1 = 90.81\%\ (CH_4) = 0.9081$
$a_2 = 2.96\%\ (C_2H_6) = 0.0296$
$a_3 = 2.7\%\ (C_3H_8) = 0.027$
$a_4 = 0.54\%\ (C_4H_{10}) = 0.0054$
$a_5 = 0.76\%\ (C_4H_{10}) = 0.0076$
$a_6 = 0.2\%\ (C_5H_{12}) = 0.002$
$a_7 = 0.16\%\ (C_5H_{12}) = 0.0016$
$a_8 = 0.26\%\ (C_6H_{14}) = 0.0026$

$a_9 = 1.49\% \ (CO_2) = 0.0149$
$a_{10} = 0.09\% \ (N_2) = 0.0009$

Mass of fuel,

$m_{fuel} = a_1 \ M_{CH4} + a_2 \ M_{C2H6} + a_3 \ M_{C3H8} + a_4 \ M_{C4H10i} + a_5 \ M_{C4H10n} + a_6 \ M_{C5H12i} + a_7 \ M_{C5H12n} + a_8 \ M_{C6H14} + a_9 \ M_{CO2} + a_{10} \ M_{N2}$

$= 18.58$ kg/ kg mol fuel

Energy of fuel,

$HHV = a_1 \ h_{fCH4} + a_2 \ h_{fC2H6} + a_3 \ h_{fC3H8} + a_4 \ h_{fC4H10i} + a_5 \ h_{fC4H10n} + a_6 \ h_{fC5H12i} + a_7 \ h_{fC5H12}n + a_8 \ h_{fC6H14} - (a_1 + 2a_2 + 3a_3 + 4a_4 + 5a_6 + 5a_7 + 6a_8 + a_9) \ h_{fCO2} - (2a_1 + 3a_2 + 4a_3 + 5a_4 + 5a_5 + 6a_6 + 6a_7 + 7a_8) \ h_{fH2OL}$

$= 970110$ kJ/kg mol $= 52202$ kJ/kg

Exergy of fuel,

$\varepsilon = a_1 \ \varepsilon_{CH4} + a_2 \ \varepsilon_{C2H6} + a_3 \ \varepsilon_{C3H8} + a_4 \ \varepsilon_{C4H10} + a_5 \ \varepsilon_{C4H10} + a_6 \ \varepsilon_{C5H12} + a_7 \ \varepsilon_{C5H12} + a_8 \ \varepsilon_{C6H14} + a_9 \ \varepsilon_{CO2} + a_{10} \ \varepsilon_{N2} + RT_0 \ (a_1 \ln(xa_1) + xa_2 \ln(xa_2) + xa_3 \ln(xa_3) + xa_4 \log(xa_4) + xa_5 \ln(xa_5) + xa_6 \ln(xa_6) + xa_7 \ln(xa_7) + xa_8 \ln(xa_8) + xa_9 \ln(xa_9) + xa_{10} \ln(xa_{10})))$

$= 922230$ kJ/kg mol $= 49626$ kJ/kg

Let us consider the initial guess of x is equal to 30
The mass balance of chemical reaction,
$b_1 = a_1 + 2a_2 + 3a_3 + 4a_4 + 4a_5 + 5a_6 + 5a_7 + 6a_8 + a_9$
$b_1 = 1.15$
$b_2 = 2a_1 + 3a_2 + 4a_3 + 5a_4 + 5a_5 + 6a_6 + 6a_7 + 7a_8$
$b_2 = 2.12$
$b_3 = a_9 + 0.21x - b_1 - 0.5b_2$
$b_4 = a_{10} + 0.79x$

$h_{fuel} = a1 \ h_{fCH4} + a_2 \ h_{fC2H6} + a_3 \ h_{fC3H8} + a_4 \ h_{fC4H10i} + a_5 \ h_{fC4H10n} + a_6 \ h_{fC5H12i} + a_7 \ h_{fC5H12n} + a_8 \ h_{fC6H14} + a_9 \ h_{fCO2}$

$= -81906$ kJ/ kg mol fuel

$h_{air} = 0.21 \ h_{T,\,O2} + 0.79 \ h_{T,\,N2}$

$= 9163.3$ kJ/kg mol air

Enthalpy of reactants,

$HR = h_{fuel} + x \ h_{air}$

Enthalpy of products,

$HP = b_1(h_{fCO2} + h_{T,\,CO2}) + b_2 \ (h_{fH2O} + h_{T,\,H2O}) + b_3 \ h_{T,\,O2} + b_4 \ h_{T,\,N2}$

Iteration of energy balance of the chemical reaction for air supply to the combustion chamber

x	HR – HP
30 (initial guess)	21578
30.21	15925
30.37	11752
30.49	8673.3
- -	- -
30.82	0.0001

At x = 30.82, HR = HP = 200540 kJ/kg mol

After the iterations,

x = 30.82

$b_3 = 4.28$

$b_4 = 24.35$

$n_{total} = b_1 + b_2 + b_3 + b_4 = 31.90$

$CO_2 = 3.60\%$

$H_2O = 6.64\%$

$O_2 = 13.41\%$

$N_2 = 76.34\%$

Air fuel ratio = 47.83:1

Following is the resulting combustion equation at pressure ratio of 10 and temperature of 1400 K.

$[0.908\ CH_4 + 0.0297\ C_2H_6 + 0.027\ C_3H_8 + 0.0054\ (C_4H_{10})_i + 0.0076\ (C_4H_{10})_n + 0.002\ (C_5H_{12})_i + 0.0016\ (C_5H_{12})_n + 0.0027\ (C_6H_{14})_n + 0.0149\ CO_2 + 0.0009\ N_2]_{NG} + [30.82\ (0.21\ O_2 + 0.79\ N_2)]_{air} \rightarrow [1.15\ CO_2 + 2.12\ H_2O + 4.28\ O_2 + 24.35\ N_2]_{exhaust}$

Isentropic efficiency of a gas turbine, $$\eta_{gt} = \frac{1-\left(\frac{P_4}{P_3}\right)^{\frac{\eta_{\infty gt}(\gamma-1)}{\gamma}}}{1-\left(\frac{P_4}{P_3}\right)^{\frac{(\gamma-1)}{\gamma}}}$$

$$= \frac{1-(10)^{\frac{0.875(1.4-1)}{\gamma}}}{1-(10)^{\frac{(1.4-1)}{1.4}}} = 0.91$$

Initial guess of $T_{4'}$

$$T_4' = T_3\left(\frac{P_4}{P_3}\right)^{\frac{\gamma-1}{\gamma}}$$

$$= 1400\left(\frac{1}{10}\right)^{\frac{1.4-1}{1.4}}$$

$$= 742.35\ K$$

For an isentropic expansion process in the gas turbine, the equation is given as:

$\Delta s = 0 = s_3 - s_{4'}$

$$s_3 = [b_1 s_{CO_2} + b_2 s_{H_2O} + b_3 s_{O_2} + b_4 s_{N_2}]_3 - (b_1 + b_2 + b_3 + b_4)\ R\ \ln\left(\frac{P_3}{P_0}\right)$$

$$s_{4'} = [b_1 s_{CO_2} + b_2 s_{H_2O} + d_3 s_{O_2} + b_4 s_{N_2}]_{4'} - (b_1 + b_2 + b_3 + b_4)\ R\ \ln\left(\frac{P_4}{P_0}\right)$$

The temperature of the gas after expansion in the gas turbine (T_4') is estimated from iterations of the equation. The actual temperature of gas at the gas turbine exit (T_4) is determined with the gas turbine's isentropic efficiency.

Iteration for turbine exit temperature after the isentropic expansion

T_4'	$s_{4'} - s_3$
742.35 (initial guess)	108.31
753.17	93.30
762.51	80.51
- - -	- - -
823.38	0.01

$s_3 = s_{4'} = 981.06$ kJ/kg mol

$T_4' = 823.28$ K

$h_3 = b_1\, h_{CO2,\,T3} + b_2\, h_{H2O,\,T3} + b_3\, h_{O2,\,T3} + b_4\, h_{N2,\,T3} = 1165400$ kJ/kg mol

$h_4 = b_1\, h_{CO2,\,T4} + b_2\, h_{H2O,\,T4} + b_3\, h_{O2,\,T4} + b_4\, h_{N2,\,T4} = 523150$ kJ/kg mol

$$\eta_T = \frac{h_3 - h_4}{h_3 - h_4'}$$

$$h_4 = h_3 - \eta_T (h_3 - h_4')$$
$$= 1165400 + 0.91(1165400 - 523150) = 582980 \text{ kJ/kg mol}$$

Initial guess of T_4 based on isentropic efficiency,

$$T_4 = T_3 - \eta_T (T_3 - T_4')$$
$$= 1400 - 0.91(1400 - 823.28) = 877.1 \text{ K}$$

Iteration for turbine exit temperature after the adiabatic expansion

T_4	h_4 (=582980 kJ/kg mol)
877.1 (initial guess)	580250
879.8	583170
879.6	582970
- - -	- - -
879.65	582980

Therefore, the exit temperature of the turbine is 879.65 K

$h_3 = b_1\, h_{CO2,\,T3} + b_2\, h_{H2O,\,T3} + b_3\, h_{O2,\,T3} + b_4\, h_{N2,\,T3} = 1165400$ kJ/kg mol

$h_4 = b_1\, h_{CO2,\,T4} + b_2\, h_{H2O,\,T4} + b_3\, h_{O2,\,T4} + b_4\, h_{N2,\,T4} = 583000$ kJ/kg mol

$h_2 = x\,(0.21\, h_{O2,\,T2} + b_2\, h_{N2,\,T2}) = 282440$ kJ/kg mol

$w_c = (h_2 - h_1) = 282440$ kJ/kg mol

$w_T = (h_3 - h_4) = 582430$ kJ/kg mol

$w_{net} = w_t - w_c = 300000$ kJ/kg mol

But the mass of air is 1 kg/s, therefore the multiplication factor for the chemical reaction is found as follows,

$$m_{a,factor} = \frac{m_a}{x(0.21M_{O_2} + 0.79M_{N_2})} \qquad \because m_a = m_{a,factor}\, x(0.21M_{O_2} + 0.79M_{N_2})$$
$$= 0.0011$$

Power generation,

$$W_c = \frac{m_{a,factor} w_c}{\eta_{m,c}}$$

$=324.1$ kW

$$W_T = \eta_{m,T} \eta_G m_{a,factor} w_T$$

$= 609.8$ kW

Net power generation $= 285.7$ kW

Thermal efficiency,

$$\eta_1 = \frac{W_{net}}{m_{a,factor} M_f \; HHV / kg} \times 100$$

$= 26.18\%$

Exergy evaluation

$$s_1 = x\left(0.21 s_{O_2,1} + 0.79 s_{N_2,1} - R \ln\left(\frac{P_1}{P_0}\right)\right)$$

$= 0$

$$s_2 = x\left(0.21 s_{O_2,2} + 0.79 s_{N_2,2} - R \ln\left(\frac{P_2}{P_0}\right)\right)$$

$= 58.9$ kJ/kg mol

$$s_3 = [b_1 s_{CO_2} + b_2 s_{H_2O} + b_3 s_{O_2} + b_4 s_{N_2}]_3 - (b_1 + b_2 + b_3 + b_4) \; R \ln\left(\frac{P_3}{P_0}\right)$$

$= 981.1$ kJ/kg mol

$$s_4 = [b_1 s_{CO_2} + b_2 s_{H_2O} + b_3 s_{O_2} + b_4 s_{N_2}]_4 - (b_1 + b_2 + b_3 + b_4) \; R \ln\left(\frac{P_4}{P_0}\right)$$

$= 1051.3$ kJ/kg mol

Mole fractions of combustion gases

$nt_{total} = b_1 + b_2 + b_3 + b_4$

$= 31.90$

$$xb_1 = \frac{b_1}{n_{total}}$$

$= 0.036$

$$xb_2 = \frac{b_2}{n_{total}}$$

$= 0.0664$

$$xb_3 = \frac{b_3}{n_{total}}$$
$$= 0.1342$$

$$xb_4 = \frac{b_4}{n_{total}}$$
$$= 0.7634$$

Chemical exergy of air,

$$\varepsilon_{ch,\, air} = x\,(0.21\, \varepsilon_{0,\, O2} + 0.79\, \varepsilon_{0,\, N2} + R\, T_0\, (0.21\, \ln(0.21) + 0.79\, \ln(0.79)))$$
$$= 3960.7 \text{ kJ/kg mol}$$

Physical exergy of air at the inlet of compressor,

$$\varepsilon_{ph,\, air\, 1} = 0$$

Physical exergy of air at the exit of the compressor:

$$\varepsilon_{ph,\, air\, 2} = (h - T_0 s)_2 = 264880 \text{ kJ/kg mol}$$

Physical exergy of gas at the exit of the combustion chamber/inlet of gas turbine:

$$\varepsilon_{ph,\, gas\, 3} = (h - T_0 s)_3 = 872900 \text{ kJ/kg mol}$$

Physical exergy of gas at the exit of the gas turbine:

$$\varepsilon_{ph,\, gas\, 4} = (h - T_0 s)_4 = 269520 \text{ kJ/kg mol}$$

Chemical exergy of gas,

$$\varepsilon_{ch,\, gas} = n_{total}\, (xb_1\, \varepsilon_{0,\, CO2} + xb_2\, \varepsilon_{0,\, H2OG} + xb_3\, \varepsilon_{0,\, O2} + xb_4\, \varepsilon_{0N2} + xb_5\, \varepsilon_{0NO2} + R\, T_0\, (xb_1\, \ln(xb_1)$$
$$+ xb_2\, \ln(xb_2) + xb_3\, \ln(xb_3) + xb_4\, \ln(xb_4) + xb_5\, \ln(xb_5)))$$
$$= 21170 \text{ kJ/kg mol}$$

Total exergy of fluid = chemical exergy + physical exergy.

Exergy of air at the inlet of the compressor,

$$\varepsilon_1 = \varepsilon_{ch,\, air} + \varepsilon_{ph,\, air\, 1} = 3960 \text{ kJ/kg mol}$$

Exergy of air at the exit of the compressor/inlet of combustion,

$$\varepsilon_2 = \varepsilon_{ch,\, air} + \varepsilon_{ph,\, air\, 2} = 268840 \text{ kJ/kg mol}$$

Exergy of gas at the exit of the combustion/inlet of gas turbine,

$$\varepsilon_3 = \varepsilon_{ch,\, gas} + \varepsilon_{ph,\, gas\, 3} = 894080 \text{ kJ/kg mol}$$

Exergy of gas at the exit of the gas turbine,

$$\varepsilon_4 = \varepsilon_{ch,\, gas} + \varepsilon_{ph,\, gas\, 4} = 290690 \text{ kJ/kg mol}$$

Irreversibility of the compressor,

$$i_{compressor} = \varepsilon_1 + w_c - \varepsilon_2 = 17554 \text{ kJ/kg mol} = 19.73 \text{ kW}$$

Irreversibility of combustion,

$$i_{combustion} = \varepsilon_{fuel} + \varepsilon_2 - \varepsilon_3 = 296990 \text{ kJ/kg mol} = 333.97 \text{ kW}$$

Irreversibility of the gas turbine,

$i_T = \varepsilon_3 - w_T - \varepsilon_4 = 20954$ kJ/kg mol = 23.56 kW

Irreversibility of exhaust gas,

$i_{exhaust} = \varepsilon_4 = 290690$ kJ/kg mol = 326.89 kW

The exergy balance applied to the gas power plant

$\varepsilon_{fuel} + \varepsilon_{air,\ in,\ 1} + W_C = W_T + \varepsilon_{air,\ out,\ 4} + I_{compressor} + I_{combustion} + I_T$

Exergy efficiency of the gas power plant,

$$\eta_2 = \frac{W_{net}}{m_{f\ factor} M_f\ \varepsilon_{fue/kg}} \times 100$$

$$= 27.55\%$$

2. Evaluate the Brayton cycle with reference to energy and exergy performance. Use the fuel air cycle concept. The gas turbine inlet temperature = 1400 K with a compressor pressure ratio of 10. The specific heat of air and gas can be taken as 1.005 kJ/kg K and 1.05 kJ/kg K respectively. The calorific value of fuel is 52204 kJ/kg. The isentropic efficiencies of the compressor and turbine are 0.88 and 0.91 respectively. The reference pressure and temperature are 1.01325 bar and 25°C respectively. The condition of air before compression is 1.01325 bar at 25°C. The pressure at gas turbine exit is 1.1 bar. Exergy of fuel is 49626 kJ/kg. Efficiency of electrical generator = 98%. Mechanical efficiency of gas turbine = 95%. Mechanical efficiency of air compressor = 98%. Consider the mass of air as 1 kg/s.

Solution:

Temperature of air after the isentropic compression of air,

$$T_2' = T_1\left(\frac{P_2}{P_1}\right)^{\frac{\gamma-1}{\gamma}}$$

$$= 298.15(10)^{\frac{1.4-1}{1.4}}$$

$$= 575.6\ K$$

$T_2' = 575.6$ K

$$\eta_c = \frac{h_2' - h_1}{h_2 - h_1}$$

$$h_2 = h_1 + \frac{h_2' - h_1}{\eta_c}$$

$$T_2 = T_1 + \frac{T_2' - T_1}{\eta_c} \quad \text{(since } c_p \text{ is constant)}$$

$$= 25 + \frac{575.6 - 25}{0.88} = 613.5\ K$$

$m_f CV = (m_f + m_a)\ c_{pg}\ (T_{comb} - T_{air})$

where m_f and m_a are mass flow rates of fuel and air respectively, kg/s. CV is the calorific value of fuel, kJ/kg. c_{pg} is the specific heat of gas, kJ/kg K. T_{comb} and T_{air} are the temperatures of the combustion chamber (products of gas) and compressed air respectively.

The above equation can be simplified as:

$$CV = \frac{(m_f + m_a)c_{pg}(T_{comb} - T_{air})}{m_f} = (1 + AFR)c_{pg}(T_{comb} - T_{air})$$

where AFR is the air fuel ratio.

Therefore,

$$AFR = \frac{CV}{c_{pg}(T_{comb} - T_{air})} - 1$$

$$= \frac{52204}{1.04(1400 - 613.5)} - 1$$

$= 62.21$

Fuel air ratio,

$$FAR = \frac{1}{AFR}$$

$$= \frac{1}{62.21} = 0.0161$$

Mass of air, $m_a = 1$ kg/s

Mass of gas,

$m_g = (1 + FAR)\, m_a$

$= (1 + 0.0161) \times 1 = 1.0161$ kg/s

Mass of fuel,

$$m_f = \frac{m_a}{AFR}$$

$$= \frac{1}{62.21} = 0.0161\ kg/s$$

The exit temperature of gas from the gas turbine,

$$T_4' = T_3\left(\frac{P_4}{P_3}\right)^{\frac{\gamma-1}{\gamma}}$$

$$= 1400\left(\frac{1}{10}\right)^{\frac{1.4-1}{1.4}}$$

$= 742.34$ K

$T_4' = 742.34$ K

Initial guess of T_4 based on isentropic efficiency,

$$\eta_T = \frac{h_3 - h_4}{h_3 - h_4'}$$

$$\eta_T = \frac{T_3 - T_4}{T_3 - T_4'} \text{ (since } c_p \text{ is constant)}$$

$$T_4 = T_3 - \eta_T (T_3 - T_4')$$

$$= 1400 - 0.91(1400 - 742.34) = 801.5 \text{ K}$$

$h_1 = c_{pa} (T_1 - T_0) = 0$

$h_2 = c_{pa} (T_2 - T_0) = 306.9$ kJ/kg

$h_3 = c_{pg} (T_3 - T_0) = 1156.9$ kJ/kg

$h_4 = c_{pg} (T_4 - T_0) = 528.6$ kJ/kg

$$s_1 = c_{pa}(T_1 - T_0) - R \ln\left(\frac{P_1}{P_0}\right)$$

$$= 0$$

$$s_2 = c_{pa}(T_2 - T_0) - R \ln\left(\frac{P_2}{P_0}\right)$$

$$= 0.0643 \text{ kJ/kg K}$$

$$s_3 = c_{pg}(T_3 - T_0) - R \ln\left(\frac{P_3}{P_0}\right)$$

$$= 0.9631 \text{ kJ/kg K}$$

$$s_4 = c_{pg}(T_4 - T_0) - R \ln\left(\frac{P_4}{P_0}\right)$$

$$= 1.0148 \text{ kJ/kg K}$$

Power supply to compressor,

$$W_c = \frac{m_a w_c}{\eta_{m,c}}$$

$$= 323.37 \text{ kW}$$

$$W_T = \eta_{m,T} \eta_G m_g w_T$$

$$= 594.43 \text{ kW}$$

Net power generation = 271.06 kW

Thermal efficiency,

$$\eta_1 = \frac{W_{net}}{m_f HHV} \times 100$$

$$= 32.20\%$$

$\varepsilon_1 = m_a (h_1 - T_0 s_1) = 0$

$\varepsilon_2 = m_a (h_2 - T_0 s_2) = 297.73$ kW

$\varepsilon_3 = m_g (h_3 - T_0 s_3) = 883.77$ kW

$\varepsilon_4 = m_g (h_4 - T_0 s_4) = 229.62$ kW

Irreversibility of compressor,

$I_{compressor} = \varepsilon_1 + W_c - \varepsilon_2 = 25.64$ kW

Irreversibility of combustion,

$i_{combustion} = \varepsilon_{fuel} + \varepsilon_2 - \varepsilon_3 = 211.64$ kW

Irreversibility of gas turbine,

$I_T = \varepsilon_3 - W_T - \varepsilon_4 = 59.71$ kW

Irreversibility of exhaust,

$I_{exhaust} = \varepsilon_4 = m_g\,(h_4 - T_0 s_4) = 229.62$ kW

Exergy efficiency of gas power plant,

$$\eta_2 = \frac{m_f \varepsilon_{sup\,ply} - \sum I}{m_f \varepsilon_{sup\,ply}} \times 100$$

$$= 33.98\%$$

3. Evaluate the energy and exergy performance of a gas power plant operating on Brayton cycle as follows. The gas turbine inlet temperature = 1400 K with a compressor pressure ratio of 10. Specific heat of air and gas can be taken as 1.005 kJ/kg K. The isentropic efficiencies of compressor and turbine are 0.88 and 0.91 respectively. The reference pressure and temperature are 1.01325 bar and 25°C respectively. The condition of air before compression is 1.01325 bar at 25°C. The pressure at gas turbine exit is 1.1 bar. Efficiency of electrical generator = 98%. Mechanical efficiency of gas turbine = 95%. Mechanical efficiency of air compressor = 98%. Consider the mass of the working fluid as 1 kg/s.

Solution:

Temperature of air after the isentropic compression of air,

$$T_2' = T_1\left(\frac{P_2}{P_1}\right)^{\frac{\gamma-1}{\gamma}}$$

$$= 298.15(10)^{\frac{1.4-1}{1.4}}$$

$$= 575.6 \text{ K}$$

$T_2' = 575.6$ K

$$\eta_c = \frac{h_2' - h_1}{h_2 - h_1}$$

$$h_2 = h_1 + \frac{h_2' - h_1}{\eta_c}$$

$$T_2 = T_1 + \frac{T_2' - T_1}{\eta_c} \quad \text{(since } c_p \text{ is constant)}$$

$$= 25 + 273.15 + \frac{575.6 - (25 + 273.15)}{0.88} = 613.5 \ K$$

The exit temperature of gas from the gas turbine,

$$T_4' = T_3\left(\frac{P_4}{P_3}\right)^{\frac{\gamma-1}{\gamma}}$$

$$= 1400\left(\frac{1}{10}\right)^{\frac{1.4-1}{1.4}}$$

$$= 742.34\ \text{K}$$

$T_4' = 742.34$ K

Initial value ascribed to T_4 based on isentropic efficiency,

$$\eta_T = \frac{h_3 - h_4}{h_3 - h_4'}$$

$$\eta_T = \frac{T_3 - T_4}{T_3 - T_4'} \text{ (since } c_p \text{ is constant)}$$

$$T_4 = T_3 - \eta_T (T_3 - T_4')$$

$$= 1400 - 0.91(1400 - 742.34) = 801.5$$

$h_1 = c_{pa} (T_1 - T_0) = 0$

$h_2 = c_{pa} (T_2 - T_0) = 306.9$ kJ/kg

$h_3 = c_{pa} (T_3 - T_0) = 11074$ kJ/kg

$h_4 = c_{pa} (T_4 - T_0) = 505.9$ kJ/kg

$$s_1 = c_{pa}(T_1 - T_0) - R\ \ln\left(\frac{P_1}{P_0}\right)$$

$$= 0$$

$$s_2 = c_{pa}(T_2 - T_0) - R\ \ln\left(\frac{P_2}{P_0}\right)$$

$$= 0.0643\ \text{kJ/kg K}$$

$$s_3 = c_{pa}(T_3 - T_0) - R\ \ln\left(\frac{P_3}{P_0}\right)$$

$$= 0.8934\ \text{kJ/kg K}$$

$$s_4 = c_{pa}(T_4 - T_0) - R\ \ln\left(\frac{P_4}{P_0}\right)$$

$$= 0.9703\ \text{kJ/kg K}$$

Power supply to compressor,

$$W_c = \frac{1 \times w_c}{\eta_{m,c}}$$

$=323.37$ kW (as the mass of working fluid is 1 kg/s)

$$W_T = \eta_{m,T}\eta_G 1 \times w_T$$

$$= 559.95 \text{ kW}$$

Net power generated = 236.58 kW

Heat supply, $Q_1 = m(h_3 - h_4)$
$= 790.45$ kJ/kg

Thermal efficiency,

$$\eta_1 = \frac{W_{net}}{Q_1} \times 100$$

$$= 29.93\%$$

$\varepsilon_1 = m\ (h_1 - T_0 s_1) = 0$

$\varepsilon_2 = m\ (h_2 - T_0 s_2) = 297.73$ kW

$\varepsilon_3 = m\ (h_3 - T_0 s_3) = 883.77$ kW

$\varepsilon_4 = m\ (h_4 - T_0 s_4) = 229.62$ kW

Exergy supply to the cycle

$$\varepsilon_{supply} = Q_1 \left(1 - \frac{T_0}{T_{max}}\right)$$

$$622.11 \text{ kW}$$

Irreversibility of the compressor,

$I_{compressor} = \varepsilon_1 + W_c - \varepsilon_2 = 25.64$ kW

Irreversibility of combustion,

$I_{combustion} = \varepsilon_{supply} + \varepsilon_2 - \varepsilon_3 = 78.89$ kW

Irreversibility of the gas turbine,

$I_T = \varepsilon_3 - W_T - \varepsilon_4 = 64.39$ kW

Irreversibility of exhaust,

$I_{exhaust} = \varepsilon_4 = 229.62$ kW

Exergy efficiency of the gas power plant,

$$\eta_2 = \frac{\varepsilon_{supply} - \sum I}{\varepsilon_{supply}} \times 100$$

$$= 38.03\%$$

Comparison of results evaluated by three methods

S. No.	Particulars	Air standard performance	Fuel-air cycle	Using JANAF tables
1.	Compressor power, kW	323.37	323.37	324.1
2.	Gas turbine, power, kW	559.95	594.43	609.8
3.	Net Power, kW	236.95	271.06	285.7
4.	Heat supply, kW		839.12	1090.9
5.	Thermal efficiency, %	29.93 (cycle)	32.20	26.18
6.	Irreversibility of compressor, kW (% of exergy supply)	25.64 (4.12%)	25.64 (3.21%)	19.74 (1.90%)
7.	Irreversibility of combustion chamber, kW (%)	78.89 (12.68%)	211.64 (26.53%)	333.97 (32.20%)
8	Irreversibility of gas turbine, kW, (%)	64.39 (10.35%)	59.71 (7.48%)	23.56 (2.27%)
9.	Irreversibility of exhaust, kW, (%)	216.61 (34.82)	229.62 (28.79%)	326.89 (31.52%)
10.	Total irreversibilities, kW (%)	385.83 (61.97%)	526.62 (66.02%)	704.16 (67.90%)
11.	Exergy supply, kW		797.68	1037.1
12.	Exergy efficiency, %	38.03 (cycle)	33.98	27.55

4. A reheat steam Rankine cycle operates between two pressure values of 100 bar and 0.1 bar respectively at a boiler and a condenser. The reheat pressure is maintained at 25% of boiler pressure. The turbine inlet temperature and reheat temperature are the same at 500°C. Consider 90% and 85% of isentropic efficiencies for turbine and pump, find the specific power, steam rate and thermal efficiency of the plant.

Solution:

Figure 8.33 and Fig. 8.34 show the schematic layout and T-s diagram for a reheat steam Rankine cycle. The properties of the reheat steam Rankine cycle can be determined based on the formulations supplied. The calculated properties are listed in the table. From the calculated enthalpies, the specific power, steam rate, and thermal efficiency of the plant can be calculated.

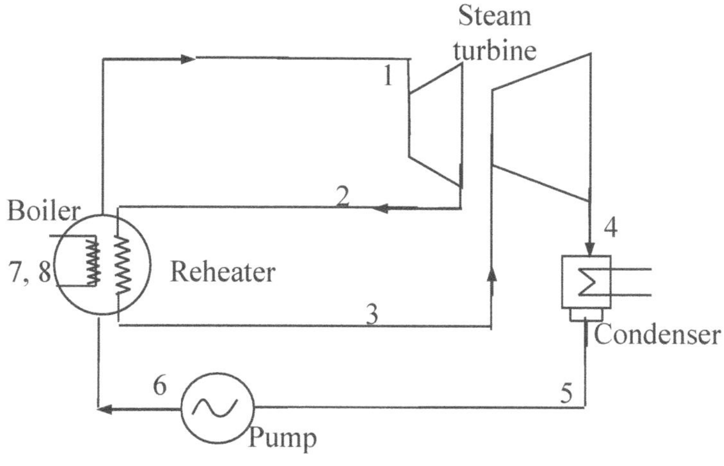

Fig. 8.33. Schematic of reheat Rankine cycle.

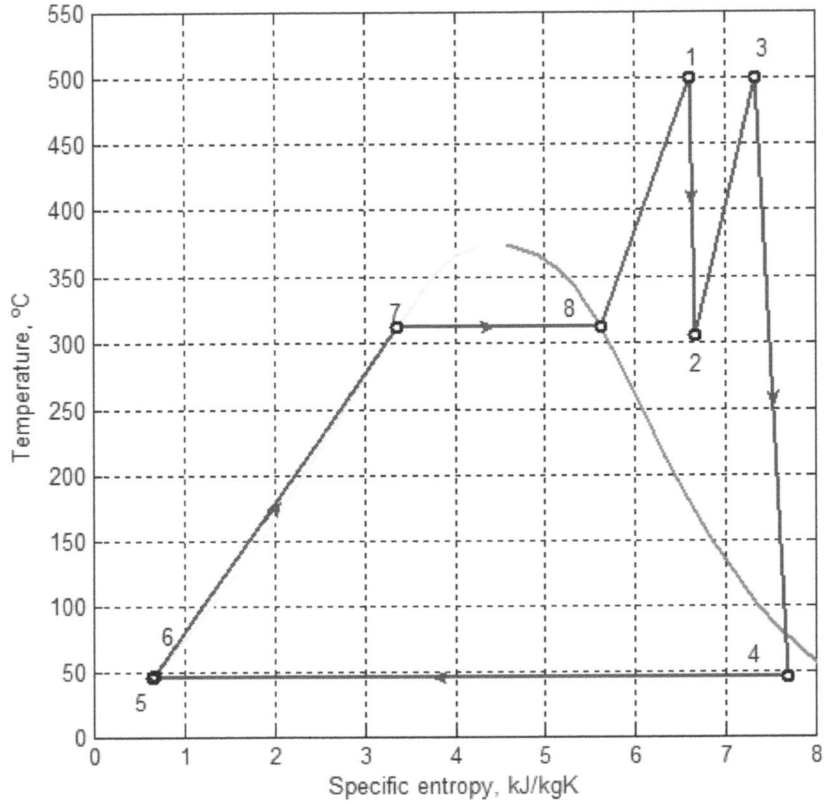

Fig. 8.34. Solution.

State	P, bar	T,°C	h, kJ/kg	s, kJ/kg K
1	100.00	500.00	3374.60	6.60
2	25.00	304.92	3022.44	6.67
3	25.00	500.00	3461.74	7.32
4	0.10	45.83	2435.03	7.68
5	0.10	45.83	191.83	0.65
6	100.00	46.73	195.59	0.66
7	100.00	310.96	1407.67	3.36
8	100.00	310.96	2728.07	5.62

5. In a steam power plant, steam has been generated from hot gas at 350°C and 300000 Nm^3/h. The steam turbine inlet temperature is 16 bar and 325°C. The condenser pressure is 0.1 bar. Consider a Rankine cycle with a deaerator at a mid-temperature of a condenser and boiler. Assume steam turbine efficiency to be 85% and pump isentropic efficiency to be 75%. Evaluate using the second law of thermodynamics. Analyze the irreversibility losses. The circulating water inlet temperature in a condenser is 25°C with a temperature rise of 8°C. Molecular weight of hot gas is 29 and its specific heat is 1.03 kJ/kg K. The hot gas exit temperature is 200°C.

Solution:

Refer to Fig. 8.35 for the plant layout of the steam power plant with a deaerator. Figure 8.36 and Fig. 8.37 detail the T-s properties.

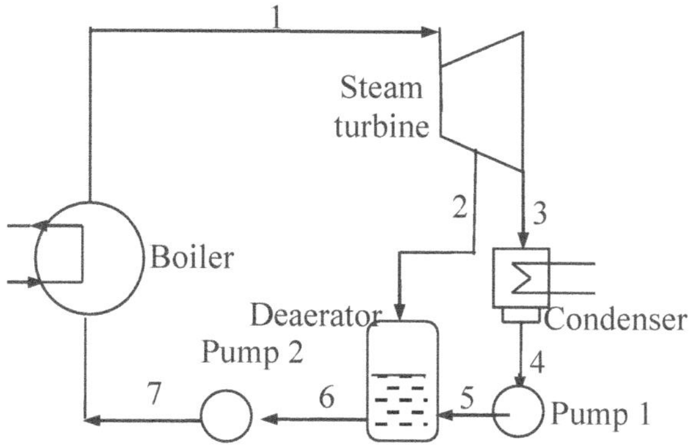

Fig. 8.35. Schematic of a Rankine cycle with a deaerator.

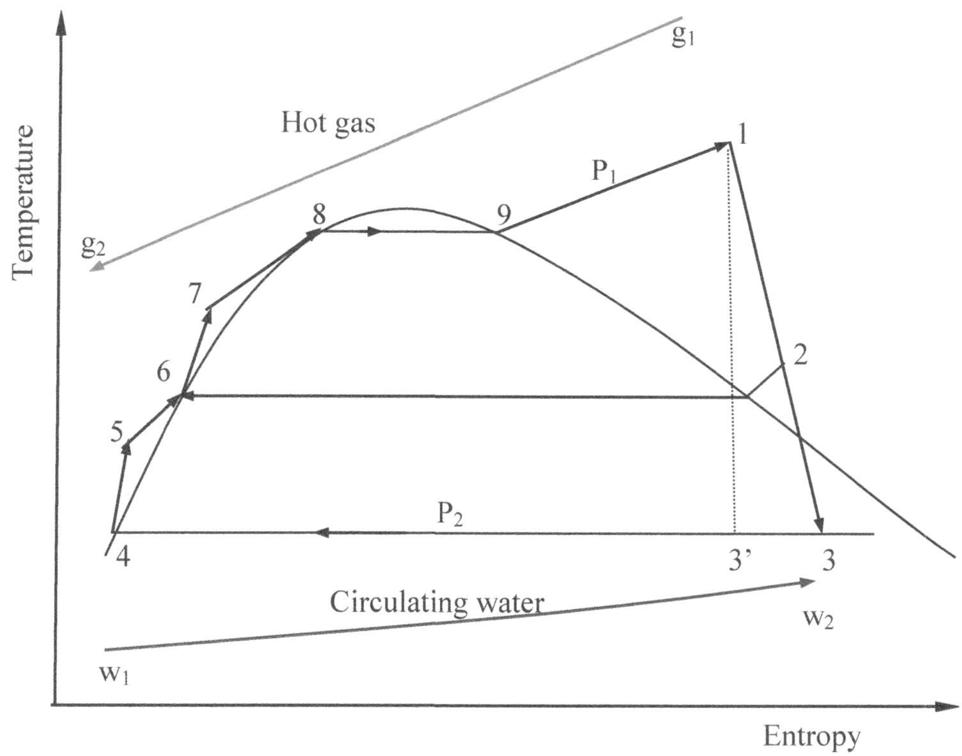

Fig. 8.36. Temperature-entropy diagram of Rankine cycle with deaerator.

$T_{g1} = 350°C$, $T_{g2} = 200°C$, $m_g = 300000\ Nm^3/h$

$P_1 = 16$ bar, $T_1 = 325°C$, $P_2 = 0.1$ bar

$$m_g = \frac{V_g M_g}{22.4 \times 3600} = \frac{300000 \times 29}{22.4 \times 3600} = 107.88\ kg/s$$

The saturation temperature of boiler, $T_{sat1} = 201.37°C$

Similarly, the saturation temperature of condenser, $T_{sat3} = 45.83°C$

Deaerator temperature, $T_{sat3} = T_{sat3} + 0.5(T_{sat1} - T_{sat3}) = 123.60°C$

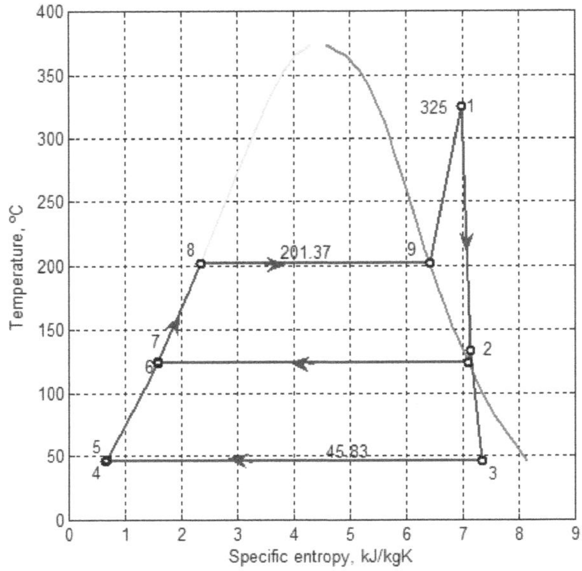

Fig. 8.37. T-s diagram.

The saturation pressure of deaerator = 2.22 bar

Using steam tables and applying first law of thermodynamics

$h_1 = 3091.9$ kJ/kg

$h_2 = 2731.5$ kJ/kg

$h_3 = 2333.2$ kJ/kg

$h_4 = 191.8$ kJ/kg

$h_5 = 196.1$ kJ/kg

$h_6 = 519.0$ kJ/kg

$h_7 = 524.3$ kJ/kg

$s_1 = 6.9825$ kJ/kgK

$s_2 = 7.1422$ kJ/kgK

$s_3 = 7.3625$ kJ/kgK

$s_4 = 0.6493$ kJ/kgK

$s_5 = 0.6626$ kJ/kg K

$s_6 = 1.5663$ kJ/kg K

$s_7 = 1.5796$ kJ/kg K

From the energy balance in heat recovery, the mass of steam:

$$m_s = \frac{m_g c_{pg} (T_{g1} - T_{g2})}{h_1 - h_7} = \frac{107.88 \times 1.03(350 - 200)}{3091.9 - 524.3} = 6.4919 \ \ kg/s$$

Bled steam,

$$m_2 = \frac{m_1(h_6 - h_5)}{h_2 - h_5} = \frac{6.4919(519.0 - 196.1)}{2731.5 - 196.1} = 0.8269 \ kg/s$$

Steam turbine output,

$$W_t = m_1(h_1 - h_2) + (m_1 - m_2)(h_2 - h_3)$$

$$= 6.4919\,(3091.9 - 2731.5) + 5.665\,(2731.5 - 2333.2) = 4596 \text{ kW}$$

$$W_p = m_4(h_5 - h_4) + m_6(h_7 - h_6) = 57 \text{ kW}$$

$$W_{net} = W_t - W_p = 4596 - 57 = 4539 \text{ kW}$$

$$Q_{supply} = m_g c_{pg}(T_{g1} - T_{g2}) = m_1(h_1 - h_7) = 16669 \text{ kW}$$

First law of efficiency:

$$\eta_1 = \frac{W_{net}}{Q_{supply}} \times 100 = \frac{4539}{16669} = 27.23\%$$

Irreversibilities:

$$I_{heat\ recovery} = T_0\left(m_g c_{pg} \ln\frac{T_{g2}}{T_{g1}} + m_1(s_1 - s_7)\right)$$

$$= 13360 \text{ kW}$$

$$I_{steam\ turbine} = T_0\left(m_1(s_2 - s_1) + m_3(s_3 - s_2)\right) = 681.25 \text{ kW}$$

$$I_{deaerator} = T_0\left(m_6 s_6 - m_2 s_2 - m_5 s_5\right) = 151.7 \text{ kW}$$

From the energy balance in the steam condenser, the circulating cooling water,

$$m_w = \frac{m_s(h_3 - h_4)}{c_{pw}(T_{w2} - T_{w1})} = 416.72 \ kg/s$$

$$I_{condenser} = T_0\left(m_w c_{pw} \ln\frac{T_{w2}}{T_{w1}} + m_s(s_4 - s_3)\right)$$

$$= 724.67 \text{ kW}$$

$$I_{pump} = T_0\left(m_4(s_5 - s_4) + m_6(s_7 - s_6)\right)$$

$$= 48.38 \text{ kW}$$

$$I_{exhaust} = m_g c_{pg}(T_{g2} - T_0) - T_0 m_g c_{pg} \ln\frac{T_{g2}}{T_0}$$

$$= 4146 \text{ kW}$$

$$I_{hot\ water} = m_w c_{pw}(T_{w2} - T_0) - T_0 m_w c_{pw} \ln\frac{T_{w2}}{T_0}$$

$$= 183.23 \text{ kW}$$

Total irreversibility losses,

$$I_{total} = I_{heat\ recovery} + I_{steam\ turbine} + I_{deaerator} + I_{condenser} + I_{pump} + I_{exhaust} + I_{hot\ water}$$

$$= 7269.2 \text{ kW}$$

Exergy of hot gas source,

$$E_g = m_g c_{pg}(T_{g1} - T_0) - T_0 m_g c_{pg} \ln\frac{T_{g1}}{T_0}$$

$$= 11691 \text{ kW}$$

$$W_{net,\ check} = E_{hot\ gas} - I_{total}$$

$$= 4421 \text{ kW (matched with } W_{net})$$

Exergy efficiency from the second law of thermodynamics,

$$\eta_2 = \frac{W_{act}}{W_{max}} \times 100 = \frac{E_g - I_{total}}{E_g} \times 100 = \frac{11691 - 7269.2}{11691} \times 100 = 38.82\%$$

Figure 8.38 summarizes the exergy losses and power generation.

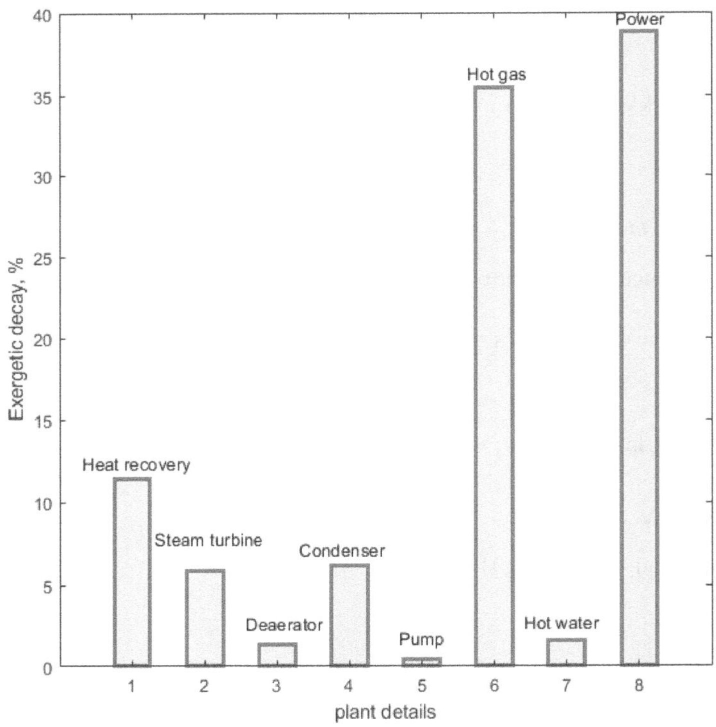

Fig. 8.38. Exergy analysis.

6. Find the evaporator pressure for a single pressure heat recovery system. Take $c_{p\ gas} = 1.03$ kJ/kg K. Pinch point, PP = 25 K. Approach Point, AP = 15 K. Terminal temperature difference, TTD = 50 K and Degree of super heat = 270 K. The gas supply temperature for heat recovery, $T_{g\ in} = 600$°C. Also find the exhaust gas temperature at evaporator outlet and water temperature at economizer exit in the water line.

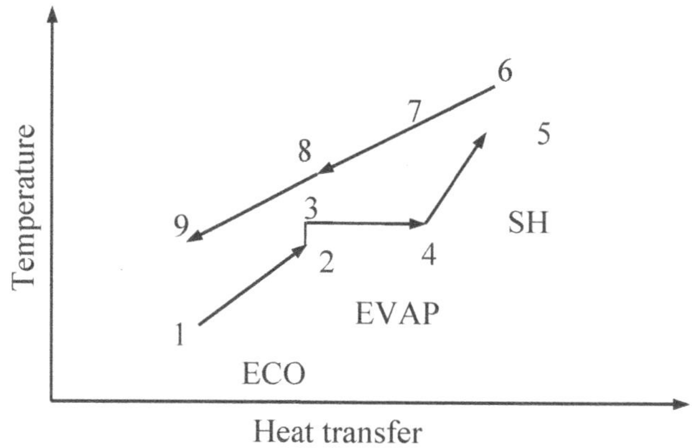

Fig. 8.39. Heat recovery with single pressure HRSG.

Solution:

Refering to Fig. 8.39, the saturation temperature of evaporator, $T_{sat} = T_3 = T_4 = T_{g\ in} - TTD - DSH$
$= 600 - 50 - 270 = 280°C$

From steam tables, the saturation pressure corresponding to 280°C = 64.20 bar

Exhaust gas temperature at evaporator outlet = $T_{sat} + PP = 280 + 25 = 305°C$

Water exit temperature at economizer outlet = $T_{sat} - AP = 280 - 15 = 265°C$

7. Find the steam generation in a single pressure heat recovery system for 650 kg/s of exhaust flow. Take steam pressure at 160 bar, $c_{p\ gas} = 1.03$ kJ/kg K, Pinch point, PP = 25 K, Approach Point, AP = 15 K and Terminal temperature difference, TTD = 50 K. The exhaust gas supply temperature to HRSG, $T_{g\ in} = 600°C$.

Solution:

From steam tables,

At 160 bar, the saturation temperature of steam, $T_{sat} = 347.4°C$

The exhaust gas temperature at evaporator exit, $T_{go} = T_{sat} + PP$
$= 347.4 + 25 = 372.4°C$

Water inlet temperature at evaporator inlet, $T_2 = T_{sat} - AP$
$= 347.4 - 15 = 332.4°C$

$T_5 = T_6 - TTD$

$= 600 - 50 = 550°C$

From steam tables, $h_5 = 3437.8$ kJ/kg

$h_2 = 1535$ kJ/kg

Heat balance in SH and evaporator:

$m_g\ c_{pg}\ (T_6 - T_8) = m_s\ (h_5 - h_2)$

$$m_s = \frac{m_g c_{pg} (T_6 - T_8)}{h_5 - h_2}$$

$$m_s = \frac{650 \times 1.03(600 - 372.4)}{3437.8 - 1535} = 78 \text{ kg/s}$$

8. Find the LP for a dual pressure heat recovery system. Assume HP = 200 bar, $c_{p\ gas}$ = 1.03 kJ/kg K. Pinch point, PP = 25 K. Approach Point, AP = 15 K. Terminal temperature difference, TTD = 25 K. Degree of super heat in LP heater = 50 K. HP feed water inlet temperature = 150°C and the exhaust gas supply temperature to HRSG, $T_{g\ in}$ = 600°C. Also find the flow rates of HP and LP steams for 650 kg/s of exhaust.

Solution:

Refer Fig. 8.40 for the arrangement of hot gas and working fluid.

At 200 bar HP, $T_{sat\ HP} = 365.7°C$

$$T_{10} = T_{11} - TTD$$
$$= 600 - 25 = 575°C$$

Water exit temperature at HP economizer, $T_7 = T_{sat\ HP} - AP$

$$= T_8 - AP$$
$$= 365.7 - 15 = 350.7°C$$

Exhaust gas temperature at HP evaporator exit, $T_{13} = T_{sat\ HP} + PP$

$$= T_8 + PP$$
$$= 365.7 + 25 = 390.7°C$$

Given $T_6 = 150°C$

From steam tables, $h_6 = 632.2$ kJ/kg

$h_7 = 1673$ kJ/kg

$h_{10} = 3461.9$ kJ/kg

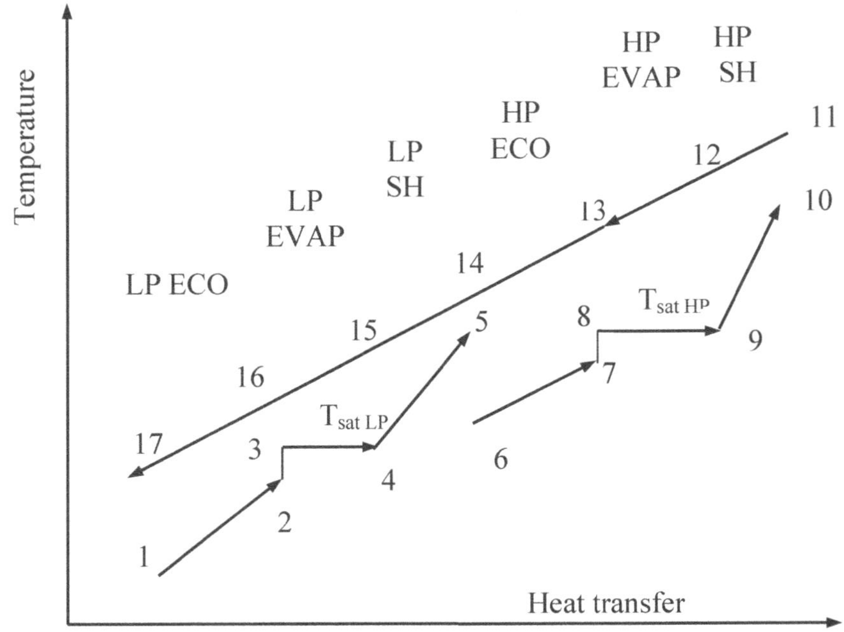

Fig. 8.40. Dural pressure HRSG.

Heat balance in HP EVAP+HPSH:

$m_{11}(h_{11} - h_{13}) = m_{HP}(h_{10} - h_7)$

$$m_{HP} = \frac{m_{11}(h_{11} - h_{13})}{(h_{10} - h_7)}$$

$$= \frac{650 \times 1.03(600 - 390.7)}{(3461.7 - 1673)} = 78.34 \text{ kg/s}$$

Heat balance in HP ECO:

$m_{13}(h_{13} - h_{14}) = m_{HP}(h_7 - h_6)$

$$m_{13}c_{pg}(T_{13} - T_{14}) = m_{HP}(h_7 - h_6)$$

$$T_{14} = T_{13} - \frac{m_{HP}(h_7 - h_6)}{m_{13}c_{pg}}$$

$$T_{14} = 390.7 - \frac{78.34(1673 - 632.2)}{650 \times 1.03}$$

$= 268.7°C$

$T_5 = T_{14} - TTD$

$= 268.7 - 25 = 243.9°C$

$T_{sat\ LP} = T_3 = T_{14} - TTD - DSH_{LP}$

$= 268.7 - 25 - 50$
$= 193.7°C$

From steam tables, the saturation pressure at 97.4°C is 13 bar

Water exit temperature at LP economizer, $T_2 = T_{sat\ LP} - AP$
$= T_3 - AP$
$= 193.7 - 15 = 178.7°C$

Exhaust gas temperature at LP evaporator exit, $T_{16} = T_{sat\ LP} + PP$

$= T_3 + PP$
$= 193.7 + 25 = 218.7°C$

From gas tables,

$h_5 = 3920$ kJ/kg

$h_2 = 760$ kJ/kg

Heat balance in LP EVAP+LPSH

$m_{14}(h_{14} - h_{16}) = m_{LP}(h_5 - h_2)$

$$m_{HP} = \frac{m_{14}(h_{14} - h_{16})}{(h_5 - h_2)}$$

$$= \frac{650 \times 1.03(268.7 - 218.7)}{(3920 - 760)} = 10.59 \text{ kg/s}$$

Chapter 9
Cooling Systems

9.1 Introduction

Cooling demands are increasing in residential, commercial, and industrial sectors. Therefore, there is a need to develop advanced cooling systems such as solar air conditioning, cooling cogeneration, cooling polygeneration, etc. The important component in a cooling system is the throttling unit. It expands the high-pressure liquid into a l ow-pressure fluid to generate cooling. To create this high pressure, a compressor is required. Since the compressed fluid is in a vapor form, it needs to condense into liquid form to suit liquid throttling. An evaporator is a place for cooling. These components together form a vapor compression refrigeration (VCR) system. In solar refrigeration, in place of a mechanical compressor, a thermal compressor is used. This cycle is called the vapor absorption refrigeration (VAR) system.

9.2 Vapor compression refrigeration (VCR) system

Figure 9.1 shows the layout of the VCR system. The main components are a vapor compressor, condenser, throttling device, and evaporator. The condenser may be an air-cooled condenser or a water-cooled condenser, depending on its capacity (Stoecker and Jones, 1981). The processes are depicted on the p-h diagram and the T-s diagram, respectively, on Fig. 9.2(a) and Fig. 9.2(b). It is

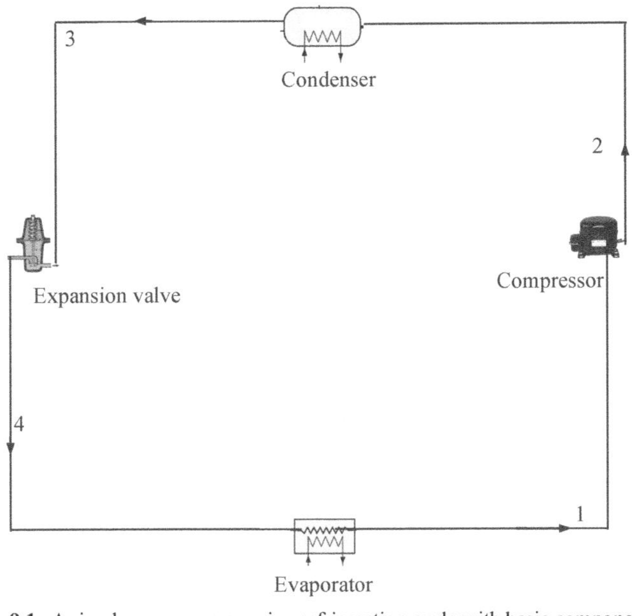

Fig. 9.1. A simple vapor compression refrigeration cycle with basic components.

DOI: 10.1201/9781032642505-9

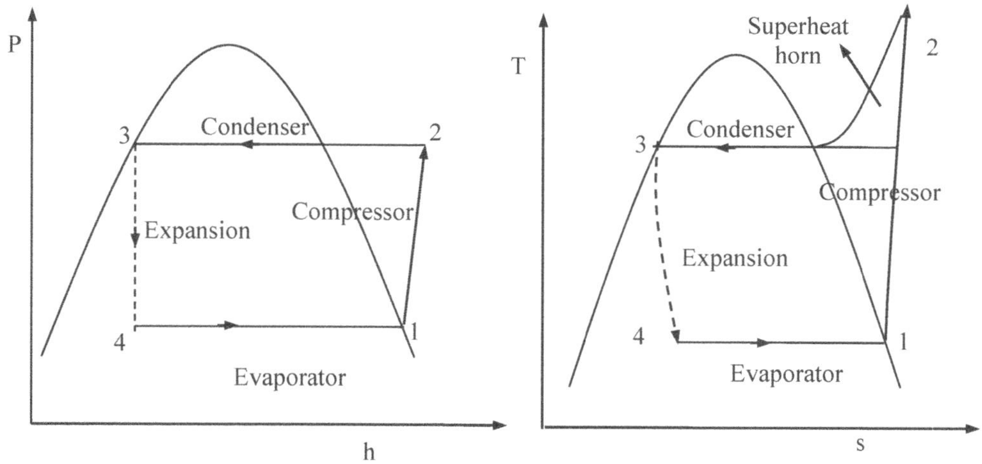

Fig. 9.2. Representation of the vapor compression refrigeration cycle on (a) P-h diagram and (b) T-s diagram.

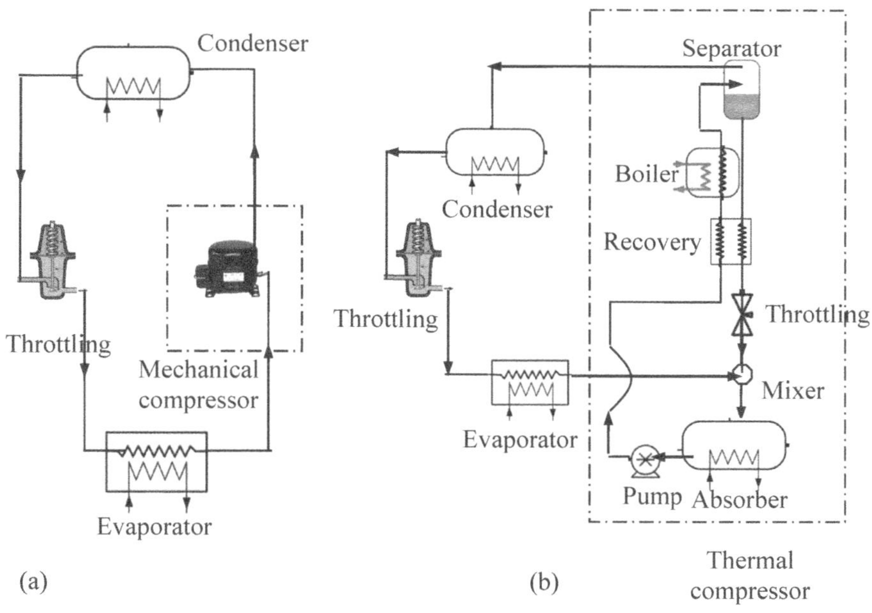

Fig. 9.3. Comparison of the vapor compression refrigeration cycle and vapor absorption refrigeration cycle with highlights in (a) Mechanical compressor and (b) thermal compressor.

assumed that the condenser exit and evaporator exit states are saturated liquid and saturated vapor conditions.

Figure 9.3 outlines the layout of the VCR cycle and the VAR cycle. The main difference is in the type of compressor used. In the VAR cycle, low-grade energy, i.e., heat, is used to increase the pressure of the vapor. The thermal compressor consists of a pump, boiler, and separator. Compared to a compressor, a pump demands a small amount of work. This pump may also be replaced with heat operations such as bubble pumps and steam jet operations.

The work supply to a compressor,

$$w_c = h_2 - h_1 \tag{9.1}$$

The condenser heat load includes the desuperheater and condensation.

$$q_{condenser} = h_2 - h_3 \tag{9.2}$$

Since throttling is an isenthalpic process,

$$h_4 = h_3$$

The heat load in the evaporator,

$$q_{evaporator} = h_1 - h_4 = h_1 - h_3 \tag{9.3}$$

The coefficient of performance (COP) is the ratio of cooling generation and work supplied.

$$COP = \frac{q_{evaporator}}{w_c} = \frac{h_1 - h_4}{h_2 - h_1} = \frac{h_1 - h_3}{h_2 - h_1} \tag{9.4}$$

Generally, the refrigerating effect or cooling output is expressed in Ton of Refrigeration (TR). One ton of refrigeration is equal to the cooling required for the generation of one ton of ice (2000 lb) from water and at 0°C in 24 hours.

The refrigerating effect in TR is,

$$TR = \frac{907 \times 334}{24 \times 3600} = 3.506 \approx 3.5 \ kW \tag{9.5}$$

Therefore, the refrigerating effect can be converted from kW to ton of refrigeration as follows,

$$q_{refrigeration,\ TR} = \frac{q_{evaporator,\ kW}}{3.5} = \frac{h_1 - h_3}{3.5} \tag{9.6}$$

Figure 9.4 shows the comparison of Carnot VCR with the actual VCR system. In the actual cycle, superheat horm with the compressor and desuperheat can be observed. The isentropic expansion in Carnot VCR differs from the throttling process in the actual cycle.

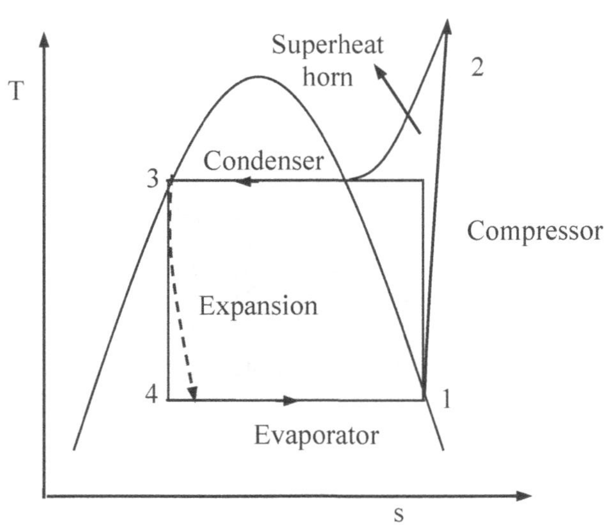

Fig. 9.4. Comparison of Carnot vapor refrigeration and vapor refrigeration.

9.3 Internal heat recovery in vapor compression refrigeration

The condensed liquid at the exit of the condenser needs to cool to evaporator temperature. Similarly, the low temperature vapor at the exit evaporator needs to increase its pressure to condenser pressure. A heat exchanger between the exit of the condenser and the inlet of the compressor provides internal heat recovery. It increases the condenser load, evaporator load, and compressor load. The increase in refrigerating effect may cause a rise in COP. But the shifting of compression towards the superheating region increases the compressor's work; hence, it may cause a drop in the COP. Therefore, the internal heat recovery may increase or decrease the COP with an increase in refrigerating effect. The arrangement of this internal heat recovery heat exchanger between these two pressure components is shown in Fig. 9.5. The thermodynamic processes are depicted in Figs. 9.6a and 9.6b, respectively, on the p-h and T-s diagrams. The process 3-4 is subcooling. The equivalent superheating at the inlet of the compressor is shown as a process 6-1.

The heat loss by condensed liquid is equal to the heat gain by the vapor at the exit of the evaporator.

$$h_3 - h_4 = h_1 - h_6$$

The practical advantage of this internal heat recovery heat exchanger is that it ensures dry vapor at the inlet of the compressor and also avoids vapor bubbles at the inlet of the throttling device due to subcooling of the liquid.

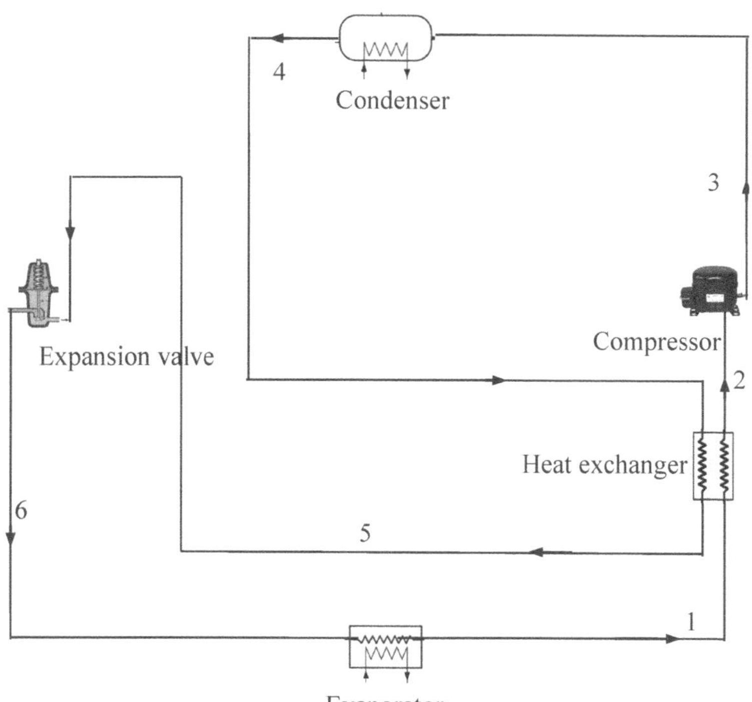

Fig. 9.5. Use of a subcooler (internal heat recovery) in a VCR.

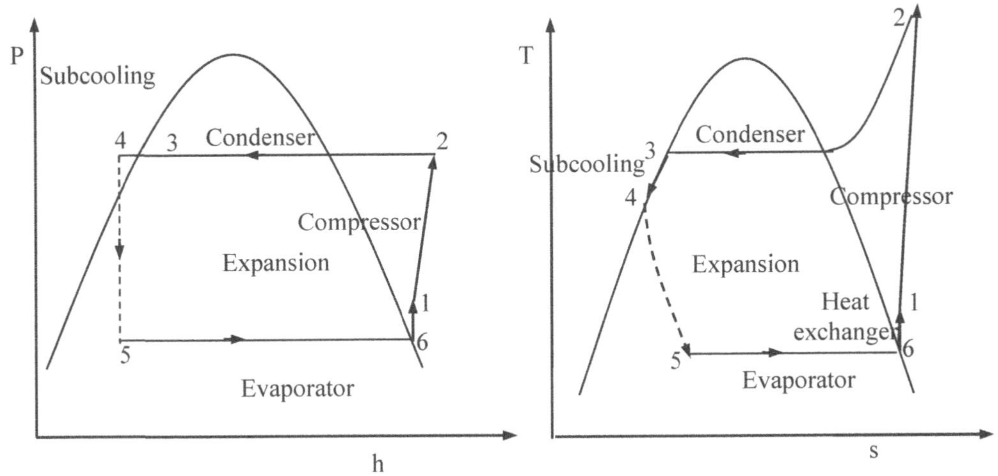

Fig. 9.6. Operation of a subcooler heat exchanger for internal heat recovery on (a) P-h diagram and (b) T-s diagram.

9.4 Flash chamber in vapor compression refrigeration

A flash chamber can be used to separate the liquid and vapor to increase the performance of the refrigeration system. There are many methods to apply the flash chamber for an increase in cooling and coefficient of performance (COP).

9.4.1 Flash chamber intercooler to superheated state

In this model, one flash chamber is associated with two expansion valves as shown in Fig. 9.7. A flash chamber with a single expansion valve is also possible. If the condensed fluid is expanded directly from the condenser pressure to evaporator pressure and the liquid and vapor are separated at the evaporator temperature or pressure, one expansion is enough. The processes in multi-stage compression with a flash chamber are shown in p-h diagram (Fig. 9.8(a)) and T-s diagram (Fig. 9.8(b)).

9.4.2 Flash chamber intercooler to a saturated state without liquid separation

Figure 9.9 shows the layout of a VCR with a flash chamber without liquid separation. The corresponding processes are detailed in Fig. 9.10(a) and Fig. 9.10(b), respectively, for the p-h diagram and the T-s diagram.

9.4.3 Flash chamber intercooler to a saturated stage with liquid separation

Figure 9.11 shows the layout of a VCR with a flash chamber with liquid separation. The corresponding processes are detailed in Fig. 9.12(a) and Fig. 9.12(b), respectively, for the p-h diagram and the T-s diagram.

9.4.4 Flash chamber with parallel operation of a multi-stage compressor

Figure 9.13 shows the layout of VCR with a flash chamber with parallel multi-stage compression. The corresponding processes are detailed in Fig. 9.14(a) and Fig. 9.14(b), respectively, with the p-h diagram and the T-s diagram.

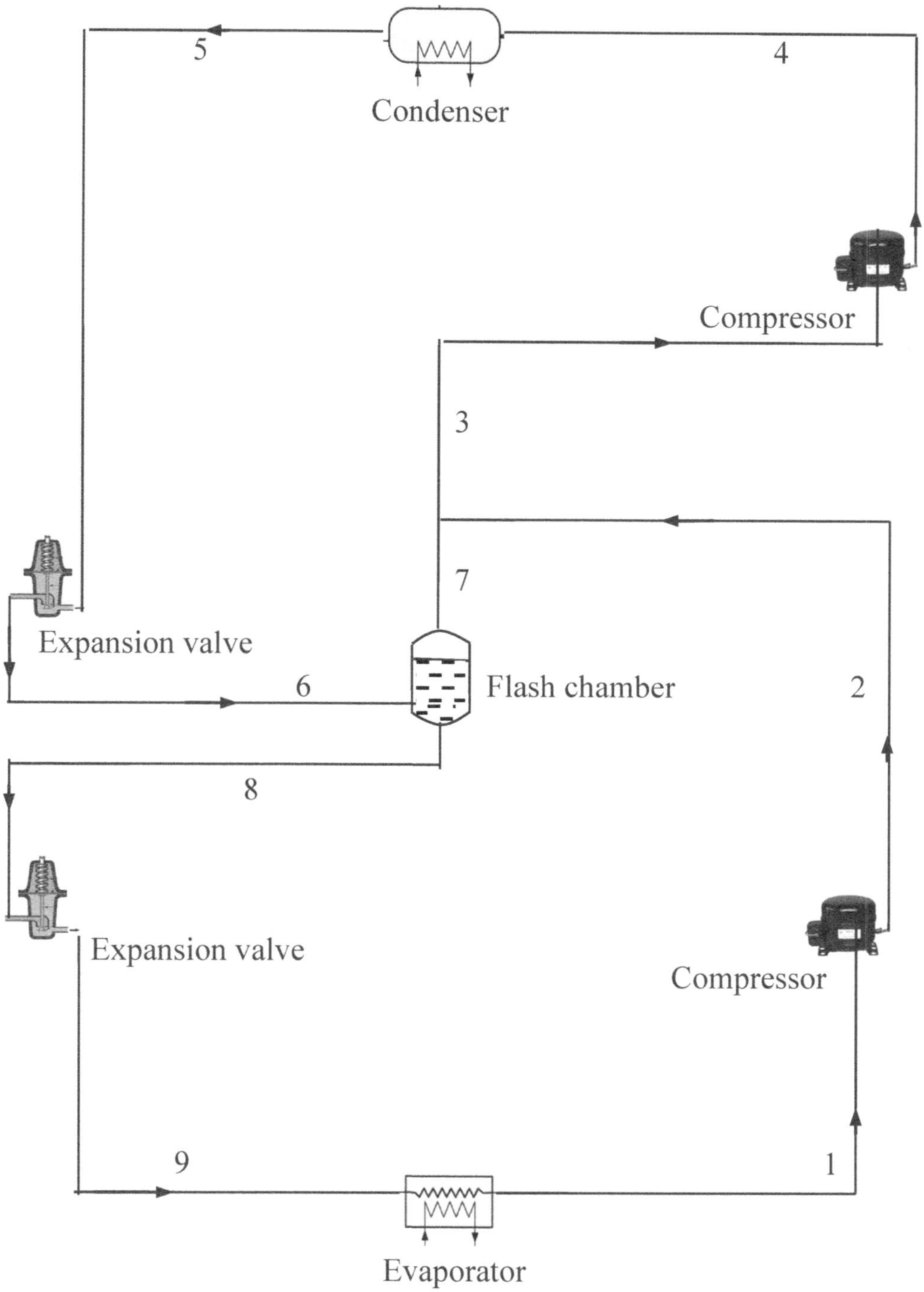

Fig. 9.7. Multi-stage compression with flash chamber intercooler to superheated state.

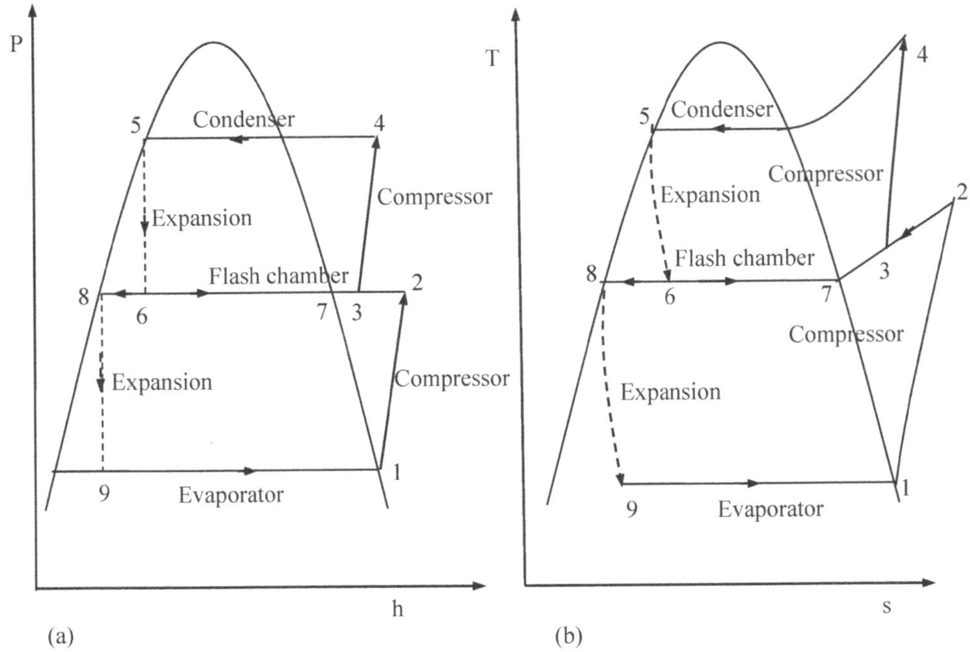

Fig. 9.8. Multi-stage compression with flash chamber intercooler to superheated state on property chart (a) pressure-enthalpy and (b) temperature-entropy diagrams.

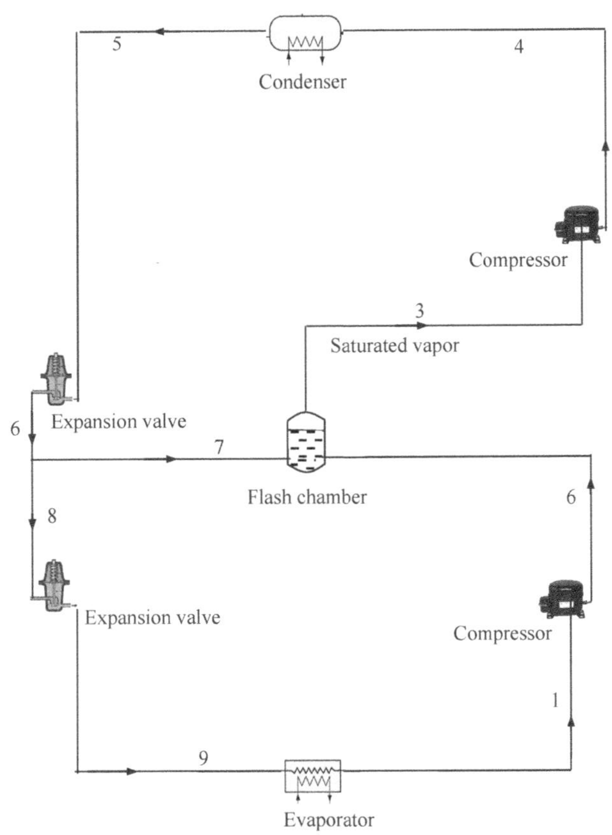

Fig. 9.9. Multi-stage compression with a flash chamber intercooler to a saturated state without liquid separation.

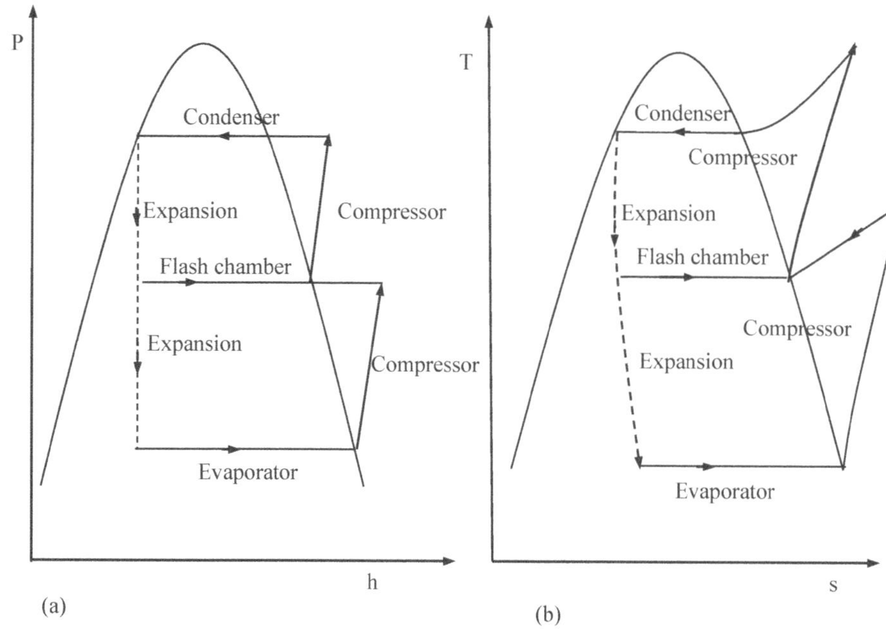

Fig. 9.10. Multi-stage compression with a flash chamber to a saturated state without liquid separation on property chart (a) pressure-enthalpy and (b) temperature-entropy diagrams.

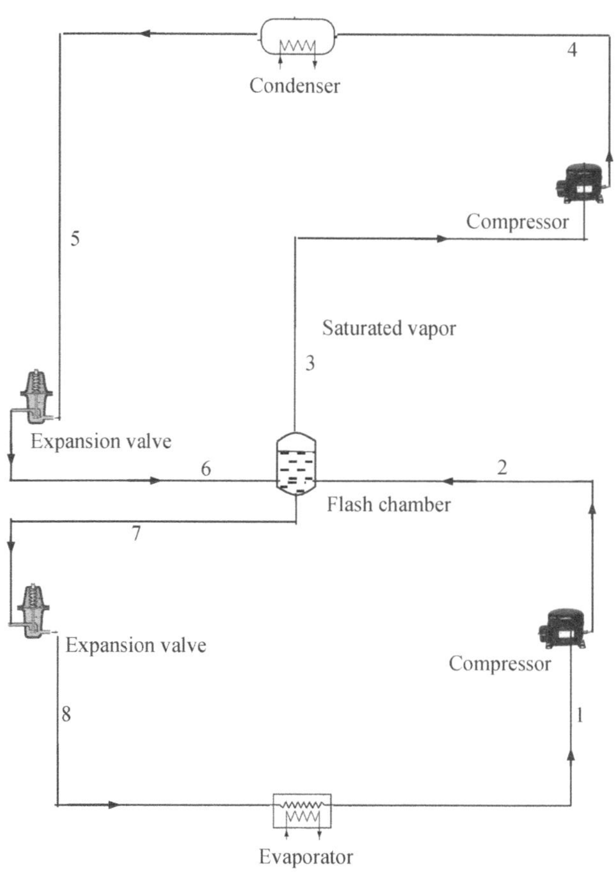

Fig. 9.11. Multi-stage compression with a flash chamber intercooler to a saturated stage with liquid separation.

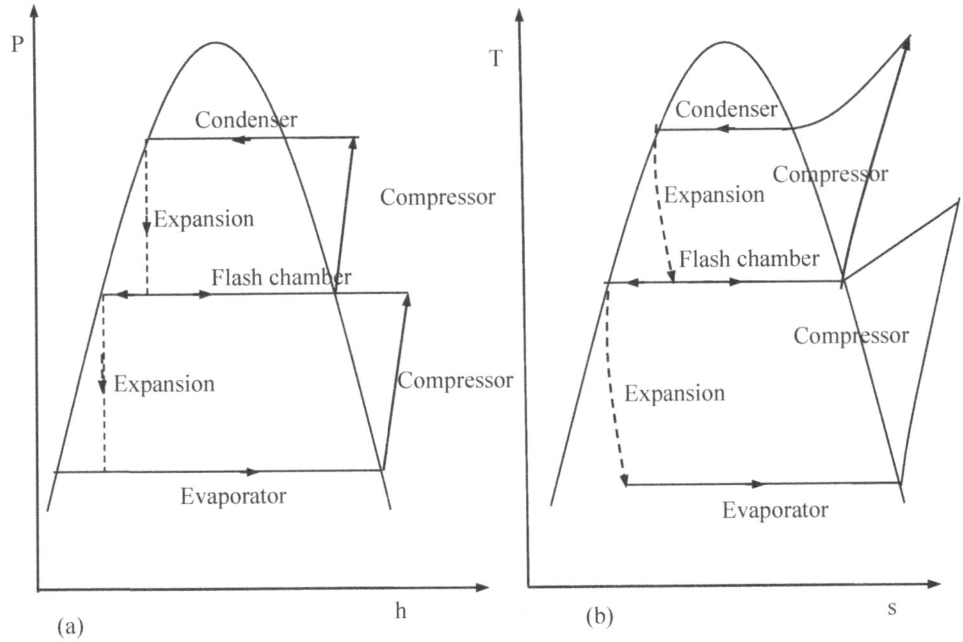

Fig. 9.12. Multi-stage compression with flash chamber intercooler to saturated stage with liquid separation on property chart (a) pressure-enthalpy and (b) temperature-entropy diagrams.

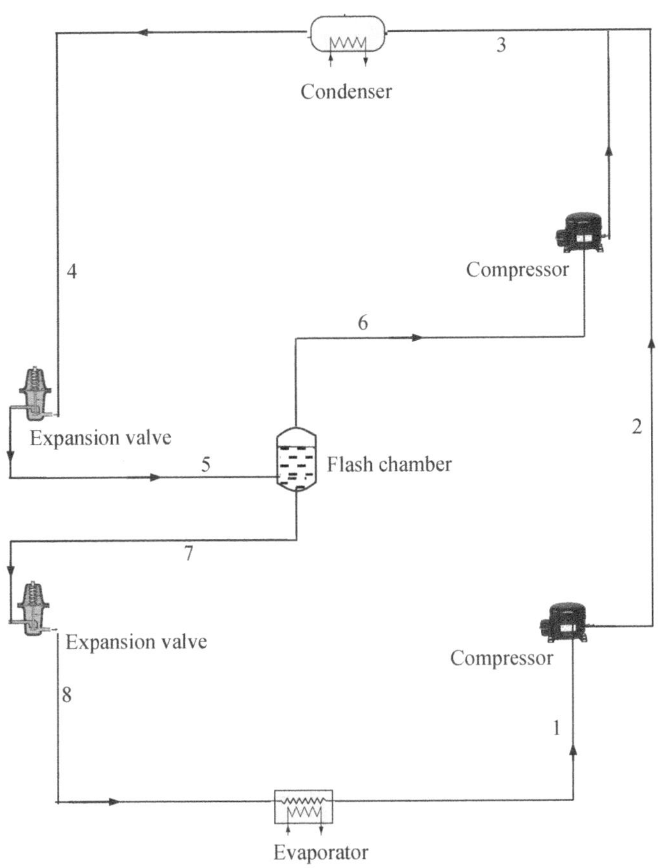

Fig. 9.13. Multi-stage compression with a flash chamber with parallel operations of a multi-stage compressor.

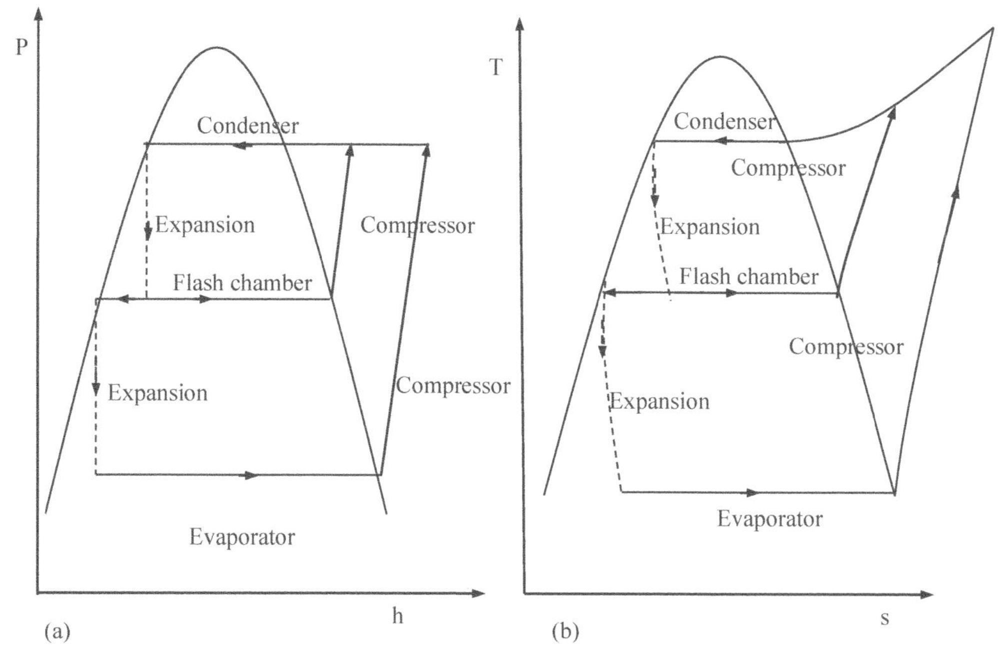

Fig. 9.14. Multi-stage compression with a flash chamber with parallel operations of a multi-stage compressor on property chart (a) pressure-enthalpy and (b) temperature-entropy diagrams.

9.5 Multi evaporator system

In refrigeration, when there are two or more than two evaporators at the same temperature or different temperatures, it is termed a multi evaporator system. Most commercial applications demand a multi evaporator system. For example, where food preservation temperature and water cooler temperature are different, multi evaporator refrigeration is needed.

Multi evaporator systems are classified into three and they are

1. Multi evaporators at the same temperature
2. Multi evaporators at different temperatures with parallel expansion valves
3. Multi evaporators at different temperatures with multi-stage expansion valves

9.5.1 Multi evaporators at the same temperature

Figure 9.15 shows the layout of VCR with multi evaporators at the same temperature. The corresponding processes are detailed in Fig. 9.16(a) and Fig. 9.16(b), respectively, for the p-h diagram and the T-s diagram.

9.5.2 Multi evaporators at different temperatures with parallel expansion valves

Figure 9.17 shows the layout of VCR with multi evaporators at different temperatures with parallel expansion valves. The corresponding processes are detailed in Fig. 9.18(a) and Fig. 9.18(b), respectively, for the p-h diagram and the T-s diagram.

9.5.3 Multi evaporators at different temperatures with multi-stage expansion valves

Figure 9.19 shows the layout of a VCR with multi evaporators at different temperatures and multi-stage expansion valves. The corresponding processes are detailed in Fig. 9.20(a) and Fig. 9.20(b), respectively, with the p-h diagram and the T-s diagram.

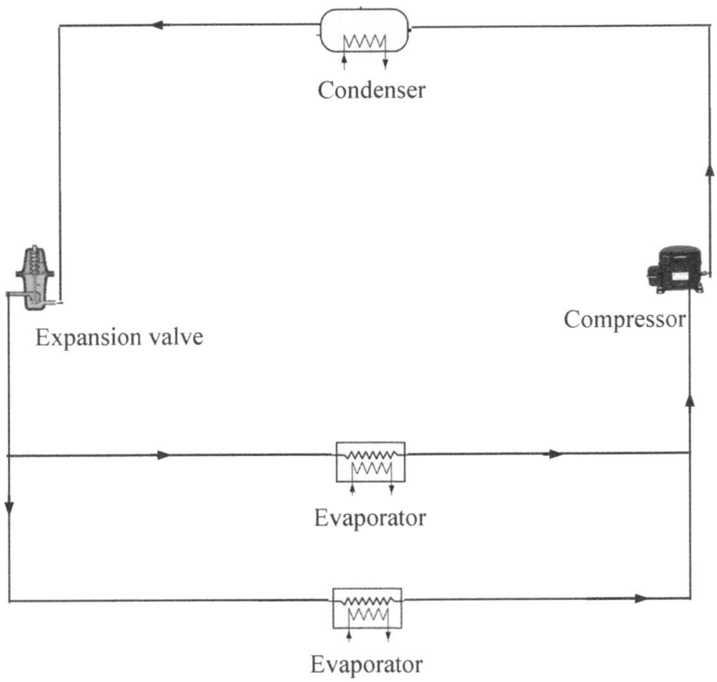

Fig. 9.15. Multi evaporators at the same temperature.

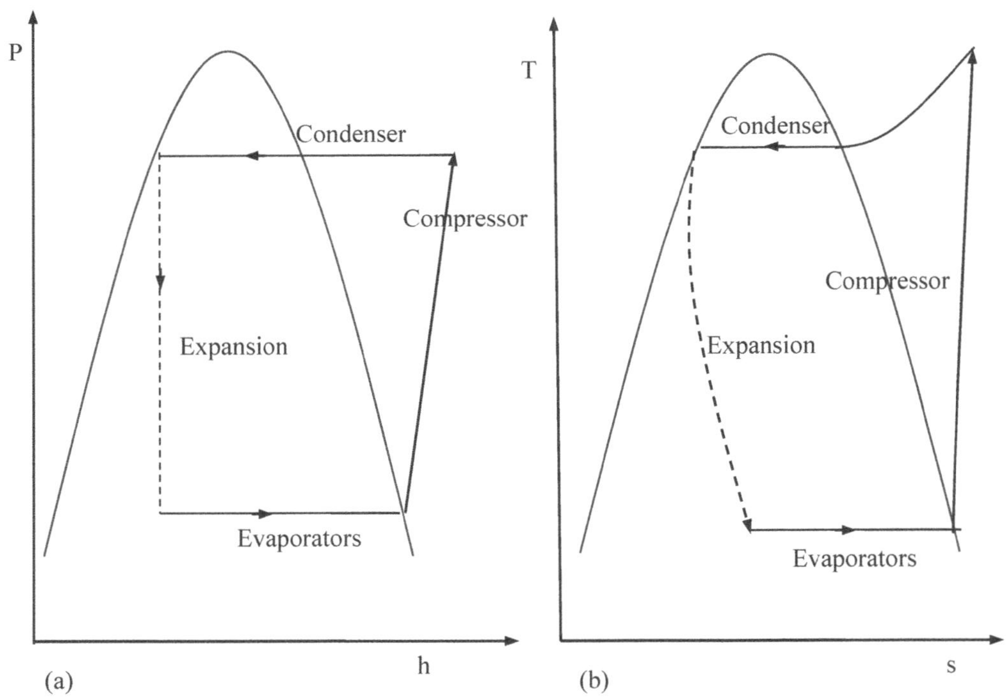

Fig. 9.16. Multi evaporators at the same temperature on property chart (a) pressure-enthalpy and (b) temperature-entropy diagrams.

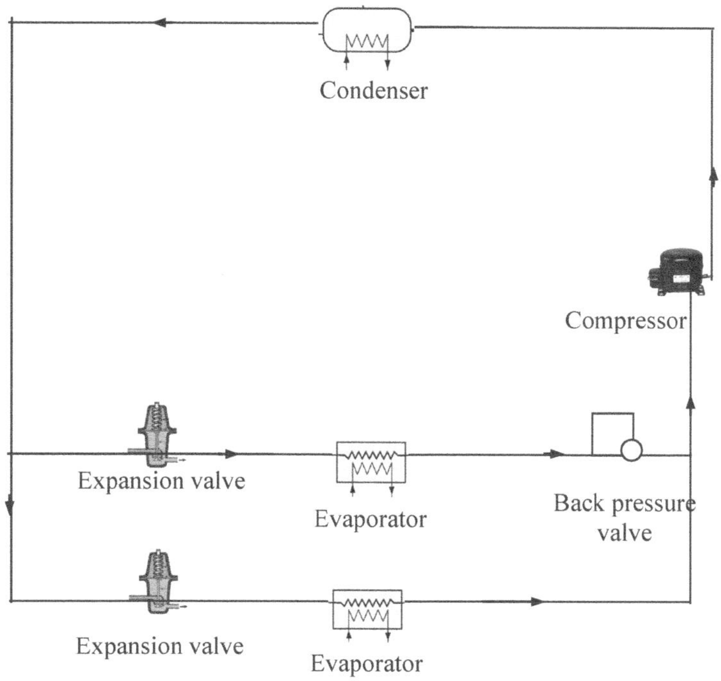

Fig. 9.17. Multi evaporators at different temperatures with parallel expansion valves.

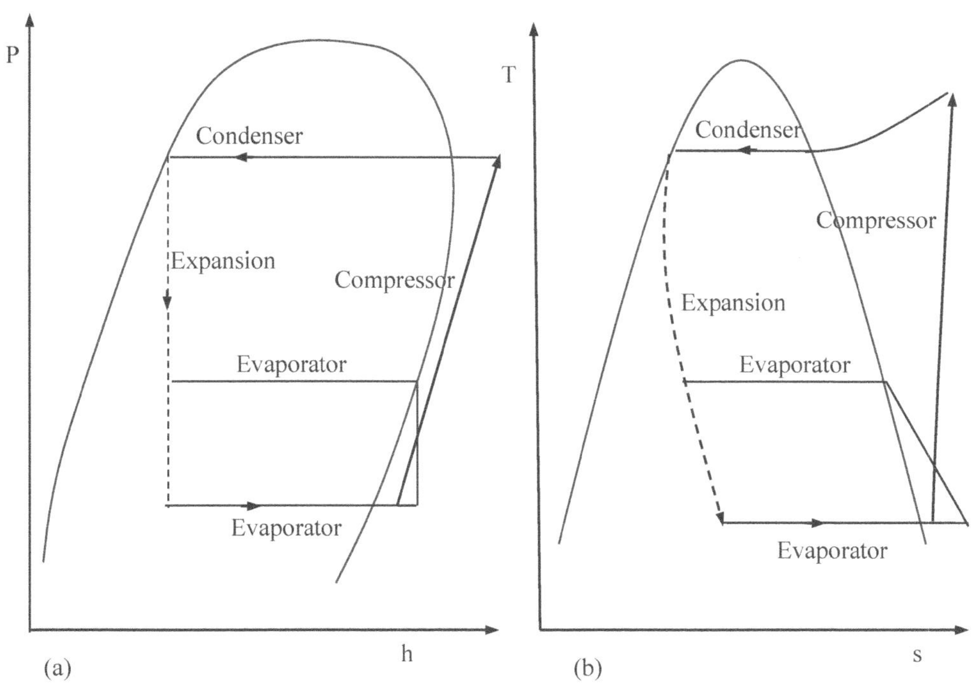

Fig. 9.18. Multi evaporators at different temperatures with parallel expansion valves on property chart (a) pressure-enthalpy and (b) temperature-entropy diagrams.

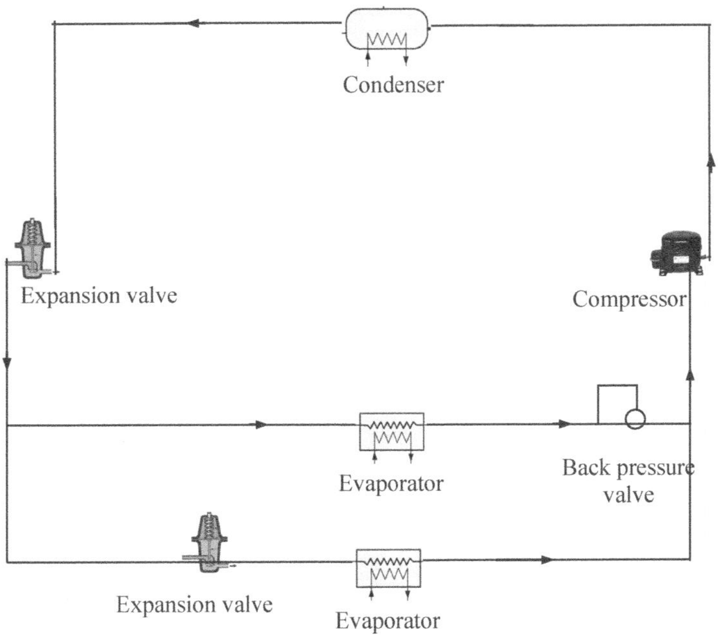

Fig. 9.19. Multi evaporators at different temperatures with multi-stage expansion valves.

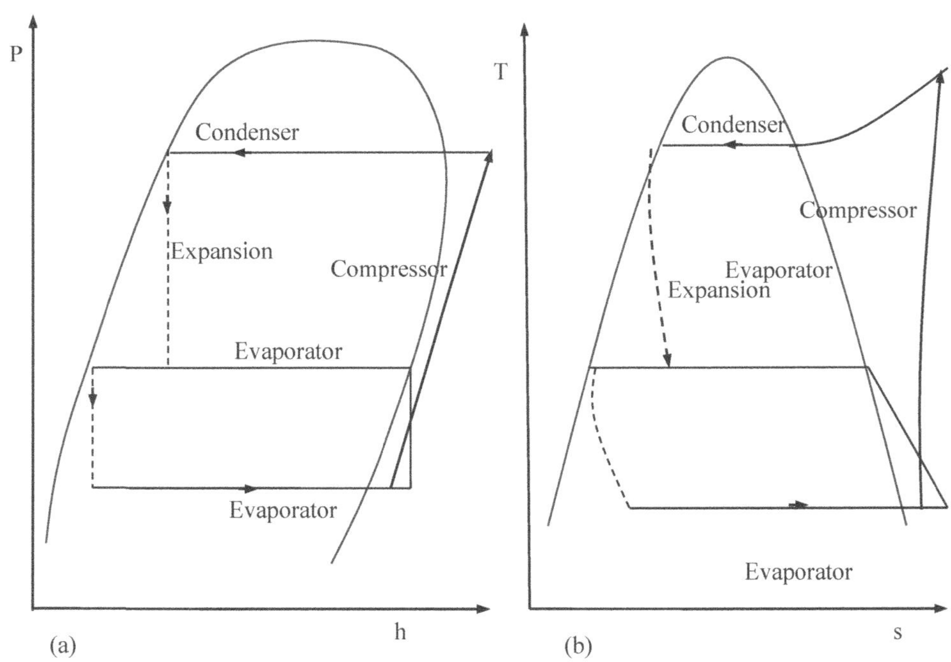

Fig. 9.20. Multi evaporators at different temperatures with multi-stage expansion valves on property chart (a) pressure-enthalpy and (b) temperature-entropy diagrams.

In addition to the above models, multi-stage compressors with multi evaporator systems can be designed to suit the applications.

9.6 Multi-purpose refrigeration systems

A refrigerator with a single compressor can provide refrigeration at several temperatures if the refrigerant is throttled in stages. Figure 9.21 shows the layout of a VCR with a multi-purpose system. The corresponding processes are detailed in Fig. 9.22(a) and Fig. 9.22(b), respectively, with the p-h diagram and the T-s diagram.

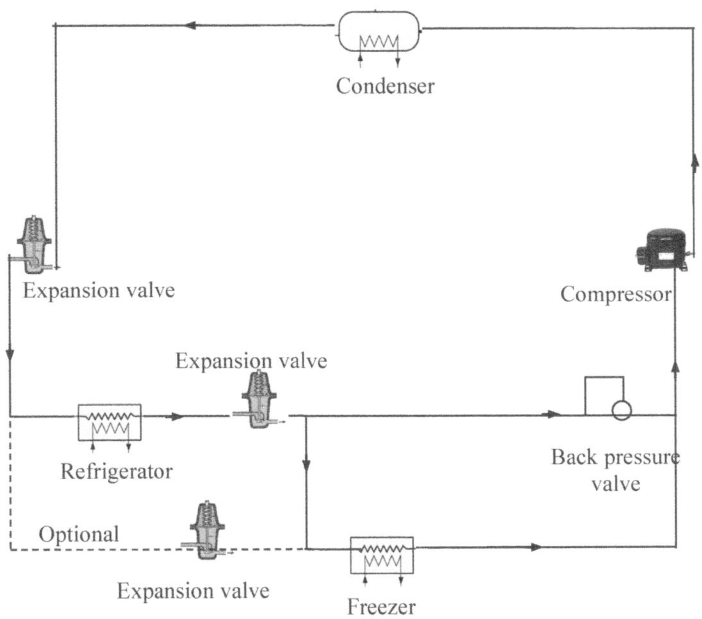

Fig. 9.21. Multi-purpose refrigeration.

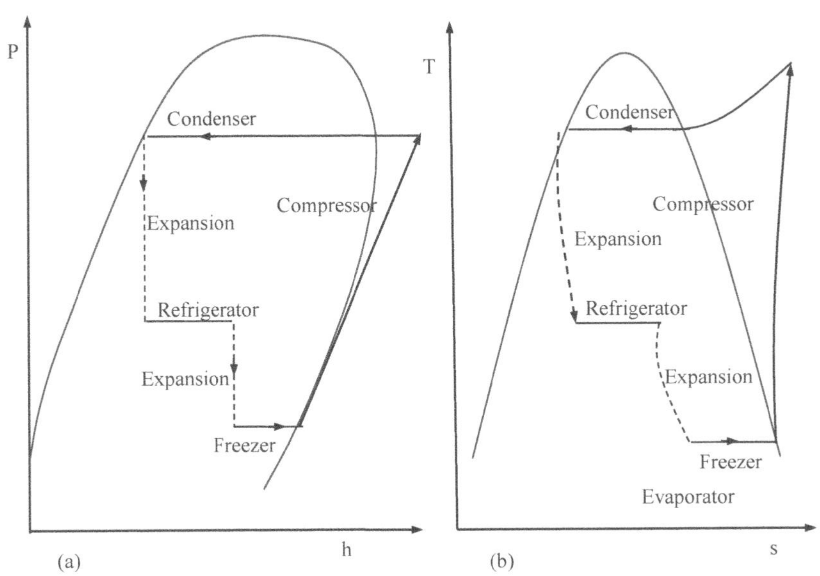

Fig. 9.22. A multi-purpose refrigeration system on property chart (a) pressure-enthalpy and (b) temperature-entropy diagrams.

9.7 Cascade refrigeration systems

Very low temperatures can be achieved by operating two or more vapor-compression systems in series, called cascading. The low temperature range focused in a separate study of cryogenics is from –100°C to –273°C. The study of low temperature refrigeration is called as cryogenics which has a great significance in fields of satellites, liquefied petroleum gas (LPG) production, hydrogen liquefaction and storage, liquefaction of gases such as air, oxygen, and nitrogen, solidification of carbon dioxide, magnetic refrigeration etc. The single stage VCR alone is not sufficient to produce low temperature. To generate a very low temperature in the evaporator, very low pressure should be maintained in the heat exchanger which demands a bigger size. The refrigeration freezing temperature also should be very low and below the required temperature. The low evaporator pressure and high condenser pressure result in a high pressure ratio and a low COP. To avoid these difficulties, cascade refrigeration is developed to generate low temperature with high COP which is more than that given by the single effect VCR.

The cascade refrigeration system looks similar to the combined cycle power plant. The cooling effect and COP of a refrigeration system also increases as a result of cascading in addition to the low temperature generation. Figure 9.23 shows the coupling of two VCR systems such that the topping VCR is absorbing heat from the condenser of the bottoming VCR. It is called two stage cascade refrigeration. It permits the use of two different refrigerants which have different temperature

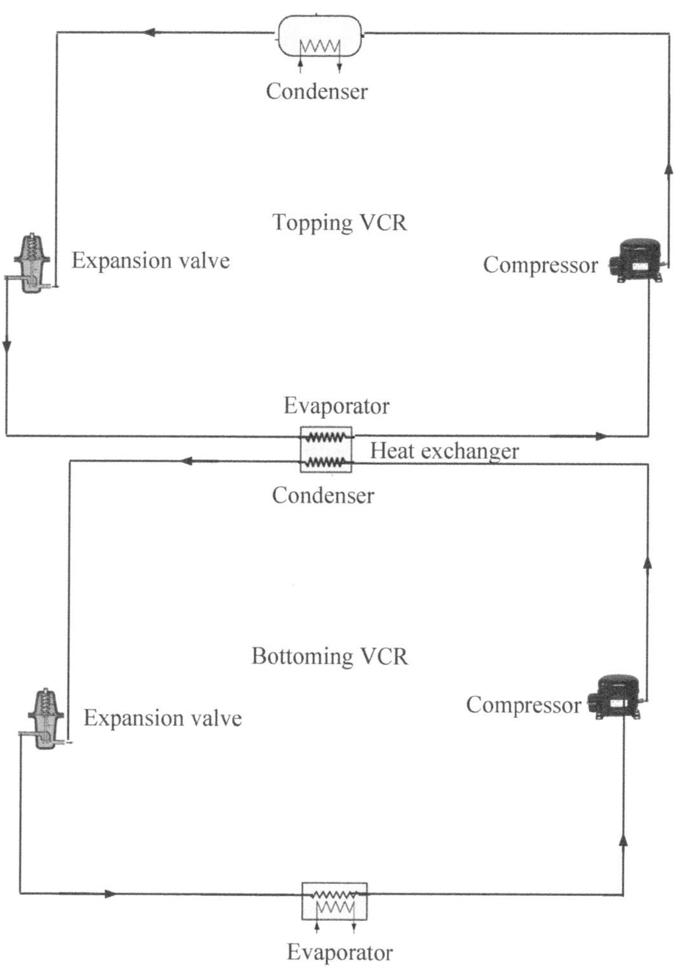

Fig. 9.23. Cascade refrigeration system.

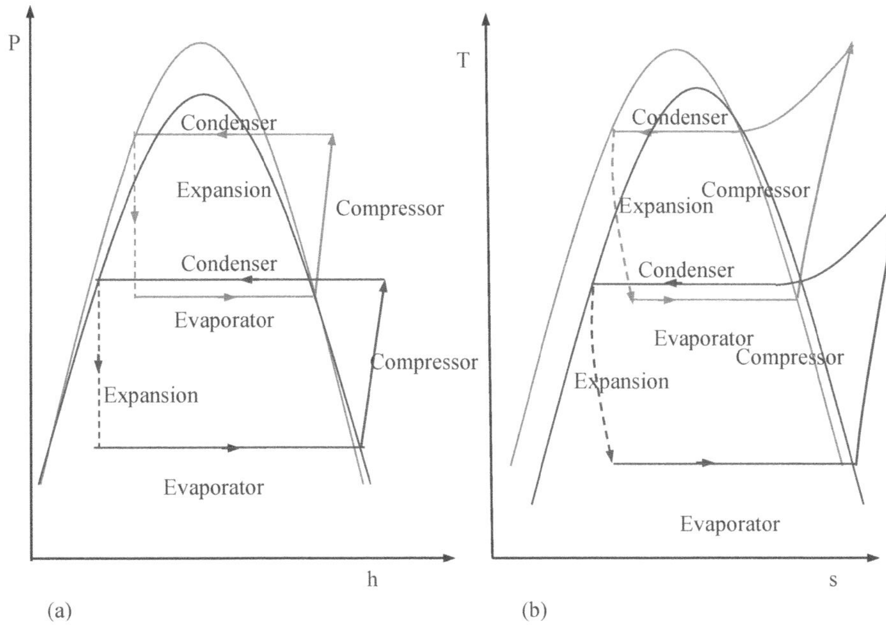

Fig. 9.24. Cascade refrigeration on property chart (a) pressure-enthalpy and (b) temperature-entropy diagrams.

ranges. Two or more refrigeration cycles are coupled with the heat exchanger(s). The hot line of heat exchangers is called the condenser of the bottoming VCR and the cold line of the heat exchanger is called as the absorber of the topping VCR. Finite temperature difference is required to transfer the heat from the bottoming VCR to the topping VCR. This temperature difference is called temperature overlap. If these two temperatures are equal then it is termed intermediate temperature.

Figure 9.24 shows the P-h and T-s diagrams to help understand the properties changes in a cascade refrigeration cycle. The temperature of fluid in the topping VCR absorber is lower than the temperature of fluid in the condenser of the bottoming VCR. It assures the heat flow from the bottoming VCR to the topping VCR.

9.8 Air conditioning – a simple system with 100% re-circulated air

In this simple system, there is no outside air, and the same air is recirculated as shown in Figs. 9.25a and 9.25b. Figure 9.25a shows the process layout, and Fig. 9.25b shows the psychrometric chart. The split air conditioning and window air conditioning systems for rooms and laboratories are examples of air conditioning with 100% recirculated air. Since fresh air is not taken from outside, there is no extra load on the cooling coil. The same cooled air is circulated through the refrigeration system again and again. The supply of air to the space should be maintained at the required temperature so must its and relative humidity. The sensible and latent heat loads of the space should be counted.

With the assumptions of no heat gains in the supply and return ducts and no energy addition from fans, using energy balance across the room, the room sensible cooling load ($Q_{s, r}$), room latent cooling load ($Q_{l, r}$), and room total cooling load ($Q_{t, r}$) are given by,

$$Q_{s,r} = m_a c_{pm} (T_1 - T_2) \tag{9.7}$$

$$Q_{l,r} = m_a (\omega_1 - \omega_2) h_{fg} \tag{9.8}$$

$$Q_{t,r} = Q_{s,r} + Q_{l,r} = m_a (h_1 - h_2) \tag{9.9}$$

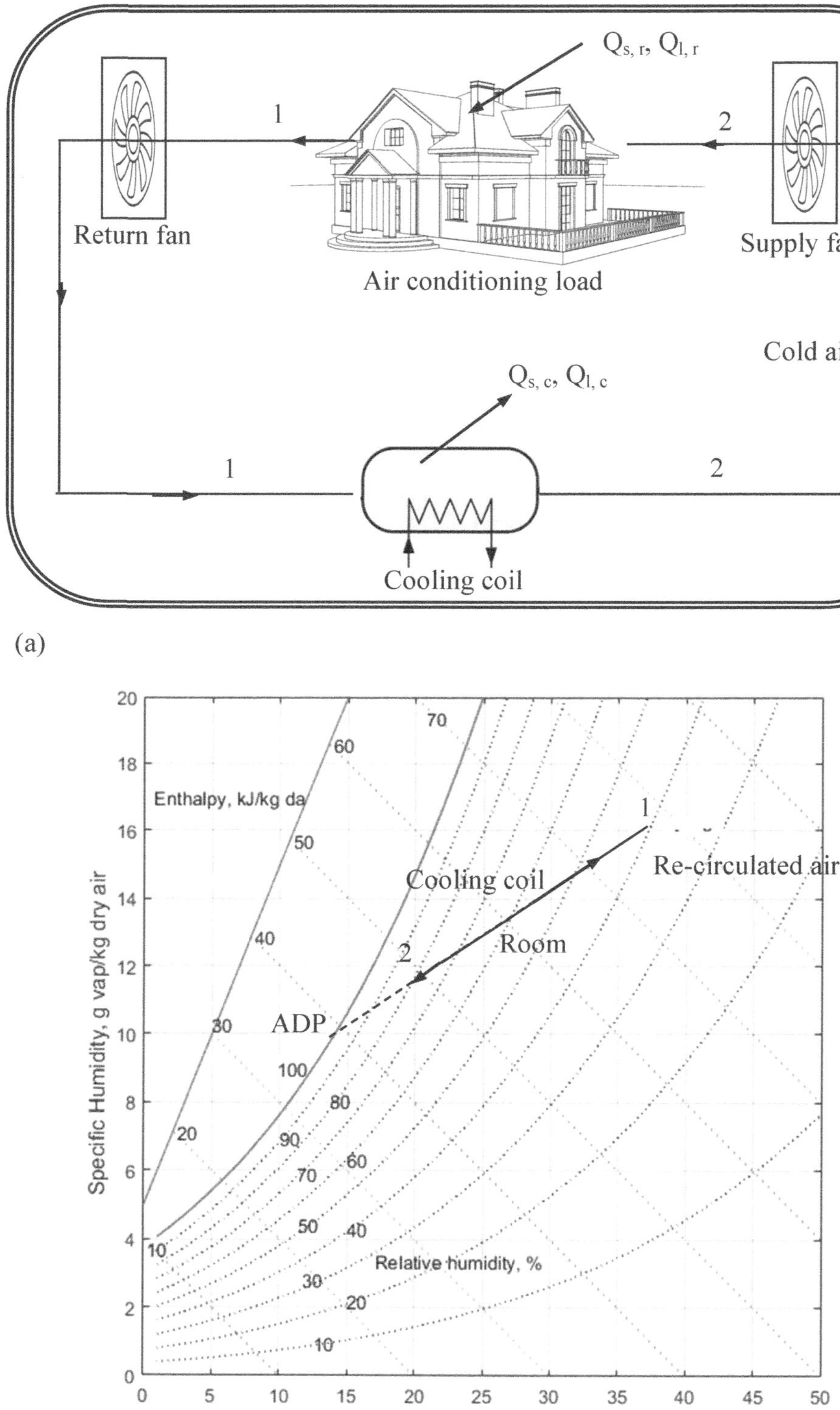

Fig. 9.25. Air conditioning with 100% re-circulated air with (a) plant layout and (b) psychrometric chart.

From cooling load calculations, the sensible, latent and total cooling loads on the room are obtained. Hence one can find the room sensible heat factor (RSHF) from the equation,

$$RSHF = \frac{Q_{s,\ r}}{Q_{s,\ r} + Q_{l,\ r}} = \frac{Q_{s,\ r}}{Q_{t,\ r}} \tag{9.10}$$

From the RSHF value one can calculate the slope of the process undergone by the air as it flows through the conditioned space (process 1-2) as,

$$\text{Slope of the line, 1-2, } \tan\theta = \frac{1}{2451} \times \left(\frac{1 - RSHF}{RSHF}\right) \tag{9.11}$$

From Fig. 9.1a, the condition '1' is a known state. We can draw a line from 2 to 1 for the process. At a given sensible heat load and latent heat load, the process should extract these loads in the required proportions. As the recirculated air is 100%, processes 1-2 and 2-1 are the same and opposite in direction. The temperature and humidity ratio decreases in the cooling coil and increases in the conditioned space. Assuming the sensible and latent heat loads of the cooling coil are equal to the sensible and latent heat loads of conditioned space, neglecting other heat transfer loads from ducts and fans,

$Q_{s,\ r} = Q_{s,\ c}$ and $Q_{l,\ r} = Q_{l,\ c}$

The supply conditions are fixed in the three equations. There are four unknowns, m_a, T_2, ω_s, and h_2, and three equations; one condition may be fixed to determine the other three unknown conditions. If the bypass factor (BF) of the cooling coil is known from room conditions, coil appatus dew point (ADP), and the bypass factor, the suppled air temperature T can be determined from the bypass factor as

$$BF = \frac{T_2 - T_{ADP}}{T_1 - T_{ADP}} \tag{9.12}$$

$$T_2 = T_{ADP} + BF(T_1 - T_{ADP}) \tag{9.13}$$

The mass flow rate of supply air,

$$m_1 = \frac{Q_{s,\ r}}{c_{pm}(T_1 - T_2)} = \frac{Q_{s,\ r}}{c_{pm}(T_1 - (T_{ADP} + BF(T_1 - T_{ADP}))}$$

$$= \frac{Q_{s,\ r}}{c_{pm}(T_1 - T_{ADP} - BF(T_1 - T_{ADP})}$$

$$m_1 = \frac{Q_{s,\ r}}{c_{pm}(T_1 - T_{ADP})(1 - BF)} \tag{9.14}$$

This is equal to the dry air, i.e., $m_a = m_1$

Specific humidity of air at state 2,

$$\omega_2 = \omega_1 - \frac{Q_{l,\ r}}{m_a h_{fg}} \tag{9.15}$$

Specific enthalpy of air at state 2,

$$h_2 = h_1 - \frac{Q_{t,\ r}}{m_a} \tag{9.16}$$

As per Eq. (9.14), the required mass at state 1 decreases with decrease in BF. It is clear that the required mass flow rate of supply air decreases as the bypass factor BF decreases.

At zero BF, the minimum amount of supply air is:

$$m_{1,\ \min} = \frac{Q_{s,\ r}}{c_{pm}(T_1 - T_{ADP})} \tag{9.17}$$

Thus with 100% re-circulated air, the room ADP, i.e., T_2 is equal to the coil ADP. Since outdoor air is not mixed, the load on the coil is equal to the load on the room. The drawback of 100% re-circulated air is that there is a loss of freshness and oxygen in the air after a long time. Therefore, for small size applications and short durations the 100% re-circulated air unit is suitable.

9.9 Air conditioning system with 100% outdoor air

In some situations, a room always needs fresh air to be air conditioned. The recirculation of conditioned air may not be possible or not advisable as per the room situation. Buses, cars, aircraft and hospitals are the examples of air conditioning systems with 100% outdoor air circulation. Figure 9.26a and Fig. 9.26b show the processes on a psychrometric chart and its plant layout respectively. The cooling coil load is greater than the room load as the cooling air is not re-circulated and sent to out. This is the price to be paid for the 100% fresh air requirement.

The energy balance equations in room sensible heat and latent heat,

$$Q_{s,\ r} = m_a\, c_{pm}\, (T_3 - T_2) \tag{9.18}$$

$$Q_{l,\ r} = m_a\, (\omega_3 - \omega_2)\, h_{fg} \tag{9.19}$$

$$Q_{t,\ r} = Q_{s.\ r} + Q_{l,\ r} = m_a\, (h_3 - h_2) \tag{9.20}$$

The room exit specific humidity,

$$\omega_3 = \omega_2 + \frac{Q_{l,\ r}}{m_a h_{fg}} \tag{9.21}$$

The room exit specific enthalpy,

$$h_3 = h_2 + \frac{Q_{t,\ r}}{m_a} \tag{9.22}$$

The room sensible heat factor,

$$RSHF = \frac{Q_{s,\ r}}{Q_{s,\ r} + Q_{l,\ r}} = \frac{Q_{s,\ r}}{Q_{t,\ r}} \tag{9.23}$$

In this case, since the air conditioning unit or cooling coil heat load includes the outdoor load in addition to room load, the air conditioning apparatus has to meet the grand total heat load. Grand sensible heat factor is defined as the ratio of sensible heat load of cooling coil to its total load.

The grand sensible heat factor,

$$GSHF = \frac{Q_{s,\ c}}{Q_{s,\ c} + Q_{l,\ c}} = \frac{Q_{s,\ c}}{Q_{t,\ c}} \tag{9.24}$$

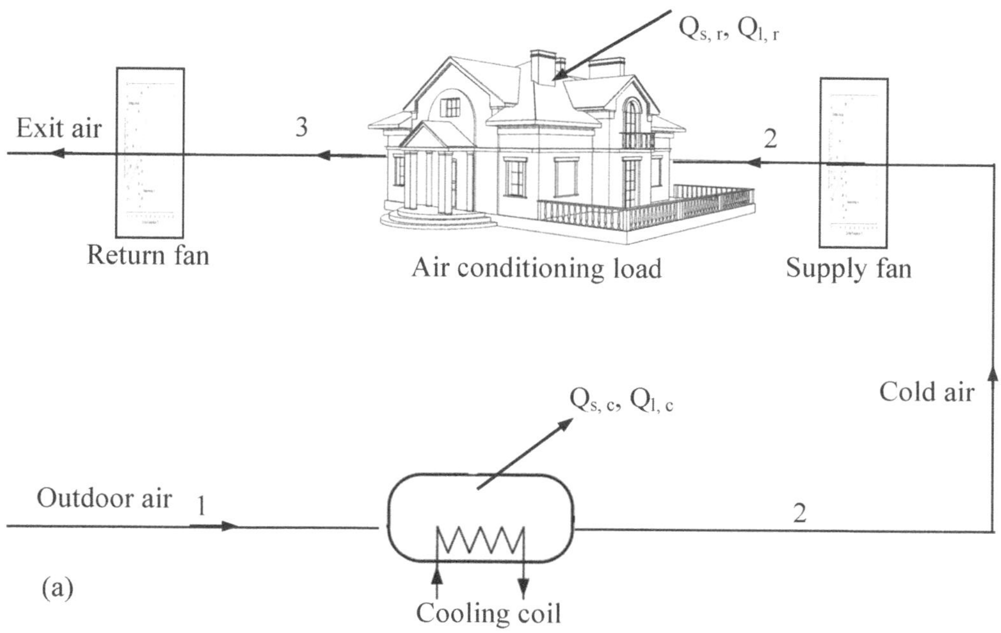

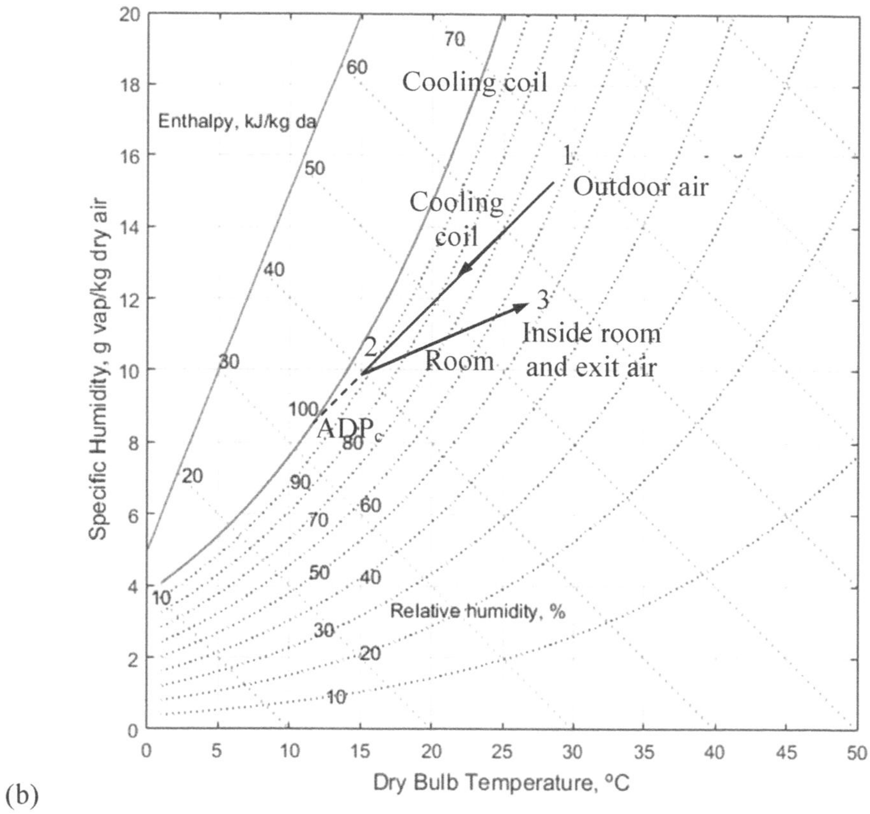

Fig. 9.26. (a) Air conditioning with 100% outdoor air on a process layout and (b) psychrometric chart.

9.10 Air conditioning system with outdoor air ventilation and zero bypass factor

For ventilation purposes, some outdoor fresh air is required for the air conditioning system. The amount of outdoor air depends on occupancy and type of application; for example, an operation theater in a hospital needs 100% outdoor air. Generally, the outdoor air is expressed in its quantity as a fraction of recirculated air. Figure 9.27a shows the layout of a summer air conditioning system with outdoor air. Figure 9.27b is the representation of these air conditioning processes on a psychrometric chart with a zero bypass factor.

The process 3-4 can be drawn on a psychrometric chart from the known sensible heat load, latent heat load, and RSHF. The cooling load calculations result in sensible heat load and latent heat load values. The intersection of this line with the saturation curve results in ADP. At zero bypass factor, the room ADP is equal to the coil ADP and temperature of supply air.

The required supply air,

$$m_2 = \frac{Q_{s,\ r}}{c_{pm}(T_4 - T_3)} = \frac{Q_{s,\ r}}{c_{pm}(T_4 - T_{ADP})} \tag{9.25}$$

From the supply mass flow rate, supply air humidity ratio and enthalpy can be evaluated.

From mass balance of air,

$$m_2 = m_1 + m_6 \tag{9.26}$$

In the above equation m_6 is the re-circulated air and m_1 is the outdoor air. From m_1 or m_1/m_6 ratio of re-circulated air can be determined.

From the energy balance across the cooling coil; the sensible, latent and total heat transfer rates, $Q_{s,\ c}$, $Q_{l,\ c}$ and $Q_{t,\ c}$ at the cooling coil are given by:

$$Q_{s,\ c} = m_2\, c_{pm}\, (T_2 - T_3) \tag{9.27}$$

$$Q_{l,\ c} = m_2\, (\omega_2 - \omega_3)\, h_{fg} \tag{9.28}$$

$$Q_{t,\ c} = Q_{s,\ c} + Q_{l,\ c} = m_2\, (h_2 - h_3) \tag{9.29}$$

where '2' refers to the mixing condition which a is a result of mixing of re-circulated air (6) and outdoor air (1). Applying mass and energy balance to the mixing process one can obtain the state of the mixed air from the equation,

$$\frac{m_1}{m_2} = \frac{\omega_2 - \omega_4}{\omega_1 - \omega_4} = \frac{h_2 - h_4}{h_1 - h_4} \approx \frac{T_2 - T_4}{T_1 - T_4} \tag{9.30}$$

Since $m_1/m_2 > 0$, from the above equation it is clear that $\omega_2 > \omega_4$, $h_2 > h_4$ and $T_2 > T_4$.

This implies that $m_2\,(h_2 - h_3) > m_2\,(h_4 - h_3)$ or the load on the cooling coil is greater than the load on the conditioned space.

During mixing, some amount of hot and humid air is added and the same cool and dry air is removed ($m_1 = m_5$).

The heat loads in the room are as follows:

$$Q_{s,\ r} = m_3\, c_{pm}\, (T_4 - T_3) \tag{9.31}$$

$$Q_{l,\ r} = m_3\, (\omega_4 - \omega_3)\, h_{fg} \tag{9.32}$$

$$Q_{t,\ r} = Q_{s,\ r} + Q_{l,\ r} = m_3\, (h_4 - h_3) \tag{9.33}$$

$$Q_{s,\ c} - Q_{s,\ r} = m_2\, c_{pm}\, (T_2 - T_4) \tag{9.34}$$

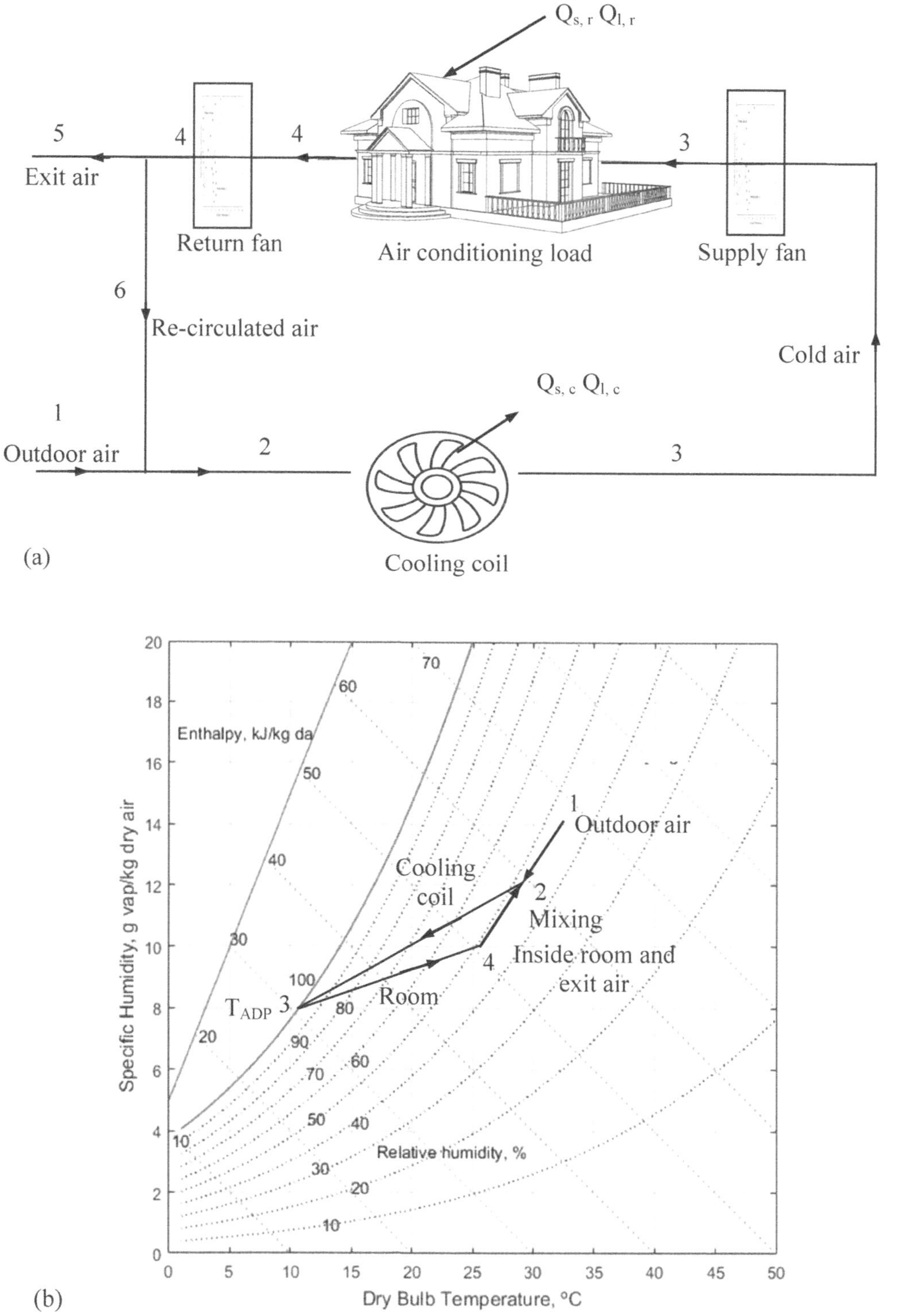

Fig. 9.27. Summer air conditioning system with outdoor air and zero bypass factor with (a) components arrangement and (b) psychrometric chart.

$$Q_{l,c} - Q_{l,r} = m_2 (\omega_2 - \omega_4) h_{fg} \tag{9.35}$$

$$Q_{t,c} - Q_{t,r} = m_2 (h_2 - h_4) \tag{9.36}$$

From the mixing of outdoor air and re-circulated air,

$$m_1T_1 + m_6T_6 = m_2T_2 \tag{9.37}$$

$$T_2 = \frac{m_1T_1 + m_6T_6}{m_2} = \frac{m_1T_1 + m_6T_4}{m_2} \tag{9.38}$$

Substituting T_2 in Eq. (9.34),

$$\begin{aligned} Q_{s,c} - Q_{s,r} &= m_2 c_{pm}\left(\frac{m_1T_1 + m_6T_4}{m_2} - T_4\right) \\ &= m_1c_{pm}T_1 + m_6c_{pm}T_4 - m_2c_{pm}T_4 \\ &= m_1c_{pm}T_1 - (m_2 - m_6)\, c_{pm}T_4 \\ &= m_1c_{pm}T_1 - m_1c_{pm}T_4 \\ &= m_1c_{pm}(T_1 - T_4) \end{aligned} \tag{9.39}$$

Similarly,

$$Q_{l,c} - Q_{l,r} = m_1 (\omega_1 - \omega_4) h_{fg} \tag{9.40}$$

$$Q_{t,c} - Q_{t,r} = m_1 (h_1 - h_4) \tag{9.41}$$

The difference in heat load, i.e., between the cooling coil and the room, increases with outdoor air (m1), its temperature, and relative humidity. The line 2-ADP is the process line that is flowing on the cooling coil. The slope of this line depends on the sensible heat factor of the cooling coil, SHF_c.

$$SHF_c = \frac{Q_{s,c}}{Q_{s,c} + Q_{l,c}} = \frac{Q_{s,c}}{Q_{t,c}} \tag{9.42}$$

The coil sensible heat factor is also called GSHF as discussed in earlier section.

9.11 Air conditioning system with outdoor air ventilation and non-zero bypass factor

Figure 9.28a shows the summer air conditioning layout with outdoor and bypass factors. The bypass factor lies between zero and one. The exit temperature of air is higher than the cooling coil temperature with the bypass factor. The psychrometric processes of this air conditioning system are shown on the psychrometric chart (Fig. 9.28b). As per the bypass factor, the room ADP and coil ADP differ from each other. This process is adequate with RSHF greater than 0.75. At the supply air temperature, T_3, or the temperature rise of air ($T_2 - T_3$), the solution can be developed with the following steps:

1. Since the supply temperature is specified one can calculate the required supply air flow rate (m_3), specific humidity (ω_3) and specific enthalpy (h_3) from sensible heat, latent heat and total heat of the room respectively.
2. Since the room and its exit condition (4), supply air temperature (T_3) and RSHF are known, one can draw the line 4-3. The intersection of this line with the saturation curve gives the room ADP.
3. The condition of air after mixing (2) is obtained from data of m_6 and m_1 with mass and energy balances combined from the lever rule of mixing.

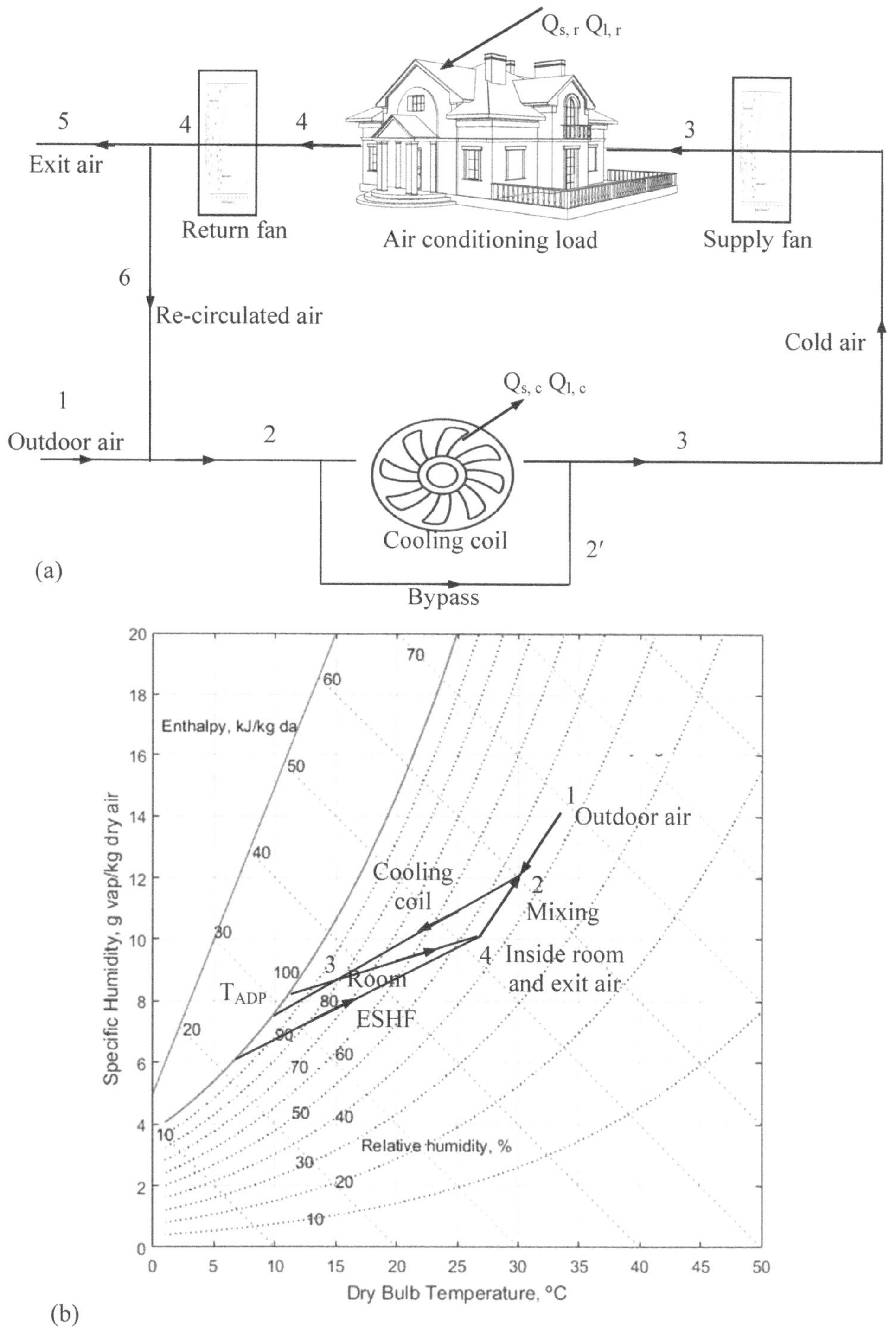

Fig. 9.28. Summer air conditioning system with outdoor air and bypass factor on (a) processes flow diagram and (b) psychrometric chart.

4. Now joining points 2 and 3 gives the process line of air as it flows through the cooling coil. The intersection of this line with the saturation curve gives the coil ADP. It can be seen that the coil ADP is lower than the room ADP.
5. The capacity of the cooling coil is obtained from the sensible heat, latent heat and total heat equations applied to the cooling coil.
6. From points 2, 3 and coil ADP, the bypass factor of the cooling coil can be determined.

If the coil ADP and coil bypass factor are given instead of the supplied air temperature, then a trial and error method has to be employed to obtain the supply condition.

In the earlier discussion, from the room load, RSHF and cooling coil load GSHF are formulated. The partial or full entry of outdoor air or fresh air increases the RSHF due to added sensible heat and latent heat from the outdoor air. It is the effective sensible heat factor (ESHF).

$$ESHF = \frac{Q_{s,\ r} + BF \times Q_{s,\ OA}}{Q_{s,\ r} + BF \times Q_{s,\ OA} + Q_{l,\ r} + BF \times Q_{l,\ OA}} \tag{9.43}$$

where OA is the outdoor air. ESHF renders the same value as GSHF.

9.12 Air conditioning system with outdoor air ventilation, non-zero bypass factor and reheating

A low coil ADP results in a high latent load, high outside humidity, large ventilation requirements (e.g., hospitals), and high internal latent loads (e.g., the presence of a kitchen or laundry). A low-coil ADP demands a low evaporator temperature in the refrigeration system. With the reheater, the difficulty of low ADP can be solved as shown in Fig. 9.29. The operation of a refrigeration system at low evaporator temperature decreases its COP. Therefore, a reheater coil permits the use of a cooling coil at high ADP, and it also addresses the high latent heat load.

Figure 9.30a shows a summer air conditioning system with a reheat coil for high latent heat load and Fig. 9.30b shows it with the psychrometric processes n. At the cooling coil, the air is cooled and dehumidified as shown in processes 2–3. Later, it is heated by the reheat coil to the required temperature. At the required supply temperature, the mass flow rate, and the state of the conditioned

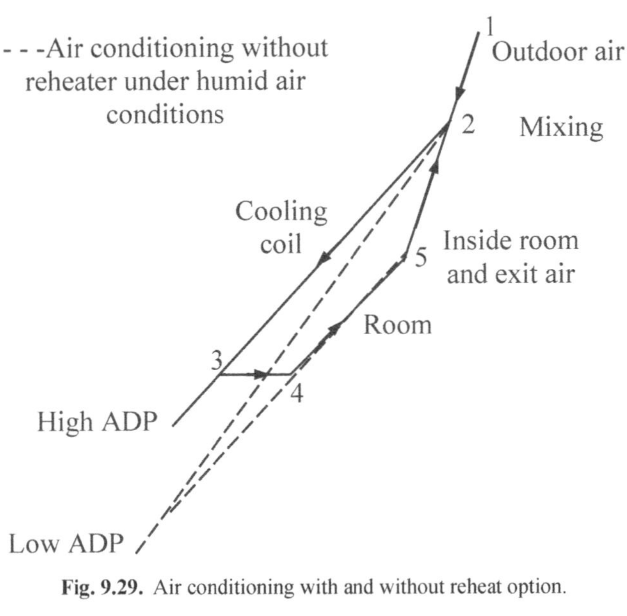

Fig. 9.29. Air conditioning with and without reheat option.

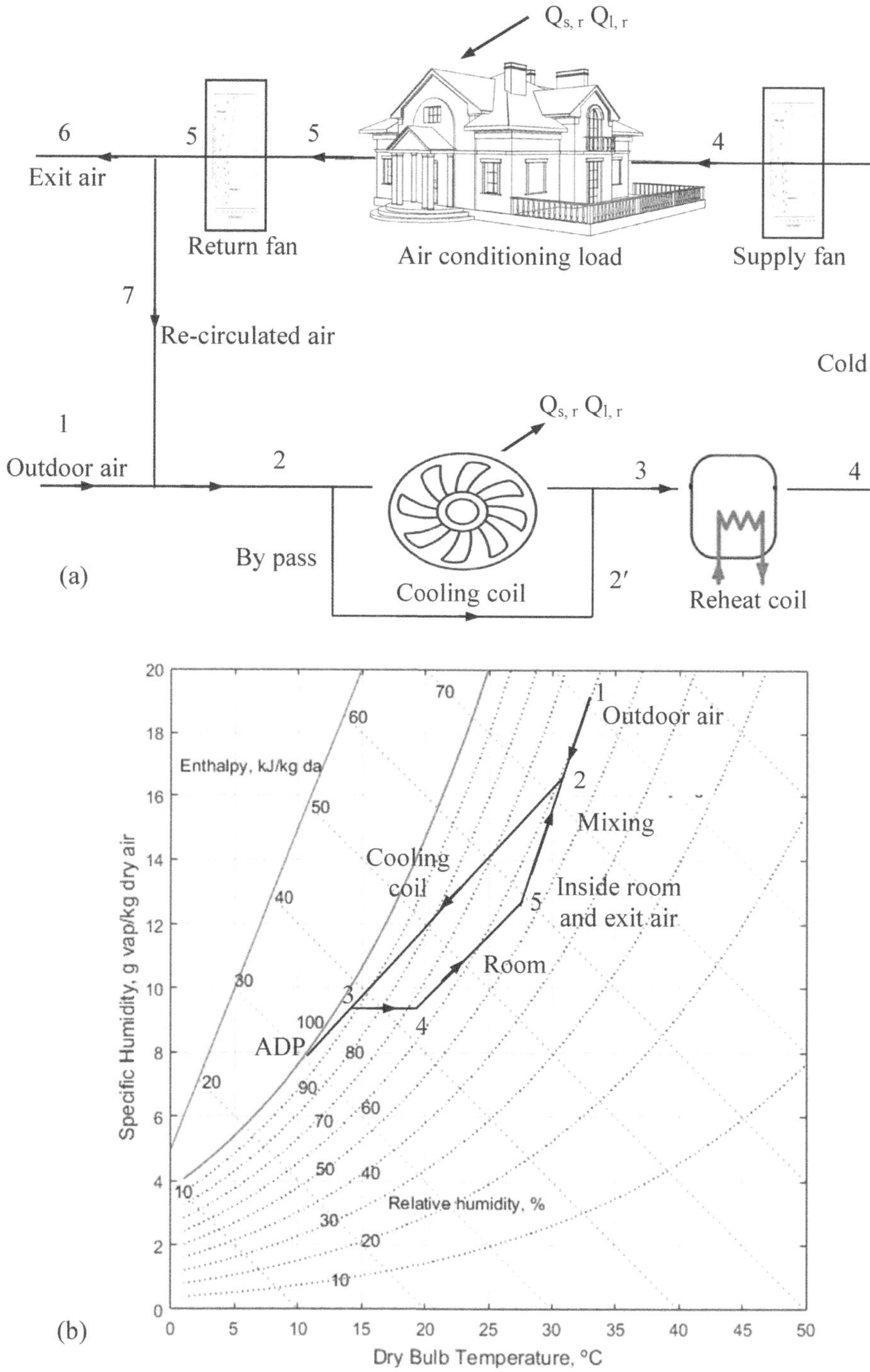

Fig. 9.30. Summer air conditioning system with reheat coil for high latent heat load on (a) processes flow diagram and (b) psychrometric chart.

air after mixing, from the energy balance equations. The process in the reheat coil is sensible heating, as shown by a horizontal line 3-4 on psychrometric chart (Fig. 9.30). If the coil ADP is known, then a coil condition line can be drawn, and the intersection of this line with the horizontal line drawn from supply state '3' gives the condition of the air at the exit of the cooling coil. From this state, the cooling coil load can be determined from the supply mass flow rate and state of air after mixing. The capacity of the reheat coil is then obtained from the energy balance across it.

The following are the advantages and disadvantages of air conditioning with a reheat coil:

1. A higher COP can be obtained with a high evaporator coil due to the use of the reheater.
2. The decrease in temperature rise ($T_5 - T_4$) increases the supplied air.
3. Cooling of air followed by heating of air, at the cooling coil and reheat coil respectively, results in some wastage of energy. By using waste heat recovery for the reheating of air, this problem can be addressed.

9.13 Numerical solutions

1. A vapor compression cycle generates 50 kW of cooling with R134a fluid at a condensing temperature of 35°C and an evaporating temperature of –10°C. Calculate (a) the cooling effect in kJ/kg, (b) the refrigerant, kg/s, (c) compressor power, kW, (d) COP, (e) the volume flow rate measured at the compressor suction, (f) the power per kW of refrigeration and (g) the compressor discharge temperature.

Solution:

Refer Fig. 9.31 for the notations.

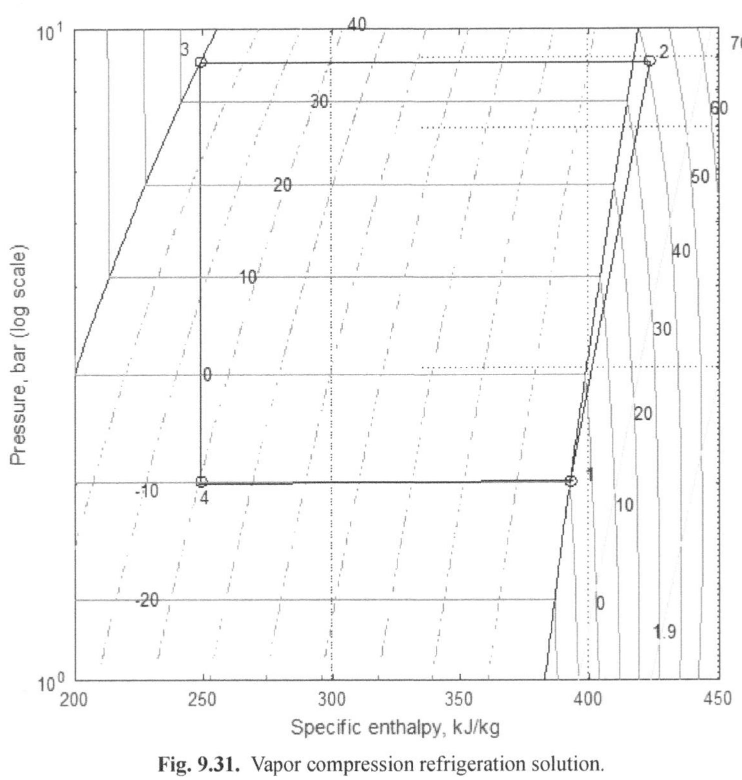

Fig. 9.31. Vapor compression refrigeration solution.

At –10°C, $h_1 = 401.6$ kJ/kg, $v_1 = 0.0651$ m³/kg

$h_3 = h_4 = 243.1$ kJ/kg

The enthalpy at the exit of the compressor can be determined in three ways, i.e., (i) using superheated vapor tables, (ii) using c_p of vapor and (iii) using a p-h chart.

The superheated vapor tables are available for a few refrigerants and not available for phased out refrigerants.

The c_p of vapor at 35°C is 0.9485 kJ/kg K

$h_2 = h_g + c_p (T_2 - T_{sat})$

To find T_2, the isentropic process can be used, i.e., $s_2 = s_1 = 1.7658$ kJ/kg K

$$s_2 = s_g + c_p \ln\left(\frac{T_2}{T_{sat}}\right)$$

$$T_2 = T_{sat} \exp\left(\frac{s_2 - s_g}{c_p}\right)$$

$$T_2 = (35 + 273.15)\exp\left(\frac{1.7658 - 1.7049}{0.9485}\right) - 273.15 = 55.43°C$$

$h_2 = h_g + c_p (T_2 - T_{sat}) = 415.34 + 0.9485(55.43 - 35) = 434.72$ kJ/kg

The actual value is:

$h_2 = 435.2$ kJ/kg at the superheated condition

a) refrigerating effect,

$q_e = h_1 - h_4 = 401.6 - 243.1 = 158.1$ kJ/kg

b) the refrigerant circulation rate,

$$m = \frac{RE}{h_1 - h_4} = \frac{50}{158.1} = 0.315 \ kg/s$$

c) The power required to compressor,

$W = m (h_2 - h_1) = 0.315 (435.2 - 401.6) = 10.6$ kW

d) coefficient of performance,

$$COP = \frac{RE}{W} = \frac{50}{10.6} = 4.72$$

e) The volume rate of flow at the compressor inlet,

$V = mv_1 = 0.315 \times 0.0654 = 0.0206$ m³/s = 20.6 lit/s

f) the compressor power per kW of refrigeration,

$$w = \frac{W}{RE} = \frac{10.6}{50} = 0.212$$

2. For a vapor compression refrigeration (VCR) system, thermodynamic properties at all the states are shown in Fig. 9.32 and Table 9.1. Find the COP, irreversibilities of the components and second law efficiency of the plant.

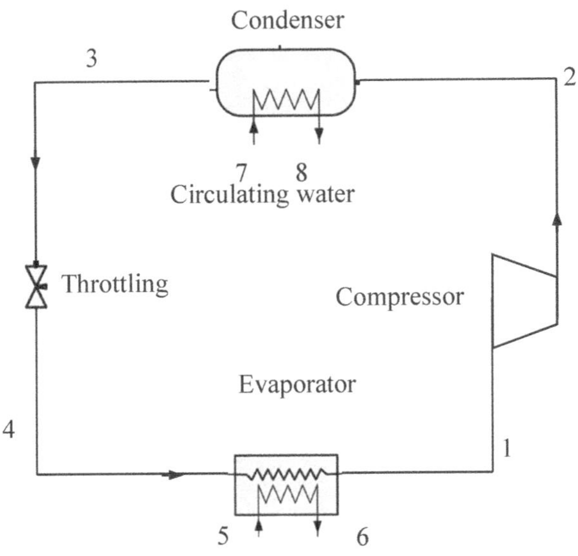

Fig. 9.32. Components arrangement in simple (ideal) vapor compression refrigeration.

Table 9.1. Thermodynamic properties of a vapor compression refrigeration system.

State	P, bar	T, °C	m, kg/s	h, kJ/kg	s, kJ/kg K
1	1.40	–18.58	0.05	387.65	1.74
2	8.00	38.92	0.05	425.77	1.75
3	8.00	31.33	0.05	243.80	1.15
4	1.40	–18.58	0.05	243.80	1.18
5	1.01	20.00	0.31	–20.90	–0.07
6	1.01	12.00	0.31	–54.34	–0.19
7	1.01	26.00	0.22	4.18	0.01
8	1.01	33.00	0.22	33.44	0.11

Solution:

$$COP = \frac{RE}{W} = \frac{h_1 - h_4}{h_2 - h_1} = \frac{387.65 - 243.8}{425.77 - 387.65} = 3.77$$

Irreversibilities:

$$I_{compressor} = T_0 m_1 (s_2 - s_1) = 298.15 \times 0.05 \times (1.75 - 1.74) = 0.15\ kW$$

$$I_{condenser} = T_0 \left(m_2 (s_3 - s_2) + m_{cw} c_w \ln \frac{T_8}{T_7} \right)$$

$$= 298.15 \left(0.05(1.15 - 1.75 + 0.2151 \times 4.18 \ln \frac{33 + 273.15}{26 + 273.15} \right)$$

$$= 0.01\ kW$$

$$I_{throttling} = T_0 m_3 (s_4 - s_3) = 298.15 \times 0.05 \times (1.175 - 1.15) = 0.38\ kW$$

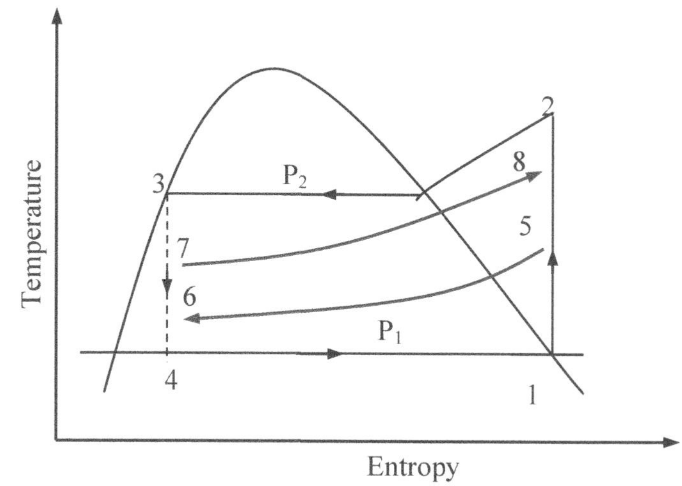

Fig. 9.33. Temperature-entropy diagram of a simple VCR.

$$I_{evaporator} = T_0\left(m_4(s_1 - s_4) + m_{chw}c_w \ln\frac{T_6}{T_5}\right) = 298.15\left(0.05(1.74 - 1.175 + 0.311 \times 4.18 \ln\frac{12 + 273.15}{20 + 273.15}\right) = 1\ kW$$

Total irreversibility losses,

$I_{total} = I_{compressor} + I_{condenser} + I_{throttling} + I_{evaporator}$

$= 0.15 + 0.01 + 0.38 + 1 = 1.55$ kW

$W_{in} = m_1 (h_2 - h_1)$

$= 0.05(425.77 - 387.65) = 1.9$ kW

Second law of performance,

$$COP_2 = \frac{W_{in} - I_{total}}{W_{in}} \times 100 = \frac{1.9 - 1.55}{1.9} \times 100 = 18.85\%$$

Figure 9.34 shows the results on p-h chart.

3. Calculate the compressor power for 1.2 kg/s of ammonia from saturated vapor at 80 kPa to 1000 kPa (a) by single stage compression and (b) by two-stage compression with intercooling by liquid refrigerant at 300 kPa.

Solution:

Without intercooling:

Cycle 1-2-3-6-7

Ammonia properties:

$h_1 = 1410$ kJ/kg

$h_2 = 1588$ kJ/kg

$h_3 = 1800$ kJ/kg

$W_{c1} = 213.6$ kW

$W_{c2} = 254.4$ kW

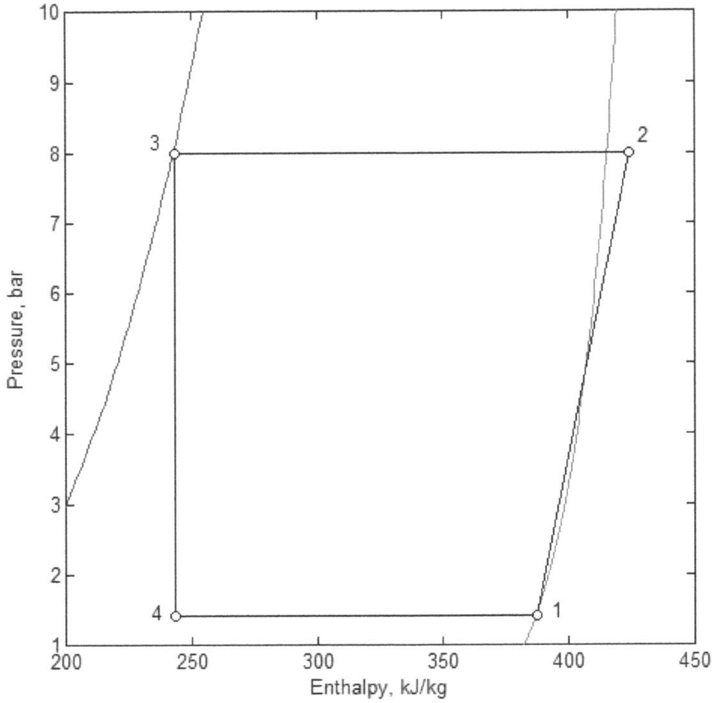

Fig. 9.34. P-h chart for a VCR system.

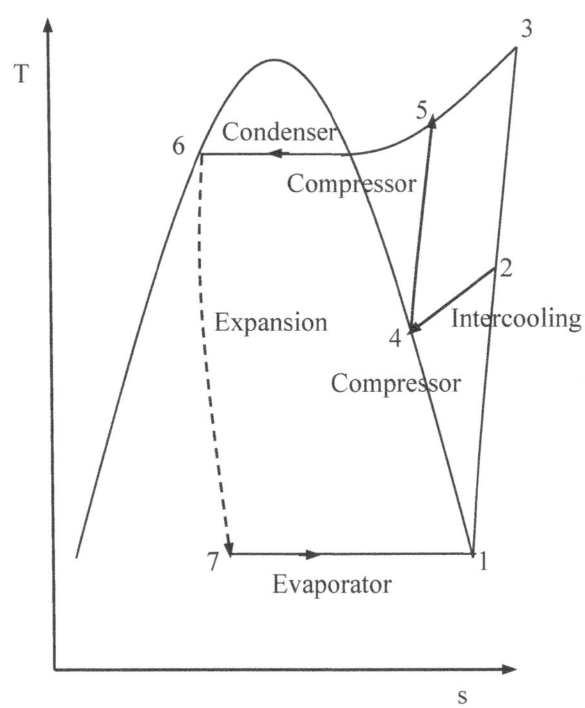

Fig. 9.35. Vapor compression with intercooler.

$W_c = W_{c1} + W_{c2} = 468$ kW

With intercooling:

Cycle 1-2-4-5-6-7

$h_1 = 1410$ kJ/kg

$h_2 = 1588$ kJ/kg

$h_4 = 1450$ kJ/kg

$h_5 = 1628$ kJ/kg

$m_1 = 1.2$ kg/s

$m_4 = 1.346$ kg/s

$W_{c1} = 213.6$ kW

$W_{c2} = 239.6$ kW

$W_c = W_{c1} + W_{c2} = 453.2$ kW

Figure 9.36 shows the results on p-h chart.

The high stage compressor in the intercooled system must compress 1.2 kg/s plus the flow rate of refrigerant that evaporates to de-superheat the gas at 2. The flow rate of ammonia compressed in the high stage can be calculated by making a heat and a mass balance about the intercooler.

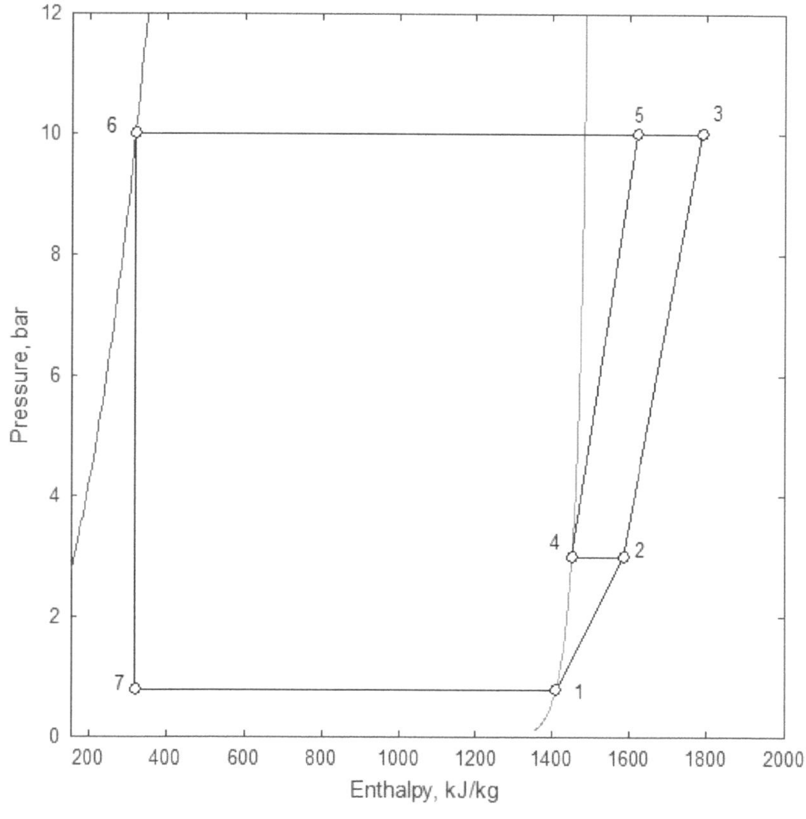

Fig. 9.36. Solution for multi-stage compression with intercooler.

4. Develop mass and energy formulation for multi-stage compression (or solar hybridized VCR with a flat plate collector and compressor connected in series) as shown in Fig. 3.35. Find the COP and specific cooling of the system and compare these results with a system without a flash chamber. The low pressure and high pressure are 1.4 bar and 8 bar respectively. Take the intermediate pressure as a geometric mean of two pressures. Consider the refrigerant flow in the evaporator as 1 kg/s.

Solution (Fig. 9.37 and Fig. 9.38):

The intermediate pressure or flash chamber's pressure is the geometric mean of the evaporator pressure and condenser pressure, i.e.,

$$P_2 = \sqrt{P_1 P_4} = \sqrt{1.4 \times 8} = 3.35 \ bar$$

From the tables and P-h graph, the specific enthalpy of the working fluid can be taken as

$h_1 = 387.41$ kJ/kg

$h_2 = 405.06$ kJ/kg

$h_5 = 243.40$ kJ/kg

$h_6 = 243.40$ kJ/kg

$h_7 = 204.12$ kJ/kg

$h_8 = 204.12$ kJ/kg

$h_9 = 400.55$ kJ/kg

$$x_6 = \frac{h_6 - h_7}{h_9 - h_7}$$

$$= \frac{243.40 - 204.12}{400.55 - 204.12}$$

$$= 0.2$$

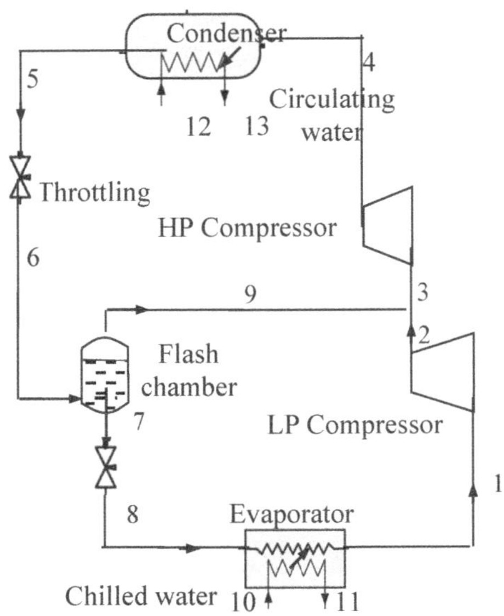

Fig. 9.37. Multi-stage compression with a flash chamber and multi expansion valves.

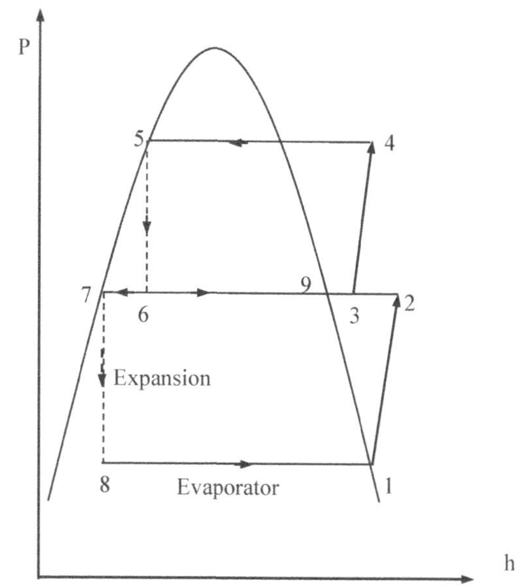

Fig. 9.38. Multi-stage compression with a flash chamber and multi expansion valves on property chart (a) pressure-enthalpy and (b) temperature-entropy diagrams (first layout).

$m_3 = m_2 + m_9$

$m_9 = x_6\, m_6 = x_6\, m_3$

$m_3 = m_2 + x_6\, m_3$

$$m_3 = \frac{m_2}{1 - x_6}$$

$$= \frac{1}{1 - 0.2}$$

$= 1.25$ kg/s

$m_6 = m_3 = 1.25$ kg/s

$m_7 = (1 - x_6)\, m_6$
$= (1 - 0.2)\; 1.25$
$= 1$ kg/s (matched with the input and verified)

$m_9 = x_6\, m_6$

$= 0.2 \times 1.25 = 0.25$ kg/s

$$h_3 = \frac{m_2 h_2 + m_9 h_9}{m_3}$$

$$= \frac{1 \times 405.06 + 0.25 \times 400.55}{1.25}$$

$= 404.06$ kJ/kg

Then from isentropic compression,

$h_4 = 423.24$ kJ/kg

Refrigerating effect,

$$RE = m_8 (h_1 - h_8)$$
$$= 1 \times (387.41 - 204.12)$$
$$= 183.29 \text{ kW}$$

Work input to a multi-stage compressor,

$$W = m_1 (h_2 - h_1) + m_3 (h_4 - h_3)$$
$$= 1 \times (405.06 - 387.41) + 1.25 \times (423.25 - 404.06)$$
$$= 41.63 \text{ kW}$$

$$COP = \frac{RE}{W} = \frac{183.29}{41.63} = 4.4$$

Figure 9.39 shows the results on p-h chart.

5. A multi load refrigeration system has three separate evaporators at temperatures –10°C, 0°C and 10°C respectively. While from the condenser the refrigerant R12 liquid is flowing to the different evaporators at a temperature of 40°C, the liquid from the condenser can be assumed to be saturated. The evaporator loads at –10°C and 10°C are 10 TR each and the load on 0°C is 20 TR. After each of the three evaporators, separate pressure reducing valves bring the pressure to the suction pressure of the compressor. A single compressor and individual expansion valves are used in the system. What is the power required for operating the system and COP?

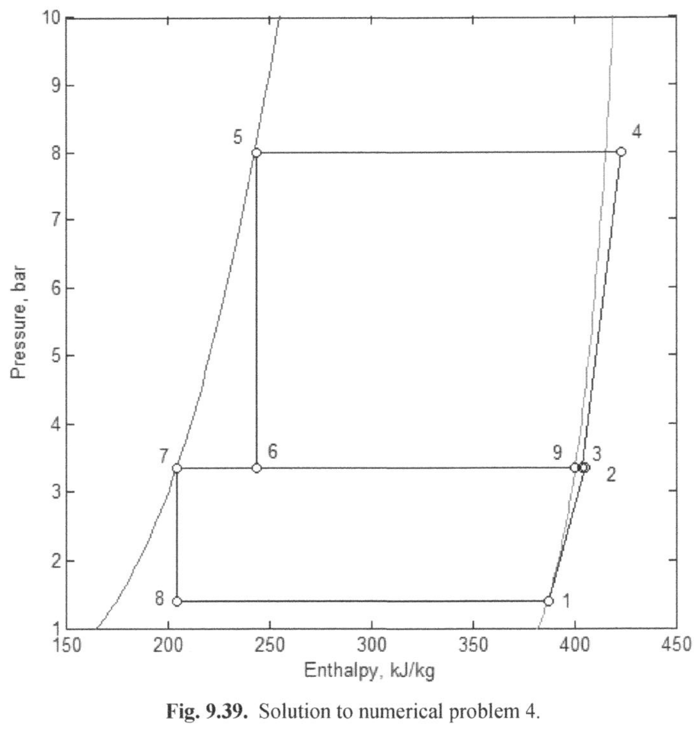

Fig. 9.39. Solution to numerical problem 4.

Solution (Fig. 9.40 and Fig. 9.41)

From the tables and P-h graph, the specific enthalpy of the working fluid can be taken as

$h_3 = h_4 = h_7 = h_{10} = 74.6$ kJ/kg

$h_5 = h_6 = 191.7$ kJ/kg

$h_8 = h_9 = 187.5$ kJ/kg

$h_{11} = 183.2$ kJ/kg

Mass flow rate in the –10°C evaporator,

$$m_{10} = \frac{10 \times 3.51}{h_{11} - h_{10}} = \frac{10 \times 3.51}{183.2 - 74.6} = 0.358 \ \ kg/s$$

Mass flow rate in the 0°C evaporator,

$$m_7 = \frac{20 \times 3.51}{h_8 - h_7} = \frac{20 \times 3.51}{187.5 - 74.6} = 0.689 \ \ kg/s$$

Mass flow rate in the 10°C evaporator,

$$m_4 = \frac{10 \times 3.51}{h_5 - h_4} = \frac{10 \times 3.51}{191.7 - 74.6} = 0.33 \ \ kg/s$$

Mixing state 1 can be obtained by solving the following energy balance equation,

$$h_1 = \frac{m_{11}h_{11} + m_9 h_9 + m_6 h_6}{m_1} = \frac{0.358 \times 183.2 + 0.689 \times 187.5 + 0.33 \times 191.7}{0.358 + 0.689 + 0.33} = 187.39 \ \ kJ/kg$$

Locate state 1 on P-h chart from h_1 and P_1 then from isentropic compression locate state 2.

$h_2 = 217$ kJ/kg

Power required for operating the system,

$W = m_1 (h_2 - h_1) = (0.358 + 0.689 + 0.33) (217 - 187.39) = 40.75$ kW

$$COP = \frac{RE}{W} = \frac{(10 + 20 + 10) \times 3.51}{40.75} = 3.82$$

Figure 9.42 shows the results on p-h chart.

6. A VCR with R12, has one compressor, one condenser and three evaporators of 10 TR, 20 TR and 30 TR capacities are maintained at 10°C, 5°C and –10°C respectively. The system comprises multiple-expansion valves and back pressure check valves. The condenser pressure is 0.609 bar (surrounding temperature = 40°C). The vapors leaving the evaporators are dry saturated. The liquid refrigerant is sub-cooled by 10°C in the condenser sub-cooler. Assume isentropic compression. Determine (a) refrigerating effect/kg in each evaporator, (b) mass flow rate in each evaporator, (c) compressor power and (d) COP of the system.

Solution (Fig. 9.43 and Fig. 9.44):

From the tables and P-h graph, the specific enthalpy of the working fluid can be taken as

$h_3 = h_4 = 64.6$ kJ/kg

$h_7 = h_8 = 45.4$ kJ/kg

$h_{11} = h_{12} = 40.7$ kJ/kg

$h_5 = 191.7$ kJ/kg

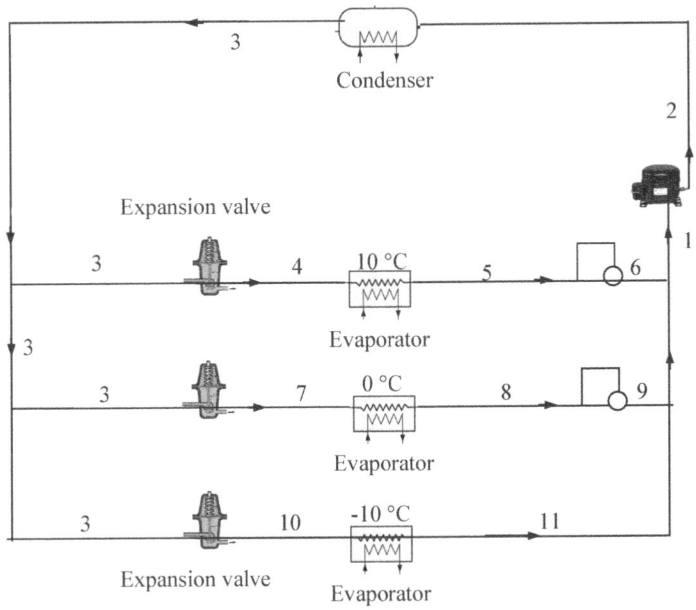

Fig. 9.40. Multi evaporators at different temperatures with back pressure valves.

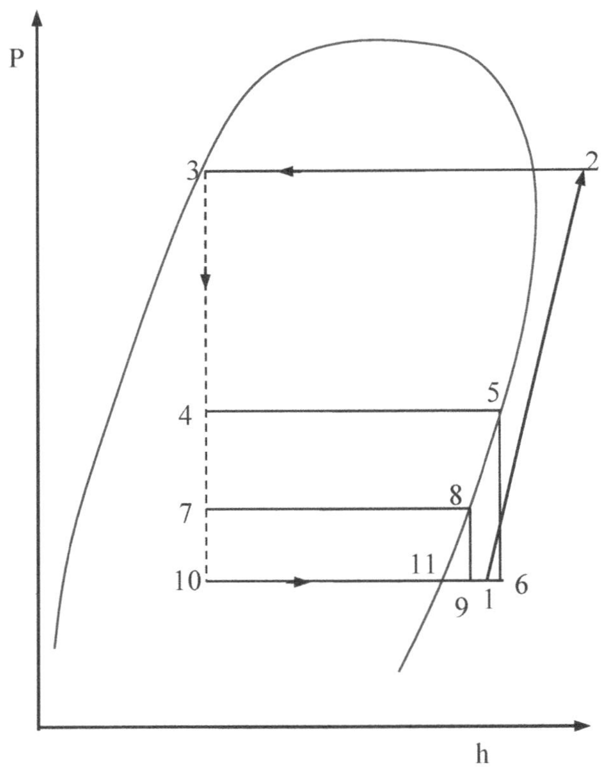

Fig. 9.41. Multi evaporators at different temperatures with back pressure valves on property chart (a) pressure-enthalpy and (b) temperature-entropy diagrams.

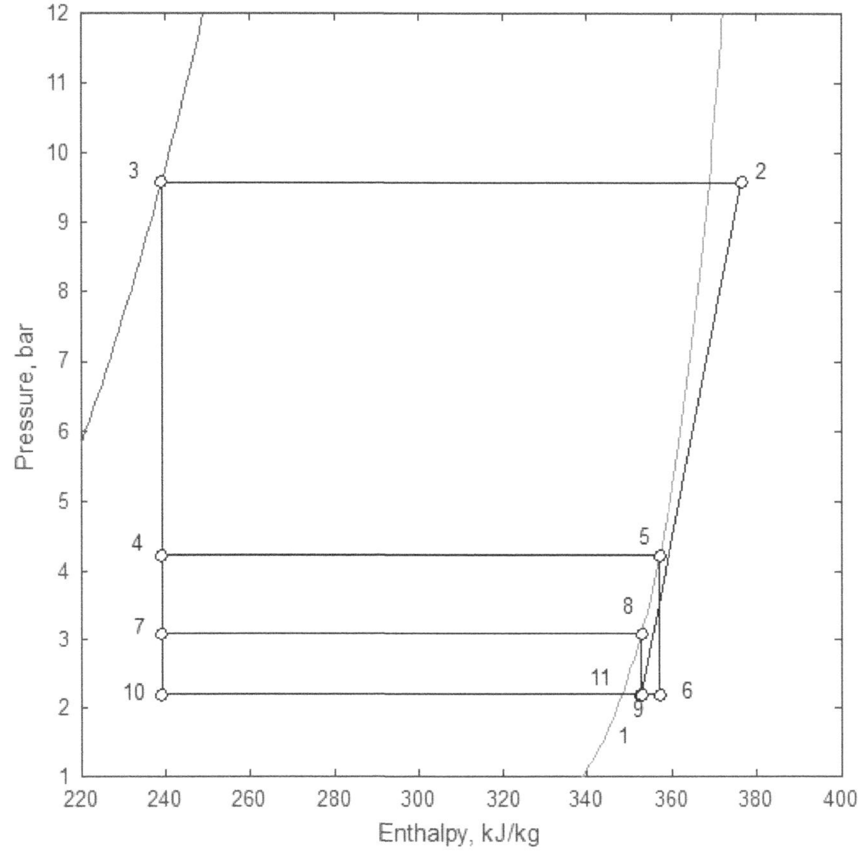

Fig. 9.42. Solution to the problem of a multi evaporator refrigeration system at different evaporator temperatures.

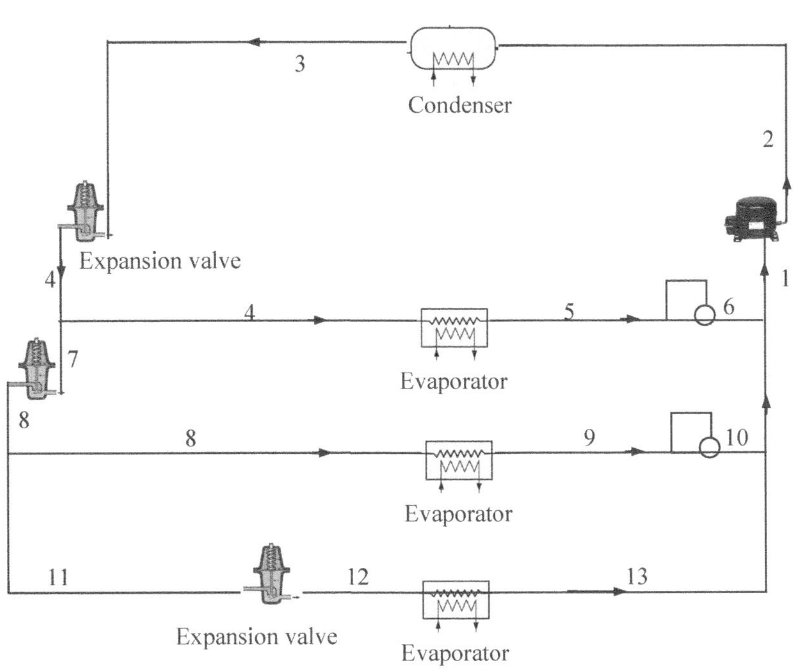

Fig. 9.43. Multi evaporators at different temperatures with back pressure valves.

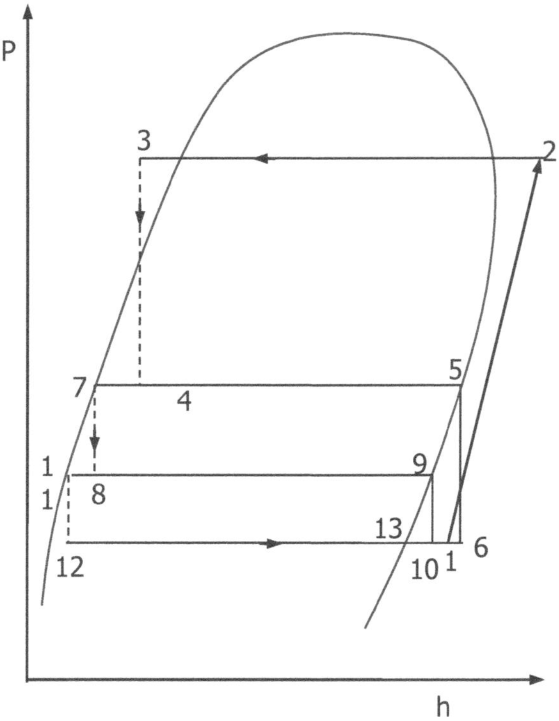

Fig. 9.44. Multi evaporators at different temperatures with back pressure valves on property chart (a) pressure-enthalpy and (b) temperature-entropy diagrams.

$h_9 = 189.7$ kJ/kg

$h_{13} = 183.2$ kJ/kg

RE in evaporator 1

$RE_1 = h_{13} - h_{12} = 183.2 - 40.7 = 142.5$ kJ/kg

Similarly,

$RE_2 = h_9 - h_8 = 189.7 - 45.7 = 142.5$ kJ/kg

$RE_3 = h_5 - h_4 = 191.7 - 64.6 = 127.1$ kJ/kg

$$x_6 = \frac{h_6 - h_7}{h_9 - h_7}$$

7. Model a two stage cascade refrigeration system with pressure limits of 8 bar and 1.4 bar. Each stage operates on a simple vapor compression refrigeration cycle with R134a fluid. The topping VCR absorbs heat from the bottoming VCR through an adiabatic counter flow heat exchanger where both streams enter at about 3.2 bar. In practice, the working fluid of the lower cycle is at a higher pressure and temperature in the heat exchanger for effective heat transfer. If the mass flow rate of the refrigerant through the upper cycle is 0.05 kg/s, determine (i) the mass flow rate of the refrigerant through the lower cycle, (ii) the rate of heat removal from the refrigerated space and the power input to the compressor, and (iii) the coefficient of performance of this cascade refrigerator.

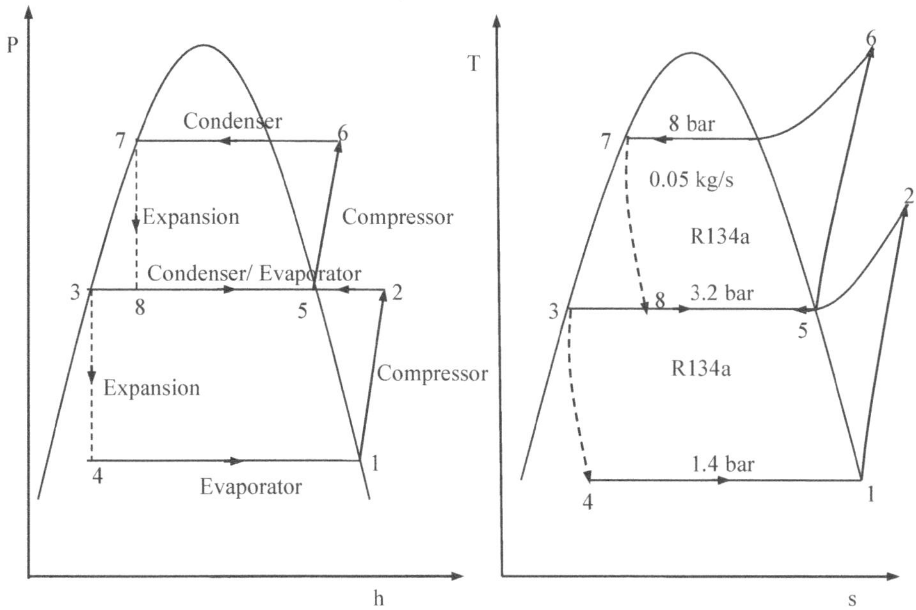

Fig. 9.45. Representation of the problem on (a) pressure-enthalpy and (b) temperature-entropy charts.

Solution (Fig. 9.45):

From the R134a tables and P-h chart the enthalpies of the working fluid are furnished as follows:

$h_1 = 239.16$ kJ/kg

$h_2 = 255.93$ kJ/kg

$h_3 = 55.16$ kJ/kg

$h_4 = 55.16$ kJ/kg

$h_5 = 251.88$ kJ/kg

$h_6 = 270.92$ kJ/kg

$h_7 = 95.47$ kJ/kg

$h_8 = 95.47$ kJ/kg

The energy balance in a heat exchanger, i.e., an evaporator condenser assembly gives the refrigeration generation in bottoming VCR,

$m_{TVCR}(h_5 - h_8) = m_{BVCR}(h_2 - h_3)$

$$m_{BVCR} = \frac{m_{TVCR}(h_5 - h_8)}{(h_2 - h_3)} = \frac{0.05(251.88 - 95.47)}{255.93 - 55.16} = 0.039 \ kg/s$$

The refrigerating effect,

$RE = m_{BVCR}(h_1 - h_4)$

$= 0.036(239.16 - 55.16)$
$= 7.18$ kW

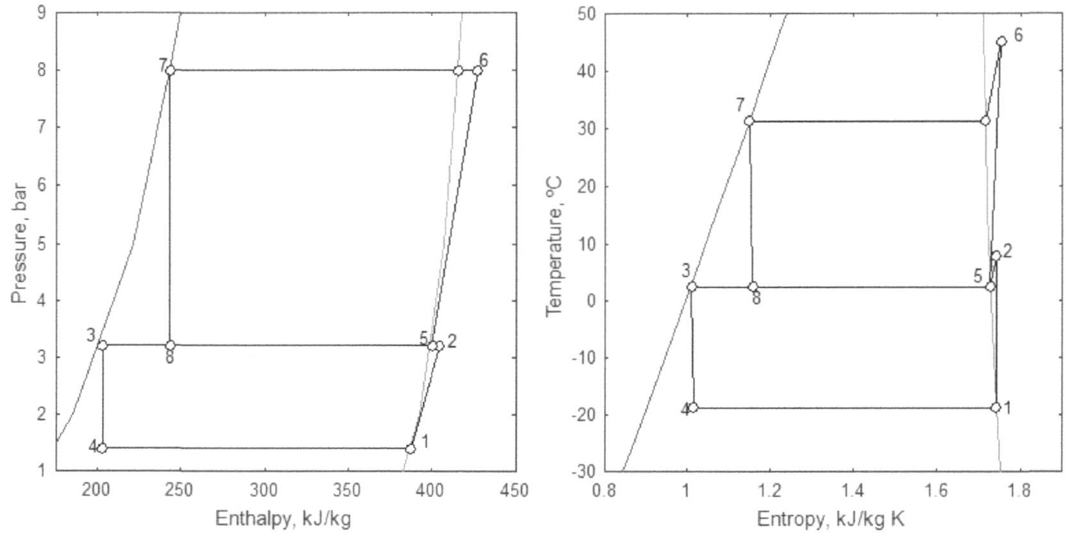

Fig. 9.46. Solution to the cascade refrigeration problem.

The work input to the two compressors,

$$W = m_{TVCR}(h_6 - h_5) + m_{BVCR}(h_2 - h_1)$$
$$= 0.05(270.92 - 251.88) + 0.036(255.93 - 239.16)$$
$$= 1.61 \text{ kW}$$

$$COP = \frac{RE}{W} = \frac{7.18}{1.61} = 4.46$$

Figure 9.46 shows the results on p-h and T-s diagrams.

8. An air conditioning system maintains 24°C dry bulb and 50% RH inside, while the outside conditions are 34°C dry bulb and 40% RH. The room sensible and latent heat gains are 60 kW and 12 kW respectively. As shown in the figure, the outside fresh air first flows over a first cooler coil and is reduced to state 2 of 10°C dry bulb and a relative humidity of 85%. It is then mixed with re circulated air, the mixture (state 3) being handled by a fan, passed over a second cooler coil and sensibly cooled to 12°C dry bulb (stage 4). The air is then delivered to the room. If the outside fresh air is used for dealing with the whole of the room's latent heat gain and if the effects of fan power and duct heat gains are ignored, find (i) mass flow rates of outside fresh air and supplied air (ii) DBT and enthalpy of the air handled by the fan (state 3) and (iii) the required cooling capacity of the first cooler coil and second sensible cooler coil.

Solution (Fig. 9.47):

The outdoor air condition:

$T_1 = 34°C$, $RH_1 = 40\%$

From psychrometric chart,

$\omega_1 = 0.01335$ kg/kg da and $h_1 = 68.21$ kJ/kg da

The room inside and its exit condition:

$T_5 = 24°C$, $RH_5 = 50\%$

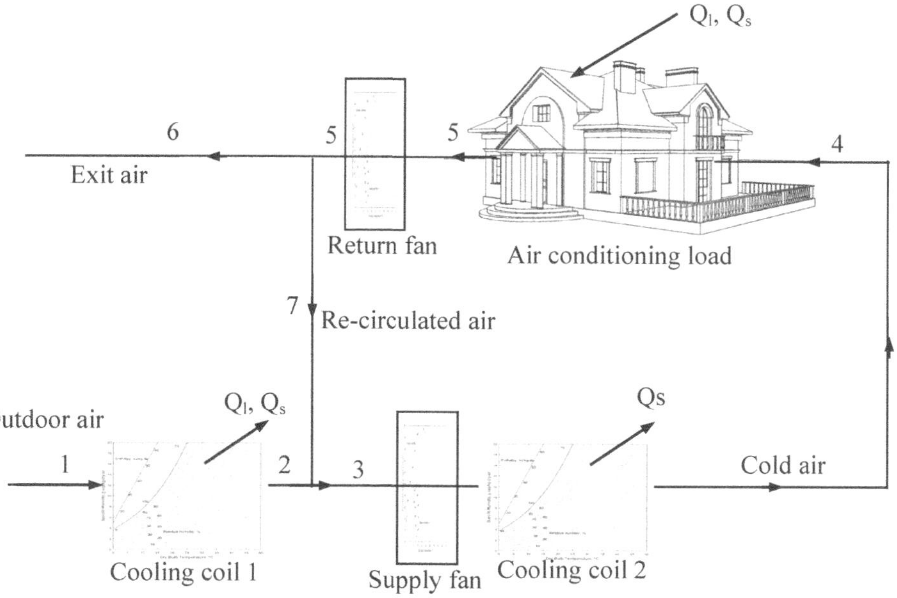

Fig. 9.47. Summer air conditioning system with outdoor air for ventilation.

From psychrometric chart,

$\omega_5 = 0.0093$ kg/kg da and $h_5 = 47.66$ kJ/kg da

$h_7 = h_6 = h_5 = 47.55$ kJ/kg da

For re-circulated air, $\omega_7 = \omega_6 - \omega_5 = 0.0093$ kg/kg da

After the first cooler, the air properties are:

$T_2 = 10°C$, $RH_2 = 85\%$

From psychrometric chart,

$\omega_2 = 0.00647$ kg/kg da and $h_2 = 26.31$ kJ/kg da

Since the air is supplied to the room at 12°C, the mass flow rate of supplied air m_4 is as follows:

$$m_4 = \frac{Q_{s,\ r}}{c_{pm}(T_5 - T_4)} = \frac{60}{1.0216(24-12)} = 4.894\ \ kg/s$$

As per the notation shown, $m_3 = m_4 = m_5 = 4.894$ kg/s (dry air)

The moisture content of supplied air is obtained from the latent heat balance of the room as follows:

$$\omega_4 = \omega_5 - \frac{Q_{l,\ r}}{m_a h_{fg}} = 0.0093 - \frac{12}{4.894 \times 2501} = 0.0083\ \ kg/kg\ \ da$$

Since the fresh air takes care of the entire latent load, the heat transfer across coil 2 is only sensible heat transfer. Therefore,

$\omega_3 = \omega_4 = 0.0083$ kg/kg da

At $T_4 = 12°C$ and $\omega_4 = 0.0083$ kg/kg da,

From psychrometric chart,

$RH_4 = 95.32\%$ and $h_4 = 33.03$ kJ/kg da

Applying mass balance across the mixing of re-circulated and fresh air,

$m_2\,\omega_2 + m_7\,\omega_7 = m_3\,\omega_3$

$m_2\,\omega_2 + (m_3 - m_2)\,\omega_7 = m_3\,\omega_3$

$m_2\,\omega_2 + m_3\,\omega_7 - m_2\,\omega_7 = m_3\,\omega_3$

$m_2\,(\omega_2 - \omega_7) = m_3\,(\omega_3 - \omega_7)$

$$m_2 = \frac{m_3(\omega_3 - \omega_7)}{(\omega_2 - \omega_7)} = \frac{4.894(0.0083 - 0.0093)}{(0.00647 - -0.0093)} = 1.73\ kg/s$$

$m_1 = m_2 = 1.73$ kg/s

The mass flow rate of re-circulated air:

$m_7 = m_5 - m_2 = 4.894 - 1.73 = 3.164$ kg/s

From the energy balance across the mixing process 2-7-3,

$$T_3 = \frac{m_2T_2 + m_7T_7}{m_3} = \frac{1.74 \times 10 + 3.164 \times 24}{4.894} = 19.05\ °C$$

Similarly,

$$h_3 = \frac{m_2h_2 + m_7h_7}{m_3} = \frac{1.74 \times 26.31 + 3.164 \times 47.55}{4.894} = 40.11\ kJ/kg$$

The cooling capacity of cooling coil 1,

$Q_{c,1} = m_1(h_1 - h_2) = 1.73\ (68.21 - 26.31) = 72.49$ kW

The cooling capacity of cooling coil 2,

$Q_{c,2} = m_3\,c_{pm}\,(T_3 - T_4) = 4.894 \times 1.0216 \times (19.05 - 12) = 35.25$ kW

Figure 9.48 shows the results on psychrometric chart.

Observation

It can be seen that the combined cooling capacity (72.49 + 35.25 = 107.74 kW) is larger than the total cooling load on the building (60 + 12 = 72 kW). The difference between these two quantities (107.74 – 72 = 35.74 kW) is equal to the cooling capacity required to reduce the enthalpy of the fresh air from outdoor conditions to the required indoor conditions. This is the cost incurred for providing fresh air to the conditioned space. Larger the fresh air requirement, larger the required cooling capacity.

9. For an air conditioned room, the sensible and latent heat loads are 10^5 kJ/h and 40,000 kJ/h, respectively. The makeup air is 40% of the total air supplied to the room. Ambient condition is $T_{db} = 40°C$ and $T_{wb} = 30°C$ and inside room condition $T_{db} = 27°C$ and $Ø = 60\%$. The conditioned air leaves the cooling coil at 18°C. The return air is mixed before the cooling coil. Determine (a) RSHF line (b) the state of the air entering the room (c) The amount of total air and make up air and (d) ADP and bypass factor.

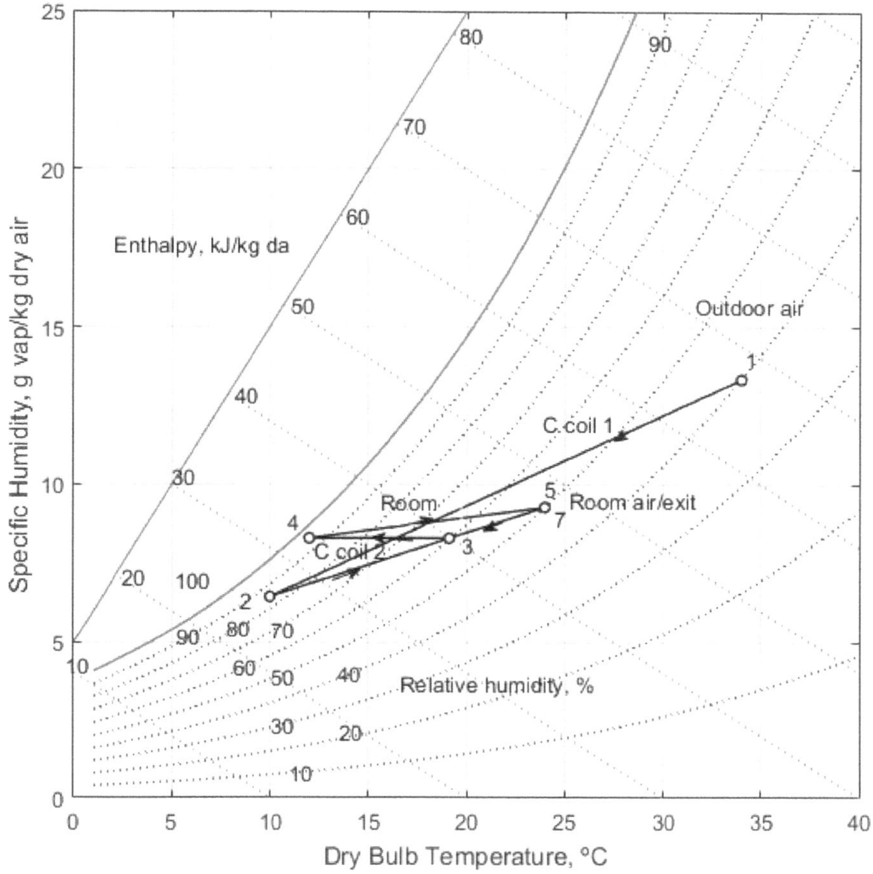

Fig. 9.48. Solution to the air conditioning problem.

Solution (Fig. 9.49):

RSH = 10^5 kJ/h

RLH =40000 kJ/h

$$RSHF = \frac{RSH}{RSH + RLH} = \frac{10^5}{10^5 + 40000} = 0.714$$

The outdoor (make up) air condition,

Share = 40%

T_{db} = 40°C and T_{wb} = 30°C

i.e., T_1 = 40°C

From the psychrometric chart,

RH_1 = 49.54%. ω_1 = 0.0233 kg/kg da and h_1 = 100.12 kJ/kg da

The room inside and its exit conditions,

T_4 = 27°C, RH_4 = 60%

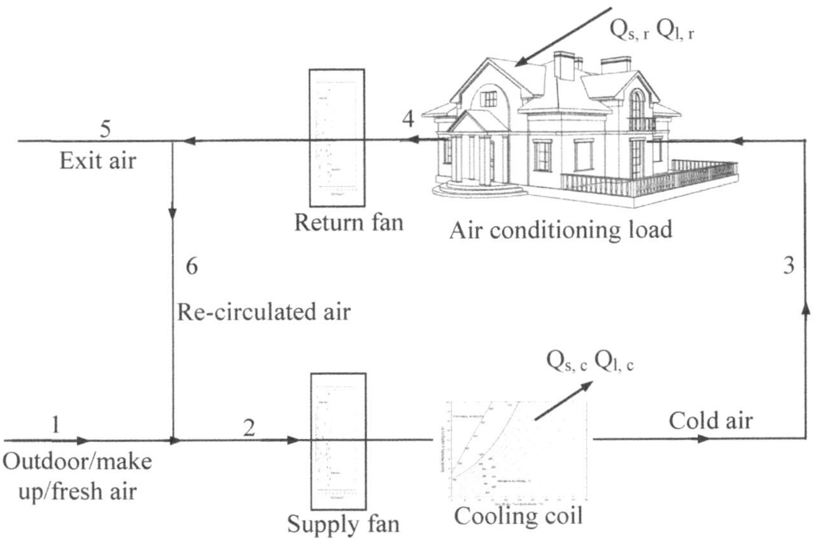

Fig. 9.49. Summer air conditioning system with makeup air and re-circulated air.

From psychrometric chart,

$\omega_4 = 0.0134$ kg/kg da and $h_4 = 61.34$ kJ/kg da

$h_5 = h_6 = h_4 = 61.34$ kJ/kg da

For re-circulated air, $\omega_6 = \omega_5 = \omega_4 = 0.0134$ kg/kg da

$T_3 = 18°C$

Since the air flow rate in the air conditioning system is:

$$m_a = m_2 = \frac{Q_{s,\ r}}{c_{pm}(T_4 - T_3)} = \frac{10^5}{1.0216(27-18)\times 3600} = 3.02\ \ kg/s$$

As per the notation, shown, $m_2 = m_3 = m_4 = 3.02$ kg/s (dry air)

$m_1 = 0.4\ m_2 = 0.4 \times 3.02 = 1.21$ kg/s

$m_6 = 0.6\ m_2 = 0.6 \times 3.02 = 1.81$ kg/s

$m_5 = m_1 = 1.21$ kg/s

The moisture content of supplied air is obtained from latent heat balance of the room as follows:

$$\omega_3 = \omega_4 - \frac{Q_{l,\ r}}{m_a h_{fg}} = 0.0134 - \frac{40000}{3.02\times 2501\times 3600} = 0.0119\ \ kg/kg\ \ da$$

At $T_3 = 18°C$ and $\omega_3 = 0.0119$ kg/kg da,

From psychrometric chart,

$RH_3 = 92.52\%$, and $h_3 = 48.35$ kJ/kg da

Applying mass balance across the mixing of re-circulated and fresh air,

$m_1\ \omega_1 + m_6\ \omega_6 = m_2\ \omega_2$

$$\varpi_2 = \frac{m_1\omega_1 + m_6\omega_6}{m_2} = \frac{1.21\times 0.0233 - 1.81\times 0.0134)}{3.02} = 0.0174\ \ kg/kg\ \ da$$

Similarly,

$$T_2 = \frac{m_1T_1 + m_6T_6}{m_2} = \frac{1.21\times 40 + 1.81\times 27}{3.02} = 32.2\ °C$$

$$h_2 = \frac{m_1h_1 + m_6h_6}{m_2} = \frac{1.21\times 100.12 + 1.81\times 61.34}{3.02} = 76.86\ kJ/kg$$

The cooling capacity of the cooling coil,

$Q_c = m_2(h_2 - h_3) = 3.02\ (76.86 - 48.35) = 86.13$ kW

The cooling coil apparatus dew point can be obtained by extending the cooling coil line on psychrometric chart up to the saturation line.

The resultant ADP_c is 15.2°C and

Bypass factor,

$$BF_C = \frac{\omega_3 - \omega_{ADP}}{\omega_2 - \omega_{ADP}} = \frac{0.0119 - 0.0107}{0.0174 - 0.0107} = 0.18$$

Figure 9.50 shows the results on psychrometric chart.

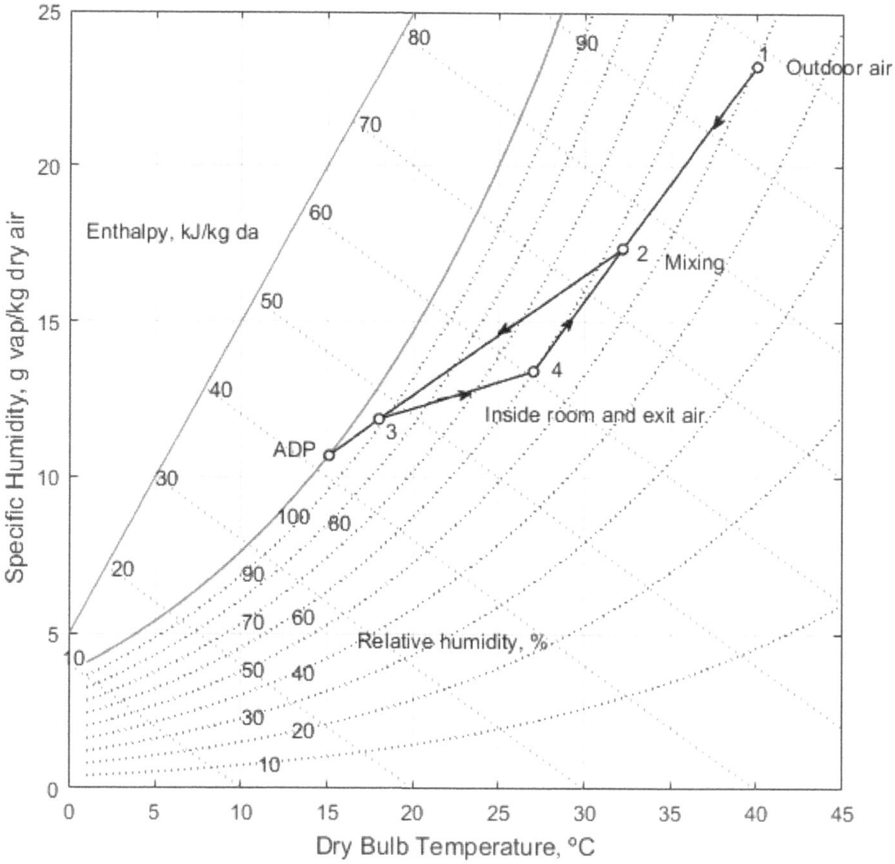

Fig. 9.50. Solution of the air conditioning problem.

Chapter 10
Solar Geometry

10.1 Introduction

The sun is the nearest star to the earth, and we are depending on the sun as a source of life. The sun occupies 99.86% of the total mass of the solar system. We cannot live or imagine the universe without the sun, as everything is dependent on the sun. Since the sun's diameter (139 × 10^4 km) is larger than the earth's diameter (1.27 × 10^4 km) and the mean distance between these two is large (14960 × 10^4 km), and the angle of vision of the sun from the earth is 0° 32¢. Therefore, it is reasonable to assume that the earth is receiving parallel rays of the sun's rays. The sun is the origin of all renewable energy sources, either directly or indirectly. For example, the temperature difference of the sun's heat at different parts of the earth causes a thermal gradient. This thermal gradient results in the density being different in air and creates wind. As shown in Fig. 10.1, solar energy sources can be grouped into direct sources and indirect sources. Applications from the direct sources of solar energy can be categorized into solar thermal technology and solar photovoltaic (PV) technology. Technologies from indirect sources are biomass, wind, etc.

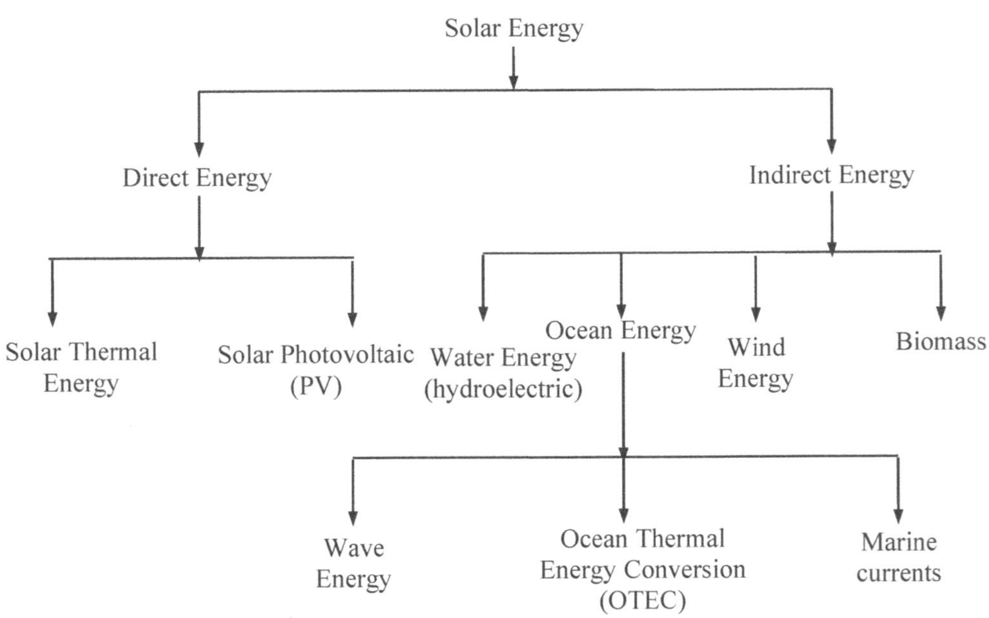

Fig. 10.1. Classification of solar energy.

DOI: 10.1201/9781032642505-10

10.2 Sun earth relation

The sun-earth relation is the basis for the design of solar tracking devices and solar radiation. To solve problems in solar geometry, first we should know its nomenclature and terminology. The geometrical terms discussed in this unit are latitude, longitude, declination angle, angle of incidence, hour angle, day length, azimuth angle, zenith angle, altitude angle, tilt angle, and air mass. The surface of the solar collector is to be faced towards the sun for effective utilization. But the position of solar rays changes daily, seasonally, and from place to place. From the top, the earth rotates in an anticlockwise direction. The place or location can be stated by geographical parameters, i.e., latitude and longitude. The earth revolves around the sun, and it will take 365.2422 days for a complete revolution. But the earth is located eccentrically in the orbit of revolution. Therefore, the distance between the sun and earth varies during the revolution of earth around the sun. This distance varies by 3%, and it will not affect the seasons much. The tilt of the earth's polar axis at 23.5° causes the seasons (Sukhatme and Nayak, 2017).

10.2.1 Angle of incidence

Angle of incidence (θ) is the angle between the normal to the tilted surface and the sun beams. It is a special focus in solar geometry, as the final objective of the study of solar geometry is the minimization of the angle of angle of incidence for the designed solar collector's surface to maximize solar collection. For horizontal surface geometry, the normal to the horizontal is called the zenith, and its angle of angle of incidence is called the zenith angle. The zenith angle is the angle between the sun's rays and the vertical to the ground. If the surface is facing the sun exactly, the rays fall on the surface normally and give the highest intensity for the heating and maximum electricity generation from a solar PV panel. To ensure this position, the surface has to continuously face the sun. If the normal of the collector's surface is passing through the sun's rays, the angle of angle of incidence is zero, which is the most wanted position. To maintain the angle of incidence at zero, solar geometry is to be studied for the effective control of the collector, also known as the tracking process.

After the derivation and simplification of solar geometry, the generalized expression to determine the angle of angle of incidence as a function of δ, ω, Ø, β and γ as follows

$$\theta = \cos^{-1}\begin{pmatrix} \sin\delta\sin\phi\cos\beta - \sin\delta\cos\phi\sin\beta\cos\gamma + \cos\delta\cos\phi\cos\beta\cos\omega \\ +\cos\delta\sin\phi\sin\beta\cos\gamma\cos\omega + \cos\delta\sin\beta\sin\gamma\sin\omega \end{pmatrix} \quad (10.1)$$

or

$$\theta = \cos^{-1}\begin{pmatrix} \sin\phi(\sin\delta\cos\beta + \cos\delta\sin\beta\cos\gamma\cos\omega) + \cos\phi(\cos\delta\cos\beta\cos\omega - \sin\delta\sin\beta\cos\gamma) \\ +\cos\delta\sin\beta\sin\gamma\sin\omega \end{pmatrix} \quad (10.2)$$

or

$$\cos\theta = \cos\theta_z \cos\beta + \sin\theta_z \sin\beta \cos(\gamma_s - \gamma) \quad (10.3)$$

For a tilted surface towards south,

$$\cos\theta = \sin\delta \sin(\varphi - \beta) + \cos\delta \cos(\varphi - \beta) \cos\omega$$

In the above equation, β and γ are derived angles and are basic angles. If the sun is behind the surface, the angle of incidence crosses 90°. Therefore, it is necessary to ensure that the earth is not blocking the sun. Therefore, the hour angle should be between sunrise and sunset. The generalized expression for the angle of incidence derived from solar geometry is a function of five variables.

The five basic variables that influence the angle of incidence are:

1. Declination, δ
2. Hour angle, ω
3. Latitude, $\emptyset$
4. Tilt angle, β
5. Surface azimuth angle, γ

Therefore, $\theta = f(\delta, \omega, \emptyset, \beta \text{ and } \gamma)$

The following sections describe these five angles and other related geometry to evaluate the angle of incidence. The declination and hour angle can be determined from the date and time of the interest. Latitude is the location or place of the solar collector. The values of the tilt angle and surface azimuth angle are obtained from the position of the collector.

10.2.2 Declination

Declination (δ) is the angle between the line joining the center of the sun and the center of the earth to the equatorial plane. It is the angle of the sun beam with the equatorial plane. The equatorial plane tilts due to the tilt of the earth's polar axis. The tilt angle of the equatorial plane to the horizontal is 23.5°. Now the sun's rays are parallel to the line joining the sun and earth. The sun's rays make an angle to the equatorial plane, which is variable as the earth revolves around the sun as shown in Fig. 10.2. If the equatorial plane projected area comes towards the sun, the declination angle becomes maximum or minimum. In the case where the equatorial plane's projected line is in front of the sun, the sun's rays pass through the equatorial plane and make a zero declination.

The declination changes from –23.5° to 23.5° and is shown in Fig. 10.3. The tilt of the earth's polar axis generates seasonal variations. Declination is the latitude of a subsolar point. The subsolar point is the place where sun's rays fall directly overhead. The subsolar point is also called the zero zenith position. At the Tropic Cancer and Tropic of Capricorn, at the declination of 23.5°, the zero zenith position will occur. Therefore, the zero zenith or subsolar point varies between the Tropic of Cancer and the Tropic of Capricorn. When the North Pole is tilted towards the sun, the opposite end, the South Pole, moves away from the sun. At this time, the northern hemisphere is exposed to summer and the southern hemisphere experiences winter, and vice versa.

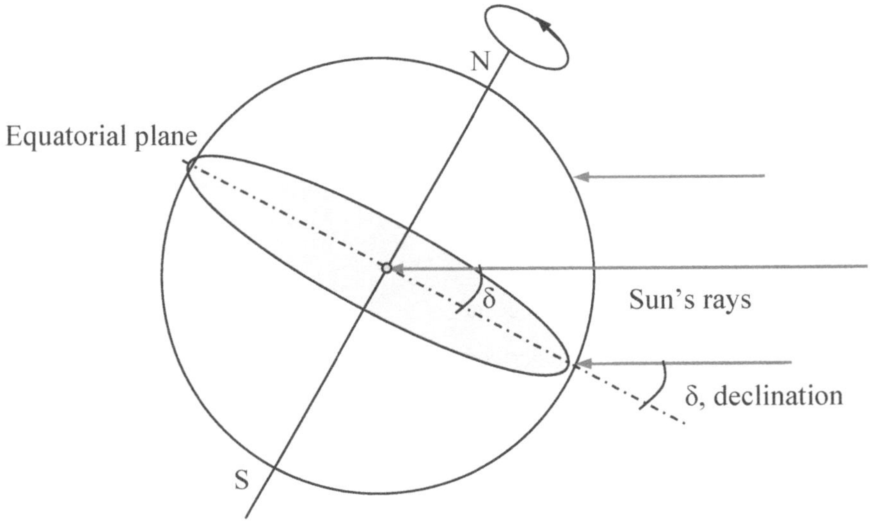

Fig. 10.2. Declination angle.

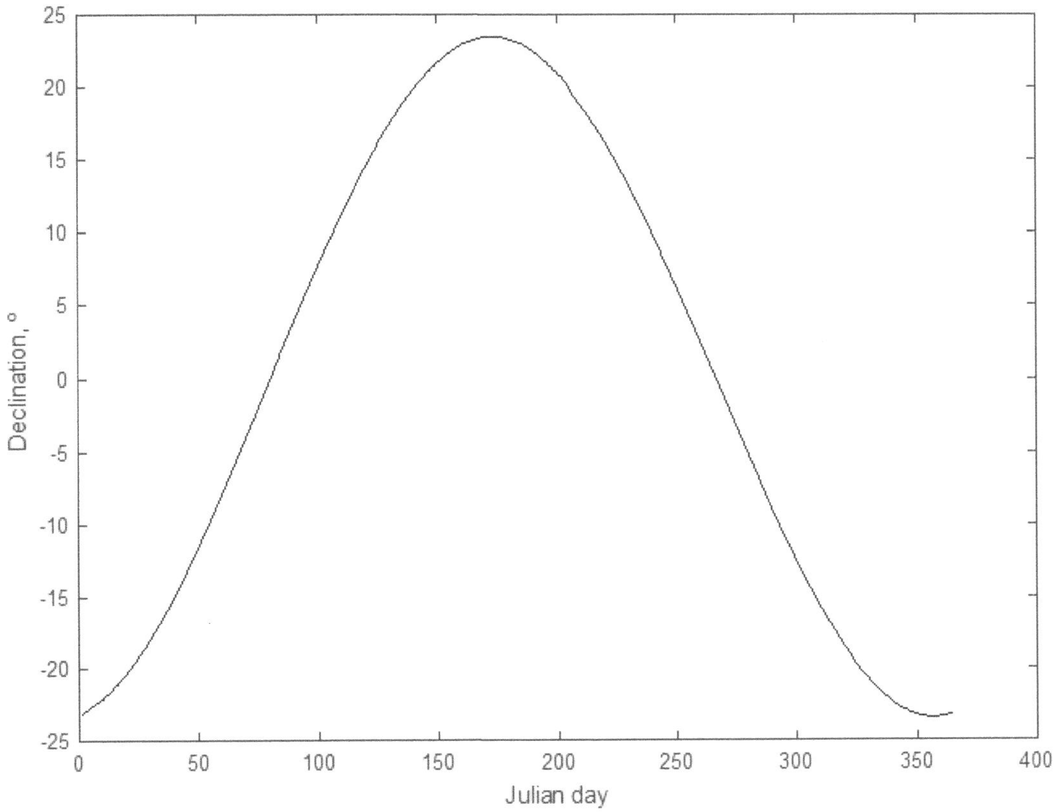

Fig. 10.3. Change in the declination angle with Julian day during a year.

Figure 10.4 shows the seasonal variation in declination angle on a daily basis. The polar axis of the earth makes an angle of 23.5° to the vertical. The resulting seasons are winter, spring, summer, and autumn. The critical declination occurs during the two equinoxes and two solstices. The northern hemisphere tilts away from the sun during the winter. On December 21, the declination is –23.5°. The earth rotates around the sun in an anticlockwise direction from the top. After the winter, the next season is spring, on March 21st, when the equinox results in a zero declination angle. At the equator, equal days and nights happen on equinox days. Yearly two equinox days appear. After the spring, summer comes, and on June 21st, the northern hemisphere moves close to the sun. At the same time, the southern hemisphere tilts away from the sun. It indicates that at the same time, if the northern sphere is in summer, the southern hemisphere is in winter. That means Christmas will be celebrated in winter in the northern hemisphere and in summer in the southern hemisphere. The declination angle on June 21st is 23.5°C. The second equinox comes on September 21st. Again, during this date, equal days and nights occur at the equator. On the autumn equinox, the declination is 0°.

Therefore, the minimum and maximum declination occur on the winter solstice and the summer solstice, respectively. During the equinox, the sun comes directly overhead, i.e., two times per year. For a location on the northern hemisphere, the sun's path is away from the equator during the winter and located south of the equator but between –23.5° and 23.5° latitude. On the other hand, the path lies on the northern side of the equator from the northern hemisphere. Thus the sun's path is on the south side for 6 months and the sun lies on the north side for the remaining 6 months. On only two days per year - during the equinox- the sun comes over head. Since the northern hemisphere moves closer to the sun during the summer, the sun looks bigger in the sky. The northern hemisphere tilts away from the sun, and a small image of the sun can be viewed on winter days in the sky. The sun

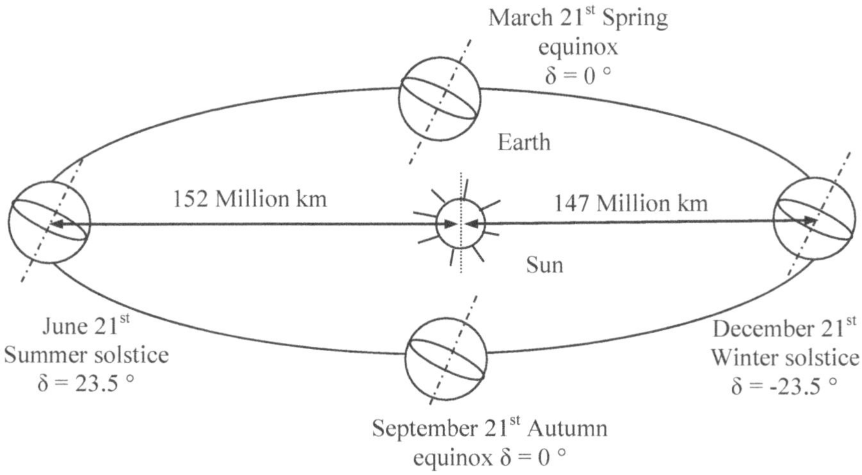

Fig. 10.4. Seasonal variation in declination angle.

is located eccentrically in the earth's orbit. Therefore, the distance traveled by the location in the northern hemisphere is 50000 km extra compared to winter travel. Therefore, the summer is longer than the winter. Similarly, the winter is long in the southern hemisphere.

Actually, the declination changes every day and every time. For simplicity, in solar geometry, the declination changes during a day are neglected. The daily variations in declinations are considered. It is assumed to be constant on a day. The declination can be determined as a function of Julian day. Julian day, n, is determined from January 1st (n = 1) to December 31st (n = 365 for non-leap years).

$$\delta = 23.45 \sin\left[\frac{360(284+n)}{365}\right] \quad ° \tag{10.4}$$

In the above equation, the declination is constant during the given Julian day, n. But actually, due to the continuous self-rotation of the earth and revolving around the sun, the declination changes from the beginning of the day till the end of the day, i.e., for 24 hours. Figure 10.5 shows the variation in declination from the midnight to midnight, i.e., 12.00 A.M. to 12.00 A.M. The declination varies linearly during the day, and its range is small in a day. The day is scaled from 0 to 1 with the steps of 0.1-day length. Midday is expressed as 0.5 days. The daily variation of declination is approximately 0.3° and may be neglected over the period of a day. It has been observed that the maximum rate of change is about 0.4° per day.

10.2.3 Hour Angle

The hour angle (ω) is designed based on the earth's self-rotation, i.e., 360° in 24 hours, i.e., 15° per hour. 15° per hour is also equal to 4 minutes per ° rotation. The reference for the hour angle is the solar noon. The hour angle is the angle between the sun's rays and the local meridian. Longitude lines are drawn vertically on the world map. The longitude, or meridian, is the angular distance between the standard meridian and local meridian. It can be determined from the given local time. Latitude lines are the horizontal lines on the map. A detailed discussion about latitude and longitude is given in the subsequent sections. Declination is determined as an angle of sun's rays with the equatorial plane. Hour angle is determined as an angle of sun's ray with the local meridian plane and shown in Fig. 10.6. The declination can be seen on the vertical plane, and the hour angle is visible on the horizontal plane. The daily variations of declination are not considered in the calculations. But the hour angle takes care of the daily variations from sunrise to sunset. The sun's rays fall east during the forenoon and west in the afternoon. As per solar geometry, the hour angle is negative

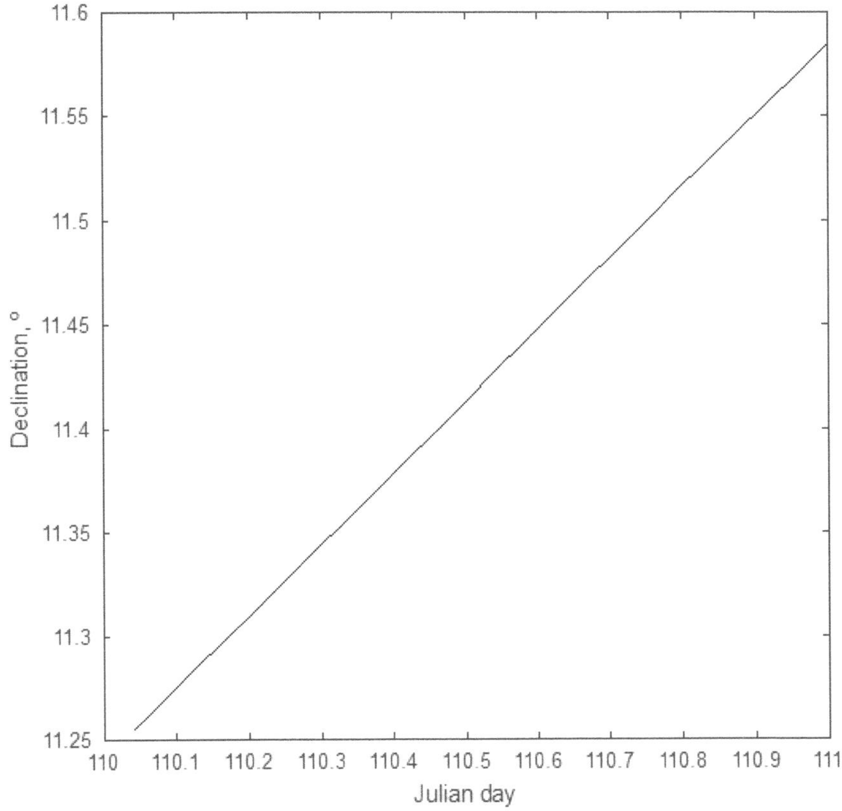

Fig. 10.5. Change in the declination from beginning of the day 12.00 A.M. to end of the day 12.00 A.M. for 24 hours on April 20 (Julian day = 110).

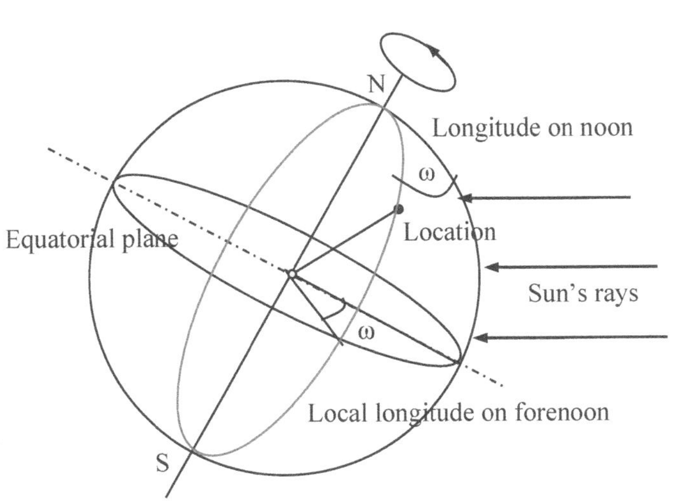

Fig. 10.6. Hour angle measured as an angle between local longitude and longitude ahead of the sun.

during the forenoon on the east and positive in the afternoon on the west side. The earth rotates 15° per hour about the polar axis. Therefore, the hour angle changes 15° per hour. This relation can also be used to convert an hour angle into time or vice versa. It also indicates how far the sun's rays are from the solar noon to local longitude (Fig. 10.7). At noon, the sun's rays fall very close to the local longitude, and the hour angle becomes zero.

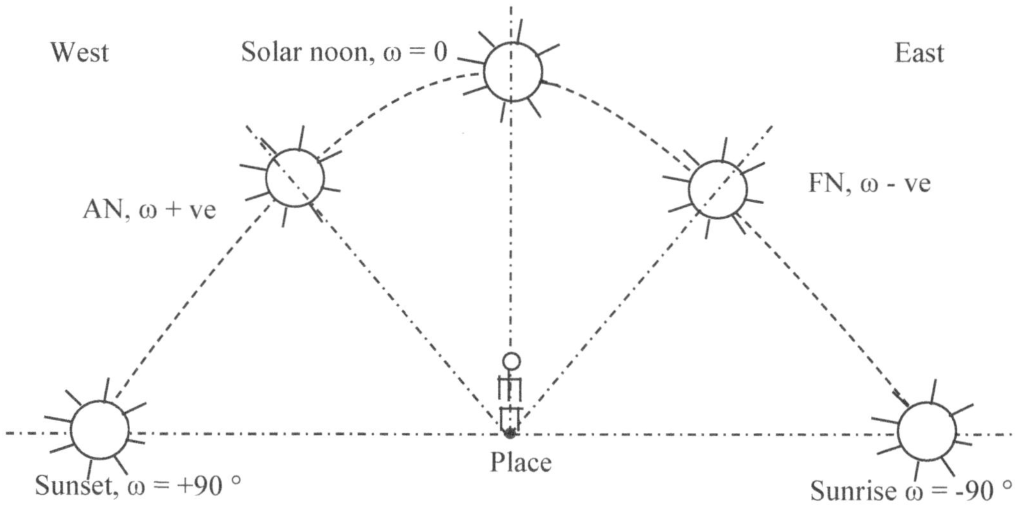

Fig. 10.7. Changes in the hour angle from sunrise to sunset during equinox.

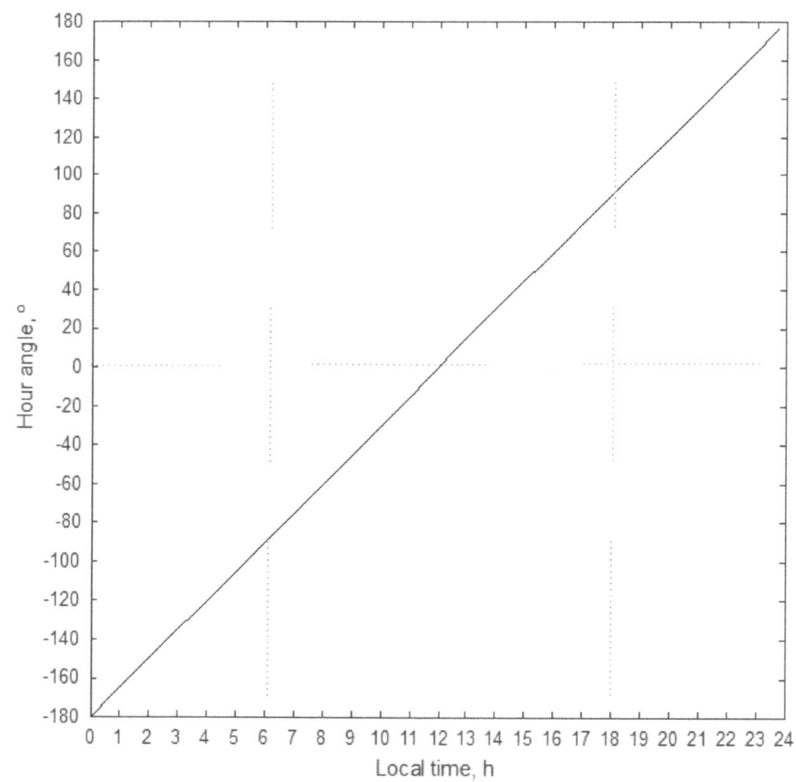

Fig. 10.8. Hour angle changes for 24 hours during the day and night.

Figure 10.8 shows the hour angle changes from the beginning of the day at 12.00 A.M. to the next midnight, i.e., for 24 hours. At midnight, the hour angle is 180°. At the equinox, the hour angle at sunrise is –90°, and it reaches 0° during the solar noon. After reaching sunset, the hour angle is 90°. The variation of hour angle with time is linear, and it is the function of time and location.

The hour angle is the function of time as follows,

$$\omega = (15^\circ / hr)\ \ hours\ \ from\ solarnoon = \frac{\text{min}\ \ from\ solarnoon}{4\,\text{min}/^\circ}$$

It also can be calculated from the following relation

$$\omega = 15^\circ\ (\text{solar time} - 12) \tag{10.5}$$

In the above equation, the time difference is determined in hours, and it is converted into degrees by multiplying with 15, as 15° is equal to one hour. The solar time should be substituted on a 24-hour scale. The sign convention of negative and positive is used in the forenoon and afternoon, respectively. The equation also results in the sign convention along with the magnitude.

10.2.4 Latitude (Ø)

Latitude (Ø) is the angular distance of a place from the equator located either on the northern hemisphere or the southern hemisphere. These are represented by horizontal circles or cross lines on the earth, as shown in Fig. 10.9. The latitude can be used to find the tilt angle for the solar collector. The solar collector is either focused or non-focused. The non-focused collectors, such as solar water heaters and solar PV panels, are generally oriented at a fixed angle without incorporating the tracking link. The non-tracking surfaces are tilted towards south (in case the location is on the northern hemisphere) with a fixed tilt angle. Theoretically, this tilt angle is equal to the latitude of the location. For the equator, the latitude is 0°. The north pole and south pole have latitudes of 90° and –90°, respectively. The Tropic of Cancer and the Tropic of Capricorn have latitudes of 23.5 N and 23.5 S, respectively. These are defined based on the polar axis at 23.5° to the vertical. Latitude information can be obtained through Google map at all places.

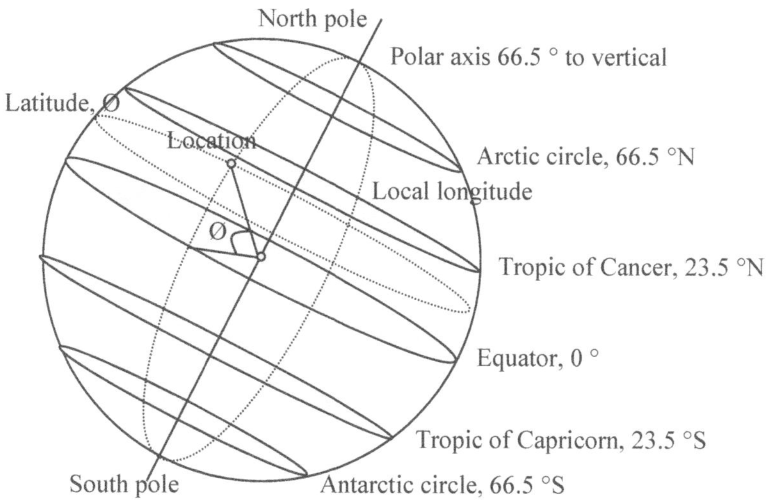

Fig. 10.9. Latitude on earth with principal meridians.

10.2.5 Tilt angle (β)

Tilt angle (β) is the angle of the collector's surface with the horizontal plane. It is also called the slope of the surface. It is one of the main tracking parameters in the zenith-axis tracking method. The zenith angle is measured on the vertical plane, as the angle is measured from the horizontal. It can be controlled by tilting the surface through the horizontal axis. In the non-tracking collectors, the tilt angle is fixed and approximately equal to the local latitude. In the zenith axis control, the tilt angle

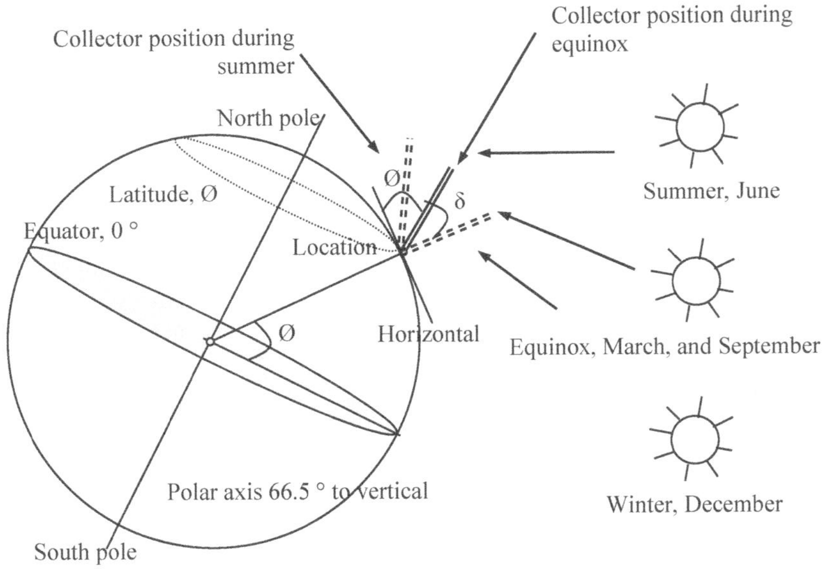

Fig. 10.10. Tilt angle of a solar collector.

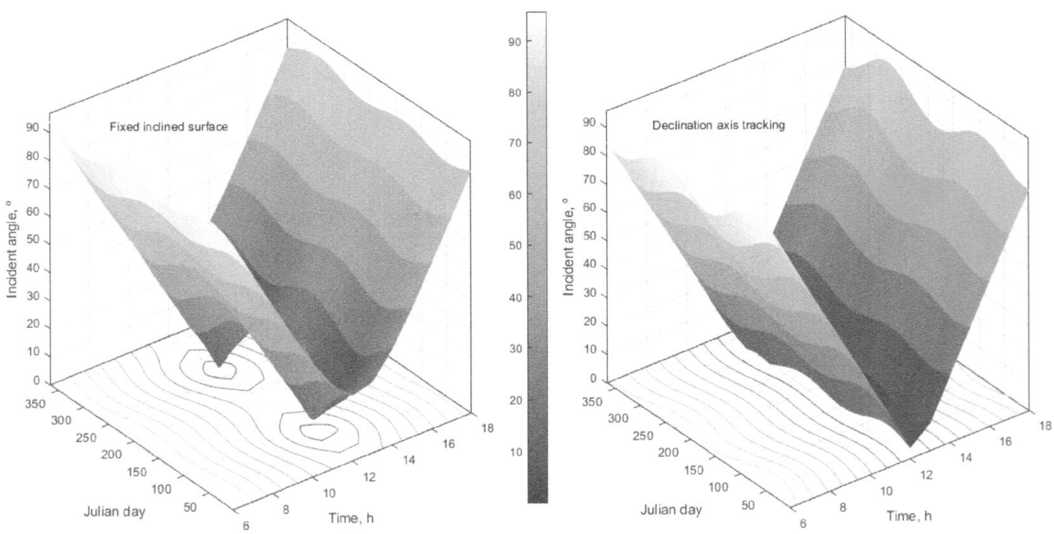

Fig. 10.11. Comparison of the angle of incidence through the tilt angle controlled by declination.

is variable and depends on tracking type. Figure 10.10 shows the nature of tilt angle at a location on a northern hemisphere. The collector needs to tilt towards the south to gain the minimum angle of incidence for places in the northern hemisphere. The surface is tilted in the opposite direction, i.e., towards the north; the sun beams pass parallel to the surface, and we cannot see the normal direction of solar rays. As per the geometry shown in Fig. 10.10, the latitude is equal to the tilt angle during the equinoxes. On the days of the equinox, the declination has no effect and the tilt is equal to the latitude. But in the summer, the sun is located nearer the north pole, and the northern part of the earth comes close to the sun. Since most of the time the sun is overhead, less tilt angle is required. Geometry says that the effective tilt angle is equal to the difference of local latitude and declination.

$$\beta = \emptyset - \delta \tag{10.6}$$

In summer, the declination is 23.5°; therefore, the tilt angle should be less than this from the local latitude. In winter, the declination is –23.5°. The effective tilt angle is equal to Ø + 23.5. On an average basis, the seasonal adjustment of the solar collector, which is a non-focusing type, can be maintained as follows:

$$\beta = Ø \pm 13° \tag{10.7}$$

Figure 10.11 compares the improvement in the angle of incidence with the tilted surface facing south, and the tilt angle is controlled by the declination, i.e., $\beta = Ø - \delta$ over the fixed surface. Instead of a fixed tilt angle, if the tilt angle is controlled with the daily declination, the angle of incidence is decreased in all the seasons from sunrise to sunset. The results are plotted for New Delhi, India (Ø = 28° 38¢ N and Longitude = 77° 12¢ E).

10.2.6 Surface azimuth angle

Surface azimuth angle (γ) is the angle of the project of normal to the surface on ground measured from south. Similar to the tilt angle, the surface azimuth angle is a controlling parameter in tracking systems. It is used in the azimuth angle tracking system. Most of the fixed tilted solar collectors, which are non-focusing, are tilted toward the south as they are located in the northern hemisphere. For this condition, the surface azimuth angle is zero. It is negative in the southeast and positive in the south-west direction as per the solar geometry. It is defined as the angle made by the horizontal projection of normal to the surface with the local longitude. The azimuth angle is measured on a horizontal plane. The azimuth angle is controlled by rotating the vertical axis. The combination of zenith angle control by horizontal axis and azimuth angle control by vertical axis rotation results in a fine tilt of the solar collector, which is required for the point-focusing collectors. Practically, the surface azimuth angle can be measured by taking the angle between the shadow of the normal line on the ground and the line joining between north and south directions. In green house design and collector's simulation, shadow analysis is required to predict the shadow falling on the buildings through the simulation of surface azimuth angle. Now some special software modules are available to perform this task.

10.2.7 Solar azimuth angle

The solar azimuth angle (γs) is the angle between the horizontal projection of sun's rays and the south line, as shown in Fig. 10.12. The definition of the solar azimuth angle is similar to the surface azimuth angle. In the solar azimuth angle, instead of normal to the surface, sun's rays are considered. The surface azimuth angle is controllable, and so is an important controlling parameter in the geometry. But the solar azimuth angle is not controllable and depends on date, time, and place. The sign convention followed in surface azimuth angle is applied in solar azimuth angle. If the projection of azimuth is on a north-south line or longitude, it is zero azimuth. For south-east projection, it is negative and south-west direction positive.

The solar azimuth angle can be determined from the zenith angle, latitude and declination, i.e., from the date, time and place.

$$\gamma_s = \cos^{-1}\left(\frac{\cos\theta_Z \sin\phi - \sin\delta}{\sin\theta_Z \cos\phi}\right) \tag{10.8}$$

The solar azimuth can angle also be derived from geometry as

$$\gamma_s = \sin^{-1}\left(\frac{\cos\delta \sin\omega}{\sin\theta_Z}\right) \tag{10.9}$$

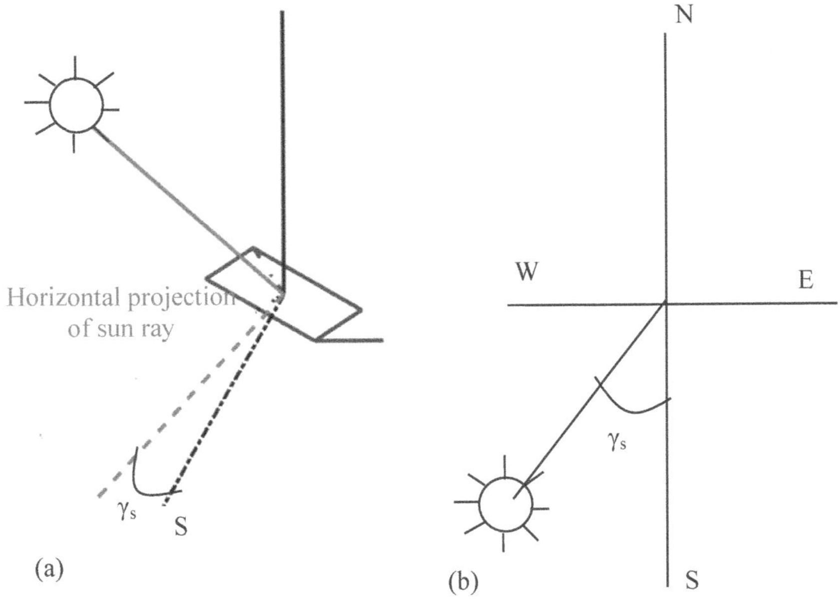

Fig. 10.12. (a) Solar azimuth angle on (a) three dimensional view and (b) plan.

For example, at the solar noon the solar azimuth angle is zero, i.e., $\gamma_s = 0$

$$\cos\gamma_s = \cos 0 = 1 = \frac{\cos\theta_Z \sin\phi - \sin\delta}{\sin\theta_Z \cos\phi}$$

$$\sin\theta_z \cos\varnothing = \cos\theta_z \sin\varnothing - \sin\delta$$

$$\sin\delta = \cos\theta_z \sin\varnothing - \sin\theta_z \cos\varnothing = \sin\varnothing \cos\theta_z - \cos\varnothing \sin\theta_z$$

$$\sin\delta = \sin(\varnothing - \theta_z)$$

$$\delta = \varnothing - \theta_z$$

$$\theta_z = \varnothing - \delta$$

It shows that at solar noon, the zenith angle is equal to the difference between latitude and declination. It also indicates the tilt angle to get the minimum angle of incidence. If the surface is tilted to this level, the normal to the surface approaches the sunbeam. The above equation gives the magnitude of the solar azimuth angle without the directions, i.e., sign. Based on the forenoon or afternoon period, we need to add the sign convention to the resulting value of the solar azimuth angle. The projection of the sun's rays falls on the ground and can be observed as a shadow of objects. Similar to the surface azimuth angle, the solar azimuth angle can also be measured from the shadow of the vertical object and the north-south line or local longitude. The duration of the sun beam projection on the ground is six months on the northern hemisphere, and another six months it will appear in the southern hemisphere.

Figure 10.13 shows the changes in solar azimuth angle and surface azimuth angle from sunrise to sunset set on a typical winter solstice in India. To differentiate these two curves, the solar azimuth angle is measured from the north, whereas the surface azimuth is from the south. As per the earlier discussion, we cannot control the solar azimuth angle, and so it results as per solar geometry but not by tracking. The surface azimuth angle can be controlled by tracking. In the plot, the surface azimuth angle is maintained by rotating the vertical axis from east to west of the collector face equal to match the solar azimuth angle. Therefore, the magnitudes of solar azimuth and surface azimuth

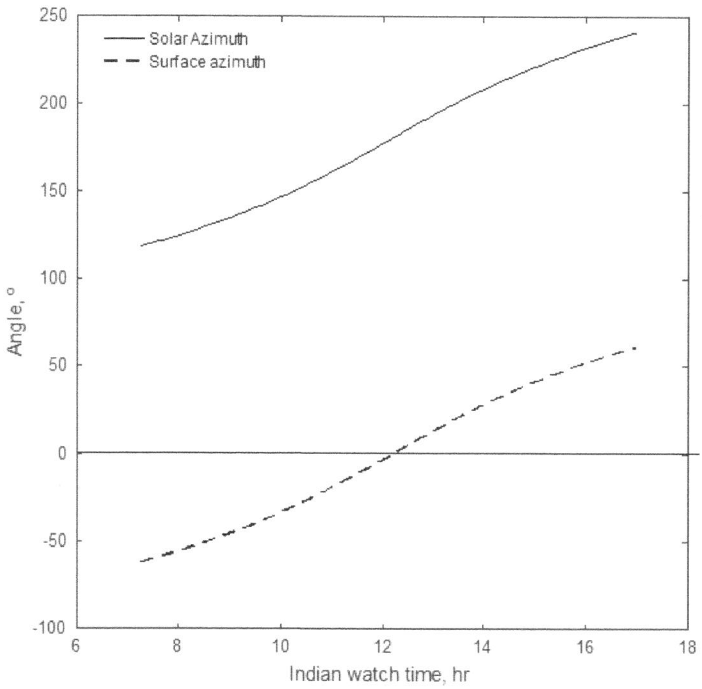

Fig. 10.13. Variation in solar azimuth angle and surface azimuth angle on winter solstice (angle change in clock direction).

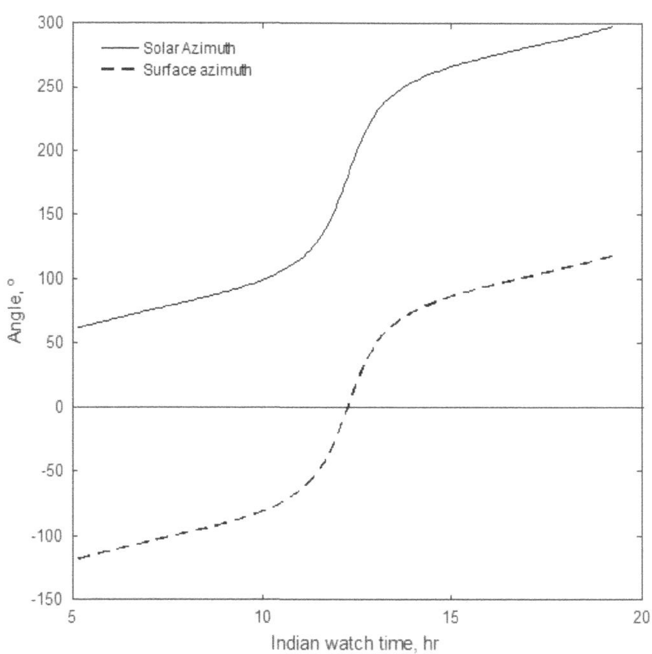

Fig. 10.14. Solar and surface azimuth variations during the summer solstice.

are equal, as shown. It can be seen that the surface azimuth is negative in the morning and zero at noon and shifts to positive during the afternoon. The surface or solar azimuth is not 90° during sunrise or sunset. Similar to the zenith angle, it depends on the seasonal tilt of the sun's path.

The changes in solar and surface azimuth are different compared in winter, as shown in Fig. 10.14. The surface rotation during noon is fast compared to the morning and evening speeds.

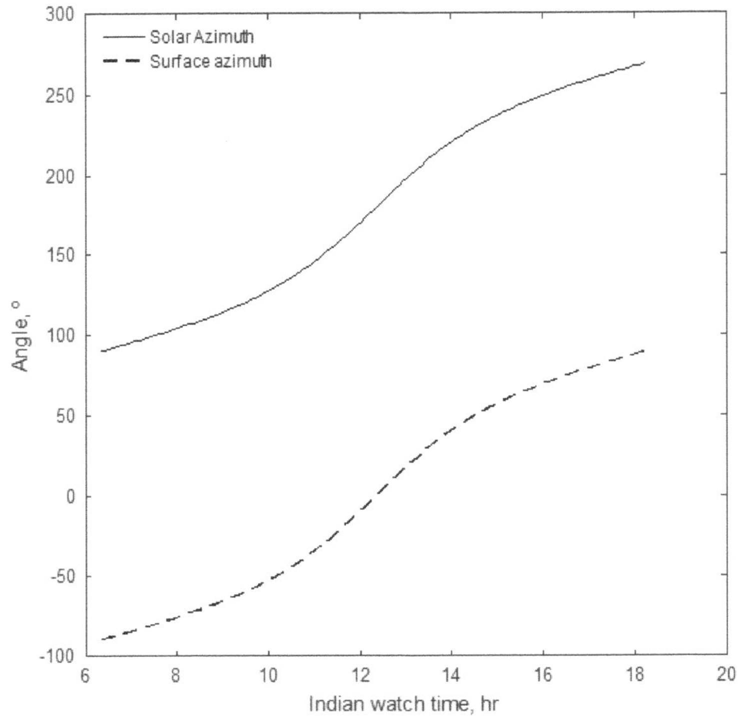

Fig. 10.15. Solar azimuth and surface azimuth on equinox date.

On the equinox day, the solar and surface azimuth changes are depicted in Fig. 10.15. During the winter, the change in azimuth angle is low compared to the changes in summer. More amount of sweep in azimuth is observed in summer. The azimuth angular changes lie between summer and winter during the equinox.

10.2.8 Zenith angle

Zenith angle (θ_z) is the angle between the sun's rays to the vertical, or zenith. It is a special case of an angle of incidence. The angle of incidence is defined for the tilted surface, and the zenith is for the horizontal surface. Zenith angle can also be referred to as the tilted surface to highlight the decrease in incidence by shifting from a horizontal surface to a tilted mode. Zenith angle and azimuth angle both refer to the sun position linked with the earth location. The combination of zenith angle and azimuth angle describes the sun path with reference to the north-south line.

The expression for the angle of incidence (Eq. 10.1) can be simplified by substituting the zero tilt angle (horizontal surface). Actually, azimuth angle cannot be defined for the horizontal surface.

$$\theta_z = \cos^{-1} (\sin\delta \sin\emptyset + \cos\delta \cos\emptyset \cos\omega) \qquad (10.10)$$

The above equation depicts the zenith angle as a function of Julian day, latitude (location), and hour angle (time). The supply of date, time, and place is enough to evaluate the zenith angle.

Figure 10.16 shows the changes in zenith angle from sunrise to sunset during a summer day in India. The zenith angle changes from 90° to 90° from sunrise to sunset for all the days in the year. The day length span changes seasonally. The minimum zenith angle can be observed during the noontime. At the equator, zero zenith results. During summer, the zenith angle at noon is small compared to the zenith in winter at noon. On the equinox, the zenith angle is in the middle of summer zenith and winter zenith at noon. The sunrise and sunset zenith will not change seasonally. The places below the Tropic of Cancer and above the Tropic of Capricorn experience zero zenith

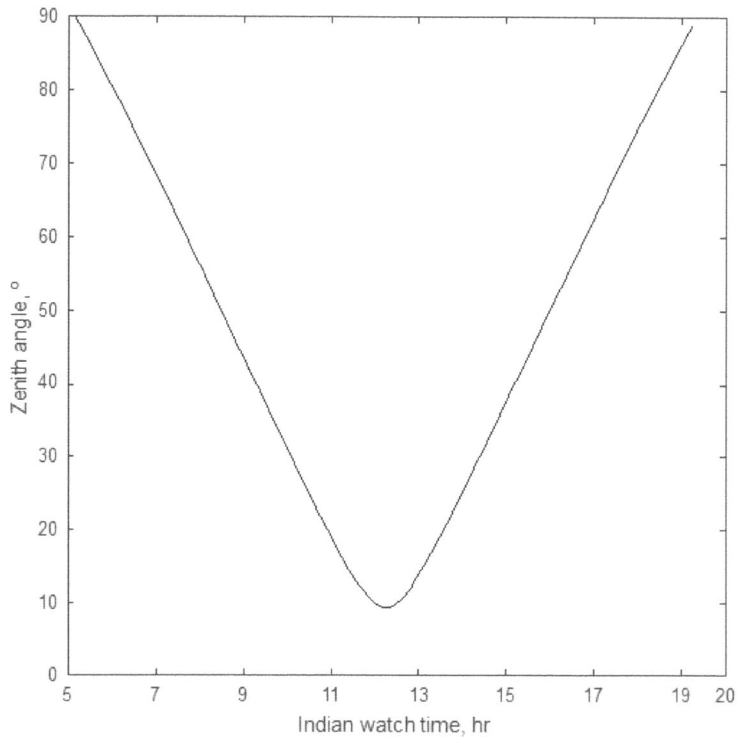

Fig. 10.16. Zenith angle changes from sunrise to sunset.

angle twice per year, as shown in Fig. 10.17. Zero zenith indicates that the sun lies exactly overhead at solar noon. Other than these places, a minimum zenith angle appears once a year instead of zero zenith.

As per Fig. 10.17a, it can be concluded that at certain places, it is possible to see the sun exactly overhead twice a year. But beyond the region of latitude 23.5°, there is no zero zenith angle as per solar geometry.

The zero zenith angle results when θ_z becomes zero.

For a horizontal surface,

$\cos\theta_z = \cos(0) = 1 = \sin\delta \sin\varphi + \cos\delta \cos\varphi \cos\omega$

The zero zenith angle results at solar noon, i.e., $\omega = 0$

$1 = \sin\delta \sin\varphi + \cos\delta \cos\varphi$

$$\sin\delta \sin\phi + \sqrt{\left(1-\sin^2\delta\right)}\cos\phi = 1$$

$$\sqrt{\left(1-\sin^2\delta\right)}\cos\phi = 1-\sin\delta\sin\phi$$

Squaring both sides

$$\left(1-\sin^2\delta\right)\cos^2\phi = 1+\sin^2\delta\sin^2\phi - 2\sin\delta\sin\phi$$

$$\cos^2\phi - \sin^2\delta\cos^2\phi = 1+\sin^2\delta\sin^2\phi - 2\sin\delta\sin\phi$$

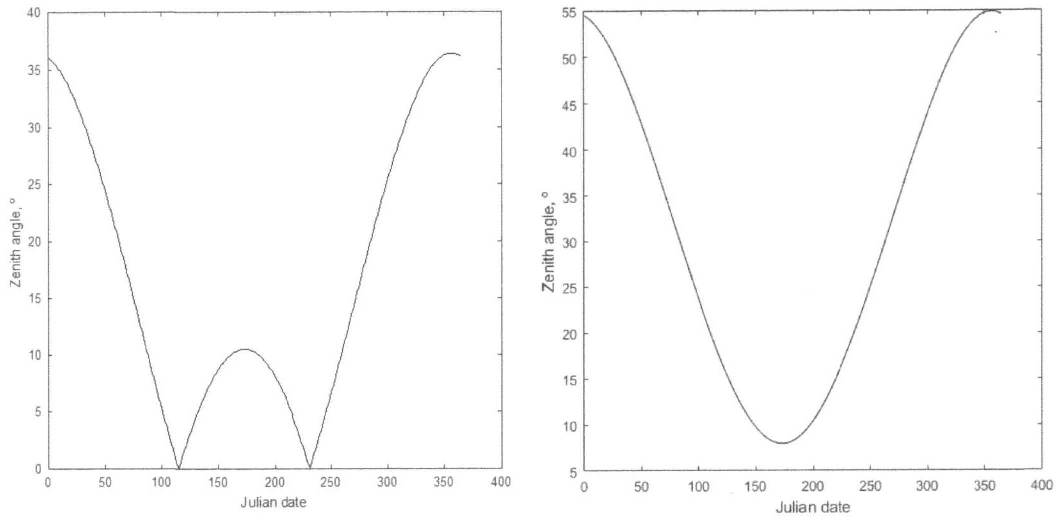

Fig. 10.17. Zero zenith angle twice per year at solar noon for the latitude of 12° (a) Ø < 28.5° (b) Ø > 23.5°.

$$\sin^2\delta\left(\sin^2\phi + \cos^2\phi\right) - 2\sin\delta\sin\phi - \cos^2\phi + 1 = 0$$

$$\sin^2\delta - 2\sin\phi\sin\delta + \sin^2\phi = 0$$

$$\left(\sin\delta - \sin\phi\right)^2 = 0$$

$\sin\delta = \sin\text{Ø}$

Therefore

$\delta = \text{Ø}$

Declination angle, $\delta = 23.45\sin\left[\frac{360(284+n)}{365}\right]\,°$

$$n = \frac{365}{360}\times\sin^{-1}\left(\frac{\delta}{23.45}\right) - 284$$

Similarly, at solar noon $\omega = 0$

$\cos\theta_z = \cos\text{Ø}\cos\delta + \sin\text{Ø}\sin\delta = \cos(\text{Ø} - \delta)$

$\theta_z = \text{Ø} - \delta$

It shows that at solar noon, the zenith angle is equal to the difference between latitude and declination. This condition was also obtained with the zero azimuth at solar noon.

10.2.9 Altitude angle

Altitude angle (α) is the complementary angle to the zenith angle. It is the angle between the sunray and horizontal line, as shown in Fig. 10.18. The zenith is measured to the vertical, and the altitude is measured to the horizontal surface. The altitude and zenith both can be read on a vertical plane.

$$\alpha = 90 - \theta_z \tag{10.11}$$

From the zenith angle,

$\cos\theta_z = \sin\delta\sin\text{Ø} + \cos\delta\cos\text{Ø}\cos\omega$

Therefore

$\cos(90-\alpha) = \sin\alpha = \sin\delta \sin\text{Ø} + \cos\delta \cos\text{Ø} \cos\omega$

Altitude angle,

$$\alpha = \sin^{-1} (\sin\delta \sin\text{Ø} + \cos\delta \cos\text{Ø} \cos\omega) \quad (10.12)$$

Figure 10.19 shows the changes in altitude angle from sunrise to sunset on a winter day in India. At sunrise and sunset, the sunrays reach horizontally to the ground. Therefore, the altitude angle

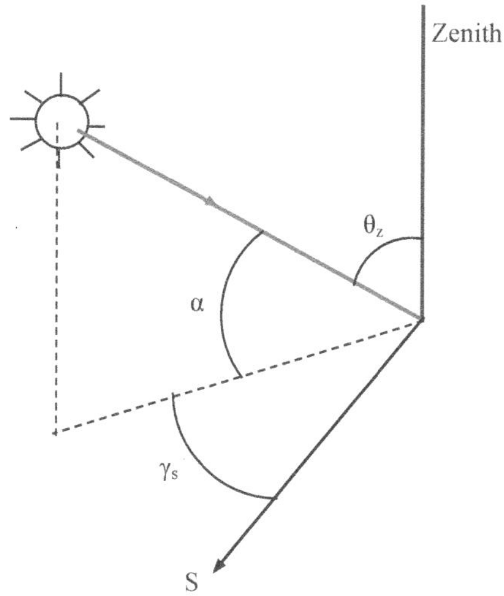

Fig. 10.18. Altitude angle.

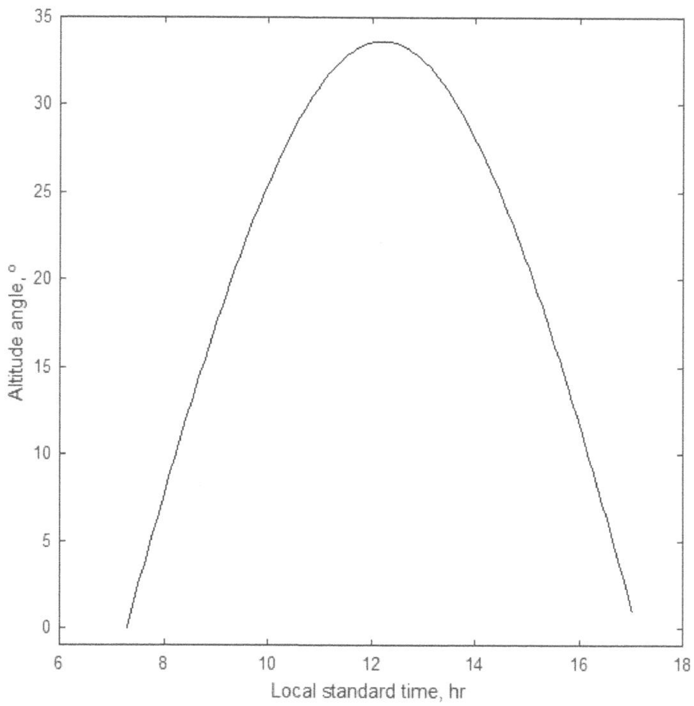

Fig. 10.19. Altitude angle from sunrise to sunset on a winter day.

is zero during these two timings. The maximum altitude angle is observed at solar noon. The 90° altitude angle indicates that the sun beam falls on the ground vertically. Some places close to the equator can experience this 90° altitude angle on days of equinox, i.e., twice per year.

10.2.10 Air mass

Air mass (m) is the ratio of the atmospheric thickness in the direction of sun's rays to the atmospheric thickness at zenith position. Figure 10.20 shows the optical thickness of the atmosphere at the actual instant (OS) and the optical thickness of the atmosphere at zenith (OZ).

As per the definition of air mass,

$$Air\ \ mass, m = \frac{OS}{OZ} = \sec\theta_z \tag{10.13}$$

If the sun is at the zenith, m = 1 at $\theta_z = 0°$, the sun beam passes through the zenith thickness. For the zenith angle of 60°, the sec (60°) results in 2. Therefore, at the angle of 60° zenith, the air mass is 2. Outside the atmosphere, the optical thickness of the atmosphere is zero, i.e., OS = 0. This instant air mass, m = 0. The values of air mass 0, 1, and 2 can be represented as air mass zero (AM0), air mass ratio one (AM1), and air mass two (AM2), respectively. If the distance traveled by the sun beam is more through the atmosphere, the intensity of radiation decreases. It happens during sunrise and sunset. At solar noon, the distance traveled by the rays is minimum, and it results in maximum solar radiation. Therefore, the air mass can be used as a multiplication factor in the solar radiation estimation other than the zenith position. The increase in air mass decreases the intensity of radiation.

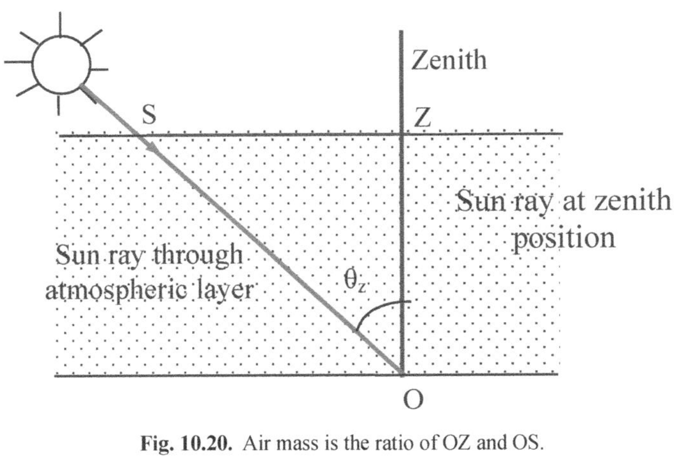

Fig. 10.20. Air mass is the ratio of OZ and OS.

10.2.11 Longitude/Meridian

Similar to latitude, longitude indicates the location of a place on earth. Now, through Google Maps, it is easy to see the latitude and longitude values of cities, villages, and almost all places. Longitude lines are the vertical circles starting from the Greenwich meridian, as shown in Fig. 10.21. The Greenwich meridian is the prime meridian of 0° passing through London, England. The longitude is defined as the angle between the prime meridian and meridian passing through the place. The longitudes continue towards east and west of the prime meridian. From the prime meridian, the longitudes extend up to 180° towards east as well as west. Therefore, the longitudes cover a total of 360° around the earth. Longitudes help calculate time by converting the local standard time into solar time or vice versa. Every country has its own standard longitude to follow its own local time. But the solar time depends on the actual fall of the solar beam at the location. This deviation can be

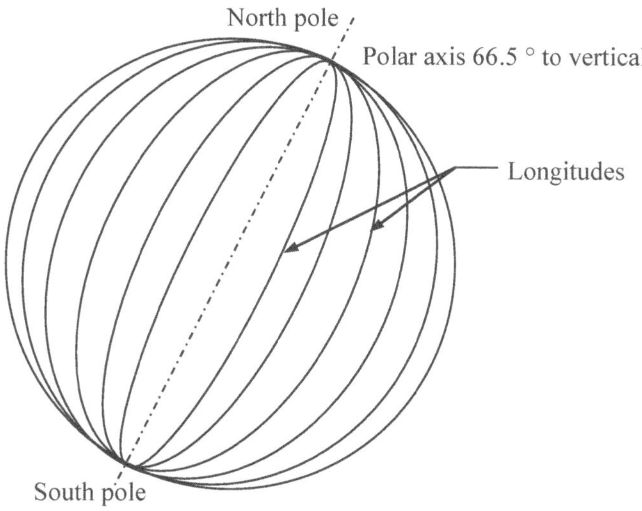

Fig. 10.21. Representation of latitude and longitude at a location on earth.

determined from the difference between the standard longitude and local longitude. The earth rotates 360° per day, i.e., 24 hours. It is equal to 15° per hour. Therefore, on a world map, 15° longitudes can be seen as an equivalent of 1 hour. The standard longitude in India passes through Prayagraj at 82.5°. India is located towards the east of Greenwich. The time difference between Greenwich and India is 85.5/15 = 5.5 hours, ahead of the Greenwich Mean Time (GMT).

10.2.12 Solar time

In ancient times, people measured time based on the sun's shadow at temples or with a sundial. These gave the solar time. Solar time is the local apparent time determined by the position of sun in the sky. All the solar geometry calculations are determined from the solar time, not with the local time. If the sun's rays are passing exactly over a local longitude, it is known as solar noon. At solar noon, the sun's rays are nearer the location, and so their intensity is maximum compared to the other times. The local time or watch time is different from the solar time. The standard longitude and the local longitude differentiate the solar time from the local standard time. We use watch time for our daily activities. But solar geometry, radiation, and all other estimations related to solar rays are found from the solar time. Therefore, as a first step, watch time has to be converted into solar time. The local standard noon is not symmetrical. But the solar noon is symmetrical in the morning and afternoon. The sunrise and sunset are also symmetrical in the solar time scale.

10.2.13 Watch time

It is difficult for the public to follow solar time as it differs with place. Using advanced mathematics and high-level computational methods, the standard time for every country has been designed. The watch time, or the local standard time (LST), is determined from the national standard longitude. For India, the standard longitude, or meridian, is 82.5° E, passes through Prayagraj. The mean or standard meridian should be selected such that it should pass through the middle of the country. The earth makes a complete rotation on its axis once in 24 hours, thereby moving across 360° longitude. The rotational speed is therefore 4 minutes per meridian (longitude) or 15° per hour. So, the time difference between the local time (based on the local meridian) and standard time (based on the standard meridian) will be 4 $(L_{st} - L_{lo})$, in minutes, where L_{st} is the standard longitude and L_{lo} is the local longitude.

$$Solar\ time - standard\ time = \frac{4\min(L_{st} - L_{lo})}{60} + \frac{EoT\ \ min}{60}$$

The eccentric location of the sun in orbit and the dynamic action of the solar system lead to an error in determining the solar time with the longitude difference hypothetically. This error is called the Equation of Time (EoT).

The EoT can be determined in minutes from the analytical formula as follows

$$\text{EoT} = 9.87 \sin 2\text{B} - 7.53 \cos \text{B} - 1.5 \sin \text{B} \tag{10.14}$$

where

$$B = \frac{360(n-81)}{364}\ ^\circ \tag{10.15}$$

The EoT is determined from the Eq. 10.13 or read from the graph Fig. 10.22.

The sign convention in time conversion for the east side of Greenwich is different compared to the west side of Greenwich. The Greenwich time is the reference time for the world, and country time is given with Greenwich time as reference. Time can be adjusted while flying from one country to another. The direction of eastern longitudes from Greenwich and sun path projection on ground are opposite.

The solar time in hours is calculated from the watch time for locations east of Greenwich (say, Asian, African, and Australian locations), as follows:

$$Solar\ time = Watch\ time - \frac{4\min(L_{st} - L_{lo})}{60} + \frac{EoT\ \ min}{60} \tag{10.16}$$

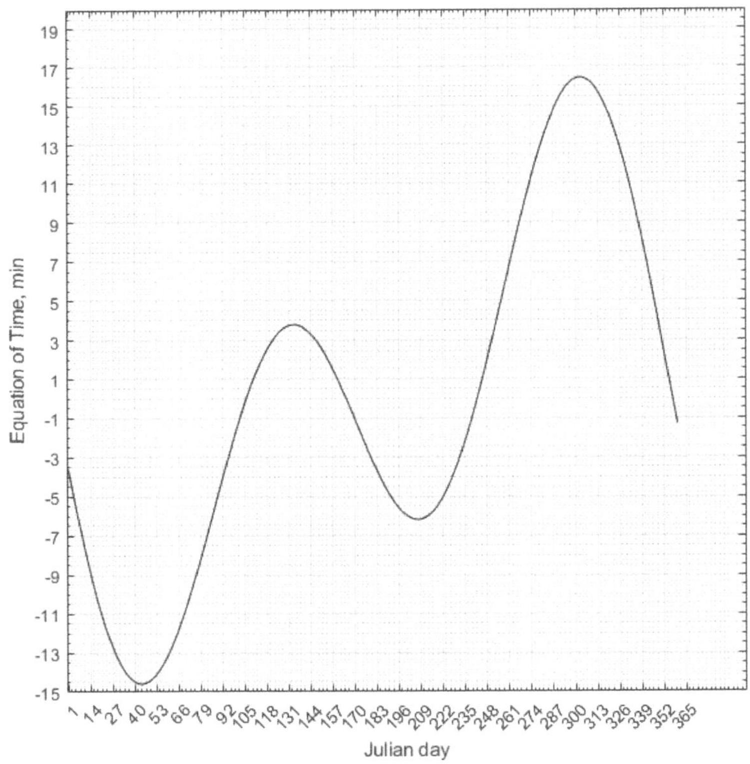

Fig. 10.22. Equation of time with Julian day.

On the west side of Greenwich, the directions of longitude markings and the sun path are in the same direction.

While for locations west of Greenwich (say, the Americas and Canada) it is

$$Solar\ time = Watch\ time + \frac{4\min(L_{st} - L_{lo})}{60} + \frac{EoT\ \ \min}{60} \quad (10.17)$$

The solar time and standard time (watch time) at Mirzapur are the same due to the fact that the difference in longitude is zero. The EoT results in some differences due to the dynamic nature of sun-earth relations. Now consider two places: one on the eastern part of the standard longitude, Itanagar, and another on the western side of the standard longitude, Jalandhar. The longitude of Itanagar is 93.6053° E, and the longitude of Jalandhar is 75.5762° E. For Itanagar, the difference in standard longitude and local longitude is (82.5°–93.6053°) –11.1053°. Since the sun passes first over Itanagar compared to Mirzapur, the solar time of Itanagar is more than Mirzapur's solar time, or standard time. So, the difference between solar time and standard time is positive. But the difference in longitudes (Lst–Llo) is negative. In the second case of Jalandhar, the difference in longitude is 82.5° (75.5762°) + 6.9238°. At Jalandhar, the sun passes after Mirzapur; hence, the solar time of Jalandhar is less than the standard time of the country. Here, too, the difference between solar time and standard time is negative. The sign convention is opposite. Therefore, the ΔL must be multiplied with a negative sign to get the correct result.

In the western part of Greenwich, the difference in longitudes is the opposite. But the difference in solar time and watch time on the east side of standard longitude and the west side of standard longitude is the same as in the eastern countries. Therefore, the sign convention for the latitude difference is positive. In other words, on the eastern side of Greenwich, it can be seen that the sun's path is from east to west, and the measurement of longitude is from west to east. The opposite direction results in a negative sign in the equation. On the western side of Greenwich, the sun's path and measurement of longitude are in the same direction, i.e., from east to west. Therefore, the same direction results in a positive sign in the equation.

Canada is the second-biggest country in the world and ranges from 52° W to 141° W. Therefore, the local time in the west is nearly 6 hours behind that of the east. Countries having a large distance from West to East are often divided into two or more time zones to adjust daytimes to the position of the sun. If Canada had chosen only a single time zone, then both the east and west parts of Canada would be offset from local time by 3 hours. As a result, eastern Canada would experience noon by 9:00 AM, while western Canada would experience it at 3:00 PM. In the world's largest country, i.e., Russia, there exists a time difference of up to 10 hours between the East and the West. Therefore, selection of multiple time zones was essential for the country. In Canada, there are 6 time zones that cover four and a half hours. From west to east, these time zones are: Pacific (Vancouver), Mountain (Edmonton), Central (Winnipeg), Eastern (Toronto), Atlantic (Halifax), and Newfoundland (St. Johns).

Daylight saving time (DST), also referred to as daylight savings time, daylight time (United States, Canada, and Australia), or summer time (United Kingdom, European Union, and others), is the practice of advancing clocks (typically by one hour) during warmer months so that darkness falls at a later clock time. The typical implementation of DST is to set clocks forward by one hour in either the late winter or spring ("spring forward") and to set clocks back by one hour in the fall ("fall back") to return to standard time. As a result, there is one 23-hour day in early spring and one 25-hour day in the middle of autumn.

10.2.14 Sunrise and sunset time

The hour angle at sunrise (ω_{sr}) and sunset (ω_{ss}) results in the time at sunrise and sunset, respectively. During the sunrise and sunset, the zenith angle is 90° and the complementary altitude angle is zero. The hour angle, ω, at sunrise and sunset is 90°. The –90° hour angle is at sunrise and 90° is at sunset.

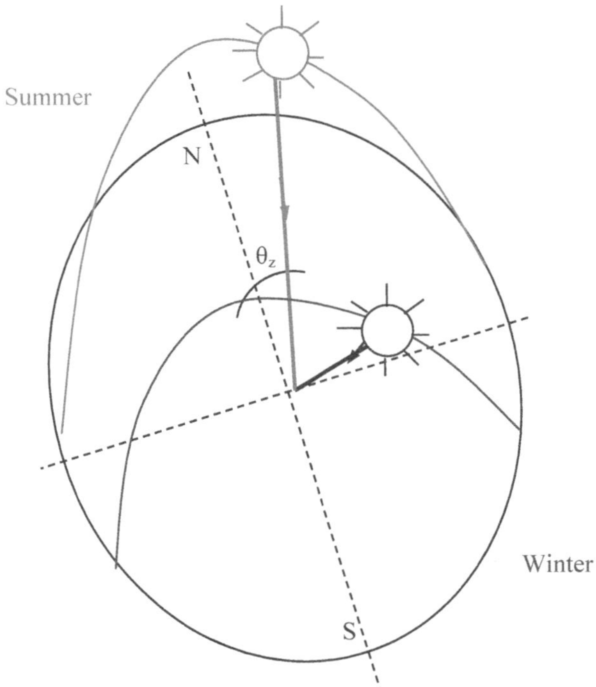

Fig. 10.23. Sunrise and sunset instants during summer and winter.

The sun path on earth during the summer lies on the northern hemisphere. The north pole tilts towards the sun and the sun path is longer, as shown in Fig. 10.23. The longer path results in longer day length in summer. The sun rises early and sets late. Most of the parts of the sun are visible in the northern hemisphere at a bigger size. Since the zenith angle during the summer is low, the shadow length at this season is short. The day length in winter is short as the sun path, as shown, is short. In winter, the sun rises late and sets early. The larger zenith angle or lower altitude angle results in longer shadows in winter.

During sunrise or sunset, the zenith angle is 90°. This condition is used to find the sunrise and sunset times. The hour angle at sunrise or sunset is ω_{sr} and ω_{st} respectively.

$$\cos\theta_z = 0 = \sin\delta \sin\emptyset + \cos\delta \cos\emptyset \cos\omega_s$$

Therefore, the hour angle at the sunrise or sunset is

$$\omega_{sr} = \omega_{ss} = \cos^{-1}(-\tan\delta \tan\emptyset) \tag{10.18}$$

Therefore, the sunrise and sunset angle can be determined from the date (declination) and place (longitude).

The sunrise or sunset time in hour,

$$sunrise = sunset = \frac{\cos^{-1}(-\tan\delta\tan\phi)}{15},\ h \tag{10.19}$$

The above equation is framed with the assumption of sunrise or sunset with a line matching with center of the sun. But the sunrise or sunset occur with the upper limb (tip) of the sun. Based on the vision angle, this error can also be adjusted by correcting the zenith angle.

10.2.15 Day length

Day length is the time between the sunrise and sunset. The sunrise and sunset formulations are prepared for a horizontal surface. Therefore, the day length is also given for the horizontal surface. For the horizontal surface, the angle of incidence is the zenith angle. From the zenith angle condition at sunrise and sunset, i.e., 90°, the expression for sunrise and sunset has been developed. For a non-horizontal surface, i.e., a tilted surface, an angle of incidence is used in place of the zenith angle to assess the sunrise and sunset. At the value of 90° zenith for horizontal surface, the sunrise hour angle ω_{sr} and sunset hour angle ω_{st} are determined.

Since the sunrise and sunset positions are symmetrical on solar time, the total hour angle for a day length is

$$\omega_{total} = \omega_{ss} - \omega_{sr}$$

The ω_{sr} is negative in forenoon.

$\omega_{total} = \omega_{ss} - (-\omega_{sr}) = 2\ \omega_{sr}$ or $2\ \omega_{ss}$ due to symmetrical

The magnitude of ω_{sr} or ω_{ss} is same, then $\omega_{sr} = \omega_{ss} = \omega_s$

The corresponding day length in hours

$$S = \frac{2\omega_s}{15} = \frac{2\cos^{-1}(-\tan\delta\tan\phi)}{15} \tag{10.20}$$

For tilted surfaces towards south, during the sunrise or sunset, the angle of incidence is 90°.

$$\cos\theta = 0 = \sin\delta\ \sin(\varphi - \beta) + \cos\delta\ \cos(\varphi - \beta)\ \cos\omega_s$$

$$S = \frac{2\omega_s}{15} = \frac{2\cos^{-1}(-\tan\delta\tan(\phi-\beta))}{15} \tag{10.21}$$

It is the maximum length of a day determined from sunrise to sunset. The actual day length is less than the day length calculated from sunrise to sunset due to cloudy hours. The sunrise and sunset hour angles do not exactly follow the time published in local newspapers. The sunrise reported in the newspaper will be earlier, and the sunset will be later. The reason for this difference is that the sunlight is refracted as it moves through the earth's atmosphere, the sun appears slightly higher in the sky than estimated by simple geometrical calculations. This is the same effect that makes a stick appear to bend when it is placed in water. During the middle of the day, effect is small, but during the sunrise or sunset periods, the effect can change the apparent solar time by about 5 minutes.

Sunrise, sunset, and day length data are required in the assessment of tracking performance. It gives the time period the collector needs to rotate. The day length information guides the designer to track the collector for a known period of time. Figure 10.24 shows the information on day length with Julian day for all locations on the world map. The latitude changes from 0 to 90° in the northern hemisphere and also in the southern hemisphere. Some interesting points can be observed: the day length at the north pole is six months, and the balance is six months; it is in night mode. During the summer in the northern hemisphere, the southern hemisphere is exposed to winter with a tilt of earth such that one pole is toward the sun and the other is away from the sun.

Table 10.1 also shows the shortest day and longest day resulted from the day length formula for the regions of tropics, temperate, and polar places with the latitude. The increase in the latitude is from equator to poles. The shortest day decreases from 12 hours to 0 hours, and the longest day decreases from 12 hours to 24 hours.

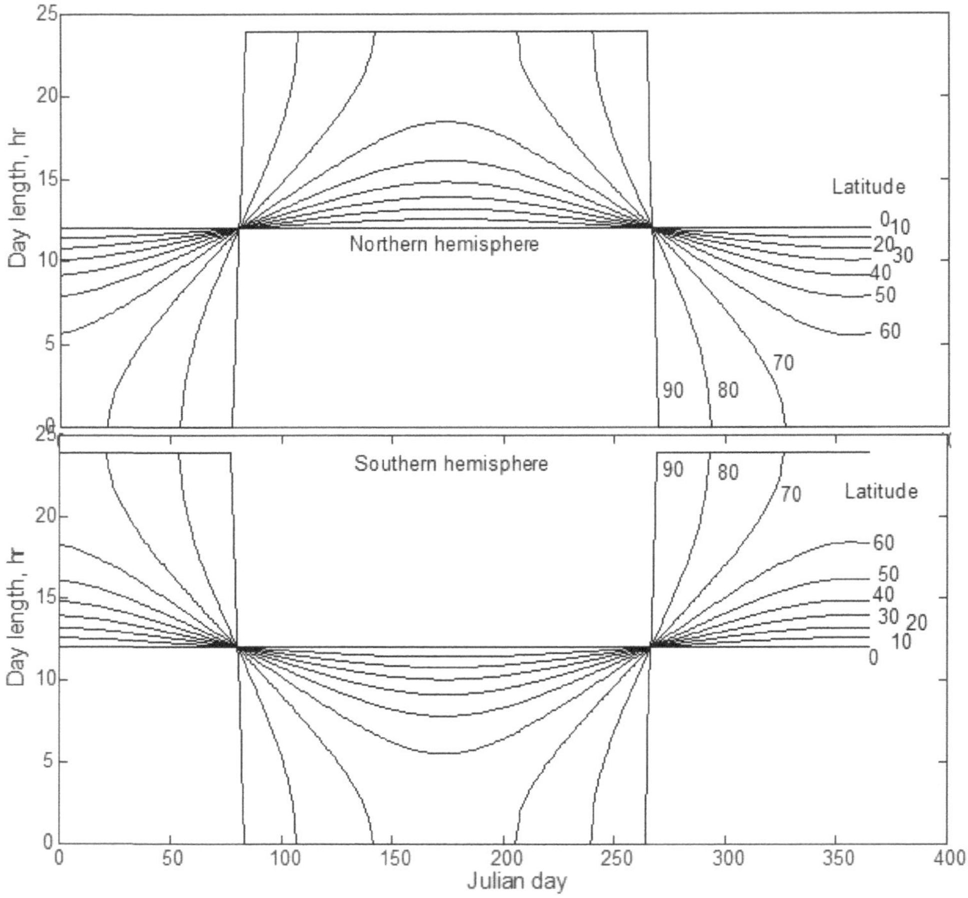

Fig. 10.24. Assessment of day length for Julian days in the northern hemisphere and southern hemisphere.

Table 10.1. Day length with longitude.

Region	Latitude, °	Shortest day, hr	Longest day, hr
Tropics	00	12.07	12.07
	10	11.32	12.42
	20	10.56	13.20
Temperate	30	10.14	14.04
	40	09.20	15.00
	50	08.05	16.21
	60	05.54	18.49
Polar	70	00.00	24.00
	80	00.00	24.00
	90	00.00	24.00

10.3 Sun path diagram

The projection of the sun on the earth appears on the north for six months and south during the next six months. The sun path is the motion of the sun's shadow on earth from sunrise to sunset per year at the given location. Figure 10.25 shows a typical sun track position at the latitude of 13° N.

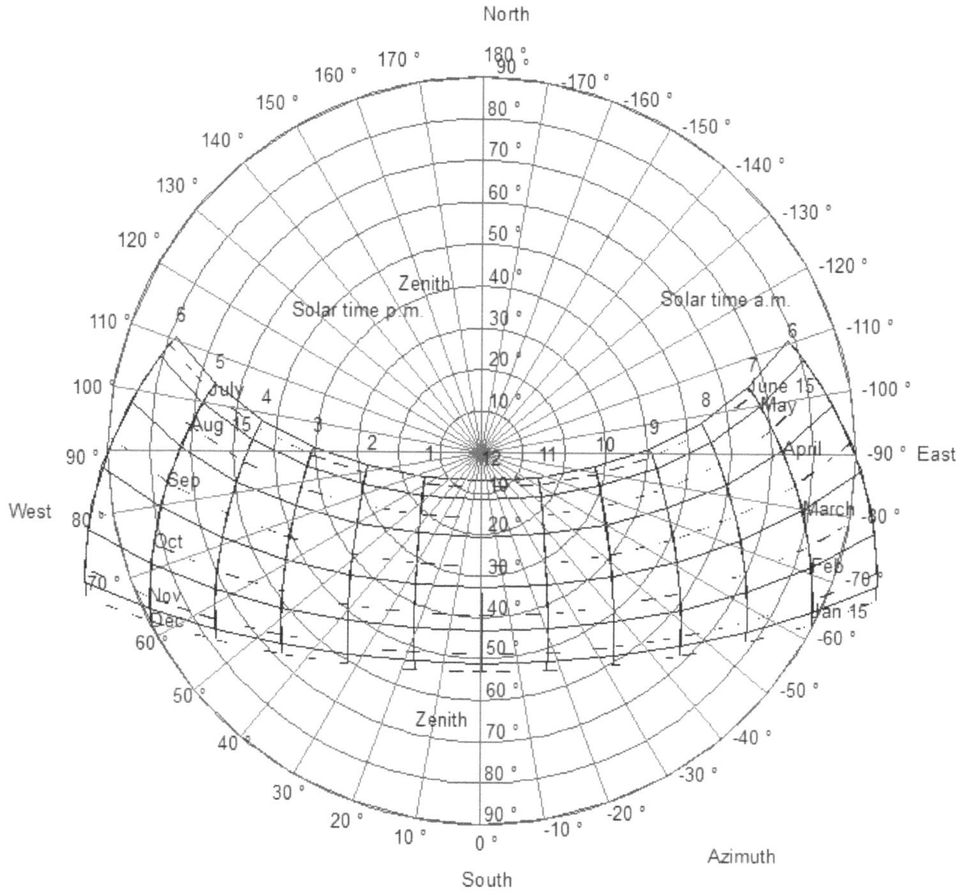

Fig. 10.25. Sun path diagram.

The zenith angle is the function of declination, hour angle, and latitude. It is the top view of sun's projections on the earth. The sun path diagram shown in Fig. 10.25 can be prepared with the function of these three variables. Since it is difficult to show the three variables on a two-dimensional plot, the declination (from January 1st to December 31st) and hour angle (from sunrise to sunset) can be varied at the selected place (which is kept fixed). The distance of the sun's projection on the earth surface is shown in the four quadrants from this place. The quadrants from 1 to 4 are respectively north-east, north-west, south-west, and south-east. The projection of the sun on the earth can be determined from the zenith angle and solar azimuth angle. The distance of the projection is drawn to scale with the apparent units from the location. The sun path diagram at a given date, time, and place is useful for shadow analysis for green building design, tracking simulations, and area optimization of solar farms.

The horizontal distance of the sun's projection on the earth towards east or west

$$X = \text{length units} \times \sin(\theta_z) \tag{10.22}$$

The vertical distance of sun projection on the earth towards north or south

$$Y = \text{length units} \times \cos(\gamma_s) \tag{10.23}$$

The most common representation of the sun path diagram is in the radial coordinates. The altitude can be read in the radial direction and the azimuth angle in the tangential direction. The sun path diagrams from 0° latitude to 90° latitude are generated and shown in Appendix. The sun path diagrams can be used to read the altitude and azimuth angle at a known place, date, and time using a graphical method without using formulae.

10.4 Solar geometry in assessment of the angle of incidence

Figure 10.26 shows solar geometry of an inclined surface. It can be seen that the normal to the tilted surface is not matching with the sunbeam. The angle between the sun beam and the surface is called tracking error. To minimize this tracking error, many tracking methods have developed. For example, to make the angle of incidence zero or zero tracking error, what is to be controlled? The answer is, two parameters control. Out of five variables in the angle of incidence, viz., declination (date), hour angle (time), latitude (place), tilt angle, and surface azimuth angle, the first three variables are not controllable as they depend on the time and place. The last two variables are independent and can be called controlling variables to result in a minimum angle of incidence. To arrive at a zero angle of incidence for the tilted surface, the tilt angle should be equal to the zenith angle, and the surface azimuth angle should be equal to the solar azimuth angle. The answer to this question is shown in Fig. 10.27. After tilting the surface equal to the zenith angle and surface azimuth angle with the solar azimuth, the resulting angle of incidence (angle between the sun beam and normal to the surface) is zero.

After observing the above solar geometry, the expression for angle of incidence can be simplified for special cases as follows.

The Eq. 10.1 can be simplified to a surface facing due south ($\gamma = 0$),

$$\cos\theta = \sin\delta\sin\phi\cos\beta - \sin\delta\cos\phi\sin\beta + \cos\delta\cos\phi\cos\beta\cos\omega + \cos\delta\sin\phi\sin\beta\cos\omega$$

$$= \sin\delta\,(\sin\varphi\cos\beta - \cos\varphi\sin\beta) + \cos\delta\,(\cos\varphi\cos\beta + \sin\varphi\sin\beta)\cos\omega$$

After simplification

$$\cos\theta = \sin\delta\sin(\varphi - \beta) + \cos\delta\cos(\varphi - \beta)\cos\omega \qquad (10.24)$$

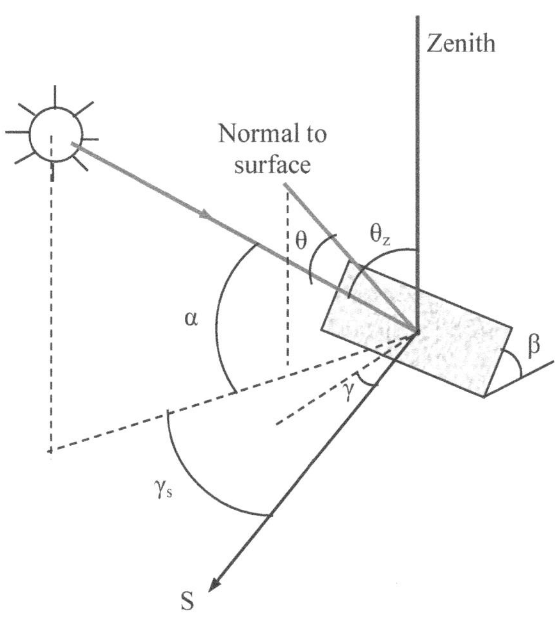

Fig. 10.26. Relation of the angle of incidence with zenith, azimuth (solar and surface) angle and tilt angle.

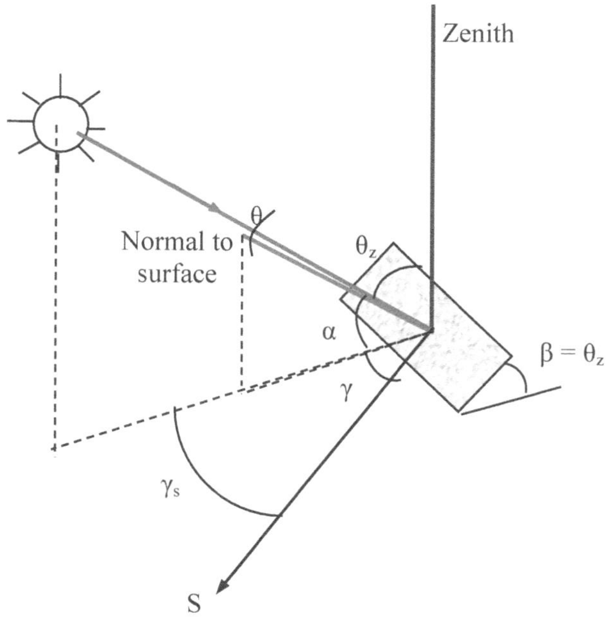

Fig. 10.27. Zero angle of incidence by tilting the surface equal to zenith angle and surface azimuth with the solar azimuth.

For a horizontal surface, the angle of incidence is equal to the zenith angle as a special case.

$$\cos\theta = \cos\theta_z = \sin\delta \sin\varphi + \cos\delta \cos\varphi \cos\omega \tag{10.25}$$

The altitude angle is the complementary angle of zenith angle

$$\sin\alpha = \cos\theta_z = \sin\delta \sin\varphi + \cos\delta \cos\varphi \cos\omega \tag{10.26}$$

For vertical surfaces (walls), the tilt angle is 90°. Therefore,

$$\cos\theta = -\sin\delta \cos\varphi \cos\gamma + \cos\delta \sin\varphi \cos\gamma \cos\omega + \cos\delta \sin\gamma \sin\omega \tag{10.27}$$

10.5 Methodology for solar geometry solutions

In solar geometry, the main task of the solar engineer, particularly in the sun-earth relations, is finding the angle of incidence at the designed orientation of the solar collector. Figure 10.28 shows the methodology of evaluation of the angle of incidence from the input of declination, hour angle, latitude, tilt angle, and surface azimuth angle. In the first step, the local standard time is to be converted into solar time, as all the numerical solutions depend on solar time. The method not only gives the angle of incidence but also other useful information such as sunrise, sunset, and day length. These are also useful data in the implementation of the design for effective utilization. Solar geometry requires adopting the method given in the flow chart (Fig. 10.28). The sequential operations are arranged logically here to simplify the method for a practicing engineer.

Figure 10.29 depicts the results of solar geometry controlling a tilted surface such that the slope is equal to the zenith and the surface and solar azimuths are the same during a summer day. By controlling these two axes, the angle of incidence can be made zero, which is the condition for perfect tracking from sunrise to sunset, i.e., the whole day. The surface azimuth is measured and shown from the south, and the solar azimuth is shown from the north line for convenience.

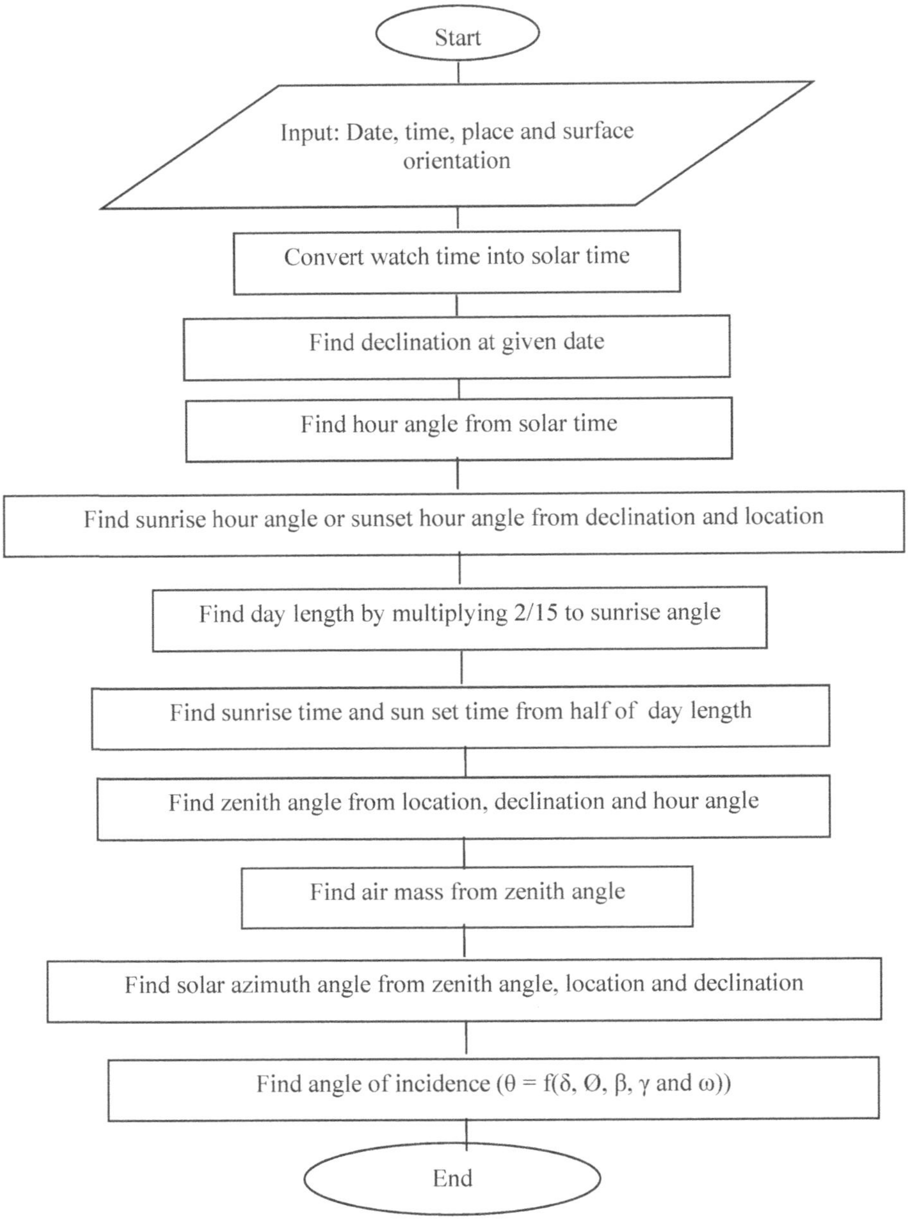

Fig. 10.28. Solar geometry methodology to find the and angle of incidence.

Zenith angle and azimuth angle show all the geometric aspects for a horizontal surface. It is interesting to know the accuracy of the mathematical model by comparing it with the data obtained from the solar observations built with the measuring instruments. Figure 10.30 shows a comparison of simulated solar geometry from sunrise to sunset on 21st March with the real values taken from the solar radiation and geometric instruments. The simulated geometry matches the laboratory values, and so the mathematical model can be used satisfactorily as it is validated with the real data.

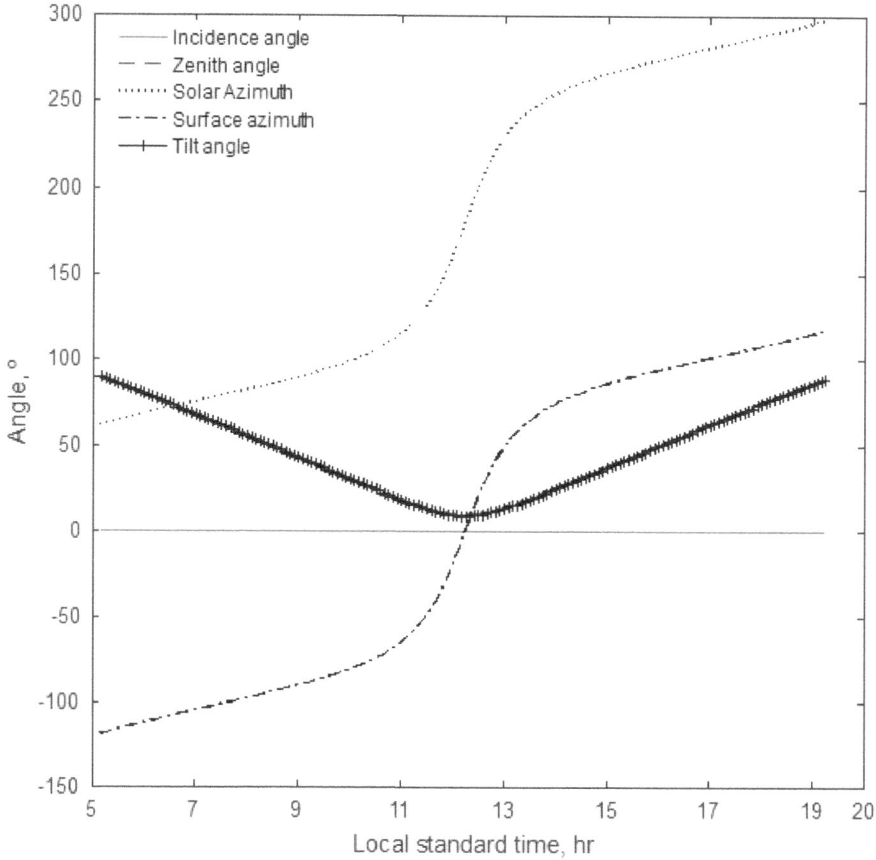

Fig. 10.29. Solar geometry to study the angle of incidence by equating tilt to zenith and surface azimuth with solar azimuth.

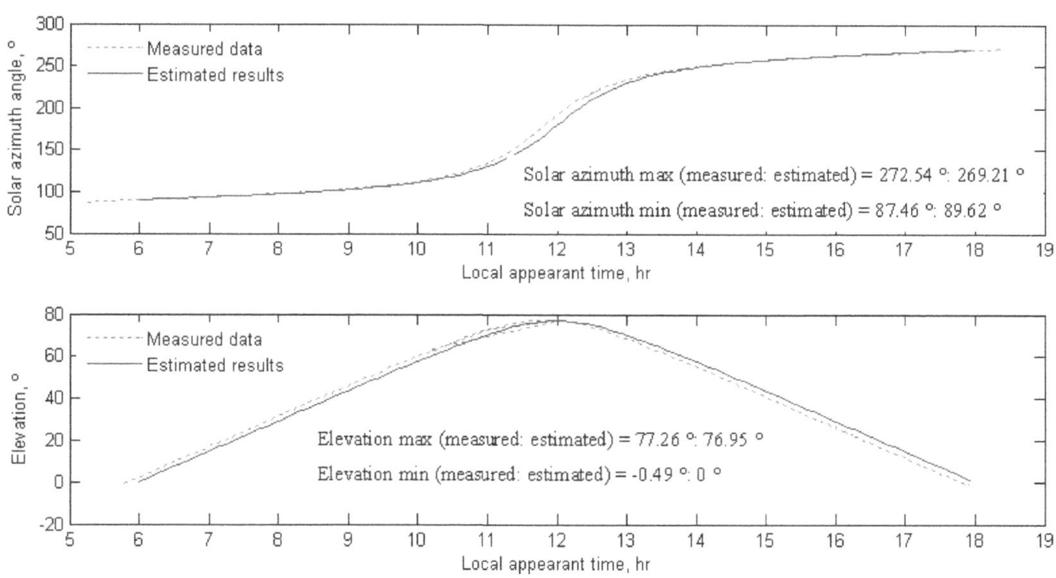

Fig. 10.30. Validation of solar geometry.

10.6 Minimum angle of incidence

Many tracking practices are developed to achieve the minimum angle of incidence through a continuous change in the orientation of the collector. The tracking of the collector may be done in single-axis mode or dual-axis mode. In this section, four methods for single-axis tracking are developed and compared to pick the best practice. The condition for perfect tracking is applied for dual-axis tracking (zenith axis and azimuth axis), and the resultant angle of incidence is observed for the period of one day and up to one year. Including non-tracking, single axis, and dual axis, a total of six methods are compared, viz. (i) fixed inclined surface (without tracking); (ii) east-west (E-W) horizontal axis rotation; (iii) north-south (N-S) horizontal axis rotation; (iv) N-S inclined axis with azimuth angle rotation; (v) N-S inclined axis with zenith angle rotation; and (vi) dual axis tracking. For the single-axis tracking system, N-S horizontal axis rotation proved better than the other cases. The accuracy of the tracking device depends on the method of tracking, rotation, and orientation. The focusing type solar collector demands accurate tracking machines for maximum capture of solar energy. This section is aimed at the comparison and development of tracking motion with time. Solar geometry is well developed, and problems are solved using spherical trigonometry.

There is a need to study tracking modes on a daily basis, seasonal basis, and yearly basis. This section is focused on analysis of the angle of incidence for tilted surfaces, horizontal axis rotation (E-W and N-S), N-S inclined axis rotation (zenith angle and azimuth angle), and two axis tracking. It highlights the best tracking practice for single-axis tracking over a year. It also defines the conditions for perfect tracking in a bi-axis tracker.

For single-axis tracking, the axis of the collector can be oriented either on the E-W or N-S axes. The main objective of any tracking device is to generate a zero tracking error with zero angle of incidence (θ). The angle of incidence depends on declination (δ) from date, hour angle (ω) from time, latitude ($\emptyset$) from place, collector's tilt angle to horizontal (β), and surface azimuth angle (γ). The solar azimuth angle (γ_s) influences the surface azimuth in dual-axis tracking. Zenith angle (θ_z) is a special case of the angle of incidence. The sunbeam angle can be measured from normal to a horizontal surface (vertical), called the zenith angle, θ_z, or from normal to an inclined surface, known as the angle of incidence. The declination, hour angle, and latitude are the basic information for the angle of incidence, and the tilt angle and surface azimuth angle can be derived for the type of tracking movement.

The following are the six cases of collector orientations in this study.

10.6.1 Fixed inclined surface

For tilted surface, $\beta = \phi$,

In the above mode, the solar collector plane is tilted towards south, i.e., $\gamma = 0$.

After substituting the condition of γ in the angle of incidence,

$\theta = \cos{-1}\ (\sin\delta \sin(\varphi - \beta) + \cos\delta \cos(\varphi - \beta) \cos\omega)$

Since $\varphi - \beta = 0$,

$$\theta = \cos^{-1}(\cos\delta \cos\omega) \tag{10.28}$$

The above equation shows that the angle of incidence is a function of date and time for the inclined surface.

10.6.2 E-W horizontal axis rotation

In this case, the focal axis of the collector is in E-W horizontal and rotated about this axis continuously so that the solar beam makes the minimum angle of incidence with the aperture plane at all times.

The Eq. 10.1 can be differentiated with respect to and equated to zero to get the condition for the minimum angle of incidence at $\gamma = 0$.

The resulting condition is

$$\beta = \phi - \tan^{-1}\left(\frac{\tan\delta}{\cos\omega}\right) \tag{10.29}$$

For $\gamma = 0$, the angle of incidence,

$\theta = \cos{-1}\,(\sin\delta \sin(\varphi - \beta) + \cos\delta \cos(\varphi - \beta)\cos\omega)$

$\phi - \beta = \tan^{-1}\left(\frac{\tan\delta}{\cos\omega}\right)$ can be substituted in the angle of incidence.

$$\theta = \cos^{-1}\left(\sin\delta \sin\left(\tan^{-1}\left(\frac{\tan\delta}{\cos\omega}\right)\right) + \cos\delta\cos\left(\tan^{-1}\left(\frac{\tan\delta}{\cos\omega}\right)\right)\cos\omega\right) \tag{10.30}$$

After substituting the $\varphi - \beta$, the minimum angle of incidence can be taken as a function of date (declination), time (hour angle). The tilt angle varies continuously with the function of declination, hour angle, and latitude (fixed).

10.6.3 North south (N-S) horizontal axis rotation

In this mode, the focal axis is N-S horizontal. The collector is rotated continuously to result in a minimum angle of incidence. In this case the surface azimuth angle, γ_s is $-90°$ before noon and $+90°$ after noon.

Therefore Eq. 10.1 can be differentiated with reference to β with the condition in surface azimuth angle,

$$\beta = \tan^{-1}\left(\frac{\pm\cos\delta\sin\omega}{\sin\phi\sin\delta + \cos\phi\cos\delta\cos\omega}\right) \tag{10.31}$$

In Eq. 10.31, the negative sign is in the forenoon and positive is in the afternoon. The angle of incidence can be determined from Eq. 10.1.

10.6.4 North south (N-S) fixed inclined axis with azimuth angle rotation

In this case, $\beta = \varphi$ and $\gamma = \gamma_s$

It is also called azimuth angle tracking.

where $\gamma_s = \cos^{-1}\left(\frac{\cos\theta_Z \sin\phi - \sin\delta}{\sin\theta_Z \cos\phi}\right)$

Angle of incidence can be determined from Eq. 10.1

10.6.5 North south (N-S) fixed inclined axis with zenith angle rotation

In this case, $\beta = \theta_z$ and $\gamma = 0$

For $\gamma = 0$, the angle of incidence,

$\theta = \cos^{-1}(\sin\delta \sin(\varphi - \beta) + \cos\delta \cos(\varphi - \beta)\cos\omega)$

After substituting the tilt angle,

$$\theta = \cos^{-1}(\sin\delta \sin(\varphi - \theta_z) + \cos\delta\cos(\varphi - \theta_z)\cos\omega) \tag{10.32}$$

10.6.6 Dual axis tracking

In this case, $\beta = \theta_z$ and $\gamma = \gamma_s$

The important solar geometry angles that influence the rotation of the solar collector are the zenith angle, solar azimuth angle, surface azimuth angle, and tilt angle. To study the seasonal changes using geometry of the angle of incidence, five cases of tracking have been developed at the winter solstice, equinox, and summer solstice.

Table 10.2 summarizes all the above modes of surface orientations with reference to the tilt angle and surface azimuth angle. The condition for the tilt angle is derived from the differentiation of the angle of incidence with respect to the tilt angle at zero. The expression for the angle of incidence can be directly determined from the generalized equation with the given conditions, or the conditions can be substituted in the angle of incidence.

Table 10.2. The conditions for minimum angle of incidence with different modes of tracking.

S. No.	Tracking mode	Tilt angle, β °	Surface azimuth angle, γ °	Minimum angle of incidence, θ °
1.	Horizontal surface,	0	0	$\cos^{-1}(\sin\delta \sin\varphi + \cos\delta \cos\varphi \cos\omega)$
2.	Fixed tilt angle	Ø	0	$\cos^{-1}(\cos\delta)$
3.	E-W horizontal axis rotation	$\beta = \phi - \tan^{-1}\left(\frac{\tan\delta}{\cos\omega}\right)$	0	$\cos^{-1}(\sin\delta \sin(\varphi - \beta) + \cos\delta \cos(\varphi - \beta) \cos\omega)$
4.	N-S horizontal axis ratation	$\beta = \tan^{-1}\left(\frac{\pm\cos\delta\sin\omega}{\sin\phi\sin\delta + \cos\phi\cos\delta\cos\omega}\right)$ –ve FN and +ve AN	$\gamma = \pm 90$	Eq. 10.1
5.	N-S fixed inclined axis and azimuth angle rotation	Ø	γ_s	Eq. 10.1
6.	N-S fixed inclined axis and zenith angle rotation	θ_z	0	$\cos^{-1}(\sin\delta \sin(\varphi - \theta_z) + \cos\delta \cos(\varphi - \theta_z) \cos\omega)$
7.	Two axis tracking, $\beta = \theta_z$, $\gamma = \gamma_s$	θ_z	γ_s	Eq. 10.1

10.6.7 Minimum angle of incidence analysis

Figure 10.31 depicts the angle of incidence variations from sunrise to sunset during the winter solstice. A fixed inclined surface (Fig. 10.31a) is tilted (equal to latitude) towards the south for a location in the northern hemisphere. The resulting angle of incidence is less than the zenith angle and shows the benefit due to tilt over the horizontal. For seasonal adjustment, the tilt may be more (approximately 10°) in winter and less in summer. The angle of incidence that results with dual axis tracking is compared with the fixed inclined surface, which is also called the 'collector's surface' without tracking. In Fig. 10.31a, the angle of incidence is decreased from 90° to a minimum, then rises to 90° from sunrise to sunset. The dual-axis tracking results in zero angle of incidence from hypothetical calculations. The solar azimuth angle and surface azimuth angles are shown from south in the results. For this, the solar and surface azimuth angles are maintained the same. Similarly, the tilt angle and zenith angles are equal. In two-axis tracking, the tilt angle (about the horizontal axis) is rotated equal to the zenith angle, and the surface azimuth angle (about the vertical axis) is rotated equal to the solar azimuth angle (Fig. 10.31b). The surface can be adjusted so that the normal to

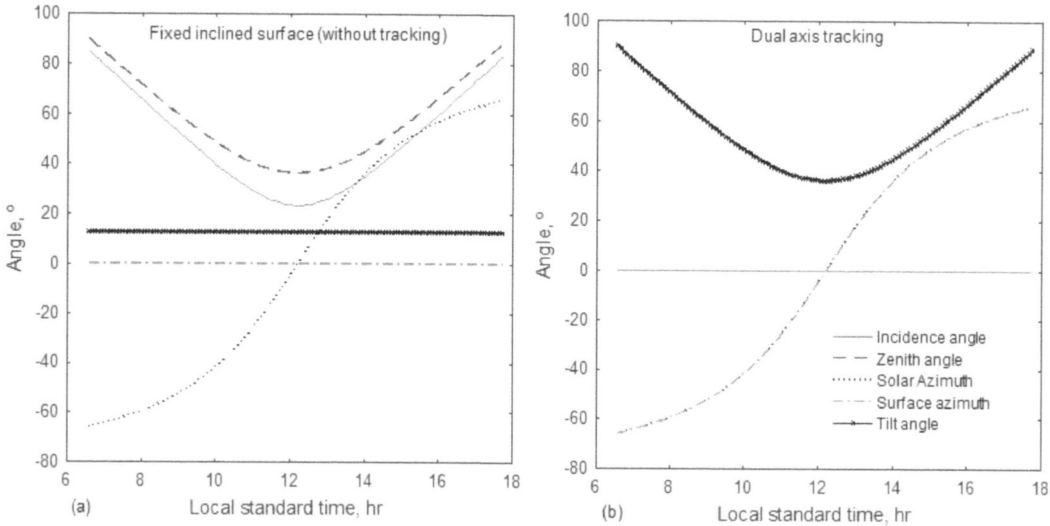

Fig. 10.31. Angle of incidence from sunrise to sunset for (a) without tracking and fixed inclined surface (b) dual axis tracking during winter solstice.

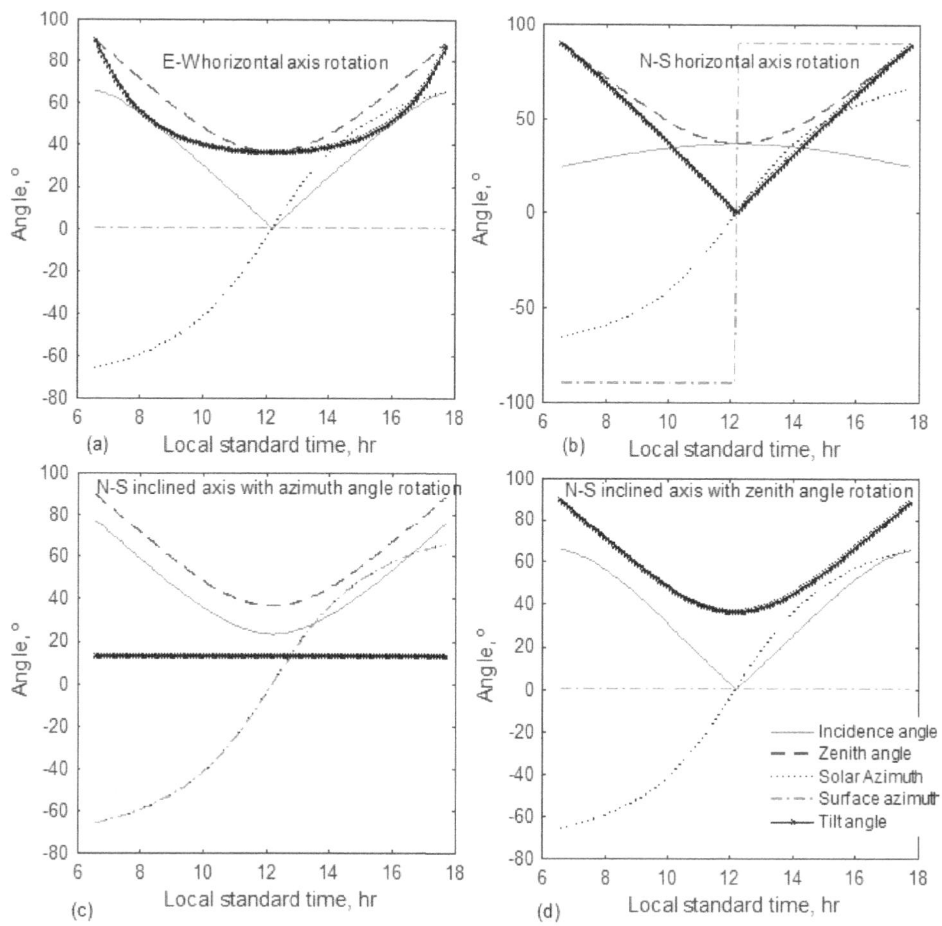

Fig. 10.32. Angle of incidence for single axis tracking from sunrise to sunset for (a) E-W horizontal axis rotation (b) N-S horizontal axis rotation (c) N-S inclined axis with azimuth angle rotation and (d) N-S inclined axis with zenith angle rotation during winter solstice.

the surface is passed through the sunbeam by tilting equal to zenith and surface azimuth to solar azimuth. At this position, the substitution of these two conditions results in zero incidence, which is perfect tracking with zero tracking error. The surface azimuth is less than 90° in dual axis tracking and falls on east-south in the morning and west-south in the afternoon.

Figure 10.32 shows the solar geometry and collector's angle of incidence on winter solstice for a single-axis tracking collector. The shown single-axis results are (a) E-W horizontal axis tracking, (b) N-S horizontal axis tracking, (c) N-S inclined axis with azimuth angle tracking, and (d) N-S inclined axis with zenith angle tracking. The change in the angle of incidence is focused on changes in zenith angle, solar azimuth angle, surface azimuth angle, and tilt angle from sunrise to sunset. The zenith angle is the angle of incidence of a horizontal surface since the normal passes through the zenith (vertical). In E-W horizontal axis rotation (Fig. 10.32a), the surface azimuth is zero since it is oriented towards north or south. At the zero surface azimuth condition, the mathematical equation of the angle of incidence can be differentiated and equated to zero to result in the angular rotation (tilt angle) of the E-W axis. The substitution of this tilt angle results in the minimum angle of incidence. The angle of incidence changes from maximum to zero and again reaches the maximum in the afternoon. During noon, the angle of incidence is zero. There is some improvement in the angle of incidence switching from a fixed inclined surface to E-W axis rotation.

The single axis can also be oriented in the N-S line. In this case, the surface azimuth angle in the forenoon is –90° with tilt towards east and 90° in the afternoon with tilt towards west. Similar to the E-W axis, the N-S axis can be found with differentiation for minimum incidence. In the E-W axis, the minimum incidence results at noon, while in the N-S axis, the minimum is at sunrise and at sunset instants. A great improvement is found in N-S axis tracking (Fig. 10.32b). The tilt angle for minimum incidence is changed from 90° (east) to 0°, then taken to 90° (west) in the period of a day length. The N-S axis can be tilted toward south for the location on the northern hemisphere with an option. For the tilted N-S axis, either the entire setup can be rotated about the vertical axis to change the azimuth values, which is called azimuth angle rotation, or the N-S axis can be rotated from east to west with a change in tilt, which is called zenith angle rotation. In the N-S inclined axis with azimuth angle rotation, the tilt angle is fixed at the local latitude and the azimuth is changed by rotating the collector from the east to west about a vertical axis. The resulting angle of incidence is not very effective compared to the other methods. The angle of incidence is decreased in the morning, increased in the afternoon, and reaches a minimum at the noon position (Fig. 10.32c). The surface azimuth measured from south varies from east (negative) to zero and reaches west (positive). This orientation is not resulting in the in the lowest angle of incidence compared to the others. In the fourth case, i.e., the N-S inclined axis with zenith angle rotation, the angle of incidence decreases in the morning and increases in the afternoon. But at noon, the incidence reaches zero value. Similar to the other N-S inclined axis tracking, this model also does not result in an effective angle of incidence.

Figure 10.33 shows the change in angle of incidence on the spring equinox from sunrise to sunset for (a) fixed surfaces and (b) dual axis tracking. On the fixed axis, the tilt angle is fixed as shown, and the resulting angle of incidence is not much improved compared to the zenith angle except at noon. At noon, it results in zero due to the equinox. As expected, the dual axis results in zero angle of incidence for the whole day. At the spring equinox also, the sun's shadow falls on an east-south direction in the morning and west-south in the afternoon as per the shown solar azimuth angle.

Figure 10.34 shows the angle of incidence from sunrise to sunset on the spring equinox with (a) E-W horizontal axis rotation (b) N-S horizontal axis rotation (c) N-S inclined axis with vertical axis rotation and (d) N-S inclined axis with horizontal axis rotation. The angle of incidence with tilted surface shown in Fig. 10.33a and E-W horizontal axis rotation (Fig. 10.34a) is nearly equal to the zenith angle. There is no improvement in the angle of incidence compared to the winter condition for tilted surfaces and E-W axis tracking. The N-S axis with vertical axis rotation results in better incidence than tilted surfaces (two cases) and E-W axis rotation. So, N-S horizontal axis tracking

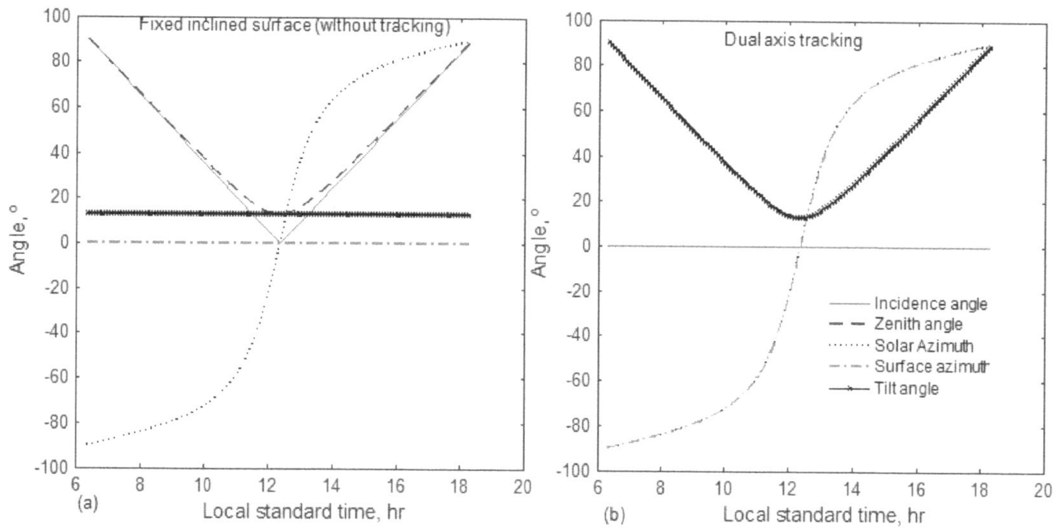

Fig. 10.33. Angle of incidence from sunrise to sunset for (a) fixed tilted surface (without tracking) and (b) dual axis rotation during spring equinox.

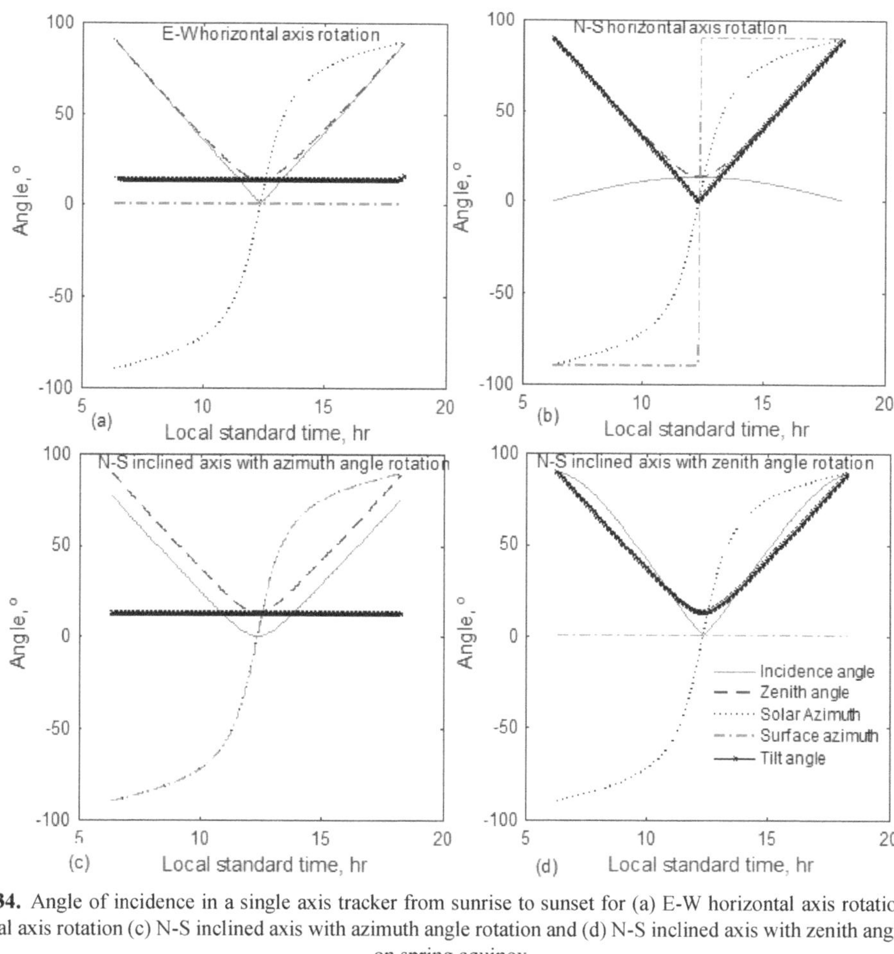

Fig. 10.34. Angle of incidence in a single axis tracker from sunrise to sunset for (a) E-W horizontal axis rotation (b) N-S horizontal axis rotation (c) N-S inclined axis with azimuth angle rotation and (d) N-S inclined axis with zenith angle rotation on spring equinox.

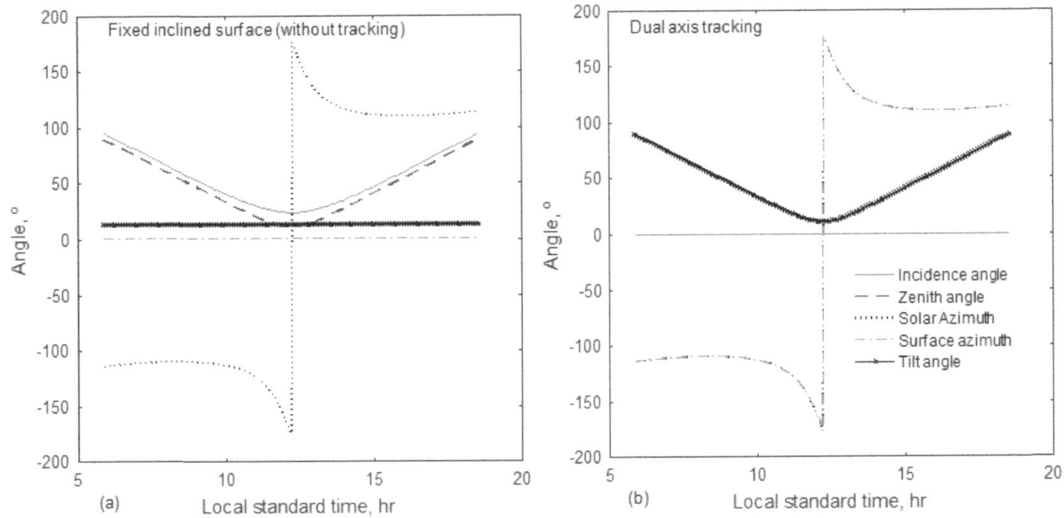

Fig. 10.35. Angle of incidence from sunrise to sunset for (a) fixed inclined surface and (b) dual axis rotation on summer solstice.

is more effective than the N-S inclined axis, either vertical axis rotation or horizontal axis rotation. Compared to the winter position, the angle of incidence is improved with the N-S horizontal axis tracking during the equinox period.

Figure 10.35 shows solar geometry results with (a) tilted surface (fixed) and (b) dual axis rotation on summer solstice. The generated angle of incidence is changed from 90° to minimum and again to 90° from sunrise to sunset, with the minimum incidence at noon. The incidence is zero with a dual-axis track. The sun's position is north-east during the forenoon (< –90°) and north-west in the afternoon (> 90°). At noon, the sun position comes on the north-south line, i.e., at midpoint.

Figure 10.36 shows the change in the angle of incidence on summer solstice with (a) E-W axis rotation, (b) N-S horizontal axis rotation, (c) N-S inclined axis with base rotation, and (d) N-S inclined axis with horizontal axis rotation. The results show that the angle of incidence is less varied in N-S horizontal axis rotation compared to the other single axis tracking methods. Compared to winter, in summer, the angle of incidence is low for the N-S horizontal axis position. With the N-S horizontal axis, the collector faces east and west; therefore, the surface azimuth is –90° and 90°, respectively. In the E-W horizontal and N-S inclined zenith axis rotation, the surface azimuth is maintained at zero for the day length. In the N-S inclined axis azimuth axis tracking, the surface azimuth is following the solar azimuth as per design.

Table 10.3 compares the minimum and maximum angle of incidences of the collector orientations in (i) horizontal surface, (ii) fixed inclined surface, (iii) E-W horizontal axis rotation, (iv) N-S horizontal axis rotation, (v) N-S tilted axis with vertical axis rotation, (vi) N-S tilted axis with horizontal axis rotation, and (vii) dual axis rotation on winter solstice, spring equinox, summer solstice, and autumn equinox. The angle of incidence resulted on the horizontal surface (zenith angle) is the primary reference for all the cases to compare the improvement. Similarly, the tilted surface (fixed) can be used as a reference to compare with tracking results and without tracking results. Except in summer, a small improvement can be observed in angle of incidence by shifting from the horizontal surface to a fixed inclined surface. There is not much benefit with E-W axis rotation compared to the inclined surface. The tracking results of the fixed inclined surface and the E-W axis are nearly the same. A great improvement with N-S horizontal axis rotation results over the other single axis tracking methods. The N-S inclined axis with rotation about vertical results in improvement over the E-W axis rotation. Even the N-S tilted axis with zenith angle rotation also did

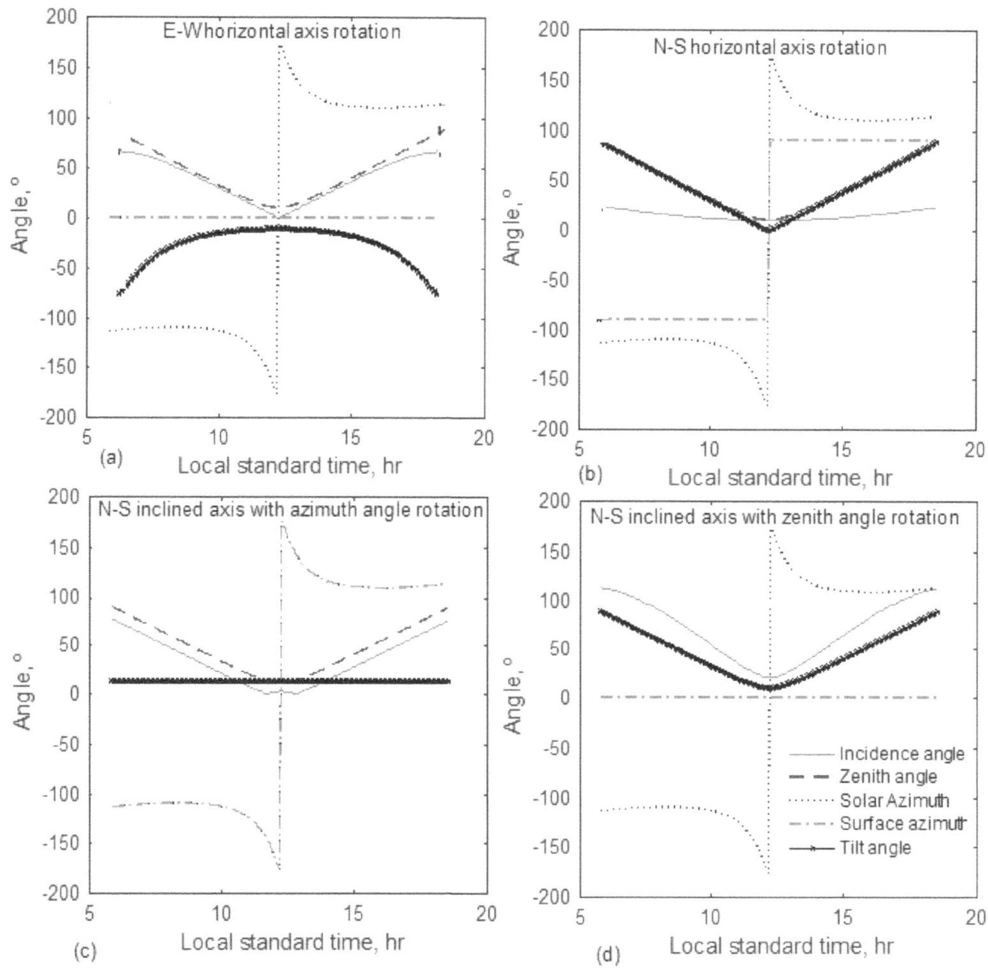

Fig. 10.36. Angle of incidence from sunrise to sunset for (a) E-W horizontal axis rotation (b) N-S horizontal axis rotation (c) N-S inclined axis with azimuth angle rotation and (d) N-S inclined axis with zenith angle rotation during summer solstice.

not prove well for angle of incidence minimization. The two-axes rotation results in zero incidence at the pre-defined conditions.

Figure 10.37 plots the angle of incidence changes on Julian day from the 7 a.m. in the morning to 5 p.m. in the evening for a tilted solar collector such as a solar water heater and solar PV panel. In the afternoon, the incidence is higher at 80°, and at noon, it decreases, and then there is a rise. The fluctuations are found with seasonal changes. The equinox results in a minimum incidence, and in the summer and winter it is maximum. The angle of incidence results in this case may be fixed as the base while comparing with tracking results.

Figure 10.38 shows the variation in angle of incidence from 7 A.M. to 5 P.M. in a year with (a) E-W horizontal axis, (b) N-S horizontal axis rotation, (c) N-S inclined axis with azimuth angle rotation, and (d) N-S fixed inclined axis with zenith angle rotation. Since the two-axis rotation always results in zero angle of incidence, it is not included in the results. In E-W horizontal mode and N-S tilted axis with azimuth angle rotation, two peaks are found during the year. Three peaks in the N-S horizontal axis and one peak in the N-S tilted axis with zenith-angle rotation are identified. There is not much difference in the angle of incidence with a fixed inclined surface to E-W horizontal rotation. The equinox period results in a lower incidence than the winter and summer. The lowest incidence is observed with N-S horizontal axis rotation compared to the other single axis tracking

Table 10.3. Minimum and maximum Angle of incidence on seasonal days at VIT University, Vellore, India.

S. No.	Tracking mode	Angle of incidence, θ							
		Winter solstice		Spring equinox		Summer solstice		Autumn equinox	
		Min. θ	Max. θ	Min. θ	Max. θ	Min. θ	Max. θ	Min. θ	Max. θ
1.	Horizontal surface, $\beta = 0$	36.39	90.00	13.03	90.00	10.49	90.00	11.94	90.00
2.	Fixed tilt angle, $\beta = \emptyset$, $\gamma = 0$	23.42	84.74	0.06	89.98	13.45	95.26	1.02	90.23
3.	E-W horizontal axis rotation, $\beta = \phi - \tan^{-1}\left(\frac{\tan\delta}{\cos\omega}\right)$, $\gamma = 0$	0.57	65.92	0.00	89.93	0.57	114.10	0.00	91.05
4.	N-S horizontal axis rotation, $\beta = \tan^{-1}\left(\frac{\pm\cos\delta\sin\omega}{\sin\phi\sin\delta + \cos\phi\cos\delta\cos\omega}\right)$, $\gamma = \pm 90$	24.27	36.39	0.06	13.03	10.47	23.82	0.07	11.94
5.	N-S fixed inclined axis and azimuth angle rotation. $\beta = \emptyset$, $\gamma = \gamma_s$	24.42	77.02	0.06	77.02	0.03	77.02	0.03	77.02
6.	N-S fixed inclined axis and zenith angle rotation. $\beta = \theta_z$, $\gamma = 0$	0.57	65.92	0.00	89.93	20.98	114.10	0.00	91.05
7.	Two axis tracking, $\beta = \theta_z$, $\gamma = \gamma_s$	0.00	0.00	0.00	0.00	0.00	0.00	0.00	0.00

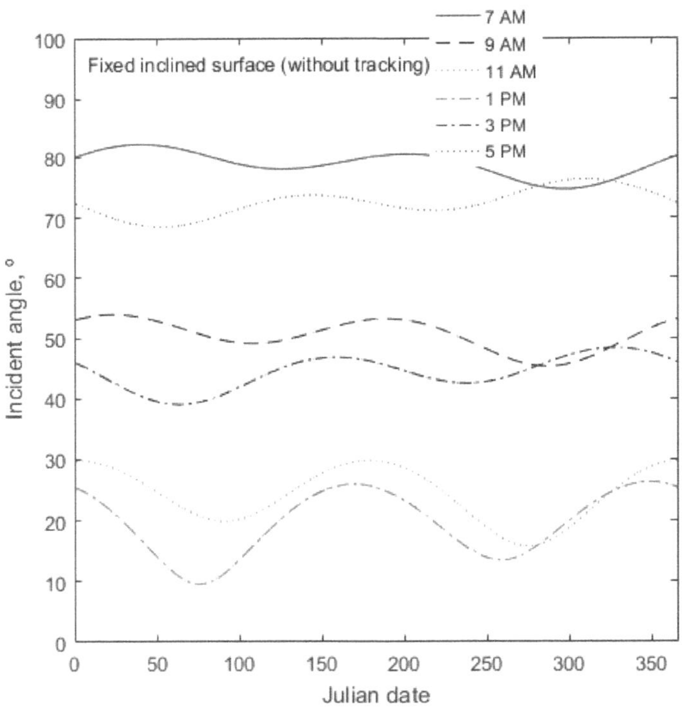

Fig. 10.37. Comparison of the angle of incidence for a fixed tilted surface from 7.00 A.M. to 5.00 P.M. in a year.

modes. Out of all the single-axis tracking models, N-S horizontal gives the greatest improvement during all times and all dates.

Figure 10.39 summarizes the angle of incidence given by tracking modes on the winter solstice, spring equinox, summer solstice, and autumn equinox. All six cases, starting from the tilted surface to dual-axis tracking, are compared with each other on seasonal dates. The angle of incidence during equinox is much lower than in the summer or winter. As per the earlier discussion, the N-S horizontal axis can be recommended for the lowest angle of incidence.

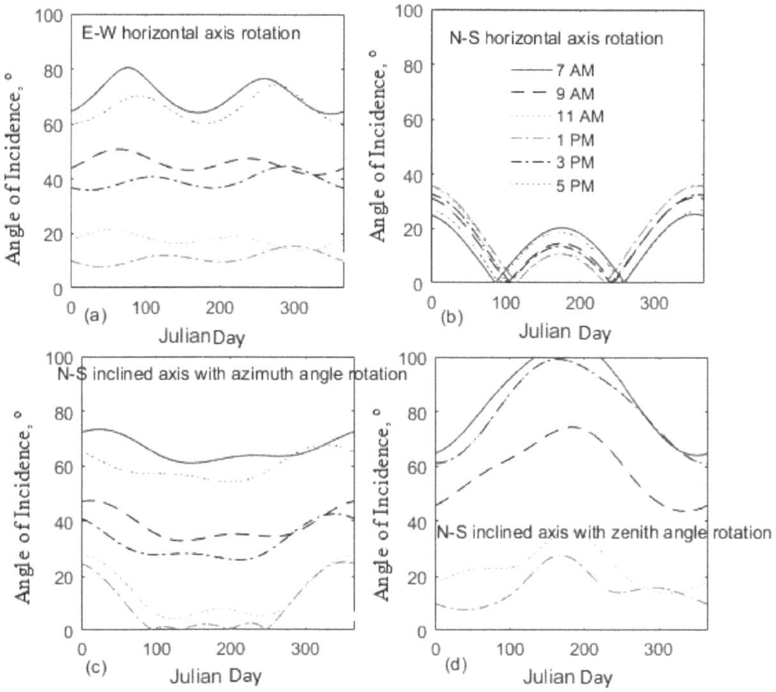

Fig. 10.38. Comparison of angle of incidence with a single axis tracking from 7.00 A.M. to 5.00 P.M. in a year for (a) E-W axis (b) N-S horizontal axis (c) N-S inclined axis with azimuth angle tracking and (d) N-S inclined axis with zenith angle tracking.

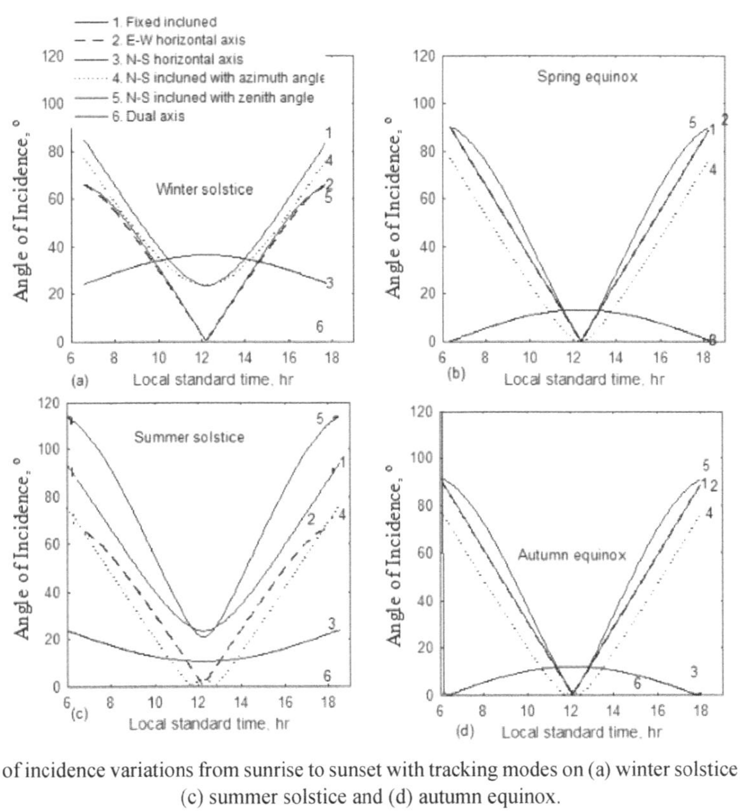

Fig. 10.39. Angle of incidence variations from sunrise to sunset with tracking modes on (a) winter solstice (b) spring equinox (c) summer solstice and (d) autumn equinox.

10.7 Numerical solutions

1. Find declination, hour angle, day length, sunrise time, sunset time, zenith angle, altitude angle, air mass, solar azimuth angle, surface azimuth angle, and angle of incidence at Vellore, India, on 20th January. Take the latitude and longitude of the place considered as 12.916° N and 79.1833° E, respectively. Indian standard longitude is 82.5° E. Time of study is 12.30 p.m. Assume the solar collector's tilt is equal to the latitude angle and due south. Consider the day length, sunrise time, and sunset time for a horizontal surface.

Solution through App Designer

Date 20 Month 1 Time, LST, h,.min (24 hours scale) 12.5

Latitude, ° 12.92 Longitude, ° 79.18 Standard Longitude, ° 82.5

Direction from Greenwich 1. Eastern 2. Westren 1

Orientation: 1. Horizontal 2. Fixed tile 3. EWH 4. NSH 5. NSIA 6. NSIZ 7. Dual 2

EWH: East West Header, NSH: North South Horizontal, NSIA: North South Inclined Axis, NSIZ: North South Inclined Zenith

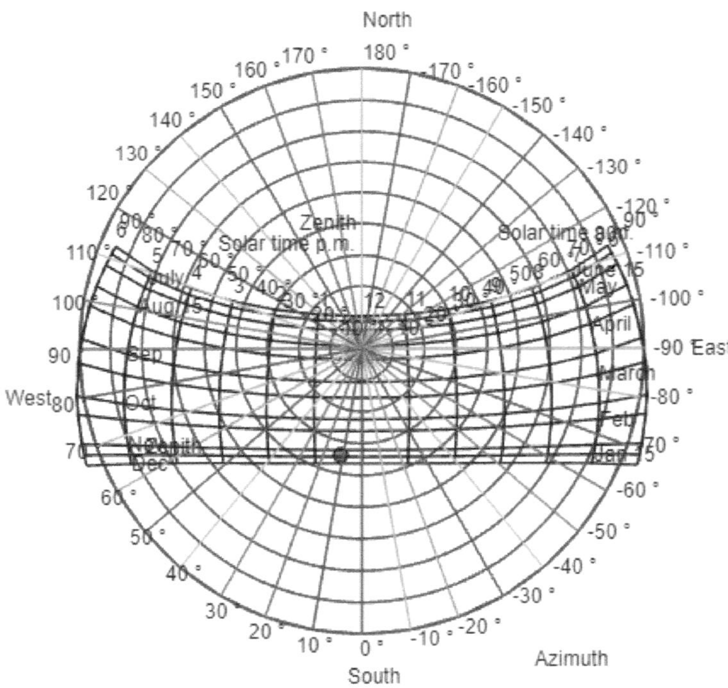

Day is January 20, the day number, n = 20

Parameter	Equation	Substitution	Result
Solar Time	$B = \frac{360(n-81)}{364}°$	$\frac{360(20-81)}{364}$	–60.33
	EoT = 9.87 sin 2B – 7.53 cos B – 1.5 sin B	9.87 sin (2 × (–60.3297°)) – 7.53 cos (–60.3297°) – 1.5 sin (–60. 3297°)	–10.907 min
	Solar time (Local Apparent Time) = Watch Time – 4 (L_{st} – L_{lo}) + EoT (in minutes)	12.5 – 4 (82.5 – 79.1833)/60 – 10.907/60	12.10 h = 12 h.06 min
Declination	$\delta = 23.45 \sin\left[\frac{360(284+n)}{365}\right]°$	$23.45 \sin\left[\frac{360(284+20)}{365}\right]$	–20.34°
Hour angle	ω = 15 (solar time – 12)°	15 (12.10 – 12)°	1.45°
Hour angle at sunrise or sunset	$\omega_s = \cos^{-1}(-\tan\delta \tan\varphi)$	$\cos^{-1}(-\tan(-20.34)\tan(12.9166))$	85.12°
Day length	$S = \frac{2}{15}\omega_s$	$\frac{2}{15} \times 85.12$	11.35 h
Sunrise time,	srt(ST) = 12 – S/2 srt(LST) = srt (ST) – ((–4 × (L_{st} – L_{lo}) + EoT)/60)	12 – 11.35/2 (ST) 6.325 – ((–4 × (82.5 – 79.1833) –10.91)/60) (LST)	6.325 h (ST) 6.73 h (LST) 6.44 am
Sunset time,	sst(ST) = S/2 = (solar time) sst(LST) = sst – ((–4 × (Lst – Llo) + EoT)/60	11.35/2 (ST) 5.675 – ((–4 × (82.5 – 79.1833) –10.907)/60) (LST)	5.675 h (ST) 6.08 h (LST) 6.05 pm
Zenith angle	$\theta_z = \cos^{-1}(\sin\varphi \sin\delta + \cos\varphi \cos\delta \cos\omega)$	$\cos^{-1}(\sin(12.9166)\sin(-20.34) + \cos(12.9166)\cos(-20.34)\cos(1.45))$	33.29°
Air mass	$\frac{1}{\cos\theta_Z}$	$\frac{1}{\cos 33.29°}$	1.20
Solar azimuth angle	$\gamma_s = \cos^{-1}\left(\frac{\cos\theta_Z \sin\phi - \sin\delta}{\sin\theta_Z \cos\phi}\right)$	$\gamma_s = \cos^{-1}\left(\frac{\cos(33.29°)\sin(12.9166°) - \sin(-20.34°)}{\sin(33.29°)\cos(12.9166°)}\right)$	2.49°
Angle of incidence	$\theta = \cos^{-1}(\sin\delta \sin(\varphi-\beta) + \cos\delta \cos(\varphi-\beta)\cos\omega)$	$\cos^{-1}(\sin(-20.34)\sin(12.9166 - 12.9166) + \cos(-20.34)\cos(12.9166 - 12.9166)\cos(1.45))$	20.9°

2. A solar collector is oriented with seven different modes, viz. (i) horizontal surface, (ii) fixed tilt without tracking, (iii) E-W horizontal axis rotation, (iv) N-S horizontal axis rotation, (v) N-S inclined axis with azimuth angle rotation, (vi) N-S inclined axis with zenith angle rotation, and (vii) dual axis tracking. For each mode, find the resulting angle of incidence at Vellore (12.916° N and 79.1833° E). The date and time of study are 24th December at 3.00 p.m. The standard Indian longitude is 82.5° E. Consolidate and analyze the results.

Solution through App Designer

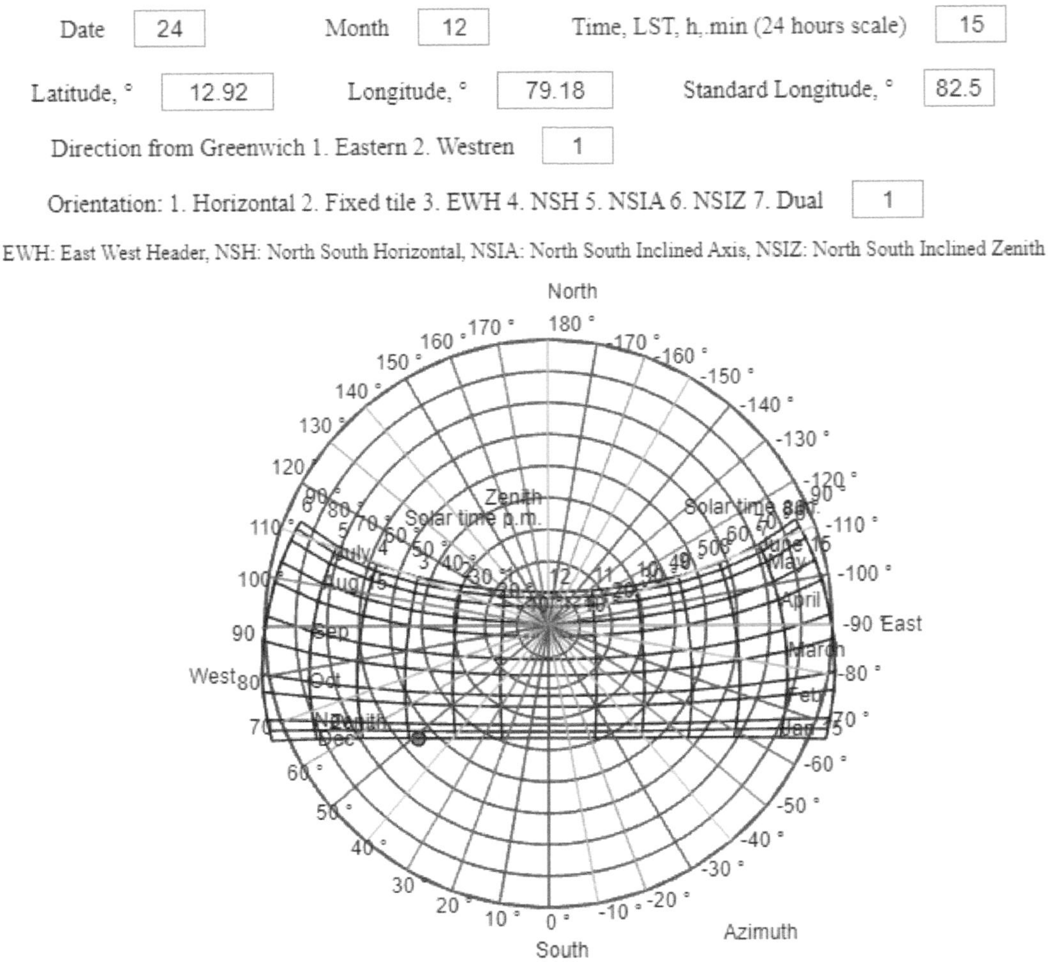

Day is December 24.

The day number, n = 358

Parameter	Equation	Substitution	Result
Solar Time	$B = \frac{360(n-81)}{364}$ °	$\frac{360(358-81)}{364}$	273.96°
	EoT = 9.87 sin 2B – 7.53 cos B – 1.5 sin B	9.87 sin (2 × 273.96°) – 7.53 cos (273.96°) – 1.5 sin (273.96°)	–0.38 min
	Solar time (Local Apparent Time) = Watch Time – 4 ($L_{st} - L_{lo}$) + EoT (in minutes)	15 – 4 (82.5 – 79.1833)/60 – 0.38/60	14.77 h = 14 h.46 min

Table contd. ...

...Table contd.

Parameter	**Equation**	**Substitution**	**Result**
Declination	$\delta = 23.45 \sin\left[\frac{360(284+n)}{365}\right]^\circ$	$23.45 \sin\left[\frac{360(284+358)}{365}\right]$	–23.41°
Hour angle	ω = 15 (solar time – 12)°	15 (14.77 – 12)°	41.59°
Hour angle at sunrise or sunset	$\omega_s = \cos^{-1}(-\tan\delta \tan\varphi)$	$\cos^{-1}(-\tan(-23.41) \tan(12.9166))$	84.30°
Day length	$S = \frac{2}{15}\omega_S$	$\frac{2}{15} \times 84.3$	11.24 h
Sunrise time,	srt(ST) = 12 – S/2 srt(LST) = srt (ST) – ((–4 × (L_{st} – L_{lo}) + EoT)/60)	12 – 11.24/2 (ST) 6.325 – 4 × (82.5 – 79.1833)/60 –0.38/60 (LST)	6.38 h (ST) 6.61 h (LST) 6.36 am
Sunset time,	sst(ST) = S/2 = (solar time) sst(LST) = sst – ((–4 × (Lst – Llo) + EoT)/60	11.24/2 (ST) 5.62 – 4 × (82.5 – 79.1833)/60 –0.38/60 (LST)	5.62 h (ST) 5.85 h (LST) 5.51 pm
Zenith angle	$\theta_z = \cos^{-1}(\sin\varphi \sin\delta + \cos\varphi \cos\delta \cos\omega)$	$\cos^{-1}(\sin(12.9166) \sin(-23.41) + \cos(12.9166) \cos(-23.41) \cos(41.59))$	54.54°
Air mass	$\frac{1}{\cos\theta_Z}$	$\frac{1}{\cos 54.54°}$	1.72
Solar azimuth angle	$\gamma_S = \cos^{-1}\left(\frac{\cos\theta_Z \sin\phi - \sin\delta}{\sin\theta_Z \cos\phi}\right)$	$\gamma_S = \cos^{-1}\left(\frac{\cos(54.54°)\sin(12.9166°) - \sin(-23.41°)}{\sin(54.54°)\cos(12.9166°)}\right)$	48.40°
Angle of incidence	$\theta = \cos^{-1}(\sin\delta \sin(\varphi - \beta) + \cos\delta \cos(\varphi - \beta) \cos\omega)$	$\cos^{-1}(\sin(-23.41) \sin(12.9166 - 0) + \cos(-23.41) \cos(12.9166 - 0) \cos(41.59))$	54.54°

i) Horizontal surface

ii) Fixed tilt angle, i.e., without tracking.

$\beta = \emptyset$

For the surface tilted due south, surface azimuth angle, $\gamma = 0°$

For surface tilted due south,

Angle of incidence, $\theta = \cos^{-1}(\sin\delta \sin(\varphi-\beta) + \cos\delta \cos(\varphi-\beta) \cos\omega)$

$= \cos^{-1}(\cos\delta \cos\omega)$ (since $\beta = \emptyset$)

$= \cos^{-1}(\cos(-23.41) \cos(41.56))$

$= 46.63°$

iii) E-W horizontal axis tracking

$$\beta = \phi - \tan^{-1}\left(\frac{\tan\delta}{\cos\omega}\right)$$

$$= 12.9166 - \tan^{-1}\left(\frac{\tan -23.41}{\cos 41.56}\right)$$

$$= 43.03°$$

$\gamma = 0$

$\delta = -23.41°$

ω = 41.56°

$$\phi - \beta = \tan^{-1}\left(\frac{\tan\delta}{\cos\omega}\right)$$

$$= \tan^{-1}\left(\frac{\tan(-23.41)}{\cos(41.56)}\right)$$

= –30.05°

After substituting the φ–β, the minimum angle of incidence can be obtained as a function of declination and hour angle.

$$\theta = \cos^{-1}\left(\sin\delta\sin\left(\tan^{-1}\left(\frac{\tan\delta}{\cos\omega}\right)\right) + \cos\delta\cos\left(\tan^{-1}\left(\frac{\tan\delta}{\cos\omega}\right)\right)\cos\omega\right)$$

$$= \cos^{-1}\left(\sin(-23.41)\sin(-30.05.92) + \cos(-23.41)\cos(-30.05)\cos(41.56)\right)$$

= 37.50°

iv) N-S horizontal axis tracking
γ = –90° in FN and
γ = +90° in AN
δ = –23.41°
ω = 41.56°

$$\beta = \tan^{-1}\left(\frac{\pm\cos\delta\sin\omega}{\sin\phi\sin\delta + \cos\phi\cos\delta\cos\omega}\right)$$

Since the give time is in afternoon,

γ = 90°

$$\beta = \tan^{-1}\left(\frac{\cos(-23.41)\sin(41.56)}{\sin(12.9166)\sin(-23.41) + \cos(12.9166)\cos(-23.41)\cos(41.56)}\right)$$

= 46.39°

Substituting these values in cos θ

Cosθ = Sinδ Sinφ Cosβ
– Sinδ Cosφ Sinβ Cosγ
+ Cosδ Cosφ Cosβ Cosω
+ Cosδ Sinφ Sinβ Cosγ Cosω
+ Cosδ Sinβ Sinγ Sinω

Cosθ = Sin(–23.41) Sin(12.916) Cos(46.39)
– Sin(–23.41) Cos(12.916) Sin(46.39) Cos(90)
+ Cos(–23.41) Cos(12.916) Cos(46.39) Cos(41.56)
+ Cos(–23.41) Sin(12.916) Sin(46.39) Cos(90) Cos(41.56)
+ Cos(–23.41) Sin(12.916) Sin(90) Sin(41.56)

θ = $\cos^{-1}$ (θ) = **32.77°**

v) N-S inclined axis with azimuth angle rotation
β = Ø = 12.9166
δ = –23.41°

$\omega = 41.56°$

Solar azimuth angle, $\gamma_s = \cos^{-1}\left(\frac{\cos\theta_Z \sin\phi - \sin\delta}{\sin\theta_Z \cos\phi}\right)$

$$= \cos^{-1}\left(\frac{\cos(54.56°)\sin(12.9166°) - \sin(-23.41°)}{\sin(54.56°)\cos(12.9166°)}\right)$$

$$= 48.35°$$

Therefore, $\gamma = \gamma_s = 48.35°$ (positive in AN, west)

substituting these values in cos θ

$$\begin{aligned}\text{Cos}\theta = {} & \text{Sin}\delta\ \text{Sin}\varphi\ \text{Cos}\beta \\ & - \text{Sin}\delta\ \text{Cos}\varphi\ \text{Sin}\beta\ \text{Cos}\gamma \\ & + \text{Cos}\delta\ \text{Cos}\varphi\ \text{Cos}\beta\ \text{Cos}\omega \\ & + \text{Cos}\delta\ \text{Sin}\varphi\ \text{Sin}\beta\ \text{Cos}\gamma\ \text{Cos}\omega \\ & + \text{Cos}\delta\ \text{Sin}\beta\ \text{Sin}\gamma\ \text{Sin}\omega\end{aligned}$$

$$\begin{aligned}\text{Cos}\theta = {} & \text{Sin}(-23.41)\ \text{Sin}(12.916)\ \text{Cos}(12.916) \\ & - \text{Sin}(-23.41)\ \text{Cos}(12.916)\ \text{Sin}(12.916)\ \text{Cos}(48.35) \\ & + \text{Cos}(-23.41)\ \text{Cos}(12.916)\ \text{Cos}(12.916)\ \text{Cos}(41.56) \\ & + \text{Cos}(-23.41)\ \text{Sin}(12.916)\ \text{Sin}(12.916)\ \text{Cos}(48.35)\ \text{Cos}(41.56) \\ & + \text{Cos}(-23.41)\ \text{Sin}(12.916)\ \text{Sin}(48.35)\ \text{Sin}(41.56)\end{aligned}$$

$\theta = \cos^{-1}(\theta) = \mathbf{41.58°}$

vi) N-S inclined axis with zenith axis rotation

$\beta = \theta_z = 54.56°$

$\gamma = 0$

$\delta = -23.41°$

$\omega = 41.56°$

Angle of incidence, $\theta = \cos^{-1}(\sin\delta \sin(\varphi - \theta_z) + \cos\delta \cos(\varphi - \theta_z) \cos\omega)$

$$\begin{aligned}&= \cos^{-1}(\sin(-23.41)\sin(12.9166 - 54.56) \\ &\quad + \cos(-23.41)\cos(12.9166 - 54.56)\cos(41.56)) \\ &= 38.98°\end{aligned}$$

vii) Two axis tracking

$\beta = \theta_Z = 54.56°$

$\gamma = \gamma_s = 48.35°$ (positive in FN, west)

$\delta = -23.41°$

$\omega = 41.56°$

Substituting these values in cos θ

$$\begin{aligned}\text{Cos}\theta = {} & \text{Sin}\delta\ \text{Sin}\varphi\ \text{Cos}\beta \\ & - \text{Sin}\delta\ \text{Cos}\varphi\ \text{Sin}\beta\ \text{Cos}\gamma \\ & + \text{Cos}\delta\ \text{Cos}\varphi\ \text{Cos}\beta\ \text{Cos}\omega \\ & + \text{Cos}\delta\ \text{Sin}\varphi\ \text{Sin}\beta\ \text{Cos}\gamma\ \text{Cos}\omega \\ & + \text{Cos}\delta\ \text{Sin}\beta\ \text{Sin}\gamma\ \text{Sin}\omega\end{aligned}$$

$$\begin{aligned}\text{Cos}\theta = {} & \text{Sin}(-23.41)\ \text{Sin}(12.916)\ \text{Cos}(54.56) \\ & - \text{Sin}(-23.41)\ \text{Cos}(12.916)\ \text{Sin}(54.56)\ \text{Cos}(48.35) \\ & + \text{Cos}(-23.41)\ \text{Cos}(12.916)\ \text{Cos}(54.56)\ \text{Cos}(41.56)\end{aligned}$$

+ Cos(–23.41) Sin(12.916) Sin(54.56) Cos(48.35) Cos(41.56)
+ Cos(–23.41) Sin(54.56) Sin(48.35) Sin(41.56)

$\theta = \cos^{-1}(\theta) = \mathbf{0°}$

Table 10.4 summarises the angle of incidence with the different modes of surface. These results show that the angle of incidence influences the collector's orientation. For the dual axis, zero incidence can be obtained. By tilting the surface from the horizontal, the angle of incidence is decreased from 54.56° to 46.63°. Out of all the single-axis tracking modes, the N-S horizontal axis results in a minimum angle of incidence. Zenith angle results are better than the azimuth angle tracking approach. But the study from sunrise to sunset will give the full result for the proper analysis.

Table 10.4. Summary of angle of incidence with mode of surface orientation.

S. No.	Orientation	Angle of incidence, θ°
1.	Horizontal surface, $\beta = 0$	54.54
2.	Fixed tilt angle, $\beta = \Phi$, $\gamma = 0$	46.63
3.	E-W horizontal axis tracking, $\beta = \theta_z$, $\gamma = 0$	37.50
4.	N-S horizontal axis tracking, $\beta = \Phi$, $\gamma = \gamma_s$	32.77
5.	N-S tilted axis with azimuth angle tracking, $\beta = \Phi$, $\gamma = \gamma_s$	41.58
6.	N-S tilted axis with zenith angle tracking, $\beta = \Phi$, $\gamma = \gamma_s$	38.98
7.	Dual axes tracking, $\beta = \theta_z$, $\gamma = \gamma_s$	00.00

3. At Vellore, Tamil Nadu, India (12.916° N and 79.1833° E), on what date will the zenith angle be zero at noon?

Solution

Zero zenith angle and solar noon imply $\theta_z = 0$ and $\omega = 0$.

After the simplification of the expression of the zenith angle,

$\delta = \varnothing$

$$\text{Declination angle, } \delta = 23.45 \sin\left[\frac{360(284+n)}{365}\right]°$$

$$n = \frac{365}{360} \times \sin^{-1}\left(\frac{\delta}{23.45}\right) - 284$$

$\delta = \varnothing = 12.916°$

$$n = \frac{365}{360} \times \sin^{-1}\left(\frac{12.916}{23.45}\right) - 284$$

$= -249.94$

In case n is negative, we need to add 365 to n

n = –250 + 365 = 115

The day numbers are 31 + 28 + 31 + 25 = 115

Now the date is April 25th.

Similarly, zero zenith occurs on one more day:

Another day = 365 – 115 = 250

The day numbers are 31 + 28 + 31 + 30 + 31 + 30 +31 + 31 + 7 = 250

The date is September 7th at the place chosen.

Chapter 11
Solar Radiation

11.1 Introduction

Solar radiation transmits from the sun to the earth through electromagnetic waves as per Maxwell's theorem. Plank's distribution law explains the nature of solar radiation as a function of the wave length and temperature of a body. The heat transfer radiation principles can be applied to determine the nature of solar radiation. Based on their emissive properties, bodies are classified as black body, grey body, white body, opaque body, reflective body, etc. Therefore, emissivity is an important property of solar radiation. In the first stage of solar radiation assessment, the amount of radiation transferred from the sun to the earth is determined. Out of the solar radiation falling on the ground, the portion of radiation that is received by the solar collector is dealt with in Chapter xx dealing with solar collectors. In this chapter, focus is given to the measurement and assessment procedure of solar radiation on earth. Solar radiation can be measured using relevant instruments after exposure to the sun. The radiation can also be noted from solar and atmosphere stations. It is practically impossible to measure solar radiation all locations on earth; however, there is a need for this data, which is why estimation techniques for solar radiation become necessary.

11.2 Solar radiation components

Solar radiation is the sun's energy that is received by the earth as electromagnetic waves, which consist of light and heat. Plank's distribution law can be used to find the monochromatic emissive power at a given wave length and body temperature, as shown in Fig. 11.1. With an increase in the body temperature, the emissive power increases proportionally.

The sun emits 3.846×10^{26} W. To find the solar flux falling on earth, we have to imagine a sphere with the sun as a center and a radius equal to the distance between the sun and earth. The distance between the sun and earth is 1.496×10^{11} m.

Therefore, the solar flux falling on earth which is also called irradiance or solar constant (SC) is,

$$I_{SC} = \frac{E}{4\pi R^2} = \frac{3.846 \times 10^{26}}{4\pi \left(1.496 \times 10^{11}\right)^2} = 1367 \ W/m^2 \tag{11.1}$$

The irradiance, or solar constant, is also called extraterrestrial radiation. Now, the portion of solar radiation towards the earth out of the total solar radiation in all directions can be determined from the ratio of the earth's half surface area to the total sphere's surface area, with radius equal to the distance between the sun and the earth.

The portion of earth receiving the radiation, out of the total solar radiation,

$$\frac{I_{earth}}{I_{sun}} = \frac{4\pi r^2/2}{4\pi R^2} = \frac{r^2}{2R^2} = \frac{1}{2}\left(\frac{6.371 \times 10^6}{1.496 \times 10^{11}}\right)^2 = 9 \times 10^{-10} \tag{11.2}$$

DOI: 10.1201/9781032642505-11

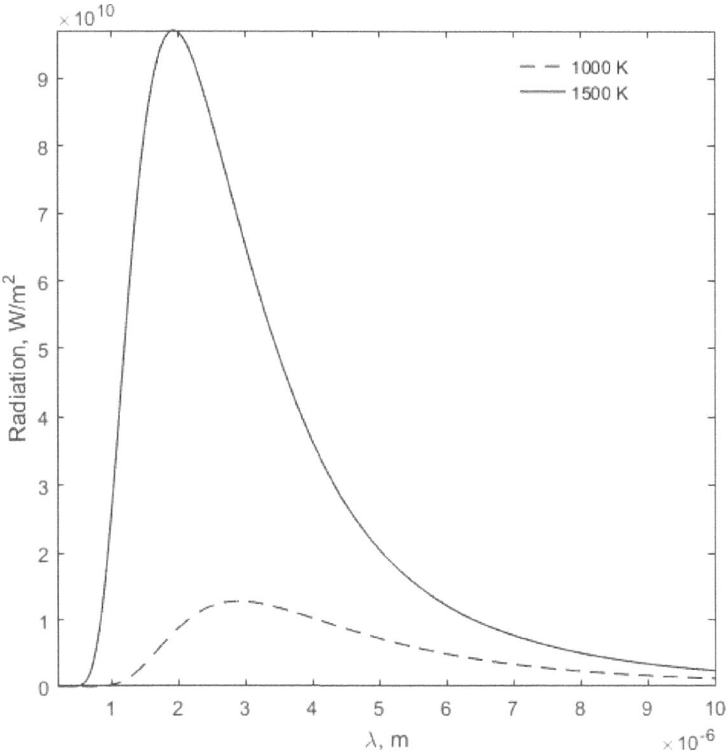

Fig. 11.1. Spectral radiation at 1000 K and 1500 K using Plank's distribution law.

Its value is nearly 0.00000009%. Therefore, out of the total solar radiation the earth's portion is very small.

Even out of the solar energy directed towards the earth, all the radiation does not reach the earth. The earth receives radiation directly (direct radiation) and indirectly (diffuse radiation). The direct radiation falls on the earth without any obstacles or disturbance. The indirect radiation, also called diffuse radiation, is radiation that scatters or diffuses due to atmospheric particles. The sum of the direct and diffuse radiation is called global or total radiation. Apart from these radiations, the surface also receives the radiation from the ground reflection. Therefore, the instrument-measured global radiation is slightly higher than the actual solar radiation falling on earth.

Irradiance, I, W/m^2, is also defined as a heat flux falling on a surface, whether it is beam, diffuse, or global radiation. Insolation, H, J/m^2, is the solar energy flux falling on the surface determined by integrating the irradiance over a period of time. Irradiance is instantaneous solar energy, and insolation is the energy content of solar energy with a specified time period.

Figure 11.2 details the solar radiation budget on earth. The solar radiation falling on earth is absorbed (10%), reflected or scattered (25%), and transmitted (17%) through the sky. Some solar radiation directly falls (24%) on earth without scattering, as shown in Fig. 11.2. The balance of 24% falls on the atmosphere, where again it is absorbed (9%), reflected (9%), and transmitted (6%). So, the total radiation falling on earth is the sum of direct radiation (24%), transmitted radiation (17%), and atmospheric transmission (6%), which is equal to 47%. The remaining 53% does not reach the earth. Overall, about half of the solar radiation is falls on earth, and the balance does not reach it. So, the collection efficiency of the earth for the solar radiation is 47%, which is nearly equal to 50%. Half of the surface area of the earth is 4πr2/2. From Eq. 11.1, the solar energy falling on earth before losses = $I_{SC} \times 2\pi\ r^2 = 1367 \times 2\pi\ (6.371 \times 106)^2 = 3.486 \times 1017$ W = 3.486 × 105 TW.

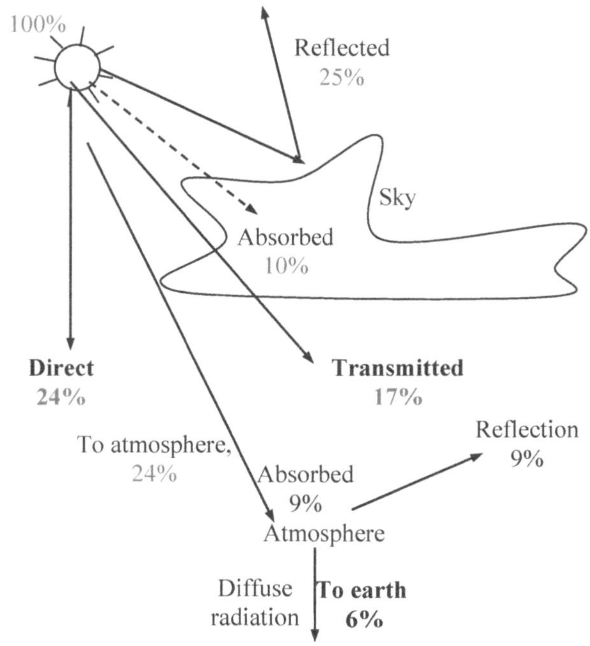

Fig. 11.2. Solar radiation budget.

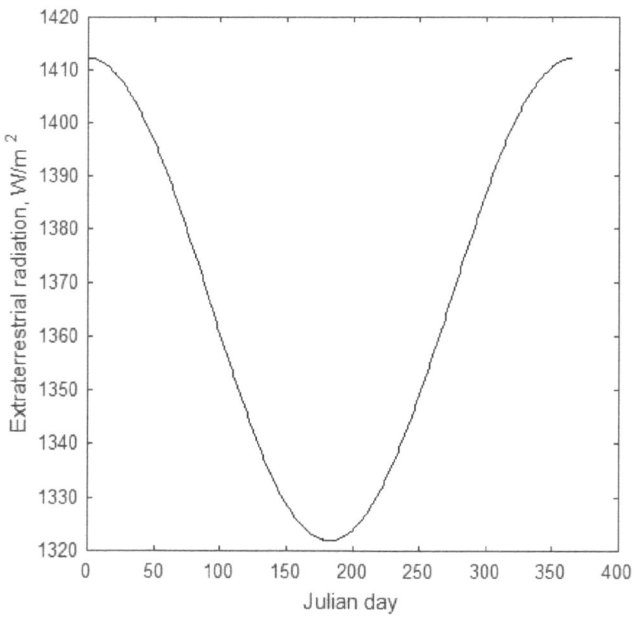

Fig. 11.3. Extraterrestrial radiation as a function of Julian day.

The global electricity demand is 18 TW, which is very low compared to the solar energy falling on earth before subtracting the losses. If 53% of this solar radiation is not received by the ground, the balanced solar energy falling on the ground is 163842 TW, i.e., nearly 164000 TW. So, it is possible to meet the total global energy demand through solar energy.

As stated in the earlier section, solar constant, or irradiance, is also called extraterrestrial solar radiation. The extraterrestrial radiation is the solar radiation available outside of the earth. It depends on the distance between the sun and earth. At the mean distance of 1.496×10^6 m, the extraterrestrial solar radiation is 1367 W/m². Since this distance is not fixed, the extraterrestrial radiation also varies with time. This variation is nearly 3%. As stated in the earlier section, 53% of this extraterrestrial radiation is not reaching the earth, and the balanced 47% is only reaching the ground through direct and indirect radiation.

The extraterrestrial radiation as a function of Julian day (n),

$$I'_{SC} = I_{SC}\left(1 + 0.033\cos\frac{360n}{365}\right) \tag{11.3}$$

Figure 11.3 shows the extraterrestrial solar radiation varying with date. Since the distance between sun and earth changes continuously it is plotted with the function of Julian date. The results show that the mean distance approaches during equinox, and the mean solar irradiance is 1367 W/m². During summer, the distance is greater and the extraterrestrial radiation is minimal. In winter, the shorter distance between the sun and earth results in maximum extraterrestrial radiation.

11.3 Prediction of solar radiation

To know the available solar energy at a site, area required for the solar collectors, capacity of the solar power plant, solar energy data either measured or predicted is required. The solar radiation data is not available for all the places on earth. Therefore, there is a need to develop empirical models to predict the components of solar radiation without much error. To predict the actual solar radiation under a cloudy sky, in addition to solar geometry, some metrological parameters such as location height from the sea level, relative humidity, rail fall, ambient temperature, sunshine hours, cloud cover, etc. are required. Figure 11.4 shows typical daily radiation, predicted hourly from sunrise to sunset, on a particular day at a location. It compares the actual and predicted data of (a) global radiation and (b) beam radiation. The randomness in the actual radiation can be observed due to cloudy conditions. The factor for the cloudy conditions can be included in hypothetical models, but the exact trace is not reproducible using the model. The predicted result should prove both magnetite and range of variations. The magnitude or the energy content from sunrise to sunset can be determined by the integration of instantaneous solar radiation over the period of selection. The matching of shape ensures the distribution and spread of the data over the period of time (Sukhatme and Nayak, 2017).

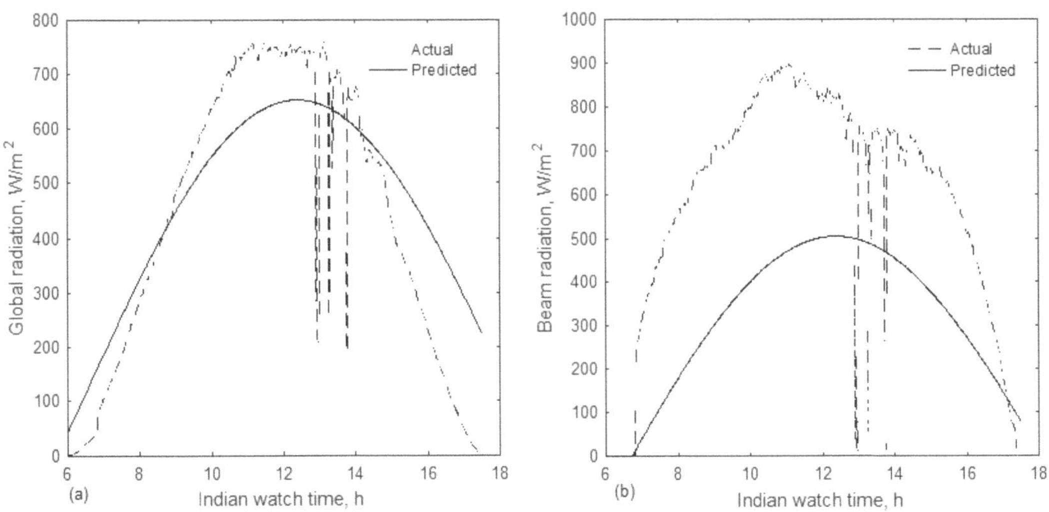

Fig. 11.4. Actual and estimated solar radiation components (a) global radiation and (b) beam radiation.

The estimation of solar radiation is either for a day or for a month expressed in kJ/m²/day. The following section is focused on estimation of monthly average daily global radiation, monthly average daily diffuse radiation, monthly average hourly global radiation, and monthly average hourly diffuse radiation.

11.3.1 Hourly global, beam and diffuse radiation under cloudless skies (clear days)

The global radiation I_g reaching a horizontal surface on the earth is given by

$$I_g = I_b + I_d$$

where I_g = hourly global radiation
I_b = hourly beam radiation
I_d = hourly diffuse radiation

Now, $I_b = I_{bn} \cos \theta_z$

where
I_{bn} = beam radiation in the direction of the rays, and
θ_z = angle of incidence on a horizontal surface, i.e., the zenith angle. Thus

$$I_g = I_{bn} \cos \theta_z + I_d$$

In the ASHRAE (American Society of Heating, Refrigerating and Air-conditioning Engineers) model, it is postulated that

$$I_{bn} = A \exp\left(\frac{-B}{\cos \theta_z}\right) \text{ and}$$

$$I_d = C\, I_{bn}$$

Where A, B, and C are constants whose values have been determined on a month-wise basis (Tables 11.1 and 11.2). These constants change during the year because of seasonal changes in the dust and water vapor content of the atmosphere and also because of the changing earth-sun distance.

Table 11.1. Values of the constants A, B and C used for predicting hourly solar radiation on clear days in India.

S. No.	Month, Day	A, W/m²	B	C
1.	January 21	610.00	0.000	0.242
2.	February 21	652.20	0.010	0.249
3.	March 21	667.86	0.036	0.299
4.	April 21	613.35	0.121	0.395
5.	May 21	558.39	0.200	0.495
6.	June 21	340.71	0.428	1.058
7.	July 21	232.87	0.171	1.611
8.	August 21	240.80	0.148	1.624
9.	September 21	426.21	0.074	0.688
10.	October 21	584.73	0.020	0.366
11.	November 21	616.60	0.008	0.253
12.	December 21	622.52	0.000	0.243

Table 11.2. Values of constants, A, B and C obtained for predicting hourly solar radiation in USA.

S. No.	Month, Day	A, W/m^2	B	C
1.	January 21	1202	0.141	0.103
2.	February 21	1187	0.142	0.104
3.	March 21	1164	0.149	0.109
4.	April 21	1130	0.164	0.120
5.	May 21	1106	0.177	0.130
6.	June 21	1092	0.185	0.137
7.	July 21	1093	0.186	0.138
8.	August 21	1107	0.182	0.134
9.	September 21	1136	0.165	0.121
10.	October 21	1136	0.152	0.111
11.	November 21	1190	0.144	0.106
12.	December 21	1204	0.141	0.103

11.3.2 Monthly average daily global radiation

The daily radiation, or monthly average daily radiation, is the total solar energy received from sunrise to sunset in a specified month and date under cloudy conditions. To determine the solar radiation under cloudy conditions, an actual atmospheric parameter or parameters are required. In this method, the actual day length measured by the sunshine recorder is considered. The solar radiation decreases from space to ground due to the resistance offered by the atmosphere layer. The Daily global radiation is recorded in dimensionless numbers. The day length is expressed as a number by the ratio of actual daylength and maximum day length on a particular date. The maximum day length is the duration between the sunrise and sunset. It can be determined from the solar geometry for the location and date. Under cloudy conditions, the actual day length is less than the clear sky day length, i.e., the maximum day length. The daily global radiation is also expressed as a number, which is the ratio of actual solar radiation and reference solar radiation, such a clear sky solar energy or extraterrestrial solar radiation energy.

To develop the empirical correlations for the solar radiation at the selected location, earlier annual solar data is required at least for a duration of 10 years. The data of global radiation, beam radiation, and diffuse radiation may be collected at regular intervals. The cumulative solar energy can be determined from this instantaneous solar radiation by the sum of the readings of the day's length. The daily solar energy can be expressed in kJ/m^2/day or kWh/m^2/day.

In 1924, Angstrom developed a linear equation to determine the daily radiation from the monthly average for a horizontal surface. The linear equation consists of two constants 'a' and 'b' determined from the actual monthly and annual average data and equation fitting (Tables 11.3 and 11.4). The constants 'a' and 'b' are determined at various cities internationally as follows.

Angstrom equation,

$$\frac{\overline{H}_g}{\overline{H}_c} = a + b\left(\frac{\overline{S}}{\overline{S}_{max}}\right) \tag{11.4}$$

The equation has the following limitations:

Different observers have obtained different values for the constant 'b' related with a clear sky. It is influenced by observer's judgment regarding a clear sky. $\overline{H}_c$ is calculated on the basis of the solar constant and geometrical calculations of solar ray angles and latitude of the place etc. It depends on a clarity factor assumed in calculations. Hence $\overline{H}_c$ has wide variation. Hence the Angstrom's equation has been modified by use of $\overline{H}_0$ in place of $\overline{H}_c$.

Table 11.3. Constants 'a' and 'b' for many cities of the world.

S. No.	Location	Range of $\frac{\bar{S}}{\bar{S}_{max}}$	a	b
1.	Albuquerque, New Mexico, USA	0.68–0.85	0.41	0.37
2.	Atlanta, Georgia, USA	0.45–0.71	0.38	0.26
3.	Blue Hill, Mass, USA	0.47–0.80	0.22	0.50
4.	Brownsville, Texas, USA	0.47–0.80	0.35	0.31
5.	Buenos Aires, Argentina	0.47–0.68	0.26	0.50
6.	Charleston, SC, USA	0.60–0.75	0.48	0.09
7.	Dairen, Manchuria	0.55–0.81	0.36	0.23
8.	EI Paso, Texas, USA	0.78–0.88	0.54	0.20
9.	Ely, Nevada, USA	0.61–0.88	0.54	0.18
10.	Hamburg, Germany	0.11–0.49	0.22	0.57
11.	Honolulu, Hawaii, USA	0.57–0.77	0.14	0.73
12.	Madison, Wisconsin, USA	0.40–0.72	0.30	0.34
13.	Malanga, Angola	0.41–0.84	0.34	0.34
14.	Miami, Florida, USA	0.56–0.71	0.42	0.22
15.	Nice, France	0.49–0.76	0.17	0.63
16.	Pune, India	0.25–0.49 0.65–0.89	0.30 0.41	0.51 0.34
17.	Stanleyville, Congo	0.34–0.56	0.28	0.39
18.	Tamanrasset, Algeria	0.76–0.88	0.30	0.43

H_0 is the daily average of extra-terrestrial radiation corresponding to the solar constant and is fairly constant during the complete year.

Modified Angstrom's equation for daily global radiation,

$$\frac{\bar{H}_g}{\bar{H}_0} = a + b\left(\frac{\bar{S}}{\bar{S}_{max}}\right) \tag{11.5}$$

$\bar{H}_0$ is the daily mean value of extra-terrestrial (beyond earth's atmosphere) radiation on a unit surface perpendicular to sun rays; corresponding to the surface on ground under consideration; the daily mean value for the month in terms of kJ/m^2.day.

H_0 is obtained by integrating the solar insolation on a horizontal surface over a period of a complete solar day (hour angle $-\omega_s$ to $+\omega_s$).

The average value of $\bar{H}_0$ is calculated by averaging the values of all days in aa month. For months having 30 days

$$\bar{H}_0 = \frac{H_1 + H_2 + ...H_{30}}{30} \tag{11.6}$$

Individual H_1, H_2, H_3.... are calculated from the equation.

Table 11.4. Constant 'a' and 'b' for Indian cities.

S. No.	Location	a	b	Mean error, %
1.	Ahmedabad	0.48	0.28	3.0
2.	Bangalore	0.64	0.18	3.9
3.	Bhavnagar	0.47	0.28	2.8
4.	Calcutta	0.42	0.28	1.3
5.	Chennai	0.44	0.30	3.5
6.	Goa	0.48	0.30	2.1
7.	Jodhpur	0.46	0.33	2.0
8.	Kodaikanal	0.55	0.32	2.9
9.	Mangalore	0.43	0.27	4.2
10.	Minicoy	0.39	0.26	1.4
11.	Nagpur	0.50	0.27	1.6
12.	New Delhi	0.57	0.25	3.0
13.	Pune	0.43	0.31	1.9
14.	Shillong	0.57	0.22	3.0
15.	Srinagar	0.40	0.35	4.7
16.	Thiruvananthapuram	0.39	0.37	2.5
17.	Vishakhapatnam	0.47	0.28	1.2

For horizontal surfaces, the tilt angle is zero. Therefore, the incidence angle is given by the zenith angle.

$\text{Cos}\theta = \text{Cos}\theta_z = \text{Sin}\varphi \ \text{Sin}\delta + \text{Cos}\varphi \ \text{Cos}\delta \ \text{Cos}\omega$

$$H_0 = I_{sc}\left(1+0.033.\cos\frac{360n}{365}\right)\int \cos\theta_z dt \tag{11.7}$$

$$H_0 = I_{sc}\left(1+0.033.\cos\frac{360n}{365}\right)\int [\sin\phi.\sin\delta + \cos\phi.\cos\delta.\cos\omega]\,dt \tag{11.8}$$

The integral $\int dt$ in the above equation is in terms of hours. It can be changed to angle $\int d\omega$ in radiations.

$$1 \text{ hour} = 15°. \text{ Hence, } dt = \frac{dw}{15}\times\frac{180}{\pi} = \frac{12}{\pi}d\omega \tag{11.9}$$

Substituting in the equation and integrating from sunrise ($-\omega_s$) to sunset ($+\omega_s$) for the total day length,

$$H_0 = \frac{12}{\pi}I_{sc}\left(1+0.033.\cos\frac{360n}{365}\right)\int_{-\omega_s}^{\omega_s}(\sin\phi.\sin\delta + \cos\phi.\cos\delta.\cos\omega)d\omega$$

$$= \frac{12}{\pi}I_{sc}\left(1+0.033.\cos\frac{360n}{365}\right)(\sin\phi.\sin\delta.\omega + \cos\phi.\cos\delta.\sin\omega)_{-\omega_s}^{\omega_s}$$

$$= \frac{12}{\pi}I_{sc}\left(1+0.033.\cos\frac{360n}{365}\right)\left(\sin\phi.\sin\delta\left(\omega_s-(-\omega_s)\right) + \cos\phi.\cos\delta.\left(\sin\omega_s - \sin(-\omega_s)\right)\right)$$

$$= \frac{12}{\pi} I_{sc}\left(1+0.033.\cos\frac{360n}{365}\right)(2\omega_s \sin\phi.\sin\delta + \cos\phi.\cos\delta.2\sin\omega_s)$$

$$H_0 = \frac{24}{\pi} I_{sc}\left(1+0.033.\cos\frac{360n}{365}\right)(\omega_s \sin\phi.\sin\delta + \cos\phi.\cos\delta.\sin\omega_s)$$

$$H_0 = \frac{24}{\pi} I_{sc} \times 3600\left(1+0.033.\cos\frac{360n}{365}\right)(\omega_s \sin\phi.\sin\delta + \cos\phi.\cos\delta.\sin\omega_s),\ kJ/m^2 - day \quad (11.10)$$

Figure 11.5 compares the extraterrestrial radiation of the northern hemisphere and the southern hemisphere. It can be observed that the summer days in the southern hemisphere are hotter than the northern hemisphere. It is because of the closeness of the sun during the summer compared to the summer in the northern hemisphere. In winter, H is much less at high latitudes because of shorter day length, more oblique incidence, and greater atmospheric attenuation. The calculation of H_0 is simplified by determining the day in each month on which the extraterrestrial radiation is nearly equal to the monthly mean value. The dates on which the value of H_0 is equal to the average value are as follows: January 17, February 16, March 16, April 15, May 15, June 11, July 17, August 16, September 15, October 15, November 14, and December 10. As expected, these dates are close to the middle of the month (Table 11.5).

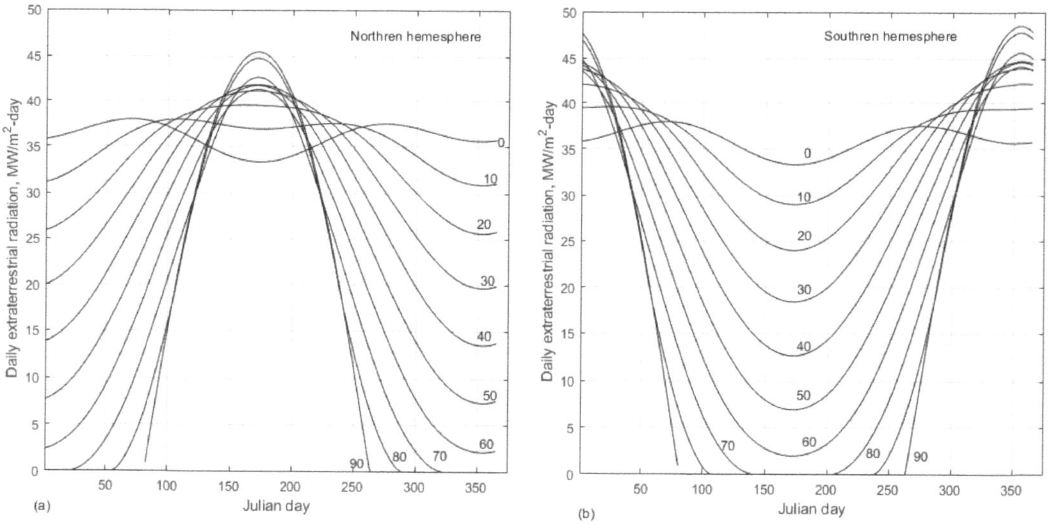

Fig. 11.5. Daily extra-terrestrial solar energy with latitude on (a) northern hemisphere and (b) southern hemisphere.

11.4 Solar radiation on tilted surfaces

Most solar equipment (e.g., flat-plate collectors) for absorbing radiation are tilted at an angle to the horizontal. It therefore becomes necessary to calculate the flux that falls on a tilted surface. This flux is the sum of the beam and diffuse radiation falling directly on the surface and the radiation reflected onto the surface from the surroundings.

11.4.1 Beam radiation

The ratio of the beam radiation flux falling on a tilted surface to that falling on a horizontal surface is called the tilt factor for beam radiation. It is denoted by the symbol r_b. For the case of a tilted surface facing south (i.e., $\gamma = 0°$),

Table 11.5. Representative dates for the months to determine the monthly average global solar radiation.

S. No.	Month	n for i^{th} day of month	Date, i
1	January	i	17
2	February	31 + i	16
3	March	59 + i	16
4	April	90 + i	15
5	May	120 + i	15
6	June	151 + i	11
7	July	181 + i	17
8	August	212 + i	16
9	September	243 + i	15
10	October	273 + i	15
11	November	304 + i	14
12	December	334 + i	10

$$\cos\theta = \sin\delta\sin(\phi-\beta)+\cos\delta\cos\omega\cos(\phi-\beta)$$

While for a horizontal surface

$$\cos\theta_z = \sin\phi\sin\delta+\cos\phi\cos\delta\cos\omega$$

Hence

$$r_b = \frac{I_{bn}\cos\theta}{I_{bn}\cos\theta_z} = \frac{\sin\delta\sin(\phi-\beta)+\cos\delta\cos\omega\cos(\phi-\beta)}{\sin\phi\sin\delta+\cos\phi\cos\delta\cos\omega} \tag{11.11}$$

Similarly, expressions for r_b can be derived for other situations in which the tilted surface is oriented in a different direction with $\gamma \neq 0°$.

11.4.2 Diffuse radiation

The tilt factor r_d for diffuse radiation is the ratio of the diffuse radiation flux falling on the tilted surface to that falling on a horizontal surface. The value of this tilt factor depends upon the distribution of diffuse radiation over the sky and on the portion of the sky dome seen by the tilted surface. Assuming that the sky is an isotropic source of diffuse radiation, we have for a tilted surface with a slope β,

$$r_d = \frac{1+\cos\beta}{2} \tag{11.12}$$

Since (1+cos β)/2 is the radiation shape factor for a tilted surface with respect to the sky.

11.4.3 Reflected radiation

Since (1+cos β)/2 is the radiation shape factor for a tilted surface with respect to the sky, it follows that (1+cos β)/2 is the radiation shape factor for the surface with respect to the surrounding ground. Assuming that the reflection of the beam and diffuse radiations falling on the ground is diffuse and isotropic, and that the reflectivity, is ρ, the tilt factor for reflected radiation is given by

$$r_r = \rho\left(\frac{1-\cos\beta}{2}\right) \tag{11.13}$$

11.4.4 Flux on tilted surface

The flux I_T falling on a tilted surface at any instant is given by

$$I_T = I_b r_b + I_d r_d + (I_b + I_d)\, r_r \tag{11.14}$$

Dividing both sides by I_g, we obtain the ratio of the flux falling on a tilted surface at any instant to that on a horizontal surface.

$$\frac{I_T}{I_g} = \left(1 - \frac{I_d}{I_g}\right) r_b + \frac{I_d}{I_g} r_b + r_r \tag{11.15}$$

This equation can be used for calculating the hourly radiation falling on a tilted surface if the value of ω is taken at the mid-point of the hour. It can also be used for calculating the monthly average hourly value $(\overline{I_T})$, if the calculations are done for the representative day of the month. The modified form of the equation is

$$\frac{\overline{I_T}}{\overline{I_g}} = \left(1 - \frac{\overline{I_d}}{\overline{I_g}}\right) \overline{r_b} + \frac{\overline{I_d}}{\overline{I_g}} \overline{r_b} + \overline{r_r} \tag{11.16}$$

where $\overline{r_b} = r_b$, $\overline{r_d} = r_d$ and $\overline{r_r} = r_r$

The daily radiation falling on a tilted surface is also of interest in many applications. The ratio of the daily radiation falling on a tilted surface (H_T) to the daily global radiation on a horizontal surface (H_g) is

$$\frac{H_T}{H_g} = \left(1 - \frac{H_d}{H_g}\right) R_b + \frac{H_d}{H_g} R_d + R_r \tag{11.17}$$

For a south facing surface ($\gamma = 0°$),

$$R_b = \frac{\omega_{st} \sin\delta \sin(\phi - \beta) + \cos\delta \sin\omega_{st} \cos(\phi - \beta)}{\omega_s \sin\phi \sin\delta + \cos\phi \cos\delta \sin\omega_s} \tag{11.18}$$

$R_d = r_d$ and $R_r = r_r$

ω_{st} and ω_s are the sunrise or sunset hour angles (expressed in radians) for the tilted surface and a horizontal surface respectively.

The equation can also be used for calculating the monthly average daily radiation falling on a tilted surface, if the values required are calculated for the representative day of the month.

$$\frac{\overline{H_T}}{\overline{H_g}} = \left(1 - \frac{\overline{H_d}}{\overline{H_g}}\right) \overline{R_b} + \frac{\overline{H_d}}{\overline{H_g}} \overline{R_d} + \overline{R_r} \tag{11.19}$$

11.5 Development of solar radiation at a typical location, Vellore, India

Assessment of solar radiation saves, time for measurement, and use of costly equipment. It is difficult to measure solar radiation in remote places such as deserts, forests, and villages. The current mathematical work has developed a new and specific forecasting model to estimate instantaneous solar heat flux and monthly average daily solar energy flux for select locations in India. Linear and quadratic fit coefficients are developed from solar radiation data collected at Vellore, India. The instantaneous solar radiation is generated as a function of the sunshine hour ratio and time for every month. Similarly, monthly average daily solar energy is the function of the sunshine hour ratio for every month. The results are compared with the available literature model and highlight the improvement. The results of the existing model deviate from the actual data due to assumptions. The proposed model is a good match with quantity and quality in variations.

The use of solar energy is increasing day by day due to the depletion of conventional sources and the global warming alarm. Instantaneous solar radiation can be determined from the existing correlations developed for India. The model does not incorporate the actual weather conditions. The geography and climatic conditions change from place to place. For example, Vellore (12.9165° N, 79.1325° E) in south India is surrounded by the hills that affect the distribution of solar radiation components, i.e., beam and diffusion. The reported monthly average daily solar radiation predictions for a few cities in India are prepared without the option of month selection. The coefficients are applied for the whole year, which leads to more error. The solar radiation coefficients are available at the nearest station, Chennai, to find the monthly average daily solar energy. So far, there is no solar model reported to predict the solar radiation at Vellore. The adoption of Chennai coefficients to predict the radiation at Vellore is not suitable for Vellore due to different climatic conditions, even though it is not far from Chennai. These issues are addressed in the proposed model by developing the solar heat flux and solar energy coefficients at specific levels. The novelty in this work is the development of empirical correlations for instantaneous solar heat flux as a function of month, date, and time at select locations. The work is focused on empirical correlations for monthly average daily solar energy as a function of a month at the selected place. The generation of coefficients in the correlation at the specific level, increases the accuracy of the model compared to those reported in literature.

In the current work, the gaps in the existing models are addressed, and a novel method is developed and applied to generate the specific solar radiation under cloudy conditions for every month. The main objective of the current work is to predict the solar radiation data (instantaneous and monthly average daily) at Vellore, month-wise. For both instantaneous solar heat flux and monthly average daily solar energy, coefficients are presented for each month.

11.5.1 Mathematical model

In this work, a novel mathematical model has been developed to predict the solar radiation for a particular location in a more refined way. The radiation models are categorized into two, viz., instantaneous solar heat flux and monthly average daily solar energy flux. Vellore is geographically located at the center of south India. In this city, many tourist attractions, hospitals, institutes, and industries are located. In the future, this place is going to develop exponentially. Solar radiation correlation studies of Vellore have not been done or reported. The current work is the first of its kind; it is initiated to generate a solar radiation prediction model for Vellore, Tamil Nadu, India.

11.5.1.1 Empirical correlation for instantaneous solar heat flux

In the existing model, two coefficients are used to predict the instantaneous solar direct beam heat flux falling on a horizontal surface. One coefficient is used to assess the instantaneous diffuse heat flux. Since global radiation is the sum of the direct beam and diffuse components, these three coefficients are involved in the prediction of the instantaneous global radiation. The instantaneous global radiation, Ig, falling on a horizontal surface is the sum of beam radiation, I_b, and diffuse radiation, I_d. The beam radiation falling on a horizontal surface can be determined by taking the solar radiation in the direction of the beam, which is transferred to the normal direction of the horizontal surface.

The beam radiation in the direction of sun rays,

$$I_b = I_{bn} \cos \theta_z \tag{11.20}$$

The zenith angle,

$$\theta_z = \cos^{-1} (\sin\delta \sin\varnothing + \cos\delta \cos\varnothing \cos\omega)$$

Zenith angle is the angle between the sun beam and normal to the horizontal surface. The zenith angle is the function of declination (δ), latitude (Ø), and hour angle (ω). Declination is the angle

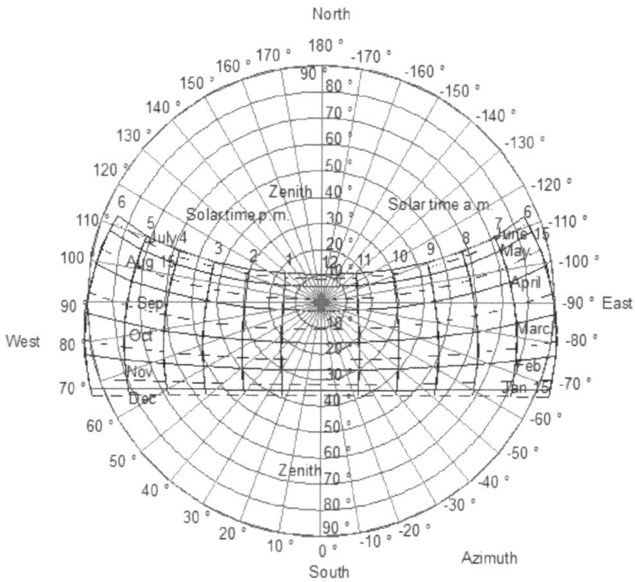

Fig. 11.6. Sun path diagram to depict the variations in the zenith angle and azimuth angle at Vellore from sunrise to sunset for a year.

between the sunbeam and the equatorial plane. The latitude of a location is its angular displacement from the equator. The hour angle shows the angle measure from the solar noon; it is negative in the forenoon and positive in the afternoon.

Figure 11.6 is the sun path diagram prepared for Vellore, Tamil Nadu, India, to show the changes in the solar beam position with time, date, and month. It can be imagined as the sun's position seen from the sky. From the figure, it can be seen that most of the time, the sun's position lies in a southerly direction. The mid date of the month is shown on the sun path diagram, as it is difficult to draw all 365 days in a year. The circles show the zenith angle, and the radial lines represent the solar azimuth angle.

Therefore, the global radiation falling on the horizontal surface is the sum of the beam radiation component on the horizontal surface and diffuse radiation.

$$I_g = I_{bn} \cos \theta_z + I_d \tag{11.21}$$

In the ASHRAE model, the correlations equations are presented to predict the direct normal beam radiation and diffuse radiation and are given in the following equations.

Beam radiation in the direction of the sun rays,

$$I_{bn} = A \exp\left(\frac{-B}{\cos \theta_z}\right) \tag{11.22}$$

Diffuse radiation falling on a horizontal surface,

$$I_d = c\, I_{bn} \tag{11.23}$$

Beam radiation falling on a horizontal surface,

$$I_b = A \exp\left(\frac{-B}{\cos \theta_z}\right) \cos \theta_z \tag{11.24}$$

Therefore, the global radiation falling on a horizontal surface,

$$I_g = A\exp\left(\frac{-B}{\cos\theta_z}\right)\cos\theta_z + CI_{bn} \tag{11.25}$$

In Eqs. A, B, and C are coefficients given for the month applicable, for all places in the country. The latitude, declination, and hour angle account for the location, date, and time, respectively. Since the coefficients are given month-wise, it also considers seasonal variations. But it does not account for the weather conditions and the local solar radiation data.

The proposed model is developed with more specific information such as location, month, date, time, atmospheric parameters, and local solar radiation data.

The correlation developed to predict the instantaneous solar heat flux, either global or as direct beam, on a particular month and location is as follows:

$$I = \left(a_1\left(\frac{S}{S_{\max}}\right)^2 + a_2\left(\frac{S}{S_{\max}}\right) + a_3\right)t^2 + \left(b_1\left(\frac{S}{S_{\max}}\right)^2 + b_2\left(\frac{S}{S_{\max}}\right) + b_3\right)t + c_1\left(\frac{S}{S_{\max}}\right)^2 + c_2\left(\frac{S}{S_{\max}}\right) + c_3 \tag{11.26}$$

Equation 11.26 shows that the instantaneous heat flux is the function of actual day length to maximum day length ratio $\left(\frac{S}{S_{\max}}\right)$ and time (t) on a specified month. Separate coefficients are solved for each month. It is a nonlinear equation to fit the continuous solar flux variations. In the literature, instantaneous heat flux can be determined with the two constants for the global component. Equation 11.26 is a function of two variables, i.e., relative sunshine hours and time. Each variable is correlated with three variables. So, these two variables account for a total of nine coefficients. These nine coefficients are developed month-wise from the annual solar radiation data at the location. For 12 months of modeling, a total of 108 coefficients were developed. Compared to the existing, the use of specific coefficients is complex. But with the recent developments in computational method and tools, it is not an issue. The reported correlation for hourly global radiation uses two variables per month, which is simple but is prone to errors in results. In this work, Eq. 11.26 is used for global radiation and beam radiation. The corresponding coefficients are developed from the local solar radiation data by incorporating the variables. The resulting coefficients for instantaneous global heat flux and instantaneous beam heat flux are tabulated in Table 11.6, which is shown in the results section.

11.5.1.2 Empirical correlations for monthly average daily solar energy

The empirical correlations for monthly average daily solar energy for the global component and beam component are generated from the processing of earlier radiation data.

Modified Angstrom's equation for daily global radiation,

$$\overline{H}_g = \left(a\left(\frac{\overline{S}}{\overline{S}_{\max}}\right) + b\right)\overline{H}_0 \tag{11.27}$$

In Eq. 11.27, $\overline{H}_0$ is daily average of extra-terrestrial radiation corresponding to the solar constant and is fairly constant during the complete year. It is the daily mean value of extra-terrestrial radiation (beyond earth's atmosphere) on a unit surface perpendicular to the sun's rays; corresponding to the surface on the ground under consideration; the daily mean value for the month in terms of kJ/m^2.day. H_0 is obtained by integrating the solar insolation on a horizontal surface over a period of complete solar day (hour angle $-\omega_s$ to ω_s).

$$H_0 = \frac{24}{\pi} I_{sc} \times 3600\left(1 + 0.033.\cos\frac{360n}{365}\right)\left(\omega_s \sin\phi.\sin\delta + \cos\phi.\cos\delta.\sin\omega_s\right) \tag{11.28}$$

Since the distance between sun and earth changes continuously every day, extra terrestrial radiations and H_0 will change accordingly.

$$\frac{\overline{H}_g}{\overline{H}_0} = a\left(\frac{\overline{S}}{\overline{S}_{\max}}\right)^2 + b\left(\frac{\overline{S}}{\overline{S}_{\max}}\right) + c \tag{11.29}$$

The ratio, $\frac{\overline{H}_g}{\overline{H}_0}$ is called monthly average clearness index.

Similarly, the monthly average daily beam energy is determined as follows.

The linear equation is

$$\overline{H}_b = \left(a\left(\frac{\overline{S}}{\overline{S}_{\max}}\right) + b\right)\overline{H}_0 \tag{11.30}$$

The non-linear equation for the monthly average daily beam energy

$$\overline{H}_b = \left(a\left(\frac{\overline{S}}{\overline{S}_{\max}}\right)^2 + b\left(\frac{\overline{S}}{\overline{S}_{\max}}\right) + c\right)\overline{H}_0 \tag{11.31}$$

11.5.2 Results and interpretation

The results focus on instantaneous and daily solar radiation with their components. This section presents the resulting coefficients from the linear and non-linear models and the generated solar radiation. The radiation models are compared with the reported model to highlight the differences. The values are validated by comparing with the actual data.

11.5.2.1 Instantaneous solar heat flux

The proposed and existing mathematical models are described in an earlier section to estimate the instantaneous solar radiation and monthly average daily solar energy flux at a particular location. This section presents the results, including the developed coefficients for estimation of solar radiation, both instantaneous and integrated energy, for the proposed correlations. The correlations consist of coefficients (constants) and variables. In instantaneous solar radiation, day length ratio and time are the controlling parameters.

In the monthly average daily radiation, day length ratio is the main input variable. These coefficients are generated month-wise and applicable at the considered location. Equations 11.24 and 11.25 are the existing correlations to find the instantaneous solar radiation, respectively, for beam and global solar radiation components. Equation 11.26 is the empirical equation proposed to generate the instantaneous solar radiation from sunrise to sunset for both beam and global radiation. Equation 11.26 has nine coefficients, viz. a_1, a_2, a_3, b_1, b_2, b_3, c_1, c_2, and c_3. These coefficients are tabulated for global radiation and beam radiation in Table 11.6 after making the curve fit from the solar radiation data. The data is selected from sunrise to sunset with a one-minute time interval. Laboratory measuring instruments and control systems are used for continuous recording of the solar radiation. The solar radiation during the night is not included in the coefficient development or cumulative solar radiation. The coefficients available in the literature are used to evaluate solar radiation. The existing instantaneous solar radiation model was also constructed with month-wise coefficients for the whole country. It does not incorporate the controlling variable from the actual weather condition. The proposed model adds the actual sunshine hours to incorporate the cloudy condition. The predicted model is constructed with a cloudy condition. Since the available model has no weather variables, it is the clear sky radiation or cloudless radiation. It is the first attempt to predict the instantaneous radiation under cloudy conditions with day-length radio at a selected place.

Table 11.6. Coefficients of empirical correlations to predict the instantaneous solar heat flux at Vellore, India.

Radiation	Coefficient	January	February	March	April	May	June	July	August	September	October	November	December
Global	a_1	4.2609	5.2995	13.5211	31.2336	–0.9710	12.4187	5.9345	21.8238	–1.6973	–21.8839	–15.1628	1.0270
	a_2	–22.6624	–22.6825	–29.3432	–44.8601	–8.3486	–25.3761	–19.0068	–31.6706	–14.1632	7.3162	–10.2349	–19.7484
	a_3	–9.0484	–12.3398	–12.9808	–10.3537	–14.7942	–10.5833	–12.3357	–12.7487	–15.2271	–18.6260	–11.0350	–9.9794
	b_1	–113.1770	–130.6633	–346.5105	–686.2771	62.7211	–281.5339	–104.8595	–423.6912	99.8966	527.4090	348.7239	–22.2118
	b_2	551.9143	538.9259	715.9900	1035.6324	166.2341	576.0545	413.8902	677.7971	280.7414	–168.9243	231.6284	475.1866
	b_3	211.7068	300.3756	307.3388	226.3699	332.4428	246.8889	292.5534	297.5503	356.8238	436.7802	245.8700	221.1939
	c_1	586.7494	899.3617	1903.1440	3251.2607	–413.5792	1290.9470	263.6095	1525.9822	–959.3804	–2667.8956	–1710.4719	–58.5574
	c_2	–2714.0348	–2875.3465	–3515.4336	–4926.4674	–677.9372	–2512.2644	–1662.0480	–2665.9358	–817.5716	882.2242	–905.2728	–2186.4302
	c_3	–979.5046	–1378.5486	–1474.0686	–913.4714	–1361.4786	–1084.7451	–1314.6175	–1396.9038	–1656.7335	–2057.1508	–1069.2326	–974.0614
Beam	a_1	–21.7220	15.3642	–10.5333	21.5130	1.2117	–9.9478	–5.5087	0.4298	–14.7403	–27.8278	–8.1084	–15.2214
	a_2	–2.3477	–53.4493	–12.1273	–42.1407	–22.4482	–11.1007	–11.4547	–19.6577	–11.5376	–6.3605	–18.4032	–9.2298
	a_3	–0.9730	13.0907	–2.7347	2.4977	1.1976	0.0223	0.4012	0.2135	–0.9860	–0.7382	0.3248	–1.0023
	b_1	510.8459	–309.2786	223.9026	–451.1015	5.1170	240.3492	151.9244	62.4332	420.6651	706.4110	226.5952	378.4734
	b_2	70.4864	1196.9600	318.2780	960.9491	508.5141	248.1016	249.7556	405.1758	210.3202	122.9280	402.2360	206.9692
	b_3	22.1073	–277.2738	64.6391	–62.9851	–32.9868	1.6297	–6.4310	0.2939	29.1851	20.3912	–7.2491	19.8192
	c_1	–2450.2177	1654.7108	–688.5288	2160.9556	–22.0341	–1090.7456	–830.1891	–591.4519	–2386.3011	–3667.1808	–1158.8920	–1971.3018
	c_2	–430.5493	–5981.4855	–1883.2017	–4608.6902	–2431.6365	–1094.4230	–1051.7539	–1637.3666	–600.0759	–398.9895	–1779.3566	–901.6447
	c_3	–97.2467	1330.1450	–299.4134	344.5489	221.8356	–20.3010	14.9943	–31.7421	–176.6414	–120.6501	37.8950	–75.6863

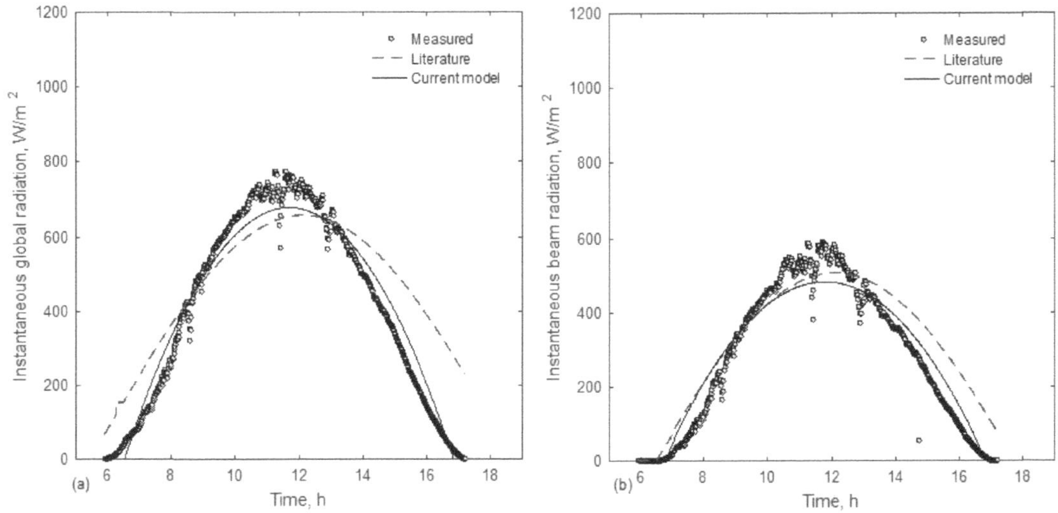

Fig. 11.7. Validation of (a) instantaneous solar global heat flux and (b) instantaneous solar beam heat flux with measured data and literature model on a typical winter day.

Figure 11.7a and Fig. 11.7b show the variation of instantaneous solar radiation from sunrise to sunset, respectively, for global radiation and beam radiation. The time shown is the local standard time, i.e., watch time. The measured data, generated data from the literature, and proposed work are compared with each other. In both diagrams, the axes are fixed to compare the magnitude of global radiation and beam radiation. It is difficult to show the solar radiation from sunrise to sunset for all the days in a year; three days are presented from: winter, equinox, and summer. It is solar radiation as on a typical winter day in December. The existing mathematical model is generalized to apply

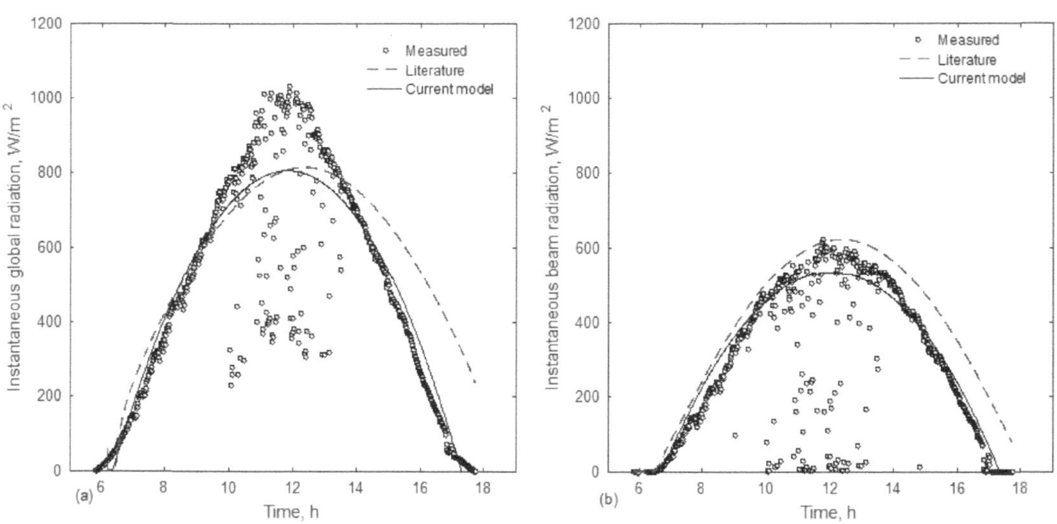

Fig. 11.8. Validation of (a) instantaneous solar global heat flux and (b) instantaneous solar beam heat flux with measured data and literature model on equinox in the month of March.

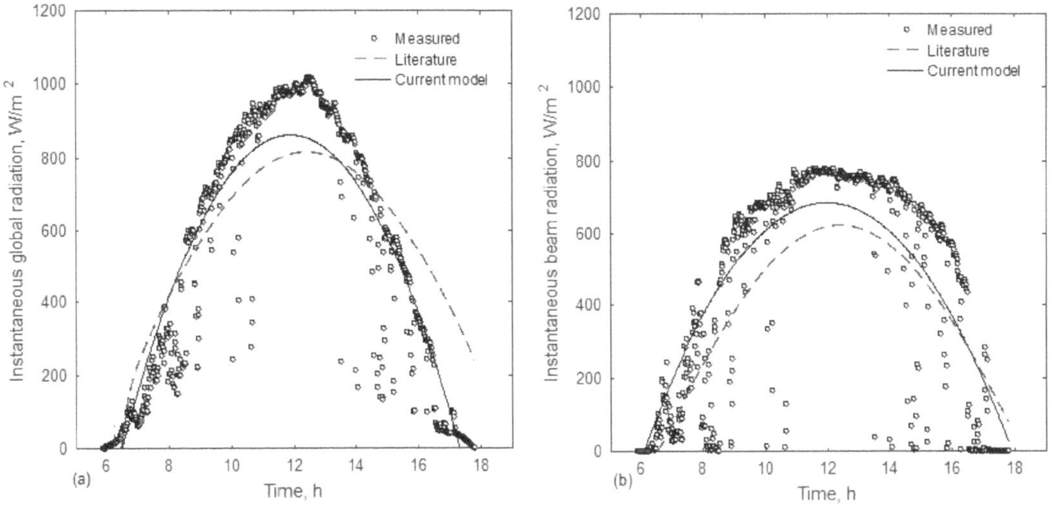

Fig. 11.9. Validation of (a) instantaneous solar global heat flux and (b) instantaneous solar beam heat flux with measured data and literature based model on a typical summer day.

at all the places in India; hence, the local sunrise and sunset are not matched exactly. Since the proposed model counts the local timings, there is no mismatch from the beginning (sunrise) to the ending (sunset) of radiation. The solar radiation evaluated from the earlier model deviates from the measured data in the morning and afternoon as it is not specific for the local time and place. More closeness has been observed between the actual data and the proposed model.

Figure 11.8 shows (a) instantaneous solar global heat flux and (b) instantaneous solar beam heat flux models on equinox. It can be seen from the comparison of predicted data from the proposed model and measured data. The daily or monthly solar radiation may not repeat every year. The mismatch between the measured data and predicted data using the existing mathematical model is a result of not incorporating local parameters. The peak is also not matched with the existing model. The proposed model results in a peak with a good match with the data obtained by the measuring instruments such as pyranometer and perheliometer. The climatic variations are not uniform every year. The proposed model's solar radiation is symmetrical to the solar noon, but the reported model does not result in symmetrical radiation. Similarly, Fig. 11.9 shows the instantaneous solar radiation in summer.

The instantaneous solar radiation components, i.e., global and beam heat fluxes, are predicted using the averaged coefficients for all the days in a year. In a month of January from the beginning of the month to the end of the month, for all the dates, solar radiation flux is predicted and shown in Fig. 11.10 from sunrise to sunset. The low peak in the prediction compensates for the lower solar radiation during the cloudy periods. The existing model does not incorporate the actual weather parameters to predict instantaneous data. Therefore, the predicted data exactly reproduces the actual data as it is limited to the considered year. The proposed model has a facility to incorporate the present climatic variables; therefore, the result will match with the actual radiation. The solar

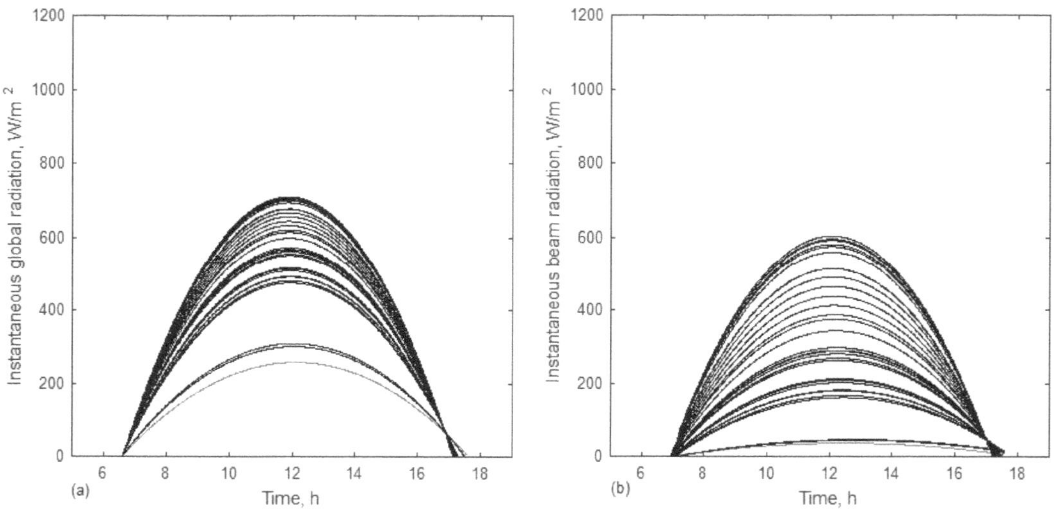

Fig. 11.10. Predicted solar radiation for the month of January from 1st to 31st (a) global radiation and (b) beam radiation.

radiation by the proposed model varies every year with reference to the actual weather conditions. In the presented graphs, the averaged sunshine hours' ratio is fed to the proposed model, resulting in the past data. In place of the past sunshine hours' ratio, the present actual sunshine hours' ratio results in the current solar radiation status.

11.5.2.2 Monthly average daily solar energy

The annual average sunshine hours and sunshine hours' ratio is tabulated in Table 11.7 for prediction of the monthly average daily solar radiation. The monthly average solar radiation is a function of sunshine hours and clearness index. Table 11.7 also shows the coefficients to generate the monthly average daily radiation. The coefficients are generated for both linear fit and non-linear fit equations. The total average values of the linear coefficients for the global radiation are 0.3726 and 0.2948, respectively, for 'a' and 'b' in Eq. 11.27. These are 0.44 and 0.30 for the nearest station in Chennai. The linear coefficients for the assessment of the monthly average beam energy are 0.4839 and –0.0451, respectively, for 'a' and 'b' in Eq. 11.30.

Table 11.8 compares the generated results of solar radiation with the measured data. It shows the comparison of the predicted data with the measured data. The mathematical models used in the work are ASHRAE standards based. The radiation measurements are counted from sunrise to sunset, and the night radiation is not accounted for, as it is minor compared to day.

Table 11.7. Coefficients of empirical relations to generate the monthly average daily global energy flux and monthly average daily beam energy flux

Radiation	Coefficient	January	February	March	April	May	June	July	August	September	October	November	December
S_{ave}		6.9661	9.0179	8.6965	8.6828	7.4409	5.4791	5.5629	4.6105	5.9442	6.1699	3.7203	6.0720
S_{ave}/S_{max}		0.6151	0.7663	0.7285	0.6808	0.6057	0.4744	0.4947	0.3930	0.4768	0.4839	0.3037	0.5262
Global, linear	a	0.3988	0.3415	0.3569	0.3568	0.3891	0.3931	0.3931	0.4645	0.3632	0.2521	0.3821	0.3799
	b	0.2333	0.2822	0.3013	0.3065	0.3314	0.4122	0.4122	0.2803	0.2838	0.2852	0.2003	0.2084
Global, non linear	a	–0.1563	0.0999	–0.2507	–0.2027	–0.1919	–0.1049	–0.1049	–0.2226	0.0660	0.3002	0.1564	–0.1896
	b	0.5532	0.1984	0.6533	0.5990	0.5721	0.4957	0.4957	0.6314	0.3039	–0.0156	0.3103	0.5505
	c	0.2077	0.3318	0.2249	0.2403	0.3022	0.3936	0.3936	0.2611	0.2935	0.3256	0.1973	0.1822
Beam, linear	a	0.4976	0.6521	0.5380	0.4929	0.4944	0.4069	0.4069	0.4152	0.4995	0.5542	0.4190	0.4306
	b	–0.0532	–0.1460	–0.0623	–0.0312	–0.0435	–0.0309	–0.0309	–0.0176	–0.0393	–0.0517	–0.0102	–0.0249
Beam, non linear	a	0.4290	0.3821	0.3677	0.0628	0.3205	0.1863	0.1863	0.2080	0.4382	0.5451	0.3063	0.2720
	b	0.0660	0.1128	0.0989	0.4223	0.1887	0.2286	0.2286	0.2435	0.1122	0.0968	0.2320	0.1860
	c	0.0227	0.0358	0.0538	–0.0137	0.0046	0.0003	0.0003	0.0055	0.0252	0.0101	0.0010	0.0127

Table 11.8. Comparison of monthly average daily solar energy flux with the measured data.

Radiation		January	February	March	April	May	June	July	August	September	October	November	December
Global radiation, kWh/m²-day	**Measured**	4.1286	5.2591	5.8624	6.0624	4.0497	4.7495	5.5205	4.6434	5.1088	4.2468	3.7233	3.7233
	Predicted from the literature	4.8471	6.2238	6.2748	6.2879	5.9197	4.6294	4.2500	4.5784	5.4157	5.3792	4.8214	4.8214
	Error, %	−17.4042	−18.3434	−7.0347	−3.7196	−46.1750	2.5287	23.0151	1.4009	−6.0073	−26.6648	−29.4927	−29.4927
	Predicted from current model	4.0611	5.3121	5.6720	5.7603	4.8430	4.6963	4.9869	4.4825	4.8374	4.2195	3.7035	3.7035
	Error, %	1.6337	−1.0068	3.2478	4.9832	−19.5891	1.1212	9.6658	3.4662	5.3124	0.6428	0.5331	0.5331
Beam radiation, kWh/m²-day	**Measured**	1.9932	3.3756	3.5009	3.5583	4.0497	1.6161	1.8141	1.6135	2.4016	2.0656	1.8303	1.8688
	Predicted	2.1483	3.4553	3.3373	3.1893	2.5062	1.7368	1.4366	1.4235	2.1089	2.2974	1.8688	1.8303
	Error, %	−7.7815	−2.3611	4.6731	10.3701	38.1127	−7.4655	20.8120	11.7756	12.1898	−11.2219	−2.1035	2.0601

11.6 Numerical solutions

1. Find the solar radiation components of direct beam, diffuse and global on February 4th at 2 p.m. at Vellore, Tamil Nadu, India (Ø = 12.9166° N, L =79.1833 E °).

Solution through App Designer

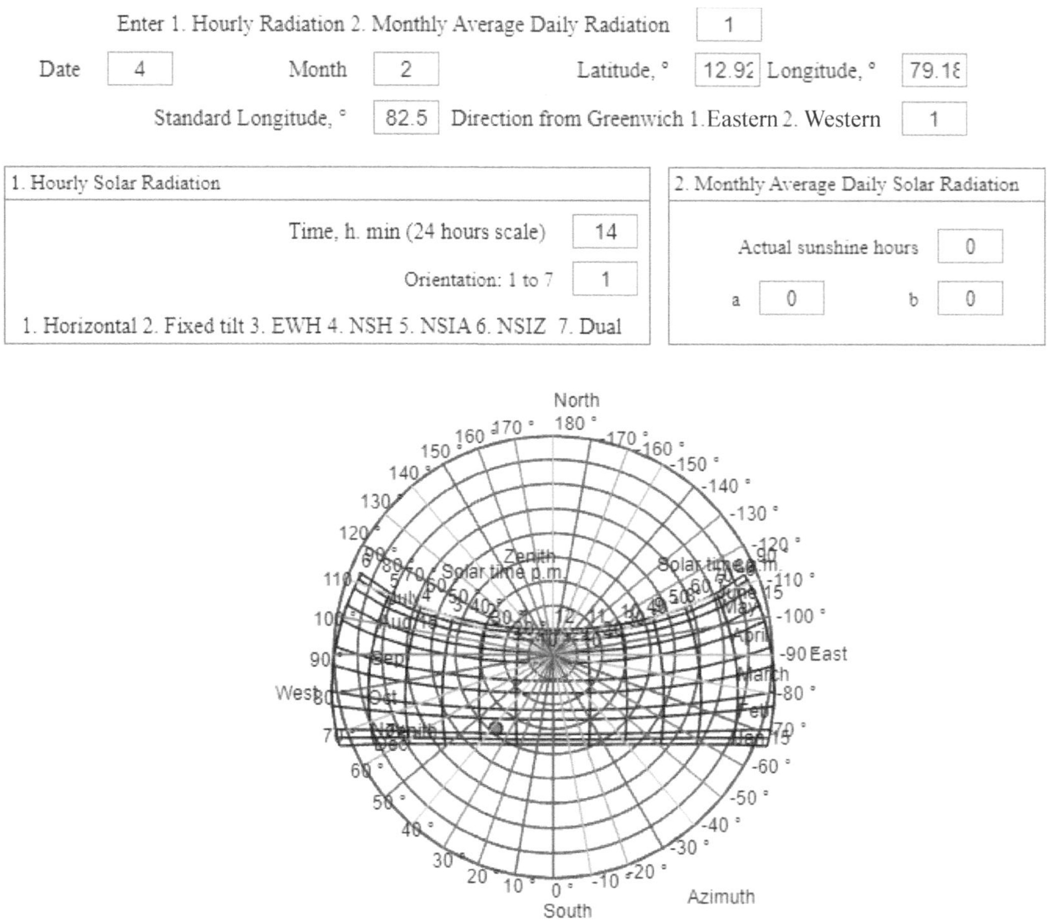

The day is February 4th, the day number, n = 35

Parameter	Equation	Substitution	Result
Solar Time	$B = \frac{360(n-81)}{364}°$	$\frac{360(35-81)}{364}$	–45.49 °
	EoT = 9.87 sin 2B – 7.53 cos B – 1.5 sin B	9.87 sin (2 × –45.49°) – 7.53 cos –45.49°) – 1.5 sin –45.49°)	–14.08 min
	Solar time (Local Apparent Time) = Watch Time – 4 ($L_{st} - L_{lo}$) + EoT (in minutes)	12.5 – 4 (82.5 – 79.1833)/60 –14.075/60	13.54 h 13 h.33 min
Declination	$\delta = 23.45 \sin\left[\frac{360(284+n)}{365}\right]°$	$23.45 \sin\left[\frac{360(284+35)}{365}\right]$	–16.69°

Table contd. ...

...Table contd.

Parameter	Equation	Substitution	Result
Hour angle	$\omega = 15$ (solar time – 12)°	15 (13.54 – 12)°	23.16°
Zenith angle	$\theta_z = \cos^{-1} (\sin\varphi \sin\delta + \cos\varphi \cos\delta \cos\omega)$	$\cos^{-1}$ (sin (12.9166) sin (–16.69) + cos (12.9166) cos (–16.69) cos (23.16))	37.42°
Normal beam radiation	$I_{bn} = A\exp\left(\dfrac{-B}{\cos\theta_z}\right)$	$652.2\exp\left(\dfrac{-0.01}{\cos 37.39}\right)$	644.04 W/m²
Direct beam radiation	$I_{bn} \cos\theta_z$	644.04 × cos (37.39)	511.57 W/m²
Diffuse radiation	$C\, I_{bn}$	0.249 × 644.04	160.36 W/m²
Total or global radiation	$I_g = I_b + I_d$	511.57 + 160.36	671.9 W/m²

2. Estimate the monthly average daily global radiation and daily diffuse radiation on a horizontal surface at Chennai (Φ = 13.08° N) during the month of January if the average actual sunshine hours is 8.6 h. Consider the constant 'a' and 'b' as 0.3 and 0.44 respectively.

Solution through App Designer

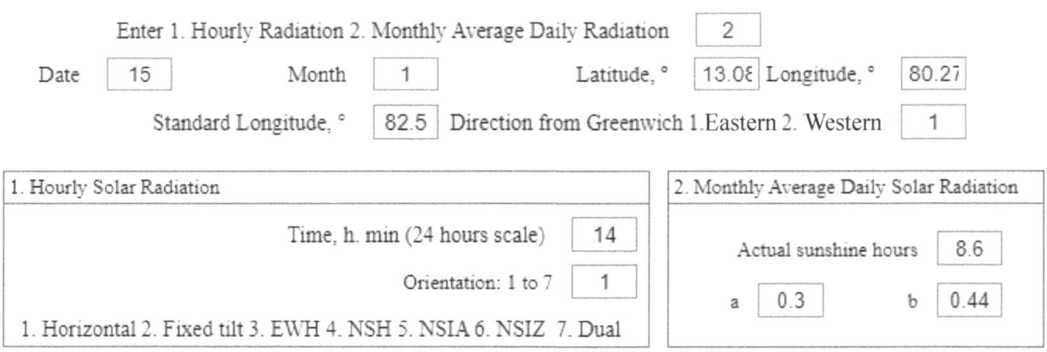

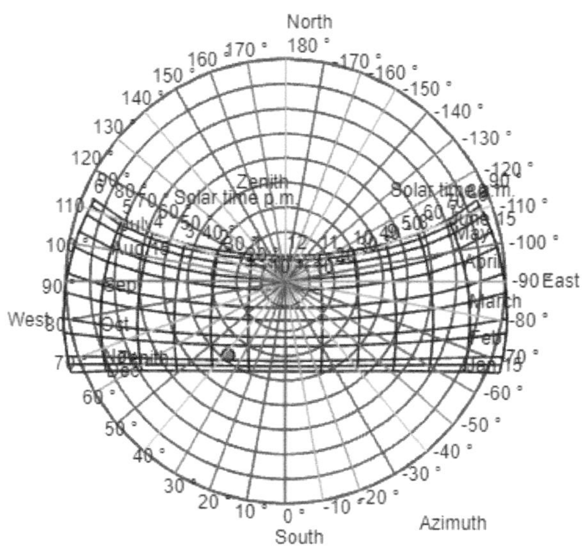

January 17, day number, n = 17

Parameter	Equation	Substitution	Result
Declination	$\delta = 23.45 \sin\left[\frac{360(284+n)}{365}\right]^{\circ}$	$23.45 \sin\left[\frac{360(284+17)}{365}\right]$	–20.92°
Hour angle at sunrise	$\omega_s = \cos^{-1}(-\tan\delta \tan\varphi)$	$\cos^{-1}(-\tan(-20.92)\tan(13.08))$	84.91°
Day length	$S_{max} = \frac{2}{15}\omega_s$	$\frac{2}{15} \times 84.91$	11.32
Extraterrestrial radiation on a horizontal surface	$H_0 = \frac{24}{\pi} I_{sc} \times 3600\left(1 + 0.033.\cos\frac{360n}{365}\right)$ $(\omega_s \sin\phi.\sin\delta + \cos\phi.\cos\delta.\sin\omega_s)$	$= \frac{24}{\pi} 1.367 \times 3600\left(1 + 0.033.\cos\frac{360 \times 17}{365}\right)$ $\times \left(\begin{array}{l} 84.91 \times \left(\frac{\pi}{180}\right)\sin(13.08). \\ \sin(-20.92) + \cos(13.08). \\ \cos(-20.92).\sin(84.91) \end{array}\right)$	30504.40 kJ/m²-day 8.47 kWh/m²-day
Monthly average daily global radiation	$\overline{H}_g = \overline{H_0}\left(a + b\left(\frac{\overline{S}}{\overline{S}_{max}}\right)\right)$	$\overline{H}_g = 30504.4 \times \left(0.3 + 0.44\left(\frac{8.6}{11.32}\right)\right)$	19347.55 kJ/m²-day 5.37 kWh/m²-day
Monthly average daily diffuse radiation	$\frac{\overline{H}_d}{\overline{H}_g} = 1.411 - 1.696\frac{\overline{H}_g}{\overline{H}_0}$	19347.55 × (1.411–1.696 × 0.63)	6487.32 kJ/m²-day 1.8 kWh/m²-day

Chapter 12
Solar Energy Systems

12.1 Introduction

The study of solar energy involves solar energy collection, solar energy storage, and solar energy utilization. Solar collectors are the main components of solar energy systems. Solar energy collectors include solar thermal collectors and solar photovoltaic (PV) collectors. The collection of solar energy on a flat plate collector, solar air heaters, solar dryers, and PV panels are examples of non-focusing collectors. The concentrating collectors are linear and point-focusing collectors, specifically parabolic trough collectors and paraboloid collectors. The design of solar thermal collectors includes optical efficiency and thermal efficiency. Determining optical efficiency involves optical properties such as reflection, refraction index, transmission, and absorption properties. Measuring thermal efficiency has to take into account optical losses and thermal losses. To estimate thermal losses, conduction, convection, and radiation models are applied with the thermodynamic and thermo-physical properties.

12.2 Flat plate collector

Figure 12.1 shows the structural details of a flat plate collector. The main components of a flat plate collector are tubes, absorber plate, insulating material, transparent cover(s), storage tank, drain plug, and plumbing. Thermal efficiency of a flat plate collector is defined as a ratio of output heat to the supplied solar radiation. It may be determined either at an instantaneous time or over a period of time (Duffie et al., 2020).

Thermal efficiency of a flat plate collector,

$$\eta = \frac{\int Q_u d\tau}{A_c \int H d\tau} = \text{(Useful Energy Delivered)/(Incident Solar Radiation)} \tag{12.1}$$

where, Q_u is the useful heat collected by a collector with an area of A_c over a period of time $d\tau$ during which it received solar radiation H (W/m^2).

$$Q_u = A_c F_R (H_a - U_l(T_{fi} - T_a)) \tag{12.2}$$

where A_c - area of the collector, m^2
F_R - heat removal factor (dimensionless)
H_a - solar radiation absorbed by the collector, W/m^2
U_L - overall heat loss coefficient, W/m^2 °C
T_{fi} - inlet fluid temperature, °C
T_a - ambient temperature, °C

DOI: 10.1201/9781032642505-12

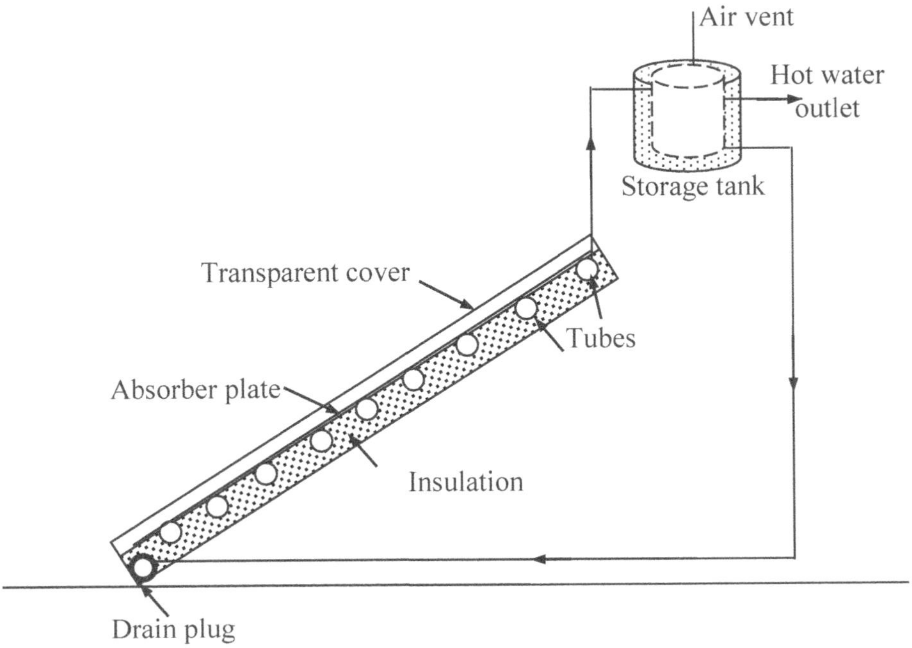

Fig. 12.1. Flat plate collector.

The above equation is known as the Hottel-Whillier-Bliss (HWB) equation. The heat loss coefficient will typically have values in the order of 5 W/m^2C at 1000 W/m^2 radiation levels. The heat removal factor expresses the relative level of effectiveness of heat transfer through the absorber. It is the ratio between the actual amount of heat removed to the maximum possible heat removal (i.e., at when the absorber is at the fluid inlet temperature).

First we can conduct steady state modelling, simulation and analysis for a flat plate collector. Later it can be extended to transient state modelling and its results. The absorber tube is fixed under the absorber plate.

The energy balance equation applied to the absorber plate is

$$Q_u = A_p\, S - Q_l \tag{12.3}$$

Generally, a time period of an hour is considered for calculations.

Where

Q_u = useful heat transfer to fluid, kJ/h

S = solar radiation falling on an absorber plate, kJ/m^2-h

A_p = area of the absorber plate, m^2

Q_l = total heat losses (conduction, convection and re-radiation), kJ/h

The solar radiation, I_T falling on top of a tilted surface is

$$I_T = I_b r_b + I_d r_d + (I_b + I_d)\, r_r \tag{12.4}$$

To get the radiation falling on absorber surface, transmissivity-absorptivity product $(\tau\alpha)$ can be multiplied on both sides of equation.

$$S = I_b r_b(\tau\alpha)_b + \{I_d r_d + (I_b + I_d)\, r_r\}(\tau\alpha)_d \tag{12.5}$$

where

τ = transmissivity of the transparent cover, which is the ratio of the solar radiation coming through after reflection at the glass-air interfaces and absorption in the glass to the radiation incident on the glass cover system,
α = absorptivity of the absorber plate,
$(\tau\alpha)_b$ = transmittivity-absorptivity product for beam radiation
$(\tau\alpha)_d$ = transmissivity-absorptivity product for diffuse radiation

Therefore, to determine the useful heat, q_u, the values of $(\tau\alpha)_b$, $(\tau\alpha)_d$ and q_l have to be derived.

Instantaneous collection efficiency,

η_i = Useful heat/Solar radiation on tilted surface

$$\eta_i = \frac{Q_u}{I_T A_p} \times 100 \tag{12.6}$$

Generally, the collector aperture area (A_a) or the collector gross area (A_c) is also used in place of A_p. The collector aperture area is the net opening in the topmost cover through which solar radiation is admitted into the collector, while the collector gross area is the area of the topmost cover (including the frame). A_a is usually about 10 to 15% more than A_p, while A_c is about 15 to 20% more than A_p. With no fluid flow rate, there is no useful heat, and the thermal efficiency of the collector is zero. At these conditions, the received radiation falling on the absorber plate will be lost through conduction, convection, and radiation losses such that $A_pS = q_l$. At this point, the plate temperature is maximum, which is called the stagnation temperature. Finding the stagnation temperature helps in designing the collecting system to select suitable material.

To find the optical efficiency of a flat plate collector, the transmissivity of the transparent cover is required. It is modelled by considering the transmissivity separately with reflection-refraction and absorption and is given by the product.

$$\tau = \tau_r \tau_a \tag{12.7}$$

where τ_r = transmissivity with only reflection and refraction
τ_a = transmissivity with only absorption

Transmissivity based on reflection-refraction can be evaluated as per the following:

Figure 12.2 shows the reflection and refraction of a transparent cover. When solar radiation, I_{bn} falling on a transparent medium 1 (air), it strikes the interface separating it into reflected beam and refracted beam. The angle of reflection is equal to the incidence angle. The refracted beam travels

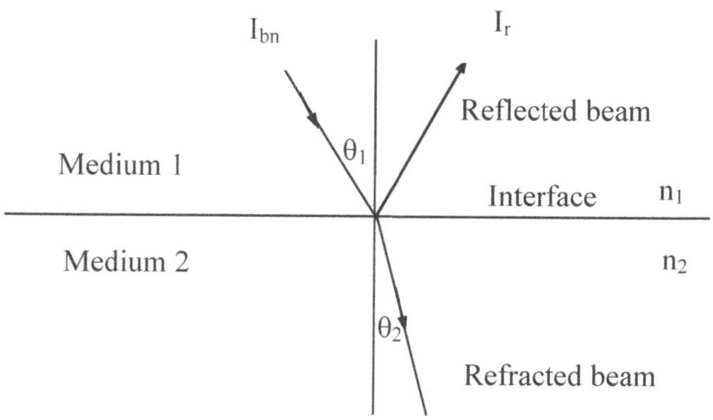

Fig. 12.2. Reflection-refraction of a transparent cover with media.

through the second medium (transparent cover) with a deviation in angle. It is called the angle of refraction. The intensity of reflected beam, I_r decreases.

The angle of refraction can be determined from Snell's law.

$$\frac{\sin\theta_1}{\sin\theta_2} = \frac{n_2}{n_1} \tag{12.8}$$

where θ_1 = angle of incidence,
θ_2 = angle of refraction
n_1 and n_2 = refractive indices of the two media.

$$\sin\theta_2 = \frac{\sin\theta_1}{n}, n = \frac{n_2}{n_1}$$

The reflectivity ρ ($=I_r/I_{bn}$) is related to the angles of incidence and refraction by the equations

$$\rho = \frac{1}{2}(\rho_I + \rho_{II}) \tag{12.9}$$

$$\rho_I = \frac{\sin^2(\theta_2 - \theta_1)}{\sin^2(\theta_2 + \theta_1)} \tag{12.10}$$

$$\rho_{II} = \frac{\tan^2(\theta_2 - \theta_1)}{\tan^2(\theta_2 + \theta_1)} \tag{12.11}$$

$$\rho = \frac{1}{2}\left[\frac{\tan^2(\theta_1 - \theta_2)}{\tan^2(\theta_1 + \theta_2)} + \frac{\sin^2(\theta_1 - \theta_2)}{\sin^2(\theta_1 + \theta_2)}\right] \tag{12.12}$$

ρ_I and ρ_{II} being the reflectivity of the two components of polarization.

For the special case of normal incidence ($\theta_1 = 0°$), it can be shown that

$$\rho = \rho_I = \rho_{II} = \left(\frac{n_1 - n_2}{n_1 + n_2}\right)^2 \tag{12.13}$$

Transmissivity τ_r is given by an expression similar to that for ρ

$$\tau_r = \frac{1}{2}(\tau_{rI} + \tau_{rII}) \tag{12.14}$$

where τ_{rI} and τ_{rII} are the transmissivities of the two components of polarization.

Consider one of the components of polarization of a beam incident on a single cover. Because of the fact that there are two interfaces, multiple reflections and refractions will occur and are shown in Fig. 12.3.

$$\begin{aligned}\text{Hence, } \tau_{rI} &= (1-\rho_I)^2 + \rho_I^2(1-\rho_I)^2 + \rho_I^4(1-\rho_I)^2 + \cdots \\ &= (1-\rho_I)^2(1+\rho_I^2+\rho_I^4+\cdots) \\ &= \frac{(1-\rho_I)^2}{1-\rho_I^2} = \frac{1-\rho_I}{1+\rho_I}\end{aligned} \tag{12.15}$$

Similarly,

$$\tau_{rII} = \frac{1-\rho_{II}}{1+\rho_{II}} \tag{12.16}$$

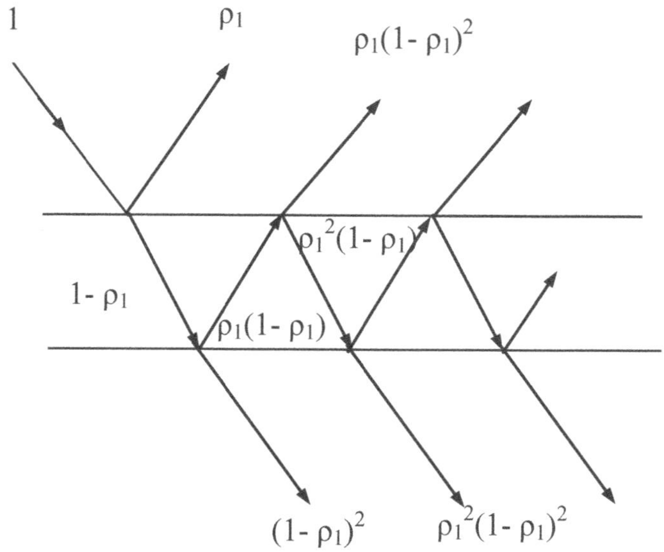

Fig. 12.3. Ray diagram showing transmission through a single cover considering reflection-refraction alone (non-absorbing surface).

These results can be readily extended to a system of M covers for which it can be shown that

$$\tau_{rI} = \frac{1-\rho_I}{1+(2M-1)\rho_I} \tag{12.17}$$

and

$$\tau_{rII} = \frac{1-\rho_{II}}{1+(2M-1)\rho_{II}} \tag{12.18}$$

$$\tau_r = \frac{1}{2}\left(\frac{1-\rho_I}{1+(2M-1)\rho_I} + \frac{1-\rho_{II}}{1+(2M-1)\rho_{II}}\right) \tag{12.19}$$

The transmissivity based on absorption is modelled with the assumption of attenuation due to absorption is proportional to the local intensity (Bouger's law). Consider a beam radiation, I_{bn} incident normally on a transparent cover of thickness δ_c and emerging with an intensity I_l as shown in Fig. 12.4.

From Bouger's law,

$$dI = -K\,I\,dx \tag{12.20}$$

where K is a constant of proportionality and is called the extinction coefficient. It will be assumed to have a value independent of wave length. Integrating over the length traversed by the beam, we have

$$\tau_a = \frac{I_l}{I_{bn}} = e^{-K\delta c} = \frac{I_{transmitted}}{I_{incident}} \tag{12.21}$$

For the beam radiation with incident angle, θ_1, the path traversed through the cover would be $\left(\frac{\delta_c}{\cos\theta_2}\right)$, where θ_2 is the angle of refraction.

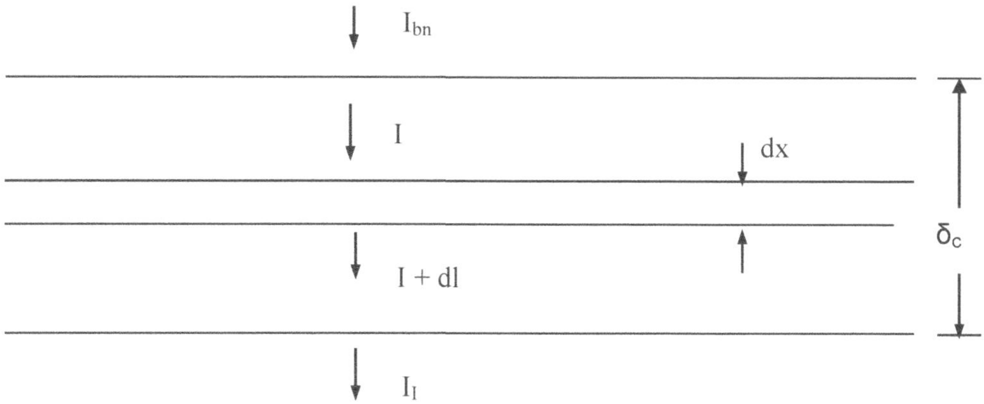

Fig. 12.4. Absorption in a transparent cover.

After the substitution,

$$\tau_a = e^{\frac{-KM\delta_c}{\cos\theta_2}} \tag{12.22}$$

The extinction coefficient K is a property of the transparent cover material. Its value varies about 5 to 25 m^{-1} for different qualities of glass. A low value is obviously desirable.

Transmittivity for beam radiation,

$$\tau_b = \tau_r \tau_a = \frac{1}{2}\left(\frac{1-\rho_I}{1+(2M-1)\rho_I} + \frac{1-\rho_{II}}{1+(2M-1)\rho_{II}}\right) \times e^{\frac{-KM\delta_c}{\cos\theta_2}} \tag{12.23}$$

Simulation of transmissivity for diffuse radiation is difficult as it scatters in all directions. Therefore, it is difficult to simulate diffuse radiation. But this difficulty can be solved by assuming an equivalence angle of 60° to beam radiation; diffuse radiation can be applied. The transmissivity-absorptivity product is defined as the radio of the flux absorbed in the absorber plate to the flux incident on the cover system and is denoted by the symbol, τα. The substitutes b and d are added to indicate the beam radiation and diffuse radiation, respectively. The transmissivity-absorptivity product is modelled as follows:

τ is the fraction transmitted through the transparent cover. This is absorbed and partly reflected diffusely, as shown in Fig. 12.5. From the reflected portion, again a part is transmitted through the transparent cover system and the remaining is reflected back towards the absorber plate. The process of reflection and absorption continue indefinitely.

The net portion of radiation absorbed on the plate,

$$(\tau\alpha) = \tau\alpha\left[1+(1-\alpha)\rho_d + (1-\alpha)^2\rho_d^2 + \cdots\right] \tag{12.24}$$

$$(\tau\alpha)_e = \frac{\tau\alpha}{1-(1-\alpha)\rho_d} \tag{12.25}$$

The symbol ρ_d represents the diffuse reflectivity of the cover system. It can be found by determining the value $\tau_a(1-\tau_r)$ for the cover system for an incidence angle of 60°.

$$\rho_d = \tau_a(1-\tau_r) \text{ at } 60° \tag{12.26}$$

$$\rho_d = \tau_a - \tau_a\,\tau_r \text{ since, } \tau_d = \tau_r\tau_a \tag{12.27}$$

$$\rho_d = \tau_a - \tau_d \tag{12.28}$$

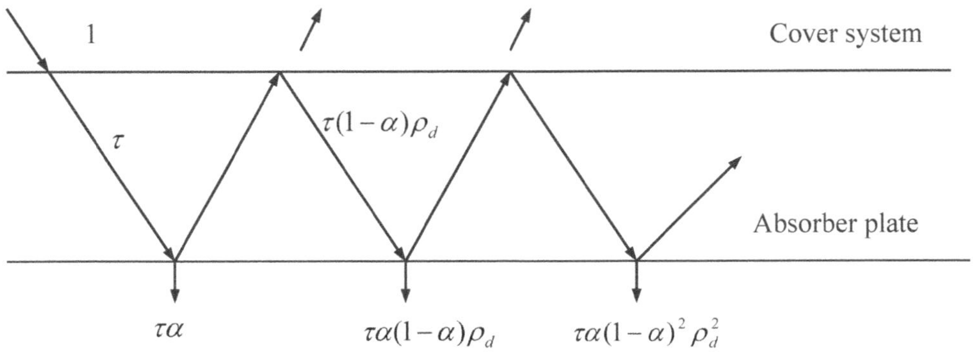

Fig. 12.5. Multiple absorption and reflection at the absorber plate which is below the transparent cover.

The above section details the determination of optical resistance and optical efficiency of a flat plate collector (Fig. 12.6). To model the thermal efficiency of a flat plate collector, the overall heat loss coefficient and transparent cover temperature are to be iterated through the numerical models or software tools.

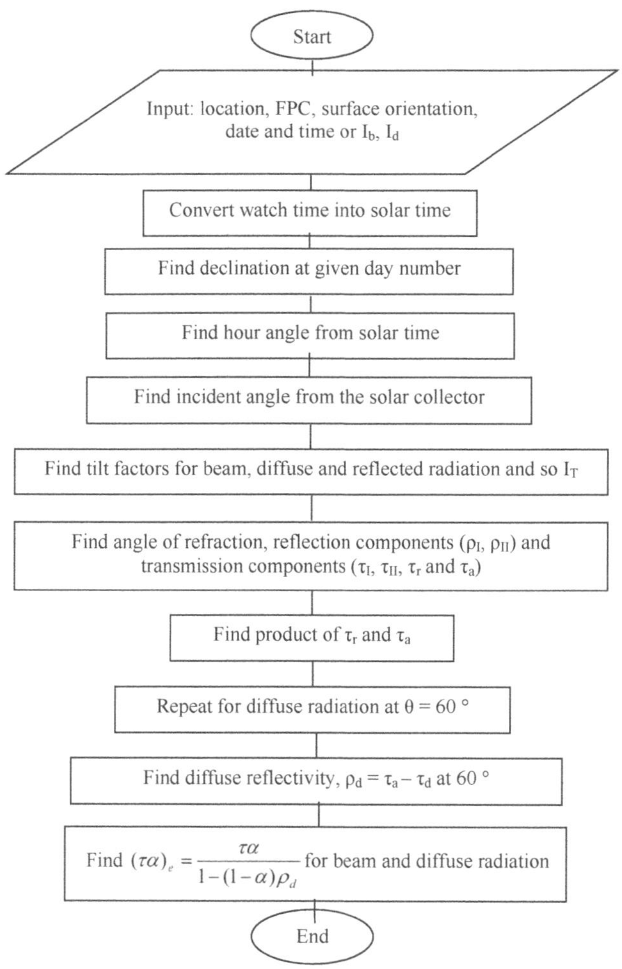

Fig. 12.6. Flow chart showing the methodology for optical efficiency of a solar flat plate collector.

$Q_l = U_l A_p (T_{pm} - T_a)$ (12.29)

where U_l = overall loss coefficient, W/m^2 K
A_p = area of the absorber plate, m^2
T_{pm} = average temperature of the absorber plate, °C
and
T_a = temperature of the surrounding air, °C

The total heat loss from a flat plate collector is the sum of top heat loss, side heat loss and bottom heat loss.

$Q_l = Q_t + Q_b + Q_s$ (12.30)

where Q_t = rate at which heat is lost from the top, W
Q_b = rate at which heat is lost from the bottom, W
Q_s = rate at which heat is lost from the sides, W

Each of these losses is also expressed in terms of coefficients called the top loss coefficient, the bottom loss coefficient and the side loss coefficient and defined by the equations

$Q_t = Ut\, A_p (T_{pm} - T_a)$ (12.31)

$Q_b = U_b A_p (T_{pm} - T_a)$ (12.32)

$Q_s = U_s A_p (T_{pm} - T_a)$ (12.33)

In all these three equations, the heat transfer is the function of area, A_p and the temperature difference, $(T_{mp} - T_a)$.

$U_l = U_t + U_b + U_s$ (12.34)

As per electrical analogy, these losses can also be expressed as a function of thermal resistances, as shown in Fig. 12.7. The overall heat loss coefficient varies from 2 to 10 W/m^{2}K. The three energy balance equations can be used to solve the variables of Q_t, T_{c1}, and T_{c2}. In addition to these equations, heat transfer correlations are required to determine the $h_{p\text{-}c1}$, $h_{c1\text{-}c2}$, h_w and T_{sky}. The top loss coefficient is a function of convective and radiative heat transfer losses. For heat transfer calculations, the absorber plate and transparent cover are assumed to be infinite parallel plates. Heat transfer is considered in a one-dimensional direction and steady state. The temperature drop in the thickness of the covers is considered negligible. The outgoing radiation losses are of large wavelengths. For these wavelengths, the transparent cover is assumed to be opaque.

Figure 12.8 shows the arrangement of the absorber plate and transparent covers to determine the top heat loss coefficient as per assumptions. The heat transfer by convection and radiation happens between the absorber plate and the first transparent cover, the first transparent cover and second transparent cover, and the last transparent cover and surroundings.

$$\frac{Q_t}{A_p} = h_p - c_1(T_{pm} - T_{c1}) + \frac{\sigma(T_{pm}^4 - T_{c1}^4)}{\left(\frac{1}{\varepsilon_p} + \frac{1}{\varepsilon_c} - 1\right)} \quad (12.35)$$

$$= h_{c1-c2}(T_{c1} - T_{c2}) + \frac{\sigma(T_{c1}^4 - T_{c2}^4)}{\left(\frac{1}{\varepsilon_c} + \frac{1}{\varepsilon_c} - 1\right)} \quad (12.36)$$

$$= h_w(T_{c2} - T_a) + \sigma\varepsilon_c(T_{c2}^4 - T_{sky}^4) \quad (12.37)$$

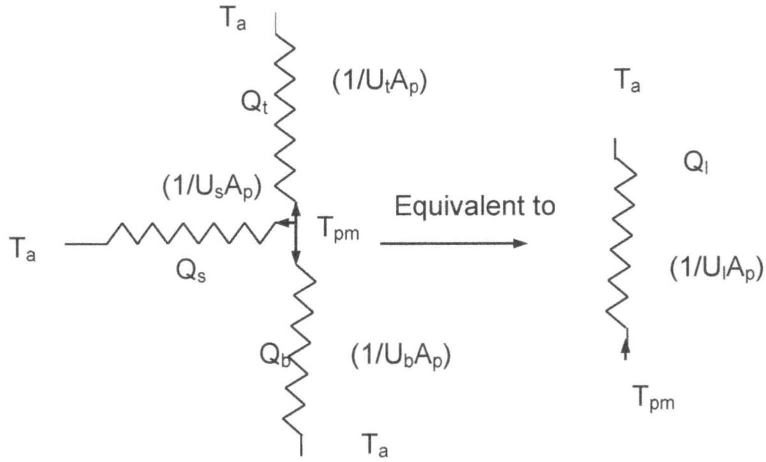

Fig. 12.7. Thermal resistance network showing collector losses.

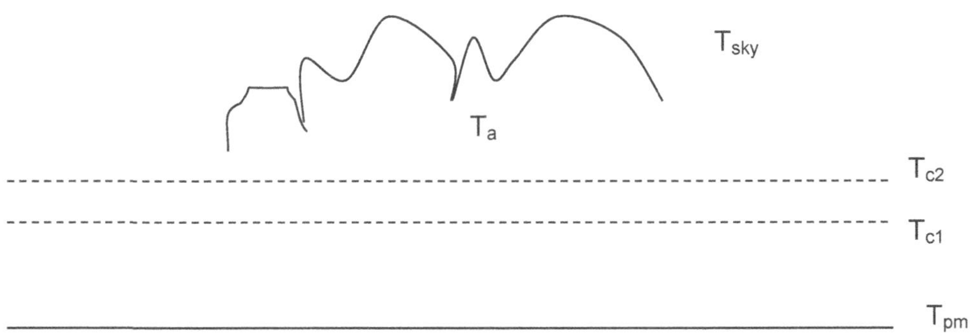

Fig. 12.8. Arrangement of covers for the calculation of top loss coefficient.

where

$h_{p\text{-}c1}$ = convective heat transfer coefficient between the absorber plate and the first transparent cover, $W/m^2\,K$

$h_{c1\text{-}c2}$ = convective heat transfer coefficient between the first and second covers, $W/m^2\,K$

h_w = convective heat transfer coefficient between the top most cover and the surrounding air, $W/m^2\,K$

T_{c1}, T_{c2} = temperatures of transparent covers, K

T_{sky} = effective temperature of the sky, K

ε_p = emissivity of the absorber plate

ε_c = emissivity of the covers

The heat transfer coefficient between inclined parallel plates can be determined from the following correlations.

$$Nu_L = 1;\ Ra_L \cos\beta < 1708 \tag{12.38}$$

$$Nu_L = 1 + 1.446\left(1 - \frac{1708}{Ra_L \cos\beta}\right) : 1708 < Ra_L \cos\beta < 5900 \tag{12.39}$$

$$Nu_L = 0.229(Ra_L \cos\beta)^{0.252};5900 < Ra_L \cos\beta < 9.23\times10^4 \tag{12.40}$$

$$Nu_L = 0.157(Ra_L \cos\beta)^{0.285}; 9.23\times10^4 < Ra_L \cos\beta < 10^6 \quad (12.41)$$

Nu_L and Ra_L are the Nusselt and Reyleigh numbers respectively. The characteristic dimension L is the space between the plates, while properties are evaluated at the mean of the surface temperatures.

Gr.Pr = Rayleigh's number, Ra.

$$Gr_L = \frac{\rho^2 \beta g \Delta T L^3}{\mu^2} = \frac{\beta g \Delta T L^3}{\nu^2} \quad (12.42)$$

and

$$\Pr = \frac{\mu C_p}{k} \quad (12.43)$$

$$Ra_L = Gr_L \Pr = \frac{\beta g \Delta T L^3}{\nu^2} \times \Pr \quad (12.44)$$

The heat transfer coefficient at the top cover can be determined from the following heat transfer correlations.

$$h_w = 5.7 + 3.8\, V_\infty \quad (12.45)$$

V_∞ = wind speed in m/s

Alternatively, the following dimensionless correlations also can be used.

$$j = 0.86(\mathrm{Re}^*_L)^{-1/2} \quad (12.46)$$

where j = j-factor given by $(h_w/\rho C_p V_\infty)\Pr^{2/3}$

Re^*_L = Reynolds number $(V_\infty L^*/\nu)$ based on the characteristic dimension

$$L^* = 4A_c/C_c \quad (12.47)$$

A_c = collector gross area,

C_c = circumference associated with the collector gross area.

The effective sky temperature,

$$T_{sky} = T_a - 6 \quad (12.48)$$

In finding the bottom loss coefficient, the convection can be considered negligible, as the heat loss occurs mainly with conduction. Figure 12.9 details the bottom and side losses for the heat transfer formulation.

$$U_b = \frac{k_i}{\delta_b} \quad (12.49)$$

where

k_i = thermal conductivity of the insulating material for bottom and sides, W/mK

δ_b = thickness of the insulation, m

The side loss coefficient, which is related to the four sides of a flat plate collector, can be determined similar to the bottom loss coefficient. Conduction is only considered with one-dimensional assumptions. If the area of the collector is $L_1 \times L_2$ with a height of L3, then the total area for heat flow is $2(L_1+L_2)L_3$. The temperature difference changes from $(T_{pm} - T_a)$ at the absorber

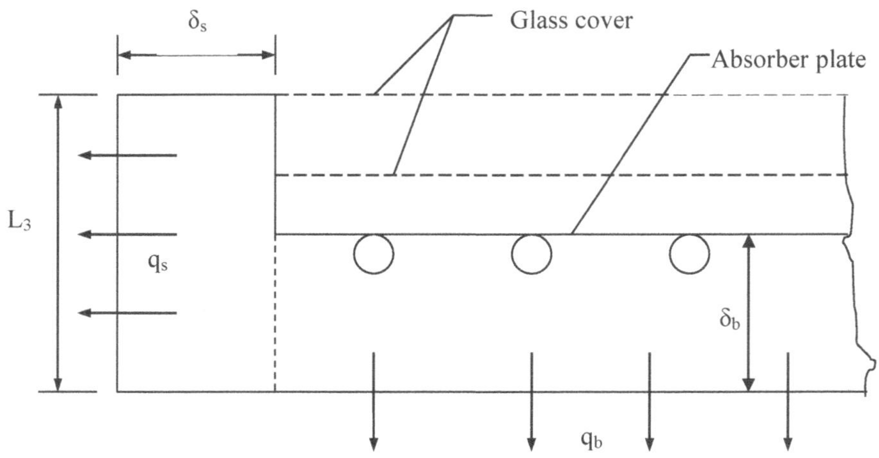

Fig. 12.9. Details of bottom and side losses from a flat plate collector.

plate to zero at the bottom. If the average temperature across the insulation is $(T_{pm} - T_a)/2$ with the thickness of insulation of δ_s,

$$Q_s = 2L_3(L_1 + L_2)k_i \frac{T_{pm} - T_a}{2\delta_s} \tag{12.50}$$

Thus, from the equation,

$$Q_s = U_s A_p (T_{pm} - T_a) \tag{12.51}$$

$$U_s = \frac{(L_1 + L_2)L_3 k_i}{L_1 L_2 \delta_s} \tag{12.52}$$

Similar to the ε-NTU method in a heat exchanger, the collector efficiency factor and heat removal factor can be used to simulate the heat load of a flat plate collector even without knowing the fluid outlet temperature. The one-dimensional heat flow at right angle to the absorber plate and in the tubes is considered to evaluate the flat plate collector. If the length and width of an absorber tube are L_1 and L_2, and N is the number of fluid tubes, then the pitch of the tubes is W ($=L_2/N$). Let D_i and D_o be the inner and outer diameters of tubes, as shown in Fig. 12.10.

As shown in Fig. 12.10, consider a section of absorber plate between two tubes. The temperature of the plate, T_p, changes in the x-direction with a one-dimensional approach. Taking a small slice dy along the direction of flow (z-axis in Fig. 12.10) and neglecting heat flow in the y direction, an energy balance can be applied for this element dx × dy on the absorber plate.

Conduction heat + solar flux = heat loss

$$k_p \delta_p \frac{d^2 T_p}{dx^2} dxdy + Sdxdy = U_l dxdy(T_p - T_a) \tag{12.53}$$

Simplifying,

$$\frac{d^2 T_p}{dx^2} = \frac{U_l}{k_p \delta_p}\left(T_p - T_a - \frac{S}{U_l}\right)$$

Let $m = \left(\dfrac{U_l}{k_p \delta_p}\right)^{1/2}$

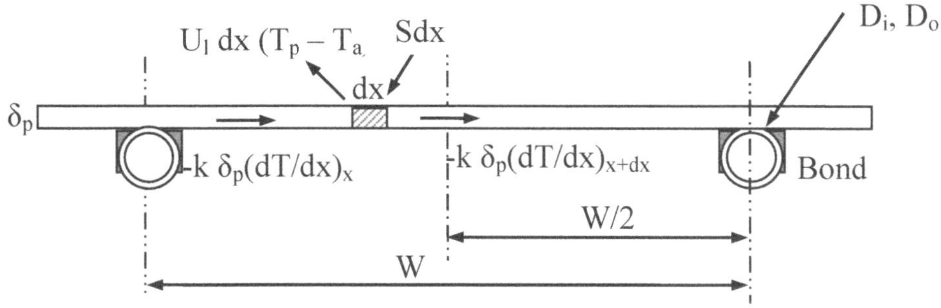

Fig. 12.10. Energy balance on a fin element.

and

$$T_p - \left(T_a + \frac{S}{U_l}\right) = \theta$$

$$\frac{d^2\theta}{dx^2} - m^2\theta = 0$$

The general solution is, $\theta = C_1.e^{-mx} + C_2.e^{mx}$

The boundary condition required to solve C_1 & C_2

The above equation is then solved using the boundary conditions

$$at\ x = 0, \frac{dT_p}{dx} = 0\ or \frac{d\theta}{dx} = 0$$

and

$$x = \left(\frac{W - D_o}{2}\right), T_p = T_{po}\ or\ \theta = \theta_0$$

$$\frac{d\theta}{dx} = -m.C_1.e^{-mx} + m.C_2.e^{mx}$$

$0 = -.C_1 + .C_2$ at $x = 0$, i.e., $C_1 = C_2$

$$\theta_0 = C_1.e^{-m\left(\frac{W-D_o}{2}\right)} + C_2.e^{m\left(\frac{W-D_o}{2}\right)}\ at\ x = \left(\frac{W - D_o}{2}\right),$$

$$\theta_0 = C_1\left(e^{-m\left(\frac{W-D_o}{2}\right)} + e^{m\left(\frac{W-D_o}{2}\right)}\right)$$

$$C_1 = C_2 = \frac{\theta_0}{\left(e^{-m\left(\frac{W-D_o}{2}\right)} + e^{m\left(\frac{W-D_o}{2}\right)}\right)}$$

Substituting C_1 and C_2 in $\theta = C_1.e^{-mx} + C_2.e^{mx}$

$$\theta = \theta_0 \frac{e^{-mx} + e^{mx}}{\left(e^{-m\left(\frac{W-D_o}{2}\right)} + e^{m\left(\frac{W-D_o}{2}\right)} \right)}$$

$$\because \cosh x = \frac{e^x + e^{-x}}{2}$$

$$\theta = \theta_0 \frac{\cosh(mx)}{\cosh\left(m\left(\frac{W-D_o}{2}\right)\right)}$$

$$\frac{T_p - \left(T_a + \frac{S}{U_l}\right)}{T_{po} - \left(T_a + \frac{S}{U_l}\right)} = \frac{\cosh mx}{\cosh\left(\frac{m(W-D_o)}{2}\right)}$$

where $m = \left(\frac{U_l}{k_p \delta_p}\right)^{1/2}$

The temperature changes are similar to the changes in a long fin.

The rate at which energy is conducted through the plate to one fluid tube from both sides

$$= -2k_p\delta_p \left(\frac{dT_p}{dx}\right)_{x=(W-D_o)/2} dy$$

$$= -2k_p\delta_p.\theta_0(m)\frac{\sinh mx}{\cosh\left(\frac{m(W-D_o)}{2}\right)}\Bigg|_{x=\frac{(W-D_o)}{2}}$$

$$= -2\left(\frac{U_l}{k_p\delta_p}\right)^{1/2} k_p\delta_p.\left(T_{po} - \left(T_a + \frac{S}{U_l}\right)\right)\tanh\left(\frac{m(W-D_o)}{2}\right) dy$$

$$= -2(k_p\delta_p U_l)^{1/2}\frac{U_l^{1/2}}{U_l^{1/2}}\left(T_{po} - \left(T_a + \frac{S}{U_l}\right)\right)\tanh\left(\frac{m(W-D_o)}{2}\right) dy$$

$$= 2\left(\frac{k_p\delta_p}{U_l}\right)^{1/2} [S - U_l(T_{po} - T_a)]\tanh\left(\frac{m(W-D_o)}{2}\right) dy$$

The rate at which energy is absorbed just above the tube = $D_0[S - U_l(T_{po} - T_a)]$ dy

Thus, the useful energy gain for all the N tubes of the collector over a length dy is given by

$$dq_u = N[S - U_l(T_{po} - T_a)]\left[2\left(\frac{k_p\delta_p}{U_l}\right)^{1/2} \tanh\frac{m(W-D_o)}{2} + D_o\right] dy$$

The equation can be simplified with the plate effectiveness, Φ, which is the ratio of heat conducted through the plate to the fluid at the thermal conductivity of the plate material.

$$\phi = \frac{\tanh\left[\frac{m(W - D_0)}{2}\right]}{\frac{m(W - D_0)}{2}}$$

Thus $dq_u = N[S - U_l(T_{po} - T_a)]\left[2\frac{1}{m}\phi\frac{m(W - D_o)}{2} + D_o\right]dy$

$$\frac{1}{N}\left(\frac{dq_u}{dy}\right) = [S - U_l(T_{po} - T_a)][\phi(W - D_o) + D_o]$$

For the heat flow from the plate to the fluid, there are three thermal resistances that exist. They are the adhesive of the tubes, tube wall resistance, and the inner heat transfer coefficient. The tube wall resistance can be ignored due to the thinness of the wall.

$$\frac{1}{N}\left(\frac{dq_u}{dy}\right) = \frac{(T_{p0} - T_f)}{\left(\frac{\delta_a}{k_a D_o} + \frac{1}{\pi D_i h_f}\right)}$$

where δ_a = average thickness of the adhesive
k_a = thermal conductivity of the adhesive material,
T_f = local fluid
and h_f = heat transfer coefficient on the inside surface of the tube.

Combining the above two equations so as to eliminate the intermediate temperature T_{po}, we have

$$\frac{1}{N}\left(\frac{dq_u}{dy}\right) = [S - U_l(T_{po} - T_a)][\phi(W - D_o) + D_o]$$

$$\frac{\frac{1}{N}\left(\frac{dq_u}{dy}\right)}{[\phi(W - D_o) + D_o]} = S - U_l T_{po} + U_l T_a$$

$$T_{po} = \frac{1}{U_l}\left(S - \frac{\frac{1}{N}\left(\frac{dq_u}{dy}\right)}{[\phi(W - D_o) + D_o]} + U_l T_a\right)$$

and from $\frac{1}{N}\left(\frac{dq_u}{dy}\right) = \frac{(T_{p0} - T_f)}{\left(\frac{\delta_a}{k_a D_o} + \frac{1}{\pi D_i h_f}\right)}$

$$T_{p0} = \frac{1}{N}\left(\frac{dq_u}{dy}\right)\left(\frac{\delta_a}{k_a D_o} + \frac{1}{\pi D_i h_f}\right) + T_f$$

Equating these two equations of T_{p0v}

$$\frac{1}{U_l}\left(\frac{S-\frac{1}{N}\left(\frac{dq_u}{dy}\right)}{[\phi(W-D_o)+D_o]}+U_lT_a\right)=\frac{1}{N}\left(\frac{dq_u}{dy}\right)\left(\frac{\delta_a}{k_aD_o}+\frac{1}{\pi D_ih_f}\right)+T_f$$

$$\left(\frac{S[\phi(W-D_o)+D_o]-\frac{1}{N}\left(\frac{dq_u}{dy}\right)+U_lT_a[\phi(W-D_o)+D_o]}{U_l[\phi(W-D_o)+D_o]}\right)=\frac{1}{N}\left(\frac{dq_u}{dy}\right)\left(\frac{\delta_a}{k_aD_o}+\frac{1}{\pi D_ih_f}\right)+T_f$$

$$\left(\frac{S[\phi(W-D_o)+D_o]-\frac{1}{N}\left(\frac{dq_u}{dy}\right)+U_lT_a[\phi(W-D_o)+D_o]-U_l[\phi(W-D_o)+D_o]T_f}{U_l[\phi(W-D_o)+D_o]}\right)$$

$$=\frac{1}{N}\left(\frac{dq_u}{dy}\right)\left(\frac{\delta_a}{k_aD_o}+\frac{1}{\pi D_ih_f}\right)$$

$$\left(\frac{(S+U_lT_a-U_lT_f)}{U_l}\right)\frac{1}{N}\left(\frac{dq_u}{dy}\right)\left(\frac{\delta_a}{k_aD_o}+\frac{1}{\pi D_ih_f}\right)+\frac{\frac{1}{N}\left(\frac{dq_u}{dy}\right)}{U_l[\phi(W-D_o)+D_o]}$$

$$\left(\frac{S-U_l(T_f-T_a)}{U_l}\right)=\frac{1}{N}\left(\frac{dq_u}{dy}\right)\left(\left(\frac{\delta_a}{k_aD_o}+\frac{1}{\pi D_ih_f}\right)+\frac{1}{U_l[\phi(W-D_o)+D_o]}\right)$$

$$\frac{1}{N}\left(\frac{dq_u}{dy}\right)=\frac{S-U_l(T_f-T_a)}{U_l\left(\frac{1}{U_l[W-D_o)\phi+D_o]}+\frac{\delta_a}{k_aD_o}+\frac{1}{\pi D_ih_f}\right)}$$

We now define a term called the collector efficiency factor (F') as follows

$$F'=\frac{1}{WU_l\left(\frac{1}{U_l[(W-D_o)\phi+D_o]}+\frac{\delta_a}{k_aD_o}+\frac{1}{\pi D_ih_f}\right)} \tag{12.54}$$

Substituting this definition into an equation, we get

$$\frac{1}{N}\left(\frac{dq_u}{dy}\right)=WF'[[S-U_l(T_f-T_a)]$$

$$F'=\frac{actual\ useful\ energy\ collected}{useful\ energy\ collected\ if\ the\ entire\ absorber\ surface\ is\ at\ the\ local\ fluid\ temperature}$$

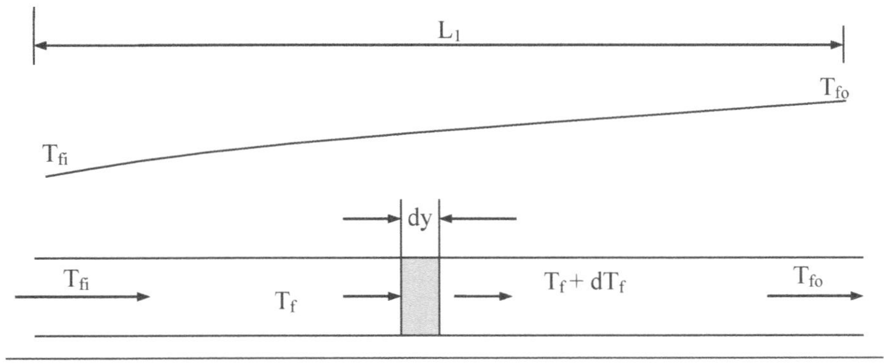

Fig. 12.11. Change of fluid temperature in the direction of fluid flow (y-axis).

$$q_u = F'A_p\left[S - U_l(T_f - T_a)\right]$$

T_f is the average of inlet and outlet temperatures.

The instantaneous efficiency of flat plate collector in terms of mean film temperature,

$$\eta_m = F'\frac{S}{I_T} - F'U_l\frac{(T_f - T_a)}{I_T}$$

Similar to the collector efficiency factor, a heat removal factor can be used in simulation of fluid outlet temperature and utilized with a flat plate collector. The one dimensional heat in the direction of fluid flow is considered as shown in Fig. 12.11.

Applying the energy balance equation to the small element,

The heat transfer in the fluid = net rate of energy transfer to the absorber plate

$$\left(\frac{\dot{m}}{N}\right)C_p dT_f = \frac{1}{N}dq_u = WF'[S - U_l(T_f - T_a)]dy$$

$$\frac{dT_f}{dy} = \frac{WF'U_l}{\left(\frac{\dot{m}}{N}\right)C_p}\left[\left(\frac{S}{U_l} + T_a\right) - T_f\right]$$

Integrating and using the inlet condition, y = 0, $T_f = T_{fi}$ the temperature distribution

$$\frac{\left(\frac{S}{U_l} + T_a\right) - T_f}{\left(\frac{S}{U_l} + T_a\right) - T_{fi}} = \exp\left(\frac{-L_2 F'U_l y}{\dot{m}C_p}\right)$$

The fluid outlet temperature T_{fo} is obtained by substituting $T_f = T_{fo}$ and $y = L_1$ in the above equation. Thus,

$$\frac{\left(\frac{S}{U_l} + T_a\right) - T_{fo}}{\left(\frac{S}{U_l} + T_a\right) - T_{fi}} = \exp\left(\frac{-F'U_l L_1 L_2}{\dot{m}C_p}\right)$$

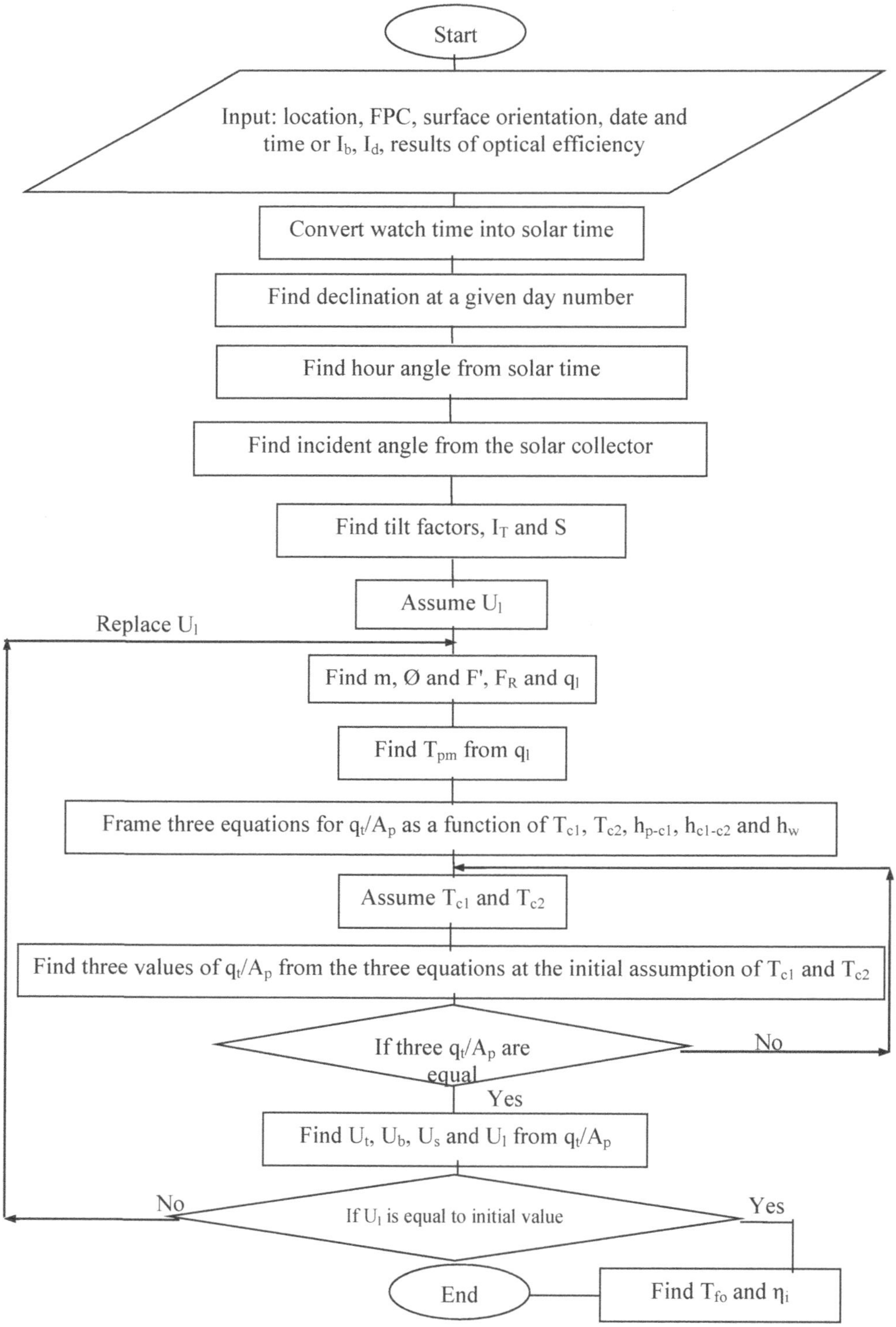

Fig. 12.12. Flow chart showing the methodology for thermal efficiency of a solar flat plate collector.

Subtracting both sides of this equation from the unit,

$$\frac{T_{fo} - T_{fi}}{\left(\frac{S}{U_l} + T_a\right) - T_{fi}} = 1 - \exp\left(\frac{-F'U_l A_p}{\dot{m}C_p}\right)$$

Thus, the useful heat gain rate for the collector

$Q_u = \dot{m}c_p\,(T_{pm} - T_a)$

$$= \dot{m}C_p\left[\left(\frac{S}{U_l} + T_a\right) - T_{fi}\right]\left[1 - \exp\left(\frac{-F'U_l A_p}{\dot{m}C_p}\right)\right]$$

$$= \frac{\dot{m}C_p}{U_l}\left[S - U_l(T_{fi} - T_a)\right]\left[1 - \exp\left(\frac{-F'U_l A_p}{\dot{m}C_p}\right)\right]$$

$Q_u = F_R\,A_p\,(S - U_l\,(T_{fi} - T_a)$

It is often referred to as the Hottel-Whillier-Bliss equation.
where,

$$F_R = \frac{\dot{m}C_p}{U_l A_p}\left[1 - \exp\left(\frac{-F'U_l A_p}{\dot{m}C_p}\right)\right] \tag{12.55}$$

$$F_R = \frac{\textit{actual useful energy collected}}{\textit{useful energy collected if the entire absorber surface is at the temperature of the fluid entering the collector}}$$

Therefore, with the use of F_R,

$Q_u = A_p\,(S - U_l\,(T_p - T_a)$ becomes

The instantaneous solar collector efficiency η is defined as

$$\eta_i = \frac{\textit{useful heat}}{\textit{solar radiation on collector}} \times 100$$

$$\eta_i = \frac{Q_u}{I_T A_p} \times 100$$

Substituting Q_u,

$$\eta = F_R\frac{S}{I_T} - F_R U_l\frac{(T_{fi} - T_a)}{I_T} \tag{12.56}$$

Figure 12.12 summarizes the modelling of thermal efficiency of a flat plat collector.

12.3 Transient analysis of flat plate collector

We will now consider a type of transient problem. On any day, a collector takes a few hours in the morning to reach its operating temperature, after which it may operate under quasi-steady conditions. Thus, a transient analysis is needed for these hours (from say 7 to 10 a.m.) to consider the heating of the collector from its low overnight temperature to its operating temperature. We will do an approximate lumped-parameter analysis for a single-cover collector for this situation. We assume that the absorber plate and tubes, the bottom and side insulation, and the water in the tubes have a heat capacity $(mC)_p$ and are at the same temperature T_{pm}, while the cover has a heat capacity $(mC)_c$ and is at a temperature T_c. Since there is no useful heat gain during this period, we obtain the following energy balance equations for the absorber plate and the cover

$$\frac{(mC)_p}{A_p}\frac{dT_{pm}}{dt} = S - \left[h_{p-c1}(T_{pm}-T_c) + \frac{\sigma(T_{pm}^4 - T_c^4)}{\left(\dfrac{1}{\varepsilon_p}+\dfrac{1}{\varepsilon_c}-1\right)}\right] - (U_b + U_s)(T_{pm} - T_a) \tag{12.57}$$

$$\frac{(mC)_p}{A_p}\frac{dT_c}{dt} = h_{p-c}(T_{pm}-T_c) + \frac{\sigma(T_{pm}^4 - T_c^4}{\left(\dfrac{1}{\varepsilon_p}+\dfrac{1}{\varepsilon_c}-1\right)} - \left[h_w(T_c - T_a) + \sigma\varepsilon_c(T_c^4 - T_{sky}^4)\right] \tag{12.58}$$

We combine the convective and radiative losses from the top by defining overall coefficients U_{t1} and U_{t2} as follows

$$U_{t1}(T_{pm}-T_c) = h_{p-c}(T_{pm}-T_c) + \frac{\sigma(T_{pm}^4 - T_c^4}{\left(\dfrac{1}{\varepsilon_p}+\dfrac{1}{\varepsilon_c}-1\right)} \tag{12.59}$$

$$U_{t2}(T_c - T_a) = h_w(T_c - T_a) + \sigma\varepsilon_c(T_c^4 - T_{sky}^4) \tag{12.60}$$

Thus, the first two equations become

$$\frac{(mC)_p}{A_p}\frac{dT_{pm}}{dt} = S - U_{t1}(T_{pm}-T_c) - (U_b + U_s)(T_{pm} - T_a) \tag{12.61}$$

$$\frac{(mC)_c}{A_p}\frac{dT_c}{dt} = U_{t1}(T_{pm}-T_c) - U_{t2}(T_c - T_a) \tag{12.62}$$

Now in a steady state

$$U_{t2}(T_c - T_a) = U_t(T_{pm} - T_a) \tag{12.63}$$

We assume that this relation holds during the transient state also and that T_a is a constant. Differentiating the above equation, we have

$$U_{t2}\frac{dT_c}{dt} = U_t\frac{dT_{pm}}{dt} \tag{12.64}$$

Substituting for U_{t2} and $\dfrac{dT_c}{dt}$ in equations,

$$\frac{1}{A_p}\left[(mC)_p + (mC)_c\frac{U_t}{U_{t2}}\right]\frac{dT_{pm}}{dt} = S - U_l(T_{pm} - T_a) \tag{12.65}$$

The quantity within the square brackets represents an effective heat capacity of the collector and will be denoted by the symbol $(mC)_e$. The differential equation can be integrated under the assumption that the values of S and T_a are constants. This is valid if the time interval of the integration is kept reasonably small. Using the initial condition

$t = 0,\ T_{pm} = T_{pi}$

we get

$$\frac{S-U_l(T_{pm}-T_a)}{S-U_l(T_{pi}-T_a)}=\exp\left(-\frac{A_pU_lt}{(mC)_e}\right) \tag{12.66}$$

The equation can be used either to find the time taken by the collector to reach the mean plate temperature corresponding to its prescribed fluid inlet temperature or to determine the mean plate temperature after a time lapse..

The efficiency values calculated from the equation are plotted against the parameter $(T_{fi}-T_a)/I_T$. The reason for doing this is apparent if one considers the Hottle-Whillier-Bliss equation. Dividing both sides of equation $q_u = F_RA_p\,[S - U_l\,(T_{fi}-T_a)]$ by A_cI_T, we have

$$\eta_i = F_R\left(\frac{A_P}{A_c}\right)\left[\frac{S}{I_T}-U_l\frac{(T_{fi}-T_a)}{I_T}\right] \tag{12.67}$$

We have

$$S = I_T(\tau\alpha)_{av} \tag{12.70}$$

where $(\tau\alpha)_{av}$ is an average transmissivity-absorptivity product for both beam and diffuse radiation. Hence

$$\eta_i = F_R\left(\frac{A_P}{A_c}\right)\left[(\tau\alpha)_{av}-U_l\frac{(T_{fi}-T_a)}{I_T}\right] \tag{12.68}$$

Since the values of F_R, $(\tau\alpha)_{av}$ and U_l are essentially constant, it is seen from equation that if η_i is plotted against $(T_{fi}-T_a)/I_T$ a straight line with a negative slope will be obtained. The intercept on the y-axis will give the value of $[F_R\,(\tau\alpha)_{av}\,A_p/A_c]$, while the slope of the line will give the value of $[F_R\,U_l\,A_p/A_c]$. In the equation, the value of η_i is based on the collector gross area. It could be based on the absorber plate area also. In that case, the term (A_p/A_c) would drop out of equations. The intercept on the y-axis would then be $F_R\,(\tau\alpha)_{av}$ and the slope of the line would be F_RU_l.

Experimental values of η_i plotted against the parameter $(T_{fi}-T_a)/I_T$ generally yield straight lines. However, the scatter of the data is always wide.

12.4 Solar air heater

A solar air heater is lighter than a solar water heater due to the density of air. There is no rust or corrosion in parts, so cheaper material can be used in an air heater. But the thermal properties of air are inferior to those of water. Solar air heaters can be used for space heating, dryers, and many other applications. In place of tubes, ducts will be used in a solar air heater. Let us take the size of the absorber plate of length L1 and width L2 with the air flows in a duct kept under the absorber plate as shown in Fig. 12.13.

The modeling of the solar air heater is similar to the solar water heater for finding the optical efficiency and thermal efficiency. Consider a small slice of width L_2 and thickness dx at a distance of x from the inlet. The energy balance equation can be developed with the assumptions that the bulk mean temperature of air changes from T_f to $(T_f + dT_f)$, and the mean temperatures of the absorber plate and the plate below are T_{pm} and T_{bm}, respectively, are constant, and side losses can be treated as negligible.

For the absorber plate

$$SL_2dx = U_tL_2dx(T_{pm}-T_a)+h_{fp}L_2dx(T_{pm}-T_f)+\frac{\sigma L_2dx(T_{pm}^4-T_{bm}^4)}{\left(\frac{1}{\varepsilon_p}+\frac{1}{\varepsilon_b}-1\right)} \tag{12.69}$$

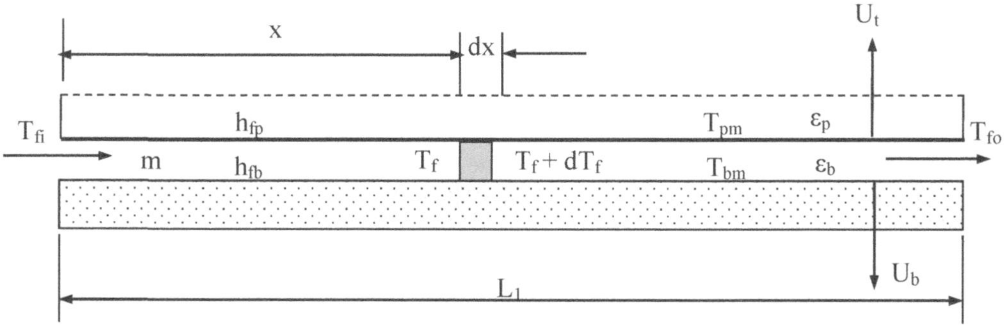

Fig. 12.13. Analysis of a conventional solar air heater.

For the bottom plate

$$\frac{\sigma L_2 dx(T_{pm}^4 - T_{bm}^4)}{\left(\frac{1}{\varepsilon_p} + \frac{1}{\varepsilon_b} - 1\right)} = h_{fb} L_2 dx(T_{bm} - T_f) + U_b L_2 dx(T_{bm} - T_a) \tag{12.70}$$

For the air stream

$$\dot{m} C_p dT_f = h_{fp} L_2 dx(T_{pm} - T_f) + h_{fb} L_2 dx(T_{bm} - T_f) \tag{12.71}$$

In the above equations,
S = flux absorbed in the absorber plate,
U_t = top loss coefficient, W/m² K
U_b = bottom loss coefficient, W/m² K
h_{fp} = convective heat transfer coefficient between the fluid and absorber plate, W/m² K
h_{fb} = convective heat transfer coefficient between the fluid and bottom plate, W/m² K
ε_p = emissivity of the absorber plate, and
ε_b = emissivity of the bottom plate.

There are not different correlations for convective heat transfer coefficients hfp and hfb; therefore, we can assume $h_{fp} = h_{fb}$.

The equivalent radiative heat transfer coefficient, h_r

$$h_r(T_{pm} - T_{bm}) \frac{\sigma(T_{pm}^4 - T_{bm}^4)}{\left(\frac{1}{\varepsilon_p} + \frac{1}{\varepsilon_b} - 1\right)} \tag{12.72}$$

For small values of the temperature difference ($T_{pm} - T_{bm}$), the term ($T_{pm}^4 - T_{bm}^4$) can be approximated to 4 $T_{av}^4 (T_{pm} - T_{bm})$ with $T_{av} = (T_{pm} + T_{bm})/2$.

Then

$$h_r = \frac{4\sigma T_{av}^4}{\left(\frac{1}{\varepsilon_p} + \frac{1}{\varepsilon_b} - 1\right)} \tag{12.73}$$

The bottom loss coefficient can be clubbed with the top loss coefficient as its value is small.

$$S = U_t(T_{pm} - T_a) + h_{fp}(T_{pm} - T_f) + h_r(T_{pm} - T_{bm}) \tag{12.74}$$

$$h_r(T_{pm} - T_{bm}) = h_{fb}(T_{bm} - T_f) \tag{12.75}$$

$$\frac{\dot{m}C_p}{L_2}\frac{dT_f}{dx} = h_{fp}(T_{pm} - T_f) + h_{fb}(T_{pm} - T_f) \tag{12.76}$$

From the equation, we get

$$T_{bm} = \frac{h_r T_{pm} + h_{fb} T_f}{h_r + h_{fb}} \tag{12.77}$$

Substituting this expression into equation, we have

$$T_{pm} = \frac{S + U_l T_a + h_e T_f}{U_l + h_e} \tag{12.78}$$

Effective heat transfer coefficient between the absorber plate and fluid,

$$\text{where } h_e = h_{fp} + \frac{h_r h_{fb}}{h_r + h_{fb}} \tag{12.79}$$

$$\text{Hence, } (T_{pm} - T_a) = \frac{S + h_e(T_f - T_a)}{U_l + h_e}$$

From the equations

$$\frac{\dot{m}C_p}{L_2}\frac{dT_f}{dx} = S - U_l(T_f - T_a) \tag{12.80}$$

Substituting the expression for $(T_{pm} - T_a)$ into the equation, we get the differential equation

$$\frac{\dot{m}C_p}{L_2}\frac{dT_f}{dx} = \frac{1}{\left(1 + \frac{U_l}{h_e}\right)}\left\{S - U_l(T_f - T_a)\right\} \tag{12.81}$$

In a manner analogous to the liquid flat-plate collector, we now define a collector efficiency factor F' given by

$$F' = \left(1 + \frac{U_l}{h_e}\right)^{-1} \tag{12.82}$$

Therefore

$$\frac{\dot{m}C_p}{L_2}\frac{dT_f}{dx} = F'\left\{S - U_l(T_{pm} - T_a)\right\} \tag{12.83}$$

This is derived first for the liquid flat plate collector. The solution now proceeds along identical lines, and we obtain the fluid temperature distribution

$$\frac{\left(\frac{S}{U_l} + T_a\right) - T_f}{\left(\frac{S}{U_l} + T_a\right) - T_{fi}} = \exp\left(\frac{-L_2 F' U_l x}{\dot{m}C_p}\right) \tag{12.84}$$

Similarly, the useful heat gain rate for the collector is given by

$$Q_u = F_R A_p [S - U_l (T_{fi} - T_a)] \tag{12.88}$$

where F_R = collector heat-removal factor

$$F_R = \frac{\dot{m}C_p}{U_l A_p}\left[1-\exp\left(\frac{-F'U_l A_p}{\dot{m}C_p}\right)\right] \tag{12.85}$$

and A_p = area of the absorber plate = L_1L_2

It is worth noting that if the simplifying assumption of deleting U_b from the equation and clubbing it with the first equation had not been made, we would have obtained the following differential equation.

$$\frac{\dot{m}C_p}{L_2}\frac{dT_f}{dx} = F'\{S - U_l''(T_f - T_a)\} \tag{12.86}$$

where U_l'' is an equivalent overall loss coefficient.

F' and U_l'' are defined as follows:

$$F' = \left(1+\frac{U_l'}{h_e}\right)^{-1} \tag{12.87}$$

$$U_l'' = U'_l + \frac{1}{F'}\frac{U_b h_{fb}}{(h_r + h_{fb} + U_b)} \tag{12.88}$$

where

$$U_l' = U_l + \frac{h_r U_b}{(h_r + h_{fb} + U_b)} \tag{12.89}$$

and

$$h_e = h_{fp} + \frac{h_r h_{fb}}{(h_r + h_{fb} + U_b)} \tag{12.90}$$

The useful heat gain rate for the collector is then given by

$$q_u = F_R A_p[S - U_l''(T_{fi} - T_a)] \tag{12.91}$$

where F_R = collector heat-removal factor

$$F_R = \frac{\dot{m}C_p}{U_l'' A_p}\left[1-\exp\left(\frac{-F'U_l'' A_p}{\dot{m}C_p}\right)\right] \tag{12.92}$$

Heat transfer and pressure drop in a parallel plate duct

Nusselt number,

$$Nu = 0.0158\ Re^{0.8} \tag{12.93}$$

Characteristic dimension,

$$d_e = \frac{4 \times cross\ sectional\ area\ of\ duct}{wetted\ \ perimeter} \tag{12.94}$$

Friction factor,

$$f = 0.079\ Re^{-0.25} \tag{12.95}$$

Pressure drop, $$\Delta P_{friction} = \frac{4 f \rho L_1 V^2}{2d_e} \tag{12.96}$$

12.5 Parabolic trough collector

Figure 12.14 shows the parts of a parabolic collector with a receiver tube. For the evaluation of a cylindrical parabolic collector, let

W = aperture, m
L = length of collector, m
Φ_r = rim angle, °
D_i = absorber tube has an inner diameter, m
D_o = absorber tube outer diameter, m
D_{ci} = concentric glass cover of inner diameter, m
D_{co} = glass cover outer diameter, m
$\dot{m}$ = fluid flow rate, kg/s
c_p = specific heat, kJ/kg K
T_{fi} = fluid inlet temperature, °C
T_{fo}.= fluid outlet temperature, °C

Generally, the axis of a cylindrical parabolic collector is oriented on the north-south axis to minimize the incidence angle and so capture maximum solar radiation. Since it is a single-axis tracking model, the focus of sun beams on the receiver is not uniform. Similar to earlier models, it can be assumed that the temperature drop in the absorber tube and transparent cover are considered negligible.

The concentration ratio of the collector,

$$C = \frac{Effective\ aperature\ area}{Abosrber\ tube\ area} = \frac{(W - D_o)L}{\pi D_o L} = \frac{(W - D_o)}{\pi D_o} \tag{12.97}$$

$$C = \frac{(W - D_o)}{\pi D_o} \tag{12.98}$$

Similar to the flat plate collector, an energy balance applied to an elementary strip dx of the absorber tube at a distance x from the inlet, results in the following steady state equation.

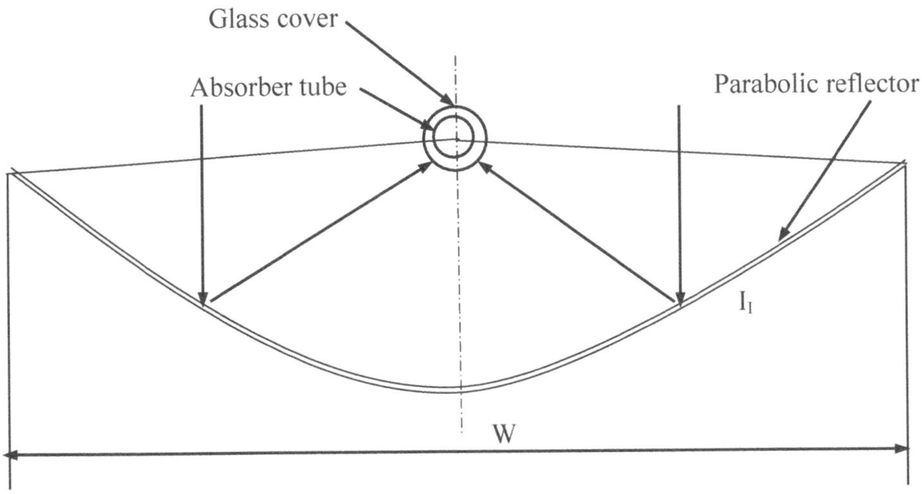

Fig. 12.14. Cylindrical parabolic collector.

$$dQ_u = [I_b r_b (W - D_o)\rho\gamma\, (\tau\alpha)_b + I_b r_b D_o\, (\tau\alpha)_b - U_l \pi D_o (T_p - T_a)]\, dx \tag{12.99}$$

where

dQ_u = useful heat gain rate for length dx,

ρ = specular reflectivity of the concentrator surface,

γ = intercept factor, the fraction of the secularly-reflected radiation intercepted by the absorber tube,

$(\tau\alpha)_b$ = average value of the transitivity-absorptivity product for beam radiation,

U_l = overall loss coefficient,

T_p = local temperature of the absorber tube, and

T_a = ambient temperature.

In the Eq. 12.99, the first term on the right-hand side represents the solar radiation falling on a reflective surface and the same focused on the absorber surface. The second term represents the direct falling of solar radiation on the absorber tube. The last term is related to the thermal losses from the receiver tube.

Absorbed solar flux per unit area,

$$S = I_b r_b \rho\gamma(\tau\alpha)_b + I_b r_b (\tau\alpha)_b \left(\frac{D_o}{W - D_o}\right) \tag{12.100}$$

Collector efficiency factor,

$$F' = \frac{1}{U_l\left(\frac{1}{U_l} + \frac{D_o}{D_i h_f}\right)} \tag{12.101}$$

h_f = heat transfer coefficient on the inside surface of the tube

Useful energy

$$Q_u = \dot{m} c_p (T_{fo} - T_{fi})$$

$$= F_R (W - D_o) L \left[S - \frac{U_l}{C}(T_{fi} - T_a)\right] \tag{12.102}$$

where F_R is the heat removal factor

$$F_R = \frac{\dot{m} C_p}{\pi D_o L U_l}\left[1 - \exp\left(\frac{-F'\pi D_o L U_l}{\dot{m} C_p}\right)\right] \tag{12.103}$$

The instantaneous collection efficiency based on global radiation,

$$\eta_i = \frac{Q_u}{I_T A_p} \times 100$$

$$\eta_i = \frac{Q_u}{(I_b r_b + I_d r_d) W L} \times 100 \tag{12.104}$$

In this equation, the ground reflection is treated as negligible.

The instantaneous efficiency based on beam radiation,

$$\eta_{ib} = \frac{Q_u}{I_b r_b W L} \times 100 \tag{12.105}$$

The overall loss coefficient and heat transfer coefficients can be determined from the energy balance and heat transfer correlations.

$$\frac{Q_l}{L} = h_{p-c}\pi D_o(T_{pm} - T_c) + \frac{\sigma\pi D_o(T_{pm}^4 - T_c^4)}{\left(\frac{1}{\varepsilon_p} + \frac{D_o}{D_{ci}}\left(\frac{1}{\varepsilon_c} - 1\right)\right)} \tag{12.106}$$

$$= h_w \pi D_{co}(T_c - T_a) + \sigma\pi D_{co}\epsilon_c(T_c^4 - T_{sky}^4) \tag{12.107}$$

where

Q_l/L = heat loss rate per unit length, W/m

h_{p-c} = convective heat transfer coefficient between the absorber tube and the glass cover, W/m²K

T_{pm} = average temperature of the absorber tube, °C

T_c = temperature attained by the cover, °C

In the above two non linear equations, the two unknowns are (Q_l/L) and T_c. In these equations h_{p-c} and h_w are determined from the heat transfer correlations.

The natural heat transfer coefficient between the absorber tube and cover, h_{p-c} is determined from the correlations.

$$\frac{k_{eff}}{k} = 0.317(Ra^*)^{\frac{1}{4}} \tag{12.108}$$

where k_{eff} = effective thermal conductivity, defined as the thermal conductivity that the motionless air in the gap must have to transmit the same amount of heat as the moving air, and

Ra^* = modified Rayleigh number related to the usual Reyleigh number by the following equation;

the characteristic dimension used for the calculation of the Rayleigh number is the radial gap $b = \frac{D_{ci} - D_o}{2}$.

Properties are evaluated at the mean temperature $\frac{T_{pm} + T_c}{2}$.

It is to be noted that the effective thermal conductivity k_{eff} cannot be less than the thermal conductivity k.

Hence $\frac{k_{eff}}{k}$ is taken as equal to unity if the use of the equation yields a value less than unity.

$$(Ra^*)^{\frac{1}{4}} = \frac{\ln(D_{ci}/D_o)}{b^{3/4}\left(\frac{1}{D_o^{3/5}} + \frac{1}{D_{ci}^{3/5}}\right)^{5/4}} Ra^{\frac{1}{4}} \tag{12.109}$$

The relationship between the heat transfer coefficient h_{p-c} and the effective thermal conductivity can be found by equating expressions for the heat exchange rate per unit length.

$$\frac{2\pi k_{eff}}{\ln(D_{ci}/D_o)}(T_{pm} - T_c) = h_{p-c}\pi D_o(T_{pm} - T_c) \tag{12.110}$$

Thus,

$$h_{p-c} = \frac{2k_{eff}}{D_o \ln(D_{ci}/D_o)} \tag{12.111}$$

The heat transfer coefficient on the outside surface of the cover is also determined from the correlations.

The wall convective heat transfer coefficient, h_w on the outside surface of the cover is determined by the following equation.

$Nu = C_1 Re^n$

where C_1 and n are constants having the following values:

for $40 < Re < 4000$, $C_1 = 0.615$, $n = 0.466$

for $4000 < Re < 40\,000$, $C_1 = 0.714$, $n = 0.618$

for $40\,000 < Re < 400\,000$, $C_1 = 0.0239$, $n = 0.805$ (12.112)

D_{co} is the characteristic dimension to be used in the equation. Properties are be evaluated at the mean temperature $\dfrac{T_c + T_a}{2}$.

Similarly, the heat transfer coefficient inside the absorber tube is determined from the correlation equation related to forced convection.

The convective heat transfer coefficient h_f on the inside surface of the absorber tube can be calculated under the assumption that the flow is fully developed. This assumption is justified because the length-to-diameter ratio (L/D_i) is large, usually greater than 20. For a Reynolds number less than 2000, the flow is laminar, and the heat transfer coefficient may be calculated from the equation

$$Nu = 3.66 \tag{12.113}$$

On the other hand, for a Reynolds number greater than 2000, the flow is turbulent, and the heat transfer coefficient may be calculated from the well-known Dittus Boelter equation

$Nu = 0.023\ Re^{0.8}\ Pr^{0.4}$

The characteristic dimension used for calculating Nu and Re in the equations is D_i. Properties are evaluated at the mean temperature $\dfrac{T_{fi} + T_{fo}}{2}$.

For augmentation of heat transfer coefficient, one of the simplest techniques is to use a twisted tape of width D_i inserted all along the inside of the absorber tube. Hong and Bergles (1976) have suggested the correlation for this case.

$$Nu = 5.172\left[1 + 0.005484\left(Pr\left(\frac{Re}{X}\right)^{1.78}\right)^{0.7}\right]^{0.5} \tag{12.114}$$

where X = tape twist ratio $= \dfrac{H}{D_i}$

and H = length over which the tape is twisted through 180°

The characteristic dimension used for calculating Nu and Re is D_i. Calculations with the equation show that the use of a twisted tape results in very significant increases in the heat transfer coefficient when the Prandtl number is high. At the same time, the pressure drop does not increase in the same proportion.

$$f\,Re = 38.4\left(\frac{Re}{X}\right)^{0.05} \quad for\ 6.7 \le \frac{Re}{X} \le 100 \tag{12.115}$$

$$= C_2\left(\frac{Re}{X}\right)^{0.3} \quad for\ \frac{Re}{X} > 100 \tag{12.116}$$

where f = friction factor,

$$C_2 = 8.8201\,X - 2.1193\,X^2 + 0.2108\,X^3 - 0.0069\,X^4 \tag{12.117}$$

12.6 Numerical solutions

1. Find a flat plate collector's optical efficiency at 11 a.m. located at a place (Φ = 12.916° N and L = 79.1833° E) on 28 February. Number of covers = one, Thickness of each cover = 4 mm, Refractive index of glass relative to air = 1.52, Extinction coefficient of glass = 15 m^{-1}. Absorptivity of the black plate = 0.95. The tilt angle = 20°. The dimensions of the flat plate collector = 1.17 m × 1 m. Fluid inlet temperature = 45°C. Reflectivity of surrounding surface = 0.2

Solution through App Designer

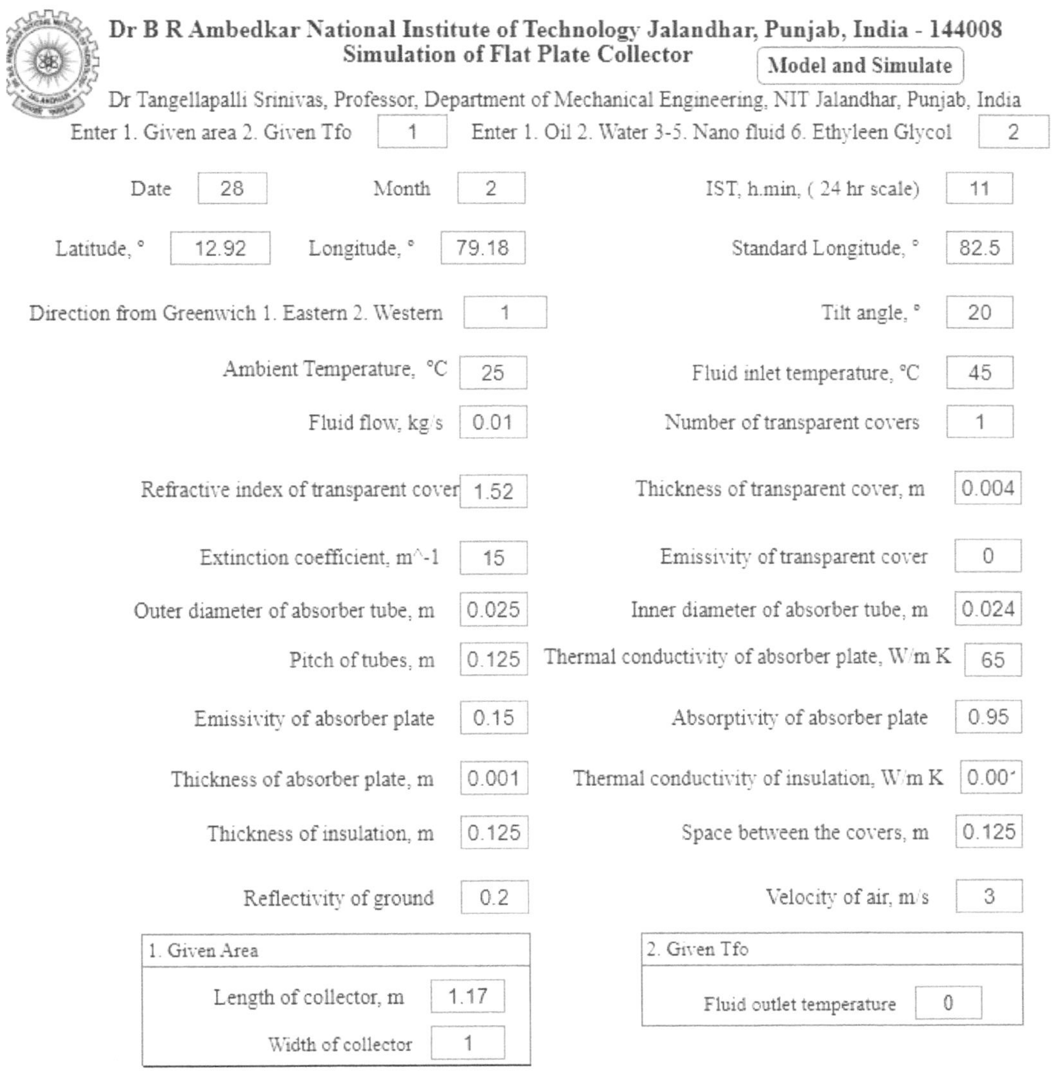

Day is February 28, the day number n = 59

Parameter	Equation	Substitution	Result
Solar Time	$B = \frac{360(n-81)}{364}{}^{\circ}$	$\frac{360(59-81)}{364}$	–21.76
	EoT = 9.87 sin 2B – 7.53 cos B – 1.5 sin B	9.87 sin (2 × (–21.76°)) – 7.53 cos (–21.76°) – 1.5 sin (–21.767°)	–13.23 min
	Solar time (Local Apparent Time) = Watch Time – 4 ($L_{st} - L_{lo}$) + EoT (in minutes)	11.00 – 4 (82.5 – 79.1833)/60 – 13.23/60	10.56 h = 10 h.33 min
Declination	$\delta = 23.45 \sin\left[\frac{360(284+n)}{365}\right]^{\circ}$	$23.45 \sin\left[\frac{360(284\quad 59)}{365}\right]$	–8.67°
Hour angle	ω = 15 (solar time – 12)°	15 (10.56 – 12)°	–21.63°
Zenith angle	$\theta_z = \cos^{-1}(\sin\varphi \sin\delta + \cos\varphi \cos\delta \cos\omega)$	$\cos^{-1}$ (sin (12.9166) sin (–8.62) + cos (12.9166) cos (–8.62) cos (–21.62))	30.45°
Incidence angle	$\theta = \cos^{-1}(\sin\delta \sin(\varphi-\beta) + \cos\delta \cos(\varphi-\beta) \cos\omega)$	$\cos^{-1}$ (sin (–8.67) sin (12.9166 – 20) + cos (–8.62) cos (12.9166 – 20) cos (–21.63))	21.47°
Normal beam radiation	$I_{bn} = A \exp\left(\frac{-B}{\cos\theta_z}\right)$	$652.2 \exp\left(\frac{-0.01}{\cos 30.45}\right)$	644.68 W/m²
Direct beam radiation	$I_b = I_{bn} \cos\theta_z$	644.68 × cos (30.45)	555.74 W/m²
Diffuse radiation	$I_d = C\, I_{bn}$	0.249 × 644.68	160.52 W/m²
Total or global radiation	$I_g = I_b + I_d$	555.74 + 160.52	716.26 W/m²
Tilt factor for beam radiation	$r_b = \frac{\sin\delta \sin(\varnothing-\beta) + \cos\delta \cos\omega \cos(\varnothing-\beta)}{\sin\delta \sin\varnothing + \cos\delta \cos\omega \cos\varnothing}$	$r_b = \frac{\sin(-8.67)\sin(12.9166-20) + \cos(-8.67)\cos(-21.63)\cos(12.9166-20)}{\sin(-8.67)\sin(12.9166) + \cos(-8.67)\cos(-21.63)\cos(12.9166)}$	1.08
Tilt factor for diffuse radiation	$r_d = \frac{1+\cos\beta}{2}$	$\frac{1+\cos(20)^{\circ}}{2}$	0.97
Tilt factor for ground reflection	$r_r = \rho\left(\frac{1-\cos\beta}{2}\right)$	$0.2\left(\frac{1-\cos(20)^{\circ}}{2}\right)$	0.01
Beam radiation on tilted surface	$I_T = I_b r_b + I_d r_d + (I_b + I_d)\, r_r$	(555.74 × 1.08) + (160.52 × 0.97) + ((555.74 + 160.52) × 0.01)	759.92 W/m²
Angle of refraction for beam radiation	$\theta_2 = \sin^{-1}\left(\frac{\sin\theta_1}{\frac{n_2}{n_1}}\right)$	$\theta_2 = \sin^{-1}\left(\frac{\sin 21.47^{\circ}}{1.52}\right)$	13.93°
Reflectivity of the two components of polarization 1 (beam)	$\rho_I = \frac{\sin^2(\theta_2-\theta_1)}{\sin^2(\theta_2+\theta_1)}$	$\frac{\sin^2(13.93^{\circ}-21.47^{\circ})}{\sin^2(13.93^{\circ}+21.47^{\circ})}$	0.05
Reflectivity of the two components of polarization 2 (beam)	$\rho_{II} = \frac{\tan^2(\theta_2-\theta_1)}{\tan^2(\theta_2+\theta_1)}$	$\frac{\tan^2(13.93^{\circ}-21.47^{\circ})}{\tan^2(13.93^{\circ}+21.47^{\circ})}$	0.03

Table contd. ...

...Table contd.

Parameter	Equation	Substitution	Result
Transmissivity of the two components of polarization 1 (beam)	$\tau_{rI} = \frac{1-\rho_I}{1+(2M-1)\rho_I}$	$\frac{1-0.0513}{1+0.0513}$	0.90
Transmissivity of the two components of polarization 2 (beam)	$\tau_{rII} = \frac{1-\rho_{II}}{1+(2M-1)\rho_{II}}$	$\frac{1-0.0346}{1+0.0346}$	0.94
Transmissivity based on reflection-refraction (beam)	$\tau_r = \frac{\tau_{rI}+\tau_{rII}}{2}$	$\frac{1}{2}(0.9025+0.933)$	0.92
Transmissivity based on absorption (beam)	$\tau_a = e^{\frac{-MK\delta_c}{\cos\theta_2}}$	$\exp[-(1\times 15\times 0.004/\cos 13.93°)]$	0.94
Transmissivity for beam radiation	$\tau_b = \tau_r\,\tau_a$	0.92×0.94	0.86
Angle of refraction for diffuse radiation (considering $\theta_1 =$ 60°)	$\theta_2 = \sin^{-1}\left(\frac{\sin\theta_1}{\frac{n_2}{n_1}}\right)$	$\sin^{-1}\left(\frac{\sin(60°)}{1.52}\right)$	34.73°
Reflectivity of the two components of polarization 1 (diffuse)	$\rho_I = \frac{\sin^2(\theta_2-\theta_1)}{\sin^2(\theta_2+\theta_1)}$	$\frac{\sin^2(34.73°-60°)}{\sin^2(34.73°+60°)}$	0.18
Reflectivity of the two components of polarization 2 (diffuse)	$\rho_{II} = \frac{\tan^2(\theta_2-\theta_1)}{\tan^2(\theta_2+\theta_1)}$	$\frac{\tan^2(34.73°-60°)}{\tan^2(34.73°+60°)}$	0.0015
Transmissivity of the two components of polarization 1 (diffuse)	$\tau_{rI} = \frac{1-\rho_I}{1+(2M-1)\rho_I}$	$\frac{1-0.18}{1+0.18}$	0.69
Transmissivity of the two components of polarization 2 (diffuse)	$\tau_{rII} = \frac{1-\rho_{II}}{1+(2M-1)\rho_{II}}$	$\frac{1-0.0015}{1+0.0015}$	0.997
Transmissivity based on reflection-refraction (diffuse)	$\tau_r = \frac{\tau_{rI}+\tau_{rII}}{2}$	$\frac{0.69+0.997}{2}$	0.84
Transmissivity based on absorption (diffuse)	$\tau_a = e^{\frac{-MK\delta_c}{\cos\theta_2}}$	$e^{\frac{-1\times 15\times 0.004}{\cos 60°}}$	0.93
Transmissivity for beam radiation (diffuse)	$\tau_d = \tau_r\,\tau_a$	0.84×0.93	0.78
Diffuse reflectivity	$\rho_d = \tau_a - \tau_d$	$0.93 - 0.78$	0.09

Table contd. ...

...Table contd.

Parameter	Equation	Substitution	Result
Net fraction of absorbed beam radiation	$(\tau\alpha)_b = \frac{\tau_b \alpha}{1-(1-\alpha)\rho_d}$	$\frac{0.86 \times 0.95}{1-(1-0.95)0.09}$	0.82
Net fraction of absorbed diffuse radiation	$(\tau\alpha)_d = \frac{\tau_d \alpha}{1-(1-\alpha)\rho_d}$	$\frac{0.78 \times 0.95}{1-(1-0.95)0.09}$	0.75
Solar flux at the absorber plate	$S = I_b r_b (\tau\alpha)_b + \{I_d r_d + (I_b + I_d)\, r_r\}(\tau\alpha)_d$	(555.74 × 1.08 × 0.82) + [(160.52 × 0.97) + ((555.74+160.52) × 0.01)] × 0.75)	613.25 W/m²
Optical efficiency	$\eta_{optical} = \frac{S}{I_T} \times 100$	$\frac{613.25}{759.92} \times 100$	80.70%

2. A flat plate collector has a copper absorber plate with one glass cover. Evaluate the collector with the following data.

Length of collector = 2.08 m
Width of collector = 1.07 m
Length of absorber plate = 2.00 m
Width of absorber plate = 0.98 m
Plate to cover spacing = 2.5 cm
Thermal conductivity of plate material = 350 W/m K
Plate thickness = 0.15 mm
Plate absorptivity for solar radiation = 0.94
Plate emissivity for re-radiation = 0.14
Outer diameter of tube = 13.7 mm
Inner diameter of tube = 12.5 mm
Tube centre-to-centre distance =11.3 cm = 113 mm
Location of collector = Pune, India (18°32' N, 73°51' E)
Date = May 15
Time = 12 noon. (IST)
Collector tilt = latitude angle
Surface azimuth angle = 0°
Solar flux available on absorber plate, S = 514.48 452.29 W/m²
Solar flux on tilted surface, I_T = 651.37 W/m²K
Adhesive resistance = Negligible
Water flow rate = 70 kg/h
Water inlet temperature = 60°C
Ambient temperature = 25°C
Wind speed = 3.1 m/s
Back insulation thickness = 5 cm
Insulation thermal conductivity = 0.04 W/m K
Reflectivity of surrounding surface = 0.2

Assume that the side loss coefficient is 10 percent of the bottom loss coefficient.

Calculate,

1. The angle of incidence of beam radiation on the collector,
2. The total solar flux incident on the collector,
3. The incident flux absorbed by the absorber plate,
4. The collector heat-removal factor and overall loss coefficient,
5. The water outlet temperature, and
6. The instantaneous efficiency.

Solution through App Designer

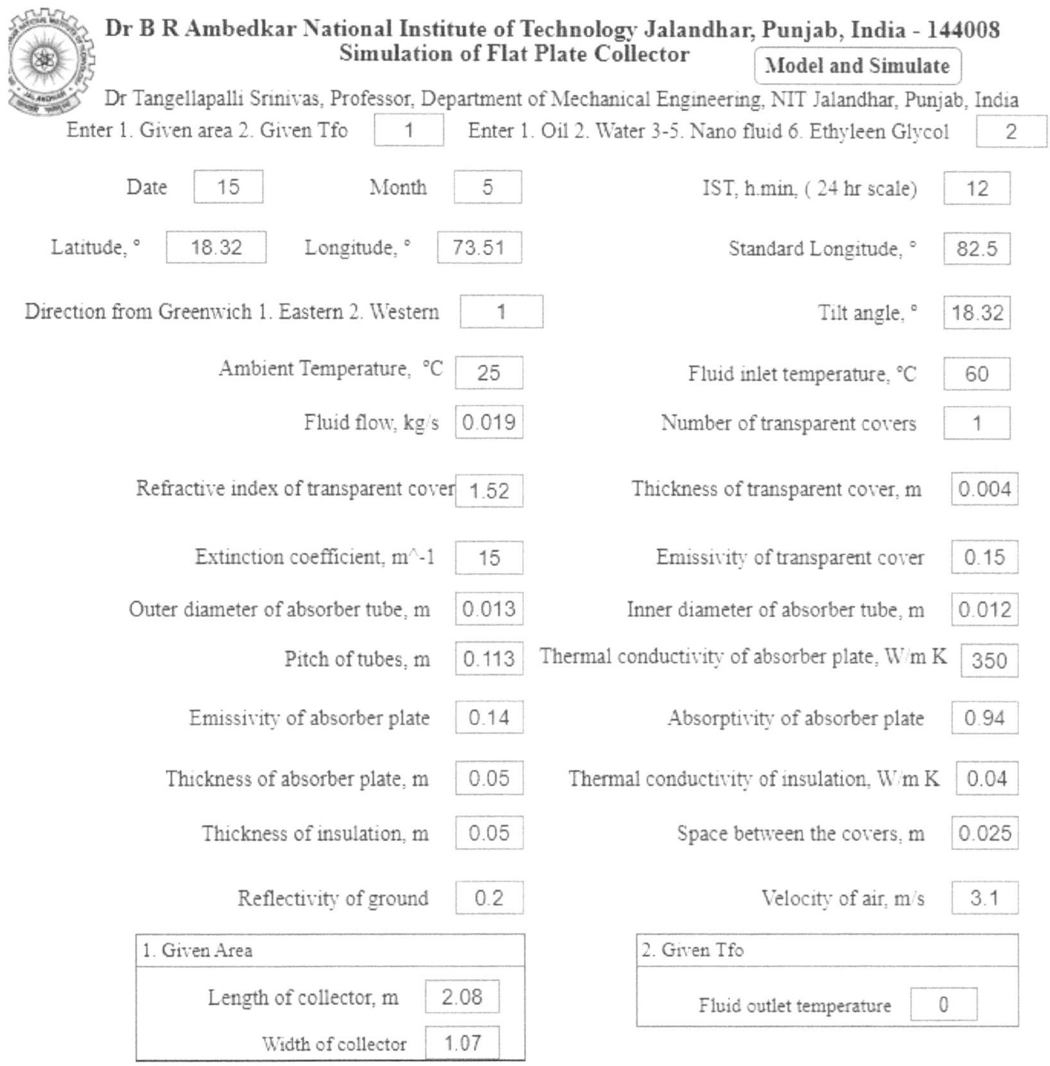

Note: The results of optical and solar radiation are furnished as input to avoid repetition in calculation.

Parameter	Equation	Substitution	Result
	For simple demonstration, the final values of U_l and T_c are given to avoid the iterations. $U_l = 4.27$ W/m²°C. $T_c = 33.29$ °C	--	--
Fin parameter	$m = \left(\frac{U_l}{k_p \delta_p}\right)^{1/2}$	$\left(\frac{4.27}{350 \times 0.0015}\right)^{1/2}$	2.85 m⁻¹
Effectiveness	$\varepsilon = \frac{m(W - D_o)}{2}$	$\frac{2.85(0.113 - 0.0137)}{2}$	0.14
Factor	$\varnothing = \frac{\tanh\left(\frac{m(W-D_o)}{2}\right)}{\left(\frac{m(W-D_o)}{2}\right)}$	$\frac{\tanh(0.14)}{(0.14)}$	0.99
Heat transfer coefficient in pipe	$h_f = 4.36\left(\frac{k_f}{D_i}\right)$	$4.36\left(\frac{0.6594}{0.0125}\right)$	230 W/m²
Collector efficiency factor	$F' = \frac{1}{WU_l\left(\frac{1}{(U_l(W-D_0)\varnothing + D_0))} + \frac{\delta_a = 0}{k_a D_0} + \frac{1}{\pi D_i h_f}\right)}$	$\frac{1}{0.113 \times 4.27\left(\frac{1}{(4.27(0.113 - 0.0137)0.99 + 0.0137))} + \frac{1}{\pi 0.0125 \times 230}\right)}$	0.94
Heat removal factor	$F_R = \frac{\dot{m}c_p}{U_l A_p}\left(1 - \exp\left(\frac{-F' U_l A_p}{\dot{m}c_p}\right)\right)$	$F_R = \frac{0.019 \times 4185}{4.27 \times 2.2256}\left(1 - \exp\left(\frac{-0.95 \times 4.27 \times 2.2256}{0.019 \times 4185}\right)\right)$	0.89
Useful heat gain	$Q_u = F_R A_P (S - U_l(T_{fi} - T_a))$	0.89 × 2.2256 × (514.48 – 4.27(60–25)	724.89 W
Heat loss	$Q_l = S A_p - Q_u$	514.48 × 2.2256 – 724.89	420.13 W
Plate mean temperature	$T_{pm} = T_a + \frac{Q_l}{U_l A_p}$	$25 + \frac{420.13}{4.27 \times 2.2256}$	69.16 °C
Sky temperature	$T_{sky} = T_a - 6$	25 – 6	19 °C
Mean film temperature	$T_{mf} = \frac{T_{pm} + T_c}{2}$	$\frac{69.16 + 33.29}{2}$	51.22 °C
Prandtl number	$\Pr = \frac{\mu c_p}{k_f}$	$\frac{(1.0915 \times 1.841 \times 10^{-5}) \times 1006.7}{0.028}$	0.72
Grashof number	$Gr = \frac{9.81\left(\frac{1}{T_{mf}}\right)(T_{pm} - T_c)L_c^3}{v^2}$	$\frac{9.81\left(\frac{1}{324.37}\right)(69.16 - 33.29)0.0250^3}{(1.8433 \times 10^{-5})^2}$	49970.28
Rayleigh number	$Ra = Gr \Pr \cos(\beta)$	4.9918 × 0.72 cos(18.32)	3.42×10^4
Nusselt number	$Nu = 0.229 Ra^{0.252}$	$0.229\ (3.42 \times 10^4)^{0.252}$	3.18

Table contd. ...

...Table contd.

Parameter	Equation	Substitution	Result
Heat transfer coefficient between plate and cover	$h_{pc} = \frac{Nuk_f}{L_c}$	$\frac{3.18 \times 0.028}{0.025}$	3.57 W/ m^2
Plate to cover heat loss	$\frac{Q_t}{A_p} = h_{pc}(T_{pm} - T_c) + \frac{\sigma(T_{pm}^4 - T_c^4)}{\left(\frac{1}{\varepsilon_p} + \frac{1}{\varepsilon_p} - 1\right)}$	$3.57(69.16 - 33.29) + \frac{5.667 \times 10^{-8}(69.16 + 273.15)^4 - (33.29 + 273.15)^4}{\left(\frac{1}{0.14} + \frac{1}{0.14} - 1\right)}$	149.71 W/m^2
Wall heat transfer coefficient	$h_w = 1.25(8.55 + 2.56\ V)$	$1.25(8.55 + 2.56 \times 3.1)$	16.48 W/ m^2
Top heat loss	$\frac{q_t}{A_p} = h_w(T_c - T_a) + \sigma\varepsilon_c(T_c^4 - T_{sky}^4)$	$16.48(33.29 - 25) + 5.667 \times 10^{-8} \times 0.14(33.29 + 273.15)^4 - 292.19^4)$	149.74 W/m^2
Top loss coefficient	$U_t = \frac{\frac{Q_t}{A_p}}{(T_{pm} - T_a)}$	$\frac{149.73}{(69.16 - 25)}$	3.39 W/ m^2 K
Bottom loss coefficient	$U_b = \frac{k_i}{\delta_b}$	$\frac{0.04}{0.05}$	0.8 W/ m^2 K
Side loss coefficient	U_s = 10% of bottom loss coefficient,	0.1×0.8	0.08 W/ m^2 K
Overall heat transfer coefficient (total)	$U_l = U_t + U_b + U_s$	$3.39 + 0.8 + 0.08$	4.27 W/ m^2 K
Fluid outlet temperature	$T_{fo} = T_{fi} + \frac{Q_u}{\dot{m}c_p}$	$60 + \frac{724.89}{0.019 \times 4180}$	69.11 ºC
Instantaneous thermal efficiency	$\eta_{fo} = \frac{Q_u}{I_T A_p} \times 100$	$\frac{724.89}{651.37 \times 2.2256} \times 100$	50%

3. Calculate the performance of a conventional solar air heater including pressure drop due to friction with the following data:

Length of absorber plate = 2 m.

Width of absorber plate = 1 m.

Space between absorber plate and bottom absorber plate = 2 cm.

Air flow rate = 200 kg/h.

Air inlet temperature = 50°C.

Ambient temperature = 30°C.

Solar flux on absorber plate = 578.13 W/m^2.

Solar radiation on tilted surface = 719.07 W/m^2.

Heat loss coefficient = 4.49 W/m^2K.

$\varepsilon_p = \varepsilon_b = 0.15.$

Neglect heat loss from the sides.

Solution through App Designer

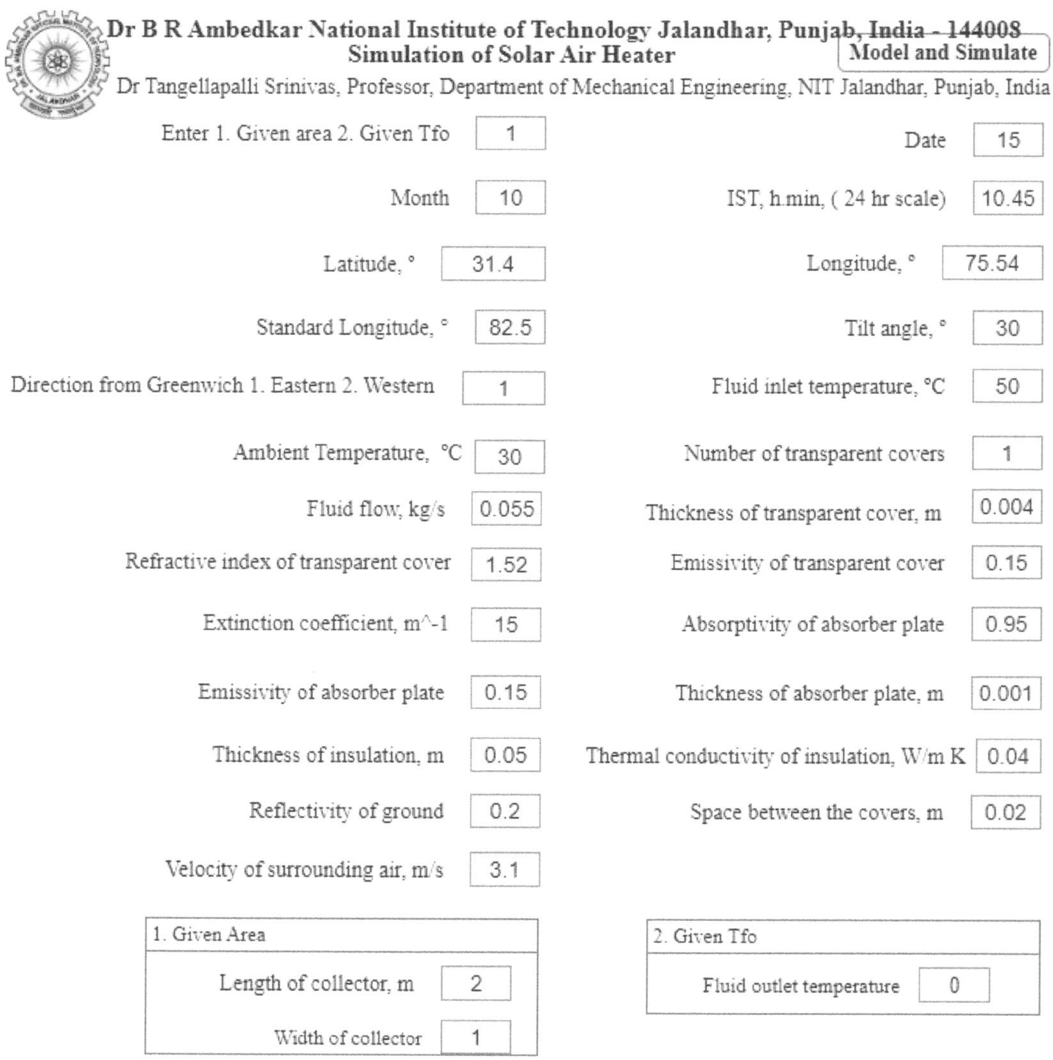

Note: The results of optical, solar radiation and heat loss coefficients have been furnished as input to avoid repetition in calculation.

Parameter	Equation	Substitution	Result
	Mean fluid temperature, T_{mf} is 55.15°C and heat loss coefficient of 4.49 W/m^2 K	--	--
Equivalent diameter	$d_e = \frac{4 \times cross\ sectional\ area\ of\ duct}{wetted\ perimeter}$	$\frac{4 \times 1 \times 0.02}{2(1+0.02)}$	0.0392 m
Velocity of air	$V = \frac{\dot{m}}{\rho A}$	$\frac{200}{3600 \times 1.0754 \times 1 \times 0.02}$	2.58 m/s
Reynolds number	$\mathrm{Re} = \frac{\rho V d_e}{\mu}$	$\frac{1.0754 \times 2.58 \times 0.0392}{19.845 \times 10^{-5}}$	5493.7
Nusselt number	$Nu = 0.0158\,\mathrm{Re}^{0.8}$	$Nu = 0.0158 \times 5493.7^{0.8}$	15.51
Convective heat transfer coefficient	$h_{fp} = h_{fb} = \frac{Nuk}{d_e}$	$\frac{15.51 \times 0.0282}{0.0392}$	11.14 W/m^2 K
Radiative heat transfer coefficient	$h_r = \frac{4\sigma T_{mf}^3}{\left(\frac{1}{\varepsilon_p} + \frac{1}{\varepsilon_p} - 1\right)}$	$\frac{4 \times 5.67 \times 10^{-8}(55.15+273.15)^3}{\left(\frac{1}{0.15} + \frac{1}{0.15} - 1\right)}$	0.65 W/m^2 K
Equivalent heat transfer coefficient	$h_e = h_{fp} + \frac{h_r h_{fp}}{h_r + h_{fp}}$	$11.14 + \frac{0.65 \times 11.14}{0.65 + 11.14}$	11.75 W/m^2 K
Collector efficiency factor	$F' = \left(1 + \frac{U_l}{h_e}\right)^{-1}$	$\left(1 + \frac{4.49}{11.75}\right)^{-1}$	0.72
Factor	$a = \frac{\dot{m} c_p}{U A_p}$	$\frac{200 \times 1.007 \times 1000}{3600 \times 1 \times 2}$	6.23
Collector heat removal factor	$F_R = a\left(1 - \exp\left(\frac{-F'}{a}\right)\right)$	$6.23\left(1 - \exp\left(\frac{-0.74}{6.23}\right)\right)$	0.68
Useful heat gain	$q_u = F_R A_P (S - U_l(T_{fi} - T_a))$	0.68 × 1 × 2 (578.13 – 4.49(50 – 30))	666.80 W
Fluid outlet temperature	$T_{fo} = T_{fi} + \frac{q_u}{\dot{m} c_p}$	$50 + \frac{666.80 \times 3600}{200 \times 1.007 \times 1000}$	61.90°C
Instantaneous thermal efficiency	$\eta_{fo} = \frac{q_u}{I_T A_p} \times 100$	$\frac{666.80}{719.07 \times 2} \times 100$	46.37%
Friction factor	$f = 0.079\ \mathrm{Re}^{-0.25}$	$0.079 \times 5493.7^{-0.25}$	0.0092
Pressure drop	$\Delta P_{friction} = \frac{4 f \rho L_1 V^2}{2 d_e}$	$\frac{4 \times 0.0092 \times 1.0754 \times 2 \times 2.58^2}{2 \times 0.0392}$	6.72 N/m^2

4. Calculate the overall loss coefficient U_l for the receiver of a cylindrical parabolic concentrating collector system. The receiver consists of a selective coated absorber tube with one glass cover around it. The following data is given:

Absorber tube, inner diameter 7.5 cm, outer diameter 8.1 cm

Glass cover, inner diameter 14.4 cm, outer diameter 15 cm

Emissivity of absorber tube surface = 0.15

Emissivity of glass cover = 0.88

Mean temperature of absorber tube = 170°C

Ambient temperature = 25°C

Wind velocity = 4 m/s

Solution with App Designer

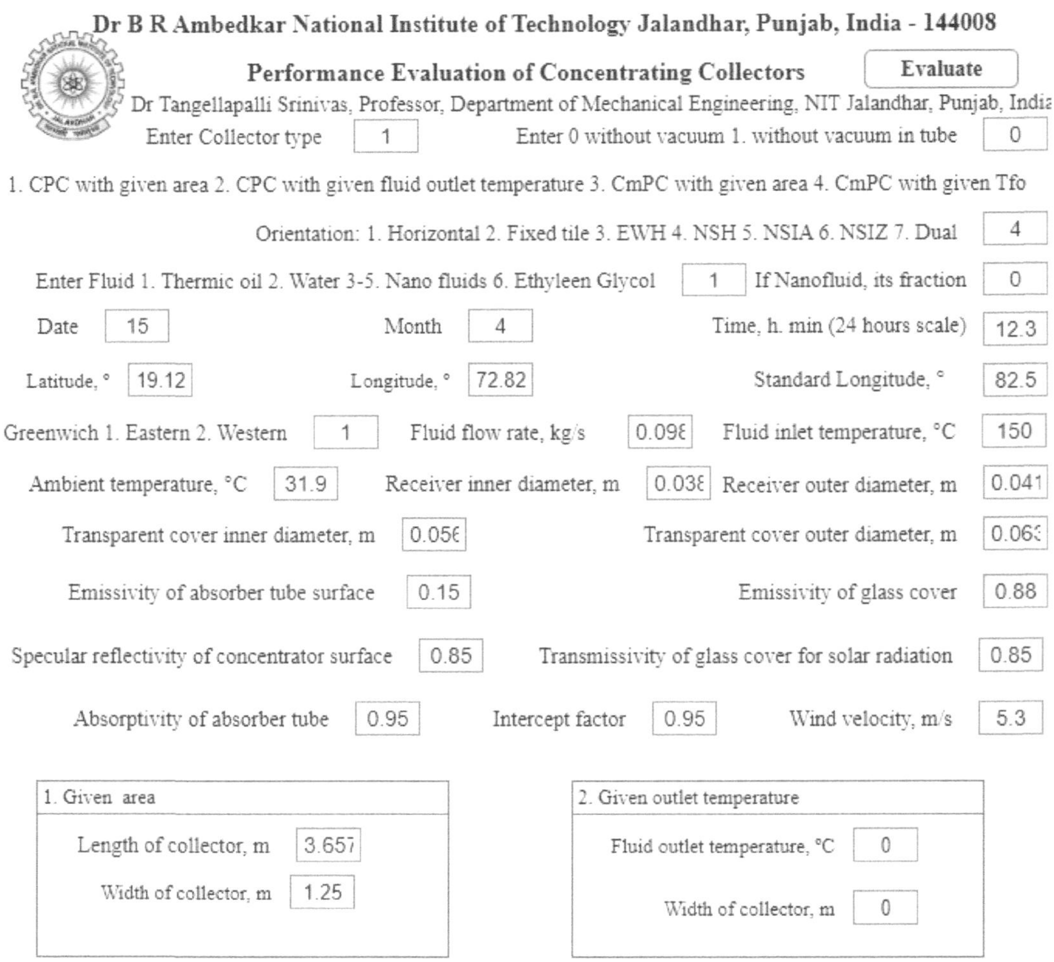

Note: The results of optical and solar radiation are furnished as input to avoid repetition in calculation.

Julian day = 105

Parameter	Equation	Substitution	Result
	The correct values of U_l and T_c are determined by doing an iterative procedure on a computer. Therefore, the solution is illustrated with the correct values of these values. $U_l = 6.43$ W/m² °C $T_c = 45.8$°C	--	--
Factor	$B = \frac{360(n-81)}{364}°$	$\frac{360(105-81)}{365}$	23.74°
Equation of time	EoT = 9.87 sin 2B – 7.53 cos B – 1.5 sin B	9.87 sin (2 × (23.74°)) – 7.53 cos (23.74°) – 1.5 sin (23.74°)	–0.223 min
Solar time	Solar time (Local Apparent Time) = Watch Time – 4 ($L_{st} - L_{lo}$) + EoT (in minutes)	12.50 – 4 (82.5 – 72.82)/60 – 0.223/60	11.85 h = h min
Declination	$\delta = 23.45 \sin\left[\frac{360(284+n)}{365}\right]°$	$23.45 \sin\left(\frac{360(284+105)}{365}\right)$	9.41°
Hour angle	ω = 15 (solar time - 12)°	15 (12.5 – 12)°	–2.23°
Zenith angle	$\theta_z = \cos^{-1}(\sin\varphi \sin\delta + \cos\varphi \cos\delta \cos\omega)$	$\cos^{-1}$ (sin (19.12) sin (9.41) + cos (19.12) cos (9.41) cos (–2.23))	9.94°
Tilt angle	$\beta = \tan^{-1}\left(\frac{\pm\cos\delta\sin\omega}{\sin\varnothing\sin\delta + \cos\varnothing\cos\delta\cos\omega}\right)$	$\tan^{-1}\left(\frac{-\cos(9.41)\sin(-2.23)}{\sin(19.12)\sin(9.41) + \cos(19.12)\cos(9.41)\cos(-2.23)}\right)$	2.24 °
Incidence angle	cosθ = sinδ sinφ cosβ – sinδ cosφ sinβ cosγ + cosδ cosφ cosβ cosω + cosδ sinφ sinβ cosγ cosω + cosδ sinβ sinγ sinω	sin(9.41) sin(19.12) cos(2.24) –sin(9.41) cos(19.12) sin(2.24) cos(–90) + cos(9.41) cos(19.12) cos(2.24) cos(–2.23) + cos(9.41) sin(19.12) sin(2.24) cos(–90) cos(–2.23) + cos(9.41) sin(2.24) sin(–90) sin(–2.23)	9.69°
Normal beam radiation	$I_{bn} = A\exp\left(\frac{-B}{\cos\theta_z}\right)$	$613.35\exp\frac{0.121}{\cos(9.94)}$	542.45 W/m²
Direct beam radiation	$I_b = I_{bn}\cos\theta_z$	542.45 × cos (9.94)	534.30 W/m²
Diffuse radiation	$I_d = C\, I_{bn}$	0.395 × 542.45	214.27 W/m²
Total or global radiation	$I_g = I_b + I_d$	534.30 + 214.27	748.56 W/m²
Tilt factor for beam radiation	$r_b = \frac{\cos\theta}{\cos\theta_Z}$	$\frac{\cos(9.69)}{\cos(9.94)}$	1.0008
Tilt factor for diffuse radiation	$r_d = \frac{1+\cos\beta}{2}$	$\frac{1+\cos(2.24)}{\cos(9.94)}$	0.99
Absorbed radiation flux	$S = I_b r_b (\tau\alpha)_b \left(\rho\gamma + \left(\frac{D_o}{W - D_o}\right)\right)$	$534.3 \times 1.0008 \times 0.85 \times 0.95\left(\frac{0.413}{1.25-0.413}\right)$	363.43 W/m²
Prandtl number	$\Pr = \frac{\mu c_p}{k_f}$	$\frac{(954.36 \times 5.9869 \times 10^{-7}) \times 1092}{0.1207}$	9.09

Table contd. ...

...Table contd.

Parameter	Equation	Substitution	Result
Average velocity	$V_{ave} = \frac{4\dot{m}}{\pi D_i^2}$	$\frac{4 \times 0.0986}{\pi 0.0381^2}$	0.0906 m/s
Reynolds number	$\text{Re} = \frac{V_{ave} D_i}{v}$	$\frac{0.0906 \times 0.0381}{5.9869 \times 10^{-7}}$	5767
Nusselt number	$\text{Nu} = 0.023\ \text{Re}^{0.8}\ \text{Pr}^{0.4}$	$0.023 \times 5567^{0.8} \times 9.09^{0.8}$	56.74
Heat transfer coefficient of fluid in pipe	$h_f = \frac{Nuk_f}{D_i}$	$\frac{356.74 \times 0.1207}{0.0381}$	179.8 W/m K
Collector efficiency factor	$F' = \frac{1}{U_l\left(\frac{1}{U_l} + \frac{D_o}{D_i h_f}\right)}$	$\frac{1}{6.43\left(\frac{1}{6.43} + \frac{0.0413}{0.0381 \times 179.8}\right)}$	0.96
Heat removal factor	$F_R = \frac{\dot{m}c_p}{\pi D_o L U_l}\left(1 - \exp\left(\frac{-F'\pi D_o L U_l}{\dot{m}c_p}\right)\right)$	$F_R = \frac{0.0986 \times 1920.9}{\pi \times 0.0413 \times 3.675 \times 6.43}$ $\left(1 - \exp 6\left(\frac{-0.96 \times \pi \times 0.0413 \times 3.675\pi \times 6.43}{0.0986 \times 1920.9}\right)\right)$	0.95
Concentration ratio	$CR = \frac{W - D}{\pi D}$	$\frac{1.25 \times 0.0413}{\pi \times 0.0413}$	9.3
Useful heat gain	$Q_u = F_R(W - D_o)L\left(S - \frac{U_l}{CR}(T_{fi} - T_a)\right)$	$0.95(1.25 - 0.0413)3.65\left(363.43 - \frac{6.43}{9.3}(150 - 31.9)\right)$	1194.3 W
Heat loss	$q_l = S\ (W - D_o)L - q_u$	363.43 (1.25 – 0.0413) 3.65 – 1194.3	412.07 W
Plate mean temperature	$T_{pm} = T_a + \frac{q_l}{\pi D_o L U_l}$	$31.9 + \frac{412.07}{\pi \times 0.0413 \times 3.65 \times 6.43}$	168.33 °C
Characteristic length	$b = \frac{D_{ci} - D_o}{2}$	$\frac{0.056 \times 0.0413}{2}$	0.0073
Prandtl number	$\text{Pr} = \frac{\mu_a c_{pa}}{k_a}$	$\frac{(0.9301 \times 2.405 \times 10^{-5}) \times 1011.8}{0.0321}$	0.7
Grashof number	$Gr = \frac{9.81\left(\frac{1}{T_{mf}}\right)(T_{pm} - T_c)b^3}{v^2}$	$\frac{9.81\left(\frac{1}{380.22}\right)(168.33 - 45.8)0.0250^3}{(2.405 \times 10^{-5})^2}$	2147.5
Rayleigh number	Ra = Gr Pr	2147.5 × 0.7	1512.4
Effective thermal conductivity	$k_{eff} = \frac{k \times 0.317 \times ln\left(\frac{D_{ci}}{D_o}\right)}{b^{3/4}\left(\frac{1}{D_o^{3/5}} + \frac{1}{D_{ci}^{3/5}}\right)^{5/4}} Ra^{1/4}$	$\frac{0.0321 \times 0.317 \times ln\left(\frac{0.056}{0.0413}\right)}{0.0073^{3/4}\left(\frac{1}{0.0413^{3/5}} + \frac{1}{0.056^{3/5}}\right)^{5/4}} 1512.4^{1/4}$	0.0331 W/m k

Table contd. ...

...Table contd.

Parameter	Equation	Substitution	Result
Heat Transfer coefficient on inside surface of absorber surface	$h_{p-c} = \frac{2k_{eff}}{D_o \ln\left(\frac{D_{ci}}{D_o}\right)}$	$\frac{2 \times 0.0331}{0.0413 \ln\left(\frac{0.056}{0.0413}\right)}$	5.27 W/m² K
Reynolds number	$Re = \frac{V_{ave} D_i}{v}$	$\frac{0.063 \times 5.3}{1.7252 \times 10^{-5}}$	19354
Nusselt number	$Nu = 0.229 Ra^{0.252}$	$0.174\ (19354)^{0.618}$	77.58
Heat transfer coefficient on the outside surface of the cover	$h_w = \frac{Nuk_a}{D_{co}}$	$\frac{77.58 \times 0.0271}{0.063}$	33.39 W/m² K
Sky temperature	$T_{sky} = T_a - 5$	$31.9 - 5$	26.9 °C
Plate to cover heat loss	$\frac{Q_t}{L} = h_{p-c}\pi D_o(T_{pm} - T_c) + \frac{\sigma \pi D_o (T_{pm}^4 - T_c^4)}{\left(\frac{1}{\varepsilon_p} + \frac{D_o}{D_{ci}}\left(\frac{1}{\varepsilon_c} - 1\right)\right)}$	$5.27\pi \times 0.0413(168.33 - 45.81) + \frac{5.667 \times 10^{-8}\pi \times 0.0413(168.33 + 273.15)^4 - (45.81 + 273.15)^4)}{\left(\frac{1}{0.15} + \frac{0.0413}{0.056} \times \left(\frac{1}{0.88} - 1\right)\right)}$	114.03 W/m
Top heat loss	$\frac{Q}{L} = h_w \pi D_{co}(T_c - T_a) + \sigma \pi D_{co} \varepsilon_c (T_c^4 - T_{sky}^4)$	$33.39\pi \times 0.063(45.80 - 31.9) + 5.667 \times 10^{-8}\pi \times 0.063 \times 0.88(45.8+273.15)^4 - (26.9 + 273.15)^4)$	114.07 W/m
Top loss coefficient	$U_t = \frac{\frac{Q_t}{L}}{\pi D_o (T_{pm} - T_a)}$	$\frac{114.05}{\pi \times 0.0413(168.34 - 31.39)}$	6.43 W/m² K
Fluid outlet temperature	$T_{fo} = T_{fi} + \frac{Q_u}{\dot{m} c_p}$	$150 + \frac{114.05}{0.0986 \times 1920.9}$	156.31 °C
Instantaneous thermal efficiency	$\eta_{ig} = \frac{Q_u}{(I_T + I_T r_d) WL} \times 100$	$\frac{114.05}{(534.7 \times 214.27 \times 0.99)1.25 \times 3.657} \times 100$	34.88%
Instantaneous thermal efficiency based on beam radiation	$\eta_{ib} = \frac{Q_u}{I_T WL} \times 100$	$\frac{114.05}{(534.7 \times 1.25 \times 3.657} \times 100$	48.86%

Chapter 13
Solar Tracking

13.1 Introduction

A motorless tracking system can be operated without the use of electricity, batteries, or such power sources. But it is not a perpetual motion machine of the first kind (PPM1). It makes use of stored energy, which is in the form of potential energy. Its energy has been stored as gravitational energy by keeping a liquid (water) in a header. There are two headers on either side of a linear concentrator, viz., the east header and the west header. The liquid in the west header is half full, whereas the east header is completely filled with liquid. The tracking of the collector starts at sunrise and driving the collector till sunset. When starting up, the operator has to open the liquid valve of the east header at the predetermined flow rate. The dripping continues till sunset. At noon, half of the liquid is consumed, and both headers have the same quantity of fluid that maintains the collector at mid-position, which is vertically upward. During the forenoon operation, the liquid in the east header is greater than the west header. Therefore, the collector is tilted towards the east with a North-South (N-S) axis orientation. In the afternoon, the west header liquid is greater (fixed quantity) than the east header liquid. It keeps the collector's tilt towards the west. At sunset, the liquid in the east header is completely exhausted and empty. Therefore, after the start of the track, the unit rotates continuously without the attention of the operator. The next day, in the morning, the process repeats as described earlier. The water can be collected in a container without wastage, and the same can be recycled by replacing the empty header with the full header.

13.2 Design of the proposed novel tracking system

Figure 13.1 shows the operation of motorless tracking with a suspension spring. The suspension spring has been connected between the center of gravity of the collector and the fixed eccentric point below the focus. The eccentricity is equal to the maximum deflection of the spring. To avoid damage to the spring, the spring can be loaded to 75% of the maximum load. It also avoids slack during noon. Since the spring is fixed below the focus, during sunrise and sunset, the spring is not horizontal. The spring is a compression spring. The compression is maximum during sunrise and sunset and minimum at noon. This minimum deflection is a setting of the spring during its fitting to avoid slack and jerk at noon. The spring has two eyes, and these two are fitted to the center of gravity of the collector, one to a fixed eccentric point at the bottom of the focus. Two springs have been used on either side of the collector. For N-S axis rotation, these springs are located towards the north and south. The collector rotates about its focal line and the spring rotates about its eccentric point. Eccentricity describes the distance between the focus and the eccentric point. The shaft is mounted between bearings. Two sprocket wheels are mounted on the shaft on either side of the collector, i.e., the north side and the south side. These two sprocket wheels carry the liquid containers of the east and west headers through the roller chains. The dripping of liquid on one header with a fixed liquid on the other side causes the weight difference between these two headers. The collector is imbalanced

DOI: 10.1201/9781032642505-13

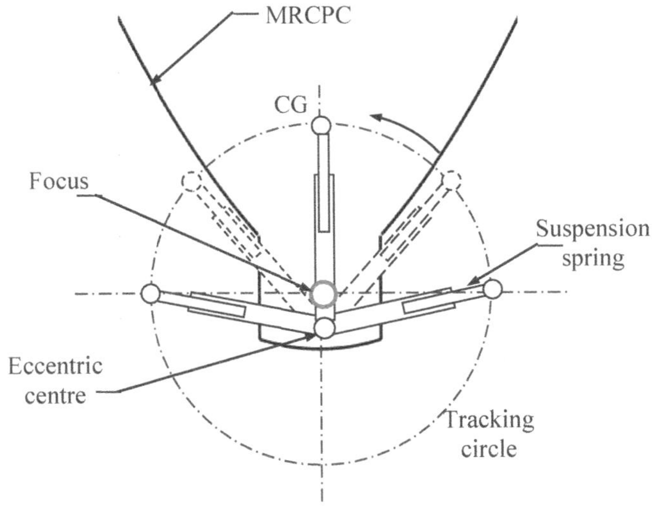

Fig. 13.1. The working principle of the suspension spring for controlling the rotation of the solar collector.

with these unequal loads. This difference in weight of the liquid drives the collector with the spring control. The weight of the collector is compensated for by the equal and opposite counter weight. Therefore, the driving force is caused by the difference in the header weight, which is controlled by the damping spring. The torque of the spring is determined by the force and perpendicular distance of spring action to the focus. Since the spring is of a compressive type, the spring force acts in the opposite direction, i.e., it is tensile in nature. The collector may be rotated about its center of gravity, a point below the collector or its focus. The absence of a counter balance is a benefit of rotating the collector about its center of gravity. The weight is evenly distributed around the center of gravity. But the focus is not a fixed member, and it rotates along with the collector. A flexible tube can be fixed for the supply and collection of the fluid. These connections also result in resistance to the track. The shaft of rotation also provides the shadow effect on the receiver tube. On the other hand, the shadow receiver from the driving shaft is placed below the collector. Here too, the receiver tube is not fixed. Since the distance from the center of gravity increases, it demands more counterbalance. It is recommended for a large size geometric factor (greater than 0.6). At these higher geometric factors, the radiation of multiple reflections decreases from the upper parabolic reflectors and flat surfaces. It decreases the effectiveness of the flat surfaces. The driving of the collector towards its focus has the benefit of a shadow-free receiver tube from the driving force. The main benefit of this method is a fixed receiver tube. But this also demands a counterbalance. But by properly selecting the geometry factor, the counter balance can be minimized. The shaft is not continuing and there are two parts to the track. The shaft is not a single piece. The collector's two plates and base should be strong enough to transmit the motion from the tracking drive, or else the collector will twist (Srinivas et al., 2024).

Figure 13.2 shows the arrangement of the spring (CB) and eccentric line (OC) geometrically. The collector is rotated about the centre of gravity, B. The eccentric point, which is fixed to the stand, connects one end of the spring. Another end of the spring eye is fixed to the centre of gravity of the collector, and it's a movable point. This side of the spring changes its position with the centre of gravity. The collector's angle with the horizontal is ζ. It changes from 0° to 90° from sunrise to noon and from 90° to 0° from noon to sunset. The radius, r_t, is the distance from focus to the centre of gravity (OB). This radial line also matches the axis of the collector. ϑ is the angle between the eccentric line (OC) and the collector's axis (OB) or radial line of tracking.

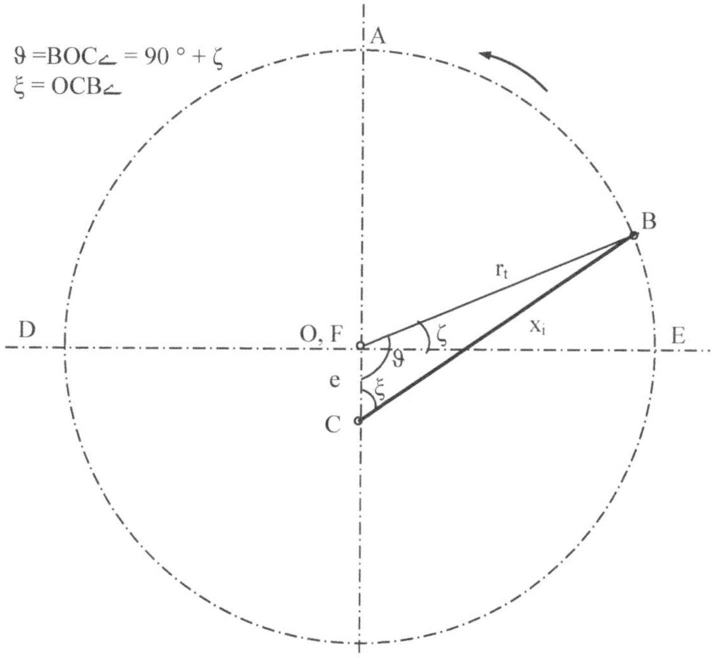

Fig. 13.2. Arrangement of suspension spring with reference to focus and eccentricity during forenoon.

From the Fig. 13.2, the angle of rotation of the collector axis with the eccentricity line (vertical),

$$\vartheta = 90 + \zeta \tag{13.1}$$

The collector is rotated as per the tracking orientation. For a higher solar energy collection, the axis of the collector is kept on the N-S axis, and it rotates from east to west. The collectors' angle, ζ changes from 0° to 90° till noon and afterwards decreases from 90° to 0° towards the west as the collector moves towards west. Therefore, the angle between the two lines, i.e., the eccentric line and the collector's axis, ϑ can be determined by adding 90° to the collector's angle.

Geometrically, the eccentric length can be fixed by the maximum allowed compression of the shock absorber. The radius of tracking is also a known value, which is equal to the distance between the focus and the centre of gravity of the collector. The spring length changes with the rotation of the collector. The instantaneous length of the spring of the shock absorber is evaluated from the eccentric length, radius of tracking and the angle between these two lines. The instantaneous length of the spring is the distance between the two eyes of the shock absorber.

Let x_i be the instantaneous length of the spring.

In Δ BOC, having sides, x, rt and e, using law of cosine,

$$x_i^2 = r_t^2 + e^2 - 2r_t e \ \cos\vartheta \tag{13.2}$$

Similarly, Fig. 13.3 shows the position of the solar collector during the afternoon. Since the collector angle and spring inclination are defined in a similar way to the forenoon, Eqs. 13.1 and 13.2 also hold here. The collector angle changes from 90° to 0°.

With reference to Fig. 13.4, the suspension spring tension acts along the length of BC. To find the moment of this force about the focus (O), the perpendicular distance of the spring force is required.

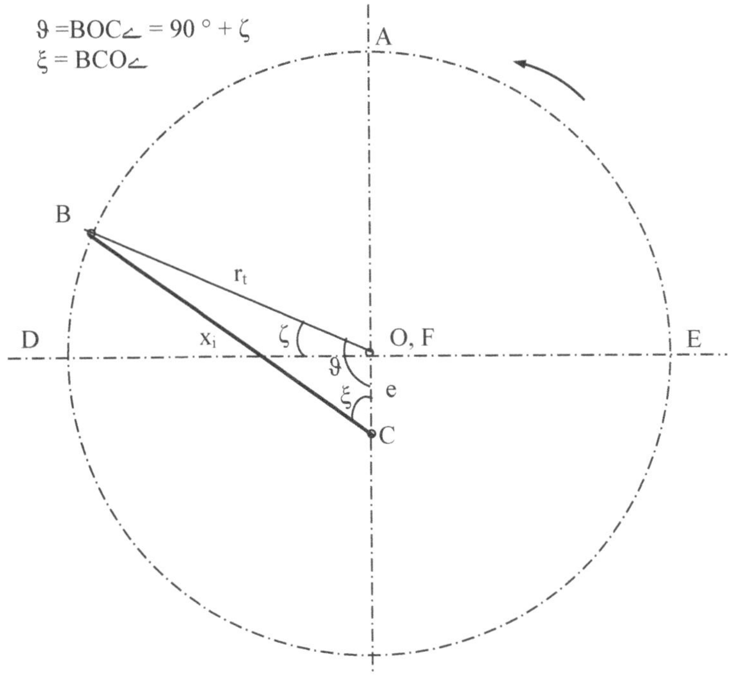

Fig. 13.3. Suspension spring arrangement with reference to focus and eccentricity during the afternoon period.

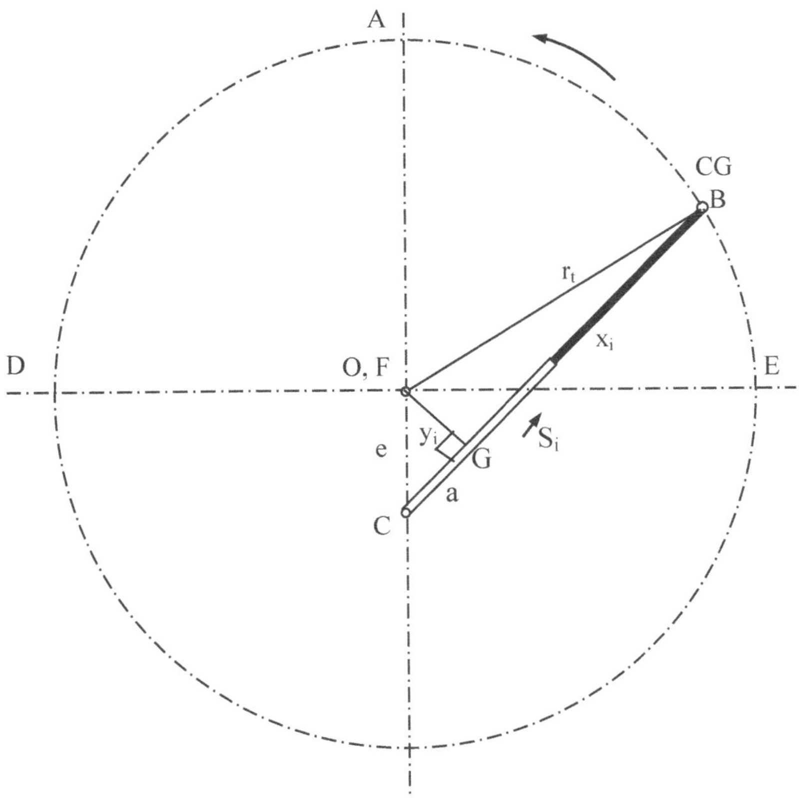

Fig. 13.4. The perpendicular distance of the spring line to focus.

Let OG be a perpendicular line drawn from O towards the spring. This perpendicular distance changes with the rotation of the solar collector. The instantaneous perpendicular distance is y_i. To find the perpendicular distance, y_i, let GC be the source.

The perpendicular distance, y_i can be determined from the two right angle triangles, Δ OGC and Δ OGB.

From Δ OGC,

$$e^2 = a^2 + y_i^2 \tag{13.3}$$

Similarly, from Δ OGB,

$$r_t^2 = (x_i - a)^2 + y_i^2 \tag{13.4}$$

$$a = \frac{x_i^2 + e^2 - r_t^2}{2x_i} \tag{13.5}$$

$$y_i = \sqrt{e^2 - \left(\frac{x_i^2 + e^2 - r_t^2}{2x_i}\right)^2} = \sqrt{r_t^2 - \left(\frac{x_i^2 + r_t^2 - e^2}{2x_i}\right)^2} \tag{13.6}$$

Therefore, the perpendicular distance of the spring to focus is a function of the tracking radius, instantaneous length of the spring, and eccentricity. The eccentricity, i.e., the distance between the focus and the eccentric point, is the sum of maximum and minimum spring compression at noon. The maximum instantaneous spring length is at noon. It is equal to the tracking radius and eccentricity. The tracking radius is the distance between the centre of gravity of the collector and the focus. The centre of gravity changes with a change in geometric factor. Therefore, the selection of the geometry factor should meet this condition.

13.3 Geometry of MRCPC

The ratio of the focus distance to cavity width is called geometry factor.

$$GF = \frac{f}{w} \tag{13.7}$$

The focus, f, is measured from the base line of the bottom small parabola. The width of the bottom small parabola or cavity is w.

If FP is the focal parameter, it is equal to twice the focal height or equal to the distance between two foci of two parabolas in this compound parabola.

$$FP = 2f \tag{13.8}$$

Therefore,

$$FP = 2\ GF\ w \tag{13.9}$$

Generally, in the design of the solar collector, the geometry should be developed as per the given width of the collector at the top, which results from the size of the solar thermal plant.

The width of the upper parabola (P_1–P_6),

$$W = 2w\left(\frac{5}{4} - \left(\frac{FP}{w}\right)^2 + \sqrt{\left(\frac{9}{4} - \left(\frac{FP}{w}\right)^2\right)^2 + \left(\frac{FP}{w}\right)^2}\right) \tag{13.10}$$

Substituting,

$$\frac{FP}{w} = 2GF \tag{13.11}$$

$$W = 2w\left(\frac{5}{4} - 4GF^2 + \sqrt{\left(\frac{9}{4} - 4GF^2\right)^2 + 4GF^2}\right) \tag{13.12}$$

From the selected GF and collector width, W, the cavity width can be determined.

The cavity width,

$$w = \frac{W}{2\left(\frac{5}{4} - 4GF^2 + \sqrt{\left(\frac{9}{4} - 4GF^2\right)^2 + 4GF^2}\right)} \tag{13.13}$$

The coordinates (X, Y) of the parabola are generated from the focal parameter and cavity width.

Left-hand curve, $Y = \frac{(X - w)^2}{2FP}$ (13.14)

Right-hand curve $Y = \frac{(X + w)^2}{2FP}$ (13.15)

13.4 Tracking of the collector during forenoon

If the west header is half filled and fixed, the liquid in the east header can be varied by either filling or dripping. In the case of dripping, the east header is filled completely at sunrise. The dripping of the east header causes an imbalance and drives the shaft and collector through the tracking mechanism. At noon, half of the liquid will have dripped from the east header, and the equal fluid in both headers keeps the collector in a horizontal position. Later, further dripping from the east header drives the collector towards the west as the fluid in the west header is greater than in the east header. Figure 13.5 shows the free body diagram to balance the moment in a clockwise and counter-clockwise direction (Natarajan and Srinivas, 2017).

For the force balance of one bearing, only half of the instantaneous fluid load (both at the east header and west header) needs to be considered as two bearings share this load. The load in the header includes the container load and the fluid load. The hydraulic suspension spring is of a compression type. This also works as a shock absorber and dampens the oscillations caused by windy periods. The deviation of the collector from its noon position to either the left or right causes compression of the spring at its support, which is eccentrically fixed to the focal point. This collector tries to come to a neutral position, i.e., the mid-position with the shock absorber's spring. The spring force acts tangentially in a counter-clockwise direction on a circle with a radius of eccentricity.

The balancing of moment on both directions for the equilibrium results in the following,

$$\left(\frac{m_{Ei}}{2}\right) r_w = \left(\frac{m_{Wi}}{2}\right) r_w + S_i y_i \tag{13.16}$$

$$m_{Ei} r_w = m_{Wi} r_w + 2 S_i y_i \tag{13.17}$$

The perpendicular distance of spring length, y_i can be determined from the tracking geometry detailed in the earlier section. This distance is a function of tracking radius, r_t, eccentricity, e, and instantaneous spring length, x_i.

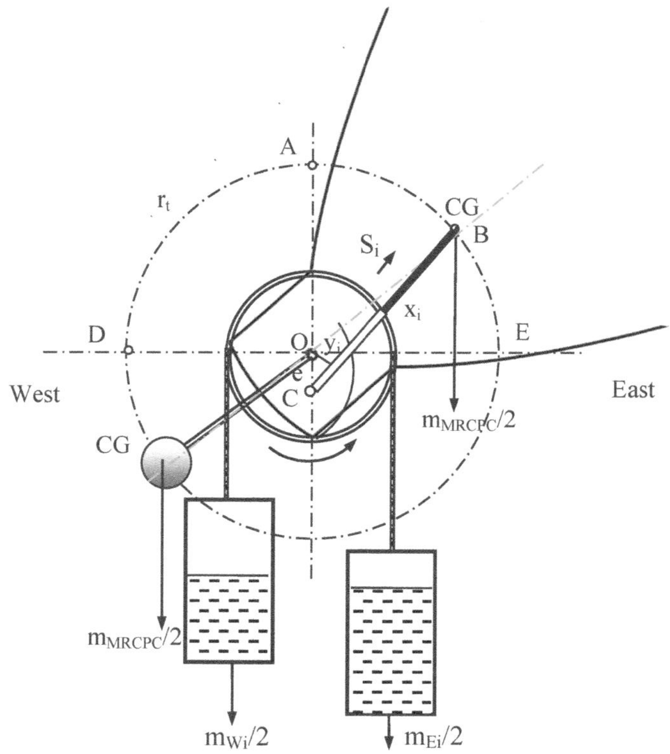

Fig. 13.5. Free body diagram of the motorless tracking mechanism from sunrise to solar noon.

The load, m_{wi} is fixed. The variable load in the east head,

$$m_{Ei} = m_{Wi} + \frac{2S_i y_i}{r_w} \tag{13.18}$$

The perpendicular distance from the focus to the spring decreases; hence, the spring force decreased to zero during the forenoon.

The spring deflection is at its maximum at the start and end of the operation.

At the initial state, the east header is full of liquid with a fixed mass in the west head, which is equal to half the full mass.

Therefore, in the initial state,

$$m_{Wi} = m_{W,\ sr} = \frac{m_{E,\ sr}}{2} \tag{13.19}$$

Substituting this initial condition, in the tracking load,

$$m_{E,\ sr} = \frac{4S_{sr} y_{sr}}{r_w} \tag{13.20}$$

and $m_{W,\ sr} = \dfrac{2S_{sr} y_{sr}}{r_w}$ (13.21)

It shows that, at the beginning of tracking, the fluid mass in the east and west headers can be determined from the spring force, perpendicular distance, and sprocket wheel radius at that point. At this state, the spring force and perpendicular distance are at their maximum. Therefore, the fluid in the east header is at its maximum. Low fluid is required for the lighter spring compared to the stiff spring. But the spring stiffness must be selected to resist the collector weight. Therefore, the collector weight acts indirectly on the tracking mechanism even though its term is cancelled in the moment equation. The increase in the sprocket wheel radius decreases the fluid mass in the header.

13.5 Tracking of the collector during the afternoon

Figure 13.6 is a free diagram of a tracking unit operating from noon to sunset. The collector weight falls to the west and the counter weight is to the east. The counterweight balances the collector weight for the dynamic balancing. The spring balances the tracking force caused by the differential weight on the east header and the west header.

After simplification of the force and moment equations,

$$m_{Ei} = m_{Wi} - \frac{2S_i y_i}{r_w} \tag{13.22}$$

In the final state, the east header is empty of liquid with a fixed mass in the west header, which is equal to half the initial full mass of the east head.

Therefore, in the final state,

$$m_{W,\ ss} = \frac{2S_{ss} y_{ss}}{r_w} \tag{13.23}$$

Since, $S_{sr} = S_{ss}$ and $y_{sr} = y_{ss}$, this result is same as the earlier result,

$$m_{W,\ sr} = m_{W,\ ss} = m_{Wi,} = \frac{2S_{sr} y_{sr}}{r_w} \tag{13.24}$$

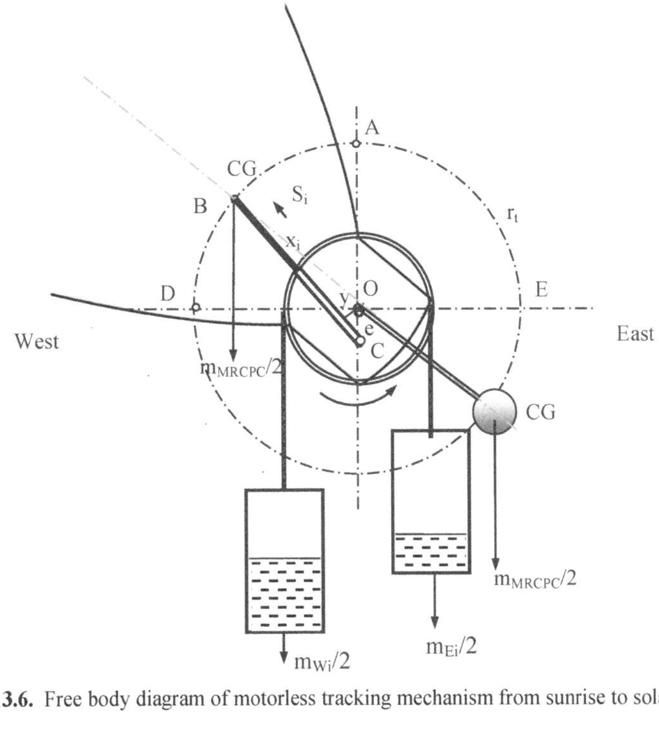

Fig. 13.6. Free body diagram of motorless tracking mechanism from sunrise to solar noon.

13.6 Evaluation of reaction on the bearings

Figure 13.7 shows the force balance applied to a simply supported beam with the tracking loads, collector mass, counter weight, and reaction at the two supports. The shaft is considered a beam. The reaction at the bearings can be calculated by the simplification of the mass balance. The small gaps between the bearing, sprocket wheel, and tracking wheel are neglected.

The spring angle to the vertical or eccentric line (ξ) is taken into account (Fig. 13.2).

$$r_t^2 = x_i^2 + e^2 - 2x_i e \ \cos\xi \tag{13.25}$$

$$\xi = \cos^{-1}\left(\frac{x_i^2 + e^2 - r_t^2}{2x_i e}\right) \tag{13.26}$$

The force balance with reference to Fig. 13.7,

$$R_i + S_i \cos\xi_i + R_i + Si \cos\xi_i = (m_{Ei} + m_{Wi} + m_C) + m_{MRCPC} \tag{13.27}$$

Due to the symmetric arrangement, the reaction at each support is the same.

$$2(R_i + S_i \cos\xi_i) = m_{Ei} + m_{Wi} + m_{MRCPC} + m_C \tag{13.28}$$

$$R_i = \frac{m_{Ei} + m_{Wi} + m_{MRCPC} + m_C}{2} - S_i \cos\xi_i \tag{13.29}$$

Moment about G,

$$m_{MRCPC}\left(\frac{z}{2}\right) + \left(\frac{m_{Ei} + m_{Wi} + m_C}{2}\right)z = \left(R_i + S_i \cos\xi_i\right)z \tag{13.30}$$

$$R_i = \frac{m_{Ei} + m_{Wi} + m_{MRCPC} + m_C}{2} - S_i \sin\xi_i \tag{13.31}$$

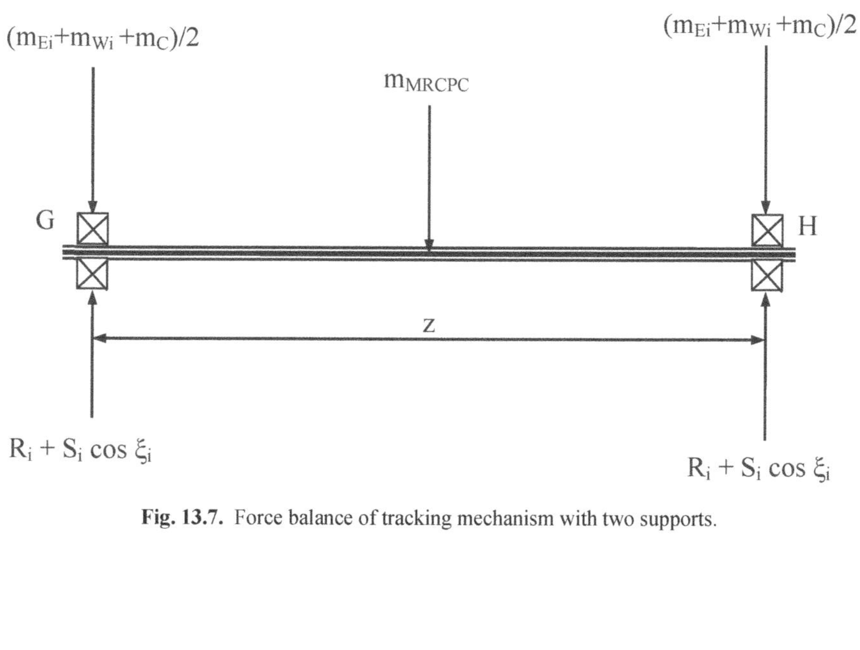

Fig. 13.7. Force balance of tracking mechanism with two supports.

13.7 Results and discussion

13.7.1 Analysis of shape and size of a solar collector

Figure 13.8 depicts the spring angle with vertical, spring length, spring deflection, and spring perpendicular distance to focus from sunrise (0°) to sunset (180°). The angle of the spring is measured with a vertical (ξ). Since the spring is mounted eccentrically below the focus, during sunrise and sunset, it is not perpendicular to the horizontal (< 90°). The spring should not be loaded fully. Underloading is ensured by rotating the collector with an angle less than 90°. Therefore, the designer can keep the maximum deflection of the spring as per the distance between the focus and the eccentric point (eccentricity). At noon, the spring is aligned with the vertical axis of the solar collector. At this time, the deflection of the spring is zero or minimum according to the initial setting (Fig. 13.8a). The spring used is a compression spring, and its length is greatest at noon and shortest at sunrise and sunset (Fig. 13.8b). The deflection of the spring has been depicted, excluding the initial compression to avoid the backlash during noon. The spring deflection is maximum during sunrise and sunset and minimum at noon. For evaluating the spring moment about the focus, its perpendicular distance is required as the moment is equal to the product of spring force and its perpendicular distance about the fulcrum (focus). This perpendicular distance is at its maximum at sunrise and sunset periods. At noon, the spring axis and collector axis are aligned and pass through the focus. The perpendicular distance is zero at this instant.

Figure 13.9 shows the changes in the size of MRCPC with the geometric factor at a fixed collector width of 2 m. The focus of the collector increases with an increase in geometric factor. It is the ratio of focus to cavity width. For example, at a geometric factor of 0.5, the focus is half the cavity width, as shown in Fig. 13.9. The geometric factor has a greater influence on the size of the collector as it increases. The height (total height, curved height, and height from the focus) decreases with an increase in the geometric factor. It indicates that the upper portion of the collector decreases and the lower portion of the collector (focus, cavity width, and cavity length) increases with an increase in the geometric factor. Therefore, the distance between the focus and the center of gravity of the collector, which is the radius of tracking, decreases with an increase in the geometric factor. Since the height of the center of gravity is decreasing, the stability of the collector increases. Therefore, the tracking of the collector at a higher geometric factor is easy compared to the lower

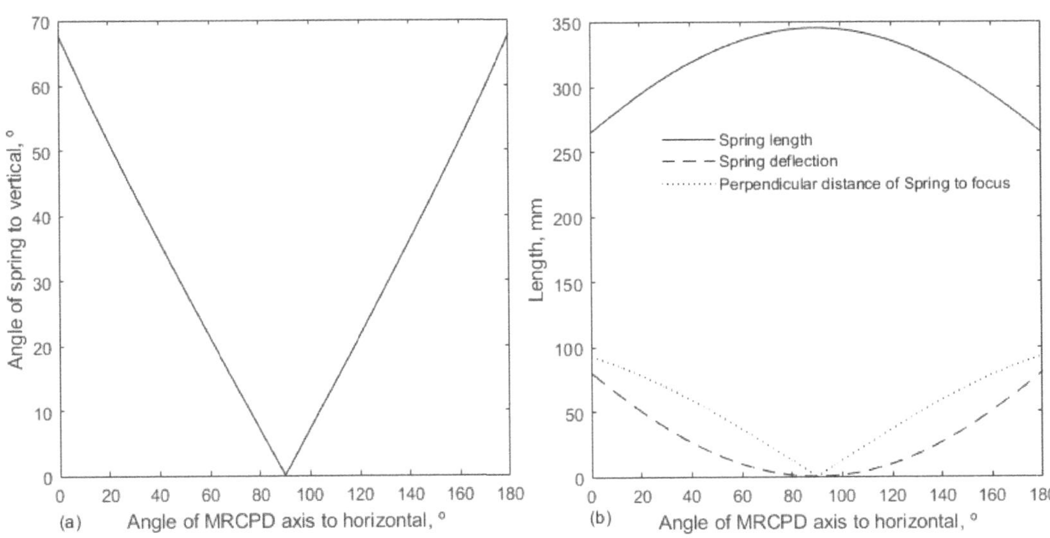

Fig. 13.8. Change in geometry of tracking mechanism for a 180° rotation of collector.

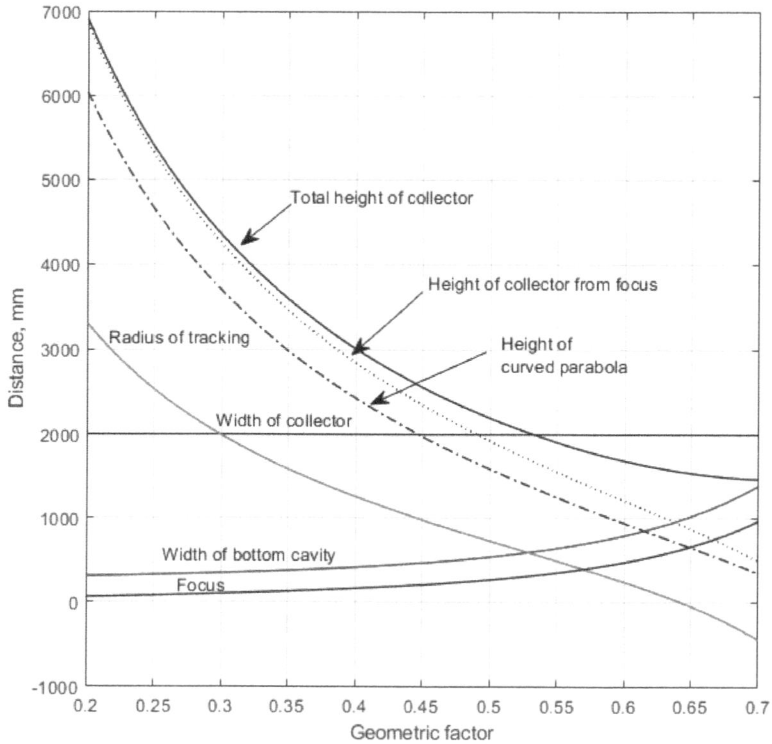

Fig. 13.9. The effect of the geometric factor (ratio of focus and cavity width) on the dimensions of the MRCPC.

geometric factor. At the lower geometric factor, the height of the collector and the height of the center of gravity are greater. Due to the lower stability, a continuous and smooth rotation of the collector is difficult. Therefore, a higher geometric factor is recommended for smooth and easy tracking with a lower tracking force. But after a certain value of the geometric factor (0.63), the center of gravity shifts below the focus and the radius of tracking becomes negative. At this critical geometric factor (CGF), the radius of tracking is zero, i.e., the center of gravity coincides with the focus of the collector. It will not permit space for the suspension spring due to a lack of a gap between the center of gravity and the focus. Therefore, the geometric factor should be below the CGF. The selected GF should provide a sufficient gap between CG and focus to accommodate the suspension spring. Therefore, for this work, the recommended GF is between 0.55 and 0.6. The spring length is measured from the center of gravity to eccentricity.

Figure 13.10 relates the geometric factor to the area of the reflective sheet. The earlier discussion reveals the relationship of the geometric factor to the size and shape of the solar collector. The upper area increases, and the lower area increases with an increase in the GF. The minimum reflective sheet results in a cost-effective solution for the solar collector as it lowers the cost of the reflective sheet. The lowest reflective area results from the GF of 0.65. But it is slightly greater than the CGF. Therefore, a value close to the CGF also ensures the accommodation of the suspension spring to be selected to satisfy an easy tracking condition with the minimum reflective area. At the CGF, the center of gravity of the collector and focus are the same and result in a minimum reflective sheet. But it is not possible for a spring to be located between CG and focus. After this CGF, the CG shifts below the focus and the lower reflective sheet increases. Therefore, the total area of the reflective sheet increases after the CGF.

Figure 13.11 shows the geometrical changes of MRCPC with a change in the geometric factor. Five significant geometric factors (0.45, 0.49, 0.53, and 0.6) have been selected to observe the changes in the shape and size of the MRCP. In this analysis, the collector width (P_1–P_6) has been

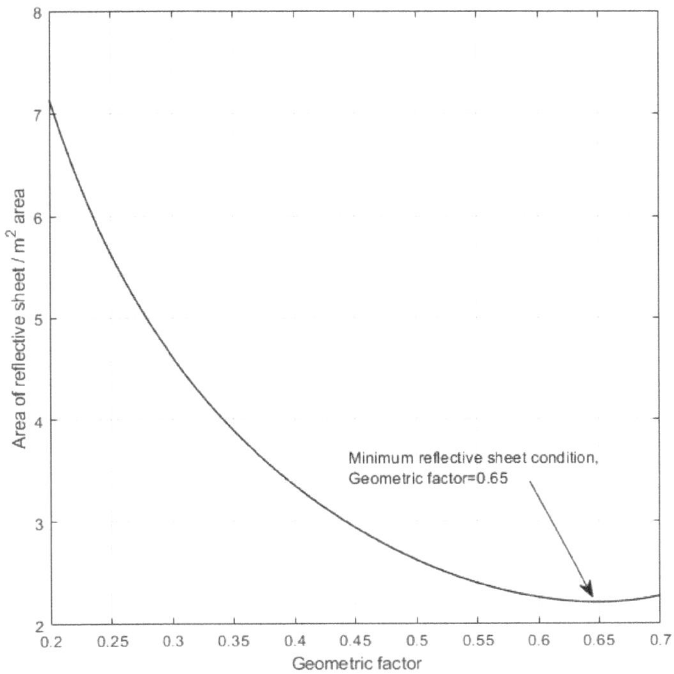

Fig. 13.10. Effect of focus to cavity width ratio for minimum reflective sheet area.

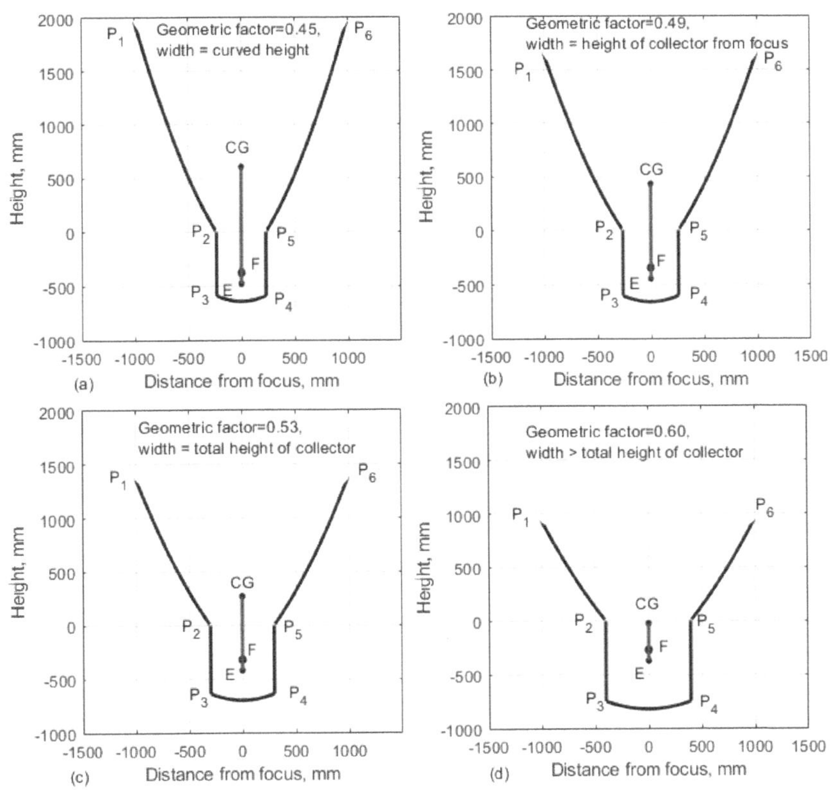

Fig. 13.11. Effect of geometric factor width ratio on the shape and size of MRCPC at a fixed collector's width of 2 m at the top.

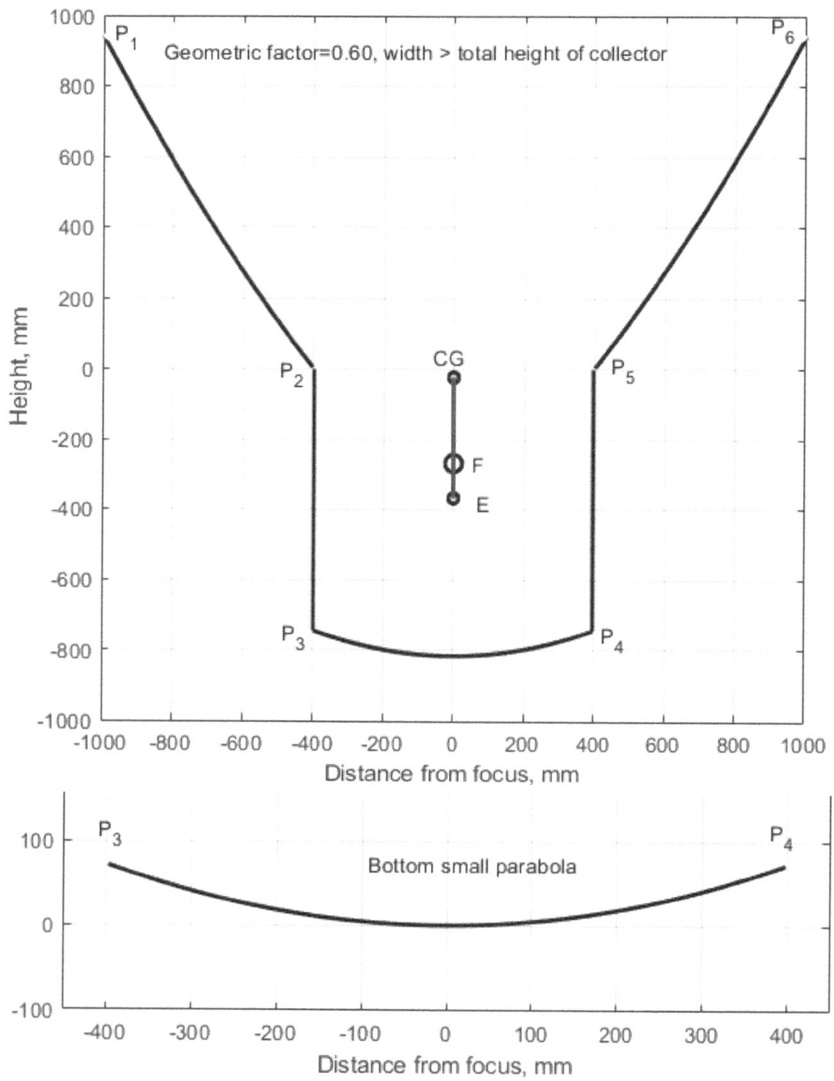

Fig. 13.12. Size of a MRCPC at a typical geometric factor of 0.60.

fixed at 2 m. With reference to Fig. 13.11, at a 0.45 geometric factor, the collector width is equal to the length of the upper parabolic curve. The 0.49 geometric factor equals the collector width equal to the collector height from the focus. The width of the collector is equal to the total height of the collector from the base at a 0.53 geometric factor. After the geometric factor of 0.53, the width of the collector is greater than the total height of the collector. For a minimum sweep area and tracking without jerks (stability), a geometric factor of more than 0.53 has been recommended.. A geometric factor of 0.6 is shown in Fig. 13.11d to study the benefits of the higher geometric factor with a feasible space for the suspension spring.

Based on the above discussion, the shape and size of a typical MRCPC have been depicted in Fig. 13.12 with a GF of 0.6. The dimensions of the bottom parabola have been enlarged for a better understanding of its size. The surface end points are shown from P_1 to P_6. The points CG, F, and E stand for the center of gravity, focus, and eccentricity. The spring is connected between the eye joints of CG and E. Because the CG is in the lower part of the collector, close to the mid-point, it provides tracking stability.

The particulars of this MRCPC have been listed in Table 13.1. The results are generated with the inputs of a collector area (2 m × 2.7 m), geometric factor (0.6), and eccentricity (0.1 m). The results are focused on the dimensions of the MRCPC. Since the geometric factor is greater than 0.53, the width of the collector (2 m) is greater than the collector's height (1.69 m). The bottom cavity width and height are 7.95 m and 7.45 m, respectively. It shows that the width of the bottom cavity is greater than its height. The height of the CG from the focus is 0.245 m. Therefore, the distance between the CG and the eccentric point (E) is 0.345 m (0.245+0.1). The length of the suspension spring should be equal to 0.345 mm. Since the collector is MRCPC, the reflective area is 12.22 m^2 for the collector's projected area of 5.4 m^2. The reflective area is nearly double the collector's area.

Table 13.1. Specifications of the MRCPC.

Input conditions		
S. No	**Description**	**Result**
1.	Width of collector at top, mm	2000.00
2.	Length of collector, mm	2700.00
3.	Geometric factor	0.60
4.	Eccentricity, mm	70.00
5.	Diameter of shaft, mm	50.00
6.	Weight of the collector per m^2 area, kg	30.00
Output conditions		
S. No	**Description**	**Result**
1.	Cavity width or width of bottom parabola, mm	795.04
2.	Focus from base, mm	477.03
3.	Height of curved parabola, mm	943.33
4.	Depth of bottom small parabola, mm	71.96
5.	Cavity height, mm	745.35
6.	Total height of collector, mm	1688.68
7.	Length of curved parabola, mm	1120.88
8.	Centre of gravity from focus, mm	245.55
9.	Radius of tracking, mm	345.55
10.	Arc length of small parabola	812.09
11.	Length of MRCPC on one side, mm	2263.76
12	Collector's projected area, m^2	5.40
13.	Reflective sheet surface area, m^2	12.22

13.7.2 Analysis of sprocket wheel diameter and focus to cavity width ratio

Figure 13.13 analyses the effect of focus, geometric factor, and sprocket wheel diameter on the full tracking load of the east header during sunrise. It is not a function of the weight of the solar collector as it is balanced by the counterweight arranged symmetrically in the opposite direction. The increase in focus to cavity width ratio decreases the height of the collector. Therefore, the center of gravity of the collector moves towards the focus and increases the stability of the collector. Therefore, the rotation of the collector around the focus with the connection at the center of gravity is easy compared to the higher center of gravity distance. The shorter center of gravity results in a higher focus to cavity width ratio, resulting in a lower tracking load for the collector rotation. The tracking load decreases at a diminishing rate with an increase in the sprocket wheel diameter. The increase in the sprocket wheel increases the driving moment and torque, which is balanced by the

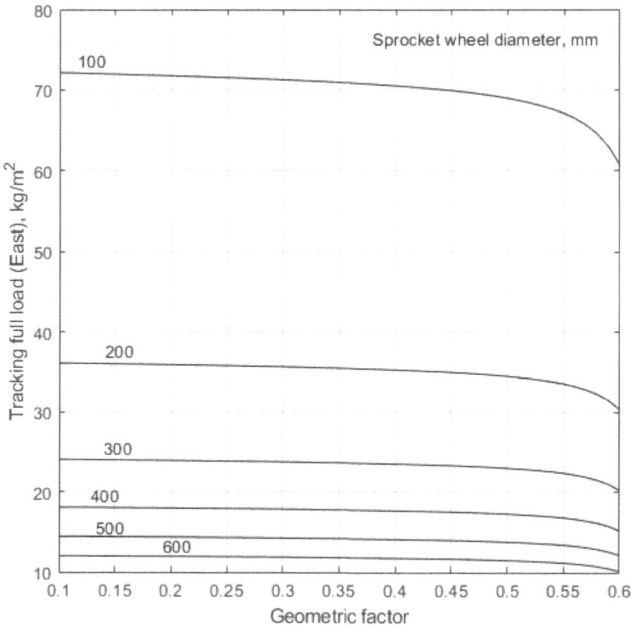

Fig. 13.13. The specific tracking load with the size of the sprocket wheel and geometric factor.

moment caused by spring force. The spring force is tensile in nature as the compression load is acting on the spring. This analysis shows that to track the collector with a minimum load, a large sprocket wheel and a higher focus on bottom cavity width need to be selected. Since the center of gravity is very close after the 0.6 ratio, the tracking load becomes negative, i.e., it is to be filled. Practically, it is difficult to have a continuous and automatic operation. Therefore, in this analysis, the ratio is restricted to 0.6. A more significant drop in the tracking load is observed from a ratio of 0.5. Therefore, the ratio can be selected between these two limits. The sprocket wheel size increases the gap between two liquid headers.

13.7.3 Tracking loads from sunrise to sunset

During summer, the tracking unit has to be operated with a maximum span. As a result, the summer solstice (21 June) has been chosen for the tracking study. Figure 13.14a shows the dynamic changes of collector rotation and the resulting incidence angle. From sunrise to noon, the collector rotates from the horizontal position to the vertical position. The operation continues during the afternoon from the vertical position to the horizontal position. Figure 13.14b focuses on the tracking load variations from sunrise to sunset. The load in the east header changes from sunrise to sunset. The load on the west header is fixed at half of the east header's full load. It has been observed that the full load in the east header is 12.12 kg/m^2 of collector area. At noon, instead of zero deflection of the spring, some compression has been created during the setting to avoid the backlash of the track. The minimum deflection in this setting is 5 mm at noon. The collector tilt angle and its surface azimuth angle are determined from the conditions of the minimum incidence angle. Two-axis tracking only results in a zero incidence angle. The incidence angle of the two-axis tracking can be designed to be as small as possible. Differentiating the incident angle function with reference to the tilt angle will result in the condition of tile angle, and the substitution of this condition results in the minimum incidence angle. Before and after the noon position (noon period), the resultant incidence angle is low compared to the sunrise and sunset periods. To meet the required minimum incidence angle, the east header's load should be changed in a non-linear way.

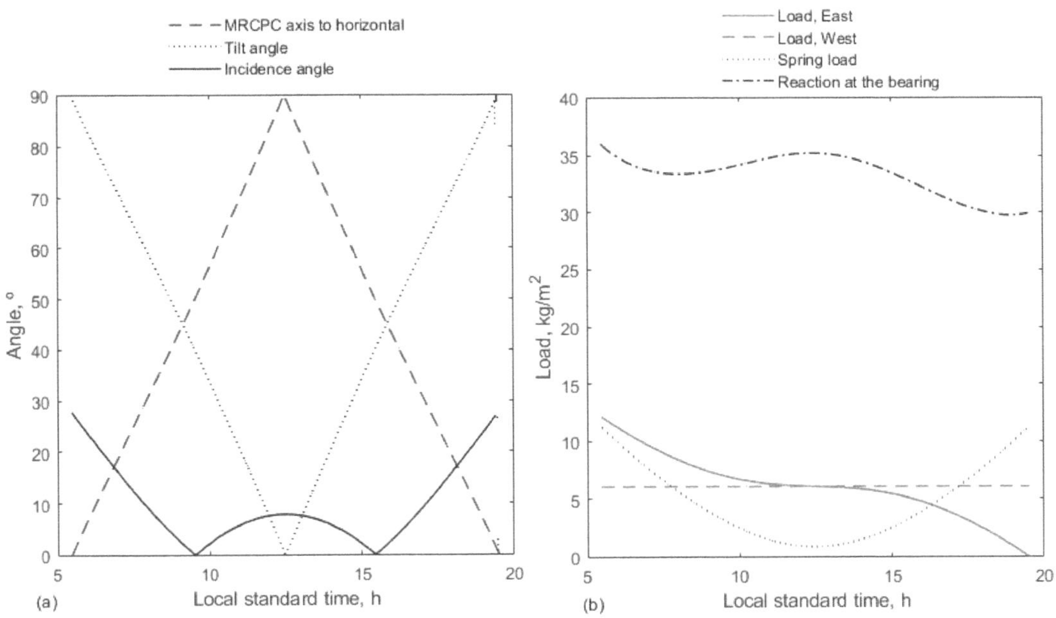

Fig. 13.14. Load tracking from sunrise to sunset using a linear variation in collector rotation.

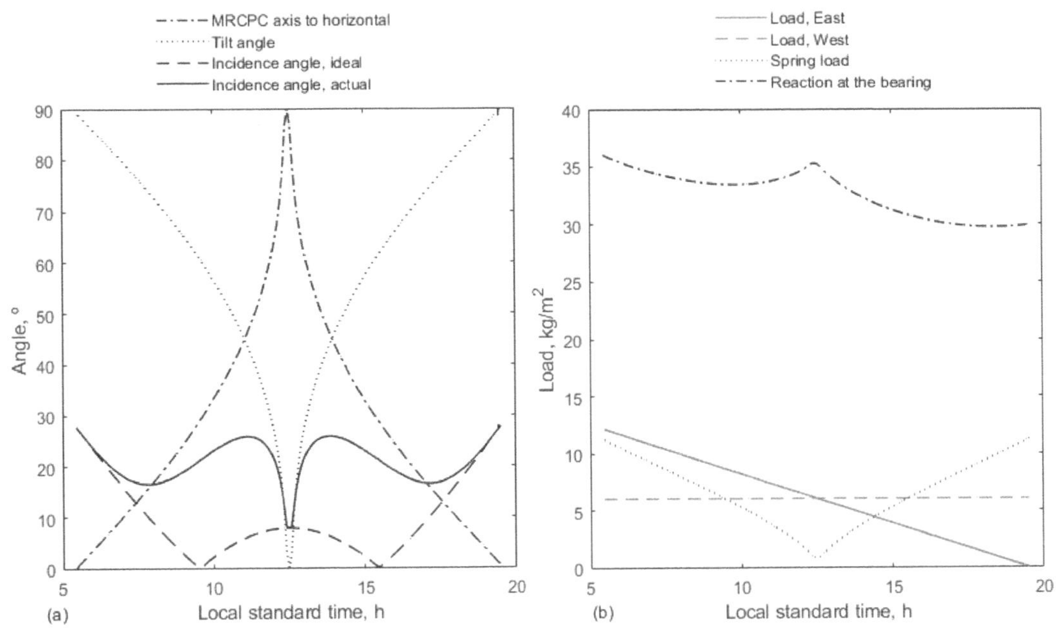

Fig. 13.15. Tracking loads from sunrise to sunset with linear variation in east header load.

Figure 13.15 shows the particulars of the (a) angular changes in the collector and (b) variations in the load from sunrise to sunset. As per the earlier discussion, to track the collector for the minimum incidence angle, the discharge in the east header should be nonlinear, i.e., maximum during sunrise and sunset and minimum at noon. However, it is difficult to maintain this control over the discharge of fluid in the east header in practice. The actual may be assumed to be linear as per Fig. 13.15. Compared to the earlier case (Fig. 13.15a), the results shown in Fig. 13.15a differ due to changes in the discharge method. The collector rotation and the resulted tilt angle are not uniform. These result in a higher incidence angle compared to the ideal incidence angle. The incidence angle is less than 30°. At the beginning of the operation, the east header load is full and the resistance in the two bearings at 36 kg/m^2 is also high.

Table 13.2 summarizes the results when the tracking conditions are as follows: sprocket wheel diameter (500 mm), spring length (260 mm), spring stiffness (1 kg/mm), size of the MRCPC (2 m × 2.7 m), geometric factor (0.6), eccentricity (70 mm) and deflection of the spring at noon setting (5 mm). The minimum and maximum changes in the tracking load in the east header are 0 and 12.12 kg/m^2 collector area, respectively. The load in the west header is fixed, which is equal to half of the load in the east header. Since the load in the west header is fixed, there is no discharge. The east header discharge at the beginning is 1.75 LPH/m^2. The N-S axis motorless tracking results in a maximum incidence angle of 30° with the uniform discharge of east header fluid.

Table 13.2. Results of the tracking unit with linear variation.

Input conditions			
S. No	**Description**		**Value**
1.	Sprocket wheel diameter, mm		500.00
2.	Spring length, mm		260.00
3.	Spring stiffness, kg/mm		1.00
4.	Size of the MRCPC, m		2 × 2.7
5.	Geometric factor		0.60
6.	Eccentricity, mm		70.00
7.	Deflection of spring at noon, mm		5.00
Output conditions			
S. No	**Description**	**Minimum**	**Maximum**
1.	Load in east header, kg/m^2	0.00	12.12
2.	Load in west header, kg/m^2	6.06	6.06
3.	Bearing resistance, kg/m^2	0.86	11.26
4.	Discharge in east header, LPH/m^2	0.00	1.75
5.	Discharge in west header, LPH/m^2	0.00	0.00
6.	Spring load, kg/m^2	4.50	79.50
7.	Spring length, m	0.25	0.32
8.	Spring deflection, m	0.00	0.06

13.7.4 Validation

A prototype experimental setup has been developed to check the feasibility and accuracy of the proposed tracking mechanism. The incidence angle of the linear concentrator with respect to the North–South horizontal axis can be minimized by differentiating it with the tilt angle. The resultant tilt angle through this simulation is compared with the experimental results of the current tracking mechanism (Fig. 13.16). The incidence angle is calculated theoretically from the minimum tilt angle. It is the ideal position as the considered tilt angle is at the minimum incidence angle. The tilt angle (collector position) and the incidence angle resulting from the experimental operation of the developed tracking mechanism are compared with the ideal position. It has been found that the proposed mechanism results in the required amount of tilt and incidence angle without much deviation. A deviation has been observed during the afternoon operation. The sliding of counter balance with the change of direction from the forenoon to afternoon has caused this deviation. Rectification has been planned and steps are in progress to redesign the counter balance and correct this deviation.

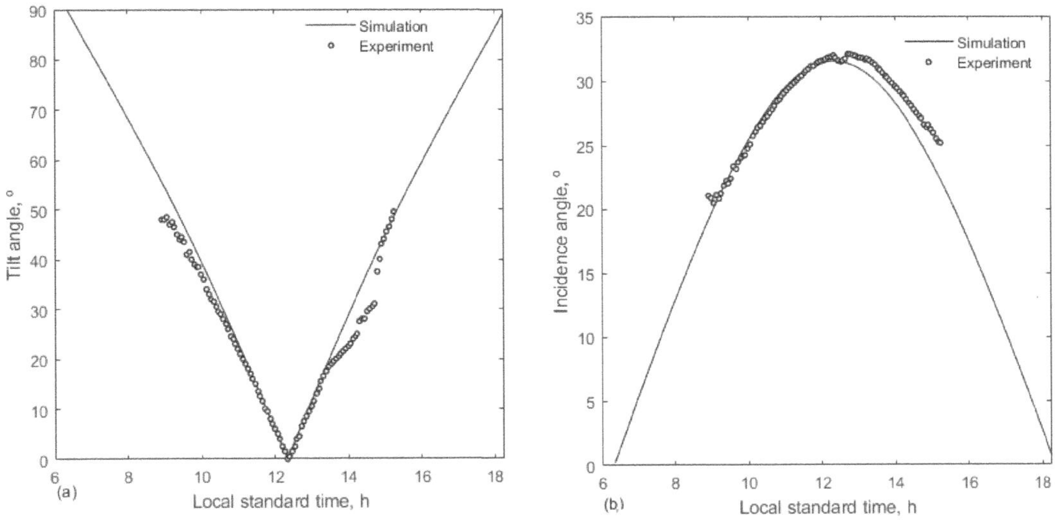

Fig. 13.16. Comparison of simulation and experimental results.

13.8 Conclusions

A novel solar tracking mechanism to drive a linear focused concentrating collector has been proposed. The unit is modelled and simulated with the target of higher dynamic stability and small unit sizes. The aim of the work is to build a compact unit with lower tracking error. The geometric factor has been identified as a critical variable in the outcome of effective tracking and minimal reflective area. The critical geometric factor at the zero tracking radius is 0.63. The geometric factor for the minimum reflective sheet area is 0.65. A large sprocket wheel and higher geometric factors result in a minimum tracking load for a motorless rotation of the collector. The minimum recommended geometric factor is 0.53; for stable tracking, a 0.6 geometric factor is recommended.

13.9 Numerical solutions

1. Find the variable instant load in the solar collector shown in Fig. 13.17 to track the collector from east to west (m_{Ei}) using water weight. The sprocket wheel radius, r_w to connect the tracking load is 0.125 m and the tracking wheel radius, r_t to connect the spring is 0.07 m. The perpendicular distance of spring line to the center of rotation is $r_t \sin(\zeta i + \xi i)$ where ζi and ξi are 18.9° and 12.6° respectively for the collector angle and spring angle with horizontal. The west load, m_{wi} is zero. The instantaneous spring load is 544.0 N. The total number of springs used at both ends of collector is 2 (The side view of the figure shows one spring only). The collector load is balanced by the counter weight.

Solution

For the total system, the moment equation is

$m_{Ei}\, r_w + m_{collector} \times$ load radius $= 2S_i\, r_t \sin(\zeta_i + \xi_i) + m_{counter}$ weight $\times$ load radius $+ m_{Wi}\, r_w$

Since $m_{collector} = m_{counter}$ weight

The moment equation can be simplified to find the tracking load at the east header,

$$m_{Ei} = 2S_i \left(\frac{r_t}{r_w}\right) \sin(\zeta_i + \xi_i) + m_{Wi}$$

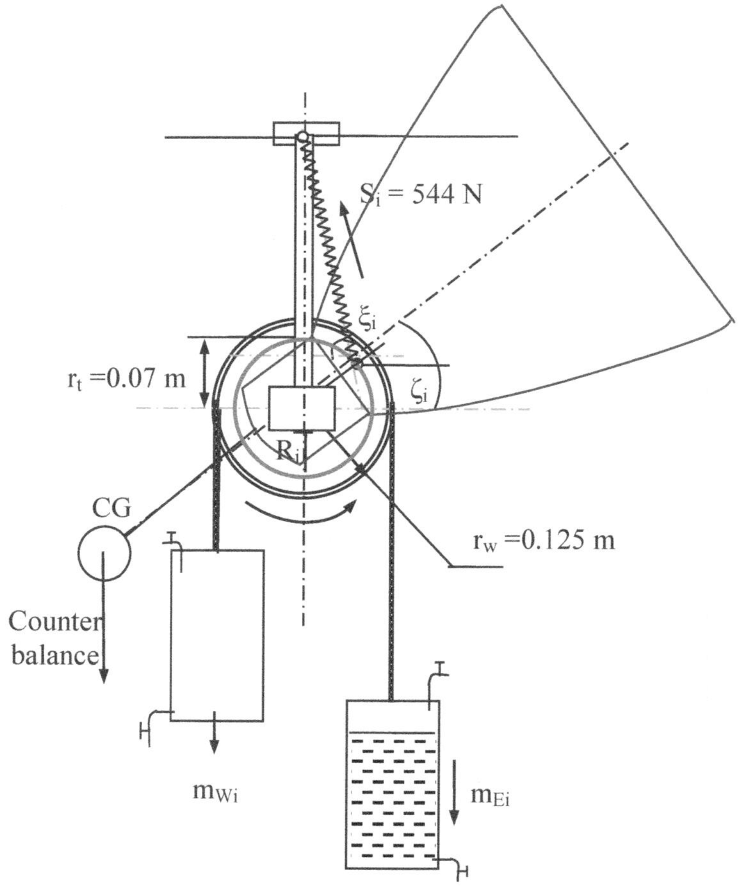

Fig. 13.17. Gravity based solar tracking system.

$$m_{Ei} = 2 \times 54.42\left(\frac{0.07}{0.125}\right)\sin(18.9 + 12.6) + 0$$

$= 31.84$ kg

2. Refer problem no 10 from chapter 1.

Chapter 14
Solar Thermal Storage

14.1 Introduction

To supply solar energy to meet load without any interruptions, a storage system is required. For example, a solar thermal power plant with the storage unit meets the electricity demand without suffering from the mismatch of solar energy collection and power plant load. The storage unit stores the excess solar radiation collected and releases it during non-available or inadequate periods. The type of the storage system required depends on the operating temperature, size of unit, heat losses, and cost of storage system.

14.2 Sensible heat storage

In sensible heat storage system, the heating and cooling medium may be a solid, (pebbles, rocks and refractories), liquid (water, heat transfer fluid, etc.) or salts. In the utilization phase of the solar energy, thermal storage has a significant role (Srinivas, 2023). Sensible heat storage systems are simple but large in size compared to the latent heat storage system. In a sensible heat storage system, the heat supply temperature is variable and in a latent heat storage system, the supply temperature is constant. A majority of sensible heat storage units use water as the storage fluid. Heat transfers between the solar collector (flat plate collector, evaluated tube collector (ETC) or concentrating collector) and the fluid may be direct or indirect. Following sections elaborate the designs, modeling, and analysis of sensible heat storage systems.

14.2.1 Direct heating without stratification

Figure 14.1 shows a sensible heat storage system with the assumption of uniform temperature of fluid. The thermal stratification considers the variation in temperature with the depth of liquid the storage tank. In this model, thermal stratification is neglected, as the temperature of the whole system is assumed constant but variable with time and usage. The tank is fitted between the solar collector (non-focusing or focusing) and the power plant load. Generally, with water as working fluid, the working temperature is below 100°C. Therefore, non-focusing collectors are used with this type of storage system.

Let the T_l be the liquid temperature in the storage tank. The energy balance equation to the storage unit is

$$\left[(\rho V c_p)_l + (\rho V c_p)_t\right]\frac{dT_l}{dt} = q_u - q_{load} - (UA)_t (T_l - T_a) \tag{14.1}$$

$(\rho V c_p)_l$ = heat capacity of the fluid in stage tank,
$(\rho V c_p)_t$ = heat capacity of the tank material,
q_u = useful heat from the collector,
q_{load} = heat load,

DOI: 10.1201/9781032642505-14

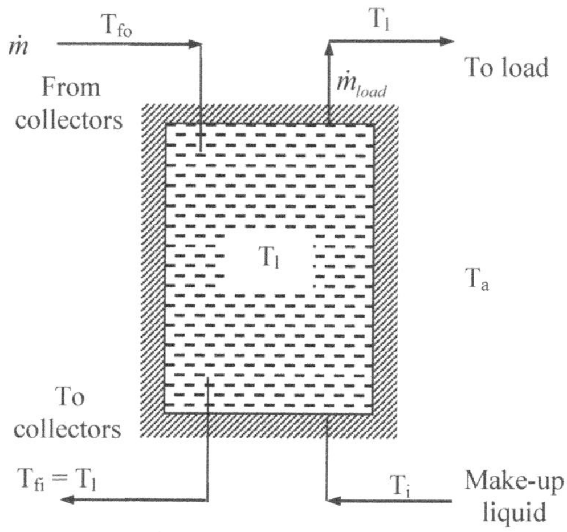

Fig. 14.1. A sensible heat storage system with direct heat without stratification in a well-mixed condition.

$(UA)_t$ = product of the overall heat transfer coefficient and surface area of the tank, and
T_a = ambient temperature around the tank
$(\rho V c_p)_t$ is significant in a small size storage system but it may be neglected in a large size storage system compared to $(\rho V c_p)_l$.

Let $(\rho V c_p)_e = (\rho V c_p)_t + (\rho V c_p)_l$

This model can be solved either using the calculus method or finite-difference technique.

Assume q_u, q_{load} and T_a are constants during the selected small time period of time.

At initial condition t = 0, $T_l = T_{li}$,

Separating the variables in energy balance equation,

$$\frac{1}{q_u - q_{load} - (UA)_t (T_l - T_a)} dT_l = \frac{dt}{(\rho V c_p)_e}$$

Integrating equation,

$$\frac{\log\left[q_u - q_{load} - (UA)_t (T_l - T_a)\right]_{T_{li}}^{T_l}}{-(UA)_t} = \left(\frac{t}{(\rho V c_p)_e}\right)_0^t$$

$$\log\left[\frac{q_u - q_{load} - (UA)_t (T_l - T_a)}{q_u - q_{load} - (UA)_t (T_{li} - T_a)}\right] = \left(\frac{-(UA)_t t}{(\rho V c_p)_e}\right)$$

$$\frac{q_u - q_{load} - (UA)_t (T_l - T_a)}{q_u - q_{load} - (UA)_t (T_{li} - T_a)} = \exp\left(\frac{-(UA)_t t}{(\rho V c_p)_e}\right) \tag{14.2}$$

$$T_l = \frac{q_u - q_{load} - \left(q_u - q_{load} - (UA)_t (T_{li} - T_a)\right)\exp\left(\frac{-(UA)_t t}{(\rho V c_p)_e}\right)}{(UA)_t} + T_a \tag{14.3}$$

Equation 14.2 can be used to simulate the temperature T_l with time as per the given changes in T_a, q_u and q_{load}.

The values of q_u and q_{load} may be calculated from the inlet and outlet temperatures and flow rates as

$q_u = m\ c_p\ (T_{fo} - T_{fi}) = m\ c_p\ (T_{fo} - T_l)$

and $q_{load} = m_{load}\ c_p\ (T_l - T_i)$

Equation 14.1 then becomes

$$\left(\rho V c_p\right)_e \frac{dT_l}{dt} = \dot{m}\, c_p (T_{fo} - T_l) - \dot{m}_{load}\ c_p (T_l - T_i) - (UA)_t (T_l - T_a)$$

q_u or q_{load} may be zero at any time. For example, if $T_{fo} < T_l$, the flow through the collectors will be stopped using a thermostat and then m = 0 and $q_{load} = 0$.

Alternatively, this can be solved using the finite difference method.

$$\left(\rho V c_p\right)_e \frac{dT_l}{dt} = \dot{m}\, c_p (T_{fo} - T_l) - \dot{m}_{load}\ c_p (T_l - T_i) - (UA)_t (T_l - T_a)$$

$$\left(\rho V c_p\right)_e \frac{T_{l,f} - T_{l,i}}{\Delta t} = \dot{m}\, c_p \left(T_{fo} - \frac{T_{l,f} + T_{l,i}}{2} \right) - \dot{m}_{load}\ c_p \left(\frac{T_{l,f} + T_{l,i}}{2} - T_i \right) - (UA)_t \left(\frac{T_{l,f} + T_{l,i}}{2} - T_a \right)$$

$$\frac{\left(\rho V c_p\right)_e}{\Delta t} T_{l,f} - \frac{\left(\rho V c_p\right)_e}{\Delta t} T_{l,i} = \dot{m}\, c_p T_{fo} - 0.5\, \dot{m}\, c_p T_{l,f} - 0.5\, \dot{m}\, c_p T_{l,i} - 0.5\, \dot{m}_{load}\ c_p T_{l,f}$$

$$-0.5\, \dot{m}_{load}\ c_p T_{l,i} + \dot{m}_{load}\ c_p T_i - 0.5 (UA)_t T_{l,f} - 0.5 (UA)_t T_{l,i} + (UA)_t T_a$$

$$\frac{\left(\rho V c_p\right)_e}{\Delta t} T_{l,f} + 0.5\, \dot{m}\, c_p T_{l,f} + 0.5\, \dot{m}_{load}\ c_p T_{l,f} + 0.5 (UA)_t T_{l,f}$$

$$= \frac{\left(\rho V c_p\right)_e}{\Delta t} T_{l,i} - 0.5\, \dot{m}\, c_p T_{l,i} - 0.5\, \dot{m}_{load}\ c_p T_{l,i} - 0.5 (UA)_t T_{l,i} + \dot{m}\, c_p T_{fo} + \dot{m}_{load}\ c_p T_i + (UA)_t T_a$$

$$\left(\frac{\left(\rho V c_p\right)_e}{\Delta t} + 0.5\, \dot{m}\, c_p + 0.5\, \dot{m}_{load}\ c_p + 0.5 (UA)_t \right) T_{l,f}$$

$$= \left(\frac{\left(\rho V c_p\right)_e}{\Delta t} - 0.5\, \dot{m}\, c_p - 0.5\, \dot{m}_{load}\ c_p - 0.5 (UA)_t \right) T_{l,i} + \dot{m}\, c_p T_{fo} + \dot{m}_{load}\ c_p T_i + (UA)_t T_a$$

$$T_{l,f} = \frac{\left(\frac{\left(\rho V c_p\right)_e}{\Delta t} - 0.5\, \dot{m}\, c_p - 0.5\, \dot{m}_{load}\ c_p - 0.5 (UA)_t \right) T_{l,i} + \dot{m}\, c_p T_{fo} + \dot{m}_{load}\ c_p T_i + (UA)_t T_a}{\left(\frac{\left(\rho V c_p\right)_e}{\Delta t} + 0.5\, \dot{m}\, c_p + 0.5\, \dot{m}_{load}\ c_p + 0.5 (UA)_t \right)} \qquad (14.4)$$

The simulated results of fluid exit temperature and heat loads are depicted in Fig. 14.2.

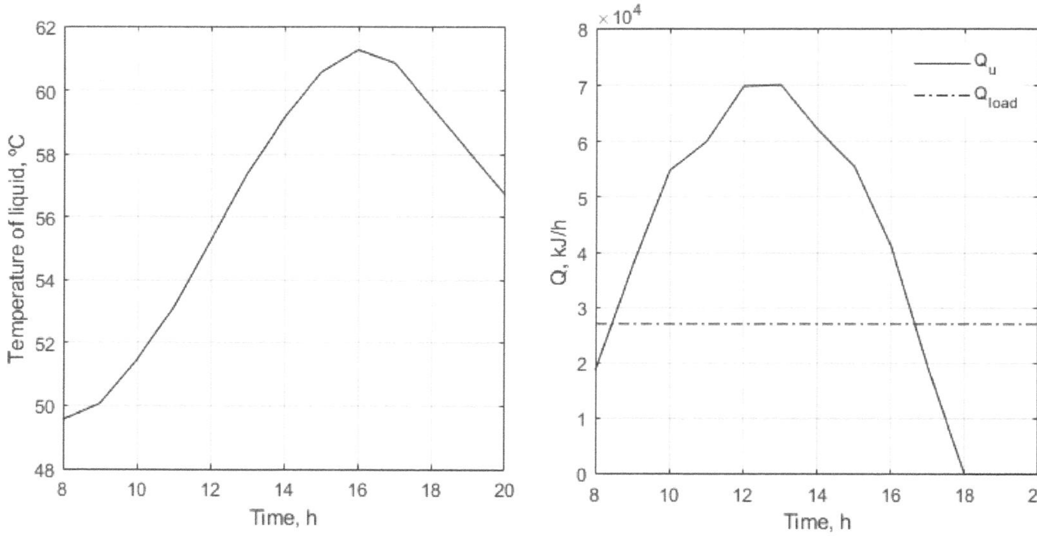

Fig. 14.2. A typical direct heating sensible heat storage system with the assumption of uniform fluid temperature.

14.2.2 Indirect heating with a heat exchanger coil for solar collector

In the indirect heating system, the fluid in a solar collector and the fluid in a heat load are different as shown in Fig. 14.3. For example, if the water is not in pure condition, this fluid may damage the solar collecting system with scale formation. To avoid this problem, a pure liquid is used in the solar collecting system. To lower the freezing temperature and increase the boiler temperature of water, glycol mixed water may be used in this type of indirect heating system. We have two ways of using a heating coil. In one option, the heating coil may be placed for the heat transfer fluid, i.e., for the collector fluid. In another option, the heating coil may be arranged on the load side. To minimize the quantity of heat transfer fluid, the heat exchanger coil is used in the collector circuit.

$$Q = \dot{m}c_{ph}(T_{fo} - T_{fi}) = (UA)_e \, LMTD = (UA)_e \left(\frac{(T_{fi} - T_l) - (T_{fo} - T_l)}{\ln \dfrac{(T_{fi} - T_l)}{(T_{fo} - T_l)}} \right) = (UA)_e \left(\frac{(T_{fi} - T_{fo})}{\ln \dfrac{(T_{fi} - T_l)}{(T_{fo} - T_l)}} \right) \quad (14.5)$$

where c_{ph} is the specific heat of hot fluid in collector.

$$\ln \frac{(T_{fi} - T_l)}{(T_{fo} - T_l)} = (UA)_e \left(\frac{(T_{fi} - T_{fo})}{\dot{m}c_{ph}(T_{fo} - T_{fi})} \right)$$

$$\frac{T_{fi} - T_l}{T_{fo} - T_l} = \exp\left(\frac{-(UA)_e}{\dot{m}c_{ph}} \right)$$

$$1 - \frac{T_{fi} - T_l}{T_{fo} - T_l} = 1 - \exp\left(\frac{-(UA)_e}{\dot{m}c_{ph}} \right)$$

$$\frac{T_{fo} - T_{fi}}{T_{fo} - T_l} = \left(1 - \exp\left(\frac{-(UA)_e}{\dot{m}c_{ph}} \right) \right)$$

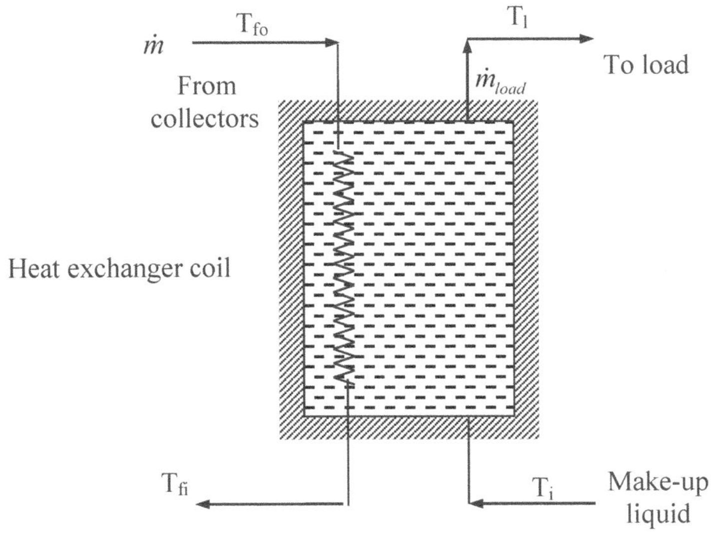

Fig. 14.3. Indirect sensible heat storage system with an immersed coil in the storage tank. The heat balance applied to a sensible heat storage system with an immersed coil.

Using the finite difference method, energy balance becomes

$$(\rho V c_w)_e \frac{dT_l}{dt} = \dot{m} c_{gw} (T_{fo} - T_l) \left(1 - \exp\left(\frac{-(UA)_e}{\dot{m} c_{gw}} \right) \right) - \dot{m}_{load} c_w (T_l - T_i) - (UA)_t (T_l - T_a) \tag{14.6}$$

let $a = 1 - \exp\left(\frac{-(UA)_e}{\dot{m} c_{gw}} \right)$

Writing the equations in finite difference form,

$$(\rho V c_w)_e \left(\frac{T_{l,f} - T_{l,i}}{\Delta t} \right) = \dot{m} c_{gw} \left(T_{fo} - \frac{T_{l,f} + T_{l,i}}{2} \right) a - \dot{m}_{load} c_w \left(\frac{T_{l,f} + T_{l,i}}{2} - T_i \right)$$
$$-(UA)_t \left(\frac{T_{l,f} + T_{l,i}}{2} - T_a \right)$$

In these equations, $T_{l,f}$ and $T_{l,i}$ represent the final and initial values of T_l over the time interval Δt.

$$\frac{(\rho V c_w)_e}{\Delta t} T_{l,f} - \frac{(\rho V c_w)_e}{\Delta t} T_{l,i} = \dot{m} c_{gw} a T_{fo} - 0.5 \dot{m} c_{gw} a T_{l,f} - 0.5 \dot{m} c_{gw} a T_{l,i}$$
$$-0.5 \dot{m}_{load} c_w T_{l,f} - 0.5 \dot{m}_{load} c_w T_{l,i} + \dot{m}_{load} c_w T_i - 0.5 (UA)_t T_{l,f} - 0.5 (UA)_t T_{l,i} + (UA)_t T_a$$

$$\frac{(\rho V c_w)_e}{\Delta t} T_{l,f} + 0.5 \dot{m} c_{gw} a T_{l,f} + 0.5 \dot{m}_{load} c_w T_{l,f} + 0.5 (UA)_t T_{l,f} = \dot{m} c_{gw} a T_{fo} - 0.5 \dot{m} c_{gw} a T_{l,i}$$
$$-0.5 \dot{m}_{load} c_w T_{l,i} + \dot{m}_{load} c_w T_i - 0.5 (UA)_t T_{l,i} + (UA)_t T_a + \frac{(\rho V c_w)_e}{\Delta t} T_{l,i}$$

$$\left(\frac{(\rho V c_w)_e}{\Delta t} + 0.5 \dot{m} c_{gw} a + 0.5 \dot{m}_{load} c_w + 0.5 (UA)_t \right) T_{l,f} = \left(\frac{(\rho V c_w)_e}{\Delta t} - 0.5 \dot{m} c_{gw} a - 0.5 \dot{m}_{load} c_w - 0.5 (UA)_t \right) T_{l,i}$$
$$+ \dot{m} c_{gw} a T_{fo} + \dot{m}_{load} c_w T_i + (UA)_t T_a$$

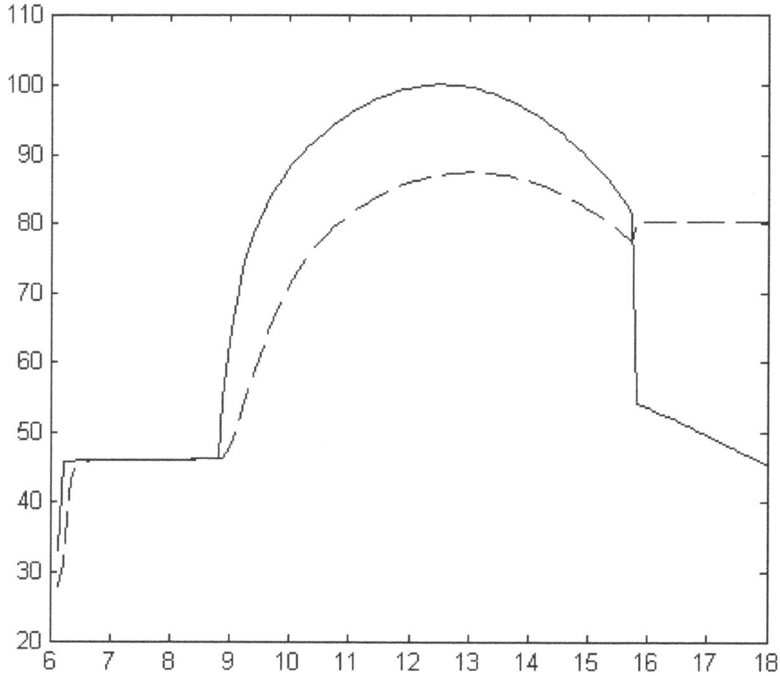

Fig. 14.4. The variations of fluid temperatures with time in an indirect sensible heating storage system.

$$T_{l,f} = \frac{\left(\frac{(\rho V c_w)_e}{\Delta t} - 0.5\left(\dot{m}c_{gw}a + \dot{m}_{load}c_w + (UA)_t\right)\right)T_{l,i} + \dot{m}c_{gw}aT_{fo} + \dot{m}_{load}c_w T_i + (UA)_t T_a}{\frac{(\rho V c_w)_e}{\Delta t} + 0.5\left(\dot{m}c_{gw}a + \dot{m}_{load}c_w + (UA)_t\right)} \tag{14.7}$$

Figure 14.4 shows the temperature distribution of a sensible storage system with an indirect heating system.

14.2.3 Indirect heating with two heat exchanger coils in storage tank for solar collector and load

Some situations need three fluids in the sensible heat storage system as shown in Fig. 14.5. If the fluid in head load system needs to be protected from freezing and boiling, a glycol mixed fluid is required. For a large sized storage tank, using glycol water is not recommended as it is not economical. Therefore, water is used in the storage tank with glycol fluid in the collector and load circuits.

For two immersed coils, the heat balance equation for the indirect heating method is

$$Q_u = \dot{m}_{sc}c_{sc}(T_{fo} - T_{fi}) = (UA)_{sc,e}\, LMTD = (UA)_{sc,e}\left(\frac{(T_{fi} - T_{st}) - (T_{fo} - T_{st})}{\ln\frac{(T_{fi} - T_{st})}{(T_{fo} - T_{st})}}\right) = (UA)_{sc,e}\left(\frac{(T_{fi} - T_{fo})}{\ln\frac{(T_{fi} - T_{st})}{(T_{fo} - T_{st})}}\right) \tag{14.8}$$

where c_{sc} is the specific heat of hot fluid in solar collector.

$$\ln\frac{(T_{fi} - T_{st})}{(T_{fo} - T_{st})} = (UA)_{sc,e}\left(\frac{(T_{fi} - T_{fo})}{\dot{m}_{sc}c_{sc}(T_{fo} - T_{fi})}\right)$$

$$\frac{T_{fi}-T_{st}}{T_{fo}-T_{st}}=\exp\left(\frac{-(UA)_{sc,e}}{\dot{m}_{sc}c_{sc}}\right)$$

$$1-\frac{T_{fi}-T_{st}}{T_{fo}-T_{st}}=1-\exp\left(\frac{-(UA)_{sc,e}}{\dot{m}_{sc}c_{sc}}\right)$$

$$\frac{T_{fo}-T_{fi}}{T_{fo}-T_{st}}=1-\exp\left(\frac{-(UA)_{sc,e}}{\dot{m}_{sc}c_{sc}}\right)$$

where $(UA)_{sc,e}$ is the product of the overall heat transfer coefficient and the area for the heat exchanger coil connected to solar collector.

Thus, the expression for the rate of useful heat gain becomes

$$Q_u=\dot{m}_{sc}c_{sc}(T_{fo}-T_{fi})=\dot{m}_{sc}c_{sc}(T_{fo}-T_{st})a_1$$

$$\text{where, } a_1=\left(1-\exp\left(\frac{-(UA)_{sc,e}}{\dot{m}_{sc}c_{sc}}\right)\right)$$

Similarly, with the heat capacity of heat exchanger coil connected to load, the temperature T_l can be related to storage tank's fluid temperature, T_{st}.

For an immersed coil connected to load

where c_l is the specific heat of hot fluid in load.

$$\ln\frac{(T_i-T_{st})}{(T_l-T_{st})}=(UA)_{l,e}\left(\frac{(T_i-T_l)}{\dot{m}_l c_l(T_l-T_i)}\right)$$

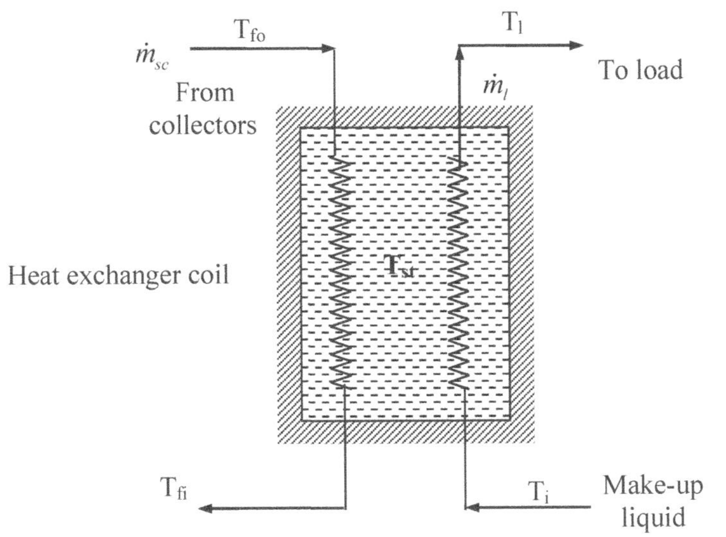

Fig. 14.5. Indirect heating sensible heat storage system with two immersed coils in storage tank.

$$\frac{T_i - T_{st}}{T_l - T_{st}} = \exp\left(\frac{-(UA)_{l,e}}{\dot{m}_l c_l}\right)$$

$$1 - \frac{T_i - T_{st}}{T_l - T_{st}} = 1 - \exp\left(\frac{-(UA)_{l,e}}{\dot{m}_l c_l}\right)$$

$$\frac{T_l - T_i}{T_l - T_{st}} = 1 - \exp\left(\frac{-(UA)_{sc,e}}{\dot{m}_{sc} c_{sc}}\right)$$

where $(UA)_{l,e}$ is the product of the overall heat transfer coefficient and the area for the heat exchanger coil connected to load.

Thus, the expression for the rate of useful heat gain becomes

$$Q_u = \dot{m}_l c_l (T_l - T_i) = \dot{m}_l c_l (T_l - T_{st})\left(1 - \exp\left(\frac{-(UA)_{l,e}}{\dot{m}_l c_l}\right)\right)$$

let, $a_2 = \left(1 - \exp\left(\frac{-(UA)_{l,e}}{\dot{m}_l c_l}\right)\right)$

$$Q_u = \dot{m}_l c_l (T_l - T_i) = \dot{m}_l c_l (T_l - T_{st}) a_2$$

The energy balance equation in differential form

$$(mc)_{st} \frac{dT_{st}}{dt} = \dot{m}_{sc} c_{sc} (T_{fo} - T_{fi}) - \dot{m}_l c_{pl} (T_l - T_i) - (UA)_t (T_{st} - T_a)$$

Substituting,

$$(T_{fo} - T_{fi}) = (T_{fo} - T_{st}) a_1$$

and

$$(T_l - T_i) = (T_l - T_{st}) a_2$$

$$(mc)_{st} \frac{dT_{st}}{dt} = \dot{m}_{sc} c_{sc} (T_{fo} - T_{st}) a_1 - \dot{m}_l c_{pl} (T_l - T_{st}) a_2 - (UA)_t (T_{st} - T_a)$$

But T_{st} and T_l are two unknown variables.

Therefore,

Another form of simplification

For an immersed coil connected to load,

$$Q_{load} = \dot{m}_l c_l (T_l - T_i) = (UA)_{l,e}\, LMTD = (UA)_{l,e}\left(\frac{(T_{st} - T_l) - (T_{st} - T_i)}{\ln \frac{(T_{st} - T_l)}{(T_{st} - T_i)}}\right) = (UA)_{l,e}\left(\frac{(T_i - T_l)}{\ln \frac{(T_{st} - T_l)}{(T_{st} - T_i)}}\right)$$

where c_l is the specific heat of hot fluid in load.

$$\ln\frac{(T_{st}-T_l)}{(T_{st}-T_i)}=(UA)_{l,e}\left(\frac{(T_i-T_l)}{\dot{m}_l c_l (T_l-T_i)}\right)$$

$$\frac{T_{st}-T_l}{T_{st}-T_i}=\exp\left(\frac{(-UA)_{l,e}}{\dot{m}_l c_l}\right)$$

$$1-\frac{T_{st}-T_l}{T_{st}-T_i}=1-\exp\left(\frac{(-UA)_{l,e}}{\dot{m}_l c_l}\right)$$

$$\frac{T_l-T_i}{T_{st}-T_i}=1-\exp\left(\frac{(-UA)_{l,e}}{\dot{m}_l c_l}\right)$$

Thus, the expression for the rate of useful heat gain becomes

$$Q_{load}=\dot{m}_l c_l (T_l-T_i)=\dot{m}_l c_l (T_{st}-T_i)\left(1-\exp\left(\frac{(-UA)_{l,e}}{\dot{m}_l c_l}\right)\right)$$

$$\text{let, } a_2=\left(1-\exp\left(\frac{(-UA)_{l,e}}{\dot{m}_l c_l}\right)\right)$$

where $(UA)_{l,e}$ is the product of the overall heat transfer coefficient and the area for the heat exchanger coil connected to load.

The final energy balance equation in differential form

$$(mc)_{st}\frac{dT_{st}}{dt}=\dot{m}_{sc}c_{sc}(T_{fo}-T_{fi})-\dot{m}_l c_{pl}(T_l-T_i)-(UA)_t(T_{st}-T_a)$$

Substituting,

$$(T_{fo}-T_{fi})=(T_{fo}-T_{st})a_1$$

and

$$T_l-T_i=(T_{st}-T_i)\,a_2$$

$$(mc)_{st}\frac{dT_{st}}{dt}=\dot{m}_{sc}c_{sc}(T_{fo}-T_{st})a_1-\dot{m}_l c_{pl}(T_{st}-T_i)a_2-(UA)_t(T_{st}-T_a)$$

$$(mc)_{st}\left(\frac{T_{st,f}-T_{st,i}}{\Delta t}\right)=\dot{m}_{sc}c_{sc}a_1\left(T_{fo}-\frac{T_{st,f}+T_{st,i}}{2}\right)$$
$$-\dot{m}_l c_{pl}a_2\left(\frac{T_{st,f}+T_{st,i}}{2}-T_i\right)-(UA)_t\left(\frac{T_{st,f}+T_{st,,i}}{2}-T_a\right)$$

In this equation, $T_{st,f}$ and $T_{st,i}$ represent the final and initial values of T_{st} over the time interval Δt.

$$\frac{(mc)_{st}}{\Delta t}T_{st,f}-\frac{(mc)_{st}}{\Delta t}T_{st,i}=\dot{m}_{sc}c_{sc}a_1T_{fo}-0.5\dot{m}_{sc}c_{sc}a_1T_{st,f}-0.5\dot{m}_{sc}c_{sc}a_1T_{st,i}$$
$$-0.5\dot{m}_l c_{pl}a_2T_{st,f}-0.5\dot{m}_l c_{pl}a_2T_{st,i}+\dot{m}_l c_{pl}a_2T_i-0.5(UA)_tT_{st,f}-0.5(UA)_tT_{st,i}+(UA)_tT_a$$

$$\frac{(mc)_{st}}{\Delta t}T_{st,f}+0.5\dot{m}_{sc}c_{sc}a_1T_{st,f}+0.5\dot{m}_lc_{pl}a_2T_{st,f}+0.5(UA)_tT_{st,f}$$
$$=\frac{(mc)_{st}}{\Delta t}T_{st,i}-0.5\dot{m}_{sc}c_{sc}a_1T_{st,i}-0.5\dot{m}_lc_{pl}a_2T_{st,i}-0.5(UA)_tT_{st,i}+\dot{m}_{sc}c_{sc}a_1T_{fo}+\dot{m}_lc_{pl}a_2T_i+(UA)_tT_a$$

$$\left(\frac{(mc)_{st}}{\Delta t}+0.5\dot{m}_{sc}c_{sc}a_1+0.5\dot{m}_lc_{pl}a_2+0.5(UA)_t\right)T_{st,f}$$
$$=\left(\frac{(mc)_{st}}{\Delta t}-0.5\dot{m}_{sc}c_{sc}a_1-0.5\dot{m}_lc_{pl}a_2-0.5(UA)_t\right)T_{st,i}+\dot{m}_{sc}c_{sc}a_1T_{fo}+\dot{m}_lc_{pl}a_2T_i+(UA)_tT_a$$

$$T_{st,f}==\frac{\left(\frac{(mc)_{st}}{\Delta t}-0.5\left(\dot{m}_{sc}c_{sc}a_1+\dot{m}_lc_{pl}a_2+(UA)_t\right)\right)T_{st,i}+\dot{m}_{sc}c_{sc}a_1T_{fo}+\dot{m}_lc_{pl}a_2T_i+(UA)_tT_a}{\frac{(mc)_{st}}{\Delta t}+0.5\left(\dot{m}_{sc}c_{sc}a_1+\dot{m}_lc_{pl}a_2+(UA)_t\right)} \quad (14.9)$$

From the earlier equations,

$$(T_{fo}-T_{fi})=(T_{fo}-T_{st})a_1$$

$$T_{fi}=T_{fo}-(T_{fo}-T_{st})a_1$$

$$(T_l-T_i)=(T_{st}-T_i)a_2$$

$$T_l=T_i+(T_{st}-T_i)a_2$$

if $(UA)_{l,e}=0$, $a_2=1-0=1$.

For $a_2=1$, $T_l=T_{st}$

14.2.4 Direct heating and thermal stratification

In thermal stratification, the variation of temperature with water level is considered. Hot water is lighter than cold water. Therefore, the temperature of fluid increases from the bottom to top. Due to temperature difference in the storage tank, the heat losses from the tank are reduced. A second advantage is that the collectors operate at a lower temperature level and deliver higher collection efficiency. In some applications, thermal stratification is obtained naturally as a matter of course. In a natural circulation water heating system, the flow rates are low and a certain degree of thermal stratification is always obtained. Thermal stratification can also be obtained in a forced circulation water heating system if the fluid inlet and outlet configurations are suitably located and shaped. Ring distributors are used so that the velocities entering and leaving the tank are low and are in such a direction as to reduce mixing. In some cases, a floating inlet, made of a wide, thin walled flexible plastic hose connected to the inlet port is used. This type of inlet delivers hot liquid from the collectors at a level at which the inlet liquid temperature is equal to that in the storage tank.

Analysis of thermal stratification is complex due to the dynamic simulation. To simplify this analysis, it may be reasonable to handle it as different sections for evaluation. By using the energy balance equation, we may predict the upper tank's fluid temperature and lower tank's fluid temperature (Fig. 14.6).

Consider the situation in which the storage tank is assumed to consist of two well mixed sections at temperatures T_{l1} and T_{l2}. Writing energy balances on each of the sections, we obtain

$$(\rho Vc_p)_{l1}\frac{dT_{l1}}{dt}=\dot{m}c_p(T_{fo}-T_{l1})-\dot{m}_{load}c_p(T_{l1}-T_{l2})-(UA)_{t1}(T_{l1}-T_a) \quad (14.10)$$

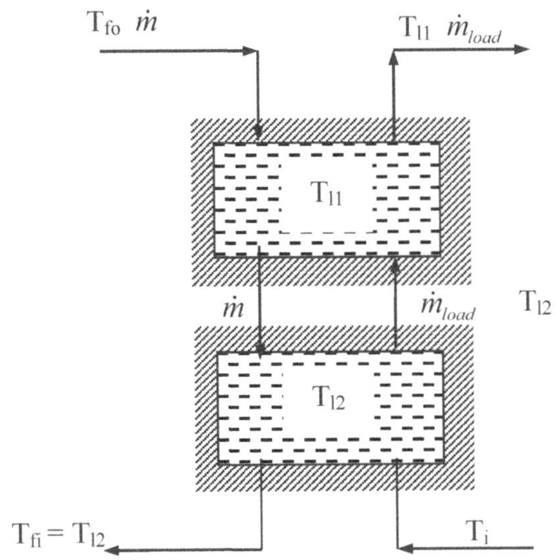

Fig. 14.6. Direct solar thermal storage system with thermal stratification.

Writing the equations in finite difference form,

$$(\rho V c_p)_{l1}\left(\frac{T_{l1,f} - T_{l1,i}}{\Delta t}\right) = \dot{m}c_p\left(T_{fo} - \frac{T_{l1,f} + T_{l1,i}}{2}\right) - \dot{m}_{load}c_p\left(\frac{T_{l1,f} + T_{l1,i}}{2} - \frac{T_{l2,f} + T_{l2,i}}{2}\right)$$
$$-(UA)_{t1}\left(\frac{T_{l1,f} + T_{l1,i}}{2} - T_a\right)$$

In these equations, $T_{l1,f}$ and $T_{l1,i}$ represent the final and initial values of T_{l1} over the time interval Δt.

$$\begin{aligned}(\rho V c_p)_{l1}\left(\frac{T_{l1,f}}{\Delta t}\right) - (\rho V c_p)_{l1}\left(\frac{T_{l1,i}}{\Delta t}\right) &= \dot{m}c_p T_{fo} - 0.5\dot{m}c_p T_{l1,f} - 0.5\dot{m}c_p T_{l1,i} \\ &-0.5\dot{m}_{load}c_p T_{l1,f} - 0.5\dot{m}_{load}c_p T_{l1,i} + 0.5\dot{m}_{load}c_p T_{l2,f} + 0.5\dot{m}_{load}c_p T_{l2,i} \\ &-0.5(UA)_{t1}T_{l1,f} - 0.5(UA)_{t1}T_{l1,i} + (UA)_{t1}T_a\end{aligned} \quad (14.11)$$

$$\left(\left(\frac{(\rho V c_p)_{l1}}{\Delta t}\right) + 0.5\dot{m}c_p + 0.5\dot{m}_{load}c_p + 0.5(UA)_{t1}\right)T_{l1,f} - 0.5\dot{m}_{load}c_p T_{l2,f}$$
$$= (\rho V c_p)_{l1}\left(\frac{T_{l1,i}}{\Delta t}\right) + \dot{m}c_p T_{fo} - 0.5\dot{m}c_p T_{l1,i} - 0.5\dot{m}_{load}c_p T_{l1,i} + 0.5\dot{m}_{load}c_p T_{l2,i} - 0.5(UA)_{t1}T_{l1,i} + (UA)_{t1}T_a$$

$$\left(\left(\frac{(\rho V c_p)_{l1}}{\Delta t}\right) + 0.5\left(\dot{m}c_p + \dot{m}_{load}c_p + (UA)_{t1}\right)\right)T_{l1,f} - \dot{m}_{load}c_p T_{l2,f}$$
$$= (\rho V c_p)_{l1}\left(\frac{T_{l1,i}}{\Delta t}\right) - 0.5\left(\dot{m}c_p + \dot{m}_{load}c_p + (UA)_{t1}\right)T_{l1,i} + 0.5\dot{m}_{load}c_p T_{l2,i} + \dot{m}c_p T_{fo} + (UA)_{t1}T_a$$

Similarly

$$(\rho Vc_p)_{l2}\frac{dT_{l2}}{dt} = \dot{m}c_p(T_{l1} - T_{l2}) - \dot{m}_{load}c_p(T_{l2} - T_i) - (UA)_{t2}(T_{l2} - T_a)$$

$$(\rho Vc_p)_{l2}\left(\frac{T_{l2,f} - T_{l2,i}}{\Delta t}\right) = \dot{m}c_p\left(\frac{T_{l1,f} + T_{l1,i}}{2} - \frac{T_{l2,f} + T_{l2,i}}{2}\right) - \dot{m}_{load}c_p\left(\frac{T_{l2,f} + T_{l2,i}}{2} - T_i\right)$$

$$-(UA)_{t2}\left(\frac{T_{l2,f} + T_{l2,i}}{2} - T_a\right)$$

$$(\rho Vc_p)_{l2}\left(\frac{T_{l2,f}}{\Delta t}\right) - (\rho Vc_p)_{l2}\left(\frac{T_{l2,i}}{\Delta t}\right) = 0.5\dot{m}c_pT_{l1,f} + 0.5\dot{m}c_pT_{l1,i} - 0.5\dot{m}c_pT_{l2,f} - 0.5\dot{m}c_pT_{l2,i}$$

$$-0.5\dot{m}_{load}c_pT_{l2,f} - 0.5\dot{m}_{load}c_pT_{l2,i} + \dot{m}_{load}c_pT_i - 0.5(UA)_{t2}T_{l2,f} - 0.5(UA)_{t2}T_{l2,i} + (UA)_{t2}T_a$$

$$\left(\left(\frac{(\rho Vc_p)_{l2}}{\Delta t}\right) + 0.5\dot{m}c_p + 0.5\dot{m}_{load}c_p + 0.5(UA)_{t2}\right)T_{l2,f} - 0.5\dot{m}c_pT_{l1,f}$$

$$= (\rho Vc_p)_{l2}\left(\frac{T_{l2,i}}{\Delta t}\right) + 0.5\dot{m}c_pT_{l1,i} - 0.5\dot{m}c_pT_{l2,i} - 0.5\dot{m}_{load}c_pT_{l2,i} + \dot{m}_{load}c_pT_i - 0.5(UA)_{t2}T_{l2,i} + (UA)_{t2}T_a$$

$$-0.5\dot{m}c_pT_{l1,f} + \left(\left(\frac{(\rho Vc_p)_{l2}}{\Delta t}\right) + 0.5\left(\dot{m}c_p + \dot{m}_{load}c_p + (UA)_{t2}\right)\right)T_{l2,f}$$

$$= (\rho Vc_p)_{l2}\left(\frac{T_{l2,i}}{\Delta t}\right) - 0.5\left(\dot{m}c_p + \dot{m}_{load}c_p + (UA)_{t2}\right)T_{l2,i} + \dot{m}_{load}c_pT_i + 0.5\dot{m}c_pT_{l1,i} + (UA)_{t2}T_a \quad (14.12)$$

Figure 14.7 shows the temperature, energy, and efficiency changes of a sensible heat storage system with thermal stratification. The two differential equations have to be solved simultaneously for the unknowns T_{l1} and T_{l2}. One way of solving them is to express them in a finite difference form and to choose a suitable time interval. Values of T_{l1} and T_{l2} are then obtained from one-time interval

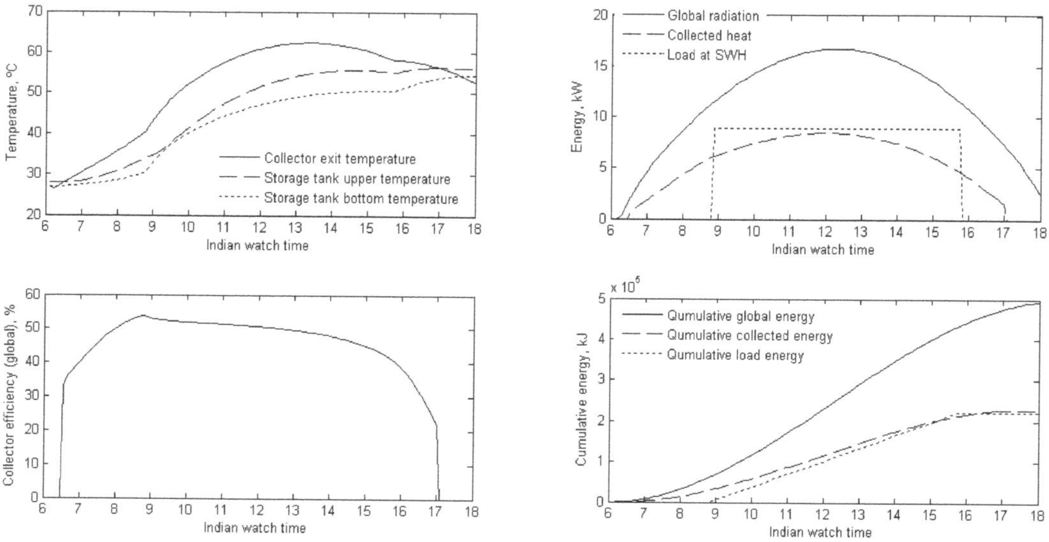

Fig. 14.7. Analysis of indirect heating with stratification.

to the next. The two equations have been written under the assumptions that the flow from the collectors can enter the storage tank only at the top of section 1. This occurs when $T_{fo} > T_{l1}$. When $T_{fo} < T_{l1}$ no flow occurs. In case the tank is also provided with an inlet between the two sections, three possibilities exist.

1) Flow $\dot{m}$ enters at the top of section 1 when $T_{fo} > T_{l1}$
2) Flow $\dot{m}$ enters between sections 1 and 2 when $T_{l1} > T_{fo} > T_{l2}$
3) There is no flow and consequently no energy to be stored when $T_{fo} < T_{l2}$

14.2.5 Indirect heating system and thermal stratification

A simplified analysis is possible by assuming that the storage tank consists of a number of well mixed sections at different temperatures and writing down energy balance equations for each section. Consider the situation in which the storage tank is assumed to consist of two well mixed sections at temperatures T_{l1} and T_{l2} (Fig. 14.8).

Writing energy balances on each of the sections, we obtain

$$q_{u1} = \dot{m}c_{gw}(T_{fo} - T_{fi,1}) = \dot{m}c_{gw}(T_{fo} - T_{l1})\left(1 - \exp\left(\frac{-(UA)_e}{\dot{m}c_{gw}}\right)\right) \tag{14.13}$$

$$(\rho V c_w)_{el1}\frac{dT_{l1}}{dt} = \dot{m}c_{gw}(T_{fo} - T_{l1})\left(1 - \exp\left(\frac{-(UA)_e}{\dot{m}c_{gw}}\right)\right) - \dot{m}_{load}c_w(T_{l1} - T_{l2}) - (UA)_{t1}(T_{l1} - T_a)$$

Writing the equations in finite difference form,

$$(\rho V c_w)_{el1}\left(\frac{T_{l1,f} - T_{l1,i}}{\Delta t}\right) = \dot{m}c_{gw}\left(T_{fo} - \frac{T_{l1,f} + T_{l1,i}}{2}\right)\left(1 - \exp\left(\frac{-(UA)_e}{\dot{m}c_{gw}}\right)\right)$$
$$-\dot{m}_{load}c_w\left(\frac{T_{l1,f} + T_{l1,i}}{2} - \frac{T_{l2,f} + T_{l2,i}}{2}\right) - (UA)_{t1}\left(\frac{T_{l1,f} + T_{l1,i}}{2} - T_a\right)$$

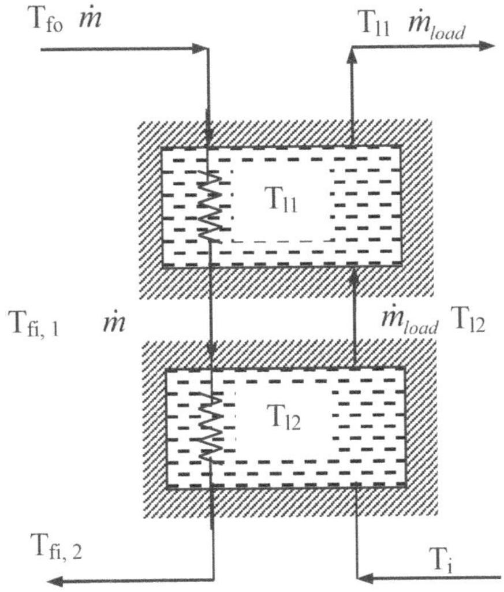

Fig. 14.8. Indirect sensible heat storage system with an immersed coil in the solar collector circuit.

In these equations, $T_{l1,f}$ and $T_{l1,i}$ represent the final and initial values of T_{l1} over the time interval Δt.

Let $a = \left(1 - \exp\left(\frac{-(UA)_e}{\dot{m}c_{gw}}\right)\right)$

$$(\rho V c_w)_{el1}\left(\frac{T_{l1,f}}{\Delta t}\right) - (\rho V c_w)_{el1}\left(\frac{T_{l1,i}}{\Delta t}\right) = a\dot{m}c_{gw}T_{fo} - 0.5a\dot{m}c_{gw}T_{l1,f} - 0.5a\dot{m}c_{gw}T_{l1,i}$$
$$-0.5\dot{m}_{load}c_wT_{l1,f} - 0.5\dot{m}_{load}c_wT_{l1,i} + 0.5\dot{m}_{load}c_wT_{l2,f} + 0.5\dot{m}_{load}c_wT_{12,i}$$
$$-0.5(UA)_{t1}T_{l1,f} - 0.5(UA)_{t1}T_{l1,i} + (UA)_{t1}T_a$$

$$\left(\left(\frac{(\rho V c_w)_{el1}}{\Delta t}\right) + 0.5a\dot{m}c_{gw} + 0.5\dot{m}_{load}c_w + 0.5(UA)_{t1}\right)T_{l1,f} - 0.5\dot{m}_{load}c_wT_{l2,f}$$
$$= (\rho V c_w)_{el1}\left(\frac{T_{l1,i}}{\Delta t}\right) - 0.5\left(a\dot{m}c_{gw} + \dot{m}_{load}c_w + (UA)_{t1}\right)T_{l1,i} + 0.5\dot{m}_{load}c_wT_{12,i} + a\dot{m}c_{gw}T_{fo} + (UA)_{t1}T_a$$

Similarly

$$q_{u2} = \dot{m}c_{gw}(T_{fi,1} - T_{fi,2}) = \dot{m}c_{gw}(T_{fi,1} - T_{l2})\left(1 - \exp\left(\frac{-(UA)_e}{\dot{m}c_{gw}}\right)\right)$$

To eliminate $T_{fi,1}$ from the above equation,

$$T_{fi,1} = T_{fo} - (T_{fo} - T_{l1})\left(1 - \exp\left(\frac{-(UA)_e}{\dot{m}c_{gw}}\right)\right) \quad \text{from } q_{u1}$$

$$\dot{m}c_{gw}(T_{fi,1} - T_{l2})\left(1 - \exp\left(\frac{-(UA)_e}{\dot{m}c_{gw}}\right)\right) = \dot{m}c_{gw}\left(T_{fo} - (T_{fo} - T_{l1})\left(1 - \exp\left(\frac{-(UA)_e}{\dot{m}c_{gw}}\right)\right) - T_{l2}\right)$$
$$\times\left(1 - \exp\left(\frac{-(UA)_e}{\dot{m}c_{gw}}\right)\right)$$

Let $a = \left(1 - \exp\left(\frac{-(UA)_e}{\dot{m}c_{gw}}\right)\right)$

$$\dot{m}c_{gw}(T_{fi,1} - T_{l2})\left(1 - \exp\left(\frac{-(UA)_e}{\dot{m}c_{gw}}\right)\right) = \dot{m}c_{gw}\left(T_{fo} - (T_{fo} - T_{l1})a - T_{l2}\right)a$$

$$= a\dot{m}c_{gw}T_{fo} - a^2\dot{m}c_{gw}T_{fo} + a^2\dot{m}c_{gw}T_{l1} - a\dot{m}c_{gw}T_{l2}$$

$$\therefore a\dot{m}c_{gw}(T_{fi,1} - T_{l2}) = a^2\dot{m}c_{gw}T_{l1} - a\dot{m}c_{gw}T_{l2} + a\dot{m}c_{gw}T_{fo}(1 - a)$$

$$(\rho V c_w)_{el2}\frac{dT_{l2}}{dt} = \dot{m}c_{gw}(T_{fi,1} - T_{l2})\left(1 - \exp\left(\frac{-(UA)_e}{\dot{m}c_{gw}}\right)\right) - \dot{m}_{load}c_w(T_{l2} - T_i) - (UA)_{t2}(T_{l2} - T_a)$$

$$(\rho V c_w)_{el2}\frac{dT_{l2}}{dt} = \left(a^2\dot{m}c_{gw}T_{l1} - a\dot{m}c_{gw}T_{l2} + a\dot{m}c_{gw}T_{fo}(1-a)\right) - \dot{m}_{load}c_w(T_{l2}-T_i) - (UA)_{t2}(T_{l2}-T_a)$$

$$(\rho V c_w)_{el2}\left(\frac{T_{l2,f}-T_{l2,i}}{\Delta t}\right) = a^2\dot{m}c_{gw}\left(\frac{T_{l1,f}+T_{l1,i}}{2}\right) - a\dot{m}c_{gw}\left(\frac{T_{l2,f}+T_{l2,i}}{2}\right) + a\dot{m}c_{gw}T_{fo}(1-a)$$
$$-\dot{m}_{load}c_w\left(\frac{T_{l2,f}+T_{l2,i}}{2} - T_i\right) - (UA)_{t2}\left(\frac{T_{l2,f}+T_{l2,i}}{2} - T_a\right)$$

$$(\rho V c_w)_{el2}\left(\frac{T_{l2,f}}{\Delta t}\right) - (\rho V c_w)_{el2}\left(\frac{T_{l2,i}}{\Delta t}\right) = 0.5a^2\dot{m}c_{gw}T_{l1,f} + 0.5a^2\dot{m}c_{gw}T_{l1,i} - 0.5a\dot{m}c_{gw}T_{l2,f}$$
$$-0.5a\dot{m}c_{gw}T_{l2,i} + a\dot{m}c_{gw}T_{fo}(1-a) - 0.5\dot{m}_{load}c_wT_{l2,f} - 0.5\dot{m}_{load}c_wT_{l2,i} + \dot{m}_{load}c_wT_i$$
$$-0.5(UA)_{t2}T_{l2,f} - 0.5(UA)_{t2}T_{l2,i} + (UA)_{t2}T_a$$

$$\left(\left(\frac{(\rho V c_w)_{el2}}{\Delta t}\right) + 0.5a\dot{m}c_{gw} + 0.5\dot{m}_{load}c_w + 0.5(UA)_{t2}\right)T_{l2,f} - 0.5a^2\dot{m}c_{gw}T_{l1,f}$$
$$= (\rho V c_w)_{el2}\left(\frac{T_{l2,i}}{\Delta t}\right) + 0.5a^2\dot{m}c_{gw}T_{l1,i} - 0.5a\dot{m}c_{gw}T_{l2,i} - 0.5\dot{m}_{load}c_wT_{l2,i} + \dot{m}_{load}c_wT_i$$
$$-0.5(UA)_{t2}T_{l2,i} + (UA)_{t2}T_a + a\dot{m}c_{gw}T_{fo}(1-a)$$

$$-0.5a^2\dot{m}c_wT_{l1,f} + \left(\left(\frac{(\rho V c_{gw})_{el2}}{\Delta t}\right) + 0.5\left(a\dot{m}c_{gw} + \dot{m}_{load}c_w + (UA)_{t2}\right)\right)T_{l2,f}$$
$$= (\rho V c_w)_{el2}\left(\frac{T_{l2,i}}{\Delta t}\right) - 0.5\left(a\dot{m}c_{gw} + \dot{m}_{load}c_w + (UA)_{t2}\right)T_{l2,i} + \dot{m}_{load}c_wT_i + 0.5a^2\dot{m}c_{gw}T_{l1,i} \qquad (14.14)$$
$$+(UA)_{t2}T_a + a(1-a)\dot{m}c_{gw}T_{fo}$$

Figure 14.9 shows the temperature, energy, and efficiency changes of a sensible heat storage system with thermal stratification and immersed coil. The two differential equations have to be

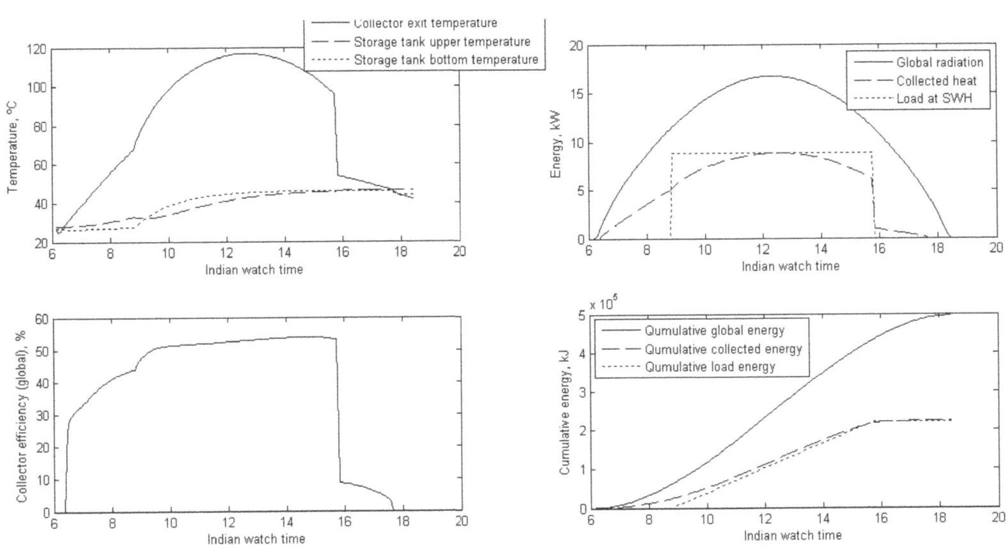

Fig. 14.9. Results of indirect heating in a sensible heat storage system with stratification and immersed coil in storage tank connected to a solar collector.

solved simultaneously for the unknown T_{l1} and T_{l2}. One way of solving them is to express them in finite difference form and to choose a suitable time interval. Values of T_{l1} and T_{l2} are then obtained from one-time interval to the next. The two equations have been written under the assumptions that the flow from the collectors can enter the storage tank only at the top of section 1. This occurs when $T_{fo} > T_{l1}$. When $T_{fo} < T_{l1}$ no flow occurs. In case the tank is also provided with an inlet between the two sections, three possibilities exist.

1) Flow $\dot{m}$ enters at the top of section 1 when $T_{fo} > T_{l1}$
2) Flow $\dot{m}$ enters between sections 1 and 2 when $T_{l1} > T_{fo} > T_{l2}$
3) There is no flow and consequently no energy to be stored when $T_{fo} < T_{l2}$

14.2.6 Indirect heating system and thermal stratification (from thermic fluid to steam generation)

A simplified analysis is possible by assuming that the storage tank consists of a number of well mixed sections at different temperatures and writing down energy balance equations for each section. Consider the situation in which the storage tank is assumed to consist of two well mixed sections at temperatures T_{l1} and T_{l2} (Fig. 14.10). Writing energy balances for each of the sections, we obtain

$$q_{u1} = \dot{m}_{load}(T_{fo} - T_{l1}) = \dot{m}c_{gw}(T_{fo} - T_{l1})\left(1 - \exp\left(\frac{-(UA)_e}{\dot{m}c_{gw}}\right)\right)$$

$$(\rho V c_w)_{el1}\frac{dT_{l1}}{dt} = \dot{m}c_{gw}(T_{fo} - T_{l1})\left(1 - \exp\left(\frac{-(UA)_e}{\dot{m}c_{gw}}\right)\right) - \dot{m}_{load}c_w(T_{l1} - T_{l2}) - (UA)_{t1}(T_{l1} - T_a)$$

Writing the equations in finite difference form,

$$(\rho V c_w)_{el1}\left(\frac{T_{l1,f} - T_{l1,i}}{\Delta t}\right) = \dot{m}c_{gw}\left(T_{fo} - \frac{T_{l1,f} + T_{l1,i}}{2}\right)\left(1 - \exp\left(\frac{-(UA)_e}{\dot{m}c_{gw}}\right)\right)$$
$$-\dot{m}_{load}c_w\left(\frac{T_{l1,f} + T_{l1,i}}{2} - \frac{T_{l2,f} + T_{l2,i}}{2}\right) - (UA)_{t1}\left(\frac{T_{l1,f} + T_{l1,i}}{2} - T_a\right)$$

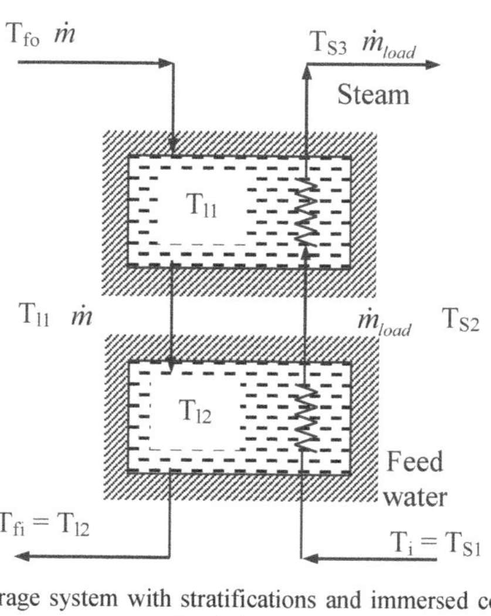

Fig. 14.10. Indirect sensible heat storage system with stratifications and immersed coil in storage tank but connected to load circuit.

In these equations, $T_{l1,f}$ and $T_{l1,i}$ represent the final and initial values of T_{l1} over the time interval Δt.

Let $a = \left(1 - \exp\left(\frac{-(UA)_e}{\dot{m}c_{gw}}\right)\right)$

$$(\rho V c_w)_{el1}\left(\frac{T_{l1,f}}{\Delta t}\right) - (\rho V c_w)_{el1}\left(\frac{T_{l1,i}}{\Delta t}\right) = a\dot{m}c_{gw}T_{fo} - 0.5a\dot{m}c_{gw}T_{l1,f} - 0.5a\dot{m}c_{gw}T_{l1,i}$$
$$-0.5\dot{m}_{load}c_w T_{l1,f} - 0.5\dot{m}_{load}c_w T_{l1,i} + 0.5\dot{m}_{load}c_w T_{l2,f} + 0.5\dot{m}_{load}c_w T_{l2,i}$$
$$-0.5(UA)_{t1}T_{l1,f} - 0.5(UA)_{t1}T_{l1,i} + (UA)_{t1}T_a$$

$$\left(\left(\frac{(\rho V c_w)_{el1}}{\Delta t}\right) + 0.5a\dot{m}c_{gw} + 0.5\dot{m}_{load}c_w + 0.5(UA)_{t1}\right)T_{l1,f} - 0.5\dot{m}_{load}c_w T_{l2,f}$$
$$= (\rho V c_w)_{el1}\left(\frac{T_{l1,i}}{\Delta t}\right) - 0.5\left(a\dot{m}c_{gw} + \dot{m}_{load}c_w + (UA)_{t1}\right)T_{l1,i} + 0.5\dot{m}_{load}c_w T_{l2,i} + a\dot{m}c_{gw}T_{fo} + (UA)_{t1}T_a$$

Similarly

$$q_{u2} = \dot{m}c_{gw}(T_{fi,1} - T_{fi,2}) = \dot{m}c_{gw}(T_{fi,1} - T_{l2})\left(1 - \exp\left(\frac{-(UA)_e}{\dot{m}c_{gw}}\right)\right)$$

To eliminate $T_{fi,1}$ from the above equation,

$$T_{fi,1} = T_{fo} - (T_{fo} - T_{l1})\left(1 - \exp\left(\frac{-(UA)_e}{\dot{m}c_{gw}}\right)\right) \text{ from } q_{u1}$$

$$\dot{m}c_{gw}(T_{fi,1} - T_{l2})\left(1 - \exp\left(\frac{-(UA)_e}{\dot{m}c_{gw}}\right)\right) = \dot{m}c_{gw}\left(T_{fo} - (T_{fo} - T_{l1})\left(1 - \exp\left(\frac{-(UA)_e}{\dot{m}c_{gw}}\right)\right) - T_{l2}\right)$$
$$\times\left(1 - \exp\left(\frac{-(UA)_e}{\dot{m}c_{gw}}\right)\right)$$

Let $a = \left(1 - \exp\left(\frac{-(UA)_e}{\dot{m}c_{gw}}\right)\right)$

$$\dot{m}c_{gw}(T_{fi,1} - T_{l2})\left(1 - \exp\left(\frac{-(UA)_e}{\dot{m}c_{gw}}\right)\right) = \dot{m}c_{gw}\left(T_{fo} - (T_{fo} - T_{l1})a - T_{l2}\right)a$$

$$= a\dot{m}c_{gw}T_{fo} - a^2\dot{m}c_{gw}T_{fo} + a^2\dot{m}c_{gw}T_{l1} - a\dot{m}c_{gw}T_{l2}$$

$$\therefore a\dot{m}c_{gw}(T_{fi,1} - T_{l2}) = a^2\dot{m}c_{gw}T_{l1} - a\dot{m}c_{gw}T_{l2} + a\dot{m}c_{gw}T_{fo}(1-a)$$

$$(\rho V c_w)_{el2}\frac{dT_{l2}}{dt} = \dot{m}c_{gw}(T_{fi,1} - T_{l2})\left(1 - \exp\left(\frac{-(UA)_e}{\dot{m}c_{gw}}\right)\right) - \dot{m}_{load}c_w(T_{l2} - T_i) - (UA)_{t2}(T_{l2} - T_a)$$

$$(\rho V c_w)_{el2} \frac{dT_{l2}}{dt} = \left(a^2 \dot{m} c_{gw} T_{l1} - a\dot{m} c_{gw} T_{l2} + a\dot{m} c_{gw} T_{fo}(1-a)\right) - \dot{m}_{load} c_w (T_{l2} - T_i) - (UA)_{t2}(T_{l2} - T_a)$$

$$(\rho V c_w)_{el2} \left(\frac{T_{l2,f} - T_{l2,i}}{\Delta t}\right) = a^2 \dot{m} c_{gw} \left(\frac{T_{l1,f} + T_{l1,i}}{2}\right) - a\dot{m} c_{gw} \left(\frac{T_{l2,f} + T_{l2,i}}{2}\right) + a\dot{m} c_{gw} T_{fo}(1-a)$$
$$-\dot{m}_{load} c_w \left(\frac{T_{l2,f} + T_{l2,i}}{2} - T_i\right) - (UA)_{t2} \left(\frac{T_{l2,f} + T_{l2,i}}{2} - T_a\right)$$

$$(\rho V c_w)_{el2} \left(\frac{T_{l2,f}}{\Delta t}\right) - (\rho V c_w)_{el2} \left(\frac{T_{l2,i}}{\Delta t}\right) = 0.5a^2 \dot{m} c_{gw} T_{l1,f} + 0.5a^2 \dot{m} c_{gw} T_{l1,i} - 0.5a\dot{m} c_{gw} T_{l2,f}$$
$$-0.5a\dot{m} c_{gw} T_{l2,i} + a\dot{m} c_{gw} T_{fo}(1-a) - 0.5\dot{m}_{load} c_w T_{l2,f} - 0.5\dot{m}_{load} c_w T_{l2,i} + \dot{m}_{load} c_w T_i$$
$$-0.5(UA)_{t2} T_{l2,f} - 0.5(UA)_{t2} T_{l2,i} + (UA)_{t2} T_a$$

$$\left(\left(\frac{(\rho V c_w)_{el2}}{\Delta t}\right) + 0.5a\dot{m} c_{gw} + 0.5\dot{m}_{load} c_w + 0.5(UA)_{t2}\right) T_{l2,f} - 0.5a^2 \dot{m} c_{gw} T_{l1,f}$$
$$= (\rho V c_w)_{el2} \left(\frac{T_{l2,i}}{\Delta t}\right) + 0.5a^2 \dot{m} c_{gw} T_{l1,i} - 0.5a\dot{m} c_{gw} T_{l2,i} - 0.5\dot{m}_{load} c_w T_{l2,i} + \dot{m}_{load} c_w T_i$$
$$-0.5(UA)_{t2} T_{l2,i} + (UA)_{t2} T_a + a\dot{m} c_{gw} T_{fo}(1-a)$$

$$-0.5a^2 \dot{m} c_w T_{l1,f} + \left(\left(\frac{(\rho V c_{gw})_{el2}}{\Delta t}\right) + 0.5\left(a\dot{m} c_{gw} + \dot{m}_{load} c_w + (UA)_{t2}\right)\right) T_{l2,f}$$
$$= (\rho V c_w)_{el2} \left(\frac{T_{l2,i}}{\Delta t}\right) - 0.5\left(a\dot{m} c_{gw} + \dot{m}_{load} c_w + (UA)_{t2}\right) T_{l2,i} + \dot{m}_{load} c_w T_i + 0.5a^2 \dot{m} c_{gw} T_{l1,i} \tag{14.15}$$
$$+(UA)_{t2} T_a + a(1-a)\dot{m} c_{gw} T_{fo}$$

Figure 14.11 shows the temperature, energy and thermal efficiency changes of a sensible heat storage system with indirect heating and thermal stratification.

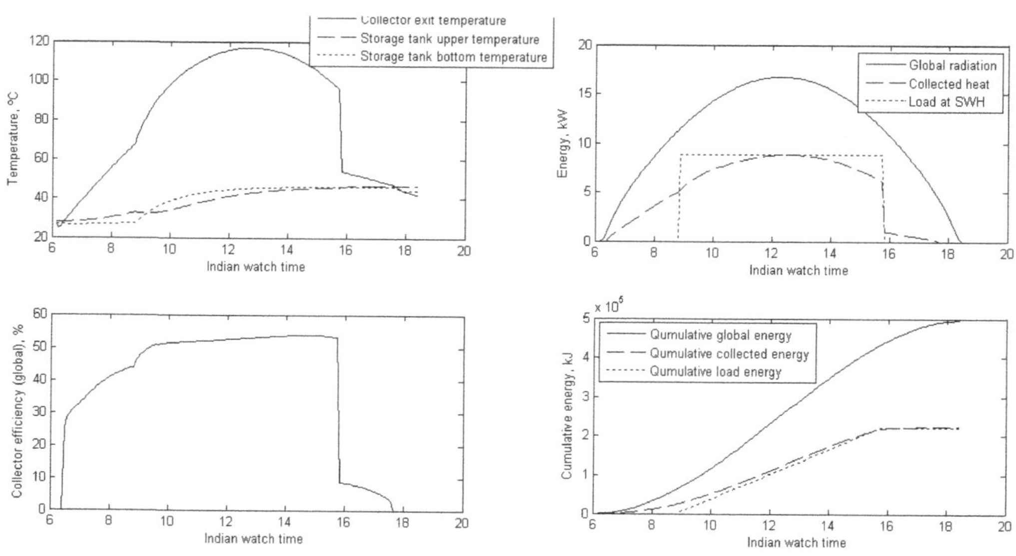

Fig. 14.11. Results of indirect heating in a sensible heat storage system with stratification.

14.3 Latent heat storage

The material used in latent heat storage melts during heating and freezes on cooling. Latent heat transfer is applied to a latent heat storage system; the material is called as phase change material (PCM). The phase change material should have a melting temperature suitable to the application, high value of latent heat transfer, high density or small volume difference, and negligible amount of sub cooling and sensible heat transfer as thermal energy storage is low.

A PCM is an organic or inorganic compound, used to store and release latent heat during phase change at constant temperature. Paraffin, salt hydrates, fatty acids, and esters are the generally used as PCMs. The organic solutions, which are water-free can be made without encapsulation, but salt based PCM solutions are encapsulated to prevent the water loss by the evaporation. Micro encapsulation has the advantage of adoption into construction material and solve the problem of poor thermal conductivity of PCMs. Molecular encapsulation is another new technology in the storage system. Recently nanoparticles also dispersed in PCM. Thermal composites also one of the advanced technology such as copper meshing in a paraffin wax.

14.4 Numerical solutions

1. A cylindrical hot water storage tank, 1.7 m in diameter and 2.1 m high, is made from a steel plate (ρ = 7800 kg/m^3, c_p = 0.46 kJ/kg K) 6 mm thick. Apart from the mass of steel required for making the surface, an additional 200 kg of steel is required in the form of angles, etc. for strengthening the tank, which is insulated all round with glass wool insulation 20 cm thick (k = 0.040 W/m-K). The initial temperature of the water in the tank is 50°C at 0700 h in the morning on a particular day and the variation of q_u and T_a up to 2000 h is shown in Table 14.1.

Table 14.1. Observations of a sensible heat transfer system.

Time	**7–8**	**8–9**	**9–10**	**10–11**	**11–12**	**12–13**	**13–14**
q_u, kJ/h	18660	37496	54890	60070	69890	70120	62205
T_a, °C	17.8	21.9	25.1	27.4	29.1	30.2	30.9
Time	**14–15**	**15–16**	**16–17**	**17–18**	**18–19**	**19–20**	
q_u, kJ/h	55490	41070	19195	0	0	0	
T_a, °C	31.2	31.3	30.8	28.9	26.0	24.0	

Refer Fig. 14.2 for the graphical representation.

2. In a domestic hot water generation system with a solar collector (non-focusing) using ETC tubes, the requirement for domestic water is 6000 liters per day (LPD). The supplied water temperature and exit water temperature of the solar water heater are 5°C and 40°C, respectively. Assume this unit is to be installed on the terrace of a building, where there is no space constraint for installation.

Figure 14.12 shows the schematic layout of a solar hot water system with evacuated tube collector (ETC) units. The system consists of an array of ETC units with thermal storage. The SHS with indirect heat transfer option is selected as per the location. Ethylene glycol in the solar collector facilitates operates the unit at a higher boiler temperature and lower freezing temperature. The fluid to load is, as usual, water, which is domestic water. The available sizes of ETC units in the market are 100 LPD, 150 LPD, 200 LPD, 250 LPD, 300 LPD, and 500 LPD. Since the required capacity is 6000 LPD, a 500 LPD unit is selected for the evaluation of the total field.
Design the system with the suitable assumptions.

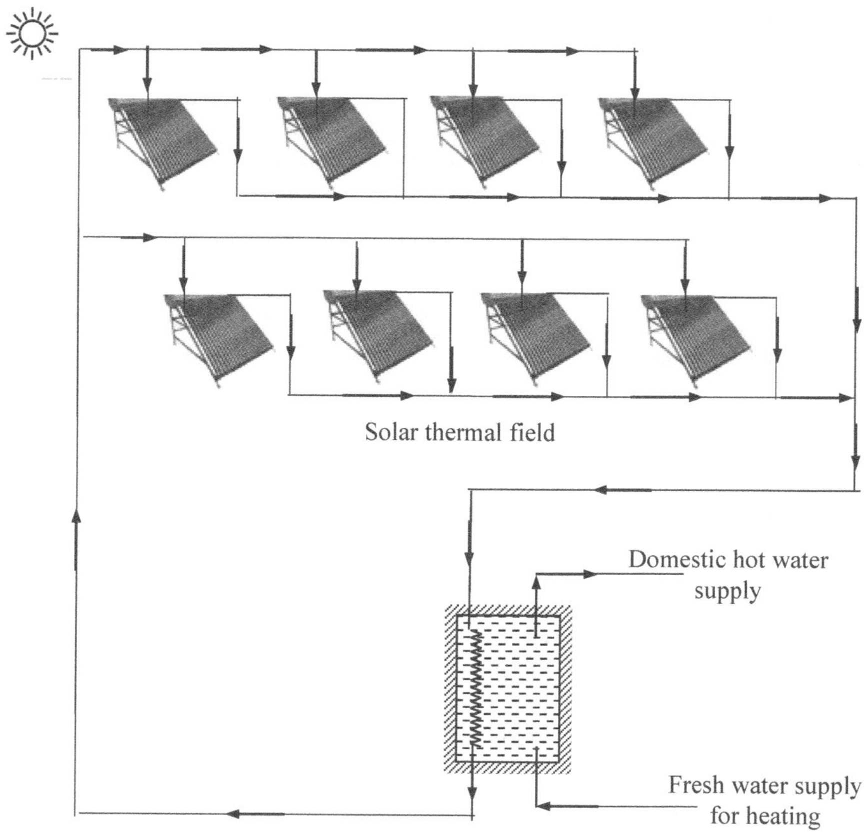

Fig. 14.12. Domestic solar hot water system.

The total energy required for domestic water is determined from the given heat load and its capacity as follows.

The energy required for domestic water per day,

$$Q = \frac{m_w c_{pw} (T_f - T_i)}{3600} \times 1hr \text{ (kWh)}$$

$$= \frac{6000 \times 4.18 \times (40 - 5)}{3600} \times 1 = 243.83 \text{ kWh}$$

This heat load for domestic water is assumed as a constant for 24 hours.

The operational conditions of ETC units with SHS are tabulated in Table 14.2. The fluid flow to the load is determined from the above calculated load. The specifications of the tube and ETC have been detailed with sizes.

Based on these conditions, a dynamic model has been created to solve the thermal storage unit connected to ETC and load. The resultant fluid temperatures are plotted and shown in Fig. 14.13 (a). Since the capacity of the system and heat load is fixed, we cannot control the fluid to load. The fluid will flow to the load as per the predetermined requirement. The iteration of the dynamic model yields the fluid temperatures, storage tank capacity, and number of ETC units required to meet the load. Figure 14.13 (b) and Fig. 14.13 (c) depict the instantaneous and cumulative heats for 24 hours. Figure 14.13 (d) shows the ETC's thermal efficiency during the day length period. The results of

Table 14.2. Operational conditions of a domestic hot water system using evacuated tube solar water heater per building.

S. No.	Description	Condition
1.	Initial temperature of Water, °C	5.00
2.	Heat load at 6000 LPD for 24 hours, kW	10.17
3.	Total energy required, kWh	244.18
4.	Fluid flow in one ETC unit, LPH	60.00
5.	Total fluid flow in the solar thermal field, LPH	1230.00
6.	Fluid flow in load, LPH	250.00
5.	Inner diameter of receiver tube, m	0.045
6.	Outer diameter of receiver tube, m	0.047
7.	Glass tube outer diameter, m	0.058
8.	Lateral distance between centers of the two tubes (pitch), m	0.076
9.	Length of evacuated tube, m	1.80
10.	Number of tubes in one unit	50.00
11.	Projected area of each unit with 50 ETC tubes, m²	6.84

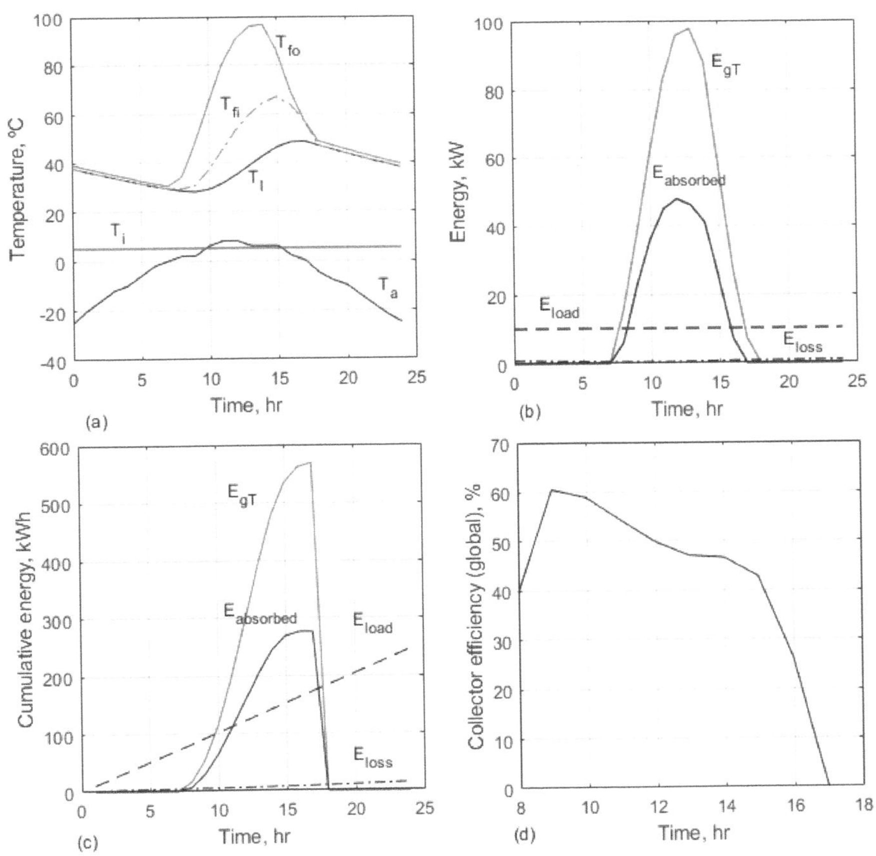

Figure 14.13. (a) Fluid temperatures in an evaluated tube solar water heater with storage unit; T_{fi}: fluid inlet temperature to collector, T_{fo}: fluid outlet temperature of collector, T_l: liquid (water) temperature in the storage tank, T_a: ambient temperature, T_i: temperature of fluid supplied to storage tank (fresh makeup water) (b) energy capacity of an evaluated tube solar water heater with storage unit for one building; gT: global radiation on tilted surface (c) cumulative energies of in an evaluated tube solar water heater with storage unit for 24 hours per building (d) thermal efficiency of an evaluated tube solar water heater with storage unit.

Table 14.3. Results of domestic hot water system using evacuated tube solar water heater per building.

S. No.	Description	Result
1.	Energy required for domestic water heating, kWh	244.18
2.	Global radiation falling on ETC units, tilted south, kWh	568.89
3.	Collected heat at ETC, tilted south, kWh	273.76
4.	Thermal energy used for load, kWh	244.18
5.	Storage tank capacity, Lit	7400.00
6.	Average temperature difference of fluid in load, K	39.00
7.	Maximum temperature of fluid in storage tank, °C	47.81
8.	Minimum temperature of fluid in storage tank, °C	29.57
9.	Temperature change in storage tank's fluid/day	18.24
10.	Average thermal efficiency of ETC with tilted global, %	52.50
11.	Number of ETC units (each unit has 50 ETCs)	20.00
12.	Total projected area of ETC units, m^2	137.00

the ETC system with the SHS for domestic water are tabulated in Table 14.3. Since the fluid inlet temperature is low (5°C) and below the normal temperature, the number of ETC units are greater compared to the units designed for the fluid supply at normal temperature. Therefore, the resultant ETC units are 20 (rounded), which is more than the nominal number. The iterative model gives the capacity of the storage tank which is 7400 liters to maintain the required water temperature for the load. As per the specifications listed in Table 14.2, the projected area of 20 ETC units is 137 m^2 at an ETC thermal efficiency of 52.5%.

The projected area of the 20 ETC units suitable for the given load (6000 LPD = 244.18 kWh) is 137 m^2 at the ETC thermal efficiency of 52.5%. The third floor area for one building is 347.13 m^2, which is equal to the terrace area. Since the required ETC area is less than the terrace area, all the ETC units can be installed without any constraint. Generally, to heat the water from its normal temperature, 12 ETC units (500 LPD each) are required for 6000 LPD. However, in the present case, the water supply temperature to the ETC units is 5°C. Therefore, the number of ETC units (20) is greater than the number at normal conditions for a 6000 LPD requirement. The exhaust gas from the boiler can be used for the partial generation of domestic hot water. It will decrease the size of the ETC field.

3. Design an indirect heating system with two heat exchanger coils in a storage tank for a solar collector and load (Fig. 14.14).

Solution

Figure 14.14 outlines the radiant heating system with a solar thermal field and storage unit. SHS system with indirect heating is selected to this application. The system has three fluids for the transfer of heat. The fluid in the solar field and radiant heating coils is ethylene glycol. The storage tank fluid is water, and this will work as a mediator between the solar field and load. A common energy source, i.e., solar thermal field with storage has been used for the two heat loads of radiant heat load and fresh air heating's load. The solar collector of a line focused unit consists of arrays of compound parabolic collectors (CmPCs). Evacuated tube with both ends open are used as a receiver tube. Five CmPCs are integrated to form an array. All the arrays are arranged in parallel with each other. The flow from the storage tank is equally divided and supplied to all these arrays. The CmPC arrays are tilted at 45° (higher than latitude) to the horizontal and towards south to capture maximum solar radiation during winter. Unlike cylindrical parabolic collector (CPC), CmPC need no tracking for the capture of solar radiation. Therefore, the tilt angle of CmPC is fixed.

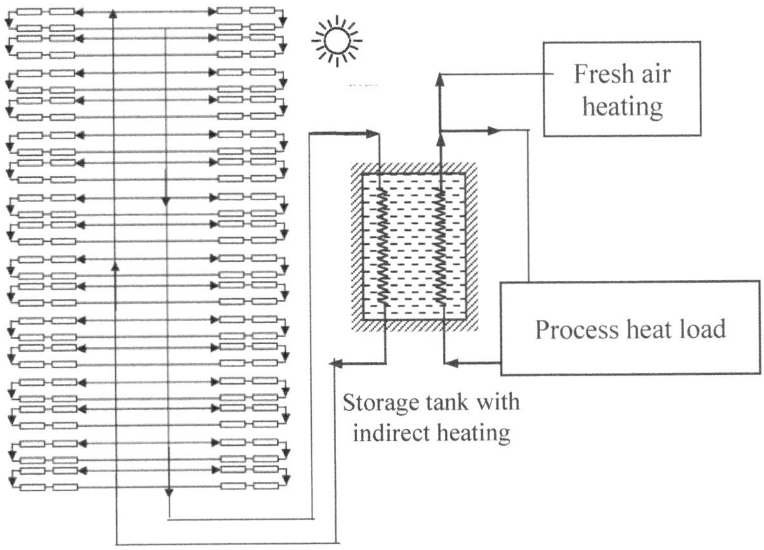

Fig. 14.14. Radiant heating system with CmPC filed and thermal energy storage.

The solar radiation model at select place in India can be evaluated as follows. The heat load calculations are conducted on the coldest day of the location. The arrays of CmPCs are tilted towards south. To evaluate the solar radiation falling on a tilted surface, the solar geometry has to be found for this day for the location. The latitude and longitude of the location are 34.1526° N, and 77.5771° E respectively. At this place minimum solar radiation falls on ground during the month of January. The solar collector has been designed for this peak winter condition. The solar geometry has been plotted for a typical winter day, 15th January. The actual solar radiation sourced from climate data for meteorological years is used for the evaluation of solar thermal systems (source: Climate. OneBuilding. Org). On this typical winter day, the sunrise and sunset are at 7.30 AM and 5.27 PM respectively with the day-length of 9.95 hrs at location. From solar geometry the incidence angle (angle between sunbeam and normal to the tilted surface) is determined to take the cosθ component to the solar radiation.

The incidence angle (θ) is the function of declination (δ), hour angle (ω), latitude (Ø), tilt angle (β) and surface azimuth angle (γ) as follows,

$$\theta = \cos^{-1}\begin{pmatrix} \sin\delta\sin\phi\cos\beta - \sin\delta\cos\phi\sin\beta\cos\gamma + \cos\delta\cos\phi\cos\beta\cos\omega \\ +\cos\delta\sin\phi\sin\beta\cos\gamma\cos\omega + \cos\delta\sin\beta\sin\gamma\sin\omega \end{pmatrix}$$

During winter, the tilt angle (β) is 45° which is more than Latitude (φ). For the surface tilted towards south, the surface azimuth angle, $\gamma = 0$.

The solar geometry was calculated using the date, time, location, and surface orientation (zenith and azimuth) of a collector. Zenith angle and azimuth angle are the functions of tilt angle and its orientation. The estimated solar geometry values are listed in Table 14.4. As shown in Fig. 14.15, the incidence angle is minimum during noon and the CmPC captures more radiation compared to sunrise and sunset.

Table 14.4. Solar geometry in January at an Indian city.

Time, hr	Tilt angle, °	Zenith, °	Solar azimuth, °	Surface azimuth, °	Incidence angle, °
8	45.00	84.62	–59.68	0.00	65.06
9	45.00	74.48	–49.89	0.00	51.09
10	45.00	65.85	–38.19	0.00	37.21
11	45.00	59.40	–24.21	0.00	23.79
12	45.00	55.93	–8.17	0.00	12.60
13	45.00	55.99	8.71	0.00	12.86
14	45.00	59.57	24.69	0.00	24.20
15	45.00	66.09	38.60	0.00	37.65
16	45.00	74.79	50.23	0.00	51.53
17	45.00	84.96	59.97	0.00	65.51

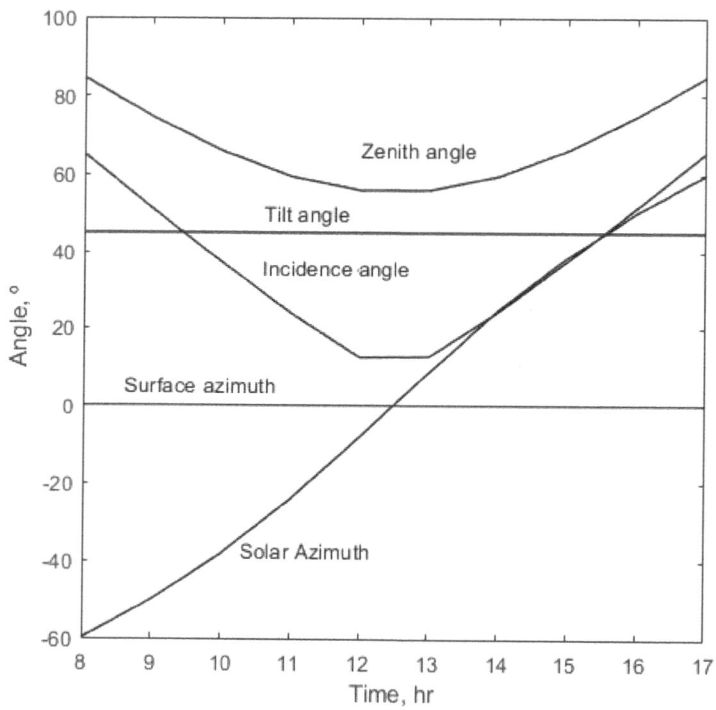

Fig. 14.15. Solar geometry with collector tilted towards south in January for the estimation of solar radiation at an Indian city.

Figure 14.16 shows the actual daily solar radiation from morning to evening. The global radiation, direct normal irradiation (DNI), and diffuse. The radiation on a tilted surface is estimated from the actual measurements and the incidence angle resulting from solar geometry as per the earlier discussion.

From the actual solar radiation data and predicted solar geometry values, the following is the result of solar radiation at a location in January (Table 14.5).
The average global radiation on horizontal surface = 336.61 W/m^2 (3.35 kWh/m^2 day)
The average DNI = 303.35 W/m^2 (3.02 kWh/m^2 day)
The average beam radiation on a tilted surface = 257.87 W/m^2 (2.56 kWh/m^2 day)

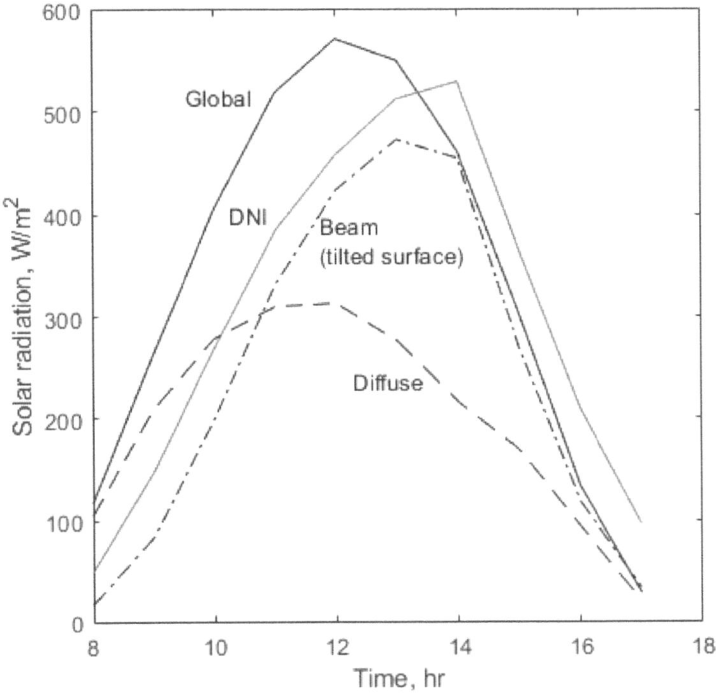

Fig. 14.16. Global radiation, DNI, beam radiation on a tilted surface, and diffuse radiation on CmPC in January.

Table 14.5. Solar radiation on a winter day (January) at a Location.

Time, hr	Horizontal global radiation, W/m^2	DNI, W/m^2	Beam radiation on tilted surface, W/m^2	Diffuse radiation, W/m^2
8	117.00	50.00	21.08	105.00
9	263.00	147.00	92.34	209.00
10	405.00	268.00	213.44	278.00
11	518.00	384.00	351.37	310.00
12	571.00	458.00	446.97	313.00
13	550.00	512.00	499.16	276.00
14	461.00	529.00	482.50	216.00
15	303.00	366.00	289.78	169.00
16	133.00	209.00	130.01	94.00
17	29.00	96.00	39.80	21.00

The average diffuse radiation = 200.06 W/m^2 (1.99 kWh/m^2 day)
The day-length on 15th January at location is 9.95 hr.

The evaluation of a solar thermal system includes collection, storage, and utilization values. For radiant heating and fresh air heating listed in Table 14.6, CmPCs are used. The evaluation of the CmPC includes finding the optical efficiency of the collector, heat transfer losses, and fluid temperatures. Thermodynamic and thermophysical properties of ethylene glycol are determined as a function of temperature at a concentration of 30%. They are used in the design of a solar thermal system.

Table 14.6. The ambient temperature and energy required for a building over the course of 24 hours.

Time, hr	Ambient temperature, °C	Process heat load, kW	Fresh air heating, kW	Total, kW
1	–20	205	91	296
2	–16	184	81	265
3	–12	161	71	232
4	–10	150	65	215
5	–6	129	55	184
6	–2	106	44	150
7	0	96	39	135
8	2	84	34	118
9	2	84	34	118
10	6	64	24	88
11	8	51	18	69
12	8	51	18	69
13	6	64	24	88
14	6	64	24	88
15	6	64	24	88
16	2	84	34	118
17	0	96	39	135
18	–5	123	52	175
19	–8	139	60	199
20	–10	150	65	215
21	–14	173	76	249
22	–18	193	86	279
23	–22	216	97	313
24	–25	233	104	337

The increase in boiling point and decrease in freezing point can be referred to as the glycol concentration in water by weight percentage. The given properties in this evaluation for the rating and sizing of the thermal equipment are density, specific heat, thermal conductivity, and viscosity, which are tabulated in Table 14.7.

In this section, solar thermal systems for radiant heating and fresh air heating have been evaluated with a single solar thermal source without keeping the constraints on available space for the erection of a solar field. Table 14.8 summarizes the assumptions used in the evaluation of a single solar energy source. It includes the size of the CmPC, the array, the receiver tube, and fluid flow rates.

The performance of the CmPC has been evaluated as per the heat transfer model and radiation losses from the evacuated tube receiver. The heat transfer coefficients in the receiver tube, between the receiver tube and transparent cover, and between the transparent cover and opaque cover have been determined from the standard heat transfer correlations. A sensible heat storage with indirect heat transfer between collector fluid and storage tank fluid is considered for the 24-hour dynamic simulation.

Table 14.7. Properties of ethylene glycol solutions.

Temperature, °C	Density, kg/m³	Specific heat, kJ/kg K	Thermal Conductivity, W/m K	Viscosity, Pa s
05	1045.10	3.77	0.47	36000.00
10	1044.40	3.79	0.48	28363.93
15	1042.20	3.80	0.48	24767.34
20	1040.01	3.81	0.48	21609.45
25	1037.81	3.82	0.48	18839.02
30	1035.62	3.83	0.49	16410.49
35	1033.42	3.84	0.49	14283.43
40	1030.90	3.85	0.49	12649.39
45	1027.96	3.86	0.49	11457.54
50	1025.02	3.87	0.50	10379.11
55	1022.07	3.88	0.50	9403.18
60	1019.13	3.89	0.50	8519.94
65	1016.19	3.90	0.50	7720.48
70	1012.79	3.91	0.50	7116.55
75	1009.33	3.93	0.50	6586.71
80	1005.87	3.94	0.50	6108.62
85	1002.41	3.95	0.50	5676.33
90	998.95	3.96	0.51	5284.68
95	995.25	3.97	0.51	4947.02
100	991.09	3.98	0.51	4667.20
105	986.92	3.99	0.51	4402.99
110	982.76	4.00	0.50	4153.55
115	978.59	4.01	0.50	3918.07
120	974.42	4.02	0.50	3695.79
125	970.06	4.03	0.50	3517.25
130	965.64	4.04	0.50	3356.70
135	961.21	4.05	0.50	3204.32
140	956.79	4.07	0.50	3059.64
145	952.37	4.08	0.50	2922.18
150	947.83	4.09	0.49	2827.73

Figure 14.17 shows the heat transfer method with SHS from the collector to the storage unit and later to the load. The heat exchanger coil provides an indirect heat transfer between the collector fluid and the fluid to load. The two heat exchanger coils facilitate a separate fluid in the storage tank without mixing with the collector's fluid or fluid in the load.

Figure 14.18(a) shows the total system temperatures for 24 hours. The required level of these temperatures can be obtained by controlling the fluid flow in the solar collector (CmPC), and load. The storage tank temperature is lower than the collector's inlet temperature and exit temperature. It is higher than the fluid supply temperature to the load as it transfers the heat from solar collectors to the load. Figure 14.18(b) shows the instantaneous energies of falling beam radiation on the collector, converted heat from the solar radiation at the receiver tube, heat used to load and heat losses for 24 hours of operation in a day. The hourly instantaneous energies of beam radiation falling on a tilted surface, energy collected at the receiver tube, energy used to load and the balanced energy

Table 14.8. Operational conditions of radiant heating and fresh air heating with a single solar thermal energy source per building.

S. No.	Description	Condition
1.	Concentration of ethylene glycol solutions, %	30.000
2.	Tilt angle of CmPC towards south, °	45.000
3.	Air exit temperature from FAH, °C	15.000
4.	Mass of fresh air, cfm	4518.000
5.	Width of CmPC, m	2.000
6.	Length of CmPC, m	2.700
5.	Effective length of CmPC, m (considering receiver length)	2.200
6.	Number of CmPCs in one array	5.000
7.	Distance between two rows of CmPC, m	1.500
8.	Actual area of one CmPC, m²	5.400
9.	Effective area of one CmPC, m²	4.400
10.	Ground area required for one array, m²	39.340
11.	Inner diameter of the receiver tube, m	0.045
12.	Outer diameter of the receiver tube, m	0.047
13.	Glass tube's outer diameter, m	0.058
14.	Fluid flow per one CmPC array, LPH	80.000
15.	Total fluid flow in a solar field, LPH	14960.000
16.	Total fluid flow in a load, LPH	80000.000

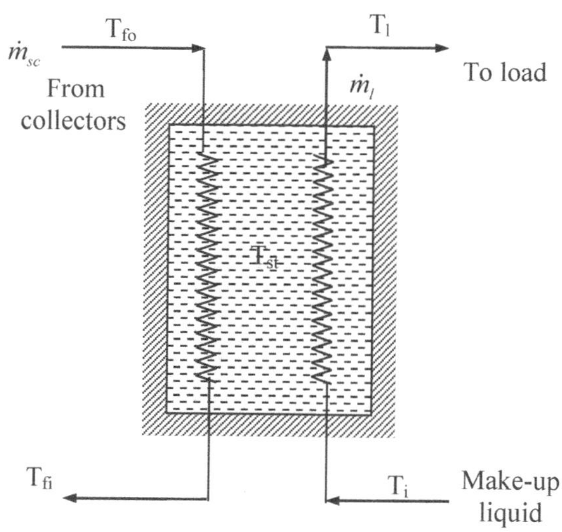

Fig. 14.17. Thermal storage with indirect heating method; m_{sc}: fluid flow in all the solar collectors, m_l: fluid flow to load, T_{fi}: fluid inlet temperature to solar collector, T_{fo}: fluid outlet temperature from the solar collector, T_i: inlet temperature of the fluid to storage tank from the load, T_l: temperature of fluid supplied to load, T_{st}: storage tank's fluid temperature.

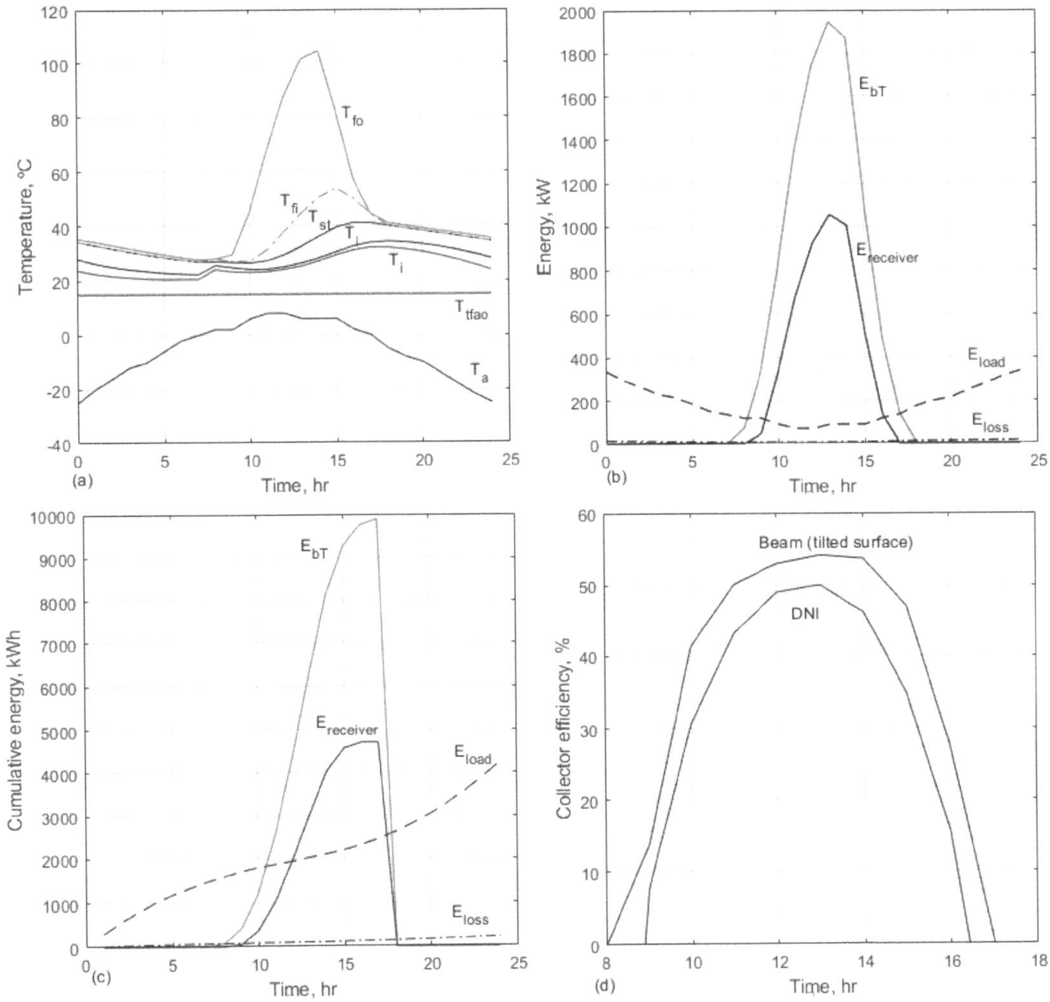

Fig. 14.18. (a) System temperatures of a single solar thermal source for radiant heating and fresh air heating, T_a: ambient temperature, T_{fi}: fluid inlet temperature to solar collector, T_{fo}: fluid outlet temperature from solar collector, T_i: fluid inlet temperature to storage tank from the load, T_l: fluid supply temperature to load, T_{tfao}: fluid outlet temperature from TFA (Treated Fresh Air) T_{st}: fluid temperature in a storage tank (b) instantaneous energies in the radiant heating's with solar collectors for one building; bT: beam radiation on tilted surface (c) cumulative energies of CmPC with storage for one building; (d) instantaneous efficiency of CmPC unit.

losses are depicted in Fig. 14.18(b). In Fig. 14.18(c), the cumulative energies of beam radiation falling on the collector, converted heat at the receiver, used heat to load and heat loss to surroundings are depicted from the 1st hour to the 24th hour (one-day cycle). The end value of the cumulative energy shows that the received energy from the solar collector is higher than the energy required to load. This difference is equal to the heat loss. The non-tracked solar collector receives less solar radiation than the tracked solar collector. Therefore, beam radiation falling on a tilted surface is lower than the DNI. The thermal efficiency of the solar collector with reference to these energy supplies is shown in Fig. 14.18(d).

Table 14.9 gives the solution of radiant heating and fresh air heating with a single energy source of solar thermal collectors. The effective area of the solar field is the collector's projected area times the effective length of the receiver, excluding the plumbing and gaps. The total area of the solar field counts the physical projected area of solar collectors without the gaps between the arrays. In the

Table 14.9. Results of radiant heating and fresh air heating with a single solar thermal energy source per building.

S. No.	Description	Result
1.	Energy required for radiant heating, kWh	2964.00
2.	Energy required for fresh air heating, kWh	1259.00
3.	Total energy requirement, kWh	4223.00
4.	Beam radiation collected on tilted CmPCs, kWh	9890.00
5.	Thermal energy collected at receiver, kWh	4686.64
6.	Thermal energy used for load, kWh	4223.00
7.	Storage tank capacity, m^3	175.00
8.	Average temperature difference of fluid in load, K	2.50
9.	Maximum temperature of fluid in storage tank, °C	40.97
10.	Minimum temperature of fluid in storage tank, °C	26.45
11.	Temperature change in storage tank's fluid/day	14.52
12.	Average thermal efficiency of CmPC, (tilted beam radiation), %	42.50
13.	Number of CmPC units	935.00
14.	Number of arrays (one array has 5 CmPCs)	187.00
15.	Number of parallel rows	187.00
16.	Total effective area of the solar field, m^2	4114.00
17.	Total area of the solar field, m^2	5049.00
18.	Total ground area required for the solar thermal field, m^2	7357.00

ground area of the solar thermal field, the gaps between the rows are also added. It is the actual land required for the erection of the solar collectors.

The required ground area for the installation of solar concentrating collectors without a boiler for the total heat load of 4224 kWh is 7357 m^2 (effective area: 4114.00 m^2) at an average solar thermal collection efficiency of 42.5%. This result was obtained at the normal component of solar radiation falling on a fixed, tilted CmPC.

References

Balaji, C. 2021. Thermal System Design and Optimization. Springer International Publishing.

Chiranjeevi, C and Srinivas, T. 2015. Experimental and simulation studies on two stage humidification-dehumidification desalination and cooling plant. Desalination 376(16): 9–16.

Duffie, J.A., Beckman, W.A. and Blair, N. 2020. Solar Engineering of Thermal Processes, Photovoltaics and Wind. John Wiley & Sons.

Ganapathy, V. 2002. Industrial Boilers and Heat Recovery Steam Generators: Design, Applications, and Calculations. CRC Press.

Hewitt, G.F., Shires, G.L. and Bott, T.R. 1995. Process heat transfer. Chemical Engineering Education 29(4): 243–275.

Horlock, J. 2013. Combined Power Plants: Including Combined Cycle Gas Turbined CCGT Plants. Elsevier.

Holman, J. and Pilling, G. 2004. Thermodynamics in context: A case study of contextualized teaching for undergraduates. Journal of Chemical Education 81(3): 373.

Kotas, T.J. 2012. The Exergy Method of Thermal Plant Analysis. Paragon Publishing.

Logan Jr, E. 1993. Turbomachinery: Basic Theory and Applications. CRC Press.

Nag, P.K. 2008. Engineering Thermodynamics. New Delhi, p.51.

Natarajan, M. and Srinivas, T. 2017. Experimental and simulation studies on a novel gravity based passive tracking system for a linear solar concentrating collector. Renewable Energy 105: 312–323.

Natarajan, M. and Srinivas, T. 2017. Design and analysis of a gravity based passive tracking mechanism to a linear solar concentrating collector. Proceedings of the Institution of Mechanical Engineers, Part C: Journal of Mechanical Engineering Science 231(13): 2503–2514.

Srinivas, T., Gupta, A.V.S.S.K.S. and Reddy, B.V. 2008. Sensitivity analysis of STIG based combined cycle with dual pressure HRSG. International Journal of Thermal Sciences 47(9): 1226–1234.

Srinivas, T., Gupta, A.V.S.S.K.S. and Reddy, B.V. 2009. Thermodynamic equilibrium model and exergy analysis of a biomass gasifier. ASME Journal of Energy Resources Technology 131(3): 1–7.

Srinivas, T., Reddy, B.V. and Gupta, A.V.S.S.K.S. 2012. Thermal performance prediction of a biomass based integrated gasification combined cycle plant. ASME Journal of Energy Resources Technology 134(2): 1–9.

Srinivas, T., Ganesh, N.S. and Shankar, R. 2019. Flexible Kalina Cycle Systems. Apple Academic Press.

Srinivas, T. 2021. Thermal Cycles of Heat Recovery Power Plants. Bentham Science Publishers.

Srinivas, T. 2023. Thermal Polygeneration. Springer Nature.

Srinivas, T., Nimesh, A.K., Tiwari, S., Kumar, R. and Singh, P. 2024. Studies on hybrid solar collector with water driven solar tracking unit. Energy Sources, Part A: Recovery, Utilization, and Environmental Effects 46(1): 8108–8131.

Stoecker, W.F. 1989. Design of Thermal Systems. McGraw Hill Education India.

Sukhatme, S.P. and Nayak, J.K. 2017. Solar Energy: Principles of Thermal Collection and Storage. McGraw-Hill Education.

Turns, S.R. 1996. Introduction to Combustion (Vol. 287, p. 569). New York, NY, USA: McGraw-Hill Companies.

W F, S. and Jones, J.W. 1981. Refrigeration and Air Conditioning. Second Edi. New York: MCGRAW-HILL Higher Education.

Yahya, S.M. 2011. Turbines Compressors and Fans (Vol. 103, No. v4, pp. 0–035). Tata Mcgraw Hill Education Pvt. Ltd., New Delhi.

Index

F

G

H

I

L

M

N

O

P

About the Author

Dr. Tangellapalli Srinivas

Dr. Tangellapalli Srinivas is a Professor and Head of the Department of Mechanical Engineering at the Dr. B.R. Ambedkar National Institute of Technology, Jalandhar, Punjab, India. Previously, he was a professor at the CO_2 Research and Green Technologies Center of the School of Mechanical Engineering at Vellore Institute of Technology, Vellore, Tamil Nadu. He received his PhD from JNT University College of Engineering, Hyderabad, Telangana, in 2008 and was a postdoctoral fellow at the University of Ontario Institute of Technology (UOIT), Canada, in 2010 and 2013. He is the principal investigator for research projects funded by CSIR, SERB, SERB (IMPRINT), and DST. He has industrial consulting experience in the areas of solar thermal energy and thermal systems. He has four granted patents and two copyrights and has published three patents. He has authored five books, 20 book chapters and 160 journal publications. He has edited three books and participated in 125 conference proceedings. His research interest includes solar thermal energy, advanced power generation, and polygeneration. He has guided six Ph.D. students, 40 masters theses, and 50 UG projects. He also has four years of industrial experience in the machine tool industry. He serves on the editorial boards of the journals *Recent Innovations in Chemical Engineering* and *Applied Solar Energy* (Springer). For teaching excellence, he received the "best teacher" award in 2021 at NIT Jalandhar. In 2023, he received the Innovative Teaching Award at the 2nd Leadership Convention on "World is One," organized by the World Leadership Academy and Kalinga Institute of Industrial Technology (KIIT), Bhubaneswar, Orissa. He is on the "Career-long Top 2% of World Scientists" list, compiled by Stanford University and Elsevier in 2024.

For Product Safety Concerns and Information please contact our EU representative GPSR@taylorandfrancis.com Taylor & Francis Verlag GmbH, Kaufingerstraße 24, 80331 München, Germany